NASA/TM–2011–216142

SPACE SHUTTLE MISSIONS SUMMARY

Robert D. Legler
Floyd V. Bennett

Mission Operations
Johnson Space Center

September 2011

THE NASA STI PROGRAM OFFICE . . . IN PROFILE

Since its founding, NASA has been dedicated to the advancement of aeronautics and space science. The NASA Scientific and Technical Information (STI) Program Office plays a key part in helping NASA maintain this important role.

The NASA STI Program Office is operated by Langley Research Center, the lead center for NASA's scientific and technical information. The NASA STI Program Office provides access to the NASA STI Database, the largest collection of aeronautical and space science STI in the world. The Program Office is also NASA's institutional mechanism for disseminating the results of its research and development activities. These results are published by NASA in the NASA STI Report Series, which includes the following report types:

- TECHNICAL PUBLICATION. Reports of completed research or a major significant phase of research that present the results of NASA programs and include extensive data or theoretical analysis. Includes compilations of significant scientific and technical data and information deemed to be of continuing reference value. NASA's counterpart of peer-reviewed formal professional papers but has less stringent limitations on manuscript length and extent of graphic presentations.

- TECHNICAL MEMORANDUM. Scientific and technical findings that are preliminary or of specialized interest, e.g., quick release reports, working papers, and bibliographies that contain minimal annotation. Does not contain extensive analysis.

- CONTRACTOR REPORT. Scientific and technical findings by NASA-sponsored contractors and grantees.

- CONFERENCE PUBLICATION. Collected papers from scientific and technical conferences, symposia, seminars, or other meetings sponsored or cosponsored by NASA.

- SPECIAL PUBLICATION. Scientific, technical, or historical information from NASA programs, projects, and mission, often concerned with subjects having substantial public interest.

- TECHNICAL TRANSLATION. English-language translations of foreign scientific and technical material pertinent to NASA's mission.

Specialized services that complement the STI Program Office's diverse offerings include creating custom thesauri, building customized databases, organizing and publishing research results . . . even providing videos.

For more information about the NASA STI Program Office, see the following:

- Access the NASA STI Program Home Page at http://www.sti nasa.gov

- E-mail your question via the internet to help@sti nasa.gov

- Fax your question to the NASA Access Help Desk at (301) 621-0134

- Telephone the NASA Access Help Desk at (301) 621-0390

- Write to:
  NASA Access Help Desk
  NASA Center for AeroSpace Information
  7115 Standard
  Hanover, MD 21076-1320

NASA/TM–2011–216142

SPACE SHUTTLE MISSIONS SUMMARY

Robert D. Legler
Floyd V. Bennett

Mission Operations
Johnson Space Center

September 2011

MOD EMBLEM DESCRIPTION

This emblem was developed during the Apollo program for the mission control team [JSC Mission Operations Directorate, MOD] to recognize their unique contribution to manned space flight since the Mercury program.

The sigma (Σ) represents the total mission team, including flight controllers, instructors, flight design and production specialists, and facility development and support teams including all engineering, scientific, operations disciplines, and supporting tasks.

The Shuttle launch represents the dynamic elements of space, the initial escape from our environment, and the thrust to explore the universe. The four stars on the Shuttle's plume represent the basic principles of the Mission Operations team: discipline, morale, toughness, and competence. Their place along the Shuttle's plume reminds us that they are the foundation upon which each mission is flown. Today's core principles include confidence, responsibility, teamwork, and vigilance. Each of these words comes into the vocabulary of Mission Operations personnel at critical points in their development. These words can never be forgotten if we are to succeed in the future.

The orbiting International Space Station symbolizes a permanent human presence in space, conducting research and developing materials leading to the commercial utilization of the space environment.

The Earth is our home and will forever be serviced by both manned and unmanned spacecrafts in order to improve our quality of life. A single star is positioned over Houston, the home of U.S. human spaceflight operations.

The comet represents those individuals who have given their lives for space exploration. The seventeen stars represent our fallen astronauts, to whom in part we dedicate our commitment to excellence. These symbols serve as a reminder of the risks inherent to space flight and recognize that we of Mission Operations provide the margin that makes the risk acceptable.

The Mercury, Gemini, Apollo, Skylab, and Apollo-Soyuz Test are represented on the bottom border. At the top of the emblem, the Moon and Mars represent our future, signifying our intent to lead the way.

The wording "RES GESTA PER EXCELLENTIAM" - "Achieve through Excellence" - is the standard for our work. It represents an individual's commitment to a belief, to craftsmanship, and to perseverance, qualities required to continue the peaceful development of space and the quest for the stars.

The original emblem was designed (at the request of White Flight, Gene Kranz) by Robert T. McCall in April 1973 and bears the inscription "For the Personnel of Mission Control with Great Respect and Admiration. Robert T. McCall." Mr. McCall died at age 90, May 5, 2010. In 1983, the original emblem was updated to support the Space Shuttle program. In 2004, with the artistic help of graphic designer Mike Okuda and participation of the Mission Operations team, the emblem was updated to recognize the achievements and contributions of the team supporting the International Space Station program as well as those that contributed to the success of the earlier Skylab and Apollo-Soyuz Test Project missions.

ABSTRACT

This document was originally produced as an informal Mission Operations book and has been updated since Space Shuttle Flight STS-1 and throughout the program. This version is a formally released NASA document. It is a handy reference guide for flight data for all Space Shuttle missions. "As-flown" data is provided as compiled from many flight support sources for ascent, on-orbit events, and descent mission phases. In addition, the specific shuttle vehicle configuration, payload, flight crew, and flight directors are identified for each flight. In the development of this book, the data for the early flights are contained on a single page per flight. For later flights, more pages per flight have been added, primarily for growth in mission complexity as noted in the "Mission Highlights" data column. This particularly applies to missions involved in the assembly of the International Space Station. Pertinent photos for each mission are also included on each mission summary page.

First Flight - 1981

THE REUSABLE SPACE SHUTTLE

The Space Shuttle Vehicle (SSV) was the world's first reusable Spacecraft. It consisted of a reusable Orbiter Vehicle with three Space Shuttle Main Engines (SSMEs), two Solid Rocket Boosters (SRBs), and an expendable External Tank (ET). The Space Shuttle System consisted of the SSV elements, Shuttle Carrier aircraft, payload accommodations, and ground support systems. The SSV was designed to perform a variety of missions to low Earth orbit with heavy payload lift capability.

SSV missions included: Manned payload bay laboratory science, deployment and servicing of payloads, and special support to space activities such as sortie missions (rescue, repair, maintenance servicing, assembly, and docking), and International Space Station (ISS) assembly, manning, and support including robotic and manned extra vehicular activities.

The SSV was flown for 30 years from 1981 to 2011. Brief mission summaries for each of these missions are provided in this document. The document contains "as flown" mission data and pertinent photographs for each flight. It was originally published as an informal document and routinely updated throughout the Shuttle era.

ABOVE: S81-30498 --- After six years of silence, the thunder of U.S. manned spaceflight is heard again, as the successful launch of the first Space Shuttle reusable vehicle, Columbia, ushers in a new concept in utilization of space - April 12,1981.

RIGHT: Thirty years later on STS-135, the Atlantis vehicle executes the final Space Shuttle landing on July 21, 2011 at KSC. With the closure of the Space Shuttle Program, the thunder of U.S. manned spaceflight is not expected to be heard again for another several years.

Last Landing - 2011

The Space Shuttle--1981 to 2011

The Space Transportation System-STS-has had a spectacular career spanning three decades of intense and productive activities in space. The Shuttle was conceived as a reusable launch system to grossly reduce the cost of transporting humans and satellites into low earth orbit and to service the entire spectrum of government and commercial space operations requirements. To accomplish this challenging task required the development of a series of new technologies in rocket engines, space systems, unique materials, highly advanced manufacturing techniques, autonomous control concepts and never before attempted flight operations maneuvers. The fact that these devices were conceived and developed and in almost all cases could be reused is a testimony to the marvelous capability of the US and allied aerospace community.

Equally significant was the ability of the government industry team to bring about the successful development of this phenomenal machine under the stringent and ever changing and fickle government budgetary process. The management team was required to continuously adjust the expenditure of funds because of both postponement and reductions in national budget that resulted in a delay in manufacturing facilities, extended testing periods and technology development which presented extraordinary circumstances regarding the ability to arrive at the first flight of the Shuttle. And although the first and subsequent STS flights were delayed by several years, the cost to build the transportation system was reasonably close to the original cost estimates. Indeed, if the effects of inflation are included, the overall cost of the program was probably within the costs estimates made almost ten years previously.

There were two devastating fatal accidents during the course of the STS time period. It should be noted that both of these accidents took place because of mismanagement. The accidents literally destroyed the user confidence in the STS and resulted in the eventual termination of the Shuttle. The Space Shuttle without these two unnecessary failures is an extremely safe space faring vehicle and it will be a long time in the future before a reusable rocket caring humans will match this accomplishment.

An overall assessment of the STS must say that history will show the accomplishments were spectacular.

Christopher C. Kraft, Jr.
First Flight Director

I look at the three decades of Space Shuttle flights with a great deal of pride. John Young and I had the privilege of flying Columbia on the initial orbital test flight. While the Shuttle didn't live up to some of the preflight hype regarding flight rate and cost, it still is the most fantastic spaceship ever built and likely will be for the foreseeable future. Yes, we had two terrible tragedies, but spaceflight is not without risk now and for the foreseeable future.

The Shuttle has accomplished many wondrous feats in its 30 years of flight. In the beginning it flew very important DOD missions that I believe played a major role in the winning of the Cold War. The payloads it has taken to orbit have revolutionized knowledge of our solar system and the universe. The Shuttle Program made possible the construction of the unbelievably complex International Space Station.

All in all, everyone associated with the Shuttle should be proud of what the program accomplished. It will be a very long time before we see a spaceship with anywhere near the Shuttle's capability.

Bob (Crip) Crippen
PLT STS-1, and CDR STS-7, STS-41C & STS-41g
KSC Center Director 1992 - 1995

Continued...

National Space Transportation System
(Space Shuttle)

Developed primarily in the 1970's, the National Space Transportation System (Space Shuttle) was, and remains to this day, the most innovative and capable human rated space launch system created by man.

As much as Apollo, the Space Shuttle established the United States as the human space flight technology leader of the world, made human access to low-Earth orbit (LEO) relatively routine, and raised the expectations of the global population in regards to the value of space to mankind. It has enabled us to learn to live and work in space to create value on Earth.

The Shuttle designers both advanced the state of technology by levying seemingly unachievable technical challenges, such as the incredibly high power density Space Shuttle Main Engine (SSME), complex redundant data processing, and reusable thermal protection systems, as well as utilizing available technology like aluminum structure and hydraulic flight control and thrust vector control systems.

The Shuttle designers both advanced the state of technology by levying seemingly unachievable technical challenges, such as the incredibly high power density Space Shuttle Main Engine (SSME), complex redundant data processing, and reusable thermal protection systems, as well as utilizing available technology like aluminum structure and hydraulic flight control and thrust vector control systems.

By advancing the state of the art in mission planning and execution, the Shuttle team took maximum advantage of the extensive capabilities available from both man and machine and the synergistic interplay between them. The results in mission accomplishments are undeniable and have forever transformed our understanding of the world in which we live.

Brewster H. Shaw, Jr.
PLT STS-9 and CDR STS-61B & STS- 28
Space Shuttle Program Mgr 1993 -1995
VP & GM Space Exploration Boeing Houston

TABLE OF CONTENTS

AIM PT	AIM POINT
AL	ASCENDING LEFT
AOA	ABORT ONCE AROUND
AR	ASCENDING RIGHT
ASC	ASCENT
ASC/ENT	ASCENT/ENTRY
AVE BRK DECEL	AVERAGE BRAKE DECELERATION
BEN	BEN GUERIRBRK INIT BRAKE INITIATION VELOCITY IN KGS
BR/UP	BREAK UP ALTITUDE OF ET IN THOUSANDS OF FEET
BYD	BANJUL
CI	CLOSEIN
CTOB	CREW TIME ON BACK
DENS ALT	DENSITY ALTITUDE
DL	DESCENDING LEFT
DOLILU	DAY OF LAUNCH I-LOAD UPDATE
DR	DESCENDING RIGHT
EDW	EDWARDS AFB
EMU	ENVIRONMENTAL MOBILITY UNIT
ET	EXTERNAL TANK
EVA	EXTRA VEHICULAR ACTIVITY
F	SS FEMALE NUMBER
FD	FLIGHT DIRECTOR
FDRD	FLIGHT DEFINITION & REQUIREMENTS DOCUMENT
FPR	FLIGHT PLANNING RESERVE
FRD	FLIGHT REQUIREMENTS DOCUMENT
GMTLO	GREENWICH MEAN TIME OF LIFTOFF
HA/HP	APOGEE AND PERIGEE IN NAUTICAL MILES
HDOT	TOUCHDOWN ALTITUDE RATE
KEAS	KNOTS EQUIVALENT AIRSPEED
KGS	KNOTS GROUND SPEED
KSC W/D	KSC WORKDAY
LD/O1	LEAD/ORBIT 1 SHIFT
LDA	LAUNCH DANGER AREA
M	SS MALE NUMBER
M 3 EOM	MACH 3 END OF MISSION
MECO	MAIN ENGINE CUT OFF
MET	MISSION ELAPSED TIME
MLGTD	MAIN LANDING GEAR TOUCHDOWN
MLP	MOBILE LAUNCH PLATFORM
MMT	MISSION MANAGEMENT TEAM
MMU	MANNED MANEUVERING UNIT/
MOD	MISSION OPERATIONS DIRECTOR MISSION OPERATIONS DIRECTORATE
MPS	MAIN PROPULSION SYSTEM
MRN	MORON
M/S	MISSION SPECIALIST
MTR	MOTOR
N	NOMINAL
NEG RET	NEGATIVE RETURN
NLGTD	NOSE LANDING GEAR TOUCHDOWN
O1, O2, O3	ORBIT 1, 2, OR 3 FLIGHT DIRECTOR SHIFTS
OFT	OFFICIAL FLIGHT DESIGNATOR
OI	OPERATIONAL INCREMENT
OMS	ORBITAL MANEUVERING SYSTEM
OPF	ORBITER PROCESSING FACILITY
ORB DIR	ORBIT DIRECTION
P	SEQUENTIAL NUMBER OF PERSON FLOWN ON SS
PAO	PUBLIC AFFAIRS OFFICE
PERF	PERFORMANCE
PERF MARGINS	PERFORMANCE MARGINS
P/L	PAYLOAD
PLNG	PLANNING SHIFT
PLS	PLANNED LANDING SITE
P/S	PAYLOAD SPECIALIST
PTA	PRESS TO ABORT ONCE AROUND
PTM	PRESS TO MECO
R	SS ROOKIE NUMBER
RECON	RECONSTRUCTED
RMS	REMOTE MANIPULATOR SYSTEM
RPT	RUPTURE OF ET IN THOUSANDS OF FEET
RSRM	REDESIGNED SOLID ROCKET MOTOR
RTLS	RETURN TO LAUNCH SITE
SEQ	SEQUENTIAL
SLS	SECONDARY LANDING SITE
SODB	SHUTTLE OPERATIONAL DATA BOOK
SS	SPACE SHUTTLE OR SUN SHIELD
SSME	SPACE SHUTTLE MAIN ENGINE
S/T	SHUTTLE TOTAL FLIGHT TIME
TAL	TRANSOCEANIC ABORT LANDING
TD NORM 195	NORMALIZED TOUCHDOWN RANGE AT 195 KEAS
TDDP	TRAJECTORY DESIGN DATA PACKAGE
TDEL	DIFFERENCE IN REFERENCE TIME FOR SSME THROTTLE ADJUSTMENT
TK	TANK
T/V	TUMBLE VALVE
V	SS VETERAN NUMBER
VAB	VEHICLE ASSEMBLY BUILDING
VEL	VELOCITY
VI	INERTIAL VELOCITY
W/D	WORKDAY
WX	WEATHER
X CG	X CENTER OF GRAVITY
XRANGE	CROSSRANGE
ZZA	ZARAGOZA (TAL SITE)

ABOUT THIS DOCUMENT

CONVERSION FROM INFORMAL DOCUMENT

Robert D. "Bob" Legler/DA8/USA was the originator of this book as an informal Mission Operations Document to provide a "handy reference guide" for "as flown" mission data, often used by JSC Flight Controllers and Mission Planners.

Mr. Legler authored the informal book from flight STS-1 through flight STS-115. After Legler's death in 2007, Floyd V. Bennett/DA8/USA/GHG took over the authorship for STS-116 and all missions to follow. In addition, a "Brief Mission Summary" statement for all ISS assembly missions and pertinent mission related photos to each summary file were incorporated.

This formal NASA document is a conversion of the informal version to provide an official historical record of pertinent Space Shuttle Missions Operational Data.

DOCUMENT FORMAT

The "as flown" operational mission data is presented in a summary table format of twelve columns. For early flights the book contains one page of data per flight. For later flights, as on-orbit activities became more and more complex, additional pages per flight were added, primarily for growth in the 12th column, "Mission Highlights".

In addition a summary table of weight data for each shuttle element and payloads for each mission is provided in Appendix A.

In Appendix B the authors acknowledge individuals for contributions to the preparation of this document and provides the data sources and Points of Contact (POCs) used in compiling flight and weight data.

Appendix C provides an historical record of JSC Flight Controllers originally compiled by Bob Legler, "History Flight". Since his death the listing has been maintained by the JSC Flight Directors Office.

And lastly, information about the authors is provided in the back of the book including an "In Memoriam" for Bob Legler.

MISSION SUMMARIES DATA DEFINITIONS

This section contains definitions of the data provided in the Mission Summaries by column number. Several entries have been assigned sequential numbers for reference purposes (e.g., # of rendezvous, # of night launches, # EVAs, etc.).

Column 1:
FLIGHT NUMBERS - The flight numbers include the official STS flight designator, followed by: the original flight designator (as applicable), the sequential flight number, the KSC launch sequential number, the OFT flight number (as applicable), the ISS flight number (as applicable), the launch pad sequential number, and MLP used.

Column 2:
ORBITER - Provides Orbiter designation, number of flights flown, & OMS PODs #'s.

Column 3:
FLIGHT CREW - Flight Crew members & titles are listed for each flight. Space shuttle flight (SS) number designators are listed for each crew member as follows:
P = sequential number of person flown on SS; R = SS rookie number; V = SS veteran number (second flight on SS); M = SS male number; F = SS female number. No attempt is made to determine which seat arrives first in orbit on the same flight. Example: P17/R2/V1/M2 - person 17, rookie 2, veteran 1, male 2. Once assigned a number, the crew member retains those R, V, & M or F numbers. Only the P number would change on subsequent flights.
EVAs - Relates to SS EVAs. Includes type of EVA, dates/times of EVAs, EVA crew member names, and sequential number of SS EVAs and EVA times.
FLIGHT DIRECTORS - The Flight Directors and Mission Operations Director are listed for each flight.
CAPCOMS - CAPCOMS are listed for missions STS-116 and all to follow.

ABOUT THIS DOCUMENT

Column 4:

LAUNCH/LIFTOFF/ASCENT DATA - Includes Pad Number, Liftoff Times [planned (P) and actual (A) in Eastern Time Zone and Greenwich Mean Time (GMT) liftoff time], Date of Launch followed by a number indicating how many SS flights have been launched on that month to date, Day-of-Week Launch followed by a number indicating how many SS flights were launched on the day of the week, Window Duration and Closure Rationale, Planned Landing Sites including those selected on Day of Launch, Ascent Events, and Abort Calls. In the later flights, there are two sets of data in the Ascent Events Column. The left set is planned METs and Velocities, and the right set is the actual METs and Velocities for the specified events.

Column 5:

LANDING DATA - Includes Landing Site/Runway followed by a Sequential Number indicating the Number of Concrete/Lakebed landings at EDW or a Sequential Number for Landings at NOR and KSC. Landing time is in local time for the landing site. The Landing Day of Week is followed by a Number indicating how many landings have been made on that day of the week. The Number after the Landing Date is the Sequential Number of Landings during that month, i.e., 4/2/92 (7), STS-45 is the seventh landing in April. Each Orbit Direction for Landing is followed by a Sequential Number. The Winds are designated in knots of head, tail and left and right crosswinds. The first listing was obtained from the MOD Descent Postflight Summary and is basically the Winds observed on a display at the touchdown time. The second listing is the "Official" Winds, which are the Two Minute Average Winds spanning the MLG Touchdown Time. The Flight Durations are determined from the time of liftoff to MLG Touchdown, specified in days, hours, minutes, and seconds.

S/T - Shuttle Total Flight Time, i.e., Accumulated Total. This is followed by an Orbiter Designator and the Accumulated Flight Time for that Orbiter.

Column 6:

SSME DATA - Includes Nominal, Abort, and Emergency Throttles, Predicted and Actual Throttle Profile, and Engine Serial Numbers followed by the Number of Flights on that engine. For a lack of space elsewhere, the Mach 3 End-of-Mission Weights and X CG and Landing Weight and X CG have been added in this column.

Column 7:

SRB/SRM/RSRM - Includes the "Build Item" Number followed by SRM/RSRM Type or Number.

ET DATA - Includes ET Numbers, ET Rupture and Breakup Altitudes and Times in MET, and Tumble Valve Use. These times and altitudes were not available for flights after STS-46. However, the time, latitude, and longitude of ET Impact are included for all missions.

Column 8 :

ORBIT INCLINATION - This is the Inclination after OMS-2 and is followed by a Sequential Number indicating how many flights were flown at that inclination. Inclinations between 28.45 and 28.55 have been considered the same for the purposes of assigning Sequential Numbers.

Column 9:

ORBIT HA/HP - Insertions were Standard Insertions unless specifically stating "Direct Insertion". Generally, Altitudes for Post OMS-2 are given, as well as Payload Deploy Altitudes and De-orbit Altitude.

Column 10:

FLIGHT SOFTWARE DESIGNATORS - OI (Operational Increment) numbers are followed by a Sequential Flight Number for that OI.

Column 11:

PAYLOAD DATA - Includes Cargo, Chargeable, Deployed, Non-Deployed, and Middeck Weights as documented in the SODB for flights STS-1 through STS-57. Effective with STS-51, the SODB data is no longer updated as flown. Therefore, the data has been obtained from the Day-of-Launch (DOL) Trajectory Design Data Package (TDDP). The following Shuttle Accumulated Weights are provided: (1) Total Payload Deployed Weights left in orbit, (2) Total Non-Deployed Payload Weights (does not include Ancillary Equipment such as ASE, cabling, etc.), and (3) Total Cargo Weights which include all Ancillary Equipment. Weights for seven DOD flights are not included. Performance Margins: Four numbers are provided - (1) Flight Planning Reserve (FPR); (2) Fuel Bias; (3) Final TDDP is margin above FPR, and Fuel Bias using mean wind and atmosphere for launch month, no unplanned drainback and final selected I-load; and (4) Recon is margin above MET wind and atmosphere, any unplanned drainback, final estimated MPS loads (a.k.a., "Reconstructed" Systems Performance). It should be noted that STS-27 Delta Margin was -295 lbs for drainback, -365 lbs for wind/atmosphere. STS-31 Delta Margin was -753 lbs for drainback, +461 lbs for wind/atmosphere. STS-41 was -358 lbs for drainback, -488 lbs for wind/atmosphere. Payloads are identified as being Primary, Payload Bay (PLB), and/or Middeck Payloads. Payload Column also contains the number of cryo Tank sets and whether a RMS was flown followed by a Sequential Number and serial number of the RMS.

Column 12:

MISSION HIGHLIGHTS/MISCELLANEOUS DATA COLUMN - Includes the Number of KSC Workdays in OPF, at VAB, at Pad, and Total Workdays. Launch Postponements may not contain early postponements. Postponements are defined as launch delays which occurred prior to call-to-stations for OMI S0007 Shuttle Countdown. Scrubs are launch date changes after the start of Shuttle countdown (countdown was terminated or recycled to a later launch date). Launch Delays are delays which occur only on the day-of-launch. Other data included are TAL Weather Data, Night Launch and Night Landing Sequential Numbers, Flight Duration Changes, Landing Site Changes, Firsts, Events, and Significant Anomalies as judged by the compiler (not all Anomalies are included). Use of Alternate and DOLILU I-loads are included with a Sequential Number for Uplinks. STS-27 was the first flight with the capability to uplink Alternate I-loads for use and STS-48 was the first flight with DOLILU capability. Rendezvous operations are identified including the Target and Sequential Number of each Space Shuttle Rendezvous. Also, a Brief Mission Summary has been added for the first ISS Assembly Mission, STS-88/2A, and all missions to follow.

SPACE SHUTTLE MISSIONS SUMMARY

FLT NO.	ORBITER	CREW (2) / TITLE, NAMES & EVA'S	LAUNCH SITE, LIFTOFF TIME, / LANDING SITES, ABORT TIMES	LANDING SITE/ RUNWAY, CROSSRANGE / LANDING TIMES FLT DURATION, WINDS	SSME-TL NOM-ABORT EMERG / THROTTLE PROFILE ENG. S.N.	SRB RSRM AND ET	ORBIT / INC	ORBIT / HA/HP	FSW	PAYLOAD WEIGHTS, / PAYLOADS/ EXPERIMENTS	MISSION HIGHLIGHTS (LAUNCH SCRUBS/DELAYS, TAL WEATHER, ASCENT I-LOADS, FIRSTS, SIGNIFICANT ANOMALIES, ETC.)
STS-1 SEQ FLT #1 KSC 1 OFT-1 PAD 39A-1	OV-102 Flight 1 Columbia OMS PODS LVO1 - 1 RVO1 - 1 FRC2 - 1	CDR: John W. Young P1/R1/M1 PLT: Robert L. Crippen P2/R2/M2 MCC FCR-1 (1) FLIGHT DIRECTORS: A/PLG - N. B. Hutchinson ORBIT - C. R. Lewis ENT/ORB - D. R. Puddy MOD - E. F. Kranz	KSC 39A 102:12:00:03.9Z 7:00:00 AM EST (P) 7:00:04 AM EST (A) Sunday 1 4/12/81 (1) WINDOW DURATION: 4.7 hours PLS - EDW SLS - NOR NO TAL AOA - EDW NOR CLS - HICKAM KADENA ROTA MAX Q = 617 M = 1.06 SRB SEP: 2:11.7 MET MECO: 8:34 MET ET SEP: 8:52.1 MET OMS-1: 10:34 MET 86.1 Seconds OMS-2: 44:02 MET 74.8 Seconds DEORBIT 148 X 146 NM VELOCITY 25731 FPS RANGE 4379 NM	EDW 23, LKBD (EDW 1, LKBD 1) 10:20:57 AM PST Tuesday 1 4/14/81 (1) XRANGE: 315 NM ORB DIR: DR (1) AIM PT: NOMINAL MLGTD: 6053 FT 104:18:20:57Z VEL: 190 KGS 183 KEAS HDOT: -1.5 FPS TD NORM 195: 4973 FT NLGTD: 9152 FT 104:18:21:07Z VEL: 156 KGS HDOT: -5.6 FPS BRK INIT:105 KGS AVE BRK DECEL: 5.9 FPS/S WHEELS STOP: 104:18:21:36Z 15046 FT ROLLOUT: 8993 FT 60 SEC WIND: 2T, 2R KNOTS OFFICIAL: 1H, 1R DENS ALT: 2200 FT FLT DURATION: 2:06:20:53 54:20:53 DISTANCE: 933,757 sm	00/100 (100) 65% 1 = 2007 (1) 2 = 2006 (1) 3 = 2005 (1) M 3 EOM WEIGHT: 195943 X CG: 1096.7 LANDING WEIGHT: 195473 X CG: 1098.1	A7/8 86-80E MTR: STD CASE: STD 168-80 SWT ET-1 START: -25.6 END: -19.9 MAX: ET BR/UP 223K 47:42 MET ET IMPACT LAT: 30.95 S LONG: 93.2 E	40.3 (1)	STANDARD INSERTION INSERTION ALTITUDE: 145 NM 152/152 172/172 SM	R16/T9	CARGO: 10823 lbs DFI: 9290 lbs SHUTTLE ACCUMULATED WEIGHTS: DEPLOYED: 0 lbs NON-DEPLOYED: 10823 lbs CARGO TOTAL: 10823 lbs PERFORMANCE MARGINS NOT AVAILABLE PAYLOADS: IECM/REM DFI NO RMS 2 CRYO TANK SETS	KSC W/D: OPF 531, VAB 33, PAD 104 =668 LAUNCH POSTPONEMENTS: Yes. LAUNCH SCRUBS: - Scrubbed 4/10/81 launch at T-18 minutes because BFS did not track PASS timing. Rescheduled launch for 4/12/81. 2-day slip. - Installed S/W patch to correct problem. LAUNCH DELAYS: 4 seconds. CONTINGENCY LANDING SITE (CLS) WX: - Rota was go. There was no TAL site for STS-1. FLIGHT DURATION CHANGES: None. FIRSTS: - First orbital flight of reusable Space Shuttle vehicle. - First manned vehicle space flgiht w/o unmanned test flight. SIGNIFICANT ANOMALIES: - SRB ignition overpressure (higher than expected) deformed FRCS oxidizer tank aft Z strut. - OMS POD tile LRSI tiles lost. - WMS problems (degraded air suction). - ET tumble system did not work. - PLBD closure overlap more than expected. - Cabin temperature controller did not maintain selected temperature. - OMS quantity gaging system was sticking during flight. - Both Radar Altimeters lost lock at 75 feet (no valid data after 75 feet). - Difficulty locking doors on two storage lockers due to misalignment. CONTINGENCY LANDING SITE: - ROTA was a contingency landing site but not required for one SSME out. S-BAND TRACKING SITES: - MIL, PDL, BDA, MAD, IOS, ORR, BUC, GDS, HAW, ACN, GWM, QUI, AGO, TUL (NOR), PTT, VDT. RADIATORS DEPLOY #1 NOTE: ON STS-1 AND STS-2, THE NOMINAL OGS AIM POINT WAS 6500 FEET (5500 FEET WAS THE CLOSE-IN AIM POINT).

...On-Orbit...
Left: CDR Young in the cockpit
Right: PLT Crippen prepares dinner on middeck

...In the MCC...
Gene Kranz/FOD, Chris Kraft/JSC Ctr Dir. & Max Faget/E&D (Father of U.S. Manned Spacecraft Design)

We Have Liftoff!
-- April 12, 1981 --
(S81-30500)

... and Touchdown at EAFB!
-- April 14, 1981 --
"That's the world's greatest flying machine"
- CDR John Young! (S81-30746)

SPACE SHUTTLE MISSIONS SUMMARY

FLT NO.	ORBITER	CREW (2) TITLE, NAMES & EVA'S	LAUNCH SITE, LIFTOFF TIME, LANDING SITES, ABORT TIMES	LANDING SITE/ RUNWAY, CROSSRANGE LANDING TIMES FLT DURATION, WINDS	SSME-TL NOM-ABORT EMERG THROTTLE PROFILE ENG. S.N.	SRB RSRM AND ET	ORBIT INC	ORBIT HA/HP	FSW	PAYLOAD WEIGHTS, PAYLOADS/ EXPERIMENTS	MISSION HIGHLIGHTS (LAUNCH SCRUBS/DELAYS, TAL WEATHER, ASCENT I-LOADS, FIRSTS, SIGNIFICANT ANOMALIES, ETC.)
STS-2 SEQ FLT # 2 KSC 2 OFT-2 PAD 39A-2	OV-102 Flight 2 Columbia OMS PODS LVO1 - 2 RVO1 - 2 FRC2 - 2	CDR: Joe H. Engle P3/R3/M3 PLT: Richard H. Truly P4/R4/M4 MCC FCR-1 (2) FLIGHT DIRECTORS: ASC - N. B. Hutchinson PLNG - T. W. Holloway ORBIT - C. R. Lewis ENT - D. R. Puddy ORB - H. M. Draughon MOD - E. F. Kranz	KSC 39A 316:15:09:59.8Z 7:20:00 AM EST (P) 10:10:00 AM EST (A) Thursday 1 11/12/81 (1) WINDOW DURATION: 4.7 hours PLS - EDW SLS - NOR TAL - ROTA (Selected) MAX Q = 640 M = 1.09 SRB SEP: 2:10 MET MECO: 8:33.8 MET ET SEP: 8:57:2 MET OMS-1: 10:33.9 MET 77 Seconds OMS-2: 41:41.7 MET 69.2 Seconds	EDW 23, LAKEBED (EDW 2, LKBD 2) 1:23:12 PM PST Saturday 1 11/14/81 (1) XRANGE: 63 NM ORB DIR: DR (2) AIM PT: NOMINAL MLGTD: 780 FT 318:21:23:12Z VEL: 186 KGS 197 KEAS HDOT: -1.0 FPS TD NORM 195: 960 FT NLGTD: 4429 FT 318:21:23:26Z VEL: 137 KGS HDOT: -5.1 FPS BRK INIT: 109 KGS AVE BRK DECEL: 6.1 FPS/S WHEELS STOP: 318:21:24:03Z 8491 FT ROLLOUT: 7711 FT 50 SEC WIND: 20H, 3R KNOTS OFFICIAL: 17H, 6L DENS ALT: 3500 FT FLT DURATION: 2:06:13:12 54:13:12 S/T: 4:12:34:05 OV-102: 4:12:34:05 DISTANCE: 933,757 sm	100/100 (107) 68% 1 = 2007 (2) 2 = 2006 (2) 3 = 2005 (2) M 3 EOM WEIGHT: 204356 X CG: 1096.6 LANDING WEIGHT: 204263 X CG: 1098.1	A9/10 MTR: STD CASE: STD 168-80 SWT ET-2 ET RPT 256K 49:20 MET ET BR/UP 219K 50:28 MET ET IMPACT LAT: 31.67 S LONG: 95.7 E	38.0 (1) 63.25 START: -53.5 END: -56.2	STANDARD INSERTION INSERTION ALTITUDE: 137 NM 120/120 137/137 NM DEORBIT 140 X 139 NM VELOCITY 25726 FPS RANGE 4474 NM	R18/T11	CARGO: 18778 lbs CHARGEABLE: SHUTTLE ACCUMULATED WEIGHTS: DEPLOYED: 0 lbs NON-DEPLOYED: 29601 lbs CARGO TOTAL: 29601 lbs PERFORMANCE MARGINS (LBS): FPR: 7057 FUEL BIAS: 1050 FINAL TDDP: 2049 RECON: 275 PAYLOADS: IECM/REM OSTA-1/PALLET MAPS SMIRR SIR-A FILE OCE DFI RMS 1 (S.N. 201) RMS CHECKOUT (UNLOADED OPS) 2 CRYO TANK SETS	KSC W/D: OPF 99, VAB 18, PAD 70 = 187 LAUNCH POSTPONEMENT: - 45-day postponement caused by FRCS N204 spill on tiles resulting in debonding of tiles. LAUNCH SCRUB: - Scrubbed 11/4/81 launch at T-31 seconds because APU's 1 & 3 lube oil outlet pressure high at 100 to 112 PSIA. Flushed APU's 1 and 3 gear boxes and changed clogged filters. Rescheduled launch for 11/12/81. 53 days total slip. LAUNCH DELAYS: - 2H40M delay MDM OF3 failure. Flew in replacement MDM which also failed. Replaced with OV-099 MDM. - 10-minute delay for KSC confidence review of systems status. - Total launch delay: 2H50M TAL WX: Rota go. FLIGHT DURATION CHANGE: - Shortened flight from 5D4H to 2D6H (priority flight after Fuel Cell 1 failed at 0/04:45 MET). FIRSTS: - First flight of RMS. SIGNIFICANT ANOMALIES: - Fuel Cell 1 failure at 0/04:45 MET resulting in priority mission. Shortened flight from planned 5D4H to 2D6H. - Icing in WSB 3 inhibited lube oil cooling, resulting in elevated APU gearbox outlet temp. - Excessive gas in drinking water. - TV camera B RMS elbow camera, PLB cameras A,B,C lenses had contamination. - CRT 1 failed due to HV power supply problem. - RH SRB lost one main chute. - RH SRM aft field joint gas leak to primary O-ring with erosion. - LH fwd windows degraded by salt spray. RADIATORS DEPLOYED #2 (port stowed last 1/2 of flight) NOTE: ON STS-1 AND STS-2, THE NOMINAL OGS AIM POINT WAS 6500 FEET (5500 FEET WAS THE CLOSE-IN AIM POINT).

Aerial view of STS-2 launch from KSC Pad 39A (S81-39840)

CDR Engle (left) & PLT Truly back at EAFB after scrub. (S81-39413)

S81-39499-- President Ronald Reagan is briefed by Dr. Christopher C. Kraft, Jr., JSC Director, pointing to MOCR screen. The President said, "Dr. Kraft, I was in the calvary, I don't understand all this." Then he talked to crew on orbit.

FLT NO.	ORBITER	CREW (2) — TITLE, NAMES & EVA'S	LAUNCH SITE, LIFTOFF TIME, LANDING SITES, ABORT TIMES	LANDING SITE/ RUNWAY, CROSSRANGE, LANDING TIMES, FLT DURATION, WINDS	SSME-TL NOM-ABORT EMERG, THROTTLE PROFILE, ENG. S.N.	SRB RSRM AND ET	INC	ORBIT HA/HP	FSW	PAYLOAD WEIGHTS, PAYLOADS/ EXPERIMENTS	MISSION HIGHLIGHTS (LAUNCH SCRUBS/DELAYS, TAL WEATHER, ASCENT I-LOADS, FIRSTS, SIGNIFICANT ANOMALIES, ETC.)
STS-3 SEQ FLT # 3 KSC 3 OFT-3 PAD 39A-3	OV-102 Flight 3 Columbia — OMS PODS LVO1-3 RVO1-3 FRC2-3	CDR: Jack R. Lousma P5/R5/M5 — PLT: C. Gordon Fullerton P6/R6/M6 — MCC FCR-1 (3) — FLIGHT DIRECTORS: ASC/PLG - T. W. Holloway LD/ORB - N. B. Hutchinson PLNG - J. T. Cox O/E - H. M. Draughon MOD - E. F. Kranz	KSC 39A 81:15:59:59.875Z 10:00:00 AM EST (P) 11:00:00 AM EST (A) Monday 1 3/22/82 (1) — WINDOW DURATION: 6.1 hours — PLS - EDW SLS - NOR TAL - ROTA (Selected) — MAX Q = 651 M = 1.04 — SRB SEP: 2:07.9 MET — MECO: 8:33 MET — ET SEP: 8:51:5 MET — OMS-1: 10:34.4 MET 85.2 Seconds — OMS-2: 40:50.4 MET 88 Seconds	WSMR 1 NORTHRUP STRIP 17 (LAKEBED) 9:04:45 AM MST Tuesday 2 3/30/82 (1) — XRANGE: 276 NM — ORB DIR: AR (1) — AIM PT: NOM — MLGTD: 1092 FT 89:16:04:44.8Z VEL: 233 KGS 220 KEAS HDOT: -5.7 FPS — TD NORM 195: 3342 FT — NLGTD: 6261 FT 89:16:04:59.7Z VEL: 176 KGS HDOT: -8.4 FPS — BRK INIT: 149 KGS — AVE BRK DECEL: 5 FPS/S — WHEELS STOP: 89:16:06.09Z 14824 FT — ROLLOUT: 13737 FT 84 SEC — WINDS: 14H, 2L KNOTS OFFICIAL: 13H, 1L — DENS ALT: 3700 FT	100/100 (107) 68% 1 = 2007 (3) 2 = 2006 (3) 3 = 2005 (3) — M 3 EOM WEIGHT: 207349 X CG: 1095.4 — LANDING WEIGHT: 207073 X CG: 1096.9 — --------- FLT DURATION: 8:00:04:45 192:04:45 — S/T: 12:12:38:50 — OV-102: 12:12:38:50 — DISTANCE: 3,900,000 sm	A11/12 MTR: STD CASE: STD 86-80E SWT ET-3 ET RPT 235K 49:18 MET ET BR/UP 210K 49:58 MET ET IMPACT LAT: 31.2 S LONG: 94.4 E	38.0 (2) 64.14 START: -33.2 END: -26.0 MAX: -36.0	STANDARD INSERTION INSERTION ALTITUDE: 130 NM 130/130 NM DEORBIT 130 X 120 NM VELOCITY 25659 FPS RANGE 4144 NM	R18/T11	CARGO: 22710 lbs — CHARGEABLE: — RETURNED: 24492.8 lbs — SHUTTLE ACCUMULATED WEIGHTS: DEPLOYED: 0 lbs NON-DEPLOYED: 52311 lbs CARGO TOTAL: 52311 lbs — PERFORMANCE MARGINS (LBS): FPR: 7444 FUEL BIAS: 1050 FINAL TDDP: 5343 RECON: 2278 — PAYLOADS: IECM/REM EEVT HBT-HEFLEX OSS-1 PDP/REM (PLASMA DIAGNOSTIC PACKAGE) DFI — RMS 2 (S.N. 201) — LOADED TESTS USING PDP — WAVE PDP OUTSIDE P/L BAY — 3 CRYO TANK SETS	KSC W/D: OPF 55, VAB 12, PAD 30=97 — LAUNCH POSTPONEMENTS: None. — LAUNCH SCRUBS: None. — LAUNCH DELAYS: - Launch delayed 1 hour. SSME GN2 purge heater temp sensor failed in GSE. — TAL WX: Rota go. — LANDING SITE CHANGE: - EDW lakebed to WSSH because EDW lakebed was wet. — FLIGHT DURATION CHANGE: - Flight extended from 7 to 8 days because of sand storm at WSSH. — FIRSTS: - First flight without white paint on ET. (800 lbs weight savings. STS-1 and STS-2 ET's were painted white.) — SIGNIFICANT ANOMALIES: - Early shutdown of APU 3 due to WSB3 freezeup causing high lube oil temp. - R ENG hydraulic lockup at 82% at To plus 8 min 12 sec due to early shutdown of APU. - RMS wrist TV camera failed causing IECM OPS to be canceled. - AFT bulkhead latch did not fully latch (top sun for 15 minutes and latches operated normally). - WMS (slinger stopped on day 5). - Missing tiles on FWD upper fuselage and upper body flap. - CCTV camera C failed, camera B zoom failed. - ARPCS GN2 usage excessive (cold soak induced anomaly). - S-Band xponder 1 failed in hi and low power modes (downlink). - S-Band xponder 2 failed in low power mode (downlink). (Contaminants in RF control relay.) - S-band Power Amp reduced power output. - VTR tape broke. - Ammonia boiler controllers A&B failed. - Cracked rotor RH outboard MLG brake. - WSMR dust storm caused significant maintenance and cleanup of orbiter (gypsum contamination). - One RH SRB main chute failure 3 seconds after deployment. — RADIATORS DEPLOYED #3

S82-28746 : First flight with ET white paint deleted for 800 lb weight savings.

NO MORE ET PAINT

CREW AT WORK ON ORBIT
ABOVE: s03-22-123 --- CDR Lousma
BELOW: s03-23-178 --- PLT Fulerton

MGR's AT WORK IN MCC-- Lt to Rt: Glynn Lunney /Mgr P/L Integ, Chris Kraft /JSC Ctr Director, a person unknown, & Aaron Cohen /Mgr Orbiter Project discuss a flight issue.

SPACE SHUTTLE MISSIONS SUMMARY

FLT NO.	ORBITER	CREW (2) — TITLE, NAMES & EVA'S	LAUNCH SITE, LIFTOFF TIME, LANDING SITES, ABORT TIMES	LANDING SITE/ RUNWAY, CROSSRANGE — LANDING TIMES FLT DURATION, WINDS	SSME-TL NOM-ABORT EMERG — THROTTLE PROFILE ENG. S.N.	SRB RSRM AND ET	ORBIT — INC	ORBIT — HA/HP	FSW	PAYLOAD WEIGHTS, PAYLOADS/ EXPERIMENTS	MISSION HIGHLIGHTS (LAUNCH SCRUBS/DELAYS, TAL WEATHER, ASCENT I-LOADS, FIRSTS, SIGNIFICANT ANOMALIES, ETC.)
STS-4 SEQ FLT # 4 KSC 4 OFT-4 PAD 39A-4	OV-102 Flight 4 Columbia — OMS PODS LVO1 - 4 RVO1 - 4 FRC2 - 4	CDR: Thomas K. Mattingly P7/R7/M7 — PLT: Henry W. Hartsfield P8/R8/M8 — MCC FCR-1 (4) — FLIGHT DIRECTORS: Asc - T. W. Holloway Ld/Orb - C. R. Lewis Plng - J. T. Cox Plng - J. H. Greene Orb/Ent - H. M. Draughon MOD - E. F. Kranz	KSC 39A 178:14:59:59.8Z 11:00:00 AM EDT (P) 11:00:00 AM EDT(A) Sunday 2 6/27/82 (1) — WINDOW DURATION: 4.4 hours — PLS - EDW SLS - KSC CLS - NOR AOA - EDW AOA WX - NOR TAL - DAKAR TAL WX - ROTA (Selected) — MAX Q = 721 M = 1.74 — SRB SEP: 2:10 MET — MECO: 8:32.7 MET — ET SEP: 8:50:4 MET — OMS-1: 10:32.6 MET 88 Seconds — OMS-2: 37:40.6 MET 104 Seconds	EDW 22, CONC (EDW 3, CONC 1) 9:09:40 AM PDT Sunday 1 7/4/82 (1) — XRANGE: 581 NM — ORB DIR: DL (1) — AIM PT: NOM — MLGTD: 948 FT 185:16:09:39.9Z VEL: 196 KGS 204 KEAS HDOT: -1.1 FPS — TD NORM 195: 1758 FT — NLGTD: 4988 FT 185:16:09:53Z VEL: 158 KGS HDOT: -3.7 FPS — BRK INIT: 133 KGS — AVE BRK DECEL: 6.4 FPS/S — WHEELS STOP: 185:16:10:44Z 10826 FT — ROLLOUT: 9878 FT 64 SEC — WIND: 15H, 7L KNOTS OFFICIAL: 12H, 1R — DENS ALT: 3563 FT — FLT DURATION: 7:01:09:40 169:09:40 — S/T: 19:13:48:30 — OV-102: 19:13:48:30 — DISTANCE: 2,900,000 sm	100/100 (107) — 100/65/ 100/65 — 1 = 2007 (4) 2 = 2006 (4) 3 = 2005 (4) — M 3 EOM — WEIGHT: 209141 — X CG: 1092.9 — LANDING — WEIGHT: 208947 — X CG: 1094.4	A13/14 — MTR: STD — CASE: STD 86-80E — SWT ET-4 — ET RPT 228K 47:19 MET — ET BR/UP 204K 47:56 MET — ET IMPACT LAT: 28.4 S LONG: 83.07 E	28.529 (1) — START: -1.2 — END: +20.5	STANDARD INSERTION — INSERTION ALTITUDE: — POST OMS-2 139.2 X 131.05 NM — DEORBIT 175 X 160 NM — VELOCITY 25800 FPS — RANGE 3810 NM	R18/T11	CARGO: 24492 lbs — PAYLOAD CHARGEABLE: 11644 lbs — PRIMARY P/L: 9800 lbs — ANCILLARY: 1844 lbs — RETURNED: 24492.8 lbs — SHUTTLE ACCUMULATED WEIGHTS: DEPLOYED: 0 lbs NON-DEPLOYED: 63955 lbs CARGO TOTAL: 76803 lbs — PERFORMANCE MARGINS (LBS): FPR: 6210 FUEL BIAS: 1474 FINAL TDDP: 4038 RECON: 1195 — PRIMARY: DOD 82-1 ICEM/REM — ANCILLARY: ACIP GAS (UTAH STATE) STUDENT EXP'S: (1) CHOLESTEROL (2) CHROMIUM LEVEL (Deficiency) MLR CFES (MID-DECK) TGE NOSL — 3 CRYO TANK SETS — RMS 3 (S.N. 201) — WAVED IECM OUTSIDE P/L BAY	KSC W/D: OPF 41, VAB 7, PAD 29=77 — LAUNCH POSTPONEMENTS: None. — LAUNCH SCRUBS: None. — LAUNCH DELAYS: None. — TAL WX: Dakar no go - crosswinds. — FLIGHT DURATION CHANGE: None. — FIRSTS: - First flight with student experiments. — SIGNIFICANT ANOMALIES: - Hail stones on tile at L-1 day (repaired tiles). - Water found in thrusters F2R & F4R. - During prelaunch rain storms, approximately 500 lbs water absorbed by tiles requiring bottom-to-sun for many hours to dry-out water (to prevent ice damage to tile). - GAS activation problems - successful workaround. - VTR would not rewind. - AFT bulkhead actuator on port PLBD stalled during latch closure. - AFT STBD, FWD port, and FWD bulkhead floodlights failed. - Thermal conditioning required to close PLBD's. - WMS slinger slowed down. - Mid-deck TV camera operation erratic. - DFI PCM recorder data lost. - Both SRB's lost (impacted water at extremely high velocity). - Right and left inboard brakes damaged. IFM - GAS EXPERIMENTS RECOVERY RADIATORS DEPLOYED #4

Columbia mission patch — MATTINGLY-HARTSFIELD

S82-31207 -- CDR Mattingly (right) & PLT Hartsfield ready to fly fourth & final Orbital Flight Test (OFT).

S82-33394: Columbia stopover at Ellington during return to KSC.

S04-28-1637: Mattingly floats in mid-deck with cameras

FLT NO.	ORBITER	CREW (4) TITLE, NAMES & EVA'S	LAUNCH SITE, LIFTOFF TIME, LANDING SITES, ABORT TIMES	LANDING SITE/ RUNWAY, CROSSRANGE LANDING TIMES FLT DURATION, WINDS	SSME-TL NOM-ABORT EMERG THROTTLE PROFILE ENG. S.N.	SRB RSRM AND ET	ORBIT INC	HA/HP	FSW	PAYLOAD WEIGHTS, PAYLOADS/ EXPERIMENTS	MISSION HIGHLIGHTS (LAUNCH SCRUBS/DELAYS, TAL WEATHER, ASCENT I-LOADS, FIRSTS, SIGNIFICANT ANOMALIES, ETC.)
STS-5 SEQ FLT # 5 KSC 5 PAD 39A-5	OV-102 Flight 5 Columbia OMS PODS LVO1 - 5 RVO1 - 5 FRC2 - 5	CDR: Vance D. Brand P9/R9/M9 PLT: Robert F. Overmyer P10/R10/M10 M/S: William B. Lenoir P11/R11/M11 M/S: Joseph P. Allen P12/R12/M12 FIRST SPACE SHUTTLE EVA SCHEDULED, BUT NOT ACCOMPLISHED BECAUSE OF EMU PROBLEMS. MCC FCR-2 (1) FLIGHT DIRECTORS: Ld/Asc/Ent - T. W. Holloway Orbit - J. T. Cox Planning - G. E. Coen MOD - E. F. Kranz	KSC 39A 315:12:18:59.997Z 7:19:00 AM EST (P) 7:19:00 AM EST (A) Thursday 2 11/11/82 (2) WINDOW DURATION: 39 Minutes (SBS Day 2 Deploy Opportunity) PLS - EDW SLS - NOR TAL - DAKAR (Selected) TAL WX - None AOA - NOR AOA WX - KSC CLS - KSC CLS WX - ROTA MAX Q = 738 M = 1.70 SRB SEP: 2:09.08 MET MECO: 8:30.68 MET ET SEP: 8:48.77 MET OMS-1: 10:30.8 MET 137.8 Seconds OMS-2: 44:40.8 MET 117.6 Seconds	EDW 22, CONC (EDW 4, CONC 2) 6:33:26 AM PST Tuesday 3 11/16/82 (2) XRANGE: 580 NM ORB DIR: DL (2) AIM PT: NOM MLGTD: 1637 FT 320:14:33:26Z VEL: 201 KGS 198 KEAS HDOT:-1.0 FPS TD NORM 195: 1907 FT NLGTD:4675 FT 320:14:33:34Z VEL:176 KGS HDOT:-4.6 FPS BRK INIT: 167 KGS AVE BRK DECEL: 6.7 FPS/S WHEELS STOP: 320:14:34:29Z 11190 FT ROLLOUT: 9553 FT 63 SEC WIND: 2 H, 0X KNOTS OFFICIAL: 2H, 0X DENS ALT: 1750 FT FLT DURATION: 5:02:14:26 122:14:26 S/T: 24:16:02:56 OV-102: 24:16:02:56 DISTANCE: 1,850,000 sm	100/100 (107) 100/85/65 1 = 2007 (5) 2 = 2006 (5) 3 = 2005 (5)	A15/16 MTR: STD CASE: STD 86-80 SWT ET-5 M 3 EOM WEIGHT: 202643 X CG: 1094.8 LANDING WEIGHT: 202480 X CG: 1096.3	28.482 (2) 89.8 START: -26.0 END: -7.2 ET RPT 236K 46:30 MET ET BR/UP 205K 47:18 MET ET IMPACT LAT: 28.3 S LONG: 82.4 E	STANDARD INSERTION INSERTION ALTITUDE: POST OMS-2 162.07 X 160.67 NM DEORBIT 154 X 148 NM VELOCITY 25758 FPS RANGE 4050 NM	R19/T12	CARGO: 32080 lbs PAYLOAD CHARGEABLE: 20830 lbs ANCILLARY P/L: 1078 lbs NON-DEPLOYED: 5167 lbs DEPLOYED: 14585 lbs RETURNED: 17495 lbs SHUTTLE ACCUMULATED WEIGHTS: DEPLOYED: 14585 lbs NON-DEPLOYED: 70200 lbs CARGO TOTAL: 108883 lbs PERFORMANCE MARGINS (LBS): FPR: 5312 FUEL BIAS:1479 FINAL TDDP:822 RECON:-1017 PRIMARY: SBS-C/PAM-D (DEPLOYED) TELESAT-E/PAM D (ANIK-D) (DEPLOYED) ANCILLARY: STUDENT EXPERIMENTS - POFERIA (SPONGE) GROWTH - SOLUTION XTAL GROWTH - CONVECTION IN ZERO-G GAS, TGE MATERIALS TEST ZERO-G DEMO 3 CRYO TK SETS NO RMS	KSC W/D: OPF 48, VAB 9, PAD 45= 102 LAUNCH POSTPONEMENTS: None. LAUNCH SCRUBS: None. LAUNCH DELAYS: None. TAL WX:DAKAR GO CLS WX: Rota go. FLIGHT DURATIONS CHANGE: None. FIRSTS: - First operational Shuttle flight. - First flight with more than 2 crewmen (4). - First flight to deploy PAM-D (SBS-C). - First OV-102 flight after Micro-Mod including disabling the two ejection seats. - First flight of OV-102 with ejection seats disabled. - First Space Shuttle IFM. IFM's: - Switched CRT-2 and CRT-4 cables on FD4 after CRT 2 failed. - Water hoses used for water dispenser failure. SIGNIFICANT ANOMALIES: - 46-hour STBD side-to-sun. - EVA canceled, EV-2 (Allen's) suit fan did not operate. EV-1 (Lenoir's) suit regulator was regulating to 3.8 psia instead of 4.3 psia. - WCCU A & B failed. - CRT-2 failed (pot in "y" deflection board). - Radar altimeter #1 failed. - FWD port & STBD PLB lights failed. - High O$_2$ flow during PCS switchover. - LHIB MLG brake locked during landing. - OMS nozzle cracks found postflight. RADIATORS DEPLOYED #5 (for SUN SIDE attitude only)

s05-07-267: First four-member crew, first operational flight, delivered by the "Ace Moving Co." Clockwise from bottom left: CDR Brand, Lenoir/MS, PLT Overmyer, & Allen/MS.

S82-28714 -- In the MOCR Lead Flight Director, Tom Holloway, surveys the room.

FLT NO.	ORBITER	CREW (4) — TITLE, NAMES & EVA'S	LAUNCH SITE, LIFTOFF TIME, LANDING SITES, ABORT TIMES	LANDING SITE/ RUNWAY, CROSSRANGE, LANDING TIMES, FLT DURATION, WINDS	SSME-TL NOM-ABORT EMERG, THROTTLE PROFILE ENG. S.N.	SRB RSRM AND ET	ORBIT — INC	ORBIT — HA/HP	FSW	PAYLOAD WEIGHTS, PAYLOADS/ EXPERIMENTS	MISSION HIGHLIGHTS (LAUNCH SCRUBS/DELAYS, TAL WEATHER, ASCENT I-LOADS, FIRSTS, SIGNIFICANT ANOMALIES, ETC.)
STS-6 SEQ FLT # 6 KSC 6 PAD 39A-6	OV-099 Flight 1 Challenger OMS PODS LPO1 - 1 RPO1 - 1 FRC9 - 1	CDR: Paul J. Weitz P13/R13/M13 PLT: Karol J. Bobko P14/R14/M14 M/S: F. Story Musgrave P15/R15/M15 M/S: Donald H. Peterson P16/R16/M16 EMU/TETHERED EVA: EVA: 4/7/83 EV1-Musgrave EV2-Peterson EVA1=3:54/4:42 Space Shuttle EVA #1 EVA HARDWARE CHECKOUT MCC FCR-2 (2) FLIGHT DIRECTORS: Ascent - J. H. Greene Orb/Ent - G. E. Coen Ld/Orb - H. M. Draughon Planning - B. R. Stone MOD - E. F. Kranz	KSC 39A 94:18:30:00.016Z 1:30:00 PM EST (P) 1:30:00 PM EST (A) Monday 2 4/4/83 (2) WINDOW DURATION: 17 Minutes (TAL Lighting) TAL - DAKAR NO TAL WX AOA - EDW AOA WX - NOR EOM - EDW MAX Q = 688 M = 1.47 SRB SEP: 2:09.4 MET MECO: 8:19.4 MET ET SEP: 8:37.55 MET OMS-1: 10:19.6 MET 139.6 Seconds OMS-2: 43:37.6 MET 119.1 Seconds	EDW 22 CONC (EDW 5, CONC 3) 10:53:42 AM PST Saturday 2 4/9/83 (2) XRANGE: 378 NM ORB DIR: AL (1) AIM PT: CLOSE IN MLGTD: 2026 FT 99:18:53:42Z VEL: 180 KGS 190 KEAS HDOT: -1.5 FPS TD NORM 195: 1576 FT NLGTD: 4970 FT 99:18:53:54Z VEL: 146 KGS HDOT: -3.9 FPS BRK INIT: 136 KGS AVE BRK DECEL: 7.3 FPS/S WHEELS STOP: 99:18:54:31Z 9270 FT ROLLOUT: 7180FT 49 SEC WIND: 21H, 5L KNOTS OFFICIAL: 12H, 3L DENS ALT: 3177 FT FLT DURATION: 5:00:23:42 120:23:42 S/T: 29:16:26:38 OV-099: 5:00:23:42 DISTANCE: 1,820,000 sm	104/104 (109) 100/104/81/ 104/65 1 = 2017 (1) 2 = 2015 (1) 3 = 2012 (1) CENTER WAS 2011	A17/18 MTR: STD CASE: LWC 86-80 231-81 LWT-1 ET-8	28.48 (3) 89.7 START: -21.6 END: -18.8 MAX: -21.9	STANDARD INSERTION INSERTION ALTITUDE: POST OMS-2 155.45 X 154.48 NM	R19/T12	CARGO: 46971 lbs CHARGEABLE: 46662 lbs DEPLOYED: 37546 lbs NON-DEPLOYED: 6853 lbs ANCILLARY P/L: 2263 lbs RETURNED: 9462 lbs SHUTTLE ACCUMULATED WEIGHTS: DEPLOYED: 52131 lbs NON-DEPLOYED: 79316 lbs CARGO TOTAL: 155854 lbs PERFORMANCE MARGINS (LBS): FPR: 5720 FUEL BIAS: 1298 FINAL TDDP: 4755 RECON: 2463 PRIMARY: TDRS-A/IUS-2 ANCILLARY: MLR CFES (MIDDECK) NOSL GAS (3) IN BAYS 3 & 4: - JAPANESE SNOWFLAKE 3 CRYO TANK SETS NO RMS	KSC W/D: OPF 123, VAB 6, PAD 115=244 LAUNCH POSTPONEMENT: - 1/20/83 launch postponed 74 days to 4/4/83 because of H2 leak in aft compartment from engine 2011 (SSME #1) during FRF 1. Post-FRF 2 found crack in MCC of 2011. 2015 and 2012 had cracked ASI fuel lines. Replaced ASI lines in all three engines. 74-day slip for engine analysis and fixes. LAUNCH SCRUBS: None. LAUNCH DELAYS: None. TAL WX: Dakar no go - haze. FLIGHT DURATION CHANGE: None. FIRSTS: - First flight of OV-099. - First flight with HUD. - First EVA on Shuttle Program. - First use of SRB LWC case. - First use of LWT ET. SIGNIFICANT ANOMALIES: - TDRS deploy at MET 10:00:01 (Rev 6). IUS problem resulted in TDRS being left in 22000 × 12000 NM orbit. TDRS was maneuvered into geosync orbit using 1 lb attitude thrusters. - IUS problem with TVC. - TPS damage AFRSI on OMS PODS, slumping tiles on nose cap and aero surfaces. - Humidity separator failed (6 wires shorted). - High flow on O2 and N2 systems. - WCCU A & B failed. - GPC 2 failed. - Teleprinter failed. - WMS slinger failed on day 5. - CRT-3 failed. - Gas path through putty on both SRM nozzle-to-case joints. IFM - Removed and stowed CCTV monitors.

Additional data (lower center columns):

M 3 EOM		ET RPT 237K 46:19 MET	DEORBIT 155 X 147 NM VELOCITY 25756 FPS RANGE 4056 NM
WEIGHT: 190627		ET BR/UP 223K 46:42 MET	
X CG: 1099.7			
LANDING		ET IMPACT LAT: 28.3 S	
WEIGHT: 190330		LONG: 83.0 E	
X CG: 1101.2			

Challenger STS-6 mission patch.

S06-10-417: First Shuttle EVA: Musgrave (left) Peterson (right) in cargo bay.

First crew to man Challenger. Seated are CDR Weitz (left) and PLT Bobko. Standing are Peterson/MS (left) and Musgrave/MS.

FLT NO.	ORBITER	CREW (5) TITLE, NAMES & EVA'S	LAUNCH SITE, LIFTOFF TIME, LANDING SITES, ABORT TIMES	LANDING SITE/ RUNWAY, CROSSRANGE LANDING TIMES FLT DURATION, WINDS	SSME-TL NOM-ABORT EMERG THROTTLE PROFILE ENG. S.N.	SRB RSRM AND ET	INC	ORBIT HA/HP	FSW	PAYLOAD WEIGHTS, PAYLOADS/ EXPERIMENTS	MISSION HIGHLIGHTS (LAUNCH SCRUBS/DELAYS, TAL WEATHER, ASCENT I-LOADS, FIRSTS, SIGNIFICANT ANOMALIES, ETC.)
STS-7 SEQ FLT # 7 KSC 7 PAD 39A-7	OV-099 Flight 2 Challenger OMS PODS LPO1 - 3 RPO1 - 3 FRC9 - 3	CDR: Robert L. Crippen (Flt 2 - STS-1) P17/R2/V1/M2 PLT: Frederick H. Hauck P18/R17/M17 M/S 1: John M. Fabian (Rt. Rear Seat) P19/R18/M18 M/S 2: Sally K. Ride (Center Seat) P20/R19/F1 M/S 3: Norman E. Thagard (Middeck Seat) P21/R20/M19 MCC FCR-2 (3) FLIGHT DIRECTORS: Ascent - J. H. Greene Ld/O1 - T. W. Holloway Orbit 2 - J. T. Cox Plng - L. S. Bourgeois Entry - G. E. Coen MOD - E. F. Kranz	KSC 39A 169:11:33:00.33Z 7:33:00 AM EDT (P) 7:33:00 AM EDT (A) Saturday 1 6/18/83 (2) PLS - KSC SLS - EDW TAL - DAKAR CLS - ROTA AOA - EDW AOA WX - KSC EOM - KSC MAX Q = 701 M = 1.56 SRB SEP: 2:06.2 MET MECO: 8:20.1 MET ET SEP: 8:38.2 MET OMS-1: 10:20.2 MET 139.5 Seconds OMS-2: 44:30.2 MET 120 Seconds	EDW 15, LAKEBED (EDW 6, LKBD 3) 6:56:59 AM PDT Friday 1 6/24/83 (1) XRANGE: 738 NM ORB DIR: DL (3) AIM PT: NOM MLGTD: 2726 FT 175:13:56:59Z VEL: 200 KGS 202 KEAS HDOT: -1.1 FPS TD NORM 195: 3356 FT NLGTD: 6843 FT 175:13:57:19Z VEL: 158 KGS HDOT: -5.1 FPS BRK INIT: 125 KGS AVE BRK DECEL: 3.6 FPS/S WHEELS STOP: 175:13:58:14Z 13176 FT ROLLOUT: 10450 FT 75 SEC WIND: 9H, 8R KNOTS OFFICIAL: 10H, 3R DENS ALT: 3000 FT FLT DURATION: 6:02:23:59 146:23:59 S/T: 35:18:50:37 OV-099: 11:02:47:41 DISTANCE: 2,220,000 sm	104/104 (109) 100/104/75/104 /65 1 = 2017 (2) 2 = 2015 (2) 3 = 2012 (2) M 3 EOM WEIGHT: 204340 X CG: 1089.8 LANDING WEIGHT: 204043 X CG: 1091.2	A51/52 MTR: STD CASE: LWC SWT ET-6 ETRPT 233K 46:20 MET ET BR/UP 188K 47:18 MET T/V OFF ET IMPACT LAT: 28.35 S LONG: 83.7 E	28.484 (4) START: +17.5 END: +41.0 MAX:	STANDARD INSERTION INSERTION ALTITUDE: POST OMS-2 161 X 159.96 NM TELESAT DEPLOY 162.21 NM PALAPA DEPLOY 162.61 NM DEORBIT 159 X 154 NM VELOCITY 25771 FPS RANGE 4042 NM	R19/T12	CARGO: 37124 lbs CHARGEABLE: 31893 lbs ANC LLARY P/L: 3942 lbs DEPLOYED: 14949 lbs NON-DEPLOYED: 13002 lbs RETURNED: 22175 lbs SHUTTLE ACCUMULATED WEIGHTS: DEPLOYED: 67080 lbs NON-DEPLOYED: 96260 lbs CARGO TOTAL: 192978 lbs PERFORMANCE MARG NS (LBS): FPR: 5539 FUEL BIAS: 1603 FINAL TDDP: 2940 RECON: 2021 PR MARY: TELESAT-F/ PAM-D (ANIK-C) DEPLOYED PALAPA-B1/PAM-D DEPLOYED SPAS-01 DEPLOYED AND RETRIEVED CFES, MLR OSTA-2: (MPE,MEA,MAUS) GAS-G002,G305, G009,G033,G088,G012 AND G345 ANC LLARY: MLR CFES (MID-DECK) GAS (7) BAYS 2-5 STUDENT EXP. 3 CRYO TK SETS RMS 4 (S.N. 201) Deployed and retrieved SPAS-01	KSC W/D: OPF 34, VAB 5, PAD 21=60 LAUNCH POSTPONEMENTS: None. LAUNCH SCRUBS: None. LAUNCH DELAYS: None. TAL WX: Dakar go. LANDING SITE CHANGE: - KSC to EDW (Poor visibility at KSC). FLIGHT DURATION CHANGE: - Extended 1 day from 5 to 6 days plus 2 revs to land at EDW. FIRSTS: - First flight with 5 crewmembers. - First US flight with female astronaut. - First payload deployed and retrieved same flight (SPAS-01). - First PROX OPS and reberthing of payload (SPAS-01). - First flight with Ku-band antenna (Ku-band not used). - First planned landing at KSC. - First PROX OPS (with SPAS-01). EVENTS: - TELESAT-F deployed on rev 4. - PALAPA-B1 deployed on rev 15. SIGNIFICANT ANOMALIES: - Reduced cabin pressure demonstration (10.2 PSIA). - Bus-tie demonstration post-landing fired one set of PYROS for MLG uplock release. - WCCU A, B and C failed. - WCCU C and E wall units failed. - Right braking system damaged. - APU 3 underspeed shutdown on-orbit. - Locker and cabin door misalignment problems. - Right inboard MLG brake damage. - Challenger window replaced after orbital debris impact.

S07-30-1574: 1st 5 member crew: In rear (lt) to (rt): CDR Crippen (1st Shuttle veteran re-flight), PLT Hauck, & Fabian/MS. Front: Ride/MS (1st U.S.Female astronaut) & Thagard/MS.

In the MCC: At Top: S83-36179 Gene Kranz, MOD Director & Cliff Charesworth/MOD in back. Left Bottom: S83-34267 Ron Epps/FIDO. Rt Bottom: S83-34270 Ed Fendell/INCO (rt) & astronaut Gordon Cooper.

SPACE SHUTTLE MISSIONS SUMMARY

FLT NO.	ORBITER	CREW (5) TITLE, NAMES & EVA'S	LAUNCH SITE, LIFTOFF TIME, LANDING SITES, ABORT TIMES	LANDING SITE/ RUNWAY, CROSSRANGE LANDING TIMES FLT DURATION, WINDS	SSME-TL NOM-ABORT EMERG THROTTLE PROFILE ENG. S.N.	SRB RSRM AND ET	INC	ORBIT HA/HP	FSW	PAYLOAD WEIGHTS, PAYLOADS/ EXPERIMENTS	MISSION HIGHLIGHTS (LAUNCH SCRUBS/DELAYS, TAL WEATHER, ASCENT I-LOADS, FIRSTS, SIGNIFICANT ANOMALIES, ETC.)
STS-8 SEQ FLT # 8 KSC 8 PAD 39A-8	OV-099 Flight 3 Challenger OMS PODS LPO1 - 2 RPO1 - 2 FRC9 - 2	CDR: Richard H. Truly (FLT 2 - STS-2) P22/R4/V2/M4 PLT: Daniel C. Brandenstein P23/R21/M20 M/S 1: Guion S. Bluford, Jr. (Center Seat) P24/R22/M21 M/S 2: Dale A. Gardner (Rt Rear Seat) P25/R23/M22 M/S 3: William E. Thornton (Middeck) P26/R24/M23 MCC FCR-2 (4) FLIGHT DIRECTORS: Asc/Plng - J. H. Greene Orbit 1 - B. R. Stone Ld/O2 - H. M. Draughon Entry - G. E. Coen MOD - E. F. Kranz	KSC 39A 242:06:32:00.009Z 2:15:00 AM EDT (P) 2:32:00 AM EDT (A) Monday 1 Tuesday 1 8/30/83 (1) LAUNCH WINDOW: 41 Minutes (INSAT Dply Rev 18) PLS - EDW SLS - KSC TAL - DAKAR NO TAL WX AOA - EDW AOA WX - NOR EOM - EDW MAX Q = 701 M = 1.53 SRB SEP: 2:04.34 MET MECO: 8:41.62 MET ET SEP: 8:59.66 MET OMS-1: 10:41.7 MET 138.8 Seconds OMS-2: 44:51.7 MET 116.5 Seconds	EDW 22, CONC (EDW 7, CONC 4) 248:07:40:43Z 12:40:43 AM PDT Monday 1 9/5/83 (1) XRANGE: 519 NM ORB DIR: DL (4) AIM PT: NOM MLGTD: 2793 FT 248:07:40:43Z VEL: 196 KGS 195 KEAS HDOT: -1.2 FPS TD NROM 195: 2793 FT NLGTD: 5515 FT 248:07:40:50Z VEL: 177 KGS HDOT: -4.3 FPS BRK INIT: 154 KGS AVE BRK DECEL: 6.9 FPS/S WHEELS STOP: 248:07:41:33Z 12164 FT ROLLOUT: 9371 FT 50 SEC WIND: 7H, 0X KNOTS OFFICIAL: 5H, 2L DENS ALT: 3600 FT FLT DURATION: 6:01:08:43 145:08:43 S/T: 41:19:59:20 OV-099: 17:03:56:24 DISTANCE: 2,220,000 sm	100/104 (104) 100/69/ 100/65 1 = 2017 (3) 2 = 2015 (3) 3 = 2012 (3)	A53/54 MTR: HPM CASE: STD LWT-2 ET-9	28.488 (5) START: -36.2 END: +29.4 MAX: +37.0	STANDARD INSERTION INSERTION ALTITUDE: POST OMS-2 161.07 X 160.14 NM INSAT DEPLOY 159.18 NM	R19/T12	CARGO: 30076 lbs PAYLOAD CHARGEABLE: 25790 lbs DEPLOYED: 7445 lbs NON-DEPLOYED: 13179 lbs ANCILLARY: 5166 lbs RETURNED: 22631 lbs SHUTTLE ACCUMULATED WEIGHTS: DEPLOYED: 74525 lbs NON-DEPLOYED: 114605 lbs CARGO TOTAL: 223054 lbs PERFORMANCE MARGINS (LBS): FPR: 6756 FUEL BIAS: 780 FINAL TDDP: 14863 RECON: 15735 PRIMARY: INSAT-1B/PAM-D (DEPLOYED) RMS/PDRS/PFTA DFI PALLET (HEAT PIPE EXPERI-MENT, 2 BOXES OF POSTAL COVERS), RME EXP, EOM ANCILLARY: CFES (MIDDECK) GAS (3) BAYS 2-8 GAS (4) BAY 5 BIO-FEEDBACK ANIMAL ENCLOSURE POSTAL COVERS 3 CRYO TK SETS RMS 5 (S.N. 201) USED FOR PFTA OPS	KSC W/D: OPF 26, VAB 4, PAD 25 = 55 LAUNCH POSTPONEMENTS: - 8/4/83 launch postponed 26 days to 8/30/83 due to removal of TDRS-B from flight (IUS not ready because of problem on STS-6) and time required to checkout TDRS-A on orbit. 26-day slip. LAUNCH SCRUBS: None. LAUNCH DELAYS: - 00H17M delay because of thunderstorms in launch area. TAL WX: Dakar go. FLIGHT DURATION CHANGE: None. FIRSTS: - First Shuttle night launch. - First Shuttle night landing. - First flight to use TDRS for communications (test mode). - First flight to use Ku-band communications. - First flight using SRM HPM. - Bluford became the first African-American to fly in space. He was selected in the first class of Space Shuttle astronauts. EVENTS: - Tile survey of Orbiter bottom made using RMS End Effector TV camera. - INSAT-1B deployed on rev 27. SIGNIFICANT ANOMALIES: - Completed all 54 DTO's and DSO's planned for flight. - Hydraulic circulation pump 2 failed - GPC-1 failed to sync (recovered OK) - WCCW A wall unit failed, B&E noisy. - CCTV C command problems & out of focus. - CCTV D failed. - TAGS failed. - Rt outboard brake had 3 cracked washers and right inboard had one cracked washer. - Nose gear thruster piston found on runway. - LH and RH SRB nozzles experienced off-nominal erosion. - SRB nozzle erosion was found after recovery. - RH mid window (W5) pitted. RADIATORS DEPLOYED #6 (for 2 days)

Additional data:

M 3 EOM	ET RPT
WEIGHT: 204141	241K 46:30 MET
X CG: 1090.4	ET BR/UP 223K 47:01 MET
LANDING WEIGHT: 203945	ET IMPACT LAT: 28.4 S LONG: 81.5 E
X CG: 1091.9	

DEORBIT
118 X
116 NM
VELOCITY
25649 FPS
RANGE
4044 NM

S83-31724 --- Crew: Front row (lt to rt) PLT Brandenstein, CDR Truly, & Bluford/MS (1st African American to fly in space). Back row (lt to rt): Gardner/MS & Thorton/MS.

At Right: S83-36307 -- INSAT P/L in prep at KSC

S83-27154- JSC Center Director Gerry Griffin visits the MOCR. Gene Kranz, Director, Mission Ops is in rear & Flight Director Jay Greene is at right. Others not identified.

FLT NO.	ORBITER	CREW (6) TITLE, NAMES & EVA'S	LAUNCH SITE, LIFTOFF TIME, LANDING SITES, ABORT TIMES	LANDING SITE/ RUNWAY, CROSSRANGE LANDING TIMES FLT DURATION, WINDS	SSME-TL NOM-ABORT EMERG THROTTLE PROFILE ENG. S.N.	SRB RSRM AND ET	ORBIT INC	ORBIT HA/HP	FSW	PAYLOAD WEIGHTS, PAYLOADS/ EXPERIMENTS	MISSION HIGHLIGHTS (LAUNCH SCRUBS/DELAYS, TAL WEATHER, ASCENT I-LOADS, FIRSTS, SIGNIFICANT ANOMALIES, ETC.)
STS-9 (STS 41-A) S/L.1 SEQ FLT # 9 KSC 9 PAD 39A-9	OV-102 Flight 6 Columbia Spacelab 1 LM (1) OMS PODS LVO1 - 6 RVO1 - 6 FRC2 - 6	CDR: John W. Young (FLT 2 - STS-1) P27/R1/V3/M1 PLT: Brewster H. Shaw, Jr. P28/R25/M24 M/S 1: Owen K. Garriott P29/R26/M25 M/S 2: Robert A. R. Parker P30/R27/M26 P/S 1: Byron K. Lichtenberg P31/R28/M27 P/S 2: Ulf Merbold (Germany) P32/R29/M28 MCC FCR-2 (5) FLIGHT DIRECTORS: Ascent - J. H. Greene Ld/Orb 1 - C. R. Lewis Orb 2 - J. T. Cox Orb 3 - L. S. Bourgeois Team4/Ent - G. E. Coen	KSC 39A 332:15:59:59.99Z 11:00:00 AM EST (P) 11:00:00 AM EST (A) Monday 3 11/28/83 (3) LAUNCH WINDOW: 14 Minutes (TAL Lighting) TAL - ZARAGOZA PLS - EDW SLS - NOR TAL - ZARAGOZA IN PLANE TAL - COLOGNE/BONN AOA - NOR AOA WX - NONE MAX Q = 676 M = 1.52 SRB SEP: 2:06.24 MET MECO: 8:29.18 MET ET SEP: 8:47.32 MET OMS-1: 10:29.3 MET 68.5 Seconds OMS-2: 40:37.4 MET 101.6 Seconds WINDS: 0 H/T, O X KNOTS OFFICIAL: 1T, OX DENS ALT: 1900 FT FLT DURATION: 10:07:47:24 247:47:24 S/T: 52:03:46:44 OV-102: 34:23:50:20 DISTANCE: 3,330,000 sm	EDW 17, LAKEBED (EDW 8, LKBD 4) 15:47:24 PM PST Thursday 1 12/8/83 (1) XRANGE: 69 NM ORB DIR: DL (5) AIM PT: NOM MLGTD: 1649 FT 342:23:47:24Z VEL: 200 KGS 185 KEAS HDOT: -1.7 FPS TD NORM 195: 749 FT NLGTD: 5897 FT 342:23:47:37Z VEL: 146 KGS HDOT: -9.9 FPS BRK INIT: 126 KGS AVE BRK DECEL: 6.8 FPS/S WHEELS STOP: 342:23:48:17Z 10105 FT ROLLOUT: 8556 FT (10105 FROM THRESHOLD) 53 SEC M 3 EOM WEIGHT: 220288 X CG: 1085.8 LANDING WEIGHT: 220027 X CG: 1087.1	104/104 (107) 100/104/ 78/104/65 1 = 2011 (1) 2 = 2018 (1) 3 = 2019 (1)	A55/60 (1) MTR: HPM CASE: STD LWT-4 ET-11 ET BR/UP 199K 1:01:00 MET ET IMPACT LAT: 59.96 S LONG: 149.9 E	57.028 (1) START: -58.0 END: -79.0 MAX: -79.9	STANDARD INSERTION INSERTION ALTITUDE: POST OMS-2 136.75 X 132.79 NM DEORBIT 129 X 124 NM VELOCITY 25696 FPS RANGE 4349 NM	OI-2 (1)	CARGO: 33264 lbs PAYLOAD CHARGEABLE: 33131 lbs PAYLOAD WEIGHT: 33,131 lbs (includes 870 lbs cryo tank) DEPLOYED: 0 lbs NON-DEPLOYED: 32261 lbs MIDDECK: 0 lbs RETURNED: 32394 lbs SHUTTLE ACCUMULATED WEIGHTS: DEPLOYED: 74525 lbs NON-DEPLOYED: 147736 lbs CARGO TOTAL: 256318 lbs PERFORMANCE MARGINS (LBS): FPR: 5404 FUEL BIAS: 1084 FINAL TDDP: 841 RECON: -411 SPACELAB-1/LM SPACELAB 1 WITH 73 EXP: - ASTRONOMY - SOLAR PHYSICS - SPACE PLASMA - ATMOSPHERIC PHYSICS - EARTH OBSERVATIONS - LIFE SCIENCES - MATERIAL SCIENCES 5 CRYO TANKS NO RMS	KSC W/D: OPF 82 (2), VAB 12 (3), PAD 34 (2) = 128 days LAUNCH POSTPONEMENTS: -10/30/83 Launch postponed 29 days to 11/28/83. Rolled back from pad and changed SRB nozzles subsequent to STS-8 excessive nozzle erosion. 29-day slip. LAUNCH SCRUBS: None. LAUNCH DELAYS: None. TAL WX: Zaragoza no go - winds, Koln-Bonn no go - clouds. FLIGHT DURATION CHANGE: - Flight extended 1 day for additional science. - Landing delay 5 revs after GPC 1 and GPC 2 hard failures - Total extension - 1 day + 5 revs. FIRSTS: - First flight with 6 crewmen. - First flight of Spacelab after Spacelab only modifications to OV-102. - First flight with non-astronauts (P/S) and first non-Americans. - First use of two shifts of 12 hours (red and blue shifts). - First flight with galley and sleep station. - First flight with 3 substack fuel cells. SIGNIFICANT ANOMALIES: - GPC SV time tag to S/L incremented by 1 day. - Ku-band TWT failed to come on (low temp problem). - Spacelab RAU 21/cooling problem. - Excessive GH₂ in water. - S-band power amp no. 2 failed. - Noises and oscillations reported by crew. - GPC 1 hard failure GPC 2 failure, re-IPL'ed, memory altered, failed again at NLG contact (delayed landing 7-3/4 hours). - IMU 1 failed (power supply failure). - APU 1 and 2 hydrazine leak/fire shutdown after landing (APU 1 and 2 damaged). - Right outboard brakes damaged. - LH OMS pod TPS damage during entry. - Mission extended one day. 8 hours extension to analyze GPC and IMU failures. - LH OMS pod removed for repair after burn-through (missing tile). RADIATORS DEPLOYED #7 (stowed for 34 hours)

S09-126-044: First 6 member crew, first non-astronauts (P/S) and first non-Americans, and 3rd Shuttle veteran (CDR Young) re-flight. Crew identified in Col 3.

s9-32-1112-- First flight of Spacelab after Spacelab only modifications to OV-

FLT NO.	ORBITER	CREW (5) TITLE, NAMES & EVA'S	LAUNCH SITE, LIFTOFF TIME, LANDING SITES, ABORT TIMES	LANDING SITE/ RUNWAY, CROSSRANGE LANDING TIMES FLT DURATION, WINDS	SSME-TL NOM-ABORT EMERG THROTTLE PROFILE ENG. S.N.	SRB RSRM AND ET	ORBIT INC	HA/HP	FSW	PAYLOAD WEIGHTS, PAYLOADS/ EXPERIMENTS	MISSION HIGHLIGHTS (LAUNCH SCRUBS/DELAYS, TAL WEATHER, ASCENT I-LOADS, FIRSTS, SIGNIFICANT ANOMALIES, ETC.)
STS-11 (STS 41-B) SEQ FLT # 10 KSC 10 PAD 39A-10	OV-099 Flight 4 Challenger OMS PODS LPO1 - 4 RPO1 - 4 FRC9 - 4	CDR: Vance D. Brand (FLT 2 - STS -5) P33/R9/V4/M9 PLT: Robert L. Gibson P34/R30/M29 M/S 1: Bruce McCandless II P35/R31/M30 M/S 2: Ronald E. McNair P36/R32/M31 M/S 3: Robert L. Stewart P37/R33/M32 UNTETHERED EVA'S MMU: EV1=McCandless EV2=Stewart EVA1=5:35/6:05 2/7/84 SS EVA #2 EVA2=6:02/6:17 2/9/84 SS EVA #3 FIRST UNTETHERED EVA's: FREE FLYER EVA's #1 & # 2 MMU CHECKOUT EVA'S	KSC 39A 34:12:59:59:998Z 8:00:00 AM EST (P) 8:00:00 AM EST (A) Friday 1 2/3/84 (1) LAUNCH WINDOW: 13 Minutes (PALAPA SUN SHIELD FAIL OPEN) PLS - KSC SLS - EDW TAL - DAKAR NO TAL WX CLS - KSC CLS - EDW AOA - EDW AOA WX - NOR EOM - KSC MAX Q = 676 M = 1.55 SRB SEP: 2:07.92 MET MECO: 8:41.42 MET ET SEP: 8:59.57 MET OMS-1: 10:41.6 MET 150 Seconds OMS-2: 45:24.6 MET 124.8 Seconds	KSC 15 (KSC 1) 7:15:55 AM EST Saturday 3 2/11/84 (1) XRANGE: 524 NM ORB DIR: DL (6) AIM PT: CLOSE IN MLGTD: 1930 FT 42:12:15:55Z VEL: 198 KGS 196 KEAS HDOT: -2.0 FPS TD NORM 195: 2020 FT NLGTD: 5789 FT 42:12:16:06Z VEL: 159 KGS HDOT: -2.8 FPS BRK INIT: 136 KGS AVE BRK DECEL: 5.1 FPS/S WHEELS STOP: 42:12:17:02 12737 FT ROLLOUT: 10,815 FEET 64 SEC WINDS: 5H, 3L KNOTS OFFICIAL: 3T, 2L DENS ALT: -200 FT FLT DURATION: 7:23:15:55 191:15:55 S/T: 60:03:02:39 OV-099: 25:03:12:19 DISTANCE: 2,870,000 sm	100/104 109 100/73/ 100/65 1 = 2109 (1) 2 = 2015 (4) 3 = 2012 (4) M 3 EOM WEIGHT: 201529 X CG: 1087.9 LANDING WEIGHT: 201239 X CG: 1089.3	A57/58 MTR: HPM CASE: MWC LWT-3 ET-10 ET RPT 231K 46:26 MET ET BR/UP 214K 46:51 MET ET IMPACT LAT: 28.3 S LONG: 80.6 E	28.486 (6) START: -26.9 END: +4.5 MAX:	STANDARD INSERTION INSERTION ALTITUDE: POST OMS-2 165.88 X 164.61 NM PALAPA DEPLOY 166.48 NM WESTAR DEPLOY 153.52 NM DEORBIT 157 X 145 NM VELOCITY 25752 FPS RANGE 4137 NM	OI-2 (2)	CARGO: 33868 lbs CHARGEABLE: 28252 LBS DEPLOYED: 15073 LBS NON-DEPLOYED 10198 lbs ANC LLARY: 2981 lbs RETURNED: 18795 LBS SHUTTLE ACCUMULATED WEIGHTS: DEPLOYED: 89598 lbs NON-DEPLOYED 160915 lbs CARGO TOTAL: 290186 lbs PERFORMANCE MARGINS (LBS): FPR: 5259 FUEL BIAS: 1038 FINAL TDDP: 12062 RECON: 6961 PR MARY: WESTAR-IV/ PAM-D (DEPLOYED) PALAPA-B2 / PAM-D (DEPLOYED) SPAS 01A MFR PLATFORM MMU (2) MMU/EMU C NEMA 360 (BAY 5) C NEMA 360 (MID-DECK) ACES EXP. EF EXP. RME EXP. ANC LLARY: RT (DEPLOYED) GAS (5) STUDENT EXP (A.E.M.) SESA+ BEAM (BAY 2) MLR EXP 4 CRYO TK SETS RMS 6 (S.N. 201) CANCELED SPAS DEPLOY (RMS PROBLEM)	KSC W/D: OPF 52, VAB 6, PAD 21=80 LAUNCH POSTPONEMENTS: - 1/24/84 launch was postponed 10 days to 2/3/84 because of ongoing analysis of APU failures on STS-9. 10-day slip. LAUNCH SCRUBS: None. LAUNCH DELAYS: None. TAL WX: Dakar no go - visibility. FLIGHT DURATION CHANGE: None. FIRSTS: - First use of Manned Maneuvering Unit (MMU) on EVA. - First untethered EVA crewman on Shuttle flight (320 foot separation from Orbiter). - First use of 10.2 PSIA cabin for EVA prep. - First use of MFR on RMS. - First landing at KSC. - First flight with spare GPC in locker (STS-9 GPC failures reaction). EVENTS: - Made Orbiter maneuver to recover foot restraint in PLB. - PALAPA-B deployed on rev 6. - WESTAR-IV deployed on rev 48. - Saw Challenger entry trail from Houston during landing at KSC. RENDEZVOUS: - Canceled planned RNDZ when IRT failed. SIGNIFICANT ANOMALIES: - RMS wrist joint failure (RMS/SPAS-01 operations canceled). RMS used for PALAPA PKM burn witness plate ops. - Left OMS POD damage from waste water dump nozzle ice (during entry). - IRT failed to inflate properly after deployment (rendezvous canceled). - Both SRB's lost one chute. - WESTAR-IV and PALAPA-B failed to achieve desired orbit due to PAM-D nozzle failure. (Both satellites were retrieved on STS 51-A). - LH SRM forward center field joint gas leak to primary O-ring with erosion. - RH SRM gas leak and erosion to primary O-ring of nozzle-to-case joint. - LH SRB main chute failed to inflate.

CHALLENGER

MCC FCR-2 (6)

FLIGHT DIRECTORS:
Asc/Ent - G. E. Coen
Orbit 1 - B. R. Stone
Ld/O2 - H. M. Draughon
Plng - L. S. Bourgeois
EVA - J. T. Cox

In back: MS's/Stewart, McNair & McCandless

First Landing at KSC

Feb. 4, 1984: McCandless performed the first untethered excursions wearing the Manned Maneuvering Unit, a rocket propelled backpack. He flew 320 ft from Obiter, further than any previous astronaut.

SPACE SHUTTLE MISSIONS SUMMARY

FLT NO.	ORBITER	CREW (5) TITLE, NAMES & EVA'S	LAUNCH SITE, LIFTOFF TIME, LANDING SITES, ABORT TIMES	LANDING SITE/ RUNWAY, CROSSRANGE LANDING TIMES, FLT DURATION, WINDS	SSME-TL NOM-ABORT EMERG THROTTLE PROFILE ENG. S.N.	SRB RSRM AND ET	ORBIT INC	HA/HP	FSW	PAYLOAD WEIGHTS, PAYLOADS/ EXPERIMENTS	MISSION HIGHLIGHTS (LAUNCH SCRUBS/DELAYS, TAL WEATHER, ASCENT I-LOADS, FIRSTS, SIGNIFICANT ANOMALIES, ETC.)
STS 41-C (STS-13) SEQ FLT # 11 KSC 11 PAD 39A-11	OV-099 Flight 5 Challenger OMS PODS LPO3 - 1 RPO1 - 5 FRC9- 5	CDR: Robert L. Crippen (Flt 3) (STS-1 & STS-7) P38/R2/V1/M2 PLT: Francis R. Scobee P39/R34/M33 M/S: Terry J. Hart P40/R35/M34 M/S: James D. van Hoften P41/R36/M35 M/S: George D. Nelson P42/R37/M36 UNTETHERED EVA'S (MMU): EV1=Nelson EV2=van Hoften EVA1=2:59/3:05 4/8/84 - SS EVA #4 SMM TPAD DOCK ATTEMPT EVA2=7:07/6:30 4/11/84 - SS EVA #5 SMM REPAIR AND RELEASE FREE FLYER EVA'S #3 AND #4	KSC 39A 097:13:57:59.999Z 8:58:00 AM EST (P) 8:58:00 AM EST (A) Friday 2 4/6/84 (3) LAUNCH WINDOW: ~3.5 MINUTES (PLANAR WINDOW/ET FOOTPRINT NEAR HAWAII) PLS - KSC SLS - EDW TAL - DAKAR TAL WX - ROTA AOA - EDW AOA WX - NOR MAX Q = 635 M = 1.03 SRB SEP: 2:05.57 MET MECO: 8:30.76 MET ET SEP: 8:48.9 MET OMS-1: NONE OMS-2: 42:54 MET 95.1 Seconds	EDW 17, LAKEBED (EDW 9, LKBD 5) 5:38:07 AM PST Friday 2 4/13/84 (3) XRANGE: 381 NM ORB DIR: DL 7 AIM PT: NOM MLGTD: 1912 FT 104:13:38:07Z VEL: 220 KGS 213 KEAS HDOT: -1.5 FPS TD NORM 195: 3505 FT NLGTD: 7167 FT 104:13:38:23Z VEL: 144 KGS HDOT: -4.6 FPS BRK INIT: 110 KGS AVE BRK DECEL: 8.4 FPS/S WHEELS STOP: 104:13:38:55Z 10628 FT ROLLOUT: 8716 FT 48 SEC WINDS: 2 H, O X KNOTS OFFICIAL: 0H, 0X DENS ALT: 1000 FT FLT DURATION: 6:23:40:07 167:40:07 S/T: 67:02:42:46 OV-099: 32:02:52:26 DISTANCE: 2,880,000 sm	104/104 (109) 100/104/ 67/104/ 65 1 = 2109 (2) 2 = 2020 (1) 3 = 2012 (5) M 3 EOM WEIGHT: 197170 LANDING WEIGHT: 196976 X CG: 1101.6	BI-012 MTR: HPM CASE: MWC ET-12 LWT-5 ET RPT 246K 1:22:15 MET ET BR/UP 228K 1:22:45 MET ET IMPACT LAT: 18.90 S LONG: 149.9 W	28.45 (7) START: -18.1 END: +12.0 MAX:	DIRECT INSERTION 252 NM DIRECT INSERTION 251.6 X 115.4 NM DEORBIT 268 X 265 NM VELOCITY 25998 FPS RANGE 4090 NM	OI-2 (3)	CARGO: 38266 lbs CHARGEABLE: 33831 lbs DEPLOYED: 21396 lbs NON-DEPLOYED: 12394 lbs MIDDECK: 41 lbs RETURNED: 16870 lbs SHUTTLE ACCUMULATED WEIGHTS: DEPLOYED: 110994 lbs NON-DEPLOYED: 173350 lbs CARGO TOTAL: 328452 lbs PERFORMANCE MARGINS (LBS): FPR: 5052 FUEL BIAS: 1038 FINAL TDDP: 995 RECON: -3322 PRIMARY: LONG DURATION EXPOSURE FACILITY (LDEF) (DEPLOYED) SMRM/FSS (RETRIEVED, REPAIRED & RELEASED) MMU (2) MMU/EMU MFR PLATFORM BAY 10 CINEMA 360 I-MAX CAMERA RME EXPERIMENT ANCILLARY: STUDENT EXPERIMENTS ACIP 4 CRYO TANK SETS RMS 7 (S.N. 302) Used for LDEF deploy, SMRM capture, berth, and deploy and water nozzle and OMS pod survey	KSC W/D: OPF 31, VAB 4, PAD 18 = 53 LAUNCH POSTPONEMENT: - 4/4/84 launch postponed 2 days to 4/6/84 to upgrade OMS pod TPS (STS 41-B problem during entry). 2-day slip. LAUNCH SCRUBS: None. LAUNCH DELAYS: None. TAL WX: Dakar no go - low clouds. FLIGHT DURATION & LANDING SITE CHANGES: - Extended flight 1 day to replan use of RMS to grapple SMM after TPAD docking failure. - Extended flight 1 rev to land at EDW because of unacceptable weather (overcast) at KSC. - Total extension: 1 day+ 1 rev. FIRSTS: - First flight to use direct insertion. - First rendezvous/satellite repair flight. - First use of TPAD. Nelson used MMU to translate to SMM and attempted to dock using TPAD. TPAD failed to fire because a thermal insulation button prevented it from firing. - First grapple of satellite using RMS. - First direct insertion (no OMS-1 burn). RENDEZVOUS 1 & 2: - To capture, repair, and release SMM. EVENTS: - Nelson held onto solar panel during MMU ops to attempt to slow SMM rotation. - Re-rendezvous with SMM on 5th day & RMS grapple of SMM. Repair and redeploy of SMM on 6th day by van Hoften & Nelson. - RMS used to survery OMS pods and monitor water dumps to ensure no ice chunks on nozzles. ET TRACKING DTO 331/318 NEAR HAWAII - ET Reentry (tumble)-KPTC RADAR poor coverage, MOTIF unusable, CAST GLANCE - LH2 rupture 264-254 Kft debris large DV, "violent rupture." SIGNIFICANT ANOMALIES: - RH SRB main parachute failure. - WCS fan SEP 1 low airflow. - WCS fan SEP 2 failed. - Brake damage similar to STS- 7 on left & right sides. - Ku-band Rndz Radar failed self test & lost lock. - RH SRB one chute failed to inflate. - RH SRM gas leak and erosion to primary O-ring (blowby) nozzle-to-case joint. RADIATORS DEPLOYED #8 (for one sleep period)

MCC FCR-2 (7)

FLIGHT DIRECTORS
Asc/Ent - G. E. Coen
Ld/O 1 - J. H. Greene
Orbit 2 - J. T. Cox
Planning - B. R. Stone
MOD - E. F. Kranz

41c-07-0262 Crew: PLT Scobee, Nelson/MS, van Hoften/MS, Hart/MS, & CDR Crippen.

STS41C-36-1618 - LDEF, Deployed by RMS, contained material samples for long term exposure to space by NASA LRC. To be retrieved by STS-32 in1990.

STS41C-38-1852 --- SSM Repair EVA

SPACE SHUTTLE MISSIONS SUMMARY

FLT NO.	ORBITER	CREW (6) TITLE, NAMES & EVA'S	LAUNCH SITE, LIFTOFF TIME, LANDING SITES, ABORT TIMES	LANDING SITE/ RUNWAY, CROSSRANGE LANDING TIMES FLT DURATION, WINDS	SSME-TL NOM-ABORT EMERG THROTTLE PROFILE ENG. S.N.	SRB RSRM AND ET	INC	ORBIT HA/HP	FSW	PAYLOAD WEIGHTS, PAYLOADS/ EXPERIMENTS	MISSION HIGHLIGHTS (LAUNCH SCRUBS/DELAYS, TAL WEATHER, ASCENT I-LOADS, FIRSTS, SIGNIFICANT ANOMALIES, ETC.)
STS 41-DR (STS-14) SEQ FLT # 12 KSC 12 PAD 39A-12	OV-103 Flight 1 Discovery OMS PODS LPO3 - 2 RPO3 - 1 FRC3 - 1	CDR: Henry W. Hartsfield (Flt 2 - STS-4) P/43/R8/V5/M8 PLT: Michael L. Coats P44/R38/M37 M/S: Steven A. Hawley P45/R39/M38 M/S: Richard m. Mullane P46/R40/M39 M/S: Judith A. Resnik P47/R41/F2 P/S: Charles Walker (MDAC) P48/R42/M40 MCC FCR-1 (5) FLIGHT DIRECTORS Asc/Ent - G. E. Coen Ld/O 1 - B. R. Stone Orbit 2 - J. T. Cox Plng - A. L. Briscoe MOD - E. E. Kranz	KSC 39A 243:12:41:50Z 8:35:00 AM EDT (P) 8:41:50 AM EDT (A) Thursday 3 8/30/84 (2) LAUNCH WINDOW: 14 minutes thermal constraint SBS-D on 5A & TELSTAR 34A EHS cutout PLS - EDW SLS - KSC TAL - DAKAR (Selected) TAL WX - MORON AOA - EDW AOA WX - NOR EOM - EDW MAX Q = 611 M = 1.26 SRB SEP: 2:04.12 MET MECO: 8:35.19 MET ET SEP: 8:53 MET OMS-1: 10:36.9 MET 159.4 Seconds OMS-2: 44:52.2 MET 126.3 Seconds	EDW 17, LAKEBED (EDW 10, LKBD 6) 6:37:54 AM PDT Wednesday 1 9/5/84 (2) XRANGE: 474 NM ORB DIR: DL 8 AIM PT: NOM MLGTD: 2510 FT 249:13:37:54Z VEL: 216 KGS 200 KEAS HDOT: -1.8 FPS TD NORM 195: 2960 FT NLGTD: 6713 FT 249:13:38:08Z VEL: 170 KGS HDOT: -5.6 FPS BRK INIT: 107 KGS AVE BRK DECEL: 5.6 FPS/S WHEELS STOP: 249:13:38:54Z 12785 FT ROLLOUT: 10270 FT 60 SEC WINDS: O H/T, O X KNOTS OFFICIAL: 2H, 2L DENS ALT: 3400 FT FLT DURATION: 6:00:56:04 144:56:04 S/T: 73:03:38:50 OV-103: 6:00:56:04 DISTANCE: 2,210,000 sm	104/104 109 100/104/ 84/65/ 104/65 1 = 2109 (3) 2 = 2018 (2) 3 = 2021 (1) M 3 EOM WEIGHT: 202317 X CG: 1090.7 LANDING WEIGHT: 201675 X CG: 1091.7	BI-011 SRM: HPM CASE: LWC LWT-6 ET-13 ET RPT 245K 45:45 MET ETBR/UP 197K 46:57 MET ET IMPACT LAT: 28.3 S LONG: 80.0 E	28.489 (8)	STANDARD INSERTION INSERTION ALTITUDE: 160 NM 160.8 X 160.8 NM POST OMS-2 161.63 X 160.95 NM SBS DEPLOY 161.43 NM (REV 6) SYNCOM DEPLOY 170.48 NM (REV 17) TELSTAR DEPLOY 174.94 NM (REV 34) DEORBIT 159 X 157 NM VELOCITY 25776 FPS RANGE 4112 NM	OI-4 (1)	CARGO: 47516 lbs CHARGEABLE: 41382 lbs DEPLOYED: 30086 lbs NON-DEPLOYED: 10122 lbs MIDDECK: 1174 lbs RETURNED: 17436 lbs SHUTTLE ACCUMULATED WEIGHTS: DEPLOYED: 141080 lbs NON-DEPLOYED: 184646 lbs CARGO TOTAL: 375968 lbs PERFORMANCE MARGINS (LBS): FPR: 4987 FUEL BIAS: 1341 FINAL TDDP:-1611 RECON: -1564 PRIMARY: SBS-D/PAM-D (DEPLOYED) TELESTAR 3-C/ PAM-D (DEPLOYED) SYNCOM-IV-2 (DEPLOYED) OAST-1/MPESS SOLAR ARRAY EXPERIMENT CFES (MIDDECK) IMAX 70MM CAMERA RME CLOUDS STUDENT EXP. SSIP-FSA EXP. 4 CRYO TANK SETS RMS 8 (S.N. 301) Used for PKM burn viewing and water dump nozzle survey and ice removal	KSC W/D: OPF 123 (2), VAB 15 (3), PAD 72 (2) = 210 LAUNCH POSTPONEMENT: - 6/22/84 launch postponed 3 days to 6/25/84 because of debonded engine shield during FRF. LAUNCH SCRUBS/PAD ABORT #1: - 6/25/84 launch scrubbed at T-20 minutes because GPC 5 (BFS) exhibited two parity errors at T-32 minutes. Rescheduled launch for 6/26/84. - 6/26/84 launch aborted at T-4 seconds when SSME #3 Main Fuel Valve failed the valve position check. (PAD abort #1.) - Rolled back to VAB and re-manifested, combining STS 41-D and STS 41-F P/L's. SSME 2021 replaced 2017. Launch slip of 63 days. - 8/29/84 launch scrubbed because MEC would not process certain critical events commands. Implemented a software patch to assure all 3 SRB fire commands are issued in proper order. 69-day total slip. LAUNCH DELAYS: - 6 M50 S delay at T-9 because of KSC GLS problems and two private planes in launch danger area. FLIGHT DURATION CHANGES: None. TAL WX: DAKAR & MORON go. FIRSTS: - First flight of Discovery - First flight to deploy 3 payloads. - First flight with commercial company P/S. SIGNIFICANT ANOMALIES: - CRT-2 failed (IFM replaced DU-2 with DU-4) - Supply/waste water nozzle iced. (12 inches in diameter by 27 inches tapered to point). - Ice from supply water nozzle removed using RMS impact . Unable to dump waste water for remainder of flight. - O₂ leak (30 lbs/hr). - Fuel cell performance monitor failed. - Vehicle pulled to right after NLGTD. Schrader valve leaking GN2 caused compressed strut. - S-band Quad antenna (ULF) (switch was R & R'ed postflight). - Five microswitch anomalies in RCS & OMS. - RH SRM forward field joint erosion. - LH SRM gas leak and erosion to primary O-ring of nozzle-to-case joint (blowby).

41D-12-034: Crew members (cc from ctr) CDR/ Hartsfield, PLT/Coats, MS/Hawley, MS/Resnik, PS/Walker, & MS/Mulane

41D-37-050 -- Telstar, last of three satellites deployed.

SPACE SHUTTLE MISSIONS SUMMARY

FLT NO.	ORBITER	CREW (7) / TITLE, NAMES & EVA'S	LAUNCH SITE, LIFTOFF TIME, LANDING SITES, ABORT TIMES	LANDING SITE/ RUNWAY, CROSSRANGE / LANDING TIMES, FLT DURATION, WINDS	SSME-TL NOM-ABORT EMERG / THROTTLE PROFILE ENG. S.N.	SRB RSRM AND ET	ORBIT / INC	HA/HP	FSW	PAYLOAD WEIGHTS, PAYLOADS/ EXPERIMENTS	MISSION HIGHLIGHTS (LAUNCH SCRUBS/DELAYS, TAL WEATHER, ASCENT I-LOADS, FIRSTS, SIGNIFICANT ANOMALIES, ETC.)
STS 41-G (STS-17) SEQ FLT # 13 KSC 13 PAD 39A-13	OV-099 Flight 6 Challenger OMS PODS LPO1 - 5 RPO1 - 6 FRC9 - 6	**CDR:** Robert L. Crippen (Flt 4 - STS-1, STS-7 & STS 41-C) P49/R2/V1/M2 **PLT:** Jon A. McBride P50/R43/M41 **M/S:** Sally K. Ride (Flt 2 - STS-7) P51/R19/V6/F1 **M/S:** Kathryn D. Sullivan P52/R44/F3 **M/S:** David C. Leestma P53/R45/M42 **P/S:** Paul D. Scully-Power (Civilian - Navy) P54/R46/M43 **P/S:** Mark Garneau (Canadian) P55/R47/M44	KSC 39A 279:11:03:00Z 7:03:00 AM EDT (P) 7:03:00 AM EDT (A) Friday 3 10/5/84 (1) **LAUNCH WINDOW:** 2 hours (EOM - LANDING KSC REV 7) PLS - KSC AOA - NOR AOA WX-NOR TAL-ZARAGOZA TAL WX-MORON (Selected) EMERGENCY COLOGNE-BONN AIRPORT **MAX Q** = 716 M = 1.42 **SRB SEP:** 2:04.5 MET **MECO:** 8:50.34 MET **ET SEP:** 9:08.41 MET **OMS-1:** 10:50.4 MET 130.6 Seconds **OMS-2:** 60:30.4 MET 144.6 Seconds	KSC 33 (KSC 2) 12:26:38 PM EDT Saturday 4 10/13/84 (1) **XRANGE:** 614 NM **ORB DIR:** DR 3 **AIM PT:** CLOSE IN **MLGTD:** 962 FT 287:16:26:38Z VEL: 209 KGS 208 KEAS HDOT: -0.5 FPS **TD NORM 195:** 2265 FT **NLGTD:** 5505 FT 287:16:26:47Z VEL: 162 KGS HDOT: -3 FPS **BRK INIT:** 113 KGS **AVE BRK DECEL:** 6.8 FPS/S **WHEELS STOP:** 287:16:27:32Z 11527 FT **ROLLOUT:** 10527 FT 54 SEC **WINDS:** 8 H, O X KNOTS OFFICIAL: 8H, 0X **DENS ALT:** 1100 FT **FLT DURATION:** 8:05:23:38 197:23:38 **S/T:** 81:09:02:28 **OV-099:** 40:08:16:04 **DISTANCE:** 3,400,000 sm	100/104 109 100/92/ 65/100/65 1 = 2023 (1) 2 = 2020 (2) 3 = 2021 (2) **M 3 EOM** **WEIGHT:** 202829 **X CG:** 1083.7 **LANDING** **WEIGHT:** 202266 **X CG:** 1084.8	A63/64 117-84 BI-013 **MTR:** HPM **CASE:** LWC 115 FT CHUTES ON SRB'S **LWT-8** ET-15 **ET BR/UP** 216K 1:01:00 MET **ET IMPACT LAT:** 57.1 S **LONG:** 150.0 E	57.08 (2) **STANDARD INSERTION** **INSERTION ALTITUDE:** POST OMS-2 191.74 X 189.06 NM **ERBS DEPLOY** 190 NM **DEORBIT** 121 X 118 NM **VELOCITY** 25684 FPS **RANGE** 4321 NM	OI-4 (2)	**CARGO:** 23465 lbs **CHARGEABLE:** 17592 lbs **DEPLOYED:** 4949 lbs **NON-DEPLOYED:** 11986 lbs **MIDDECK:** 657 lbs **RETURNED:** 18484.8 lbs **SHUTTLE ACCUMULATED WEIGHTS:** **DEPLOYED:** 146029 lbs **NON-DEPLOYED:** 197289 lbs **CARGO TOTAL:** 399433 lbs **PERFORMANCE MARGINS (LBS):** FPR: 4594 FUEL BIAS: 1152 FINAL TDDP: 2194 RECON: 3375 EARTH RADIATION BUDGET SATELLITE (ERBS) DEPLOYED OSTA-3 (SIR-B) MAPS, FILE LFC-MPESS ORS IMAX, RME CANEX (Canadian) APE, TLD GAS (8) G038, G032, G518, G013, G007, G469, G074 4 CRYO TANK SETS RMS 9 (S.N. 302) Used for ERBS deploy, TPS survey, water nozzle survey, SIR-B antenna latching assist	**KSC W/D:** OPF 53, VAB 5, PAD 22 = 80 **LAUNCH POSTPONEMENT:** - 10/1/84 launch postponed 4 days to 10/5/84 to replace SSME #2012 with #2021 from OV-103 in slot #3. Engine 2012 had non-flight HPOTP and HPFTP. 4-day slip. **LAUNCH SCRUBS:** None. **LAUNCH DELAYS:** None. **FLIGHT DURATION CHANGES:** None. **TAL WX:** ZZA no go - winds, Moron go. **FIRSTS:** - First flight with seven crewmembers. - First EVA by a female astronaut. - First use of PSA. - First Flight with 360 degree saddle brakes. - First flight with wing moment ties. - First transfer of hydrazine in space. **EVENTS:** - Used RMS to latch SIR-B antenna. - Solar heating used to free ERBS solar array when -Y solar array stuck during deploy attempt. MS2 tried deploy using SSP appendage arm and deploy switches, tb's functioned nominally but array did not deploy. Could not shake array loose using RMS back-drive procedure. ERBS was positioned to direct sun on array deploy mechanism. Array deployed approximately 15 minutes later. **SIGNIFICANT ANOMALIES:** - Found TPS screed problem postflight. Tile waterproofing caused screed deterioration requiring approx 4000 tiles to be replaced. Schedule impacted and OV-103 replaced OV-099 on STS 51-A. - FES shutdown by both controllers, probably icing in FES CORE. - DEU 2 Failed. - TPS damage on ROMS pod, approx 40-inch strip of FRSI peeled off. - Ku-Band antenna gimbal failure (beta angle motor short). EVA IFM to stow antenna. - R & R brakes post-flight. - R & R MLG tires (damaged by rough runway).

MCC FCR-2 (8)

FLIGHT DIRECTORS
Ascent - G. E. Coen
O 1/Ent - T. C. Lacefield
Ld/O 2 - J. T. Cox
Plng - G. A. Pennington
MOD - E. F. Kranz

EMU/TETHERED EVA:
EV1=Leestma
EV2=Sullivan
EVA1=3:29/3:27
10/11/84 - SS EVA #6
DEMO ON ORBIT
REFUELING SYSTEM
UNSCHEDULED
KU-BAND ANTENNA STOW

S84-43433 --- EVA: Leestma, left, & Sullivan, 1st U.S. woman to conduct EVA.

41-G-19-006 --- Crew: CDR Crippen (center back row); front row l.to.r. are: PLT McBride, Ride/MS, Sullivan/MS, and Leetsma/MS; and back row (left) Scully-Power/Civillian Oceanographer and (right) Garneau/Canadian Reseacher.

FLT NO.	ORBITER	CREW (5) / TITLE, NAMES & EVA'S	LAUNCH SITE, LIFTOFF TIME, LANDING SITES, ABORT TIMES	LANDING SITE/ RUNWAY, CROSSRANGE / LANDING TIMES FLT DURATION, WINDS	SSME-TL NOM-ABORT EMERG THROTTLE PROFILE ENG. S.N.	SRB RSRM AND ET	ORBIT INC	ORBIT HA/HP	FSW	PAYLOAD WEIGHTS, PAYLOADS/ EXPERIMENTS	MISSION HIGHLIGHTS (LAUNCH SCRUBS/DELAYS, TAL WEATHER, ASCENT I-LOADS, FIRSTS, SIGNIFICANT ANOMALIES, ETC.)
STS 51-A (STS-19) SEQ FLT # 14 KSC 14 PAD S84-40082 (August) MCC FCR-1 (6) **FLIGHT DIRECTORS** Ascent - J. H. Greene Ld/O 1 - L. S. Bourgeois Orbit 2 - B. R. Stone Plng - W. D. Reeves Entry - T. C. Lacefield MOD - E. F. Kranz	OV-103 Flight 2 Discovery OMS PODS LPO3 - 3 RPO3 - 2 FRC3 - 2	CDR: Frederick H. Hauck (Flt 2 - STS-7) P56/R17/V7/M17 PLT: David M. Walker P57/R48/M45 M/S: Joseph P. Allen (Flt 2 - STS-5) P58/R12/V8/M12 M/S: Anna L. Fisher P59/R49/F4 M/S: Dale A. Gardner (Flt 2 - STS-8) P60/R23/V9/M22 **UNTETHERED EVA'S (MMU):** EV1=Allen EV2=Gardner EVA1-6:13 11/12/84 - SS EVA #7 EVA2-6:01 11/14/84 - SS EVA #8 CAPTURE AND STOW OF PALAPA-B & WESTAR-IV FREE FLYER EVA'S #5 & #6	KSC 39A 313:12:15:00Z 7:15:00 AM EST (P) 7:15:00 AM EST (A) Thursday 4 11/8/84 (4) **LAUNCH WINDOW:** 18 Minutes PLANAR WINDOW (MAX YAW STEERING MPS LIMIT 1000 LBS FOR RENDEZVOUS) PLS - KSC TAL - DAKAR (Selected) TAL WX - MORON AOA - EDW AOA WX-NOR,KSC MAX Q = 651 M = 1.10 SRB SEP: 2:05.72 MET MECO: 8:33.16 MET ET SEP: 8:51.29 MET OMS-1: 10:33.3 MET 150.7 Seconds OMS-2: 44:43 MET 114.8 Seconds	KSC 15 (KSC 3) 6:59:56 AM EST Friday 3 11/16/84 (3) XRANGE: 486 NM ORB DIR: DL9 AIM PT: CLOSE IN MLGTD: 2724 FT 321:11:59:56Z VEL: 194 KGS 192 KEAS HDOT: -1.0 FPS TD NORM 195: 2454 FT NLGTD: 6380 FT 321:12:00:09Z VEL: 160 KGS HDOT: -4.6 FPS BRK INIT: 142 KGS AVE BRK DECEL: 6.5 FPS/S WHEELS STOP: 321:12:00:54Z 12178 FT ROLLOUT: 9461FT 58 SEC WINDS: 4 H, O X KNOTS OFFICIAL: 2T, 1R DENS ALT: -100 FT FLT DURATION: 7:23:44:56 191:44:56 S/T: 89:08:47:24 OV-103: 14:00:41:00 DISTANCE: 2,870,000 sm	104/104 109 100/89/ 67/104/ 65 1 = 2109 (4) 2 = 2018 (3) 3 = 2012 (6) M 3 EOM WEIGHT: 207983 X CG: 1081.4 LANDING WEIGHT: 207506 X CG: 1082.6	BI-014 61-84 SRM: HPM LWC 136 FT Chutes LWT-9 ET-16 ET RPT 226K 47:06 MET ET IMPACT LAT: 27.7 S LONG: 82.0 E	28.487 (9)	STANDARD INSERTION INSERTION ALTITUDE: POST OMS-2 161.22 X 151.17 NM TELESAT DEPLOY 163.48 NM SYNCOM DEPLOY 168.14 NM PALAPA RETRIEVE 194.44 NM WESTAR RETRIEVE 189.55 NM DEORBIT 191 X 188 NM VELOCITY 25870 FPS RANGE 4141 NM	OI-4 (3)	CARGO: 45306 lbs PAYLOAD CHARGEABLE: 38003 lbs DEPLOYED: 22764 lbs NON-DEPLOYED: 15052 lbs MIDDECK: 187 lbs RETRIEVED: 2381 lbs RETURNED: 24883 lbs SHUTTLE ACCUMULATED WEIGHTS: DEPLOYED: 168793 lbs NON-DEPLOYED: 212528 lbs CARGO TOTAL: 444739 lbs PERFORMANCE MARGINS (LBS): FPR: 4633 FUEL BIAS: 1566 FINAL TDDP: 281 RECON: 1003 SYNCOM IV-1 (DEPLOYED) TELESAT-H/ ANIK-D2/PAM-D (DEPLOYED) PALAPA-B2- (RETRIEVED & RETURNED) WESTAR-IV - (RETRIEVED & RETURNED) RME DMOS-3M EXP. MMU (2), EMU (3) 4 CRYO TK SETS RMS 10 (S.N. 301) Used for PALAPA/ WESTAR capture and berth, waste water dump monitor, and SYNCOM and TELESAT PKM viewing	KSC W/D: OPF 34, VAB 5, PAD 17 = 56 VEHICLE CHANGE: - OV-103 replaced OV-099 (TPS screed deterioration cased by waterproofing). LAUNCH POSTPONEMENT: None. LAUNCH SCRUBS: - 11/7/84 launch scrubbed because winds aloft exceeded Orbiter structural limits (excessive wind shear) - 1-day slip. LAUNCH DELAYS: None. TAL WX: - Dakar GO, Moron NO GO - low clouds. FLIGHT DURATION CHANGES: None. FIRSTS: - First retrieval and return of satellites. PALAPA-B AND WESTAR-IV were deployed on STS 41-B but PAM Upper Stages failed. - EVA crewmen captured spacecrafts using MMU/Stinger and stowed in payload bay. RENDEZVOUS 3 & 4: - To capture and return PALAPA & WESTAR. SIGNIFICANT ANOMALIES: - APU 2 water spray valve system A failed. - CRT 4 failed. - RCS F4R fuel leak. - Both left side EMU helmet lights failed (Bad Batteries). - Arriflex 16mm camera failed (IFM bypassed failed microswitch). - FWD RCS Manifold 3 fuel and oxidizer Iso valves lost open indications. - LRCS Sys B Fuel tank Iso Valve for manifold 3/4/5 lost open indication. - PLB blankets and metal discolored. - Brake hydraulic pressure increased when Iso valves opened at 200K (Iso valve leak). IFM's - Arriflex camera repaired, EVA helmet light repaired and DAP key changeout

S84-40082 -- CDR Hauck, seated, PLT Walker, stands next to the Eagle, 51-A mascot. Others on back row, l. to r., are Gardner/MS, Fisher/MS & Allen/MS.

51A-104-0046: Gardner donned MMU for traverse to Westar VI for first satellite retrieval, by he and Allen, for return to Earth.

FLT NO.	ORBITER	CREW (5) TITLE, NAMES & EVA'S	LAUNCH SITE, LIFTOFF TIME, LANDING SITES, ABORT TIMES	LANDING SITE/ RUNWAY, CROSSRANGE LANDING TIMES FLT DURATION, WINDS	SSME-TL NOM-ABORT EMERG THROTTLE PROFILE ENG. S.N.	SRB RSRM AND ET	ORBIT INC	HA/HP	FSW	PAYLOAD WEIGHTS, PAYLOADS/ EXPERIMENTS	MISSION HIGHLIGHTS (LAUNCH SCRUBS/DELAYS, TAL WEATHER, ASCENT I-LOADS, FIRSTS, SIGNIFICANT ANOMALIES, ETC.)
STS 51-C (STS-20) SEQ FLT # 15 KSC 15 PAD 39A-15	OV-103 Flight 3 Discovery OMS PODS LPO3 - 4 RPO3 - 3 FRC3 - 3	CDR: Thomas K. Mattingly (Flt 2 - STS-4) P61/R7/V10/M7 PLT: Loren J. Shriver P62/R50/M46 M/S: Ellison S. Onizuka P63/R51/M47 M/S: James F. Buchli P64/R52/M48 P/S: Gary E. Payton P65/R53/M49	KSC 39A 24:19:50:00Z 2:50:00 PM EST Thursday 5 1/24/85 (1) PLS - KSC SLS - EDW TAL - DAKAR TAL ALT: Zaragoza (Selected) TAL WX - MORON	KSC 15 (KSC 4) 4:23:23 PM EST Sunday 2 1/27/85 (1) XRANGE: 380 NM ORB DIR: DL 10 AIM PT: CLOSE IN MLGTD: 2753 FT 27:21:23:23Z VEL: 179 KGS 185 KEAS HDOT: -1FPS NLGTD: 5752 FT 27:21:23:35Z VEL: 146 KGS HDOT: -3.9 FPS TD NORM 195: 1853 FT BRK INIT: 117 KGS AVE BRK DECEL: 8.9 FPS/S WHEELS STOP: 27:21:24:13Z 10105 FT ROLLOUT: 7370 FT 50 SEC WINDS: 8H, 0 X KNOTS OFFICIAL: 8H, 1R DENS ALT: -100 FT FLT DURATION: 3:01:33:23 73:33:23 S/T: 92:10:20:47 OV-103: 17:02:14:23 DISTANCE: 1,242,566 sm	100/92/ 65/104/ 65 1 = 2109 (5) 2 = 2018 (4) 3 = 2012 (7) M 3 EOM WEIGHT: X CG: LANDING WEIGHT: 197700 X CG: 1091.8	BI-015 MTR: HPM CASE: LWC 115 FT Chutes LWT-7 ET-14 ET RPT 239K 46:11 MET ET BR/UP 227K 46:31 MET ET IMPACT LAT: 28.1 S LONG: 78.3 E	28.45 (10)	DEORBIT 185 X 185 NM VELOCITY 25855 FPS RANGE 4144 NM	OI-4 (4)	DOD PERFORMANCE MARGINS (LBS): FPR: FUEL BIAS: FINAL TDDP: -- RECON: -1457 ARC SFMD TRE VISION FLUID SHIFT OCEANS OASIS-1 CLOUDS AFT-T IOCM RMS 11 (S.N. 301) Used to monitor IUS/SRM burn	KSC W/D: OPF 31, VAB 5, PAD 20 = 50 LAUNCH POSTPONEMENT: None. LAUNCH SCRUBS: - 1/23/85 launch was scrubbed prior to ET tanking due to cold weather with potential for acreage ice on ET. 1-day slip. LAUNCH DELAY: Launch delay caused by right I/B elevon not in expected position. TAL WX: - Dakar & Moron NO GO - haze. Zaragoza GO. FLIGHT DURATION CHANGES: Yes. SIGNIFICANT ANOMALIES: - Right inboard elevon CH4 secondary delta pressure force flight prelaunch (cleared when APU's to full pressure). - IMU 1 and 3 excessive bias. - GHE leak in T-O umbilical. - FWD RCS dilemma during deorbit. - BFS did not proceed to MM104 after ET sep. - BFS deorbit ignition time was 8 seconds late. - TACAN 3 did not lock up. - RA2 erratic at high altitude. - TPS had long gouge under left wing. - RH SRM primary O-ring gas leak and erosion at center field joint (blowby). - LH SRM forward field joint gas leak and erosion to primary O-ring (blowby).

S84-43708: STS-51C Crew & Patch

MCC FCR-2 (9)

FLIGHT DIRECTORS
Ascent - J. H. Greene
Ld/Orb - T.W. Holloway
Plng - C. W. Shaw
Orb/Ent - T. C. Lacefield
MOD - E. F. Kranz

51C-08-023: Onizuki (left) & Shriver give thumbs up from Mid-Deck for first Department of Defense Shuttle mission.

FLT NO.	ORBITER	CREW (7) TITLE, NAMES & EVA'S	LAUNCH SITE, LIFTOFF TIME, LANDING SITES, ABORT TIMES	LANDING SITE/ RUNWAY, CROSSRANGE LANDING TIMES FLT DURATION, WINDS	SSME-TL NOM-ABORT EMERG THROTTLE PROFILE ENG. S.N.	SRB RSRM AND ET	ORBIT INC	HA/HP	FSW	PAYLOAD WEIGHTS, PAYLOADS/ EXPERIMENTS	MISSION HIGHLIGHTS (LAUNCH SCRUBS/DELAYS, TAL WEATHER, ASCENT I-LOADS, FIRSTS, SIGNIFICANT ANOMALIES, ETC.)
STS 51-E (STS-22) SEQ FLT # PAD	OV-099 Flight Challenger	CDR: Karol J. Bobko PLT: Donald E. Williams M/S: M. Rhea Seddon M/S: S. David Griggs M/S: Jeffrey A. Hoffman P/S: Patrick Baudry (French) P/S: Jake Garn (U.S. Senator from Utah) FLIGHT DIRECTORS: Asc/Ent - T. C. Lacefield Orbit 1 - C. W. Shaw Ld/Orb 2 - B. R. Stone Planning - J. M. Heflin				MTR: CASE: STD ET-17			OI-5	CARGO: CHARGEABLE : TDRS-B/IUS-2 TELESAT-I/PAM-D FEE FPE PPE	KSC W/D: OPF 57, VAB 8 (2), PAD 17 (2) = 82 days total LAUNCH POSTPONEMENT: - Launch rescheduled from 2/20/85 to 2/27/85 due to tile replacement caused by deteriorated screed on OV-099. - Launch rescheduled to 3/3/85 due to LH2 primary seal leak (17" ET/Orbiter) but decision was made that secondary seal would hold. LAUNCH SCRUBS: - **Flight canceled on 3/7/85 due to a TDRS-B problem and TELESAT-I was remanifested on OV-103 STS-51D. (Challenger was destacked.)** - ROLLED BACK TO VAB, CHANGED PAYLOAD TO SPACELAB 3 FOR STS 51-B. - THESE DATA ARE INCLUDED BECAUSE THE FLIGHT WAS SCRUBBED AFTER GOING THROUGH ALL OF THE FLIGHT REVIEWS, ETC. - 17-INCH LH2 PRIMARY SEAL REDESIGNED REDUCING WIDTH & DEPTH WITH STS 61-A AS FIRST FLIGHT.

JSC Flight Directors of 1984

(Left to right) Front row: Milt Heflin, Bill Reeves, Chuck Lewis, Al Pennington, & Cleon Lacefield.
Middle row: Jay Greene, Gary Coen, John Cox, & Harold Draughon.
Back row: Randy Stone, Chuck Shaw, Tommy Holloway, Chuck Knarr, Larry Bourgeois, & Lee Briscoe.

FLT NO.	ORBITER	CREW (7) TITLE, NAMES & EVA'S	LAUNCH SITE, LIFTOFF TIME, LANDING SITES, ABORT TIMES	LANDING SITE/ RUNWAY, CROSSRANGE LANDING TIMES FLT DURATION, WINDS	SSME-TL NOM-ABORT EMERG THROTTLE PROFILE ENG. S.N.	SRB RSRM AND ET	ORBIT INC	ORBIT HA/HP	FSW	PAYLOAD WEIGHTS, PAYLOADS/ EXPERIMENTS	MISSION HIGHLIGHTS (LAUNCH SCRUBS/DELAYS, TAL WEATHER, ASCENT I-LOADS, FIRSTS, SIGNIFICANT ANOMALIES, ETC.)
STS 51-D (STS-23) SEQ FLT # 16 KSC 16 PAD 39A-16	OV-103 Flight 4 Discovery OMS PODS LPO3 - 5 RPO3 - 4 FRC3 - 4	CDR: Karol J. Bobko (Flt 2 - STS-6) P66/R14/V11/M14 PLT: Donald E. Williams P67/R54/M50 M/S: M. Rhea Seddon P68/R55/F5 M/S: S. David Griggs P69/R56/M51 M/S: Jeffrey A. Hoffman P70/R57/M52 P/S: Jake Garn (U.S. Senator from Utah) P71/R58/M53 P/S: Charles Walker (MDAC) (Flt 2 - STS 41-DR) P72/R42/V12/M40 EVA CREWMEN: EV1= Hoffman EV2= Griggs UNSCHEDULED EVA: 4/16/85 - 3:10/3:07 (ATTACHED "FLY SWATTER" TO RMS.) SS EVA #9 SS Unscheduled EVA#1	KSC 39A 102:13:59:05Z 8:04:00 AM EST (P) 8:59:05 AM EST (A) Friday 4 4/12/85 (4) LAUNCH WINDOW: 1 Hour, 11 Minutes (ANIK SS FAIL OPEN) PLS - KSC SLS - EDW TAL - DAKAR TALWX - MORON (Selected) AOA - EDW AOA WX - NOR/KSC MAX Q = 666 M = 1.25 SRB SEP: 2:06.84 MET MECO: 8:51.96 MET ET SEP: 9:10 MET OMS-1: NONE OMS-2: 43.15 MET 143 Seconds	KSC 33 (KSC 5) 8:54:28 AM EST Friday 4 4/19/85 (4) XRANGE: 518 NM ORB DIR: DL 11 AIM PT: NOM MLGTD: 1639 FT 109:13:54:28Z VEL: 209 KGS 200 KEAS HDOT: -3.2 FPS TD NORM 195: 2089 FT NLGTD: 4303 FT 109:13:54:36Z VEL: 182 KGS HDOT: -5.9 FPS BRK INIT: 156 KGS AVE BRK DECEL: 8 FPS/S WHEELS STOP: 109:12:55:31Z 11937 FT ROLLOUT: 10,430 FT 63 SEC WINDS: 3T,5R KNOTS OFFICIAL: 4T, 7R DENS ALT: 1100 FT FLT DURATION: 6:23:55:23 167:55:23 S/T: 99:10:16:10 OV-103: 24:02:09:46 DISTANCE: 2,500,000 sm	100/104 109 100/90/ 65/100/ 65 1 = 2109 (6) 2 = 2018 (5) 3 = 2012 (8) ET-18 LWT-11 M 3 EOM WEIGHT: 198167 X CG: 1092.7 LANDING WEIGHT: 198014 X CG: 1094.3	BI-018 MTR: HPM CASE: LWC 136 Ft Chutes ET RPT ETBR/UP ET IMPACT LAT: 20.24 N LONG: 149.37 W	28.511 (11) START: END: MAX:	DIRECT INSERTION POST OMS-2 249.0 X 160.68 NM TELESAT DEPLOY 221.09 NM (REV 5) SYNCOM DEPLOY 213.16 NM (REV 15) DEORBIT 249 X 180 NM VELOCITY 25954 FPS RANGE 4064 NM	OI-5 (1)	CARGO: 35794 lbs PAYLOAD CHARGEABLE: 28747 lbs DEPLOYED: 22,576 lbs NON-DEPLOYED: 5092 lbs MIDDECK: 1079 lbs RETURNED: 13248 lbs SHUTTLE ACCUMULATED WEIGHTS: DEPLOYED: 191369 lbs NON-DEPLOYED: 218699 lbs CARGO TOTAL: 480533 lbs PERFORMANCE MARGINS (LBS): FPR: 4732 FUEL BIAS: 883 FINAL TDDP: 1243 RECON: 1957 SYNCOM IV-3 (DEPLOYED) TELESAT-I/ ANIK C-1/PAM-D (DEPLOYED) GAS(2) CFES-III, APE, PPE SSIP(2) 2 - MINIATURE COPPER STATUES OF LIBERTY MADE FROM "SOL" FRAMEWORK SKIN CLAMP (12 LBS) 4 CRYO TANK SETS RMS 12 (S.N. 301) Used for flyswatter snag of SYNCOM arm switch, PKM monitor, ET door survey, and water dump survey	KSC W/D: OPF 53, VAB 5, PAD 15 = 73 LAUNCH POSTPONEMENTS: - 3/19/85 launch postponed 9 days to 3/28/85 to remanifest TELESAT-1 from STS 51-E. - 3/28/85 launch postponed to 4/12/85 when PLBD was damaged by OPF bucket (access platform dropped on PLBD). 24-day slip. LAUNCH SCRUBS: None. LAUNCH DELAYS: - 55M5S delay - Ship in SRB recovery area. TAL WX: Dakar no go - haze, Moron go. FLIGHT DURATION CHANGES: - Extended flight from 5 to 7 days for attempt to operate SYNCOM IV-3 arming switch using IFM "Fly Swatter" (SYNCOM failed to maneuver to altitude because of defective mechanical arming switch. Crew re-rendezvoused with SYNCOM and snagged switch but switch was a single point failure and did not operate. - Landing at KSC was extended 1 rev because of KSC weather. - Extension: 2 days + 1 rev. RNDZ 5: To attempt to arm SYNCOM IV-3. ET TRACKING DTO 331/318: - ET Reentry (tumble) KPTC RADAR events detected at 245K and 232K, benign rupture. AWAC RADAR and Doppler conflicting data. MOTIF unusable/cloud coverage. CAST GLANCE no coverage/engine failure. SIGNIFICANT ANOMALIES: - Brake/tire problems resulted in programmatic decision to land at EDW lakebed until Nose Wheel Steering is used during landing at EDW. - Cryo 02 tank 1 htr ctlr auto mode failed. - Right ET door latches A and B indicated off (Thermal barrier pinned between door and sill). - Ku-band antenna motion erratic. - Hydraulic Sys 3 accum rapid pressure decay. - APU 3 shutdown load abnormal. - Right MLG inboard tire burst. - Right MLG brakes damaged (locked up). - Left OB elevon TPS damaged/skin burn. - Right RCS thruster R2U oxidizer leak. IFM: Developed and used "flyswatter" to snag SYNCOM arm switch.

MCC FCR-2 (10)

FLIGHT DIRECTORS
Asc/Ent - T. C. Lacefield
Orbit 1 - J. T. Cox
Ld/Orb 2 - B. R. Stone
Planning - J. M. Heflin
MOD - E. F. Kranz

PLT Williams --- CDR Bobko --- Griggs/MS --- Sen. Garn/PS

Hoffman/MS --- Seddon/MS --- Walker/PS

51D-09-014: First sitting member of Congress, Senator Garn/PS (left) & CDR Bobko with Doonesbury comic strip: Sen. Garn was subject of author Trudeau's creations prior to the mission.

SPACE SHUTTLE MISSIONS SUMMARY

FLT NO.	ORBITER	CREW (7) TITLE, NAMES & EVA'S	LAUNCH SITE, LIFTOFF TIME, LANDING SITES, ABORT TIMES	LANDING SITE/ RUNWAY, CROSSRANGE, LANDING TIMES FLT DURATION, WINDS	SSME-TL NOM-ABORT EMERG THROTTLE PROFILE ENG. S.N.	SRB RSRM AND ET	ORBIT INC	HA/HP	FSW	PAYLOAD WEIGHTS, PAYLOADS/ EXPERIMENTS	MISSION HIGHLIGHTS (LAUNCH SCRUBS/DELAYS, TAL WEATHER, ASCENT I-LOADS, FIRSTS, SIGNIFICANT ANOMALIES, ETC.)
STS 51-B (STS-24) SEQ FLT # 17 KSC 17 PAD 39A-17	OV-099 Flight 7 Challenger Spacelab 3 SECOND SPACELAB FLIGHT LM (2) OMS PODS LPO1 - 6 RPO4 - 1 FRC9 - 7	CDR: Robert F. Overmyer (Flt 2 - STS-5) P73/R10/V13/M10 PLT: Frederick D. Gregory P74/R59/M54 M/S: Don L. Lind P75/R60/M55 M/S: Norman E. Thagard (Flt 2 - STS-7) P76/R20/V14/M19 M/S: William E. Thornton (Flt 2 - STS-8) P77/R24/V15/M23 P/S: Taylor Wang P78/R61/M56 P/S: Lodewijk Van den Berg P79/R62/M57 MCC FCR-1 (7) FLIGHT DIRECTORS: Asc/Ent - T. C. Lacefield Ld/O 1 - G. E. Coen O 2 - W. D. Reeves O 3 - G. A. Pennington MOD - E. F. Kranz	KSC 39A 119:16:02:18Z 12:00:00 PM EDT (P) 12:02:18 PM EDT (A) Monday 4 4/29/85 (5) LAUNCH WINDOW: 3 Hours (CREW WORKDAY) PLS-EDW SLS-KSC TAL-ZARAGOZA (Selected) TAL WX-MORON MANUAL TAL-BONN MAX Q = 700 M = 1.31 SRB SEP: 2:05.88 MET MECO: 8:34.96 MET ET SEP: 8:53.05 MET OMS-1: 10:35 MET 132 Seconds OMS-2: 46.15 MET 147.5 Seconds	EDW 17,LAKEBED (EDW 11, LKBD 7) 9:11:04 AM PDT Monday 2 5/6/85 (1) XRANGE: 274 NM ORB DIR: AL 2 AIM PT: NOM MLGTD: 1576 FT 126:16:11:04Z VEL: 209 KGS 204 KEAS HDOT: -2 FPS TD NORM 195: 2386 FT NLGTD: 5528 FT 126:16:11:16Z VEL: 159 KGS HDOT: -7.1 FPS BRK INIT: 106 KGS AVE BRK DECEL: 7.1 FPS/S WHEELS STOP: 126:16:12:03Z 9893 FT ROLLOUT: 8317 FT 59 SEC WIND: 5H, 0 X KNOTS OFFICIAL: 5H, 2R DENS ALT: 3400 FT FLT DURATION: 7:00:08:46 168:08:46 S/T: 106:10:24:56 OV-099: 47:08:24:50 DISTANCE: 2,900,000 sm	104/104 109 100/94/ 65/104/ 103/72/ 65 1 = 2023 (2) 2 = 2020 (3) 3 = 2021 (3) M 3 EOM WEIGHT: 213795 X CG: 1084.1 LANDING WEIGHT: 213499 X CG: 1085.4	BI-016 MTR: HPM CASE: LWC ET-17 LWT-10 ET RPT 220K 1:01:12 MET ET BR/UP 195K 1:01:42 MET ET IMPACT LAT: 57. 1 S LONG: 150.8 E	57.004 (3) START: END: MAX:	STANDARD INSERTION INSERTION ALTITUDE: POST OMS-2 191.74 X 189.37 NM DEORBIT 192 X 189 NM VELOCITY 25857 FPS RANGE 4264 NM	OI-4 (5)	CARGO: 31377 lbs CHARGEABLE: 30748 lbs DEPLOYED: 105 lbs (NUSAT) NON-DEPLOYED: 30341 lbs MIDDECK: 302 lbs RETURNED: 30,427 lbs SHUTTLE ACCUMULATED WEIGHTS: DEPLOYED: 191474 lbs NON-DEPLOYED: 249342 lbs CARGO TOTAL: 511910 lbs PERFORMANCE MARGINS (LBS): FPR: 4887 FUEL BIAS: 849 FINAL TDDP: 2536 RECON: 3609 SPACELAB 3/LM: MPESS VWFC AFT ATMOS BTS DEMS FES GFFC IONS MICG RAHF-VT (Monkeys & Rats) UMI VCGS GAS (Deployable): - NUSAT (deployed) - GLOMR (failed to deploy) UMS 4 CRYO TANK SETS NO RMS	KSC W/D: OPF 31, VAB 4, PAD 15 = 50 AFTER STS 51-E (TDRS-B/TELESAT-1) WAS SCRUBBED, CHALLENGER WAS ROLLED BACK TO THE VAB AND PAYLOAD WAS CHANGED TO SPACELAB 3. LAUNCH POSTPONEMENT: None. LAUNCH SCRUBS: None. LAUNCH DELAYS: - 2M18S delay due to an LPS failure at T-4 minutes (lost GPC FEP). TAL WX: Zaragoza and Moron go. FLIGHT DURATION CHANGES: None. SIGNIFICANT ANOMALIES: - WSB 3 controller A inoperative. - Right ET door motor B inoperative. - SM onboard display exhibited erratic values. - Right OMS pod TPS protrusion (AFRSI). - Galley did not dispense water. - APU 3 seal cavity drain line heater 3A failed. - Smoke detector in avionics bay 2A failed self test. - Right RCS thruster R4D heater failed. - S-Band upper right antenna reflected power high and upper left antenna reflected power erratic. - APU 1 fuel by-pass line heater B failed on. - Mid MCA 2 OPS status 5 indicated zero. - PLBD close sequence failed on port aft latches. - MLG brakes damaged (LH inboard rotors destroyed). - MLG dump valve leaked 3 days after landing (power left on 3 hydraulic valves which had to be replaced). - Left OB elevon tile slumping and gap filler breach. - GLOMR failed to deploy (150 lbs). - Gas leaks and erosion in both SRM nozzle-to-case joints. - Erosion to secondary O-ring on LH SRM (blowby). IFM's: S/L drop dynamics godule experiment recovered. Spacelab ION experiment recovered.

Gregory-Lind-Thagard-Wang-van den Berg

CDR Overmyer --- Thornton

51B-116-005: CDR Overmyer captured this auroral observation in southern hemisphere halfway between Australia & Antarctic continent. There are moonlit clouds on Earth. The blue-green band and the tall red rays are aurora. Brown Streak is atmospheric luminescence.

SPACE SHUTTLE MISSIONS SUMMARY

FLT NO.	ORBITER	CREW (7) TITLE, NAMES & EVA'S	LAUNCH SITE, LIFTOFF TIME, LANDING SITES, ABORT TIMES	LANDING SITE/ RUNWAY, CROSSRANGE LANDING TIMES FLT DURATION, WINDS	SSME-TL NOM-ABORT EMERG THROTTLE PROFILE ENG. S.N.	SRB RSRM AND ET	ORBIT INC	ORBIT HA/HP	FSW	PAYLOAD WEIGHTS, PAYLOADS/ EXPERIMENTS	MISSION HIGHLIGHTS (LAUNCH SCRUBS/DELAYS, TAL WEATHER, ASCENT I-LOADS, FIRSTS, SIGNIFICANT ANOMALIES, ETC.)
STS 51-G (STS-25) SEQ FLT # 18 KSC 18 PAD 39A-18	OV-103 Flight 5 Discovery OMS PODS LPO4 - 1 RPO3 - 5 FRC3 - 5	CDR: Daniel C. Brandenstein (Flt 2 - STS-8) P80/R21/V16/M20 PLT: John O. Creighton P81/R63/M58 M/S: John M. Fabian (Flt 2 - STS-7) R82/R18/V17/M18 M/S: Steven R. Nagel P83/R64/M59 M/S: Shannon W. Lucid P84/R65/F6 P/S: Patrick Baudry (France) P85/R66/M60 P/S: Sultan S. Al-Saud (Saudia Arabia) P86/R67/M61	KSC-39A 168:11:33:00Z 7:33:00 AM EDT (P) 7:33:00 AM EDT (A) Monday 5 6/17/85 (3) LAUNCH WINDOW: 4 minutes (CLOSE ON MORELOS EARTH HORIZON SENSOR CUTOUT - 10 MINUTES WITH WAIVER OF CUTOUT) NEOM - EDW EOM WX - KSC RTLS - KSC TAL - DAKAR (Selected) TAL WX - MORON AOA - EDW AOA WX - NOR/KSC MAX Q = 648 M = 1.24 SRB SEP: 2:04.68 MET MECO: 8:35.77 MET ET SEP: 8:53.93 MET OMS-1: NONE OMS-2: 40:29 MET 179.4 Seconds	EDW 23, LAKEBED (EDW 12, LKBD 8) 6:11:52 AM PDT Monday 3 6/24/85 (2) XRANGE: 694 NM ORB DIR: DL 12 AIM PT: CLOSE IN MLGTD: 1117 FT 175:13:11:52.4Z VEL: 202 KGS 198 KEAS HDOT: -2 FPS TD NORM 195: 1387 FT NLGTD: 4990 FT 175:13:12:05Z VEL: 163 KGS HDOT: -8 FPS BRK INIT: 154 KGS AVE BRK DECEL: 8.8 FPS/S WHEELS STOP: 775:13:12:33Z 8550 FT ROLLOUT: 7433 FT 36 SEC WIND: 2H,11L KNOTS OFFICIAL: 2H, 11L DENS ALT: 3727 FT FLT DURATION: 7:01:38:52 169:38:52 S/T: 113:12:03:48 OV-103: 31:03:48:38 DISTANCE: 2,500,000 sm	104/104 109 % 100/104/ 83/65/ 104/65 1 = 2109 (7) 2 = 2018 (6) 3 = 2012 (9) M 3 EOM WEIGHT: 204321 X CG: 1082.1 LANDING WEIGHT: 204169 X CG: 1083.7	BI-019 MTR: HPM CASE: MWC ET-20 LWT-13 ET RPT 233K 1:19:15 MET ET BR/UP 219K 1:19:38 MET ET IMPACT LAT: 14.89 N LONG: 159.5 W	28.487 (12) START: END: MAX:	DIRECT INSERTION POST OMS-2 192.37 X 190.37 NM MORELOS DEPLOY 191.1 NM ARABSAT DEPLOY 193.81 NM TELSTAR DEPLOY 196.35 NM SPARTAN DEPLOY 210.3 NM DEORBIT 191 x 150 NM VELOCITY 25850 FPS RANGE 4050 NM	OI-6 (1)	CARGO: 44477 lbs. CHARGEABLE: 38258 lbs DEPLOYED: 22832 lbs NON-DEPLOYED: 14866 lbs MIDDECK: 560 lbs RETURNED: 21310 lbs SHUTTLE ACCUMULATED WEIGHTS: DEPLOYED: 214306 lbs NON-DEPLOYED: 264768 lbs CARGO TOTAL: 556387 lbs PERFORMANCE MARGINS (LBS): FPR: 5088 FUEL BIAS: 849 FINAL TDDP: 160 RECON: -1664 PRIMARY: TELSTAR-3D/ PAM-D DEPLOYED MORELOS-A/ PAM-D DEPLOYED ARABSAT-A/ PAM-D DEPLOYED SPARTAN-101DH (DEPLOYED & RETRIEVED) FEE, ADSF, FPE, HPTE, ASE GAS: G027-OFVLR G028-OFVLR G471-GSFC OLLENDORF G025-ERNO G034-EL PASO/YSLETA G314-USAF/NRL 4 CRYO TNK SETS RMS 13 (S.N. 301) Used for SPARTAN deploy, retrieve, and berth, water dump survey, PKM monitoring, and ARABSAT solar array survey	KSC W/D: OPF 37, VAB 7, PAD 14 = 58 LAUNCH POSTPONEMENTS: - 6/12/85 launch postponed to 6/14/85 due to late OPF start. - 6/14/85 launch postponed to 6/17/85 because STS 51-D landed at EDW not KSC. - 2 day extension - 5-day total slip. LAUNCH SCRUBS: None. LAUNCH DELAYS: None. TAL WX: Dakar & Moron go. FLIGHT DURATION CHANGES: None. EVENTS: - MORELOS deployed orbit 6D. - ARABSAT deployed orbit 18D. - TELSTAR deployed orbit 32D. - SPARTAN deployed orbit 51D. - Rendezvous with SPARTAN. - Wheels dug into lakebed » 6 inches at end of rollout. RENDEZVOUS 6: - With SPARTAN for retrieval and return. SIGNIFICANT ANOMALIES: - WCS Fan Separator 1 motor current high. - RCS microswitch problems. - Right RCS fuel x-feed valve 3/4/5. - Left RCS OX or Fuel Tank Iso Valve. - Right RCS OX Tank Iso Valve 3/4/5. - S-Band lower left antenna beam switch intermittent. - MDM FA3 failure (Intermittent output from secondary core power supply). - WOW dilemma (wheel off ground 800 ft). - RA2 late acquisition. - TPS debris hits. - Gas leaks and erosion on both SRM nozzle-to-case joints (blowby).

S85-32877: STS-51G Crew & Patch

MCC FCR-2 (11)

FLIGHT DIRECTORS
Asc/Ent - T. C. Lacefield
Ld/O 1 - L. S. Bourgeois
O 2 - J. M. Heflin
Plng - C. R. Knarr
MOD - T. W. Holloway

51g-s-225: Landing at EDW Lakebed.
Wheels dug into lakebed » 6 inches at end of rollout.

FLT NO.	ORBITER	CREW (7) TITLE, NAMES & EVA'S	LAUNCH SITE, LIFTOFF TIME, CROSSRANGE / LANDING SITES, ABORT TIMES	LANDING SITE/ RUNWAY, CROSSRANGE / LANDING TIMES FLT DURATION, WINDS	SSME-TL NOM-ABORT EMERG / THROTTLE PROFILE ENG. S.N.	SRB RSRM AND ET	ORBIT INC	ORBIT HA/HP	FSW	PAYLOAD WEIGHTS, / PAYLOADS/ EXPERIMENTS	MISSION HIGHLIGHTS (LAUNCH SCRUBS/DELAYS, TAL WEATHER, ASCENT I-LOADS, FIRSTS, SIGNIFICANT ANOMALIES, ETC.)
STS 51-F (STS-26) SEQ FLT # 19 KSC-19 PAD 39A-19	OV-099 Challenger (Flight 8) Spacelab 2 (IGLOO + 3 PALLETS) THIRD SPACELAB FLIGHT OMS PODS LPO1 - 7 RPO4 - 2 FRC9 - 8	CDR: C. Gordon Fullerton (Flt 2 - STS-3) P87/R6/V18/M6 PLT: Roy D. Bridges P88/R68/M62 M/S: F. Story Musgrave (Flt 2 - STS-6) P89/R15/V19/M15 M/S: Anthony W. England P90/R69/M63 M/S: Karl G. Henize P91/R70/M64 P/S: Loren W. Acton P92/R71/M65 P/S: John-David F. Bartoe P93/R72/M66 MCC FCR-1 (8) FLIGHT DIRECTORS Asc/Ent - T. C. Lacefield O 1 - G. A. Pennington Ld/O 2 - J. T. Cox O 3 - A. L. Briscoe MOD - E. F. Kranz	KSC 39A 210:21:00:00Z 3:23:00 PM EDT (P) 5:00:00 PM EDT (A) Monday 6 7/29/85 (1) LAUNCH WINDOW: 2 Hours, 25 Minutes CREW WORKDAY 3 Hours, 50 Minutes launch clearance and service window PLS - EDW SLS - KSC AOA - NOR AOA WX - KSC TAL - ZARAGOZA (Selected) TAL WX - MORON MAX Q = 762 M = 1.63 SRB SEP: 2:05.24 MET MECO: 9:41.24 MET ET SEP: 9:59.29 MET ABORT-TO-ORBIT OMS-1: 11:41 MET 106.4 Seconds OMS-2: 33:00 MET 121.8 Seconds	EDW 23, LAKEBED (EDW 13, LKBD 9) 12:45:26 PM PDT Tuesday 4 8/6/85 (1) XRANGE: 603 NM ORB DIR: AL 3 AIM PT: NOM MLGTD: 3713 FT 218:19:45:26Z VEL: 204 KGS 199 KEAS HDOT: -0.7 FPS TD NORM 195: 4073 FT NLGTD: 6412 FT 218:19:45:35Z VEL: 168 KGS HDOT: -7.1 FPS BRK INIT: 126 KGS AVE BRK DECEL: 8 FPS/S WHEELS STOP: 218:19:46:21Z 12282 FT ROLLOUT: 8569 FT 55 SEC WINDS: 10H, 1L KNOTS OFFICIAL: 9H, 3L DENS ALT: 5610 FT FLT DURATION: 7:22:45:26 190:45:26 S/T: 121:10:49:14 OV-099: 55:07:10:16 DISTANCE: 2,850,000 sm	104/104 109 % 100/104/ 97/65/ 104/91 1 = 2023 (3) 2 = 2020 (4) 3 =2021 (4) M 3 EOM WEIGHT: 216894 X CG: 1079.8 LANDING WEIGHT: 216735 X CG: 1081.3	BI-017 SRM: HPM CASE: MWC ET-19 LWT-12 ET RPT 211K 1:03:35 MET ET BR/UP 193K 1:03:58 MET ET IMPACT LAT: 48.9 S LONG: 159.0 E	49.491 (1)	142.9 X 108.7 NM STANDARD INSERTION WAS PLANNED ATO AFTER SSME #1 SHUT DOWN DEORBIT 174 X 164 NM VELOCITY 25814 FPS RANGE 4221 NM	OI5-24 (2)	CARGO: 34400 lbs CHARGEABLE: 33012 lbs DEPLOYED: 0 lbs NON-DEPLOYED: 31257 lbs MIDDECK: 1755 lbs RETURNED: 33555 lbs SHUTTLE ACCUMULATED WEIGHTS: DEPLOYED: 214306 lbs NON-DEPLOYED: 297780 lbs CARGO TOTAL: 590787 lbs PERFORMANCE MARGINS: NOT AVAILABLE SPACELAB 2 WITH 13 INVESTIGATIONS IN 7 SCIENTIFIC DISCIPLINES: SOLAR, ATMOSPHERIC, PLASMA, HIGH-ENERGY ASTRO-PHYSICS, IR ASTRONOMY, TECHNOLOGY RESEARCH, AND LIFE SCIENCES PDP, VCAP, IRT, CRNE, XRT, SOUP CHASE, HRTS, SUSIM,PGU, SUPERFLUID HELIUM, PLASMA DEPLETION PDP PROX OPS SAREX, SLSTP, CBDE PROX OPS WITH FREE FLYING PDP 4 CRYO TANK SETS RMS 14 (S.N. 302) Used for PDP deploy and retrieve, waste water dump monitor, and belly tile survey	KSC W/D: OPF 39, VAB 5, PAD 31 = 75 LAUNCH POSTPONEMENT: None. LAUNCH SCRUBS/PAD ABORT #2: - 7/12/85 launch aborted at T-4.2 seconds when SSME #2 (2020) chamber coolant valve (CCV) failed to ramp to 70% open by "CMD A", resulting in an MCF, causing shutdown. (pad abort #2). Recycled engine 2020 at pad. - 17-day launch slip. LAUNCH DELAYS: - 1H37M delay because of an error in a TMBU CMD to BFS. BFS was Re-IPL'ed and IMU's were realigned. TAL WX: Zaragoza go, Moron no go. FLIGHT DURATION CHANGES: - Extended flight 1 day (+ 1 rev) to provide additional Spacelab experiment time. FIRSTS: - First flight of Spacelab pallet only. - First flight of IPS. PROX OPS: With PDP. SIGNIFICANT ANOMALIES: - ROMS primary pitch TVC failed to respond properly to cmds on 7/10/85. - EXP computer failed prelaunch, ECOS loaded in B/U computer. - SSME #1 auto shut down at 5:43 MET. (HPFTP discharge temp B Xducer failed at 3:31 MET & Xducer A failed at 5:43) resulting in an ATO call. OMS dump (burn) of 106 seconds (4134 lbs. Prop). - SSME #3 HPFTP temp B failed at 8:12 MET, inhibited limits and accomplished ATO. - Recycled SSME 2020 at pad. - RMS tile scan to check for ET SOFI damage to Orbiter bottom TPS (100 tiles scrapped). - GPC body rate data transfer incompatible with Spacelab. - Left SRB yaw axis rate Gyro assy 3 failed hardover prelaunch (GMEM patch). - BFS logged "Stored Protect" after TMBU uplinked. - SSME 2 GH₂ Pressure Xducer failed. - No damage to brakes (runway inspection). RADIATORS DEPLOYED #9 - (port side stowed 3 hours for tile survey).

SPACELAB 2 mission patch

STS-51F Flight Crew

51-F-33-005: Experiments & IPS for Spacelab 2 are backdropped against the Libya/Tunisia Mediterranean coast.

FLT NO.	ORBITER	CREW (5) TITLE, NAMES & EVA'S	LAUNCH SITE, LIFTOFF TIME, LANDING SITES, ABORT TIMES	LANDING SITE/ RUNWAY, CROSSRANGE LANDING TIMES FLT DURATION, WINDS	SSME-TL NOM-ABORT EMERG THROTTLE PROFILE ENG. S.N.	SRB RSRM AND ET	ORBIT INC	HA/HP	FSW	PAYLOAD WEIGHTS, PAYLOADS/ EXPERIMENTS	MISSION HIGHLIGHTS (LAUNCH SCRUBS/DELAYS, TAL WEATHER, ASCENT I-LOADS, FIRSTS, SIGNIFICANT ANOMALIES, ETC.)
STS 51-I (STS-27) SEQ FLT 20 KSC-20 PAD 39A-20	OV-103 Discovery (Flight 6) OMS PODS LPO4 - 2 RPO3 - 6 FRC3 - 6	CDR: Joe H. Engle (Flt 2 - STS-2) P94/R3/V20/M3 PLT: Richard O. Covey P95/R73/M67 M/S: James D. Van Hoften (Flt 2-STS 41-C) P96/R36/V21/M35 M/S: John M. Lounge P97/R74/M68 M/S: William F. Fisher P98/R75/M69	KSC-39A 239:10:58:01Z 6:55:00 AM EDT (P) 6:58:01 AM EDT (A) Tuesday 2 8/27/85 (3) LAUNCH WINDOW: 54 Minutes (PLANAR/ET IMPACT AREA) PLS-EDW SLS-KSC ALS-NOR AOA-EDW AOA WX-NOR,KSC TAL-DAKAR TAL WX-MORON (SELECTED) MAX Q = 735 PSF M = 1.61 SRB SEP: 2:01 MET MECO: 8:27.59 MET ET SEP: 8:45.77 MET OMS-1: NONE OMS-2: 40:28 MET 183.2 Seconds	EDW 23, LAKEBED (EDW 14, LKBD 10) 6:15:43 AM PDT Tuesday 5 9/3/85 (3) XRANGE:692 NM ORB DIR: DL 13 AIM PT: NOM MLGTD: 2101 FT 246:13:15:43Z VEL: 175 KGS 191 KEAS HDOT: -0.5 FPS TD NORM 195: 1741 FT NLGTD: 4384 FT 246:13:15:51Z VEL: 144 KGS HDOT: -5.6 FPS BRK INIT: 114 KGS AVE BRK DECEL 7.3 FPS/S WHEELS STOP: 246:13:16:30Z 8201 FT ROLLOUT: 6100 FT 47 SEC WINDS: 19H, 0 X KNOTS OFFICIAL: 18H, 0X DENS ALT: 2982 FT FLT DURATION: 7:02:17:42 170:17:42 S/T: 128:13:06:56 OV-103: 38:06:06:20 DISTANCE: 2,500,000 sm	104/104 109% 100/104/ 70/67/ 104/103/ 73/67 1 = 2109 (8) 2 = 2018 (7) 3 = 2012 (10) BI-STABLE HPOTP (1) M 3 EOM WEIGHT: 196856 X CG: 1092.4 LANDING WEIGHT: 196674 X CG: 1094.2	BI-020 MTR: HPM CASE: LWC ET-21 LWT-14 ET RPT 232K 1:19:03 MET ET BR/UP 216K 1:19:29 MET ET IMPACT LAT: 11.5 N LONG: 157.6 W	28.541 (13)	DIRECT INSERTION POST OMS-2 190.51 X 190.2 NM AUSSAT DEPLOY 190.23 NM ASC DEPLOY 191.6 NM SYNCOM-F4 DEPLOY 194.6 NM DEORBIT 242 X 178 NM VELOCITY 25829 FPS RANGE 4004 NM	OI6-27 (2)	CARGO: 43988 lbs CHARGEABLE: 38884 lbs DEPLOYED: 30289 lbs NON-DEPLOY: 8221 lbs MIDDECK: 374 lbs RETURNED: 13478 lbs SHUTTLE ACCUMULATED WEIGHTS: DEPLOYED: 244595 lbs NON-DEPLOYED: 306375 lbs CARGO TOTAL: 634775 lbs PERFORMANCE MARGINS (LBS): FPR: 4983 FUEL BIAS: 839 FINAL TDDP: 176 RECON: -1145 PRIMARY: ASC-1/PAM-D DEPLOYED AUSSAT-1/PAM-D DEPLOYED SYNCOM IV-4 UNQ (LEASAT) DEPLOYED MIDDECK: PVTOS PFR/APC MFR 4 CRYO TK SETS RMS 15 (S.N. 301) Used for LEASAT capture, repair, and release, waste water dump monitor, and to open AUSSAT sunshield	KSC W/D: OPF 27, VAB 7, PAD 22 = 56 LAUNCH POSTPONEMENTS: None. LAUNCH SCRUBS: - 8/24/85 launch scheduled for 8:38 AM EDT scrubbed because of thunderstorms in launch area and ship in LDA. - 8/25/85 launch scrubbed because of GPC-5 failure. Re-IPL's GPC-5 and fault repeated 11 minutes later. Replaced GPC-5. - 3-day total slip. LAUNCH DELAYS: - 3M1S delay awaiting clearing in cloud cover and ship in SRB recovery area. TAL WX: Dakar no go - clouds, Moron go. FLIGHT DURATION CHANGES: - Shortened flight 1 day because AUSSAT was deployed early. EVENTS: - Deployed AUSSAT-1 on orbit 5 instead of 17 because of sunshield damage by RMS camera. - Deployed ASC-1 on orbit 7 at 239:22:07:32Z. - Deployed SYNCOM IV-4 on orbit 32 at 241:10:47:55z. (Failed to operate after achieving operational altitude.) - Rendezvous and EVA repair of LEASAT salvage (SYNCOM IV-3) on days 5 and 6. (Deployed on STS 51-D.) - Bi-Stable Pump - HPOTP minimum throttle of 67 percent (first flight.) RENDEZVOUS 7: To repair SYNCOM IV-3 . SIGNIFICANT ANOMALIES: - Tank A water flow rate to galley low. - Hydraulic System 3 accumulator bootstrap pressure low. - RMS elbow joint failed to respond to computer commands in primary. - Potable water nozzle temp dropped to 58 F during supply water dump. - BFS OMS 2 out-of-plane velocity computation 12.5 FPS higher than PASS. - FES topping duct zone H heater B failed. - FRCS thruster FIF chamber pressure failure. - Rt OMS fuel tank isol vlv A barber pole. - Galley water flow did not shut off. - Right OMS pod AFRSI strip loose. RADIATORS DEPLOYED #10 (one sleep period for DTO)

STS-51I Flight Crew

MCC FCR-2 (12)

FLIGHT DIRECTORS:
Asc/Ent - G. E. Coen
Ld/O 1 - J. H. Greene
O 2 - W. D. Reeves
Plng - C. R. Knarr
MOD - E. F. Kranz

EMU/TETHERED EVA'S:
EV1 - Van Hoften
EV2- Fisher

EVA1 = 8/31/85
7:20/7:07
SS EVA #10

EVA2 = 9/1/85
EV1 = 4:31/4:12
EV2 = 4:31/4:28
SS EVA #11
CAPTURE, REPAIR, AND RELEASE OF LEASAT/SYNCOM IV-4

51I-S-237: Syncom IV-3 after shove-off by Hofton/MS. Errant satellite was earlier captured & repaired by Shuttle.

SPACE SHUTTLE MISSIONS SUMMARY

FLT NO.	ORBITER	CREW (5) TITLE, NAMES & EVA'S	LAUNCH SITE, LIFTOFF TIME, LANDING TIMES ABORT TIMES	LANDING SITE/ RUNWAY, CROSSRANGE LANDING TIMES FLT DURATION, WINDS	SSME-TL NOM-ABORT EMERG THROTTLE PROFILE ENG. S.N.	SRB RSRM AND ET	INC	HA/HP	FSW	PAYLOAD WEIGHTS, PAYLOADS/ EXPERIMENTS	MISSION HIGHLIGHTS (LAUNCH SCRUBS/DELAYS, TAL WEATHER, ASCENT I-LOADS, FIRSTS, SIGNIFICANT ANOMALIES, ETC.)
STS 51-J (STS-28) SEQ. FLT # 21 KSC-21 PAD 39A-21	OV-104 Atlantis (Flight 1) OMS PODS LPO3 - 6 RPO1 - 7 - 1	CDR: Karol J. Bobko (Flt 3 - STS-6 & STS 51-D) P99/R14/V11/M14 PLT: Ronald J. Grabe P100/R76/M70 M/S: Robert L. Stewart (Flt 2 - STS 41-B) P101/R33/V22/M32 M/S: David C. Hilmers P102/R77/M71 P/S: William A. Pailes (USAF) P103/R78/M72 MCC FCR-2 (13) FLIGHT DIRECTORS: Asc/Ent - G. E. Coen O 1 - C. W. Shaw Ld/O 2 - B. R. Stone Plng - J. M. Heflin MOD - T. W. Holloway	KSC-39A 276:15:15:30Z 11:15:30 AM EDT Thursday 6 10/3/85 (2) PLS - EDW SLS - KSC TAL - Dakar TAL WX - Moron (SELECTED) TAL WX - Zaragoza	EDW 23, LAKEBED (EDW 15, LKBD 11) 10:00:08 AM PDT Monday 4 10/7/85 (2) XRANGE: 432 NM ORB DIR: DL 14 A/IM PT: CLOSE IN MLGTD: 2476 FT 280:17:00:08Z VEL: 187 KGS 192 KEAS HDOT: -2 FPS TD NORM 195: 2206 FT NLGTD: 4873 FT 280:17:00:15Z VEL: 155 KGS HDOT: -5.6 FPS BRK INIT: 117 KGS AVE BRK DECEL: 7.3FPS/S WHEELS STOP: 280:17:01:13Z 10532 FT ROLLOUT: 8056 FT 65 SEC WINDS: 14H, 1R KNOTS OFFICIAL: 11H, 4R DENS ALT: 3622 FT FLT DURATION: 4:01:44:38 97:44:38 S/T: 132:14:51:34 OV-104: 4:01:44:38 DISTANCE: 1,682,641 sm	104/104 109 100/104/ 68/65/ 104/102/ 74/65 1 = 2011 (2) 2 = 2019 (2) 3 = 2017 (4) M 3 EOM WEIGHT: X CG: LANDING WEIGHT: 190765 X CG: 1101.2	BI-021 MTR: HPM CASE: LWC ET-25 LWT-18 ET RPT 230K 1:23:04 MET ET BR/UP 215K 1:23:25 MET ET IMPACT LAT: 20.6 N LONG: 148.26 W	28.5 (14)	DEORBIT 254 X 254 NM VELOCITY 26023 FPS RANGE 3986 NM	OI6-28 (3)	DOD NO RMS OASIS-2 CLOUDS RME MARC-DN RTPA OCEANS VFT-1 VFT-2 CST AMOS WINCON	KSC W/D: OPF 84, VAB 14 PAD 34 = 132 LAUNCH POSTPONEMENTS: None. LAUNCH SCRUBS: None. FLIGHT DURATION CHANGES: None. LAUNCH DELAY: - Launch delayed because of MPS PV# 6 RPCA erratic. (LH2 prevalve close indicator.) SIGNIFICANT ANOMALIES: - Port MPM shoulder "A" pyro initiator circuit failed self test. - APU Exhaust Gas temp 2 failed. - WSB 2 regulator pressure decayed. - OPS Recorder 2 tracks 7,8, & 9 intermittent. - ROMS fuel total quantity reading offset. - TPS damage on left inboard elevon leading edge and in nose cap area. - Fuel Cell 3 O2 flowmeter failed. - SSME 1 and 2 pitch and yaw actuator secondary delta pressures high. - PLB camera "B" difficult to focus and camera "Ç" Azimuth and elevation failed. - Airlock hatch "A" tapered pin did not latch in open position. - Side hatch "T" handle difficult for crew to operate.

Hilmers/MS --- Pailes/PS

Stewart/MS -- CDR Bobko -- PLT Grabe

51J-143-126: Atlantis' vertical stabilizer (North side of photo) partially frames over-flight scene of Metropolitan Houston, muddy Galveston & Trinity Bays, Galveston Island, & Coastline of Gulf of Mexico.

FLT NO.	ORBITER	CREW (8) / TITLE, NAMES & EVA'S	LAUNCH SITE, LIFTOFF TIME, LANDING SITES, ABORT TIMES	LANDING SITE/ RUNWAY, CROSSRANGE / LANDING TIMES FLT DURATION, WINDS	SSME-TL NOM-ABORT EMERG / THROTTLE PROFILE ENG. S.N.	SRB RSRM AND ET	ORBIT INC	HA/HP	FSW	PAYLOAD WEIGHTS, PAYLOADS/ EXPERIMENTS	MISSION HIGHLIGHTS (LAUNCH SCRUBS/DELAYS, TAL WEATHER, ASCENT I-LOADS, FIRSTS, SIGNIFICANT ANOMALIES, ETC.)
STS 61-A (STS-30) SEQ FLT # 22 KSC-22 PAD 39A-22	OV-99 Challenger (Flight 9) Spacelab D-1 Flight 4th Spacelab Flight LM (3) OMS PODS LPO1 - 8 RPO3 - 7 FRC9 - 9	CDR: Henry W. Hartsfield (Flt 3 - STS-4 & STS 41-D) P104/R8/V5/M8 PLT: Steven R. Nagel (Flt 2 - STS 51-G) P105/R64/V23/M59 M/S: James F. Buchli (Flt 2 - STS 51-C) P106/R52/V24/M48 M/S: Guion S. Bluford (Flt 2 - STS-8) P107/R22/V25/M21 M/S: Bonnie J. Dunbar P108/R79/F7 P/S: Reinhard Furrer (Germany) P109/R80/M73 P/S: Ernst Messerschmid (Germany) P110/R81/M74 P/S: Wubbo J. Ockels (Netherlands) P111/R82/M75	KSC 39A 303:17:00:00Z 12:00:00 PM EST (P) 12:00:00 PM EST (A) Wednesday 1 10/30/85 (3) LAUNCH WINDOW: 180 Minutes (CREW WORKDAY) PLS - EDW SLS - KSC ALS - NOR AOA - NOR AOA WX - NONE TAL - ZARAGOZA (SELECTED) TAL WX - MORON MANUAL TAL - KOLN/BONN MAX Q = 665 PSF M = 1.25 SRB SEP: 2:05 MET MECO: 8:34.96 MET ET SEP: 8:53.05 MET OMS-1: 10:35 MET 121.4 Seconds OMS-2: 44.40 MET 132.7 Seconds	EDW 17, LAKEBED (EDW 16, LKBD 12) 9:44:51 AM PST Wednesday 2 11/06/85 (4) XRANGE: 69 NM ORB DIR: AR 2 AIM PT: NOM MLGTD: 1829 FT 310:17:44:51Z VEL: 210 KGS 203 KEAS HDOT: -1.2 FPS TD NORM 195: 2549 FT NLGTD: 4767 FT 310:17:44:59Z VEL: 178 KGS HDOT: -7.8 FPS BRK INIT: 111 KGS AVE BRK DECEL: 7.5 FPS/S WHEELS STOP: 310:17:45:40Z 10133 FT ROLLOUT: 8304 FT 49 SEC WINDS: OH, 1R KNOTS OFFICIAL: 0H, 0X DENS ALT: 2539 FT FLT DURATION: 7:00:44:51 168:44:51 S/T: 139:15:36:25 OV-099: 62:07:55:07 DISTANCE: 2,501,290 sm	104/104 109% 100/89/ 65/104/ 102/73/ 67 1 = 2023 (4) 2 = 2020 (5) 3 = 2021 (5) M 3 EOM WEIGHT: 214325 X CG: 1083.8 LANDING WEIGHT: 214171 X CG: 1085.2	BI-022 (4) MTR: HPM CASE: LWC ET-24 LWT-17 ET BR/UP 188K 1:00:57 MET ET IMPACT LAT: 59.97 S LONG: 147.96 E	56.998	STANDARD INSERTION POST OMS-2 178.99 X 175.51 NM GLOMR DEPLOY 179.62 NM DEORBIT 180 X 174 NM VELOCITY 25829 FPS RANGE 4353 NM	OI6-29 (4)	CARGO: 31911 lbs CHARGEABLE: 30519 lbs DEPLOYABLE: 150 lbs GLOMR GAS NON-DEPLOY: 27330 lbs MIDDECK: 2164 lbs RETURNED: 30732 lbs SHUTTLE ACCUMULATED WEIGHTS: DEPLOYED: 244745 lbs NON-DEPLOYED: 335869 lbs CARGO TOTAL: 666686 lbs PERFORMANCE MARGINS (LBS): FPR: 4897 FUEL BIAS: 851 FINAL TDDP: 6222 RECON: 6219 PAYLOAD: Spacelab D-1/LM (Germany) EXPERIMENTS: WL- 6 Material Science Exps PK - 3 optical diagnostic facilities (process chamber) MD - Media (material science) facility, elliptical mirror heating facility, high precision thermostat facility BW - Life Sciences VS - Vestibular sled BR - Biorack NAVES - (Nav Exp) ME - Materials Exp GLOMR (DPLY) 4 CRYO TANK SETS RMS 16 (S.N. 302) Used for waste water dump monitor	KSC W/D: OPF 35, VAB 4, PAD 14 = 53 LAUNCH POSTPONEMENTS: None. LAUNCH SCRUBS: None. LAUNCH DELAYS: None. TAL WX: Zaragoza, Moron, and Ben Guerir go. FLIGHT DURATION CHANGES: None. FIRSTS: - First flight with redesigned MPS 17" disconnect primary seal. - First flight with full nosewheel steering. - First flight with 8 crewmembers. - First flight with POCC overseas (Munich). Spacelab D-1 flight with objective science and implications of microgravity. EVENTS: - GLOMR deployed at 12:34:00 MET (rev 9). - Long-duration gravity gradient attitude (9 - 12 hours per day). SIGNIFICANT ANOMALIES: - Fuel cell 1 condenser exit temperature oscillated. - Cryo hydrogen tank 1 control pressure failed. - RRCS helium leg A operated on secondary. - RRCS helium leg B failed closed. - APU 1 gearbox GN₂P high. - Smoke detector B in avionics bay triggered false alarms. - S-Band antenna switched late. - Primary L RCS thruster L2L injector heater failed on. - RMS deploy microswitches for shoulder manipulator positioning pedestal went to zero. - Stream of particulate matter hit Orbiter. - WCS fan separator 1 fails. - LH SRM center and aft field joint gas leaks to primary O-rings (blowby). RADIATORS DEPLOYED #11 (stowed for 23 hours in -ZLV +YVV)

MCC FCR-1 (9)

FLIGHT DIRECTORS:
Asc/Ent - G. E. Coen
Ld/O 1 - L. S. Bourgeois
O 2 - G. A. Pennington
O 3 - C. R. Knarr
MOD - D. R. Puddy

STS61A-45-0098: One of many Earth views.

First 8-Member Crew

S85-40783 --- Front row (left to right) Furrer/PS (Germany), Dunbar/MS, Buchli/MS, & CDR Hartsfield. Back row (left to right) PLT Nagel, Bluford/MS, Messerschmid/PS (German), & Ockels/PS (Dutch).

FLT NO.	ORBITER	CREW (7) TITLE, NAMES & EVA'S	LAUNCH SITE, LIFTOFF TIME, LANDING SITES, ABORT TIMES	LANDING SITE/ RUNWAY, CROSSRANGE LANDING TIMES FLT DURATION, WINDS	SSME-TL NOM-ABORT EMERG THROTTLE PROFILE ENG. S.N.	SRB RSRM AND ET	INC	ORBIT HA/HP	FSW	PAYLOAD WEIGHTS, PAYLOADS/ EXPERIMENTS	MISSION HIGHLIGHTS (LAUNCH SCRUBS/DELAYS, TAL WEATHER, ASCENT I-LOADS, FIRSTS, SIGNIFICANT ANOMALIES, ETC.)
STS 61-B (STS-31) SEQ FLT #23 KSC-23 PAD 39A-23	OV-104 Atlantis (Flight 2) OMS PODS LPO3 - 7 RPO1 - 8 FRC4 - 2	CDR: Brewster H. Shaw, Jr. (Flt 2 - STS-9) P112/R25/V26/M24 PLT: Bryan D. O'Connor P113/R83/M76 M/S: Sherwood C. Spring P114/R84/M77 M/S: Mary L. Cleave P115/R85/F8 M/S: Jerry L. Ross P116/R86/M78 P/S: Charles Walker (Flt 3 - STS 41-D & STS 51-D) P117/R42/V12/M40 P/S: Rudolpho Neri Vela (Mexico) P118/R87/M79	KSC 39A 331:00:29:00Z 7:29:00 PM EST (P) 7:29:00 PM EST (A) Tuesday 3 11/26/85 (5) LAUNCH WINDOW: 9 Minutes KU-SAT B/U DPLY-AUSSAT SUN SHIELD FAIL PLS - EDW SLS - KSC ALS - NOR AOA - EDW AOA WX - NOR, KSC TAL - DAKAR (SELECTED) TAL WX - MORON MAX Q = 723 PSF M = 1.16	EDW 22, Concrete (EDW 17, CONC 5) 1:33:49 PM PST Tuesday 6 12/03/85 (2) XRANGE: 533 NM ORB DIR:AL 4 AIM PT: NOM MLGTD: 2386 FT 337:21:33:49Z VEL: 201 KGS 191 KEAS HDOT: -1.0 FPS TD NORM 195: 2026 FT NLGTD: 5909 FT 337:21:34:00Z VEL: 160 KGS HDOT: -3.6 FPS BRK INIT: 126 KGS	104/104 109% 100/104/ 65/104/ 103/74/ 65 1 = 2011 (3) 2 = 2019 (3) 3 = 2017 (5) M 3 EOM WEIGHT: 205880 X CG: 1084.4 LANDING: WEIGHT: 205732 X CG: 1085.9	BI-023 MTR: HPM CASE: LWC ET-22 LWT- 15 ET RPT 231 K 1:19:20 MET ET BR/UP 207 K 1:19:56 MET ET IMPACT LAT: 17.31 N LONG: 156.69 W	28.454 (15)	DIRECT INSERTION POST OMS-2 191.33 X 190.12 NM MORELOS DEPLOY 192.71 NM AUSSAT DEPLOY 196.43 NM SATCOM DEPLOY 197.17 NM DEORBIT 209 X 172 NM VELOCITY 25882 FPS RANGE 4099 NM	OI6-30 (5)	CARGO: 47509 lbs CHARGEABLE: 42788 lbs DEPLOYABLE: 27465 lbs NON-DEPLOY: 13986 lbs MIDDECK: 1337 lbs RETURNED: 20074 lbs SHUTTLE ACCUMULATED WEIGHTS: DEPLOYED: 272210 lbs NON-DEPLOYED: 351192 lbs CARGO TOTAL: 714195 lbs PERFORMANCE MARGINS (LBS): FPR: 5284 FUEL BIAS: 849 FINAL TDDP: 874 RECON: 2332	KSC W/D: OPF 27, VAB 4, PAD 14 = 46 LAUNCH POSTPONEMENTS: None. LAUNCH SCRUBS: None. LAUNCH DELAYS: None. NIGHT LAUNCH: Shuttle #2 TAL WX: Dakar go, Moron no-go - clouds. FLIGHT DURATION CHANGES: - EDW lakebed wet, changed to EDW 22 and landed one rev early due to lighting conditions on EDW 22. - Shortened flight by one rev. EVENTS: - OMS-1 not performed. - MORELOS deployed 331:07:46:50Z (rev 6). - AUSSAT deployed 332:01:21Z (rev 17). - SATCOM deployed 332:21:57:31Z (rev 31). - EVA 1 - Assembled/disassembled - ACCESS ten bays and six EASE assembly/disassembly cycles. - EVA 2 - Completed all tasks.

MCC FCR-2 (14)

FLIGHT DIRECTORS:
Asc/Ent - G. E. Coen
O 1 - W. D. Reeves
Ld/O 2 - J. T. Cox
Plng - C. W. Shaw
MOD - D. R. Puddy

EMU/TETHERED EVA'S:
EV1 - Jerry Ross
EV2 - Woody Spring
EVA 1 - 11/29/85
5:34 -SS EVA#12
EVA 2 - 12/1/85
6:46 - SS EVA #13
DEMO SPACE
STATION ASSEMBLY
TECHNIQUES

SRB SEP:
2:03.56 MET

MECO:
8:31.29 MET

ET SEP:
8:49.45 MET

OMS-1:
NONE

OMS-2:
40:25 MET
180.4 Seconds

AVE BRK DECEL:
7 FPS/S

WHEELS STOP:
337:21:35:07Z
13145 FT

ROLLOUT:
10759 FT
78 SEC

WINDS:
8T, 2R KNOTS
OFFICIAL:4T, 4R

DENS ALT: 2551 FT

FLT DURATION:
6:21:04:49
165:04:49

S/T: 146:12:41:14

OV-104:
10:22:49:27

DISTANCE:
2,466,956 sm

SIGNIFICANT ANOMALIES:
- Excess helium in cryo 02 fans 1 and 2.
- Fuel cell 2 performance degraded and CPM hung up.
- OMS XFD OX Center Heater failed.
- WSB #3 Reg. pressure decay.
- Port PLS R-T-L CLOSE A failed.
- Port PLBD aft.
- NLG Strut 3" low.
- Volume H locker had to be pried open.
- GSE side hatch "T" handle broke.
- Gas leaks and erosion to both nozzle-to-case joints (blowby on LH SRM).
- Radiators deployed #12 (deployed for 10-hour DTO)

PAYLOADS:
SATCOM KU-2/
PAM D-2
DEPLOYED

MORELOS-B/
PAM-D DEPLOYED

AUSSAT-2/PAM-D
DEPLOYED

SKT
EASE/ACCESS/MP
ESSIMAX
CFES
DMOS
GAS(1)
MPSE

4 CRYO TANK
SETS

RMS 17 (S.N. 303)
Used for
EASE/ACCESS
assembly, PKM
monitors, waste
water dump monitor

S85-38825 --- STS-61-B Crew Portrait

61B-41-019: During 2nd EVA Ross (above) & Spring erected a Tower known as Assembly Concept for Construction of Erectable Space Structures.

FLT NO.	ORBITER	CREW (7) / TITLE, NAMES & EVA'S	LAUNCH SITE, LIFTOFF TIME, LANDING SITES, ABORT TIMES	LANDING SITE/ RUNWAY, CROSSRANGE / LANDING TIMES, FLT DURATION, WINDS	SSME-TL NOM-ABORT EMERG / THROTTLE PROFILE ENG. S.N.	SRB RSRM AND ET	INC	ORBIT HA/HP	FSW	PAYLOAD WEIGHTS, PAYLOADS/ EXPERIMENTS	MISSION HIGHLIGHTS (LAUNCH SCRUBS/DELAYS, TAL WEATHER, ASCENT I-LOADS, FIRSTS, SIGNIFICANT ANOMALIES, ETC.)
STS 61-C (STS-32) SEQ FLT #24 KSC-24 PAD 39A-24	OV-102 Columbia (Flight 7) OMS PODS LPO4 - 3 RPO4 - 3 FRC2 - 7	CDR: Robert L. Gibson (Flt 2 - STS 41-B) P119/R30/V27/M29 PLT: Charles F. Bolden P120/R88/M80 M/S: George D. Nelson (Flt 2 - STS 41-C) P121/R37/V28/M36 M/S: Steven A. Hawley (Flt 2 STS 41-DR) P122/R39/V29/M38 M/S: Franklin Chang-Diaz P123/R89/M81 P/S: C. W. Nelson (Congressman) P124/R90/M82 P/S: R. J. Cenker (RCA) P125/R91/M83 MCC FCR-1 (10) FLIGHT DIRECTORS: AscEnt - G. E. Coen Ld/O 1 - J. H. Greene O 2 - J. M. Heflin Plng - G. A. Pennington MOD - T. W. Holloway	KSC 39A 12:11:55:00Z 6:55:00 AM EST (P) 6:55:00 AM EST (A) Sunday 3 1/12/86 (2) LAUNCH WINDOW: 49 mins SATCOM KU THERMAL CONSTR ORBIT 8A PLS - KSC SLS - EDW ALS - NOR AOA - EDW AOA WX - NOR,KSC TAL - DAKAR TAL WX - MORON (SELECTED) MAX Q = 696 PSF M = 1.13 SRB SEP: 2:07.23 MET MECO: 8:21.29 MET ET SEP: 8:39.77 MET OMS-1: 10:51 MET 164.03 Seconds ΔV = 265.8 FPS OMS-2: 46.05 MET 136.38 Seconds ΔV = 216.9 FPS	EDW 22, Concrete (EDW 18, CONC 6) 5:58:51 AM PST Saturday 5 1/18/86 (2) XRANGE: 661 NM ORB DIR: DL 15 AIM PT: NOM MLGTD: 1530 FT 18:13:58:51Z VEL: 217 KGS 212 KEAS HDOT: -2 FPS TD NORM 195: 2970 FT NLGTD: 6300 FT 18:13:59:07Z VEL: 160 KGS HDOT: -3.1 FPS BRK INIT: 138 KGS AVE BRK DECEL: 7.2 FPS/S WHEELS STOP: 18:13:59:50Z 11727 FT ROLLOUT: 10202 FT 59 SEC WINDS: 2T, 0X KNOTS OFFICIAL: 1H, 1R DENS ALT: 1088 FT FLT DURATION: 6:02:03:51 146:03:51 S/T: 152:14:45:05 OV-102: 41:01:54:11 DISTANCE: 2,197,305 sm	104/104 109% 100/104/ 85/69/ 104 1 = 2015 (5) 2 = 2018 (8) 3 = 2109 (9) BI-STABLE HPOTP (2) M 3 EOM WEIGHT: 210325 X CG: 1083.6 LANDING: WEIGHT: 210161 X CG: 1085.1	BI-024 MTR: HPM CASE: LWC ET-30 LWT- 23 ET RPT 239K 46:25 MET ET BR/UP 192K 47:41 MET ET IMPACT LAT: 28.3 S LONG: 81.3 E	28.448 (16)	STANDARD INSERTION POST OMS-2 176.13 X 175.14 NM SAT COM DEPLOY 182.63 NM DEORBIT 184 X 173 NM VELOCITY 25815 FPS RANGE 4154 NM	OI7-32 (1)	CARGO: 32733 lbs PAYLOAD CHARGEABLE: 28625 lbs DEPLOYABLE: 12351 lbs NON-DEPLOY: 15837 lbs MIDDECK: 437 lbs RETURNED: 20111 lbs SHUTTLE ACCUMULATED WEIGHTS: DEPLOYED: 284561 lbs NON-DEPLOYED: 367466 lbs CARGO TOTAL: 746928 lbs PERFORMANCE MARGINS (LBS): FPR: 5407 FUEL BIAS: 840 FINAL TDDP: 10754 RECON: 11127 PAYLOADS: SATCOM KU- 1/ PAM D2 DEPLOYED MSL-2 HITCHHIKER INFRARED - IMAGINING EXP 13 GAS CANS CHAMP IBSE HPCG STUDENT EXP (3) NORMS ACIP AADS 4 CRYO TK SETS NO RMS	KSC W/D: OPF 101, VAB 8, PAD 34 = 143 LAUNCH POSTPONEMENTS: None. LAUNCH SCRUBS: - 12/18/85 launch scrubbed to complete RCS crossfeed work in aft compartment (rescheduled before PRSD loading). 1-day slip. - 12/19/85 launch scrubbed after autohold at T-14 seconds due to RH SRB tilt HPU exceeding RPM redline (oversensitivity in control circuit). Launch rescheduled after holidays for 1/6/86. 18-day slip. - 1/6/86 launch scrubbed at T-31 seconds when GSE LO$_2$ replenish valve failed to close. Wrong manual command sequence resulted in TSM vent and drain valves opening without closing Orbiter fill/drain valve causing off-loading of approximately 18,000 lbs LO$_2$ via F/D valve. LO$_2$ SSME temperature dropped below redline limit and count recycled to T-20 minutes. Did an IMU alignment; however, launch was scrubbed when SATCOM launch window expired. Detanked and found a broken GSE LOX temperature probe lodged in SSME #2 prevalve (would have precluded full prevalve closure). Launch rescheduled for 1/7/86. 1-day slip. - 1/7/86 launch was scrubbed at T-9 hold due to bad weather at TAL sites (Dakar & Moron) and marginal KSC weather. Forty-eight hour turnaround for ovality check on MPS low pressure fuel duct. Rescheduled launch for 1/9/86. 2-day slip. - 1/9/86 launch was scrubbed on 1/8/86 because of predicted bad weather at KSC. and temperature GSE probe found in SSME #2 prevalve. Rescheduled launch for 1/10/86. 1-day slip. - 1/10/86 launch scrubbed due to rain showers at KSC with 45 minutes remaining in window. Rescheduled launch for 1/12/86). 2-day slip. - 25-day total slip. LAUNCH DELAYS: None. TAL WX: Dakar no-go - dust, Moron go. FIRSTS - First flight of OV-102 after major mod (included removal of ejection seats and modifying display panels). Continued . . .

61c-005-0036 - US Rep. C.W. Nelson, from Flordia, at work in space.

SPACE SHUTTLE MISSIONS SUMMARY

FLT NO.	ORBITER	CREW (7) TITLE, NAMES & EVA'S	LAUNCH SITE, LIFTOFF TIME, LANDING SITES, ABORT TIMES	LANDING SITE/ RUNWAY, CROSSRANGE LANDING TIMES FLT DURATION, WINDS	SSME-TL NOM-ABORT EMERG THROTTLE PROFILE ENG. S.N.	SRB RSRM AND ET	ORBIT		FSW	PAYLOAD WEIGHTS, PAYLOADS/ EXPERIMENTS	MISSION HIGHLIGHTS (LAUNCH SCRUBS/DELAYS, TAL WEATHER, ASCENT I-LOADS, FIRSTS, SIGNIFICANT ANOMALIES, ETC.)
							INC	HA/HP			
STS 61-C											

61C-14-0008 Crew in middeck; CDR Gibson (lower right corner), others counter-clockwise from upper right: PLT Bolden, U.S. Representative C.W. Nelson/PS, Cenker/RCA-PS, Hawley/MS, Chang-Diaz/MS, G.D. Nelson/MS

ABOVE: 61C-005-0036 -- SATCOM Ku-1 Communications Satellite deployed from Columbia.
BELOW: 61C-S-050 (18 January 1986) --- Second Shuttle night landing. View is of the Shuttle's main landing gear touching down at EAFB with streams of light trailing behind the orbiter.

AT LEFT: 61C-13-005 -- The crew, having received excellent service from the Waste Management System, showed this photo at their Jan. 23, 1986 Post-Flight Press Conference.

Continued . . .

FLIGHT DURATION CHANGES:
- Management decision made to change flight duration to 4 days from 5 days.
- Extended flight from 4 to 5 days due to bad weather at KSC (was 1/16/86).
- Extended flight from 5 to 6 days due to bad weather at KSC (was 1/17/86).
- Waved off KSC landing on 1/18/86 due to bad weather and landed at EDW (one rev extension).
- Flight extensions, 2 days + 1 rev.

LANDING SITE CHANGE:
- KSC to EDW.

NIGHT LANDING:
- Second Shuttle night landing.

EVENTS:
- SATCOM deployed at 9:32 MET (REV 7).
- Bi-stable Pump - HPOTP required minimum throttle of 67 percent (second flight).

SIGNIFICANT ANOMALIES:
- Fuel cell power source to essential bus 1 BC erratic.
- APU 1 gearbox GN_2 pressure high.
- APU's 1 and 3 isolation valve temperatures low.
- APU 3 fuel line system B heater failed .
- Vernier RCS jets fired excessively .
- S-band U/L and L/R antenna performance erratic.
- ECLSS pressure control system 2 oxygen flow transducer read low.
- WSB 3 System "A" heater operation erratic.
- Left RCS Helium Reg "B" leaked.
- WSB 1 system "A" cooling water use high.
- Gas leak in LH SRM nozzle-to-case joint (blowby).
- Gas leak and erosion in RH SRM nozzle-to-case joint.

FLT NO.	ORBITER	CREW (7) TITLE, NAMES & EVA'S	LAUNCH SITE, LIFTOFF TIME, LANDING SITES, ABORT TIMES	LANDING SITE/ RUNWAY, CROSSRANGE LANDING TIMES FLT DURATION, WINDS	SSME-TL NOM-ABORT EMERG THROTTLE PROFILE ENG. S.N.	SRB RSRM AND ET	INC	HA/HP	FSW	PAYLOAD WEIGHTS, PAYLOADS/ EXPERIMENTS	MISSION HIGHLIGHTS (LAUNCH SCRUBS/DELAYS, TAL WEATHER, ASCENT I-LOADS, FIRSTS, SIGNIFICANT ANOMALIES, ETC.)
STS 51-L (STS-33) SEQ FLT #25 KSC-25 PAD 39B-1	OV-099 (Flight 10) Challenger OMS PODS LVO1 - 7 RVO1 - 7 FRC9 - 10	CDR: Francis R. Scobee (Flt 2 - STS 41-C) P126/R34/V30/M33 PLT: Michael J. Smith P127/R92/M84 M/S: Judith A. Resnik (Flt 2 - STS 41-D) P128/R41/V31/F2 M/S: Ronald E. McNair (Flt 2 - STS 41-B) P129/R32/V32/M31 M/S: Ellison S. Onizuka (Flt 2 STS 51-C) P130/R51/V33/M47 P/S: Gregory Jarvis (HAC) P/131/R93/M85 P/S: Christa McAulliffe (Civilian Teacher) P132/R94/F9	KSC 39B 28:16:38:00.1Z 9:38:00 AM EST (P) 11:38:00 AM EST (A) Tuesday 4 1/28/86 LAUNCH WINDOW: 3 Hours TAL SUNSET (CASABLANCA) PLS - KSC SLS - EDW TAL - CASABLANCA TAL WX - DAKAR MAX Q = 720 PSF M = 1.35	FLT DURATION: 00:00:01:14 S/T: 152:14:46:19 OV-099: 62:07:56:21	104/104 109% 1 = 2023 (5) 2 = 2020 (6) 3 = 2021 (6)	BI-026 MTR: HPM CASE: LWC ET-26 LWT-19	28.45	PLANNED STANDARD INSERTION 153.5 NM	OI7-26 (2)	CARGO: 52685 lbs CHARGEABLE: 48633 lbs DEPLOYABLE: 37636 lbs NON-DEPLOYED: 10167 lbs MIDDECK: 830 lbs PRIMARY: TDRS-B/IUS-3 SPARTAN - HALLEY/MPESS ANCILLARY: CHAMP FDE RME TISP PPE SSIP (3) ACIP 3 CRYO TANK SETS RMS 18 (S.N. 302)	KSC W/D: OPF 30, VAB 5, PAD 28 = 63 LAUNCH POSTPONEMENTS: - On 12/23/85, the 1/22/86 launch was postponed 1 day to 1/23/86 to accommodate an integrated simulation (STS 61-C launch delay impact). 1-day slip. - On 1/22/86, the 1/23/86 launch was postponed 2 days to 1/25/86 because of KSC work schedule being impacted by STS 61-C landing delays. 2-day slip. LAUNCH SCRUBS: - 1/25/86 launch scrubbed early in count by MMT due to forecast of unacceptable weather at KSC throughout launch window. Launch rescheduled for 1/27/86. - 1/27/86 launch scrubbed. Countdown halted at T-9 minutes when a GSE hatch fixture could not be removed from exterior of side hatch, followed by a problem with a portable drill. Handling tool attach screw was drilled out. One hour and 20 minutes later, when the hatch problem was resolved, the winds at KSC RTLS runway had increased and exceeded the maximum allowable crosswind velocity. Launch rescheduled for 1/28/86. 6-day total slip. - During the night, the temperature at KSC dropped to the low twenties. Ice had accumulated in the pad area and ice inspections were made during night and morning of 1/28. LAUNCH DELAYS: - 1H00M delay during T-3 hour hold due to late ET tanking start caused by a GSE H₂ fire alarm detector problem in LH₂ ground storage tank. - 1H00M additional delay after ice team inspection of ice formed by leaking H₂O hoses. The decision was made to allow additional time for ice on pad to melt. - 2H00M launch delay total. LAUNCH: - Launch occurred at 11:38:00.010 a.m. EST on January 28, 1986. - Explosive burn at MET of 74 seconds. FIRSTS: - First Shuttle launch from pad 39B. - First flight to use Casablanca as TAL site. - First flight to use DIAL-A-TAL site. - First Shuttle failure in flight. Destroyed Vehicle and Crew.

MCC FCR-2 (15)
FLIGHT DIRECTORS:
Asc - J. H. Greene
Ent - A. L. Briscoe
Ld/O 1 - B. R. Stone
O 2 - C. W. Shaw
Plng - C. R. Knarr
MOD - D. R. Puddy

STS-51L Crew photo with Commander Francis R. Scobee, Pilot Michael J. Smith, Mission Specialists Judith A. Resnik, Ellison S. Onizuka, Ronald E. McNair and Payload Specialists Gregory B. Jarvis and Sharon Christa McAuliffe. (S85-44253)

IN MEMORIAM

Shuttle Legacy Mural - In KSC LCC Firing Room

CHALLENGER TRIBUTE

KSC-2010-4451 (http://mediaarchive.ksc.nasa.gov/index.cfm). This Tribute Display features Challenger, which blazed a trail for other vehicles with the first night landing (STS-8) and also the first landing at Kennedy Space Center (STS-41B). The spacewalker represents Challenger's role in the first spacewalk during a space shuttle mission (STS-6) and the first untethered spacewalk (STS-41B). Crew-designed patches for each of Challenger's missions lead from earth toward our remembrance of the STS-51L crew. Other significant accomplishments include the first night launch with STS-8; the first in-flight capture, repair, and redeployment of an orbiting satellite during STS-41C; the first American woman in space (Sally Ride on STS-7); the first African-American in space (Guion Bluford on STS-8); and the first American woman to walk in space (Kathryn Sullivan during STS-41G). By Mike Leinbach/Launch Director & Amy Simpson/KSC PH-2 in May 2010

SPACE SHUTTLE MISSIONS SUMMARY

FLT NO.	ORBITER	CREW (5) TITLE, NAMES & EVA'S	LAUNCH SITE, LIFTOFF TIME, LANDING SITES, ABORT TIMES	LANDING SITE/ RUNWAY, CROSSRANGE LANDING TIMES FLT DURATION, WINDS	SSME-TL NOM-ABORT EMERG THROTTLE PROFILE ENG. S.N.	SRB RSRM AND ET	INC	ORBIT HA/HP	FSW	PAYLOAD WEIGHTS, PAYLOADS/ EXPERIMENTS	MISSION HIGHLIGHTS (LAUNCH SCRUBS/DELAYS, TAL WEATHER, ASCENT I-LOADS, FIRSTS, SIGNIFICANT ANOMALIES, ETC.)
STS-26 (STS-26R) SEQ FLT #26 KSC-26 PAD 39B-2	OV-103 (Flight 7) Discovery OMS PODS LPO4 - 4 RPO3 - 8 FRC3 - 7	CDR: Frederick H. Hauck (Flt 3 - STS-7 & STS 51-A) P133/R17/V7/M17 PLT: Richard O. Covey (Flt 2 - STS 51-I) P134/R73/V34/M67 M/S 1: John M. Lounge (Flt 2 - STS 51-I) P135/R74/V35/M68 M/S 2: George D. Nelson (Flt 3 - STS 41-C & STS 61-C) P136/R37/V28/M36 M/S 3: David C. Hilmers (Flt 2 - STS 51-J) P137/R77/V36/M71 MCC FCR-1 (11) FLIGHT DIRECTORS: Asc/Ent - G. E. Coen O 1 - J. M. Heflin O 2 - C. W. Shaw Ld/Plg - L.S.Bourgeois MOD - T. W. Holloway MDR - B. R. Stone MDR - R. M. Kelso S26-09-008 --- Crew (No caption available) - see who you recognize from list above.	KSC 39B 273:15:37:00Z 9:59:00 AM EDT (P) 11:37:00 AM EDT (A) Thursday 7 9/29/88 (1) WINDOW DURATION: 3 HOURS (CREW CONSTRAINT) PLS - EDW SLS - NOR AOA- EDW - NOR TAL - BEN GUERIR TAL WX - MORON (SELECTED) AUGMENTED CTG: BANJUL MAX Q = 707 M = 1.16 SRB SEP: 2:04.8 MET MECO: 8:33.43 MET ET SEP: 8:50.5 MET OMS-1: NONE OMS-2: 39.55 MET 141.6 Seconds 222 FPS	EDW 17L (EDW 19, LKBD 13) 9:37:11 AM PDT Monday 5 10/3/88 (3) DEORBIT BURN: 277:15:34:44Z XRANGE: 383 NM ORB DIR: DL 16, REV 64 AIM PT: NOM MLGTD: 2569 FT 277:16:37:11Z VEL: 196 KGS 187 KEAS HDOT: -0.5 FPS (SR + 11 MIN) TD NORM 195: 1849 FT NLGTD: 5671 FT 277:37:16:18Z VEL: 150 KGS HDOT: -5.8 FPS BRK INIT: 127 KGS AVE BRK DECEL: 7.2 FPS/S WHEELS STOP: 277:16:37:57Z 10020 FT ROLLOUT: 7451 FEET 50 SECONDS WINDS: 3T, 0X KNOTS OFFICIAL: 5H, 1L DENS ALT: 3445 FT FLT DURATION: 4:01:00:11 97:00:11 S/T: 156:15:46:30 OV-103: 42:07:06:31 DISTANCE: 1,430,505 sm	104/104 109% 104/102/ 65/104/ 65 1 = 2019 (4) 2 = 2022 (1) 3 = 2028 (1) M 3 EOM WEIGHT: 194347 X CG: 1096.6 LANDING: WEIGHT: 194184 X CG: 1098.3	BI-029 RSRM 1 360L 001 ET-28 LWT-21 ET RPT 231K 1:17:18 MET ETBR/UP 211K 1:17:51 MET ET IMPACT LAT: 12.58 N LONG: 164.04 W	28.46 (17)	DIRECT INSERTION POST OMS-2 162.61 X 169.02 NM TDRS-C DEPLOY 165.88 NM DEORBIT 177 X 163 NM VELOCITY 25790 FPS RANGE 4117 NM	OI-8B (1)	CARGO: 46448 lbs PAYLOAD CHARGEABLE: 44601 lbs DEPLOYABLE: 37514 lbs NON-DEPLOYED: 5928 lbs MIDDECK: 1159 lbs RETURNED: 8964 lbs SHUTTLE ACCUMULATED WEIGHTS: DEPLOYED: 322075 lbs NON-DEPLOYED: 374553 lbs CARGO TOTAL: 793376 lbs PERFORMANCE MARGINS (LBS): FPR: 5169 FUEL BIAS: 949 FINAL TDDP: 1546 RECON: 624 PAYLOADS: PLB: TDRS-C/IUS DEPLOYED OASIS-1 MIDDECK: PVTOS-2 ADSF, IRCFE PCG IEF PPE ARC MLE ELRAD SSIP(2) SE84-4 SE84-5 3 CRYO TANK SETS NO RMS	KSC W/D: OPF 221, VAB 13, PAD 88 = 322 LAUNCH POSTPONEMENTS: - 9/26/88 launch postponed 3 days to 9/29/88 for Orbiter aft critical path. 3-day slip. LAUNCH SCRUBS: None. LAUNCH DELAYS: - 1H38M delay from 9:59 a.m. EDT due to: (1) winds aloft differed from planned autumn winds with exceedences of WLE-14R and WLE-14L, and (2) PLT and M/S 1 suit fan fuses blew (replaced with 10A fuses but intended 5 amp fuses). FLIGHT DURATION CHANGES: None. TAL WX: - Alternate TAL Moron selected due to rain showers and crosswind violations at Ben Guerir (Prime). FIRSTS: - Return to flight 2 yrs 8 mos after STS 51-L. EVENTS: - TDRS-C deployed at 06:13:05 MET (rev 3). - Two engines OMS SEP burn at 06:28:03 MET (16.6 sec, 30.85 FPS). - Deorbit burn 168 secs, 324.86 FPS. - ET Reentry (tumble) - CAST GLANCE violent rupture. SIGNIFICANT ANOMALIES: - Prelaunch H$_2$ leak at 4"disc. - RCS dynatube repair early in flow using clamshell. - OMS gimbal standby enable 1 fail. - FES high load evap freezing during ascent. FES shutdown during entry after OMS deorbit burn (rust/ contamination). - Ku-Band failed self test. Antenna would not follow pointing commands. (Had to use alternate stow procedure.) - GOX flow control valves 1 and 2 operated sluggish on first cycle. - WCS fan separator 1 flooded exhibiting stall currents for 80 secs. - STBD PLBD Forward R-T-L "A" Talkback failed to function. - APU#3 chamber pressure low. - Rt wing TPS damage. - 4" LH$_2$ ET/Orbiter disconnect leak. - Radar altimeter failed at 50 feet. - Video cassette tapes jammed (4 tapes).

Return - To - Flight

In MCC: G. Kranz, T. Holloway, A. Cohen, & unidentified.

FLT NO.	ORBITER	CREW (5) TITLE, NAMES & EVA'S	LAUNCH SITE, LIFTOFF TIME, LANDING SITES, ABORT TIMES	LANDING SITE/ RUNWAY, CROSSRANGE LANDING TIMES FLT DURATION, WINDS	SSME-TL NOM-ABORT EMERG THROTTLE PROFILE ENG. S.N.	SRB RSRM AND ET	ORBIT		FSW	PAYLOAD WEIGHTS, PAYLOADS/ EXPERIMENTS	MISSION HIGHLIGHTS (LAUNCH SCRUBS/DELAYS, TAL WEATHER, ASCENT I-LOADS, FIRSTS, SIGNIFICANT ANOMALIES, ETC.)
							INC	HA/HP			
STS-27 (STS-27R) SEQ FLT #27 KSC-27 PAD 39B-3	OV-104 (Flight 3) Atlantis OMS PODS LPO1 - 9 RPO1 - 9 FRC4 - 3	CDR: Robert L. Gibson (Flt 3 - STS 41-B & STS 61-C) P138/R30/V27/M29 PLT: Guy S. Gardner P139/R95/M86 M/S 1: Richard M. Mullane (Flt 2 - STS 41-DR) P140/R40/V37/M39 M/S 2: Jerry L. Ross (Flt 2 - STS 61-B) P141/R86/V38/M78 M/S 3: William M. Shephard P142/R96/M87 MCC FCR-2 (16) FLIGHT DIRECTORS: Asc - G. E. Coen O1/Ent - A. L. Briscoe Ld/O 2 - B. R. Stone Plng - C. R. Knarr MOD - T. W. Holloway	KSC 39B 337:14:30:34Z 9:30:34 AM EST Friday 5 12/2/88 (1) PLS - EDW AOA - NOR AOA WX: TAL - ZARAGOZA (SELECTED) TAL WX - MORON BEN GUERIR	EDW 17L (EDW 20, LKBD 14) 3:36:11 PM PST Tuesday 7 12/6/88 (3) DEORBIT BURN: 341:22:29:34Z CROSSRANGE: 520 NM ORBIT DIR: DR 4 AIM PT: NOM MLGTD: 1469 FT 341:23:36:11Z VEL: 204 KGS 194 KEAS HDOT: -1.0 FPS TD NORM 195: 1523 FT NLGTD: 4423 FT 341:23:36:18Z VEL:164 KGS HDOT: -4.9 FPS BRK INIT: 132 KGS AVE BRK DECEL: 9.8 FPS/S WHEELS STOP: 341:23:36:52Z 8592 FEET ROLLOUT: 7123 FEET 41 SECONDS WINDS: 0H, 2L KNOTS OFFICIAL: 0H, 0X DENS ALT: 3047 FT FLT DURATION: 4:09:05:37 105:05:37 S/T: 161:00:52:05 OV-104: 15:07:55:04 DISTANCE: 1,812,075 sm	100/104/ 96/65/ 104/65 1 = 2027 (1) 2 = 2030 (1) 3 = 2029 (1) M 3 EOM WEIGHT: X CG: LANDING: WEIGHT: 190956 X CG: 1095.1	BI-030 RSRM 2 360L 002 ET-23 LWT-16 ET RPT 236K 1:24:30 MET ET BR/UP 216K 1:25:03 MET ET IMPACT LAT: 2.86 S LONG: 123.48 W	57 (5)	DEORBIT 244 X 239 NM VELOCITY 25956 FPS RANGE 4220 NM	OI-8B (2)	DOD FLIGHT PERFORMANCE MARGINS (LBS): FPR: 4698 FUEL BIAS: 968 FINAL TDDP: 2905 * RECON: -286 SECONDARY PAYLOADS: OASIS-II AMOS APE CLOUDS CRUX RME-III VFT-2 RMS 19 (S.N. 201) Used for belly tile damage survey	KSC W/D: OPF 196, VAB 10, PAD 30 = 236 LAUNCH POSTPONEMENTS: None. LAUNCH SCRUBS: - 12/1/88 launch scrubbed due to winds aloft exceedances. Launch rescheduled for 12/2/88. 1-day slip. LAUNCH DELAYS: - Countdown held at T-9 due to winds aloft and at T-31 seconds for TAL weather. TAL WX: - Zaragoza (prime) selected, alternate sites were no go - low ceilings at Moron and Ben Guerir. ALTERNATE ASCENT I-LOADS: - LSEAT selected nominal ascent I-loads, no uplink required. FIRSTS: - First flight with alternate ascent I-loads capability. - First flight using East and West TDRS. - First flight with no communications blackout during entry (due to favorable comm look angle to West TDRS). - First flight of PDRS console position. SIGNIFICANT ANOMALIES: - Left inboard tire leaking since OPF (over-inflation plug seal). - APU #2 GG heater system malfunction. - Humidity separator B flooded. - TAGS paper jam. - TPS damage worst to date (707 hits, 298 hits > 1", most on right side bottom of wing and fuselage). - Tile survey conducted using RMS end effector camera. - R RCS Oxidizer B He regulator slow response. - Cabin temp controller #2 non-responsive. - L OMS GN_2 Isolation valve coil failure. - Engine #3 HPOTP #3 bearing inner race crack due to stress corrosion. Liquid stains, pitting, spalling - chlorine contaminant.

STS027-11-012 --- Crew on flight deck: Left to right CDR Gibson, Mullane/MS, Ross/MS, Shepherd/MS, & PLT Gardner. Floating football was presented to the NFL at the Super Bowl in Miami.

After his smooth landing at EDW, Gibson and others were astonished at severity of tile damage.

SPACE SHUTTLE MISSIONS SUMMARY

FLT NO.	ORBITER	CREW (5) / TITLE, NAMES & EVA'S	LAUNCH SITE, LIFTOFF TIME, LANDING SITES, ABORT TIMES	LANDING SITE/ RUNWAY, CROSSRANGE / LANDING TIMES FLT DURATION, WINDS	SSME-TL NOM-ABORT EMERG / THROTTLE PROFILE ENG. S.N.	SRB RSRM AND ET	ORBIT INC	ORBIT HA/HP	FSW	PAYLOAD WEIGHTS / PAYLOADS/ EXPERIMENTS	MISSION HIGHLIGHTS (LAUNCH SCRUBS/DELAYS, TAL WEATHER, ASCENT I-LOADS, FIRSTS, SIGNIFICANT ANOMALIES, ETC.)
STS-29 (STS-29R) SEQ FLT #28 KSC-28 PAD 39B-4	OV-103 (Flight 8) Discovery OMS PODS LP04 - 5 RP03 - 9 FRC3 - 8	CDR: Michael L. Coats (Flt 2 - STS 41-DR) P143/R38/V39/M37 PLT: John E. Blaha P144/R97/M88 M/S: James F. Buchli (Flt 3 - STS 51-C & STS 61-A) P145/R52/V24/M48 M/S: Robert C. Springer P146/R98/M89 M/S: James P. Bagian P147/R99/M90 MCC FCR-1 (12) FLIGHT DIRECTORS: Asc/Ent - A. L. Briscoe O 1 - G. A. Pennington Ld/O 2 - C. W. Shaw Plng - R. D. Dittemore MOD - T. W. Holloway MDR - B. R. Stone MDR - R. M. Kelso	KSC 39B 72:14:57:00Z 8:07:00 AM EST (P) 9:57:00 AM EST (A) Monday 7 3/13/89 (2) PLS - EDW AOA - NOR TAL - BEN GUERIR (Selected) TAL WX - MORON CLS - BANJUL LAUNCH WINDOW: 2.5 HOURS (CREW TIME ON BACK) MAX Q =710 M = 1.44 SRB SEP: 2:04.5 MET MECO: 8:30.8 MET ET SEP: 8:50 MET OMS-1: NONE OMS-2: 39:58 MET 141.4 Seconds 221.8 FPS	EDW 22 (EDW 21, CONC 7) 6:35:50 AM PST Saturday 6 3/18/89 (2) DEORBIT BURN: 77:13:35:15Z XRANGE: 384 NM ORB DIR: AL 5, ORBIT 79, REV 80 AIM PT: NOM MLGTD: 1195 FT 77:14:35:50Z VEL: 204 KGS 205 KEAS HDOT: -3 FPS TD NORM 195: 2085 FT NLGTD: 5027 FT 77:14:36:01Z VEL:162 KGS HDOT: -1.9 FPS BRK INIT: 129 KGS AVE BRK DECEL: 8 FPS/S WHEELS STOP: 77:14:36:41Z 10534 FT ROLLOUT: 9339 FEET 51 SECONDS WINDS: 4.4H,4.1L KNOTS OFFICIAL: 6H, 1L DENS ALT: 1853 FT FLT DURATION: 4:23:38:50 119:38:50 S/T: 166:00:30:57 OV-103: 47:06:45:21 DISTANCE: 1,800,000 sm	104/104 109% 100/104/ 66/104/ 65 1 = 2031 (1) 2 = 2022 (2) 3 = 2028 (2) M 3 EOM WEIGHT: 194940 X CG: 1093.7 LANDING: WEIGHT: 194790 X CG: 1095.3	BI-031 RSRM 3 360L 003 ET-38 LWT- 29 ET RPT 240K 1:17:11 MET ET BR/UP 217K 1:17:50 MET ET IMPACT LAT: 13.20 N LONG: 162.65 W	28.45 (18)	DIRECT INSERTION POST OMS-2 162.59 X 160.27 NM TDRS-D DEPLOY 162.63 NM DEORBIT 178 X 164 NM VELOCITY 25787 FPS RANGE 4163 NM	OI-8B (3)	CARGO: 47394 lbs PAYLOAD CHARGEABLE: 45316 lbs DEPLOYABLE: 37640 lbs NON-DEPLOYED: 6727 lbs MIDDECK: 949 lbs RETURNED: 9784 lbs SHUTTLE ACCUMULATED WEIGHTS: DEPLOYED: 359715 lbs NON-DEPLOYED: 382229 lbs CARGO TOTAL: 840770 lbs PERFORMANCE MARGINS (LBS): FPR: 4698 FUEL BIAS: 968 FINAL TDDP: 3772 RECON: 2995 PAYLOADS: PLB: TDRS-D/IUS DEPLOYED SHARE OASIS-1 MIDDECK: IMAX PCG AMOS CHROMEX SSIP (2): SE 82-08 GAS: SE 82-08 CHIX 3 CRYO TK SETS NO RMS	KSC W/D: OPF 94, VAB 11, PAD 39 = 144 LAUNCH POSTPONEMENTS: - 3/11/89 launch postponed 1 day to 3/12/89 to replace MEC #2. - 3/12/89 launch postponed 1 day to 3/13/89 to replace FPOV actuator. 2-day total slip. LAUNCH SCRUBS: None. LAUNCH DELAYS: - 1H50M launch delay due to winds aloft and ground fog at KSC. TAL WX: - Ben Guerir (prime) selected - weather good throughout. ALTERNATE ASCENT I-LOADS: - LSEAT selected YAW negative which was uplinked (first uplink). FLIGHT DURATION CHANGES: None. FIRSTS: - First flight with corner alternate I-load capability. - First flight alternate ascent I-load uplinked. EVENTS: - TDRS-D/IUS deployed at 06:12:48 MET (rev 5). - SEP burn at 06:27:48 MET, 16.48 seconds, 31.1 FPS - OASIS-1 performed nominally. - DTO 0517 NWS Runway Evaluation. - DTO 0518 Revised System Braking Test. - Deorbit burn 162 seconds, 313.2 FPS. ET ENTRY (TUMBLE) CAST GLANCE: - Tumble rate 62 deg/sec prior to rupture, max DV - 552 FPS, number of pieces-30. Continued . . .

STS029-71-000AE---IUS/TDRS-D deployment from Discovery payload bay.

FLT NO.	ORBITER	CREW (5) TITLE, NAMES & EVA'S	LAUNCH SITE, LIFTOFF TIME, LANDING SITES, ABORT TIMES	LANDING SITE/ RUNWAY, CROSSRANGE LANDING TIMES FLT DURATION, WINDS	SSME-TL NOM-ABORT EMERG THROTTLE PROFILE ENG. S.N.	SRB RSRM AND ET	ORBIT INC	HA/HP	FSW	PAYLOAD WEIGHTS, PAYLOADS/ EXPERIMENTS	MISSION HIGHLIGHTS (LAUNCH SCRUBS/DELAYS, TAL WEATHER, ASCENT I-LOADS, FIRSTS, SIGNIFICANT ANOMALIES, ETC.)

STS-29

Continued

ABOVE: S89-28089 & KSC-89PC-26---OV-103,suspended by overhead crane hooked to support structure attached at four points, is lowered for mating to ET & SRBs at KSC VAB Bay 1. SSMEs are covered with protective red shields
BELOW: STS029-04-029---CDR Coats on OV-103's forward flight deck

STS029-78-003--- IUS / TDRS-D after deployment from Discovery

BELOW: STS029-S-066--- Post Landing: Crew pose with NASA officials. Left to right: PLT Blaha, Bagian/MS, Rear Adm. Richard H. Truly/NASA Associate Administrator for Space Flight, Dr. James C. Fletcher/NASA Administrator, CDR Coats, Buchli/MS and Springer/MS.

Continued . . .

SIGNIFICANT ANOMALIES:
- RCS jet R1U failed off at ET Sep.
- Excessive vapor at H_2 ET/Orbiter umbilical area prelaunch and tower clear.
- TAGS developer overtemp; however, best TAGS performance with more than 660 pages processed.
- Sluggish GOX FCV'S system 1 and 3.
- LH2 disconnect slow to close.
- FES shutdown during deorbit prep switch reconfiguration.
- Unable to dump ops 2 track 4.
- R OMS regulator "A" anomaly (OX & FU tank pressures approx 245 psi).
- SHARE operations had problems due to vapor bubbles in liquid channels.
- IMAX camera drive mechanism problem (belt jumped off track).
- CHROMEX not cooling properly.
- PLBD PORT B CLOSED indicator failed.
- TPS 132 debris hits, 23 greater than 1"

s29-s-0041 -- Flight Directors Lee Briscoe and Ron Dittemore on console in MCC Flight Control Room.

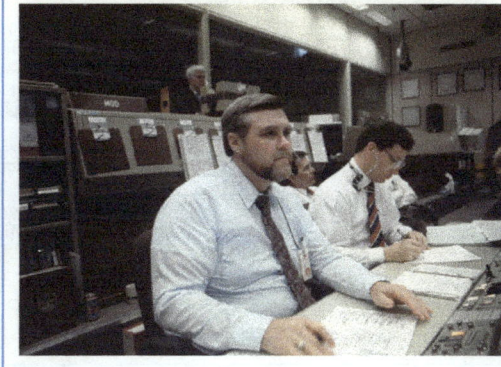

SPACE SHUTTLE MISSIONS SUMMARY

FLT NO.	ORBITER	CREW (5) / TITLE, NAMES & EVA'S	LAUNCH SITE, LIFTOFF TIME, LANDING SITES, ABORT TIMES	LANDING SITE/ RUNWAY, CROSSRANGE / LANDING TIMES FLT DURATION, WINDS	SSME-TL NOM-ABORT EMERG / THROTTLE PROFILE ENG. S.N.	SRB RSRM AND ET	INC	ORBIT HA/HP	FSW	PAYLOAD WEIGHTS, PAYLOADS/ EXPERIMENTS	MISSION HIGHLIGHTS (LAUNCH SCRUBS/DELAYS, TAL WEATHER, ASCENT I-LOADS, FIRSTS, SIGNIFICANT ANOMALIES, ETC.)
STS-30 (STS-30R) SEQ FLT #29 KSC-29 PAD 39B-5	OV-104 (Flight 4) Atlantis OMS PODS LPO1 - 10 RPO1 - 10 FRC4 - 4	CDR: David M. Walker (Flt 2 - STS 51-A) P148/R48/V40/M45 PLT: Ronald J. Grabe (Flt 2 - STS 51-J) P149/R76/V41/M70 M/S 1: Mark C. Lee R150/R100/M91 M/S 2: Norman E. Thagard (Flt 3 - STS-7 & STS 51-B) P151/R20/V14/M19 M/S 3: Mary L. Cleave (Flt 2 - STS 61-B) P152/R85/V42/F8 MCC FCR-1 (13) FLIGHT DIRECTORS: Asc - A. L. Briscoe O 1/E - R. D. Dittemore Ld/O 2 - J. M. Heflin Plng - W. D. Reeves MOD - L. S. Bourgeois MDR - C. W. Shaw	KSC 39B 124:18:46:58.975Z 1:48:00 PM EDT (P) 2:46:59 PM EDT (A) Thursday 8 5/4/89 (1) WINDOW DURATION: 64 Minutes (TAL LIGHTING) PLS - EDW AOA - EDW TAL - BEN GUERIR (SELECTED) TAL WX - MORON CTG - BANJUL RTLS 15 MAX Q = 676 M = 1.07 SRB SEP: 2:05.26 MET MECO: 8:29.37 MET ET SEP: 8:46.67 MET OMS-1: 10:29 MET 141.72 Seconds 226.29 FPS OMS-2: 44:27 MET 125.32 Seconds 197.03 FPS	EDW 22, CONC (EDW 22, CONC 8) 12:43:26 PM PDT Monday 6 5/8/89 (2) DEORBIT BURN: 128:18:40:49Z 165.7, DV 326 XRANGE: 350 NM ORB DIR: AL6, AIM PT: NOM MLGTD: 1314 FT 128:19:43:26Z VEL: 204 KGS 196 KEAS HDOT: -1.5 FPS TD NORM 195: 1354 FT NLGTD: 5088 FT 128:19:43:38Z VEL:163 KGS HDOT: -1.7 FPS BRK INIT: 128 KGS AVE BRK DECEL: 6.2 FPS/S WHEELS STOP: 128:19:44:30Z 11609 FEET ROLLOUT: 10295 FEET 64 SECONDS WINDS: VARIABLE 290/12G20 11 TO 19 KNOTS RIGHT XWIND OFFICIAL: 5H, 11R DENS ALT: 4900 FT FLT DURATION: 4:00:56:27 96:56:27 S/T: 170:01:27:24 OV-104: 19:08:51:31 DISTANCE: 1,477,500 sm	104/104 109% 100/104/ 102/65/ 104/65 1 = 2027 (2) 2 = 2030 (2) 3 = 2029 (2) M 3 EOM WEIGHT: 192558 X CG: 1097.4 LANDING WEIGHT: 192460 X CG: 1099.1	BI-027 (1) RSRM 4 360L 004 ET-39 LWT-22 ET RPT 243K 46:50 MET ET BR/UP 212K 47:40 MET T/V ET IMPACT LAT: 28.85 S LONG: 86.89 E	28.871	STANDARD INSERTION POST OMS-2 160.98 X 159.35 NM MAGELLAN DEPLOY 161.84 NM DEORBIT 176 X 160 NM VELOCITY 25788 FPS RANGE 4147 NM	OI-8B (4)	CARGO: 47783 lbs CHARGEABLE: 45823 lbs DEPLOYABLE: 40118 lbs NON-DEPLOYED: 5540 lbs MIDDECK: 165 lbs RETURNED: 7724 lbs SHUTTLE ACCUMULATED WEIGHTS: DEPLOYED: 399833 lbs NON-DEPLOYED: 387934 lbs CARGO TOTAL: 888553 lbs PERFORMANCE MARGINS (LBS): FPR: 4698 FUEL BIAS: 968 FINAL TDDP: 4709 RECON: 2650 PAYLOADS: PLB: MAGELLAN/IUS (VENUS PROBE) DEPLOYED MID-DECK: AMOS FEA MLE CRYO TK SETS - 3 NO RMS	KSC W/D: OPF 79, VAB 11, PAD 43 = 133 LAUNCH POSTPONEMENTS: None. LAUNCH SCRUBS: 4/28/89 Launch scrubbed at T-31 seconds due to an SSME 1 LH2 recirc pump failure at T-55 seconds. Launch rescheduled for 5/4/89. 6-day total slip. LAUNCH DELAYS: - 00H43M delay with hold at L-16 minutes due to RTLS ceiling violation. (1:48 PM EDT planned launch). Picked up at 2:15 PM EDT, counted down to T-5 minutes and held. Picked up count at 2:42 PM EDT when RTLS runway 15 was go (33 was no go due to broken ceiling and excessive tailwind). Total launch delay: 58M59S. TAL WX: - Ben Guerir (prime) selected - Good weather at Ben Guerir and Moron. I-LOADS: LSEAT selected nominal ascent I-loads - no uplink required. FLIGHT DURATION CHANGE: None. FIRSTS: - First interplanetary payload launch by Shuttle. First crosswind landing test. EVENTS: - Uplinked launch targeting command load ly and del Psi (inertial plane and first stage yaw steering). - Uplinked OMS targeting command load for OMS-1 and OMS-2. - IUS/Magellan deployed at 6:14:33 MET (rev 5). - Sep burn at 6:27:22 MET, 16 secs, 31.6 FPS. ET REENTRY (NO TUMBLE): - CAST GLANCE, poor quality, tumble rate not discernible. SIGNIFICANT ANOMALIES: - SSME 1 LH2 Recirc pump failure. - GPC 4 quit (poll fail on SM CRT when GPC was taken to standby). IFM replaced GPC. - Cabin P Xducer test port left on during first launch attempt. - Excess water from galley H2O dispenser. - TAGS jam on 19th page. - Teleprinter character tops illegible. - Camera A spots on image. - ARRIFLEX 16MM camera operate lever failure (crew performed IFM). - Thruster R1U failed off at ET Sep. - R RCS OX Helium P A valve failed open. - FEA problems. - WONG dilemma.

STS030-72-046   1989-05-08 --- First interplanetary payload -Magellan/IUS -launched by Shuttle.

STS030-21-013 --- Crew: Clockwise from upper right: CDR Wa ker, Cleave/MS, Lee/MS, Thagard/MS & PLT Grabe.

FLT NO.	ORBITER	CREW (5) TITLE, NAMES & EVA'S	LAUNCH SITE, LIFTOFF TIME, LANDING SITES, ABORT TIMES	LANDING SITE/ RUNWAY, CROSSRANGE LANDING TIMES FLT DURATION, WINDS	SSME-TL NOM-ABORT EMERG THROTTLE PROFILE ENG. S.N.	SRB RSRM AND ET	ORBIT INC / HA/HP	FSW	PAYLOAD WEIGHTS, PAYLOADS/ EXPERIMENTS	MISSION HIGHLIGHTS (LAUNCH SCRUBS/DELAYS, TAL WEATHER, ASCENT I-LOADS, FIRSTS, SIGNIFICANT ANOMALIES, ETC.)
STS-28 (STS-28R) SEQ FLT #30 KSC-30 PAD 39B-6	OV-102 (Flight 8) Columbia OMS PODS LPO3 - 8 RPO4- 4 FRC2 - 8	CDR: Brewster H. Shaw, Jr. (Flt 3 - STS-9 & STS 61-B) P153/R25/V26/M24 PLT: Richard N. Richards P154/R101/M92 M/S 1: James C. Adamson P155/R102/M93 M/S 2: David C. Leestma (Flt 2 - STS 41-G) P156/R45/V43/M42 M/S 3: Mark N. Brown P157/R103/M94 MCC FCR-2 (17) FLIGHT DIRECTORS: Asc/Ent-R. D. Dittemore O 1 - G. A. Pennington Ld/O 2 - C. R. Knarr Plng - N. W. Hale MOD - T. W. Holloway	KSC 39B 220:12:37:00Z 8:37:00 AM EDT Tuesday 5 8/8/89 (4) LANDING SITE PRIORITIES: 1. EDW LAKEBED 2. EDW CONC 3. NOR LAKEBED 4. KSC TAL: Zaragoza TAL WX: Moron (Selected) CLS: Banjul RTLS: KSC 33 AOA: NOR MAX Q = 679 M = 1.12 00:59.3 MET SRB SEP: 2:04 MET MECO: 8:15 MET ET SEP: 8:53 MET OMS-1: NONE OMS-2: 37:52:23 MET 106 Seconds	EDW 17 LEFT (EDW 23, LKBD 15) 6:37:09 AM PDT Sunday 3 8/13/89 (2) DEORBIT BURN: 225:12:36:57Z XRANGE: 186 NM ORB DIR: AL 7 AIM PT: NOM MLGTD: 5311 FT 225:13:37:09Z VEL: 157 KGS 155 KEAS HDOT: -1 FPS TD NORM 195: 2545 FT NLGTD: 7393 FT 225:13:37:14Z VEL:125 KGS HDOT: -9.5 FPS BRK INIT: 79 KGS AVE BRK DECEL: 6.3 FPS/S WHEELS STOP: 225:13:37:52Z 11326 FEET ROLLOUT: 6015 FEET 46 SECONDS WINDS: 160 @ 6 KTS 5.8H, 1.6 L KTS OFFICIAL: 1H, 6L DENS ALT: 3670 FT FLT DURATION: 5:01:00:09 121:00:09 S/T: 175:02:27:33 OV-102: 46:02:54:20 DISTANCE: 2,070,943 sm	104/104 109% 100/104/ 97/65/ 104/65 1 = 2019 (5) 2 = 2022 (3) 3 = 2028 (3) M 3 EOM WEIGHT: X CG: LANDING WEIGHT: 200214 X CG: 1089.4	BI-028 RSRM 5 360L 005 ET-31 LWT-24 ET BR/UP 220K 1:11:44 MET ET IMPACT LAT: 36.64 S LONG: 149.65 W	57 (6) DEORBIT 166 X 160 NM VELOCITY 25803 FPS RANGE 4332 NM	OI-8B (5)	DOD PERFORMANCE MARGINS (LBS): FPR: 4698 FUEL BIAS: 968 FINAL TDDP: 409 RECON: 158 3 CRYO TK SETS AMOS HEIN-LO IOCM/APM CLOUDS CRUX RME-III LLL SAM VFT-2	KSC W/D: OPF 190, VAB 11, PAD 25 = 227 LAUNCH POSTPONEMENTS: - 8/7/89 launch postponed to 8/8/89 due to MPS He system. 1-day slip. LAUNCH SCRUBS: None. LAUNCH DELAYS: - Launch delay at T-9 due to an NSP frame sync error and MMU 1 read problem during G9 to OPS 101 transition. - Launch delay due to KSC ground fog. TAL WX: - Zaragoza (prime) NO GO - thundershowers, Ben Guerir NO GO - crosswinds. - Moron (selected) - GO throughout. I-LOADS: - LSEAT selected nominal ascent I-loads - no uplink required. EVENTS: - No blackout during entry, comm via TDRS-W. SIGNIFICANT ANOMALIES: - Prelaunch problem, one of nose gear WOW proximity sensors began indicating weight on nose gear. Indication went away after insertion but returned later in flight causing a WOW dilemma during landing. NWS was enabled by crew by depressing SRB SEP pushbutton. - MMU input/output error on OPS-1 transition. - Pilot's seat moved aft during ascent. - Vernier thruster F5R annunciated "fail leak." - NLG WOW indication failed off. - Forward RCS F5L thruster heater failed on. - S-band PA2 power output degraded to 60 watts. - Potable water dump valve failed open. - Teleprinter cable shorted causing a 1.5-second short of 51A. - Freon coolant loop 2 flow degraded about 100 lbs/hr & FCL 1 about 50 lb/hr. - Radar altimeter 1 and 2 lost attitude reading at 26 feet. - Hydraulic system 2 unloader valve operation out-of-spec. - Body flap excessive deflection during ascent. - NSP frame sync errors prelaunch. - SSME 1 GH_2 flow control valve sluggish.

STS-28 crew portrait on middeck: Clockwise starting with Adamson/MS (mustache) are Leestma/MS, Brown/MS, PLT Richards, and CDR Shaw. In center is tail end of stuffed toy animal.

S89-41096,1989-08-09--- STS-28 Columbia, OV-102, is left at KSC LC Pad 39B by crawler transporter. Crawler transporter pulls out from under mobile launcher platform. View provided by KSC with alternate number KSC-89PC-684.

FLT NO.	ORBITER	CREW (5) TITLE, NAMES & EVA'S	LAUNCH SITE, LIFTOFF TIME, LANDING SITES, ABORT TIMES	LANDING SITE/ RUNWAY, CROSSRANGE LANDING TIMES FLT DURATION, WINDS	SSME-TL NOM-ABORT EMERG THROTTLE PROFILE ENG. S.N.	SRB RSRM AND ET	INC	ORBIT HA/HP	FSW	PAYLOAD WEIGHTS, PAYLOADS/ EXPERIMENTS	MISSION HIGHLIGHTS (LAUNCH SCRUBS/DELAYS, TAL WEATHER, ASCENT I-LOADS, FIRSTS, SIGNIFICANT ANOMALIES, ETC.)
STS-34 (STS-34R) SEQ FLT #31 KSC-31 PAD 39B-7 MLP 1 (WAS STS 61-G)	OV-104 (Flight 5) Atlantis OMS PODS LPO1 - 11 RPO3- 10 FRC4 - 5	CDR: Donald E. Williams (Flt 2 - STS 51-D) P158/R54/V44/M50 PLT: Michael J. McCulley P159/R104/M95 M/S 1: Shannon W. Lucid (Flt 2 - STS 51-G) P160/R65/V45/F6 M/S 2: Franklin Chang-Diaz (Flt 2 - STS 61-C) P161/R89/V46/M81 M/S 3: Ellen S. Baker P162/R105/F10 MCC FCR-1 (14) FLIGHT DIRECTORS: A/E/O1 - R. D. Dittemore Ld/O 2 - J. M. Heflin Plng - R. E. Castle MOD - G. E. Coen MDR - C. W. Shaw	KSC 39B 291:16:53:40Z 12:50:00 PM EDT (P) 12:53:40 PM EDT (A) Wednesday 2 10/18/89 (4) LAUNCH WINDOW: 27 Minutes (GALILEO RAAN) LANDING SITE PRIORITIES: 1. EDW LAKEBED 2. EDW CONCRETE 3. NOR 4. KSC EOM RUNWAY: Based on DTO priority: 1. Xwind DTO 2. NWS DTO EDW Concrete & Lakebed acceptable xwind < 15 knots RTLS: KSC 15 TAL: Ben Guerir TAL Wx: Zaragoza 30 (Selected) AOA: EDW 17 MAX Q = 687.9 M = 1.63 SRB SEP: 2:04.98 MET MECO: 8:31.88 MET ET SEP: 8:50 MET OMS-1: NONE OMS-2: 39:55 MET 140.64 Seconds 218.98 FPS	EDW 23L, LKBD (EDW 24, LKBD 16) 296:16:33:00Z 9:33:00 AM PDT Monday 7 10/23/89 (4) XRANGE: 496 NM DEORBIT BURN: 296:15:31:45Z 166.4 secs, 321.48 FPS ORB DIR: AL8 AIM PT: CLOSEIN MLGTD: 1871 FT 296:16:33:00Z VEL: 206 KGS 195 KEAS HDOT: -2 FPS TD NORM 195: 1880 FT NLGTD: 5355 FT 296:16:33:11Z VEL:158 KGS HDOT: -3.9 FPS BRK INIT: 77 KGS AVE BRK DECEL: 5.8 FPS/S WHEELS STOP: 296:16:34:01Z 11548 FEET ROLLOUT: 9677 FEET 61 SECONDS WINDS: 190 @ 8 KTS 1H, 4L KTS OFFICIAL: 2H, 3L DENS ALT: 2680 FT FLT DURATION: 4:23:39:20 119:39:20 S/T: 180:02:06:53 OV-104: 24:08:30:51 DISTANCE: 1,800,000 sm	104/104 109% 100/104/ 100/65/ 104/65 1 =2027 (3) 2 =2030 (3) 3 =2029 (3) M 3 EOM WEIGHT: 196112 X CG: 1093.1 LANDING WEIGHT: 195954 X G: 1094.7	BI-032 RSRM 6 ET-34 LWT-20 ET RPT 245K 1:19:00 MET ET BR/UP 228K 1:19:37 MET T/V OFF ET IMPACT LAT: 3.4 N LONG: 147.6 W	34.327 (1)	DIRECT INSERTION POST OMS-2 161.73 X 161.35 NM GALILEO DEPLOY 163.61 NM DEORBIT 177 X 162 NM VELOCITY 25784 FPS RANGE 4156 NM	OI-8C (1)	CARGO: 48613 lbs PAYLOAD CHARGEABLE: 45905 lbs DEPLOYABLE: 38323 lbs NON-DEPLOYED: 6696 lbs MIDDECK: 886 lbs RETURNED: 10320 lbs SHUTTLE ACCUMULATED WEIGHTS: DEPLOYED: 438156 lbs NON-DEPLOYED: 395516 lbs CARGO TOTAL: 937166 lbs PERFORMANCE MARGINS (LBS): FPR: 4698 FUEL BIAS: 968 FINAL TDDP: 2103 RECON: -132 PAYLOADS: PLB: GALILEO/IUS (JUPITER PROBE) (DEPLOYED) SSBUV MID-DECK: SSP (1) PM MLE GHCO STEX AMOS IMAX 3 CRYO TANKS NO RMS	KSC W/D: OPF 95, VAB 8, PAD 50 = 153 LAUNCH POSTPONEMENTS: None. LAUNCH SCRUBS: - 10/12/89 launch scrubbed during T-19 hold to replace SSME #2 controller. 5-day slip. - 10/17/89 launch scrubbed while holding at T-5 minutes due to bad RTLS weather when 27-minute window expired. Rescheduled launch for 10/18/89. 6-day total slip. LAUNCH DELAYS: - 3M40S delay into 27-minute window after reconfiguration to Zaragoza for TAL at T-5 minutes (Ben Guerir had rain showers). TAL WX: - Ben Guerir (prime) - NO GO - rain showers - Zaragoza 30 (alt) selected. I-LOADS: LSEAT selected nominal ascent I-loads, no uplink required. FLIGHT DURATION CHANGE: None. EVENTS: - Galileo/IUS deployed on rev 5. - Sep burn 06:36:23, 16.64 secs, 31.31 FPS - No blackout during entry, comm via TDRS-W. ET TRACKING DTO (NO TUMBLE): - CAST GLANCE, daylight entry, unsuccessful track. SIGNIFICANT ANOMALIES: - SRB C-Band transponders first flight. - APU 1 fault to high speed during ascent. - APU Heater GG/Fuel Pump 2-A failure. - WSB #2 Steam Vent Heater A failure. - MDM FA1 Primary Port failure. - Cryo O$_2$ manifold valve tank 2 failed to close. - Erratic waste water quantity transducer. - HSI primary miles erroneous indication. - TAGS overtemp indication. - S-Band beam control assy failed to select URF antenna. - S-Band antenna elect. 1 failed to select ULF antenna. - CCTV camera C image degraded. - R OME Cover B heater failure.

STS-34 crew portrait from left to right: CDR Williams (holding mission insignia), MS/Baker, MS/Chang-Diaz (holding stuffed toy), MS/Lucid, and PLT McCulley.

S92-52043, Alternate JPL number is P-41508 1992-12-30 --- Three years after deploy and eight days after its encounter with Earth's orbit, Galileo views the Moon and Earth from 3.9 million miles. The Moon is in the foreground, moving left to right. Antarctica is visible through clouds (bottom). The Moon's far side is seen; the shadowy indentation in the dawn terminator is the south-Pole/Aitken Basin.

FLT NO.	ORBITER	CREW (5) TITLE, NAMES & EVA'S	LAUNCH SITE, LIFTOFF TIME, LANDING SITES, ABORT TIMES	LANDING SITE/ RUNWAY, CROSSRANGE LANDING TIMES FLT DURATION, WINDS	SSME-TL NOM-ABORT EMERG THROTTLE PROFILE ENG. S.N.	SRB RSRM AND ET	ORBIT INC	ORBIT HA/HP	FSW	PAYLOAD WEIGHTS, PAYLOADS/ EXPERIMENTS	MISSION HIGHLIGHTS (LAUNCH SCRUBS/DELAYS, TAL WEATHER, ASCENT I-LOADS, FIRSTS, SIGNIFICANT ANOMALIES, ETC.)
STS-33 (STS-33R) SEQ FLT #32 KSC-32 PAD 39B-8 MLP-2	OV-103 (Flight 9) Discovery OMS PODS LPO4 - 6 RPO1- 11 FRC3 - 9	CDR: Frederick D. Gregory (Flt 2 - STS 51-B) P163/R59/V47/M54 PLT: John E. Blaha (Flt 2 - STS-29) P164/R97/V48/M88 M/S 1: Manley L. Carter, Jr. P165/R106/M96 M/S 2: F. Story Musgrave (Flt 3 - STS-6 & STS 51-F) P166/R15/V19/M15 M/S 3: Kathryn C. Thornton P167/R107/F11 MCC FCR-2 (18) FLIGHT DIRECTORS: Asc/Ent - A. L. Briscoe O 1 - N. W. Hale Ld/O 2 - C. W. Shaw Plng - R. M. Kelso MOD - T. W. Holloway	KSC 39B 327:00:23:29.98Z 7:23:30 PM EST Wednesday 3 11/22/89 (6) LANDING SITE PRIORITIES: 1. EDW LAKEBED 2. EDW CONCRETE 3. NOR 4. KSC RTLS: KSC 15 TAL: Ben Guerir 36 (Selected) CTGY: Banjul AOA: EDW 22 MAX Q = 729.3 M = 1.5 1:02.1 MET SRB SEP: 2:06.77 MET MECO: 8:26.9 MET ET SEP: 8:44 MET OMS-1: 10:25 MET 66 Seconds OMS-2: 35:16 MET 95.2 Seconds	EDW 04, CONC (EDW 25,CONC9) (04 - 1ST FLIGHT) 4:30:19 PM PST Monday 8 11/27/89 (5) DEORBIT BURN: 331:23:10:51Z 181.9 Seconds XRANGE:226 NM ORB DIR: AL 9 AIM PT: CLOSEIN MLGTD: 740 FT 332:00:30:19Z VEL: 196 KGS 199 KEAS HDOT: -1 FPS TD NORM 195: 1042 FT NLGTD: 3982 FT 332:00:30:26Z VEL:161 KGS HDOT: -2.2 FPS BRK INIT: 145 KGS AVE BRK DECEL: 8.5 FPS/S WHEELS STOP: 332:00:30:02Z 8504 FT ROLLOUT: 7764 FEET 46 SECONDS WINDS: 070 @ 8 KTS GUSTS TO 19 KTS 7.2H, 3.5R KTS OFFICIAL: 8H, 2R DENS ALT: 2302 FT FLT DURATION: 5:00:06:49 120:06:49 S/T: 185:02:13:42 OV-103: 52:06:52:10 DISTANCE: 2,045,056 sm	104/104 109% 100/104/ 97/65/ 104/65 1 = 2011 (4) 2 = 2031 (2) 3 = 2107 M 3 EOM WEIGHT: X CG: LANDING WEIGHT: 194282 X CG: 1094.8	BI-034 (19) RSRM 7 ET-38 LWT-31 ET RPT 237K 46:55 MET ET BR/UP 217K 47:26 MET T/V OFF ET IMPACT LAT: 28.57 S LONG: 86.4 E	28.45 (19)	DEORBIT 302 X 126 NM VELOCITY 25998 FPS RANGE 4068 NM	OI-8B (6)	DOD PERFORMANCE MARGINS (LBS): FPR: 4698 FUEL BIAS: 968 FINAL TDDP:1157 RECON: 653 3 CRYO TK SETS AMOS VFT-1 APE-B RME-III CLOUDS-1A	KSC W/D: OPF 114, VAB 21, PAD 27 = 162 LAUNCH POSTPONEMENTS: - 11/21/89 launch postponed to 11/22/89 due to SRB IEA cable replacement. 1-day total slip. LAUNCH SCRUBS: None. LAUNCH DELAYS: - Launch held at T-5 because of a ground purge problem for GLS confirmation of Shuttle purge flow rate and completion of APU prestart. TAL WX: - Ben Guerir 36 (prime selected - good weather after marginal ceiling earlier in day. - Banjul contingency site. I-LOADS: - LSEAT selected nominal ascent I-loads, no uplink required. NIGHT LAUNCH: Third Shuttle night launch. WAVEOFFS: - Waved off landing on fourth day due to high winds at EDW and landed one day later. FIRST SHUTTLE CREWMEMBER REPLACEMENT: - David Griggs died in private aircraft accident while in training in June 1989. He was replaced by Blaha. (This was first US spaceflight crewmember changeout since Ken Mattingly was exposed to measles 3 days before Apollo 13 launch on April 11, 970. Jack Swigert was his replacement.) EVENTS: - No entry blackout, comm via TDRS-W. SIGNIFICANT ANOMALIES: - APU 1 lube oil outlet pressure high during ascent. - Cabin leak through WCS. - TAGS jam (did not work during flight). - Galley rehydration station failed to dispense hot or cold water. - FES primary B shut down (overtemped during deorbit prep). - +X COAS line of sight shift. - CDR AMI M/VEL error. - MSBLS BITE indication. - WCCS short battery life. - Ku-Band radar self test failure. - Hydraulic system 1 and 2 accumulator pressure locked up low. - Cryo oxygen tank 2 check valve stuck twice. IFM: - Broken shear pin on WCS so crew used vice grips to drive valve linkage.

STS033-22-035, 1989-11-27 On-orbit crew portrait. Clockwise (starting at left) are CDR Gregory, Thorton/MS, PLT Blaha, Carter/MS, and Musgrave/MS.

STS033-82-071,1989-11-27 The island of Timor, Indonesia (9.0S, 125.0E) illustrates the volcanic origin of the over 1500 islands of Indonesia. The linear alignment of the volcanoes indicates the edges of the tectonic plates of the Earth's crust.

FLT NO.	ORBITER	CREW (5) — TITLE, NAMES & EVA'S	LAUNCH SITE, LIFTOFF TIME, LANDING SITES, ABORT TIMES	LANDING SITE/ RUNWAY, CROSSRANGE — LANDING TIMES, FLT DURATION, WINDS	SSME-TL NOM-ABORT EMERG — THROTTLE PROFILE ENG. S.N.	SRB RSRM AND ET	INC	ORBIT — HA/HP	FSW	PAYLOAD WEIGHTS, PAYLOADS/ EXPERIMENTS	MISSION HIGHLIGHTS (LAUNCH SCRUBS/DELAYS, TAL WEATHER, ASCENT I-LOADS, FIRSTS, SIGNIFICANT ANOMALIES, ETC.)
STS-32 (STS-32R) SEQ FLT #33 KSC-33 PAD 39A-25 MLP-3	OV-102 Columbia OMS PODS LPO3 - 9 RPO4- 5 FRC2 - 9	CDR: Daniel C. Brandenstein (Flt 3 - STS-8 & STS 51-G) P168/R21/V16/M20 PLT: James D. Wetherbee P169/R108/M97 M/S 1: Bonnie J. Dunbar (Flt 2 - STS 61-A) P170/R79/V49/F7 M/S 2: Marsha S. Ivins P171/R109/F12 M/S 3: G. David Low P172/R110/M98 MCC FCR-1 (15) FLIGHT DIRECTORS: Asc/Ent - A. L. Briscoe L/O1 - G. A. Pennington O 2 - W. D. Reeves PIng - R. E. Castle MOD - B. R. Stone	KSC 39A 09:12:35:00Z 7:35:00 AM EST (P) 7:35:00 AM EST (A) Tuesday 6 1/9/90 (4) LAUNCH WINDOW: 62 Minutes (PLANAR/PHASE/ ET IMPACT AREA) RUNWAY PRIORITIES: EDW (PLS) HEAVY WEIGHT/ FWD CG (LDEF RETURN) EOM: EDW 22/CONC NOR KSC EDW LAKEBED RTLS: KSC 33 TAL: Ben Guerir 36 AOA: EDW 22 X-WIND LIMIT > 9 DAYS, 12 KNOTS MAX Q = 641.1 M = 1.05 00:52 MET SRB SEP: 2:05 MET MECO: 8:33 MET ET SEP: 8:50 MET OMS-1: NONE OMS-2: 40:25.6 MET 140 Seconds 218 FPS	EDW 22, CONC (EDW 26,CONC 10) 20:09:35:36.2Z 1:35:36 AM PST Saturday 7 1/20/90 (3) DEORBIT BURN: 20:08:30:22Z 299.5 Seconds DV 489.7 FPS XRANGE:372 NM ORB DIR: AL10 AIM PT: NOM MLGTD: 1804 FT 20:09:35:36.2Z VEL: 209 KGS 207 KEAS HDOT: -1 FPS TD NORM 195: 3100 FT NLGTD: 6676 FT 20:09:35:51.5Z VEL:160 KGS HDOT: -2.7 FPS BRK INIT: 141 KGS AVE BRK DECEL: 6.3 FPS/S WHEELS STOP: 20:09:35:39.3Z 12495 FEET ROLLOUT: 10731 FEET 64 SECONDS WINDS: 1.9H, 3.5R KTS OFFICIAL: 1H, 4R DENS. ALT: 923 FT FLT DURATION: 10:21:00:36 261:00:36 S/T: 195:23:14:18 OV-102: 56:23:54:56 DISTANCE: 4,509,972 sm	104/104 109% 100/104/ 102/65/ 104/65 1 = 2024 (1) 2 = 2022 (4) 3 = 2028 (4) M 3 EOM WEIGHT: 228523 X CG: 1078.2 LANDING WEIGHT: 228335 X CG: 1079.6	BI-035 RSRM 8 ET-32 LWT-25 ET RPT 228K 1:18:32 MET ET BR/UP 189K 1:19:35 MET T/V OFF ET IMPACT LAT: 10.44 N LONG: 157.2 W	28.5 (20)	DIRECT INSERTION POST OMS-2 193.48 X 155.76 NM SYNCOM DEPLOY 169.09 NM LDEF RETRIEVE 178.3NM DEORBIT 178 X 173 NM VELOCITY 25823 FPS RANGE 4317 NM	OI-8C (2)	CARGO: 26458 lbs PAYLOAD CHARGEABLE: 18317 lbs DEPLOYABLE: 15316 lbs NON-DEPLOYED: 1962 lbs MIDDECK: 1039 lbs RETRIEVED (LDEF) 21393 lbs RETURNED: 32565 lbs SHUTTLE ACCUMULATED WEIGHTS: DEPLOYED: 453472 lbs NON-DEPLOYED: 398517 lbs CARGO TOTAL: 963624 lbs PERFORMANCE MARGINS (LBS): FPR: 4698 FUEL BIAS: 968 FINAL TDDP: 1956 RECON: 992 PAYLOADS: PLB: LONG DURATION EXPOSURE FACILITY (LDEF) RETRIEVAL AND RETURN SYNCOM IV-5 (DEPLOYED) MIDDECK IOCM IMAX CNCR, PCG (2) FEA, AFE, MLE L3 (LLL) AMOS ACIP AADS 5 CRYO TK SETS RMS 20 (S.N. 201) Used for LDEF capture and berth, and PKM burn monitor	KSC W/D: OPF 86, VAB 10, PAD 33 = 129 LAUNCH POSTPONEMENTS: - 12/18/89 launch postponed 21 days to 1/8/90 due to delays in readiness of pad 39A after pad modification, holidays, and Orbiter aft PCA R&R. LAUNCH SCRUBS: - 1/8/90 launch scrubbed after holding at T-9 minutes, then counting down to T-5 minutes and holding until launch window expired when RTLS weather did not improve (low ceiling/fog). Rescheduled launch for 1/9/90. - 22-day total slip. LAUNCH DELAYS: None. TAL WX: - Ben Guerir 36 (prime) - selected - good weather. I-LOADS: - LSEAT selected yaw positive I-Load - alternate I-Load uplink 2. LAUNCH TARGETING COMMAND LOAD: - Uplinked load for inertial plane of LDEF. FLIGHT DURATION CHANGE: - Extended 1 day due to fog at PLS (EDW) and unacceptable weather at NOR and KSC. - Plus One rev to reload BFS into extended GPC2. NIGHT LANDING: Third Shuttle night landing. FIRSTS: - First flight from pad 39A since STS 61-A. EVENTS: - SYNCOM-IV-F5 deployed at 1:00:43:39 MET (rev 17). - Rendezvous with Long Duration Exposure Facility (LDEF) as planned, with grapple at 3:02:41:05 MET (rev 50). LDEF was deployed on STS 41-C. - No blackout during entry, comm via TDRS-W. - Deorbit burn O-O-P component of 51 with longest OMS burn time of 299.5 seconds. RENDEZVOUS 8: With LDEF for capture and return. Continued . . .

STS032-57-006   1990-01-20 STS-32
Crew portrait with a SNOOPY stuffed toy:
CDR Brandenstein (right, rear), PLT
Wetherbee (left, rear), & front row (l to r)
MS/Ivins, MS/Dunbar, andMS/Low during
a record setting 11-day stay in Earth-orbit

STS-32 Liftoff (Wikipedia, the free encyclopedia) --- First flight from pad 39A since STS 61-A on 10/30/85.

FLT NO.	ORBITER	CREW (5) TITLE, NAMES & EVA'S	LAUNCH SITE, LIFTOFF TIME, LANDING SITES, ABORT TIMES	LANDING SITE/ RUNWAY, CROSSRANGE LANDING TIMES FLT DURATION, WINDS	SSME-TL NOM-ABORT EMERG THROTTLE PROFILE ENG. S.N.	SRB RSRM AND ET	ORBIT		FSW	PAYLOAD WEIGHTS, PAYLOADS/ EXPERIMENTS	MISSION HIGHLIGHTS (LAUNCH SCRUBS/DELAYS, TAL WEATHER, ASCENT I-LOADS, FIRSTS, SIGNIFICANT ANOMALIES, ETC.)
							INC	HA/HP			
STS-32 Continued											Continued . . .

STS032-87-030,1990-01-20 --- SYNCOM IV-5 is deployed from Columbia's payload bay.

STS032-85-051, 1990-01-20---LDEF Retrieval over South America. LDEF proposed by NASA LRC was deployed by STS-41C on 04/13/1984.

STS032-15-022   STS032-15-022   STS-32 Commander Brandenstein celebrates birthday on OV-102's aft flight deck.

S89-48717 1989-11-07 STS-32 Flight Directors in MCC standing in front of the flight director's consoles are (l. to r.) Alan L. Briscoe, Granvil A. Pennington, and Robert E. Castle, Jr.

SIGNIFICANT ANOMALIES:
- GPC 5 (BFS) registered illegal engage input/output term B during final entry checks. BFS was loaded into GPC2, GPC set restrung and GPC5 powered off. (Landing was delayed one revolution.)
- FM transmitter failed.
- APU 3 lubrication oil outlet pressure high (90 psi)
- TAGS paper jammed.
- GO_2 FCV 2 open cycle sluggish.
- Humidity separator water bypass anomalies (free water from SEP B and SEP A).
- Waste water dump line blockage at 18:13:29:00Z, no dumps performed subsequently.
- FES topping duct B string heater failure.
- IMU 1 RM failed (transient 4-axis accel-bias.
- Hydraulic systems 1 and 2 circ pump unloader valves excessive leakage.
- BFS GPC errors.
- At 17:23:46:51Z during sleep period, a bad state vector was uplinked just prior to LOS, Orbiter rotated 3 /sec.
- WSB sys 2 and 3 excessive regulator pressure decay.
- RMS was used to conduct external survey (TPS).
- Multiple S-Band dropouts.
- Smoke detector 3A transient alarm.
- WBS 3 controller A over controlling.
- Ku-band antenna feed heater erratic.
- MPS LH_2 F&D (outboard) relief valve leak.
- Pilot seat would not drive down.
- CCTV camera problems.
- Heaviest landing at 228,335 lbs.

FLT NO.	ORBITER	CREW (5) TITLE, NAMES & EVA'S	LAUNCH SITE, LIFTOFF TIME, LANDING SITES, ABORT TIMES	LANDING SITE/ RUNWAY, CROSSRANGE LANDING TIMES, FLT DURATION, WINDS	SSME-TL NOM-ABORT EMERG THROTTLE PROFILE ENG. S.N.	SRB RSRM AND ET	ORBIT INC	HA/HP	FSW	PAYLOAD WEIGHTS, PAYLOADS/ EXPERIMENTS	MISSION HIGHLIGHTS (LAUNCH SCRUBS/DELAYS, TAL WEATHER, ASCENT I-LOADS, FIRSTS, SIGNIFICANT ANOMALIES, ETC.)
STS-36 (STS-36R) SEQ FLT #34 KSC-34 PAD 39A-26 MLP-1	OV-104 (Flight 6) Atlantis OMS PODS LPO1 - 12 RPO3- 11 FRC4 - 6	CDR: John O. Creighton (Flt 2 - STS 51-G) P173/R63/V50/M58 PLT: John H. Casper P174/R111/M99 M/S 1: David C. Hilmers (Flt 3 - STS 51-J & STS-26) P175/R77/V36/M71 M/S 2: Richard M. Mullane (Flt 3 - STS 41-DR & STS-27) P176/R40/V37/M39 M/S 3: Pierre J. Thuot P177/R112/M100 MCC FCR-2 (19) FLIGHT DIRECTORS: A/E - R. D. Dittemore Ld/O 1 - L. S. Bourgeois O 2 - R. M. Kelso Plng - C. R. Knarr MOD - T. W. Holloway	KSC 39A 59:07:50:22Z 2:50:22 AM EST Wednesday 4 2/28/90 (2) LANDING SITE PRIORITIES: 1. EDW LAKEBED 2. EDW CONCRETE 3. NOR 4. KSC 1. X-WIND FIRST PRIORITY 2. NWS SECOND PRIORITY RTLS: KSC 15 TAL: Zaragoza 30 (Selected) TAL WX: Moron AOA: NOR 17 MAX Q = 743.9 M = 1.49 00:53 MET SRB SEP: 2:05.8 MET MECO: 8:30 MET ET SEP: 8:48 MET OMS-1: NONE OMS-2: 32:58.1 MET 105.4 Seconds	EDW 23L, LKBD (EDW 27, LKBD 17) 63:18:08:44Z 10:08:44 AM PST Sunday 4 3/4/90 (3) DEORBIT BURN: 63:17:11:17.24Z 125.48 Seconds 256.4 FPS XRANGE: 255 NM ORB DIR: DR 5 AIM PT: CLOSEIN MLGTD: 1622 FT 63:18:08:44Z VEL: 193 KGS 199 KEAS HDOT: -1 FPS TD NORM 195: 1959 FT NLGTD: 4862 FT 63:18:09:37.32Z VEL:145 KGS HDOT: -4.4 FPS BRK INIT: 99 KGS AVE BRK DECEL: 5.5 FPS/S WHEELS STOP: 63:18:09:37.3Z 9522 FEET ROLLOUT: 7900 FEET 53 SECONDS WINDS: 15.9H, 1.6R KTS OFFICIAL: 16H, 3R DENS ALT: 3017 FT FLT DURATION: 4:10:18:22 106:18:22 S/T: 200:09:32:40 OV-104: 28:18:49:13 DISTANCE: 1,837,962 sm	104/104 109% 100/104/ 98/75/ 104/65 1 =2019 (6) 2 =2030 (4) 3 =2027 (4) M 3 EOM WEIGHT: X CG: LANDING WEIGHT: 187200 X CG: 1096.4	BI-036 RSRM 9 ET-36 LWT-26 ET RPT 228K 1:00:35 MET ET BR/UP 217K 1:00:53 MET T/V ACTIVE LAST FLIGHT ET IMPACT LAT: 61.40 S LONG: 145.1 E	62 (1)	DEORBIT 132 X 115 NM VELOCITY 25713 FPS RANGE 4338 NM	OI-8C (3)	DOD PERFORMANCE MARGINS (LBS): FPR: 4652 FUEL BIAS: 999 FINAL TDDP: 881 RECON: 930 MIDDECK RME-III VFT-I VFT-II	KSC W/D: OPF 69, VAB 6, PAD 35 = 110 LAUNCH POSTPONEMENTS: None. LAUNCH SCRUBS: - 2/22/90 launch was scrubbed while counting from T-11 hours to T-6 hours for CDR's health (48-hour slip). - 2/24/90 launch scrubbed because of predicted bad weather at KSC. - 2/25/90 launch scrubbed due to a Range Safety backup computer problem. Count held at T-31 seconds, and during hold, the LO₂ inlet temps on all 3 engines exceeded LCC lower limit. Rescheduled launch for 2/26/90. - 2/26/90 launch scrubbed at T-9 minutes due to bad RTLS weather (cloudy). Rescheduled launch for 2/28/90. 48-hour delay to allow launch team rest. 6 days total slip. LAUNCH DELAYS: - Delay at T-9 minutes due to predicted rain in RTLS area. Resumed count to T-5 minutes, held for launch pad, RTLS, and TAL weather. TAL WX: Zaragoza 30 (prime) - Some delay waiting for STA go (until STA could see landing strip). - Moron - NO GO - ceiling. I-LOADS: - LSEAT selected yaw positive, alternate I-load uplink 3. NIGHT LAUNCH: Fourth Shuttle night launch. EVENTS: - No entry blackout - comm via TDRS-W. - Last flight with ET tumble valve active. SIGNIFICANT ANOMALIES: - AC2 Phase 2 Inverter failure. - RCS valve position indications intermittent. - WSB 2 Vent System A heater failed. - CRT 4 screen went blank. - SSME post powerdown hard failure ID. - O₂ leak into cabin. - FES overtemp shutdown. - Humidity separator A degraded operation (found 1 quart of water below middeck floor). - Supply H2O tank A-B check valve failure. - PLB floodlight failure (2). - SPOC H/W and S/W problems. - Volume H latch jammed. - TAGS paper folding. - WSB 2 vent temp heater A failure. - Hyd system leak into aft compartment. - R3D fail-off at ET SEP. - R4R jet fail-off during RCS hot fire.

STS036-151-225 1990-03-04 ---Cape Cod, MA (42.0N, 70.0W) as seen from Shuttle. Geologically, the cape is a deposit of earth and stone called a terminal moraine, left by the great Pleistocene glaciers of about 20,000 years ago.

Pilgrims first stepped ashore November 1620

STS036-21-024 1990-03-03 Atlantis crew, pose clockwise from top center: CDR Creighton, Mullane/MS, PLT Casper, Hilmers/MS, & Thuot/MS while conducting a DOD-dedicated mission.

FLT NO.	ORBITER	CREW (5) TITLE, NAMES & EVA'S	LAUNCH SITE, LIFTOFF TIME, LANDING SITES, ABORT TIMES	LANDING SITE/ RUNWAY/ CROSSRANGE LANDING TIMES FLT DURATION, WINDS	SSME-TL NOM-ABORT EMERG THROTTLE PROFILE ENG. S.N.	SRB RSRM AND ET	INC	ORBIT HA/HP	FSW	PAYLOAD WEIGHTS, PAYLOADS/ EXPERIMENTS	MISSION HIGHLIGHTS (LAUNCH SCRUBS/DELAYS, TAL WEATHER, ASCENT I-LOADS, FIRSTS, SIGNIFICANT ANOMALIES, ETC.)
STS-31 (STS-31R) SEQ FLT #35 KSC-35 PAD 39B-9 MLP-2 (WAS STS 61-J)	OV-103 (Flight 10) Discovery OMS PODS LPO4 - 7 RPO1 - 12 FRC3 - 10	CDR: Loren J. Shriver (Flt 2 - STS 51-C) P178/R50/V51/M46 PLT: Charles F. Bolden (Flt 2 - STS 61-C) P179/R88/V52/M80 M/S 1: Steven A. Hawley (Flt 3 - STS 41-DR & STS 61-C) P180/R39/V29/M38 M/S 2: Kathryn D. Sullivan (Flt 2 - STS 41-G) P181/R44/V53/F3 M/S 3: Bruce McCandless II (Flt 2 - STS 41-B) P182/R31/V54/M30 MCC FCR-1 (16) FLIGHT DIRECTORS: Asc - R. D. Dittemore Ent - N. W. Hale Ld/O 1 - W. D. Reeves O 2 - J. M. Heflin PIng - A. L. Briscoe MOD - B. R. Stone	KSC 39B 114:12:33:51Z 8:31:00 AM EDT (P) 8:33:51 AM EDT (A) Tuesday 7 4/24/90 (6) LAUNCH WINDOW: 2H30M (CREW TIME ON BACK) LANDING SITE PRIORITIES: NOEM: EDW LKBD - Prime RTLS: KSC 15 TAL: Banjul (PRI) (Planned) ALT TAL: Ben Guerir 36 (Selected) AOA or P/L Return: 1. EDW 22/04 2. EDW LKBD 3. NOR 4. KSC AOA: NOR 23 MAX Q = 656.3 M = 1.08 00:52 MET SRB SEP: 2:05.75 MET MECO: 8:30 MET ET SEP: 8:48 MET OMS-1: NONE OMS-2: 42.36 MET 305 Seconds	EDW 22, CONC (EDW 28,CONC 11) 119:13:49:57Z 6:49:57 AM PDT Sunday 5 4/29/90 (5) DEORBIT BURN: 119:12:37:36Z XRANGE: 420 NM ORB DIR: DL 17 AIM PT: NOM MLGTD: 1176 FT 119:13:49:57Z VEL: 180 KGS 177 KEAS HDOT: -4 FPS TD NORM 195: - 130 FT NLGTD: 4560 FT 119:13:50:09Z VEL:144 KGS HDOT: -3.3 FPS BRK INIT: 120 KGS AVE BRK DECEL: 5.9 FPS/S WHEELS STOP: 119:13:50:58Z 10065 FEET ROLLOUT: 8874 FEET 61 SECONDS WINDS: 180 @ 7 KTS GUSTS TO 10 KTS 4.1H, 5.7L KTS OFFICIAL: 7H, 5L DENS. ALT:2993 FT FLT DURATION: 5:01:16:06 121:16:06 S/T: 205:10:48:46 OV-103: 57:08:08:16 DISTANCE: 2,068,213 sm	104/104 109% 100/104/97/ 67/104/65 1 = 2011 (5) 2 = 2031 (3) 3 = 2107 (2) M 3 EOM WEIGHT: 189309 X CG: 1087.9 LANDING WEIGHT: 189118 X CG: 1089.7	BI-037 RSRM 10 ET-34 LWT-27 ET RPT 251K 1:24:18 MET ET BR/UP 215K 1:25:14 MET T/V OFF ON ALL SUBS FLTS ET IMPACT LAT: 19.95 N LONG: 150.0 W	28.453 (21)	DIRECT INSERTION POST OMS-2 330.63 X 310.80 NM HST DEPLOY 333.06 NM DEORBIT 333 X 327 NM VELOCITY 26120 FPS RANGE 4121 NM	OI-8C (4)	CARGO: 28643 lbs PAYLOAD CHARGEABLE: 25517 lbs DEPLOYABLE: 23095 lbs NON-DEPLOYED: 960 lbs MIDDECK: 652 lbs RETURNED: 4768 lbs SHUTTLE ACCUMULATED WEIGHTS: DEPLOYED: 476567 lbs NON-DEPLOYED: 400129 lbs CARGO TOTAL: 992267 lbs PERFORMANCE MARGINS (LBS): FPR: 4652 FUEL BIAS: 994 FINAL TDDP: 2861 * RECON: 1352 PAYLOADS: PLB: HUBBLE SPACE TELESCOPE (HST) (DEPLOYED) ICBC (IMAX) APM MIDDECK: SE-82-16 (ION ARC) IMAX RME-III AMOS IPMP PCG-III 3 CRYO TK SETS RMS 21 (S.N. 301) USED FOR HST DEPLOY	KSC W/D: OPF 78, VAB 9, PAD 39 = 126 LAUNCH POSTPONEMENTS: None. LAUNCH SCRUBS: - 4/10/90 launch scrubbed during hold at T-4 minutes due to APU anomalies. Rescheduled launch for 4/24/90 (APU 1 R&R). 14 days total slip. LAUNCH DELAYS: - 2M51S delay during hold at T-31 seconds to manually close F&D valve after failure to close by GLS (procedural enhancement problem). TAL WX: - Banjul (prime) - NO GO because redundant TACAN's down, WX marginal but acceptable. - Ben Guerir 36 (alternate) selected - marginal but GO. I-LOADS: - LSEAT selected nominal I-loads, no uplink required. FLIGHT DURATION CHANGE: None. FIRSTS/RECORDS: - First planned use of Banjul at primary TAL. - First flight with carbon brakes. - Highest Shuttle altitude to date - 333 NM. - Longest OMS burn - 305 seconds. EVENTS: - HST deployed on rev 20 (1 rev later than planned). - No entry blackout. ET REENTRY (NO TUMBLE): - ARGUS - Rupture altitude 246K feet. - AMOS/MOTIF - Tumble rate 7 deg/second. - KPTC RADAR - Max. DV 670 FPS. - VHF RADAR - Number of pieces > 3 feet - 68. - Debris scatter: 200 NM (UR/DR) 40 NM CR. SIGNIFICANT ANOMALIES: - Cabin depressed to 10.2 PSIA for approximately 72 hours. - Supply water tank C bellows stuck. - Fuel cell 2 purge anomaly. - SPOC failures. - ADTA 3 CB contamination. - TAGS problems. - WSB 2 steam vent heater A failure. - 70 mm camera jam. - L3A jet failed off, L3A fail leak. - Erratic ROMS fuel engine inlet pressure. - HST solar array deploy problem.

STS031-12-031 1990-04-29 STS-31 -- Crew: lt. to rt. PLT Bolden (top left), CDR Shriver, Sullivan/MS, McCandless/MS, and Hawley/MS

TOP: S90-32805STS31--Bill Reeves, Lead Orbit Flight Director, briefs media at preflight conference.
BOTTOM: STS031-76-026 1990-04-29 -- HST is grappled by RMS during predeployment checkout.

FLT NO.	ORBITER	CREW (5) / TITLE, NAMES & EVA'S	LAUNCH SITE, LIFTOFF TIME, LANDING SITES, ABORT TIMES	LANDING SITE/ RUNWAY, CROSSRANGE / LANDING TIMES, FLT DURATION, WINDS	SSME-TL NOM-ABORT EMERG / THROTTLE PROFILE ENG. S.N.	SRB RSRM AND ET	ORBIT / INC	HA/HP	FSW	PAYLOAD WEIGHTS, PAYLOADS/ EXPERIMENTS	MISSION HIGHLIGHTS (LAUNCH SCRUBS/DELAYS, TAL WEATHER, ASCENT I-LOADS, FIRSTS, SIGNIFICANT ANOMALIES, ETC.)
STS-41 SEQ FLT #36 KSC-36 PAD 39B-10 MLP-2 (Was STS 61-F)	OV-103 (Flight 11) Discovery OMS PODS LPO4 - 8 RPO1- 13 FR C3 - 11	CDR: Richard N. Richards (Flt 2 - STS-28) P183/R101/V55/M92 PLT: Robert D. Cabana P184/R113/M101 M/S 1: Bruce E. Melnick P185/R114/R102 M/S 2: Thomas D. Akers P186/R115/M103 M/S 3: William M. Shepherd (Flt 2 STS-27) P187/R96/V56/M87 MCC FCR-1 (17) FLIGHT DIRECTORS: A/E/O1 - R. D. Dittemore Ld/O - J. M. Heflin Plng - G. E. Coen MOD - T. W. Holloway MDR - R. M. Kelso	KSC 39B 279:11:47:14.98Z 7:35:00 AM EDT (P) 7:47:15 AM EDT (A) Saturday 2 10/6/90 (5) LAUNCH WINDOW: 2H17M (ULYSSES UPPER STAGE PERFORMANCE) LANDING SITE PRIORITIES: NOEM: EDW Lakebed - Prime RTLS: KSC 33 TAL: Banjul TAL WX: Ben Guerir 36 (Selected) AOA: NOR 17 MAX Q = 665 M = 00:49 MET SRB SEP: 2:06 MET MECO: 8:28 MET ET SEP: 8:46 MET OMS-1: None OMS-2: 39:53.3 MET 144 Seconds (223.3 FPS)	EDW 22, CONC (EDW 29,CONC 12) 283:13:57:19Z 6:57:19 AM PDT Wednesday 3 10/10/90 (5) DEORBIT BURN: 283:13:00:05Z (150 Seconds DV 286.6) XRANGE: 492 NM ORB DIR: DL 18 AIM PT: NOM MLGTD: 2295 FT 283:13:57:19Z VEL: 193 KGS 192 KEAS HDOT: -1 FPS TD NORM 195: 2315 FT NLGTD: 6359 FT 283:13:57:31Z VEL: 154 KGS HDOT: -2.7FPS BRK INIT: 135 KGS AVE BRK DECEL: 9 FPS/S WHEELS STOP: 283:13:58:08Z 10827 FEET ROLLOUT: 8478 FEET 49 SECONDS WINDS: Light & Variable Peak 3 Kts 2.3H, 2 R KNOTS OFFICIAL: 2H, 2R DENS. ALT:1308 FT FLT DURATION: 4:02:10:04 98:10:04 S/T: 209:12:58:50 OV-103: 61:10:18:20 DISTANCE: 1,707,445 sm	100/100/ 109% ACTUAL: 100/104/ 101/67/ 104/65 1 = 2011 (6) 2 = 2031 (4) 3 = 2107 (3) M 3 EOM WEIGHT: 196982 X CG: 1089.4 LANDING WEIGHT: 196869 X CG: 1091.2	BI-040 RSRM 13 ET-39 ET RPT 239K 1:16:20 MET ET BR/UP 177K 1:17:50 MET ET IMPACT LAT: 12.52 N LONG: 164.1 W	28.45 (22)	DIRECT INSERTION POST OMS-2 160.2 X 159.5 NM ULYSSES DEPLOY 160 X 159 NM POST SEP BURN 177.9 X 160 NM DEORBIT 162.4 X 151.4 NM VELOCITY 25762 FPS RANGE 4147 NM	OI-8D (1)	CARGO: 49969 LBS PAYLOAD CHARGEABLE: 46173 LBS DEPLOYABLE: 38604 LBS NON-DEPLOYED: 6732 LBS MIDDECK: 837 LBS SHUTTLE ACCUMULATED WEIGHTS: DEPLOYED: 515171 LBS NON-DEPLOYED: 407698 LBS CARGO TOTAL: 1042236 LBS PERFORMANCE MARGINS (LBS): FPR: 4652 FUEL BIAS: 994 FINAL TDDP: 1270 RECON: -152 PAYLOADS: PLB: ULYSSES/IUS/ PAM-S (SOLAR ORBIT) DEPLOYED SSBUV ISAC MID-DECK: CHROMEX VCS SSCE IPMP PSE RME-III AMOS 3 CRYO TK SETS RMS 22 (S.N. 301) Used for INTELSAT solar array coupon (witness plate) exposure	KSC W/D: OPF 109, VAB 8, PAD 32 = 149 LAUNCH POSTPONEMENTS: Launch postponed from 10/5/90 to 10/6/90 in late September. LAUNCH SCRUBS: None. LAUNCH DELAYS: - 10M43S delay at T-9 minutes due to rain showers 14 miles north of RTLS runway. - Countdown held at T-5 minutes for 10 seconds to mask GLS WSB 2 indication. - 1M22S delay at T-31 seconds due to P/L- Orbiter I/F and duct pressures out of limits. - 12M15S total delay. TAL WX: - Banjul (prime) - Marginal WX, recent rain. - Ben Guerir (alt) selected - solid GO WX. I-LOADS: LSEAT selected nominal I-loads, no uplink required. FIRSTS: - First flight with all 3 Orbiters in vertical; OV-103/STS-41 on pad B, OV-102/STS-35 on pad A, OV-104/STS-38 in VAB. - First flight after MPS LH2 leaks found in STS-35 and STS-38. - First flight using fixed (shimmed) GOX FCV's (step 1). - First flight with SRB using redesigned field joint protection system. EVENTS: - RMS parked at 1:03:35 MET with INTELSAT solar array coupon in velocity vector to witness potential solar array damage. - ULYSSES deployed at 06:01:06 MET. - No entry blackout. - Conducted RCS Hot Fire using extended firing durations (640 msecs) to attempt nitrate removal. SIGNIFICANT ANOMALIES: - MC4 (SM2) NBAT had GPC 2 assigned to FC string 3. - IMU 1 RM fail (experiencing transient 2 axis accelerometer shifts). - APU 1 GG/fuel pump heater B failed on. - Ammonia boiler PRI A controlled low, 31.6 evap out temp. - Hydraulic Sys #2 priority valve sluggish at startup. - Debris plunger (EO-2) fail to seat/ ordnance pieces found on runway. - Crescent shaped debris (22") in video camera views during Ulysses deploy. - Haz gas grab bottles indicated max 37,000 SCIM's during ascent (upward trend).

STS041-26-007 1990-10-10 **Crew in** middeck (front, lt. to rt.) CDR Richards & PLT Cabana; (rear ,lt. to rt.) Akers/MS, Melnick/MS, & Shepherd/MS.

STS041-61-009 Ulysses Deployed

S90-47615   FLT DIR's: left, Milt Heflin & right, Ron Dittemore

FLT NO.	ORBITER	CREW (5) TITLE, NAMES & EVA'S	LAUNCH SITE, LIFTOFF TIME, LANDING SITES, ABORT TIMES	LANDING SITE/ RUNWAY, CROSSRANGE LANDING TIMES FLT DURATION, WINDS	SSME-TL NOM-ABORT EMERG THROTTLE PROFILE ENG. S.N.	SRB RSRM AND ET	ORBIT		FSW	PAYLOAD WEIGHTS, PAYLOADS/ EXPERIMENTS	MISSION HIGHLIGHTS (LAUNCH SCRUBS/DELAYS, TAL WEATHER, ASCENT I-LOADS, FIRSTS, SIGNIFICANT ANOMALIES, ETC.)
							INC	HA/HP			
STS-38 SEQ FLT #37 KSC-37 PAD 39A-27 MLP-1	OV-104 (Flight 7) Atlantis OMS PODS LPO1 - 13 RPO3 - 12 FRC4 - 7	CDR: Richard O. Covey (Flt 3 - STS 51-I & STS-26) P188/R73/V34/M67 PLT: Frank L. Culbertson P189/R116/M104 M/S 1: Carle J. Meade P190/R114/M105 M/S 2: Robert C. Springer (Flt 2 - STS-29) P191/R98/V57/M89 M/S 3: Charles D. Gemar P192/R118/M106 MCC FCR-2 (20) FLIGHT DIRECTORS: Asc/Ent - A. L. Briscoe O 1 - R. M. Kelso Ld/O 2 - C. R. Knarr Plng - C. W. Shaw MOD - B. R. Stone	KSC 39A 319:23:48:15Z 6:48:15 PM EST Thursday 9 11/15/90 (7) PLS: EDW RTLS: KSC TAL: Banjul (Selected) TAL WX: Ben Guerir SELECTED: RTLS: KSC 15 TAL: BYD 32 AOA: EDW 22 MAX Q: 00:49 MET SRB SEP: 2:03 MET MECO: 8:29 MET ET SEP: 8:47 MET OMS-1: 10:30 MET OMS-2: 47:43 MET	KSC 33 (KSC 6) 324:21:42:42Z 4:42:42 PM EST Tuesday 8 11/20/90 (6) DEORBIT BURN: 324:20:46:15Z XRANGE: 3 NM ORB DIR: DL 19 AIM PT: CLOSEIN MLGTD: 1414 FT 324:21:42:42Z VEL: 195 KGS 199 KEAS HDOT: -1 FPS TD NORM 195: 1850 FT NLGTD: 4600 FT 324:21:42:52Z VEL: 162 KGS HDOT: -3.1 FPS BRK INIT: 127 KGS AVE BRK DECEL: 7 FPS/S WHEELS STOP: 324:21:43:39Z 10417 FEET ROLLOUT: 9003 Feet 57 Seconds WINDS: 4H, 4.4R KTS OFFICIAL: 4H, 4R DENS. ALT: 387 FT FLT DURATION: 4:21:54:27 117:54:27 S/T: 214:10:53:17 OV-104: 33:16:43:40 DISTANCE: 2,045,056 sm	104/104/ 109% ACTUAL: 100/104/ 104/72/ 104/65 1 = 2019 (7) 2 = 2022 (5) 3 = 2027 (5) M 3 EOM WEIGHT: X CG: LANDING WEIGHT: 191091 X CG: 1098.6	BI-039 RSRM 12 ET-40 LWT-33 ET RPT 222K 47:10 MET ET BR/UP 181K 47:56 MET ET IMPACT LAT: 28.52 S LONG: 84.9 W	28.45 (23)	STANDARD INSERTION DEORBIT BURN 114.9 SECS 228.5 FPS DEORBIT 142 X 115 NM VELOCITY 25729 FPS ENTRY RANGE 4146 NM OMS BURN 114.9 SECS 228.5 FPS	OI-8D (2)	DOD PERFORMANCE MARGINS (LBS): FPR: 4652 FUEL BIAS: 994 FINAL TDDP: 863 RECON: 474 SECONDARY PAYLOADS: APE VFT-1 RME-III AMOS APM S-BAND XPONDERS ON SRB'S	KSC W/D: OPF 134 (2), VAB 26 (3), PAD 85 (2) = 245 LAUNCH POSTPONEMENTS: - As of Jan 1990, launch date was 7/9/90. On 5/29/90, OV-102/STS-35 launch was scrubbed because of excessive H2 leak in aft compartment. Special H2 tanking tests were performed on OV-104/STS-38. - 6/18/90 - STS-38 rolled out to Pad A. Scheduled launch 7/9. - 6/29/90 - LH2 Tanking Test #1 - Excessive H2 leak detected in umbilical area. - 7/13/90 - LH2 Tanking Test #2 - Excessive H2 leak detected in umbilical and plate gap areas. - 7/25/90 - LH2 Tanking Test #3 - Excessive H2 leak ET 17" disconnect flange area. Decision made to roll back and fix leak. - 8/9/90 - Rolled stack back to VAB. - 8/15/90 - OV-104 to OPF. Umbilical removed from ET-37 and sent to MSFC and RI-D for tests. Subsequently, found follower arm seal and shaft seal leaks in tests. Decision to use ET-40 after replacing LH2 umbilical. - 10/13/90 - Rolled out to Pad A. - 10/24/90 - LH2 Tanking Test #4 successful. - Launch scheduled for 11/15/90. 129-day slip. LAUNCH SCRUBS: None during second time at pad. LAUNCH DELAY: Launch delayed because Range Bermuda command link out of service. TAL WX: - Banjul - GO (weather good). - Ben Guerir - GO (weather good). I-LOADS: - Due to seasonal slip in launch, pitch negative became pitch nominal, which LSEAT selected, and was uplinked (Uplink 4). NIGHT LAUNCH: Fifth Shuttle night launch. WAVEOFFS: - Waved off on fourth day because of excessive head and crosswinds on all three landing opportunities at EDW. - Extended one rev to land at KSC because of high winds predicted at EDW. Continued . . .

STS038-28-016   1990-11-20 Crew on Atlantis' middeck: (right to left) Springer/MS, PLT Culbertson, CDR Covey, Gemar/MS, and Meade/MS. First flight with Air Force, Navy, Army, and Marine Corps crewmembers. DOD Mission.

FLT NO.	ORBITER	CREW (5) TITLE, NAMES & EVA'S	LAUNCH SITE, LIFTOFF TIME, LANDING SITES, ABORT TIMES	LANDING SITE/ RUNWAY, CROSSRANGE LANDING TIMES FLT DURATION, WINDS	SSME-TL NOM-ABORT EMERG THROTTLE PROFILE ENG. S.N.	SRB RSRM AND ET	ORBIT INC HA/HP	FSW	PAYLOAD WEIGHTS, PAYLOADS/ EXPERIMENTS	MISSION HIGHLIGHTS (LAUNCH SCRUBS/DELAYS, TAL WEATHER, ASCENT I-LOADS, FIRSTS, SIGNIFICANT ANOMALIES, ETC.)
STS-38 Continued										Continued . . .

STS-38: LAUNCH -------- A VARIETY OF EARTH VIEWS -------- LANDING
(Captions Not Available)

s38-s-027

s38-82-093

sts038-74-086

sts038-92-077

s38-78-090

S38-86-044

sts038-86-104

s38-86- 016

s38-s041

LANDING SITE CHANGE:
- Changed from EDW to KSC landing because of predicted unfavorable winds.

FIRSTS:
- First flight with Air Force, Navy, Army, and Marine Corps crewmembers. All 4 hymns were used as wakeup music on one day.
- First flight of GOX FCV's in step 2 position.

SIGNIFICANT ANOMALIES:
- WSB 2 not cooling on controller A.
- FES water supply accumulator heater biased low.
- Vacuum cleaner short, CB 29 opened.
- CCTV monitor 2 fault light on - powered down.
- APU 2 EGT and APU 2 and 3 injector tube temps interacting.
- Right vent door 1 and 2 purge position dropped to closed position instead of purge position
- RIU PC low.
- Continuous 'Tire 'Press' FDA messages post landing.
- Several smoke detectors had event indicators go high but not high enough to trigger alarm.
- GPC mode switch found in STDBY and power switch in off.

SPACE SHUTTLE MISSIONS SUMMARY

FLT NO.	ORBITER	CREW (7) TITLE, NAMES & EVA'S	LAUNCH SITE, LIFTOFF TIME, LANDING SITES, ABORT TIMES	LANDING SITE/ RUNWAY, CROSSRANGE LANDING TIMES FLT DURATION, WINDS	SSME-TL NOM-ABORT EMERG THROTTLE PROFILE ENG. S.N.	SRB RSRM AND ET	INC	ORBIT HA/HP	FSW	PAYLOAD WEIGHTS, PAYLOADS/ EXPERIMENTS	MISSION HIGHLIGHTS (LAUNCH SCRUBS/DELAYS, TAL WEATHER, ASCENT I-LOADS, FIRSTS, SIGNIFICANT ANOMALIES, ETC.)
STS-35 (STS 61-E) SEQ FLT #38 KSC-38 PAD 39B-11 MLP-3	OV-102 (Flight 10) Columbia Fifth Spacelab Flight ASTRO-1 IGLOO + 2 PALLETS (2nd IGLOO) OMS PODS LP03 - 10 RP04 - 6 FRC2 - 10	CDR: Vance D. Brand (Flt 3 - STS-5 & STS 41B) P193/R9/V4/M9 PLT: Guy S. Gardner (Flt 2 - STS-27) P194/R95/V58/M86 M/S 1: John M. Lounge (Flt 3 - STS 51-I & STS-26) P195/R74/V35/M68 M/S 2: Jeffrey A. Hoffman (Flt 2 - STS 51-D) P196/R57/V59/M52 M/S 3: Robert A. R. Parker (Flt 2 - STS-9) P197/R27/V60/M26 P/S 1: Ronald A. Parise (CSC) P198/R119/M107 P/S 2: Samuel T. Durrance John Hopkins University P199/R120/M108 MCC FCR-1 (18) FLIGHT DIRECTORS: Asc/Ent - N. W. Hale Ld/O 1 - G. E. Coen O 2 - G. A. Pennington O 3 - R. E. Castle MOD - T. W. Holloway	KSC 39AB 336:06:49:01Z 1:28:00 AM EST (P) 1:49:01 AM EST (A) Sunday 4 12/02/90 (2) LAUNCH WINDOW 2H30M (CTOB) RTLS: KSC-15 TAL: Banjul 32 TAL WX: Ben Guerir Moron SELECTED: TAL: BYD 32 RTLS: KSC 15 AOA: EDW 22 PLS: EDW22 AOA: EDW 22 MAX Q: 696 PSF 00:50 MET SRB SEP: 2:06 MET MECO: 8:32 MET ET SEP: OMS-1: NONE OMS-2: 40:24.7 MET 180.3 SECS 179.1 FPS	EDW 22 CONC (EDW 30,CONC 13) 345:05:54:09Z 9:54:09 PM PST Monday 9 12/10/90 (4) DEORBIT BURN: 345:04:48:31Z 230.5 SECS,383 FPS XRANGE: 426 NM ORB DIR: DL 20 AIM PT: CLOSEIN MLGTD: 1535 FT 345:05:54:09Z VEL: 208 KGS 201 KEAS HDOT: -1 FPS TD NORM 195: 2247 FT NLGTD: 5559 FT 345:05:54:20Z VEL:168 KGS HDOT: -3.9 FPS BRK INIT: 136 KGS AVE BRK DECEL: 7.2 FPS/S WHEELS STOP: 345:05:55:06Z 12101 FEET ROLLOUT: 10450 Feet 58 Seconds WINDS: 0.7 T, 0.7 R KTS OFFICIAL: 1T, 1R DENS ALT: 1143 FT FLT DURATION: 8:23:05:08 215:05:08 S/T: 223:09:58:25 OV-102: 65:23:00:04 DISTANCE: 3,728,636 sm	104/104/ 109% 100/104/ 71/104/65 1 = 2024 (2) 2 = 2012 (11) 3 = 2028 (5) M 3 EOM WEIGHT: 225531 X CG: 1079.1 LANDING WEIGHT: 225329 X CG: 1080.5	BI-038 RSRM 11 ET-35 LWT-27 ET RPT 233K 1:18:39 MET ET BR/UP 203K 1:19:27 MET ET IMPACT LAT: 15.09 N LONG: 159.0 W	28.457 (24)	DIRECT INSERTION INSERTION ALTITUDE: 190.4 X 188.2 NM DEORBIT 195.2 X 180.3 NM VELOCITY 25858 FPS ENTRY RANGE 4266 NM	OI-8D (3)	CARGO: 33037 LBS CHARGEABLE: 27760 LBS DEPLOYED: 0 LBS NON-DEPLOYED: 25968 LBS MIDDECK: 1792 LBS RETURNED: SHUTTLE ACCUMULATED WEIGHTS: DEPLOYED: 515171 LBS NON-DEPLOYED: 435458 LBS CARGO TOTAL: 1075273 LBS PERFORMANCE MARGINS (LBS): FPR: 4652 FUEL BIAS: 994 FINAL TDDP: 4131 RECON: 3812 PAYLOADS: PLB: ASTRO-1: IPS, HUT, WUPPE, UIT, BBXRT (ASTRONOMY) MIDDECK: AMOS SAREX-II UVPI 5 CRYO TK SETS NO RMS	KSC W/D: OPF 126 (2), VAB 16 (3), PAD 153 (3) = 295 LAUNCH POSTPONEMENT: - As of 1/90, launch date was 5/9/90. Post-poned to 5/30/90 due to P/L argon servicing, LO₂ system leak, and FCL coolant valve contamination (low flow). 21-day slip. LAUNCH SCRUBS: - Scrubbed 5/29/90 launch during tanking due to excessive H₂ leak in aft compartment. - Failed 6/6/90 special LH2 tanking test, excessive H₂ leak in aft compartment. - 6/13/90 - Rolled back from Pad A to VAB. - 6/15/90 - OV-102 to OPF. Both OV-102 and ET-35 LH2 umbilicals sent to RI-D for special LH2 leak tests. R&R'ed ET-35 and OV-102 umbilicals (used OV-105 umbilical). - 8/2/90 - Rolled out to VAB for restacking. - 8/9/90 - Rolled to Pad A. - Scheduled launch for 9/1/90. - Scrubbed 9/1/90 launch before tanking because of BBXRT TLM problem. Rescheduled launch for 9/6/90. - Scrubbed 9/6/90 launch during tanking due to H₂ leak in aft compartment. (Estimated 30,000 SCIM's/6000 PPM.) Replaced crushed PV6 detent cover seal on SSME 3 and recirc pump package before 9/17/90 scheduled launch. - Scrubbed 9/17/90 launch during tanking at L-7 hrs due to H2 leak in aft compartment (4300 PPM). - Rescheduled launch for 10/2/90. - 10/8/90 - Rolled to Pad B after STS-41 launch (did not hard down). - 10/8-9/90 - Rolled back to VAB because of Tropical Storm Klaus threat. Replaced crushed PV5 detent seal in SSME 2. - 10/14/90 - Rolled to Pad B. MPS troubleshooting found several small H₂ leaks exceeding specs. - 10/30/90 - Instrumented LH₂ Tanking Test, successful with only 150 PPM concentration in aft compartment. - 12/2/90 - Launch successful on fifth launch attempt. 170-day launch slip. - 207-day total slip. Continued...

STS035-503-007 1990-12-11 Crew in Columbia's middeck, clockwise from bottom center, CDR Brand, Parker/MS, Parise/PS, Hoffman/MS, PLT Gardner, Lounge/MS, & Durrance/PS.

FLT NO.	ORBITER	CREW (7) TITLE, NAMES & EVA'S	LAUNCH SITE, LIFTOFF TIME, LANDING SITES, ABORT TIMES	LANDING SITE/ RUNWAY, CROSSRANGE LANDING TIMES FLT DURATION, WINDS	SSME-TL NOM-ABORT EMERG THROTTLE PROFILE ENG. S.N.	SRB RSRM AND ET	ORBIT INC	HA/HP	FSW	PAYLOAD WEIGHTS, PAYLOADS/ EXPERIMENTS	MISSION HIGHLIGHTS (LAUNCH SCRUBS/DELAYS, TAL WEATHER, ASCENT I-LOADS, FIRSTS, SIGNIFICANT ANOMALIES, ETC.)

STS-35

Continued

s35-13-008 -- Wisconsin Ultaviolet photo-Polarrimeter Experiment (WUPPE) on Spacelab pallet. The Broad Band X-Ray Telescope (BBXRT) is behind this pallet and is not visible.

STS035-28-022 1990-12-10 Astronomy Laboratory 1 (ASTRO-1) telescopes in the PL/Bay. At right is the Orion nebula. The three ultraviolet telescopes are mounted and coaligned on a common structure and attached to the Instrument Pointing System (IPS).

S88-54116 1988-11-30 Official insignia for the Johnson Space Center's (JSC's) Amateur Radio Club

S90-32048 1990-03-16 Shuttle Amateur Radio Experiment (SAREX) equipment held by R. Parise/PS at the JSC Full Fuselage Trainer. SAREX is used to conduct shortwave radio transmissions between ground amateur radio operators and a licensed onboard operator (in this case, Parise).

STS035-05-036 1990-12-11 STS-35 Commander Brand talks to family using SAREX on Columbia's middeck

Continued ...

LAUNCH DELAYS:
- 21M1S delay while Range Safety had helicopter verify 8000 foot minimum optical coverage.

TAL WX:
- Weather good at Banjul and Ben Guerir.

I-LOADS:
- Launch delayed to new season and pitch negative became pitch nominal which LSEAT selected and was uplinked (uplink 5).

NIGHT LAUNCH: Space Shuttle #6.

NIGHT LANDING: Space Shuttle #4.

EVENTS:
- Most people in Earth orbit at the same time - 12 (7 Americans and 5 Soviets).

SIGNIFICANT ANOMALIES:
- FCL-1 degraded flowrate noticed before first launch attempt. Did not affect mission and performed as predicted.
- S/L DDS 1 (DDU) failed on FD1. Crew smelled smoke.
- S/L DDS 2 failed after 4 days. Crew smelled smoke. (Crew did IPS pointing and ground sent commands to operate experiments.)
- S/L subsystem computer failed due to a command problem caused by error in workstation program, recovered by IPL.
- Degraded waste water flow, virtual blockage at 152 hours. Filled CWC with 92 lbs, wastewater transferred to 15 female UCD's and 18 male UCD's.
- TAGS jam, TAGS tool broke.
- OPS 1 track 2 and OPS 2 track 5 problems.
- P/L recorder poor data quality.
- HDRR failed after 2 days of operations.
- Cameras B, C, & D problems.
- Several software patches were required to correct experiment/IPS target tracking.
- S-band UL and LR antenna problems.
- Several payload experiment problems.
- WSGT control computer failure.
- APU 2 lube oil pressure high during ascent & entry (wax formation caused by hydrazine contamination).
- No blackout during entry.

FLT NO.	ORBITER	CREW (5) TITLE, NAMES & EVA'S	LAUNCH SITE, LIFTOFF TIME, LANDING SITES, ABORT TIMES	LANDING SITE/ RUNWAY, CROSSRANGE LANDING TIMES FLT DURATION, WINDS	SSME-TL NOM-ABORT EMERG THROTTLE PROFILE ENG. S.N.	SRB RSRM AND ET	ORBIT		FSW	PAYLOAD WEIGHTS, PAYLOADS/ EXPERIMENTS	MISSION HIGHLIGHTS (LAUNCH SCRUBS/DELAYS, TAL WEATHER, ASCENT I-LOADS, FIRSTS, SIGNIFICANT ANOMALIES, ETC.)
							INC	HA/HP			
STS-37 SEQ FLT #39 KSC-39 PAD 39B-12 MLP-1	OV-104 (Flight 8) Atlantis OMS PODS LPO1 - 14 RPO1 - 14 FRC4 - 8	CDR: Steven R. Nagel (Flt 3 - STS 51-G & STS 61-A) P200/R64/V23/M59 PLT: Kenneth D. Cameron P201/R121/M109 M/S 1: Linda M. Godwin P202/R122/F13 M/S 2: Jerry L. Ross (Flt 3 - STS 61-B & STS-27) P203/R86/V38/M78 M/S 3: Jay Apt P204/R123/M110 EMU/TETHERED EVA:* EV1 - Jerry Ross EV2 - Jay Apt EVA 1 - 4/7/91 SS EVA #14 3:40/4:32 SS UNSCHED EVA #2 RELEASE STUCK GRO HI GAIN ANTENNA EVA 2 - 4/8/91 SS EVA #15 5:47/5:57 DEMO SPACE STATION (CREW & EQUIPMENT TRANSLATION AID) Continued...	KSC 39B 95:14:22:44.98Z 9:18:00 AM EST (P) 9:22:45 AM EST (A) Friday 6 4/5/91 (7) LAUNCH WINDOW: 2H30M (CTOB) PLS: EDW LKBD TAL: BANJUL TAL ALT: BEN SELECTED: RTLS: KSC 33 TAL: BEN 36 AOA: EDW 22 TDEL: -0.16 -0.118 MAX Q: 676 681 SRB STG: 2:04.8 PERF: NOM 2 ENG TAL (BEN) 2:59 2:58 NEG RETURN: 4:04 4:07 PTA: 4:46 4:42 PTM: 5:51 5:45 MECO CMD: 8:34 8:33.3 VI: 26010 26005 OMS-2: Tig = DV=369 FPS	EDW 33, LAKEBED (EDW 31, LKBD 18) 5:55:29 AM PST Thursday 2 4/11/91 (6) XRANGE: 375 NM ORB DIR: AL 11 AIM PT: CLOSEIN MLGTD:-623 FT 101:13:55:29Z VEL: 156 KGS 168 KEAS HDOT: -2 FPS TD NORM 195: -2384 FT NLGTD: 1200 FT 101:13:55:35Z VEL:130 KGS HDOT: -8.4 FPS BRK INIT: 93 KGS AVE BRK DECEL: 4.8 FPS/S WHEELS STOP: 101:13:56:25Z 5741 FT ROLLOUT: 6364 FEET 56 SECS WINDS: 14.1H, 9.6 R KTS OFFICIAL: 15H, 8R DENS. ALT: 1732 FT FLT DURATION: 5:23:32:44 143:32:44 S/T: 229:09:31:09 OV-104: 39:16:16:24 DISTANCE: 2,487,075 sm	104/104/ 109% ACTUAL: 100/104/ 87/67/ 104/65 1 = 2019 (8) 2 = 2031 (5) 3 = 2107 (4) M 3 EOM WEIGHT: 190266 X CG: 1087.4 LANDING WEIGHT: 190098 X CG: 1089.2	BI-042 RSRM 14 ET-37 LWT-30 ET RPT 237K 1:22:20 MET ET BR/UP 195K 1:23:25 MET ET IMPACT LAT: 20.23 N LONG: 149.3 W	DIRECT INSERTION INSERTION ALTITUDE: 244.2 X 241.2 NM GRO DEPLOY HO = 246.6 NM DEORBIT 248 X 239 NM VELOCITY 24612 FPS ENTRY RANGE 4175 NM	28.453 (25)	OI-8F (1)	CARGO: 40561 LBS PAYLOAD CHARGEABLE: 36800 LBS NON-DEPLOYED: 1615 LBS DEPLOYABLE: 34442 LBS MIDDECK: 743 LBS SHUTTLE ACCUMULATED WEIGHTS: DEPLOYED: 549613 LBS NON-DEPLOYED 437816 LBS CARGO TOTAL: 1115834 LBS PERFORMANCE MARGINS (LBS): FPR: 4652 FUEL BIAS: 994 FINAL TDDP:1116 RECON: 525 PAYLOADS: PLB: GAMMA RAY OBSERVATORY (GRO) DEPLOYED APM CETA MIDDECK: PCG, BLOCK II RME-III SAREX AMOS BIMDA 3 CRYO TK SETS RMS 23 (S.N. 303 USED FOR GRO DEPLOY)	KSC W/D: OPF 97, VAB 6, PAD 22 = 125 days LAUNCH POSTPONEMENT: - On 8/2/90, launch date was 3/27/91. - 4-day postponement prior to 10/90 (launch 4/1/91). - 7-day postponement in 11/90, STS-38 launch delay, launch date 4/8/91 (under review). - On 2/28/91, decision made to rollback STS-39 from pad to repair ET door hinge cracks. OV-104 ET doors repaired before OPF rollout. OV-103 rollback caused STS-39 to be launched after STS-37. - At LSFR, launch date 4/4/91 (under review). - Postponed 1 day to 4/5/91 (tile and FRT). - 9-day total slip from 8/90. LAUNCH SCRUBS: None. LAUNCH DELAYS: - 4M45S delay due to violation of RSO 8000-foot ceiling requirement at T-9 and range "B LAST" prediction (Counted to T-5 and held for waiver.) TAL WX: - Banjul no go because of tail winds (brake energy). - Ben Guerir 36 go (selected). RTLS: - Forecast NO GO RW & ceiling, observed NO GO at T- 22 mins. Selected KSC NOM 33. I-LOADS: - LSEAT select nominal I-loads, no uplink required. FLIGHT DURATION CHANGES: - EDW 15 was first priority. Waved off one rev then extended flight 1 day due to winds/turbulence. - Extended one rev due to winds at EDW. Extension total, 1 day + 1 rev. GRO DEPLOY: 2:08:14:02 MET Unscheduled EVA to release GRO antenna. FIRSTS: - First flight of new GPC's (AP-101S). - First flight of OI-8F. - First EVA since STS 61-B on 12/01/85. Continued ...

STS037-30-024   1991-04-11 STS-37
Crew on Atlantis' middeck: Back row: CDR Nagel and PLT Cameron. Front row, left to right: Ross/MS, Godwin/MS, and Apt/MS. Cards refer to astronauts' "ACE Moving Company".

* TWO EVA TIMES ARE PROVIDED: (1) OLD DEFINITION - STARTED WHEN EMU WENT TO BAT POWER AND ENDED WHEN SWITCHED TO ORBITER POWER
(2) NEW DEFINITION - STARTS WHEN EMU GOES TO BAT POWER AND ENDS WHEN AIRLOCK REPRESS STARTS

FLT NO.	ORBITER	CREW (7) TITLE, NAMES & EVA'S	LAUNCH SITE, LIFTOFF TIME, LANDING SITES, ABORT TIMES	LANDING SITE/ RUNWAY, CROSSRANGE LANDING TIMES FLT DURATION, WINDS	SSME-TL NOM-ABORT EMERG THROTTLE PROFILE ENG. S.N.	SRB RSRM AND ET	ORBIT		FSW	PAYLOAD WEIGHTS, PAYLOADS/ EXPERIMENTS	MISSION HIGHLIGHTS (LAUNCH SCRUBS/DELAYS, TAL WEATHER, ASCENT I-LOADS, FIRSTS, SIGNIFICANT ANOMALIES, ETC.)
							INC	HA/HP			
STS-37 Continued...		Continued... MCC FCR-1 (19) FLIGHT DIRECTORS: Asc/Ent - N. W. Hale Ld/O 1 - C. W. Shaw O 2 - J. M. Heflin Plng - P. L. Engelauf MOD - G. E. Coen									Continued ... SIGNIFICANT ANOMALIES: - Thruster R1U failed off 32 seconds after MECO. - WSB 2A temporary spray bar freeze up during ascent. - WSB 2A and 3A lube oil overcooling during entry. - PRSD O_2 manifold valve failed to close. - EVA glove palm bar penetrated restraint and glove bladder. - Prelaunch BFS navigation anomaly. - Ku-band antenna erratic in ant mode. - EMU-1 failed to charge battery post EVA-1. - Abnormal O_2 concentration in aft compartment (220 PPM) - Unscheduled EVA required to deploy GRO high gain antenna. - Scheduled EVA.

STS037-99-089 1991-04-11 Deployed Gamma Ray Observatory (GRO) over Baja California, Mexico (31.5N, 113.0W), the Salton Sea and Imperial Valley region of California where the mouth of the Colorado River empties into the Sea of Cortez are clearly visible.

At Right: STS037-55-012 1991-04-11 Ross/MS drifts outside P/L Bay as he attaches a tether to a port side guidewire during EVA.

TOP: STS037-52-013 1991-04-11 Apt/MS, suited in Extravehicular Mobility Unit (EMU), tests Crew and Equipment Translation Aid (CETA) electrical hand pedal cart during EVA in P/L Bay).

FLT NO.	ORBITER	CREW (7) — TITLE, NAMES & EVA'S	LAUNCH SITE, LIFTOFF TIME, LANDING SITES, ABORT TIMES	LANDING SITE/ RUNWAY, CROSSRANGE — LANDING TIMES, FLT DURATION, WINDS	SSME-TL NOM-ABORT EMERG — THROTTLE PROFILE ENG. S.N.	SRB RSRM AND ET	ORBIT — INC — HA/HP	FSW	PAYLOAD WEIGHTS, PAYLOADS/ EXPERIMENTS	MISSION HIGHLIGHTS (LAUNCH SCRUBS/DELAYS, TAL WEATHER, ASCENT I-LOADS, FIRSTS, SIGNIFICANT ANOMALIES, ETC.)

STS-39
 	
SEQ FLT #40
 	
KSC-40
 	
PAD
39A-28
MLP-2
	OV-103
	Discovery
	(Flight 12)
	
	OMS PODS
	LPO4 - 9
	RPO3 - 13
	FRC3 - 12
		CDR:
		Michael L. Coats
		(Flt 3 - STS 41-DR & STS-29)
		P205/R38/V39/M37
		PLT:
		L. Blaine Hammond
		P206/R124/M111
		M/S 1:
		Gregory J. Harbaugh
		P207/R125/M112
		M/S 2:
		Donald McMonagle
		P208/R126/M113
		M/S 3:
		Guion S. Bluford
		(Flt 3 - STS-8 & STS 61-A)
		P209/R22/V25/M21
		M/S 4:
		Charles Lacy Veach
		P210/R127/M114
		M/S 5:
		Richard J. Hieb
		P211/R128/M115
		
		MCC FCR-1 (20)
		FLIGHT DIRECTORS:
		Asc/Ent - A. L. Briscoe
		Ld/O2 - R. D. Dittemore
		O 1 - R. E. Castle
		O 3 - R. M. Kelso
		MOD - T. W. Holloway
			KSC 39A
			118:11:33:14Z
			7:01:00 AM EDT (P)
			7:33:14 AM EDT (A)
			Sunday 5
			4/28/91 (8)
			
			LAUNCH WINDOW
			3H20M
			(AURORA CONSTR)
			PLS: EDW LKBD
			TAL: ZZA (8)
			TAL ALT:
			  BEN GUERIR
			  MORON
			
			SELECTED:
			RTLS: KSC 33/CI
			TAL: BEN 36/CI
			AOA: EDW 22
			
			TDEL:
			-0.64 -0.55
			
			MAX Q:
			709 707
			
			SRB STG:
			2:03.4 2:05
			
			PERF: Nominal
			
			2 ENG TAL (BEN):
			2:49 2:55
			
			NEG RETURN:
			4:06 4:08
			
			PTA: (ATO)
			4:56 5:10
			
			PTM:
			6:09 6:22
			VI:
			25804 25850
			
			OMS-2:
			Tig =36:08
			DV=209.6 FPS
				KSC 15 (KSC-7)
				126:18:55:35Z
				
				2:55:35 PM EDT
				Monday 10
				5/6/91 (3)
				
				DEORBIT BURN:
				126:17:53:34Z
				
				XRANGE: 616 NM
				
				ORB DIR: DL 21
				
				AIM PT: CLOSEIN
				
				MLGTD: 169 FT
				126:18:55:35
				VEL : 210 KGS
				         218 KEAS
				HDOT: - 2 FPS
				
				TD NORM 195:
				2771 FT
				
				NLGTD: 4700 FT
				126:18:55:49
				VEL: 157KGS
				HDOT: - 2.9 FPS
				
				BRK INIT:136 KGS
				
				AVE BRK DECEL:
				9.5 FPS/S
				
				WHEELS STOP:
				126:18:56:31
				9403 FT
				
				ROLLOUT:
				9234 FT
				56 Seconds
				
				WINDS:
				12H, 1R KTS
				OFFICIAL: 14H, 2R
				
				DENS ALT:1723FT
				
				FLT DURATION:
				8:07:22:21
				199:22:21
				
				S/T: 237:16:53:30
				
				OV-103:
				69:17:40:41
				
				DISTANCE:
				3,475,000 sm
					104/104/
					109%
					
					ACTUAL:
					100/100/
					94/70/
					104/67
					
					ET
					RPT
					1 = 2026 (1)
					2 = 2030 (5)
					3 = 2029 (4)
					
					M 3 EOM
					
					WEIGHT:
					211673
					
					X CG: 1080.3
					
					LANDING
					
					WEIGHT:
					211512
					
					X CG: 1082.0
					
					DEORBIT
					140 X
					138 NM
					
					VELOCITY
					25765 FPS
					
					ENTRY
					RANGE
					4502 NM
						BI-043
						
						RSRM
						15K
						
						ET-46
						LWT-39
						
						ET
						RPT
						249K
						1:09:34
						MET
						
						ET
						BR/UP
						215K
						1:10:34
						MET
						
						ET
						IMPACT
						LAT:
						43.82 S
						LONG:
						156.3 W
							DIRECT
							INSERTION
							
							INSERTION
							ALTITUDE:
							140.02 X
							138.22 NM
							
							SPAS
							DEPLOY:
							137.37 X
							136.55 NM
							
							CRO-C
							DEPLOY:
							136.4 X
							134.7 NM
							
							CRO-B
							DEPLOY:
							136.7 X
							132.7 NM
							
							SPAS RNDZ:
							135.5 X
							132.8 NM
							
							CRO-A
							DEPLOY:
							140.96 X
							138.6 NM
							
							MPEC
							DEPLOY:
							141.55 X
							139.46 NM
								57.007
								(7)
									OI-8F
									(2)
										CARGO:
										26294 LBS
										
										PYLD CHARGABLE:
										21413 LBS
										
										DEPLOYABLE:
										827 LBS
										
										NON-DEPLOYED:
										16046 LBS
										
										RETURNED:
										
										MIDDECK:
										494 LBS
										
										SHUTTLE ACCUM WEIGHTS:
										DEPLOYED:
										550440 LBS
										NON-DEPLOYED:
										454356 LBS
										CARGO TOTAL:
										1142128 LBS
										
										PERFORMANCE MARGINS (LBS):
										FPR: 4653
										FUEL BIAS: 994
										FINAL TDDP:1054
										RECON: 2768
										
										PAYLOADS:
										PLB:
										Infrared Background
										Signature Survey
										(IBSS) (SPAS-II (IV
										+ 3 GAS DEPLOY
										CRO-A, CRO-B,
										CRO-C, CIV)
										
										AF-675 (CIRRIS,
										FAR-UV, URA, HUP,
										QINMS)
										
										STP-1 (ALFE, APM,
										SKIRT, UVIM, DSE)
										
										MPEC - GAS DPLY
										
										MIDDECK:
										CLOUDS-1A
										RME-III
										UVPI
										
										4 CRYO TK SETS
										
										RMS 24 (S.N. 301)
										USED FOR
										SPAS/IBSS DPLY,
										CAPTURE, AND
										BERTH
											KSC W/D: OPF 116 (2), VAB 17 (3), PAD 47 (2) = 180
											
											LAUNCH POSTPONEMENTS:
											- As of 8/21/90, launch date is 2/26/91.
											- 2/26/91 launch postponed to 3/9/91 due to OMS pod work. (Swapped RP-03 from OV-104 for RP-01.)
											- On 2/15/91, cracks found in OV-103 ET door hinge brackets. On 2/28/91, decision made to roll back and repair ET doors resulting in STS-39 launch being scheduled after STS-37. Launch rescheduled for 4/23/91.
											- 56 days total slip based on 8/21/90 schedule.
											
											LAUNCH SCRUBS:
											- 4/23/91 launch scrubbed at L-6 hours due to SSME #3 HPOTP secondary seal pressure xducer problem and P/L servicing. Rescheduled launch for 4/28/91.
											- 5-day slip. (Total slip - 61 days.)
											
											LAUNCH DELAYS:
											- 32M14S delay caused by review of OPS 2 recorder uncommanded switching of tracks and going to run at approximate time of BFS 101 PRO.
											
											TAL WX:
											- Zaragoza and Moron no go - ceilings (broken < 8000 feet).
											
											I-LOADS:
											- LSEAT selected nominal, no uplink.
											
											FLIGHT DURATION/LANDING SITE CHANGES:
											- Landed at KSC on same rev as planned for EDW because unfavorable winds predicted at EDW.
											
											EVENTS:
											- SPAS deploy - rev 46, SPAS RNDZ - rev 72, MPEC deploy - rev 127.
											- 16 OMS burns.
											
											RENDEZVOUS 9: With Infrared Background Signature Survey (IBSS) (SPAS-II) for retrieval and return.
											
											FIRSTS:
											- First flight with 67% as standard 3g throttling.
											
											SIGNIFICANT ANOMALIES:
											- ROB tire outboard shoulder damaged during landing (3 cords).
											- OPS 2 recorder uncommanded switching of tracks and tape speed prelaunch.
											- FES feedline A system 2 heater failure.
											- APU 2 fuel pump/GGVM coolant sys A valve did not operate.
											- GFE tread mill excessive resistance.

039-07-017 STS-39 Crew On-Orbit

STS039-17-017 1991-05-06, Shuttle Pallet Satellite II (SPAS-II)/Infrared Background Signature Survey (IBSS) released by RMS.

FLT NO.	ORBITER	CREW (7) TITLE, NAMES & EVA'S	LAUNCH SITE, LIFTOFF TIME, LANDING SITES, ABORT TIMES	LANDING SITE/ RUNWAY, CROSSRANGE LANDING TIMES FLT DURATION, WINDS	SSME-TL NOM-ABORT EMERG THROTTLE PROFILE ENG. S.N.	SRB RSRM AND ET	ORBIT INC	HA/HP	FSW	PAYLOAD WEIGHTS, PAYLOADS/ EXPERIMENTS	MISSION HIGHLIGHTS (LAUNCH SCRUBS/DELAYS, TAL WEATHER, ASCENT I-LOADS, FIRSTS, SIGNIFICANT ANOMALIES, ETC.)
STS-40 SEQ FLT #41 KSC-41 PAD 39B-13 MLP-3	OV-102 Columbia (Flight 11) Sixth Spacelab Flight LM (4) First Life Sciences Flight OMS PODS LPO3 - 11 RPO4 - 7 FRC2 - 11	CDR: Bryan D. O'Connor (Flt 2 - STS 61-B) P212/R83/V61/M76 PLT: Sidney M. Gutierrez P213/R129/M116 M/S 1: James P. Bagian (Flt 2 - STS-29) P214/R99/V62/M90 M/S 2: Tamara E. Jernigan P215/R130/F14 M/S 3: Rhea Seddon (Flt 2 - STS 51-D) P216/R55/V63/F5 P/S 1: F. Drew Gaffney P217/R131/M117 P/S 2: Millie Hughes-Fulford U of Cal/VA Center P218/R132/F15 MCC FCR-1 (21) FLIGHT DIRECTORS: Asc/Ent - N. W. Hale Ld/O2 - G. A. Pennington O 1 - R. E. Castle Plng - J. W. Bantle MOD - B. R. Stone	KSC 39B 156:13:24:51Z 8:00:00 AM EDT (P) 9:24:51 AM EDT (A) Wednesday 5 6/5/91 (4) LAUNCH WINDOW: 2H00M (MAND SLS-1 SCIENCE) PLS: EDW LKBD TAL: BEN GUERIR TAL ALT: MORON ZARAGOZA SELECTED: RTLS: KSC 33/CI/N TAL: BEN 36/N//N AOA: EDW 22 PLS: EDW 22 TDEL: -0.32 +0.402 MAX QNAV: 681 689 SRB STG: 2:04.2 PERF: NOMINAL 2 ENG TAL: 2:57 3:01 NEG RETURN: 4:02 4:03 PTA: 5:15 5:18 PTM: 5:45 5:49 MECO CMD: 8:31.2 8:30.4 VI: 25850 25868 OMS-2: Tig = 2:05 DV= 199 FPS	EDW 22, CONC (EDW 32, CONC 14) 165:15:39:11Z 8:39:11 AM PDT Friday 5 6/14/91 (3) XRANGE: 211 NM ORB DIR: DR 6 AIM PT: NOMINAL MLGTD: 1485 FT 165:15:39:11Z VEL: 199 KGS 203 KEAS HDOT: -2 FPS TD NORM 195: 2202 FT NLGTD: 5914 FT 165:15:39:25Z VEL: 153 KGS HDOT: -4 FPS BRK INIT: 134 KGS AVE BRK DECEL: 6.8 FPS/S WHEELS STOP: 165:15:40:06Z 10923 FT ROLLOUT: 9438 FT 55 SECONDS WINDS: 10.4H, 6 L KTS OFFICIAL: 12H, 3L DENS ALT: 3739 FT FLT DURATION: 9:02:14:20 218:14:20 S/T: 246:19:07:50 OV-102: 75:01:14:24 DISTANCE: 3,290,226 sm	104/104/ 109% PREDICTED: 100/100/ 92/67/ 104/67 ACTUAL: 100/100/ 98/71/ 104/67 1 = 2015 (6) 2 = 2022 (6) 3 = 2027 (6) M 3 EOM WEIGHT: 226737 X CG: 1279.6 LANDING WEIGHT: 226535 X CG: 1080.9	BI-044 RSRM 16W ET-41 LWT-34 ET RPT 244K 1:19:40 MET ET BR/UP 197K 1:20:52 MET ET IMPACT LAT: 1.05 N LONG: 146.06 W	39.0156 (1)	DIRECT INSERTION POST OMS-2: 161.16 X 149.84 NM DEORBIT 157 X 146 NM VELOCITY 25772 FPS ENTRY RANGE 4339 NM	OI-8D (4)	CARGO: 33707 LBS PAYLOAD CHARGEABLE: 28114 LBS DEPLOYED: 0 LBS NON-DEPLOYED: 26237 LBS RETURNED: MIDDECK: 1877 LBS SHUTTLE ACCUMULATED WEIGHTS: DEPLOYED: 550440 LBS NON-DEPLOYED: 482470 LBS CARGO TOTAL: 1175835 LBS PERFORMANCE MARGINS (LBS): FPR: 4671 FUEL BIAS: 983 FINAL TDDP:3037 RECON: 4212 PAYLOADS: PLB: Spacelab Life Sciences-1 (SLS-1)/LM Cardiovascular, Cardiopulmonary Metabolic, Musculoskeletal, and Neurovestibular Systems Experiments GBA With 12 GAS MIDDECK: MODE-0 5 CRYO TK SETS NO RMS	KSC W/D: OPF 74, VAB 6, PAD 34 = 114 days LAUNCH POSTPONEMENT: - 1/9/91 launch date as of 8/21/90. Launch order was STS-35, STS-41, STS-38, STS-40, STS-39, and STS-37. Launch postponed due to STS-35 and STS-38 H₂ leaks. Program manifest in March set tentative schedule of 5/22/91 with STS-37 and STS-39 moved ahead of STS-40. - 129-day slip. LAUNCH SCRUBS: - 5/22/91 launch scrubbed at approximately L-1 day (during T-11 hr hold) due to (1) MDM FA2 problem, (2) GPC4 failure, and (3) SSME cryo temp probes analysis received stating probes could break and enter HP turbopumps. Changed LO₂ and LH₂ temperature transducers. Launch rescheduled for 6/1/91. 10-day turnaround. - 6/1/91 launch scrubbed at T-20 minute hold due to IMU 2 failing calibration. 96-hour turnaround. LAUNCH DELAYS: - 1H24M51S delay at T-9 minute hold due to RSO no-go for ceiling at 12K. (Moisture in middle clouds and greater than 4500 feet thick.) TAL WX: - Ben Guerir (P) go throughout (selected). - Moron go throughout - Zaragoza go. RTLS: - KSC 15/33 ceiling 12K with middle clouds thicker than 4500 ft caused delay. I-LOADS: - LSEAT selected nominal, no uplink required. SIGNIFICANT ANOMALIES: - Two ECOS failures. - Hum sep A speed sensor wire break. - PRSD H₂ tank 3 heater failure. - MECO velocity error (explained condition). - KSC wind tower data false wind gusts. - S-band degraded performance on lower antennas. - TAGS hardcopier jam. - PLBD seal section missing and 1307 bulkhead blankets unfastened. - LiOH door stuck closed (IFM freed door). - Camcorder adapter cable failure. - APU 1 fuel line heater failure. - Vernier jet L5L fail off. - S/L audio problem. - Orbiter freezer and L9I ref/freezer Freon freezeup.

STS040-610-010
1991-06-14
Spacelab Life Sciences-1 (SLS-1 in P/L Bay

STS040-605-009 1991-06-14 STS-40
Crew: Front row (lt to rt) Gaffney/PS, PLT Gutierrez, Seddon/MS, & Bagian/MS. Back row (lt to rt) CDR O'Connor, Jernigan/MS, & Hughes-Fulford/MS.

FLT NO.	ORBITER	CREW (5) TITLE, NAMES & EVA'S	LAUNCH SITE, LIFTOFF TIME, LANDING SITES, ABORT TIMES	LANDING SITE/ RUNWAY, CROSSRANGE LANDING TIMES FLT DURATION, WINDS	SSME-TL NOM-ABORT EMERG THROTTLE PROFILE ENG. S.N.	SRB RSRM AND ET	ORBIT INC	HA/HP	FSW	PAYLOAD WEIGHTS, PAYLOADS/ EXPERIMENTS	MISSION HIGHLIGHTS (LAUNCH SCRUBS/DELAYS, TAL WEATHER, ASCENT I-LOADS, FIRSTS, SIGNIFICANT ANOMALIES, ETC.)
STS-43 SEQ FLT #42 KSC-42 PAD 39A-29 MLP-1	OV-104 (Flight 9) Atlantis OMS PODS LPO1 - 15 RPO1 - 15 FRC4 - 9	CDR: John E. Blaha (Flt 3 - STS-29 & STS-33) P219/R97/V48/M88 PLT: Michael A. Baker P220/R133/M118 M/S 1: Shannon W. Lucid (Flt 3 - STS 51-G & STS-34) P221/R65/V45/F6 M/S 2: G. David Low (Flt 2 - STS-32) P222/R110/V64/M98 M/S 3: James C. Adamson (Flt 2 - STS-28) P223/R102/V615M93	KSC 39 214:15:02:00Z 11:02:00 AM EDT (P) 11:02:00 AM EDT(A) Friday 7 08/02/91 (5) LAUNCH WINDOW: 2H30M (CTOB) PLS: KSC TAL: BANJUL (P) TAL WX: BEN GUERIR MORON SELECTED: RTLS: KSC 15/CI/N TAL: BEN 36/N/N AOA: EDW 22/N/N PLS: EDW 22/N/N TDEL: 0.00 0.562 MAX QNAV: 714 PSF 718PSF SRB STG: 2:04.3 2:02.9 PERF: NOM 2 ENG TAL BEN: 3:13 3:12 NEG RETURN: 3:53 3:54 PTA (U/S 245): 5:13 5:09 PTM (U/S 245): 5:50 5:49 MECO CMD: 8:27.7 8:27.6 VI: 25875 25873 OMS-2 TIG: 39:50.09 222.2 FPS	KSC-15 (KSC-8) 223:12:23:25Z 6:23:25 AM EDT Sunday 6 08/11/91 (3) XRANGE: 180NM ORBIT DIR: DL 22 AIM PT: CLOSE IN MLGTD: 1986 FT 223:12:23:25Z VEL: 202 KGS 197 KEAS HDOT: -1 FPS TD NORM 195: 2152 FT NLGTD:5517 FT 223:12:23:36Z VEL: 165 KGS HDOT: -2.7 FPS BRK INIT: 132 KGS AVE BRK DECEL: 6.1 FPS/S WHEELS STOP: 223:12:24:24Z 11876 FT ROLLOUT: 9890 FT 59 SEC WINDS: 0.5T, 4R KTS OFFICIAL: 0T, 3R DENS ALT: 1602 FT FLT DURATION: 8:21:21:25 213:21:25 S/T: 255:16:29:15 OV-104: 48:13:37:49 DISTANCE: 3,700,400 sm	104/104/ 109% PREDICTED: 100/104/ 80/67/104 ACTUAL: 100/104/ 84/67/104 1 = 2024 (3) 2 = 2012(12) 3 = 2028 (6) M 3 EOM WEIGHT: 196353 WEIGHT: 196088 X CG: 1087.4 X CG: 1089.7 LANDING: WEIGHT: 196088 X CG: 1089.7	BI-045 RSRM 17W ET-47 LWT-40 ET RPT 234K 1:17:35 MET ET BR/UP 186K 1:18:15 MET ET IMPACT LAT: 13.47 N LONG: 162.2 W	28.46 (26)	DIRECT INSERTION 158/35 POST OMS-2: 161.3 X 160.3 NM TDRS DEPLOY: 161.2 X 159.8 NM OMS SEP MAN: 177.9 X 161.2 NM DEORBIT 174 X 161 NM VELOCITY 25794 FPS ENTRY RANGE 4312 NM	OI-20 (1)	CARGO: 49325 LBS PAYLOAD CHARGABLE: 46712 LBS DEPLOYED: 37575 LBS NON-DEPLOYED: 8146 LBS MIDDECK: 991 LBS SHUTTLE ACCUMULATED WEIGHTS: DEPLOYED: 588015 LBS NON-DEPLOYED: 491607 LBS CARGO TOTAL: 1225160LBS PERFORMANCE MARGINS (LBS): FPR: 4653 FUEL BIAS: 994 FINAL TDDP:2656 RECON: 2593 PAYLOADS: PLB: TDRS-E/IUS SSBUV SHARE-II OCTW TCPE MIDDECK: SSCE SAMS BIMDA IPMP PLG-III UVPI AMOS APE-B 4 CRYO TK SETS NO RMS	KSC W/D: OPF 60, VAB 6, PAD 35 = 101 days LAUNCH POSTPONEMENT: - 7/23/91 launch postponed on 7/19/91 to 7/24/91 due to SRB sep motor PIC wire replacement. LAUNCH SCRUBS: - 7/24/91 launch scrubbed at approximately L-6 hours (during tanking) due to SSME 3 MEC DCU "A" parity error, MCF was set. Launch rescheduled for 8/1/91. - 8/1/91 launch scrubbed at L+1H24M while holding at T-9 min. Did not get cabin vent close indication but counted down to T-20 and ran cabin pressurization test (valve was closed) but by the time cabin was vented and cabin closed out, WX at KSC was bad. Scrubbed because T-showers within 20 nm, Xwinds > 15 kts @ SLF & convection present. Rescheduled launch for 8/2/91. 10 days total slip. LAUNCH DELAYS: None. TAL WX: Ben Guerir and Moron go, Banjul late go after T-showers and ceiling no go. Selected BEN 36. I-LOADS: LSEAT selected nominal, no uplink required (uplink 6). FIRSTS: First flight of OI-20. SIGNIFICANT ANOMALIES: - Cabin vent valve failed to indicate "closed." - No cooling on WSB2 during ascent. - PDI decom problems with SHARE data. - PRSD H$_2$ tank 1 heater failed off. - APU 1 FP/GGVM overcooling. - S-band power amp 2 degradation. - PPO$_2$ sensor "C" failed. - APU 1 S/N 305 anomalous chamber pressure during entry. - PLB floodlight problems, mid-STBD RPC trip. - BIMDA cell syringe problems. - PRSD tank H$_2$ manifold valve failed to close. DISCUSSION ITEM: - LIB MLG tire rib 2 tire wear (scuffing of two cords).

Artist concept TDRS Comm Network

S90-41340 1990-06-22

STS043-601- 033 1991-08-11 TDRS-E/IUS deploy over Pacific Ocean.

MCC FCR-1 (22) FLIGHT DIRECTORS: Asc - R. D. Dittemore Ent - J. W. Bantle Ld/O 1 - R. M. Kelso O 2 - P. L. Engelauf Plng - G. E. Coen Plng - J. M. Heflin MOD - T. W. Holloway MOD - G. E. Coen MDR - B. R.Stone MDR - J. M. Heflin

STS043-40-029,1991-08-11--- Crew on Middeck: (Lt to Rt) Low/MS, Lucid/MS, Adamson/MS, CDR Blaha, & PLT Baker.

FLT NO.	ORBITER	CREW (5) TITLE, NAMES & EVA'S	LAUNCH SITE, LIFTOFF TIME, LANDING SITES, ABORT TIMES	LANDING SITE/ RUNWAY, CROSSRANGE LANDING TIMES FLT DURATION, WINDS	SSME-TL NOM-ABORT EMERG THROTTLE PROFILE ENG. S.N.	SRB RSRM AND ET	ORBIT HA/HP	FSW	PAYLOAD WEIGHTS, PAYLOADS/ EXPERIMENTS	MISSION HIGHLIGHTS (LAUNCH SCRUBS/DELAYS, TAL WEATHER, ASCENT I-LOADS, FIRSTS, SIGNIFICANT ANOMALIES, ETC.)
STS-48 SEQ FLT #43 KSC-43 PAD 39A-30 MLP-3	OV-103 (Flight 13) Discovery OMS PODS LPO4 - 10 RPO3 - 14 FRC3 - 13	CDR: John O. Creighton (Flt 3 - STS 51-G, & STS-36) P224/R63/V50/M58 PLT: Kenneth S. Reightler P225/R134/M119 M/S 1: James F. Buchli (Flt 4 - STS 51-C, STS 61-A, & STS-29) P226/R52/V24/M48 M/S 2: Mark N. Brown (Flt 2 - STS-28) P227/R103/V66/M94 M/S 3: Charles D. (Sam) Gemar (Flt 2 - STS-38) P228/R118/V67/M106 MCC FCR-1 (23) FLIGHT DIRECTORS: Asc/Ent - J. W. Bantle Ld/O1 - G. A. Pennington O 2 - R. M. Kelso Plng - P. L. Engelauf MOD - G. E. Coen	KSC 39A 255:23:11:04Z 6:57:00 PM EDT (P) 7:11:04 PM EDT (A) Thursday 10 9/12/91 (2) LAUNCH WINDOW: 2H57M (UARS RAAN & CTOB) PLS: KSC TAL: ZARAGOZA TAL ALT: MOR, BEN SELECTED: RTLS: KSC33/NOM NOM 2400 FT TAL: ZZA30/CI NOM 2900 FT AOA:NOR 17/NOM/ NOM 2900 FT PLS: EDW22/NOM/ NOM 2700 FT TDEL: -0.16 0.162/0.2 MAX Q NAV: 670 708 SRB STG: 2:04 2:05.23 PERF: NOMINAL 2 ENG TAL ZZA: 2:19 2:22 NEG RETURN: 4:09 4:14 PTA (U/S 518): 4:23 4:23 PTM (U/S 1124): 6:44 6:50 MECO CMD: 8:36 8:36 VI: 26087 26083 OMS-2 TIG: 43:39 43:40 448 FPS 450 FPS	EDW 22 NOM (EDW 33,CONC 15) 261:07:38:42Z 00:38:42 AM PDT Wednesday 4 09/18/91 (4) XRANGE: 690 NM ORBIT DIR: DR 7 AIM PT: NOMINAL MLGTD 1235 FT 261:07:38:42Z VEL: 213 KGS 203 KEAS HDOT: -1 FPS TD NORM 195: 2015 FT NLGTD: 4882 FT 261:07:38:53Z VEL: 171 KGS HDOT: -2.1 FPS BRK INIT: 145 KGS AVE BRK DECEL: 8.2 FPS/S WHEELS STOP: 10619 FT ROLLOUT: 9384 FT 49 SECS WINDS: 2.9H, 0.8 L KTS OFFICIAL: 4H, 4L DENS ALT: 3503 FT FLT DURATION: 5:08:27: 38 128:27:38 S/T: 261:00:56:53 OV-103: 75:02:08:19 DISTANCE: 2,193,670 sm	104/104/ 109% PREDICTED: 100/100/ 89/67/ 104/67 ACTUAL: 100/100/ 89/67/ 104/67 1 = 2019 (9) 2 = 2031 (6) 3 = 2107 (5) M 3 EOM WEIGHT: 192925 X CG: 1096.0 LANDING: WEIGHT: 192780 X CG: 1097.8	BI-046 RSRM 18W ET-42 LWT-35 ET RPT 229K 1:25:46 MET ET BR/UP 194K 1:26:47 MET ET IMPACT LAT: 0.26 N LONG: 121.9 W	DIRECT INSERTION 288 X 36 NM POST OMS-2: 291.5 X 289.9 NM RCS-1: 306.9 X 290.9 NM RCS-2: 308.1 X 207.9 NM UARS DEPLOY: 308.9 X 305.3 NM ENTRY: Ha/Hp: 313 X 302 NM VELOCITY 26077 FPS RANGE 4194 NM	OI-20 (2) 57.00 (8)	CARGO: 21564 LBS PAYLOAD CHARGABLE: 17144 LBS DEPLOYED: 14388 LBS NON-DEPLOYED: 2066 LBS MIDDECK: 690 LBS SHUTTLE ACCUMULATED WEIGHTS: DEPLOYED: 602403 LBS NON-DEPLOYED: 494363 LBS CARGO TOTAL: 1246729 LBS PERFORMANCE MARGINS (LBS): FPR: 4671 FUEL BIAS: 983 FINAL TDDP: 510 RECON: - 562 PAYLOADS: PLB: Upper Atmosphere Research Satellite (UARS) with 10 experiments deployed: SUSIM, SOLSTICE, PEM, CLAES, ISAMS, MLS, HALOE, HRDI, WIND II, ,ACRIM-II, APM MIDDECK: PCG-II-2 RME-III MODE IPMP AMOS PARE SAM CREAM 4 CYRO TK SETS RMS 25 (S.N. 301) used for UARS deploy	KSC W/D: OPF 78, VAB 8, PAD 27 = 101 days LAUNCH ADVANCEMENT: - Launch advanced 9 days from 9/21/91 to 9/12/91, which was the earliest date to complete crew training LAUNCH SCRUBS: None. LAUNCH DELAYS: - 14M4S because of motor boating noise on A/G voice caused by glitch on RF to MILA resulting in Delta Modulation System (DMS) false frame lock. Counted to T-5 mins, held and cleared by CDR keying A/G voice. TAL WX: Zaragoza, Moron, and Ben Guerir - all go. DOLILU/ALT I-LOADS: - First availability of DOLILU which was uplinked and used (uplink 7). DUSK LAUNCH: - Launch was planned during daylight but 14 minute delay slipped to dusk launch, RTLS would have been night. FLIGHT DURATION CHANGES: - Waved off planned rev at KSC because STA observed clouds developing south of SLF. - Flight extended one rev when STA spotted clouds forming south of SLF. Clouds were not observed on radar. FIRSTS: - First flight of enhanced MDM (OA1 only). LANDING SITE CHANGE: - Changed from KSC to EDW because of the dynamic conditions with clouds and convection observed by STA. - One rev extension. EVENTS: UARS deployed at MET 2:05:12:09. SEP 1 burn at 2:05:12:40. NIGHT LANDING: Space Shuttle #5 SIGNIFICANT ANOMALIES: - ET door centerline latch 1 motor 2 phase B failure. - Fuel cell 1 O_2 reactant valve closed indication. - Supply water dump valve leaking. - Hydraulic system 2 unloader valve leakage. - Supply water nozzle temperature temporary decrease. - APU 1 seal cavity drain pressure delay. - LINHOF camera failed.

STS048-05-024 1991-09-18 Upper Atmosphere Research Satellite (UARS)

STS048-21-004 1991-09-18 Crew on middeck: (front lt to rt) PLT Reightler, CDR Creighton, Buchli/MS and (back lt to rt) Brown/MS & Gemar/MS.

FLT NO.	ORBITER	CREW (6) — TITLE, NAMES & EVA'S	LAUNCH SITE, LIFTOFF TIME, LANDING SITES, ABORT TIMES	LANDING SITE/ RUNWAY, CROSSRANGE — LANDING TIMES, FLT DURATION, WINDS	SSME-TL NOM-ABORT EMERG — THROTTLE PROFILE ENG. S.N.	SRB RSRM AND ET	INC	ORBIT — HA/HP	FSW	PAYLOAD WEIGHTS, PAYLOADS/ EXPERIMENTS	MISSION HIGHLIGHTS (LAUNCH SCRUBS/DELAYS, TAL WEATHER, ASCENT I-LOADS, FIRSTS, SIGNIFICANT ANOMALIES, ETC.)
STS-44 SEQ FLT #44 KSC-44 PAD 39A-31 MLP-1	OV-104 (Flight 10) Atlantis OMS PODS LPO1-16 RPO1-16 FRC4-10	CDR: Frederick D. Gregory (Flt 3 - STS 51-B & STS-33) P229/R59/V47/M54 PLT: Terence (Tom) Henricks P23/0R135/M120 M/S 1: James S. Voss P231/R136/M121 M/S 2: F. Story Musgrave (Flt 4 - STS-6, STS 51-F & STS-33) P232/R15/V19/M15 M/S 3: Mario Runco, Jr P233/R137/M122 P/S: Thomas J. Hennen CWO-3, U.S. Army P234/R138/M123 MCC FCR-1 (24) FLIGHT DIRECTORS: Asc/Ent - R.D.Dittemore Ld/O 2 - J. M. Heflin O 1 - P. L. Engelauf Plng - C. W. Shaw MOD - T. W. Holloway	KSC 39, PAD A 328:23:44:00Z 6:31:00 PM EST (P) 6:44:00 PM EST (A) Sunday 6 11/24/91 (8) LAUNCH WINDOW 1H59M (DSP RAAN) EOM PLS: KSC TAL: BYD 32 TAL WX: BEN , MRN SELECTED: RTLS: KSC 33/CI/N TAL: BYD 32/N/SF AOA & PLS: EDW 22/N/N TDEL: -0.16 0.442/0.48 MAX QN: 719 PSF 728 PSF SRB STG: 2+05 2+05 PERF: NOM 2 ENG TAL BYD: 2+41 2+40 NEG RETURN: 3+57 4+00 PTA (U/S 315): 5+06 5+09 PTM (U/S 315): 5+57 6+00 MECO CMD: 8+28.5 8+30 VI: 25934 25928 OMS-2 TIG: 4+49 4+48	EDW 05 (EDW 34, LKBD 19) 335:22:34:43Z 2:34:43 PM PST Sunday 7 12/1/91 (5) XRANGE: 379 NM ORBIT DIR: AL 12 AIM PT: CLOSEIN MLGTD: 2607 FT 335:22:34:43Z VEL: 182 KGS 189 KEAS HDOT: -1 FPS TD NORM 195: 2127 FT NLGTD: 5077 FT 335:22:34:51Z VEL: 149 KGS HDOT: -5.2 FPS BRK INIT: 15 KGS AVE BRK DECEL: 1.8 FPS/S WHEELS STOP: 335:22:36:29Z 13798 FT ROLLOUT: 11191 FT 106 SEC WINDS: H12.8 KTS R2.2 KTS OFFICIAL: 13H, 0L DENS ALT: 2284 FT FLT DURATION: 6:22:50:43 166:50:43 S/T: 267:23:47:36 OV-103: 55:12:28:32 DISTANCE: 2,890,067 sm	104/104/ 109% PREDICTED 100/104/ 104/70/ 104/67 ACTUAL 100/104/ 104/73/ 104/67 1 = 2015 (7) 2 = 2030 (6) 3 = 2029 (5) M 3 EOM WEIGHT: 195047 X CG: 1090.8 LANDING WEIGHT: 194818 X CG: 1092.5	BI-047 RSRM 19W ET-53 LWT-46 ET RPT 235K 1:19:55 MET ET BR/UP 207K 1:20:38 MET ET IMPACT LAT: 17.01 N LONG: 154.05 W	28.45 (27)	DIRECT INSERTION POST OMS-2 195.0 X 194.3 NM DEPLOY: 195.5 X 194.9 NM SEP BURN: 212.4 X 195.4 NM RCS-2 195.9 X 195.3 NM COLLISION AVOIDANCE 195.9 X 195.0 NM DEORBIT 197 X 194 NM VELOCITY 25868 FPS ENTRY RANGE 4195 NM	OI-20 (3)	CARGO: 47235 LBS PAYLOAD CHARGEABLE: 44637 LBS DEPLOYED: 37588 LBS NON-DEPLOYED: 5809 LBS MIDDECK: 1240 LBS SHUTTLE ACCUMULATED WEIGHTS: DEPLOYED: 639991 LBS NON-DEPLOYED: 501412 LBS CARGO TOTAL: 1293964 LBS PERFORMANCE MARGINS (LBS): FPR: 4356 FUEL BIAS: 1337 FINAL TDDP: 565 RECON: 1025 PAYLOADS: PLB: DEFENSE SUPPORT PROGRAM (DSP)/IUS (DEPLOYED) IOCM MIDDECK: MSS-1 AMOS CREAM SAM RME-III VFT-1 TERRA-SCOUT UVPI 4 CRYO TK SETS NO RMS	KSC W/D: OPF 67, VAB 5, PAD 31 = 103 days LAUNCH POSTPONEMENTS: - As of 8/21/90, launch date was 7/5/91. - Postponed launch date to 11/15/91 caused by STS-38 and STS-35 H_2 leaks. Postponed to 11/19/91 due to STS-43 delays impacted MLP availability and WLE tee splice replacement. LAUNCH SCRUB: - Scrubbed 11/19/91 launch at T-9 hours because one IMU in IUS RIMU experienced anomalous BITE indications. Rescheduled launch for 11/24/91 to replace IUS RIMU. 5-day slip. 142 days total slip. LAUNCH DELAYS: - 11/24/91 launch was delayed 13M0S at T-9 minutes to torque down packing in a leaking LO_2 replenish valve and to avoid a COLA at 6:38 pm EST. TAL WX: Banjul (prime) and Ben Guerir were go. Moron predicted no go (ceiling) but was observed go. ALT I-LOADS: - Second flight with DOLILU capability. Nominal selected. No uplink required. NIGHT LAUNCH: Shuttle night launch #7. LANDING SITE CHANGE: Loss of one IMU caused MDF and lakebed landing, hence changed to EDW from KSC. FLIGHT DURATION CHANGES: - Extended one rev at EDW because of predicted high winds. - Flight shortened nearly 3 days due to IMU 2 failure. FIRSTS: - First flight of HAINS ALT IMU (IMU-1 only). - First flight of color CCTV monitors. SECOND SHUTTLE CREWMEMBER REPLACEMENT: David Walker was replaced by Gregory in 1990. (First Shuttle crewmember replacement occurred on STS-33.) SIGNIFICANT ANOMALIES: - Left SSME MCC P Xducer B BIAS ~30 PSIA high. - Supply water dump valve leaking after water dump. - HUMIDITY SEP B leaking water. - IMU 2 FAIL (Z AXIS ACCEL) - caused MDF and lakebed landing. - Left AIR DATA PROBE single motor deploy. - VCR tape door problem. - TREADMILL failed. - 16 mm ARRIFLEX malfunctioned. - APU 2 FUEL PUMP seal cavity drain line valve failure.

STS044-71-011 1991-12-01--- DSP/IUS Spacecraft in P/L Bay tilted for deploy.

STS044-17-030 1991-12-01 Crew: featuring "Trash Man" Hennen/PS (front ctr) star of onboard video on disposal of trash. Others (front row) CDR Gregory (left) & Voss/MS and (back row lt to rt) Runco/MS, Musgrave/MS, & PLT Henricks.

FLT NO.	ORBITER	CREW (7) TITLE, NAMES & EVA'S	LAUNCH SITE, LIFTOFF TIME, LANDING SITES, ABORT TIMES	LANDING SITE/ RUNWAY, CROSSRANGE LANDING TIMES FLT DURATION, WINDS	SSME-TL NOM-ABORT EMERG THROTTLE PROFILE ENG. S.N.	SRB RSRM AND ET	ORBIT INC	ORBIT HA/HP	FSW	PAYLOAD WEIGHTS, PAYLOADS/ EXPERIMENTS	MISSION HIGHLIGHTS (LAUNCH SCRUBS/DELAYS, TAL WEATHER, ASCENT I-LOADS, FIRSTS, SIGNIFICANT ANOMALIES, ETC.)
STS-42 SEQ FLT #45 KSC-45 PAD 39A-32 MLP-3	OV-103 (Flight 14) Discovery Seventh Spacelab Long Module (5) OMS PODS LPO4-11 RPO3-15 FRC3-14	CDR: Ronald J. Grabe (Flt 3 - STS 51-J & STS-30) P235/R76/V41/M70 PLT: Steven S. Oswald P236/R139/M124 M/S 1 (P/L CDR): Norman E. Thagard (Flt 4 - STS-7, STS 51-B, STS-30) P237/R20/V14/M19 M/S 2: William F. Readdy P238/R140/M125 M/S 3: David C. Hilmers (Flt 4 - STS 51-J, STS-26, STS-36) P239/R77/V36/M71 P/S 1: Roberta L. Bondar (Canada) P240/R141/F16 P/S 2: Ulf D. Merbold (Germany) (Flt 2 - STS-9) P241/R29/V68/M28 MCC FCR-1 (25) FLIGHT DIRECTORS: Asc/Ent - N. W. Hale Ld/O 2 - R. E. Castle O 1 - J. W. Bantle O 3 - C. W. Shaw MOD - T. W. Holloway	KSC 39, PAD A 22:14:52:33Z 8:53:00 AM EST (P) 9:52:33 AM EST (A) Wednesday 6 01/22/92 (5) LAUNCH WINDOW 2H49M (EOM/ TAL LIGHTING) PLS: EDW TAL: ZZA (P) TAL WX: MRN, BEN SELECTED: RTLS: KSC 33/N/N TAL: ZZA 30/CI/N AOA: N/A PLS: EDW 22/N/N (REV 3) EDW 04/CI/N (REV 7) TDEL: 0.00 0.562/0.6 MAX QN: 692 PSF 708 PSF SRB STG: 2+06.6 2+08 PERF: NOMINAL 2 ENG TAL ZZA: 2+51 2+48 NEG RETURN: 4+05 4+05 PTA (U/S 290): 5+20 5+10 PTM (U/S 290): 5+52 5+42 MECO CMD: VI: 25934 25928 OMS-2 TIG: 36+12.8 36+08	EDW 22 (EDW 35,CONC 16) 30:16:07:17Z 8:07:17 AM PST Thursday 3 01/30/92 (4) XRANGE: 536 NM ORBIT DIR: AR 3 AIM PT: NOMINAL MLGTD: 2835 FT 30:16:07:17Z VEL: 198 KGS 196 KEAS HDOT: -1.5 FPS TD NORM 195: 2868 FT NLGTD: 5901 FT 30:16:07:27Z VEL: 168 KGS HDOT: -4.3 FPS BRK INIT: 133 KGS AVE BRK DECEL: 6.3 FPS/S WHEELS STOP: 30:16:08:16Z 12676 FT ROLLOUT: 9841 FT 59 SEC WINDS: H 0.4 KTS R 2.0 KTS OFFICIAL: 1H, 2R DENS ALT: 670 FT FLT DURATION: 8:01:14:44 193:14:44	104/104/ 109% PREDICTED 100/100/ 100/70/ 104/67 ACTUAL 100/100/ 100/75/ 104/67 1 = 2026 (2) 2 = 2022 (7) 3 = 2027 (7) M 3 EOM WEIGHT: 218159 X CG: 1080.6 LANDING WEIGHT: 218089 X CG: 1082.2	BI-048 RSRM 20W ET-52 LWT-45 ET RPT 243K 1:09:33 MET ET BR/UP 222K 1:10:08 MET ET IMPACT LAT: 44.7 S LONG: 157.9 W	57 (9)	DIRECT INSERTION POST OMS-2 162 NM X 160 NM DEORBIT 160 X 157 NM VELOCITY 25785 FPS ENTRY RANGE 4358 NM	OI-20 (4)	CARGO: 32364 LBS PAYLOAD CHARGEABLE: 28663 LBS DEPLOYED: 0 LBS NON-DEPLOYED: 26453 LBS MIDDECK: 2210 LBS SHUTTLE ACCUMULATED WEIGHTS: DEPLOYED: 639991 LBS NON-DEPLOYED: 530075 LBS CARGO TOTAL: 1326328 LBS PERFORMANCE MARGINS (LBS): FPR: 4339 FUEL BIAS: 1394 FINAL TDDP:2511 RECON: 2716 PAYLOADS: PLB: INTERNATIONAL MICROGRAVITY LABORATORY MATERIALS SCIENCE AND LIFE SCIENCES EXPERIMENTS (IML-1/LM) GBA (12 GAS) MIDDECK: GOSAMR-1 SE 83-02 SE 81-9 IPMP RME-111 UVPI 4 CRYO TK SETS NO RMS	KSC W/D: OPF 75, VAB 6, PAD 24 = 105 days LAUNCH POSTPONEMENTS: - As of 12/19/90, launch date was 11/15/91. - Postponed to 1/13/92 as of 3/15/91. 26-day slip. - Postponed to 1/22/92 as of 8/21/91. 9-day slip. - 35 days total launch slip. LAUNCH SCRUB: None. LAUNCH DELAYS: - 1/22/92 launch was delayed 59M33S at T-9 minutes caused by: (1) Paper closure of FC2 H₂ Pump/AC2 Bus anomaly, (2) KSC field mills read >1 KVOLT/meter (determined to be caused by salt fog), (3) Excessive O₂ in mid-body, (4)"BLAST" program violation, and (5) KSC field mills read >1 KVOLT/meter (STA confirmed moisture in cloud passing over field mills). TAL WX: Zaragoza (prime), Moron, and Ben Guerir forecast and observed GO. LAKEBEDS: EDW and NOR lakebeds NO GO (WET for L&L). ALT I-LOADS: - Nominal selected. No uplink required. FLIGHT DURATION CHANGE: - Flight extended 1 day from 7 to 8 days to get additional Spacelab science data. LANDING SITE CHANGE: None. SIGNIFICANT ANOMALIES: - MIDDS computer not transferring all winds data to FDCF. - FC2 H₂ motor status/AC glitch prelaunch. - MVI CB trip during pitch operations. - Waste water dump rate degraded. - White Sands central computer failure. - WCS commode control valve linkage failure. (IFM to use vice grips to open/close.) - TAGS jam/imaging failure. - GAS can G-609 motorized door did not open. - WCCSfailures and battery shortened life - RCS jet L3A fail leak (oxidizer). - Crew reported plume from right pod, powered up MDM FA4 and confirmed R4U oxidizer leak. - SRB - Gas path in RH & LH nozzle-to-case joint polysulfide with eroded wiper O-ring. - ET - two large TPS divots on the ET intertank. Radiators Deployed #13

STS042-201-009, 1992-01-3 At work in IML-1: Bondar (left) & Oswald.

STS042-35-011 1992-01-30 Crew portrait in IML-1: Top row (lt to rt) Merbold/PS, CDR Grabe, Thagard/MS, & Bondar/PS; and bottom row (lt to rt) PLT Oswald, Hilmers/MS, Readdy/MS.

S/T: 276:01:02:20

OV-103:
83:03:23:03

DISTANCE:
3,349,830 sm

FLT NO.	ORBITER	CREW (7) TITLE, NAMES & EVA'S	LAUNCH SITE, LIFTOFF TIME, LANDING SITES, ABORT TIMES	LANDING SITE/ RUNWAY, CROSSRANGE LANDING TIMES FLT DURATION, WINDS	SSME-TL NOM-ABORT EMERG THROTTLE PROFILE ENG. S.N.	SRB RSRM AND ET	ORBIT		FSW	PAYLOAD WEIGHTS, PAYLOADS/ EXPERIMENTS	MISSION HIGHLIGHTS (LAUNCH SCRUBS/DELAYS, TAL WEATHER, ASCENT I-LOADS, FIRSTS, SIGNIFICANT ANOMALIES, ETC.)
							INC	HA/HP			
STS-45 SEQ FLT #46 KSC-46 PAD 39A-33 MLP-1	OV-104 (Flight 11) Atlantis Eighth Spacelab Flight (2 Pallets) IGLOO (3) OMS PODS LPO1-17 RPO1-17 FRC4-11	CDR: Charles F. Bolden, Jr. (Flt 3 - STS 61-C & STS-31) P242/R88/V52/M80 PLT: Brian Duffy P243/R142/M126 M/S 1: Kathryn D. Sullivan (Flt 3 - STS 41-G & STS-31) P244/R44/V53/F3 M/S 2: David C. Leestma (Flt 3 - STS 41-G & STS-28) P245/R45/V43/M42 M/S 3: C. Michael Foale P246/R143/M127 P/S 1: Dirk Frimout (Belgium) P247/R144/M128 P/S 2: Bryon Lichtenberg (Flt 2 - STS-9) P248/R28/V69/M27 MCC FCR-1 (26) FLIGHT DIRECTORS: Asc/Ent - J. W. Bantle Ld/O 2 - R. M. Kelso O 1 - R. E. Castle O 3 - L. J. Ham MOD - T. W. Holloway	KSC 39, PAD A 84:13:13:39.96Z 8:00:00 AM EST (P) 8:13:40 AM EST (A) Tuesday 8 3/24/92 (3) LAUNCH WINDOW 2H30M (CTOB) EOM PLS: KSC TAL: ZZA (P) TAL WX: MRN, BEN SELECTED: RTLS: KSC 33/CI/N TAL: ZZA 30/ CI/N AOA: NOR 17/N/N PLS: EDW 22/N/N TDEL: 0.64 0.882/0.92 MAX Q NAV: 671 PSF 678 PSF SRB STG: 2:07.7 2:07.9 PERF: NOMINAL 2 ENG TAL ZZA: 2:23 2:22 NEG RETURN: 4:11 4:13 PTA (U/S 285): 4:16 4:13 PTM (U/S 285): 4:48 4:51 MECO CMD: 8:30.9 8:31 VI: 25830 25823 OMS-2: 37:08 36:20 253.5 252.8	KSC 33 (KSC-9) 93:11:23:06Z 6:23:06 AM EST Thursday 4 4/2/92 (7) XRANGE: 679 NM ORBIT DIR: AR 4 AIM PT: CLOSE IN MLGTD: 1765 FT 93:11:23:06Z VEL: 186 KGS 192 KEAS HDOT: -1.9 FPS TD NORM 195: 1481 FT NLGTD: 4393 FT 93:11:23:14Z VEL: 161 KGS HDOT: -4.1 FPS BRK INIT: 134 KGS AVE BRK DECEL: 5.6 FPS/S WHEELS STOP: 10992 FT 93:11:24:04Z ROLLOUT: 9227 FT 56 SECS WINDS: H 5.1 KTS L 3.2 KTS OFFICIAL: 5H, 3L DENS ALT: 224 FT FLT DURATION: 8:22:09:26 214:09:26 S/T: 284:23:11:46 OV-104: 64:10:37:58 DISTANCE: 3,274,946 sm	104/104/ 109% PREDICTED 100/100/ 89/74/ 104/67 ACTUAL: 100/100/ 89/74/ 104/67 1 = 2024 (4) 2 = 2012(13) 3 = 2028 (7) M 3 EOM WEIGHT: 205672 LBS X CG: 1085.4 LANDING WEIGHT: 205588 LBS X CG: 1087.2	BI-049 RSRM 21W ET-44 LWT-37 ET RPT 249K 1:10:00 MET ET BR/UP 219K 1:10:50 MET ET IMPACT LAT: 42.7 LONG: 155.0 W	DIRECT INSERTION POST OMS-2 159.8 X 153.0 NM OMS-3: (CIRC BURN) 12.5 FPS @ 2:50:13 MET 160.5 X 159.3 NM DEORBIT 159.5 X 151.8 NM VELOCITY 25785 FPS ENTRY RANGE 4231 NM	57.02 (10)	OI-20 (5)	CARGO: 20341 LBS PAYLOAD CHARGABLE: 17683 LBS DEPLOYED: 0 LBS NON-DEPLOYED: 15538 LBS MIDDECK: 2145 LBS SHUTTLE ACCUMULATED WEIGHTS: DEPLOYED: 639991LBS NON-DEPLOYED: 547758 LBS CARGO TOTAL: 1346669 LBS PERFORMANCE MARGINS (LBS): FPR: 4671 FUEL BIAS: 983 FINAL TDDP:11017 RECON: 10427 PAYLOADS: PLB: ATLAS-1: ATMOPHERE SCIENCE: ALAE, MAS, ISO, ATMOS, GRILLE, SSBUV/A SOLAR SCIENCE: ACR, SOLCON, SOLSPEC, SUSIM SPACE PLASMA SCIENCE: AEPI, SEPAC, ENAP ASTRONOMY: FAUST GAS G-229 MIDDECK: STL-01, RME-III, VPT-2, CLOUDS-1A, SAREX-2, IPMP, UVPI 4 CRYO TK SETS NO RMS	KSC W/D: OPF 55, VAB 6, PAD 27 = 88 days LAUNCH POSTPONEMENTS: - Launch date was 3/10/92 as of 3/15/91. Postponed to 3/14/92 on 8/21/91. 4 days slip. - Postponed to 3/23/92 on 1/23/92. 9 days slip with decision made to launch during a full moon. LAUNCH SCRUB: - 3/23/92 launch was scrubbed at L-5.5 hours (fast fill + 3.5 minutes) because of H_2 and O_2 concentrations in aft compartment exceeding LCC limits (LH$_2$=750 PPM & LO$_2$=850 PPM). Could not repeat leaks during troubleshooting but scrubbed launch because could not make launch window. LAUNCH DELAYS: - 13M40S delay at T-9 minutes because of RTLS ceiling violations (cloud deck at approximately 6K feet). BLAST violations occurred during hold period. TAL WX: Zaragoza and Moron weather was GO, Moron was NO GO for runway margins, and Ben Guerir NO GO for weather (ceiling). ALT I-LOADS: - LSEAT selected YAW NEG, which was uplinked (uplink 8). DOLILU was NO GO because of greenline exceedance. FLIGHT DURATION CHANGE: - 3/29/92 MMT made decision that consumables supported an extension from 8+2 days to 9+2 days to get more science. FIRSTS: - First flight of an improved APU (APU 2 only). - First flight with a female flight director (Linda J. Ham). SIGNIFICANT ANOMALIES: - Fuel Cell 3 cell performance monitor D volts remained at self test value. - Ku-Bd power output TLM intermittent fail. - Ku-Bd auto track problem,similar to STS-37. - CCTV cameras A & C degraded. - TAGS OHC jam, cleared by crew. - APU 1 GG bed heater B intermittent. - Arriflex camera operate lever intermittent. - SEPAC electron beam accelerator operations were terminated on day 2 because 30 amp fuse between SEPAC battery and charger blew. - Lost all power to FAUST.

Linda Ham - 1st Female Flight Director

STS-45 ATLAS-1 in P/L Bay

STS045-15-003 1992-04-02

STS045-38-004 1992-04-02 Crew on Forward Flight Deck: In front are Sullivan/MS/PLC (left) & CDR Bolden. In rear are (lt to rt) Leestma/MS, PLT Duffy, Lichtenberg/PS, Frimout/MS, & Foale/MS. (The "headpieces" worn by Sullivan and Bolden are actually shadows.)

SPACE SHUTTLE MISSIONS SUMMARY

FLT NO.	ORBITER	CREW (7) TITLE, NAMES & EVA'S	LAUNCH SITE, LIFTOFF TIME, LANDING SITES, ABORT TIMES	LANDING SITE/ RUNWAY, CROSSRANGE LANDING TIMES FLT DURATION, WINDS	SSME-TL NOM-ABORT EMERG THROTTLE PROFILE ENG. S.N.	SRB RSRM AND ET	ORBIT INC	HA/HP	FSW	PAYLOAD WEIGHTS, PAYLOADS/ EXPERIMENTS	MISSION HIGHLIGHTS (LAUNCH SCRUBS/DELAYS, TAL WEATHER, ASCENT I-LOADS, FIRSTS, SIGNIFICANT ANOMALIES, ETC.)
STS-49 SEQ FLT #47 KSC-47 PAD 39B-14 MLP-2	OV-105 (Flight 1) Endeavour OMS PODS LPO3-12 RPO4-8 FRC5-1	CDR: Daniel C. Brandenstein (Flt 4 - STS-8, STS 51-G & STS-32) P249/R21/V16/M20 PLT: Kevin P. Chilton P250/R145/M129 M/S 1, EV2: Richard J. Hieb (Flt 2 - STS-39) P251/R128/V70/M115 M/S 2: Bruce E. Melnick (Flt 2 - STS-41) P252/R114/V71/M102 M/S 3, EV1: Pierre J. Thuot (Flt 2 - STS-36) P253/R112/V72/M100 M/S 4, EV3: Kathryn C. Thornton (Flt 2 - STS-33) P254/R107/V73/F11 M/S 5, EV4: Thomas D. Akers (Flt 2 - STS-41) P255/R115/V74/M103 MCC FCR-1 (27) FLIGHT DIRECTORS: Asc/Ent - N. W. Hale Ld/O 1 - G. A. Pennington O 2 - P. L. Engelauf Plng - J. M. Heflin MOD - B. R. Stone	KSC 39, PAD B 128:23:39:59.98Z 7:06:00 PM EDT (P) 7:40:00 PM EDT (A) Thursday 11 5/7/92 (2) LAUNCH WINDOW 47 Minutes (in 2 panes) EOM PLS: EDW TAL: BYD TAL WX: BEN SELECTED: RTLS: KSC 33/CI/N TAL: BEN 36/CI/N AOA: EDW 22/N/N PLS: EDW 22/N/N TDEL: 0.64 0.782/0.800 MAX Q NAV: 716 PSF 712 PSF SRB STG: 2:00.64 2:08 PERF: NOMINAL 2 ENG TAL BEN: 2:52 2:52 NEG RETURN: 4:00 4:03 PTA (U/S 285): 4:39 4:40 PTM (U/S 285): 5:53 5:43 MECO CMD: 8:28.5 8:29.8 VI: 25906 25900 OMS-2: 39:58.2 39:57.6 186.2FPS 187.97FPS	EDW 22 CONC (EDW 36,CONC 17) 137:20:57:39Z 1:57:39 PM PDT Saturday 8 5/16/92 (4) DEORBIT BURN: 137:19:55:15Z XRANGE: 411 NM ORBIT DIR: AL 14 AIM PT: NOMINAL MLGTD: 2156 FT 137:20:57:39Z VEL: 209 KGS 194 KEAS HDOT: -1.0 FPS TD NORM 195: 2329 FT NLGTD: 5770 FT 137:20:57:48Z VEL: 173 KGS HDOT: -3.5 FPS DRAG CHUTE DEPLOY: 165 KEAS 137:20:57:49Z BRK INIT: 94 KGS DRAG CHUTE JETTISON: 48 KGS 137:20:58:17Z AVE BRK DECEL: 8.0 FPS/S WHEELS STOP: 137:20:58:34Z 11646 FT ROLLOUT: 9490 FT 55 SECS WINDS: H2.0 KTS, X0.0 KTS OFFICIAL: 4H, 0L Continued. . .	104/104/ 109% PREDICTED 100/104/ 89/72/ 104/67 ACTUAL 100/104/ 89/73/ 104/67 1 = 2030 (7) 2 = 2015 (8) 3 = 2017 (6) M 3 EOM WEIGHT: 201400 LBS X CG: 1084.4 LANDING WEIGHT: 201235 LBS X CG: 1086.2	BI-050 RSRM 22K ET-43 LWT-36 ET RPT 238K 1:16:47 MET ET BR/UP 206K 1:17:45 MET ET IMPACT LAT: 12.17 S LONG: 163.6 W	28.32 (1)	DIRECT INSERTION POST OMS-2 182.5 X 139.8 NM INTELSAT RNDZ: 198 X 194 NM ORBITS: 46, 62, & 95 DEORBIT 195 X 184 NM VELOCITY 25841 FPS ENTRY RANGE 4162 NM	OI-21 (1)	CARGO: 37444 LBS PAYLOAD CHARGEABLE: 32809 LBS DEPLOYED: 23346 LBS NON-DEPLOYED: 8766 LBS MIDDECK: 697 LBS SHUTTLE ACCUMULATED WEIGHTS: DEPLOYED: 636337 LBS NON-DEPLOYED: 557221 LBS CARGO TOTAL: 1384113 LBS PERFORMANCE MARGINS (LBS): FPR: 4671 FUEL BIAS: 983 FINAL TDDP:3351 RECON: 3206 PAYLOADS: PLB: INTELSAT REBOOST (CRADLE & PERIGEE STAGE) PERIGEE STAGE ATTACHED TO INTELSAT WHICH WAS REDEPLOYED MIDDECK: CPCG BLOCK II AMOS UVPI 4 CRYO TK SETS RMS 26 (S.N. 303) Used to berth, repair, & deploy INTELSAT & monitor simulta- neous waste and supply water dump	KSC W/D: OPF 217, VAB 6, PAD 49=272 days LAUNCH POSTPONEMENTS: - Launch date was 4/16/92 as of 3/21/91. - Postponed launch to 4/30/92, then 5/4/92 on 4/23/92 at FRR because of sheer volume of work including aft ET attach point liner repair. - Postponed launch to 5/7/92 to allow a daylight launch. - 21-day total slip. LAUNCH SCRUB: None. LAUNCH DELAYS: - Launch delayed because of RTLS ceiling violations (5K-7K bkn), then TAL WX (BYD NO GO visibility/haze, BEN NO GO occasional 4K bkn and rain). MEC BITE indication and an aircraft in launch area. Counted to T-9 minutes then T-5 minutes. Switched to second pane of launch window and uplinked new launch and OMS target loads. - 34-minute total delay. TAL WX: - Banjul was NO GO - visibility, Ben Guerir late GO after occasional ceiling violation and rain. ASCENT I-LOADS: - Nominal I-loads were NO GO and DOLILU was uplinked (second DOLILU uplink and 9th total uplink). Launch and OMS targets loads uplinked for both window panes. FLIGHT DURATION CHANGE: - Flight was extended 2 days to allow the third EVA for the hand grab of INTELSAT after capture bar failed on two EVA's. RENDEZVOUS 10, 11, AND 12: - With INTELSAT for capture, berthing, AKM mounting, and deploy. FIRSTS: - First flight with drag chute. - First flight with Improved Nose Wheel Steering. - First flight of Collins TACAN, SS STAR-TRACKER, redesigned MPS 750 PSIG He Reg, MPS 850 PSIG He relief valve redesign, IAPU iso valve, redundant WOW det, brake press iso valve, improved RA antennas, deletion of vent doors 4 & 7, fourth EMU stowage, and improved PPO₂ sensor and 3 IAPU's. - First flight with 4 EVA's and first flight with 3 crewmemebers on same EVA. First flight with 4 different EVA crewmen. - First hand capture of satellite by EVA crewmen (Hieb, Thuot, and Akers), then RMS grapple of INTELSAT on capture bar. - First flight of OI-21. - First flight of Block II SSME Controller. Continued. . .

Below: Replicas of Christopher Columbus' sailing ships Santa Maria, Nina, and Pinta sail by Pad 39B in honor of Endeavour's maiden voyage.

S92-39074/KSC- 92PC-967 1992-06-18

STS049-21-005 1992-05-16 Middeck crew portrait - front row, left to rightt, Thornton/MS & Heib/MS, middlle row, left to right, Thuot/MS & Akers/MS, back row, left to rightt, CDR Brandenstein, PLT Chilton & Melnick/MS.

FLT NO.	ORBITER	CREW (7) TITLE, NAMES & EVA'S	LAUNCH SITE, LIFTOFF TIME, LANDING SITES, ABORT TIMES	LANDING SITE/ RUNWAY, CROSSRANGE LANDING TIMES FLT DURATION, WINDS	SSME-TL NOM-ABORT EMERG THROTTLE PROFILE ENG. S.N.	SRB RSRM AND ET	ORBIT INC	HA/HP	FSW	PAYLOAD WEIGHTS, PAYLOADS/ EXPERIMENTS	MISSION HIGHLIGHTS (LAUNCH SCRUBS/DELAYS, TAL WEATHER, ASCENT I-LOADS, FIRSTS, SIGNIFICANT ANOMALIES, ETC.)
STS-49 Continued		Continued. . . EMU/TETHERED EVA'S: EVA 1 - 5/10/92 SS EVA #16 BY EV1 & EV2 INTELSAT CAPTURE BAR - NO GO 3H43M EVA 2 - 5/11/92 SS EVA #17 UNSCHEDULED EVA #3 BY EV1 & EV2 INTELSAT CAPTURE BAR - NO GO 5H30M EVA3 - 5/13/92 SS EVA #18 UNSCHEDULED EVA #4 BY EV1, EV2 & EV4 INTELSAT HAND CAPTURE, REPLACED UPPER STAGE AND RELEASED 8H29M EVA4 - 5/14/92 SS EVA #19 BY EV3 AND EV4 ASEM - 7H45M		Continued. . . DENS ALT: 4664 FT FLIGHT DURATION: 8:21:17:39 213:17:39 S/T: 293:20:29:35 OV-105 TOTAL: 8:21:17:39 DISTANCE: 3,969,019 sm							Continued. . . RECORDS: - Longest ever EVA (8H29M), second longest EVA (7H45M). - Longest EVA by female astronaut (7H45M). - Four EVA's on one flight. SIGNIFICANT ANOMALIES: - Av Bay 3 high delta pressure. - O2 manifold valve 1 failed open (failed to close) - TDRSS state vector propagation errors in MCC. - Orbit Target Terminal Initiation Computation failure on third rendezvous (used D/L state vectors in Ground Computations). - WCS fan sep 1 failure. - Four floodlights failed. - RCS jet L4L fail leak. - Ku-band beta gimbal failure - IFM EVA stow of antenna similar to STS 41-G. - PLBD port aft bulkhead latch failed to reach latch position. - SSME 2 HPFT TD temp sensor failed offscale high. - GPC AP101S microcode error.

STS049-91-020 1992-05-16 STS-49 crewmembers complete successful capture of the International Telecommunications Organization Satellite (INTELSAT VI) during EVA3. Left to right, Hieb/MS, Akers/MS, & Thuot/MS, on RMS, have handholds on the satellite and prepare to attach capture bar (tethered to Hieb). Two earlier grapple attempts on two-person EVA's were unsuccessful.

S92-36605 1992-05-20 STS-49 Orbit Team 1 (O1) poses in JSC FCR with O1 Lead FD Al Pennington (left of model of James Cook's ship Endeavour) and CAPCOM, John Casper (right of model).

S93-36604   1993-06-18 Oribt 2 (O2) Flight Control Team in JSC FCR poses with O2 FD Philip Engelauf (center front, right of Endeavour model).

S92-36606 1992-05-20 Milt Heflin/FD (front right next to ship model) with STS-49 Planning Team in JSC Flight Control Room.

SPACE SHUTTLE MISSIONS SUMMARY

FLT NO.	ORBITER	CREW (7) TITLE, NAMES & EVA'S	LAUNCH SITE, LIFTOFF TIME, LANDING SITES, ABORT TIMES	LANDING SITE/ RUNWAY, CROSSRANGE LANDING TIMES FLT DURATION, WINDS	SSME-TL NOM-ABORT EMERG THROTTLE PROFILE ENG. S.N.	SRB RSRM AND ET	ORBIT INC	ORBIT HA/HP	FSW	PAYLOAD WEIGHTS, PAYLOADS/ EXPERIMENTS	MISSION HIGHLIGHTS (LAUNCH SCRUBS/DELAYS, TAL WEATHER, ASCENT I-LOADS, FIRSTS, SIGNIFICANT ANOMALIES, ETC.)
STS-50 SEQ FLT #48 KSC-48 PAD 39A-34 MLP-3	OV-102 (Flight 12) Columbia 9th Spacelab Flight Long Module (6) EDO 1 OMS PODS LP05-1 RP05-1 FRC2-12	CDR: Richard N. Richards (Flt 3 - STS-28 & STS-41) P256/R101/V55/M92 PLT: Kenneth D. Bowersox P257/R146/M130 M/S 1 (PYLD CDR): Bonnie J. Dunbar (Flt 3 - STS 61-A & STS-32) P258/R79/V49/F7 M/S 2: Ellen S. Baker (Flt 2 - STS-34) P259/R105/V75/F10 M/S 3: Carl J. Meade (Flt 2 - STS-38) P260/R117/V76/M105 P/S 1: Larry DeLucas P261/R147/M131 (U OF ALA, BIRM) P/S 2: Gene Trinh P262/R148/M132 (JPL) MCC FCR-1 (28) FLIGHT DIRECTORS: Asc/Ent - J. W. Bantle Ld/O 2 - R. E. Castle O 1 - R. D. Jackson O 3 - G. E. Coen Team 4 - R. M. Kelso MOD - A. L. Briscoe	KSC 39, PAD A 177:16:12:23Z 12:07:00 PM EDT (P) 12:12:23 PM EDT (A) Thursday 12 6/25/92 (5) LAUNCH WINDOW 2H 30M CTOB EOM PLS: EDW TAL: BYD TAL WX: BEN, ROTA SELECTED: RTLS: KSC 15/CI/N TAL: BEN 36/N/N AOA: EDW 22/N/N PLS: EDW 22/N/N TDEL: 0.48 0.682/0.72 MAX Q NAV: 688 PSF 690 PSF SRB STG: 2:05.9 2:05.9 PERF: NOMINAL 2 ENG TAL (BEN): 3:01 3:00 NEG RETURN: 3:57 4:00 PTA (U/S 235): 4:57 4:54 PTM (U/S 235): 5:58 5:40 MECO CMD: 8:26.9 8:27.6 VI: 25875 25870 OMS-2: 39:56 39:51 222.3 FPS222.6 FPS	KSC 33 (KSC 10) 191:11:42:27Z 7:42:27 AM EDT Thursday 5 7/9/92 (1) DEORBIT BURN: 191:10:41:38Z XRANGE: 389 NM ORBIT DIR: DL 23 AIM PT: NOMINAL MLGTD: 2321 FT 191:11:42:27Z VEL: 208 KGS 203 KEAS HDOT: -2 FPS TD NORM 205: 2122 FT NLGTD: 7832 FT 191:11:42:45Z VEL: 149 KGS HDOT: -5.1 FPS DRAG CHUTE DEPLOY: 136 KEAS 191:11:42:47Z BRK INIT: 111 KGS DRAG CHUTE JETTISON: 55 KGS 191:11:43:11Z AVE BRK DECEL: 6.6 FPS/S WHEELS STOP: 191:11:43:25Z 12996 FT ROLLOUT: 10675 FT 58 SECS WINDS: H 1.6 KTS L 4.8 KTS OFFICIAL 1H, 5L Continued. . .	104/104/ 109% PREDICTED: 100/104/ 104/72/104 ACTUAL: 100/104/ 104/74/104 1 = 2019 (10) 2 = 2031 (7) 3 = 2011 (7) M 3 EOM WEIGHT: 225865 LBS X CG: 1077.7 LANDING WEIGHT: 225615 LBS X CG: 1079.1	BI-051 RSRM 24W ET-50 LWT-43 ET RPT 247K 1:17:12 MET ET BR/UP 216K 1:18:03 MET ET IMPACT LAT: 13.28 N LONG: 162.64 W	28.46 (28)	DIRECT INSERTION POST OMS-2 163.5 X 159.7 NM ORBIT ADJ 1: 159.9 X 159.2 NM 04/00:23:18 ORBIT ADJ 2: 163.0 X 129.1 NM DEORBIT 163 X 130 NM VELOCITY 25786 FPS ENTRY RANGE 4347 NM	OI-21 (2)	CARGO: 32447 LBS PAYLOAD CHARGEABLE: 24305 LBS DEPLOYED: 0 LBS NON-DEPLOYED: 22126 LBS MIDDECK: 2179 LBS SHUTTLE ACCUMULATED WEIGHTS: DEPLOYED: 663337 LBS NON-DEPLOYED: 581526 LBS CARGO TOTAL: 1416560 LBS PERFORMANCE MARGINS (LBS): FPR: 4671 FUEL BIAS: 983 FINAL TDDP:2940 RECON: 3276 PAYLOADS: PLB: UNITED STATES MICROGRAVITY LABORATORY (USML-1/LM) MATERIALS SCIENCE, FLUID PHYSICS, COMBUSTION SCIENCE, BIO-TECHNOLOGY MIDDECK: IPMP UVPI SAREX-II 4 + 4 EDO CRYO TK SETS NO RMS	KSC W/D: OPF 108, VAB 5, PAD 23=136 days LAUNCH POSTPONEMENTS: - Launch date was 5/11/92 as of 7/10/91. - Launch postponed to 6/3/92. Weather delayed OV-102 delivery to KSC after major mod period at Palmdale. - Launch postponed to 6/25/92 because of Ku-Band comm work, RSB corrosion repair, and LiOH canister locker interference. LAUNCH SCRUB: None. LAUNCH DELAYS: - 5M 23S delay during T-9 hold due to a concern about a cirrus layer at 28K-33K with a detached anvil (potential lightning in launch area). WX STA PLT reported it was not a problem because he could see through it. TAL WX: - Banjul forecast and observed NO GO - ceiling. Ben Guerir forecast and observed GO (selected). Rota forecast NO GO - Vis (Haze), observed GO. ASCENT I-LOADS: - Nominal selected, no uplink required. FLIGHT DURATION/LANDING SITE CHANGE: - Extended 1 day because of forecasted rain at EDW. - Changed landing site to KSC and landed one rev early because EDW had forecast of rain in clouds. FIRSTS: - First flight of OV-102 after OMDP (Major Mods at Palmdale). - First EDO flight and EDO pallet. - First flight of RCRS (Regenerable CO2 Removal System). - First flight of OV-102 with drag chute, INWS, etc. (Second flight of drag chute - deployed after NLGTD). - First flight to exceed GEMINI VII flight duration (by 54:33). Only 3 SKYLAB flights exceed STS-50 duration. DRAG CHUTE STRATEGY: Second drag chute deploy with NLG on ground. Continued. . .

STS050-291-006 1992-07-09 In orbit crew portrait in the spacelab (USML-1/LM.

FLT NO.	ORBITER	CREW (7)	LAUNCH SITE, LIFTOFF TIME,	LANDING SITE/ RUNWAY, CROSSRANGE	SSME-TL NOM-ABORT EMERG	SRB RSRM	ORBIT		FSW	PAYLOAD WEIGHTS,	MISSION HIGHLIGHTS (LAUNCH SCRUBS/DELAYS,
		TITLE, NAMES & EVA'S	LANDING SITES, ABORT TIMES	LANDING TIMES FLT DURATION, WINDS	THROTTLE PROFILE ENG. S.N.	AND ET	INC	HA/HP		PAYLOADS/ EXPERIMENTS	TAL WEATHER, ASCENT I-LOADS, FIRSTS, SIGNIFICANT ANOMALIES, ETC.)

STS-50

Continued

STS050-291-027 1992-07-09 Dunbar/MS/PYLD CDR (rt) and DeLucas/PS in SL with Lower Body Negative Pressure Study.

Continued. . .

DENS ALT: 1423 FT

FLT DURATION:
13:19:30:04
331:30:04

S/T: 307:15:59:29

OV-102:
88:20:44:28

DISTANCE:
5,758,332 sm

EARTH VIEWS

Top Lt to Rt: Canary Islands & ocean wakes (STS050-82-002) and Dust Storm, Red Sea, & Saudi Arabia (STS050-85-037). Bottom Lt to Rt: Mt. Pinatubo Volcano - Post Eruption, Luzon, Philippines (STS050-52-026) and Andes Mountains, Chile and Argentina (STS50-112-060).

STS050-81-027 STS050-81- First U.S. Microgravity Laboratory (USML-1) module is pictured in the P/L Bay in this scene over the southern two-thirds of the Florida peninsula. KSC is just above Columbia's starboard wing.

Continued. . .

SIGNIFICANT ANOMALIES:
- RCRS shutdown due to a short in the controller, hence LiOH canisters used until IFM required use at 5 days MET.
- SL/Orbiter air not mixing properly. Found a removable inline redundant seal was not removed from tunnel air ducting as should be for on-orbit operations.
- Waste water dump line blockage causing reduction in dump rate.
- Cryo O_2 tank 2 had a 1 lb/hr leak.
- Cryo O_2 tank 2 heater A2 experienced intermittent power dropouts.
- Fuel cell 3 O_2 purge valve did not close completely. Manually closed, did not purge again for remainder of flight.
- Cryo O_2 tank 7 check valve failed in open position.
- SS inverter overvolt shut down when SL H_2O loop was turned on.
- FWD starboard floodlight did not come on.
- R OMS yaw TVC excessive movement during ascent.
- Aileron trim deflected to 2.2 at M=10.1, preflight predicted was maximum of 0.80 deflection.
- TAGS jam on day 2, used teleprinter.
- Flight deck Canon A1, Mark II camcorder failure.
- ROB brake pressure low.
- APU 1 gearbox N_2 pressure decay/ transducer erratic.
- L1U jet heater fail on.
- F2F jet fail off.

STS050-S-106 - First flight of OV-102 with drag chute, INWS, etc. (Second flight of drag chute - deployed after NLGTD).

STS50-s-084 -- Unidentified Flight Controller hangs mission plaque in FCR

SPACE SHUTTLE MISSIONS SUMMARY

FLT NO.	ORBITER	CREW (7) TITLE, NAMES & EVA'S	LAUNCH SITE, LIFTOFF TIME, LANDING SITES, ABORT TIMES	LANDING SITE/ RUNWAY, CROSSRANGE LANDING TIMES FLT DURATION, WINDS	SSME-TL NOM-ABORT EMERG THROTTLE PROFILE ENG. S.N.	SRB RSRM AND ET	INC	ORBIT HA/HP	FSW	PAYLOAD WEIGHTS, PAYLOADS/ EXPERIMENTS	MISSION HIGHLIGHTS (LAUNCH SCRUBS/DELAYS, TAL WEATHER, ASCENT I-LOADS, FIRSTS, SIGNIFICANT ANOMALIES, ETC.)
STS-46 SEQ FLT #49 KSC-49 PAD 39B-15 MLP-1	OV-104 (Flight 12) Atlantis OMS PODS LPO1-18 RPO1-18 FRC4-12	CDR: Loren J. Shriver (Flt 3 - STS 51-C & STS-31) P263/R50/V51/M46 PLT: Andrew M. Allen P264/R149/M133 M/S 1: Claude Nicollier (Switzerland) P265/R150/M134 M/S 2: Marsha S. Ivins (Flt 2 - STS-32) P266/R109/V77/F12 PYLD CDR, M/S 3: Jeffrey A. Hoffman (Flt 3 - STS 51-D & STS-35) P267/R57/V59/M52 M/S 4: Franklin R. Chang-Diaz (Flt 3 - STS 61-C & STS-34) P268/R89/V46/M81 P/S 1: Franco Malerba (Italy) P269/R151/M135 MCC FCR-1 (29) FLIGHT DIRECTORS: A/E/O 1 - R. D. Dittemore Ld/O 2 - C. W. Shaw O 3 - P. L. Engelauf MOD - B. R. Stone	KSC 39, PAD B 213:13:56:48Z 9:56:00 AM EDT (P) 9:56:48 AM EDT (A) Friday 8 7/31/92 (2) LAUNCH WINDOW 2H 30M CTOB EOM PLS: KSC TAL: BYD TAL WX: BEN, ROTA SELECTED: RTLS: KSC 15/CI/N TAL: BEN 36/N/N AOA: EDW 22/CI/N PLS: EDW 22/CI/N TDEL 0.0 0.332/0.36 MAX Q NAV: 709 PSF 718 PSF SRB STG: 2:04.2 2:06 PERF: NOMINAL 2 ENG TAL (BEN): 2:51 2:54 NEG RETURN: 3:59 4:02 PTA (U/S 285): 4:23 4:22 PTM (U/S 285): 5:29 5:29 MECO CMD: 8:29 8:29.8 VI: 25987 25985 OMS-2: 41:23.6 41:23.6 351.2 FPS 351.4 FPS	KSC 33 (KSC 11) 221:13:11:50Z 9:11:50 AM EDT Saturday 9 8/8/92 (4) DEORBIT BURN: 221:12:17:10Z XRANGE: 499 NM ORBIT DIR: DL 24 AIM PT: NOMINAL MLGTD: 1866 FT 221:13:11:50Z VEL: 202 KGS 195 KEAS HDOT: -1 FPS TD NORM 195: 1891 FT NLGTD: 6501 FT 221:13:12:05Z VEL: 154 KGS HDOT: -4.3 FPS BRK INIT: 131 KGS AVE BRK DECEL: 5.9 FPS/S WHEELS STOP: 221:13:12:55Z 12726 FT ROLLOUT: 10840 FT 55 SECS WINDS: T 0.4, L 0.9 KTS OFFICIAL 3H, 1R DENS ALT: 1834 FT FLT DURATION: 7:23:15:02 191:15:02 S/T: 315:15:14:31 OV-104: 72:09:53:00 DISTANCE: 3,321,007 sm	104/104/ 109% PREDICTED: 100/104/ 80/67/104 ACTUAL: 100/104/ 82/67/104 1 = 2032 (1) 2 = 2033 (1) 3 = 2027 (8) M 3 EOM WEIGHT: 209851 LBS X CG: 1078.2 LANDING WEIGHT: 209532 LBS X CG: 1179.6	BI-052 RSRM 25W ET-48 LWT-41 ET RPT 239K 1:21:02 MET ET BR/UP 217 K 1:21:39 MET ET IMPACT LAT: 17.86 N LONG: 153.0 W	28.46 (29)	DIRECT INSERTION POST OMS-2 230.4 X 228.3 NM EURECA DEPLOY: 231.3 X 227.8 NM TSS DEPLOY: 161.0 X 158.5 NM TSS DOCK: 161.0 X 157.8 NM DEORBIT 121 X 121 NM VELOCITY 25698 FPS ENTRY RANGE 4397 NM	OI-21 (3)	CARGO: 34060 LBS PAYLOAD CHARGEABLE: 28585 LBS DEPLOYED: 9901 LBS NON-DEPLOYED: 16094 LBS MIDDECK: 1104 LBS SHUTTLE ACCUMULATED WEIGHTS: DEPLOYED: 673238 LBS NON-DEPLOYED: 598724 LBS CARGO TOTAL: 1450620 LBS PERFORMANCE MARGINS (LBS): FPR: 4671 FUEL BIAS: 983 FINAL TDDP:2825 RECON: 1942 PAYLOADS: PLB: European Retrievable Carrier (EURECA) (Deployed) Tethered Satellite System (TSS-1) (Deployed and Retrieved) EOIM-III TEMP 2A-3 ICBC CONCAP-II CONCAP-III LDCE MIDDECK: PHCF UVPI 4 CRYO TK SETS RMS 27 (S.N. 201) USED FOR EURECA DEPLOY	KSC W/D: OPF 61, VAB 5, PAD 45=111 days LAUNCH POSTPONEMENTS: - Launch date 6/26/92 as of 6/5/91. - Launch postponed to 7/2/92 because of STS-45 launch and landing delays. - Launch postponed to 7/21/92 because of MOD STS-50 landing to launch 8-day constraint and range interference. - Launch postponed to 7/31/92 to allow additional flightcrew and flight controller training. LAUNCH SCRUB: None. LAUNCH DELAYS: - 0M 48S delay at APU startup (approxi-mately L-5 minutes). Crew did not open APU #3 fuel isolation valve within GLS window. KSC cleared hold and count continued. TAL WX: - Banjul (prime) NO GO - ceiling, Ben Guerir GO (selected), Rota (2nd flight as substitute for Moron) NO GO - visibility (haze). ASCENT I-LOADS: - DOLILU I-Load uplinked to increase margin for green squatcheloid at M=1.53. Third DOLILU uplink, total uplink #10. FLIGHT DURATION/LANDING SITE CHANGE: - Extended 1 day because of TSS deploy problems. - Waved off first landing opportunity at KSC because of scattered showers within 30 miles. Total extension, 1 day plus 1 rev. FIRSTS: - First flight of a deployment and retrieval of a tethered satellite. NOTE: TSS deployed weight of 1040 lbs plus 90 lbs prop is not included in 9901 lbs deployed. LASTS: - Last flight of fleet without drag chute, INWS, and other improvements first used on STS-49. These modifications will be made before the next flight of OV-104. THIRD SHUTTLE CREWMEMBER REPLACEMENT: - Robert "Hoot " Gibson was replaced by Shriver in 1990. (Second Shuttle crewmember replacement occurred on STS-44.) EVENTS: - EURECA deploy at 1/17:10 MET. - TSS deploy at 4/08:57:22 MET. - TSS dock at 5/08:56:12 MET. Continued. . .

STS046-12-009 1992-08-08 Crew poses in middeck. In rear (lt to rt) CDR Shriver, PLT Allen, & Chang-Diaz/MS. In front (lt to rt) Nicollier/MS CSA, Hoffman/MS PLC, Ivins/MS, and Malerba/MS (Italy). Note the crew are positioned parallel to middeck floor with sleep station in background.

FLT NO.	ORBITER	CREW (7) TITLE, NAMES & EVA'S	LAUNCH SITE, LIFTOFF TIME, LANDING SITES, ABORT TIMES	LANDING SITE/ RUNWAY, CROSSRANGE LANDING TIMES FLT DURATION, WINDS	SSME-TL NOM-ABORT EMERG THROTTLE PROFILE ENG. S.N.	SRB RSRM AND ET	ORBIT INC	HA/HP	FSW	PAYLOAD WEIGHTS, PAYLOADS/ EXPERIMENTS	MISSION HIGHLIGHTS (LAUNCH SCRUBS/DELAYS, TAL WEATHER, ASCENT I-LOADS, FIRSTS, SIGNIFICANT ANOMALIES, ETC.)
STS-46 Continued										

At left: STS046-17-017 1992-08-08 Ivins/MS (left) and Hoffman/MS and PLC are conducting the Tether Optical Phenomena (TOP) experiment.

STS046-102-021 1992-08-08 OV-104's RMS grapples EURECA-1L and holds it in deployment position above PLB

STS-46 Tethered Satellite System 1 (TSS-1) satellite is reeled out via its thin Kevlar tether into the blackness of space during deployment operations from Atlantis payload bay (PLB).

Continued. . .

SIGNIFICANT ANOMALIES:
- MPS GH2 FCV erratic pressure.
- Fan Sep 1 flooded, indicated stall currents and CB opened. Fan Sep 2 temporarily flooded.
- P/L EURECA RF data handling problem (PSP lost lock due to excessive zeros in payload bit stream).
- Flight deck speaker failed.
- TSS U2 umbilical retractions failed when commanded by crew.
- TSS deployer reel stalling at 179 and 251 meters.
- TSS upper tether control mechanism jam at 224 meters.
- Postflight investigation found the TSS level wind mechanism was jammed by a structural reinforcement bolt which was added based on late loads analysis.

SPACE SHUTTLE MISSIONS SUMMARY

FLT NO.	ORBITER	CREW (7) TITLE, NAMES & EVA'S	LAUNCH SITE, LIFTOFF TIME, LANDING SITES, ABORT TIMES	LANDING SITE/ RUNWAY, CROSSRANGE LANDING TIMES FLT DURATION, WINDS	SSME-TL NOM-ABORT EMERG THROTTLE PROFILE ENG. S.N.	SRB RSRM AND ET	ORBIT INC	HA/HP	FSW	PAYLOAD WEIGHTS, PAYLOADS/ EXPERIMENTS	MISSION HIGHLIGHTS (LAUNCH SCRUBS/DELAYS, TAL WEATHER, ASCENT I-LOADS, FIRSTS, SIGNIFICANT ANOMALIES, ETC.)
STS-47 SEQ FLT #50 KSC-50 PAD 39B-16 MLP-2	OV-105 (Flight 2) Endeavour Spacelab-J (Japan) Tenth Spacelab Flight Long Module (7) OMS PODS: LPO3 - 13 RPO4 - 9 FR5 - 2	CDR: Robert L. Gibson (Flt 4 - STS 41B, STS 61-C, STS-27) P270/R30/V27/M29 PLT: Curtis L. Brown P271/R152/M136 M/S 1: Mark C. Lee Payload CDR (Flt 2 - STS-30) P272/R100/V78/M91 M/S 2: Jay Apt (Flt 2 - STS-37) P273/R123/V79/M110 M/S 3: N. Jan Davis P274/R153/F17 M/S 4: Mae C. Jemison P275/R154/F18 P/S 1: Mamoru Mohri (Japan) P276/R155/M137 MCC FCR-1 (30) FLIGHT DIRECTORS: Asc/Ent - N. W. Hale Ld/O 2 - J. M. Heflin O 1 - G. A. Pennington O 3 - L. J. Ham MOD - G. E. Coen	KSC 39, PAD B 256:14:23:00Z 10:23:00 AM EDT (P) 10:23:00 AM EDT (A) Saturday 3 9/12/92 (3) LAUNCH WINDOW: 2H 30M CTOB EOM PLS: KSC TAL: ZZA TAL WX: ROTA, BEN SELECTED: RTLS: KSC 33/CI/N TAL: ZZA 30/N/SF AOA: NOR 17/N/N PLS: EDW 22/CI/N TDEL: -0.16 -0.118/-0.08 MAX Q NAV: 679 PSF ~682 PSF SRB STG: 2:04 PERF: NOMINAL 2 ENG TAL (ZZA): 3:05 3:07 NEG RETURN: 4:04 4:04 PTA (U/S 285): 5:22 5:22 PTM (N/A): SE PTM (U/S 476) 7:07 7:08 MECO CMD: 8:31 8:34 MECO VI: 25830 25827 OMS-2: 36:11 36:12 262 FPS 262 FPS	KSC 33 (KSC 12) 264:12:53:22Z 8:53:22 AM EDT Sunday 8 9/20/92 (5) DEORBIT BURN: 264:11:52:20Z XRANGE: 669 NM ORBIT DIR: AR 5 AIM PT: CLOSEIN MLGTD: 2458 FT 264:12:53:22Z VEL: 209 KGS 202 KEAS HDOT: 0 FPS TD NORM 205: 2367 FT DRAG CHUTE DEPLOY: 176 KEAS 264:12:53:30.9Z NLGTD: 7651 FT 264:12:53:39Z VEL: 135 KGS HDOT: -2.2 FPS BRK INIT: 114 KGS AVE BRK DECEL: 6.9 FPS/S CHUTE JETTISON: 264:12:53:57Z 55 KGS WHEELS STOP: 264:12:54:11Z 11025 FT ROLLOUT: 8567 FT 49 SECS WINDS: H 0.9, L 1.8 KTS OFFICIAL: H2, L3 DENS ALT: 1805 FT FLT DURATION: 7:22:30:22 190:30:22 S/T: 323:13:44:53 OV-105: 16:19:48:01 DISTANCE: 3,310,922 sm	104/104/ 109% PREDICTED: 100/100/ 100/67/104 ACTUAL: 100/100/ 100/67/104 1 = 2026 (3) 2 = 2022 (8) 3 = 2029 (6) M 3 EOM WEIGHT: 220325 LBS X CG: 1083.7 LANDING WEIGHT: 220195 LBS X CG: 1085.3	BI-053 RSRM 26W ET-45 LWT-38 ET RPT ET BR/UP ET IMPACT LAT: 43.99 S LONG: 158.8 W	57.02 (11)	DIRECT INSERTION POST OMS-2 163.1 X 162.7 NM DEORBIT 166 X 161 NM VELOCITY 25803 FPS ENTRY RANGE 4341 NM	OI-21 (4)	CARGO: 32480 LBS PAYLOAD CHARGEABLE: 28092 LBS DEPLOYED: 0 LBS NON-DEPLOYED: 26247 LBS MIDDECK: 1845 LBS SHUTTLE ACCUMULATED WEIGHTS: DEPLOYED: 673238 LBS NON-DEPLOYED: 626816 LBS CARGO TOTAL: 1483100LBS PERFORMANCE MARGINS (LBS): FPR: 4671 FUEL BIAS: 983 FINAL TDDP: 1348 RECON: 2887 PAYLOADS: PLB: SPACELAB-JAPAN MATERIALS SCIENCE AND LIFE SCIENCES EXPERIMENTS (SL-J/LM) GBA-12 GAS MIDDECK: ISAIAH SSCE SAREX-II 4 CRYO TK SETS RMS 28 (S.N. 303) (NOT USED PER PLAN))	KSC W/D: OPF 77, VAB 5, PAD 17=99 days LAUNCH POSTPONEMENTS: - Launch date 8/12/92 as of 8/21/91. - Launch postponed to 9/1/92 due to STS-49, STS-50, and STS-46 delays. - Launch postponed to 9/11/92 because of DFRF work and ferry to KSC being delayed. LAUNCH SCRUB: None. LAUNCH DELAYS: None. TAL WX: - Zaragoza (prime) - GO (selected), Rota - GO. Ben Guerir - GO. DOLILU/NOMINAL I-LOADS: - Nominal I-loads selected, no uplink required. FLIGHT DURATION CHANGE: - Extended one day for science gain/enhancement. - Extended one rev because rain forecast within 30 nm at KSC. FIRSTS: - First flight with married couple as crew members (M/S 1 and M/S 3). - First flight to deploy drag chute with nose in air. Deploy was at 185 KGS at 8 seconds after MLGTD. Chute pulled right 8 ± 2 causing nose to move left 27 feet. SIGNIFICANT ANOMALIES: - RCS JET L3A failed off. - L5D low chamber pressure. - DDS 1 H/W transient, screen blank and display overwrites. - Condensation on H2O loop lines. - Transient WCS fan separator stall currents. - Cryo O2 tank 4 controller problem. - H2O relief line temperature problem. - Ku-band range rate /Azimuth display failure. - APU 1 and 3 drain line temps cycling low. - RMLG line temperature high. - Loss of MCC power buses B1 and B2.

STS047-09-009 1992-09-20 Crew in Spacelab Japan (SLJ) science module Endeavour PLB: Lt to Rt, back row CDR Gibson & PLT Brown; middle row, Davis/MS, Apt/MS, & Jemison/MS; and front row, Lee/MS PLC & Mohri/PS (Japan) NASDA.

STS047-76-078

SPACE SHUTTLE MISSIONS SUMMARY

FLT NO.	ORBITER	CREW (6) TITLE, NAMES & EVA'S	LAUNCH SITE, LIFTOFF TIME, LANDING SITES, ABORT TIMES	LANDING SITE/ RUNWAY, CROSSRANGE LANDING TIMES FLT DURATION, WINDS	SSME-TL NOM-ABORT EMERG THROTTLE PROFILE ENG. S.N.	SRB RSRM AND ET	INC	ORBIT HA/HP	FSW	PAYLOAD WEIGHTS, PAYLOADS/ EXPERIMENTS	MISSION HIGHLIGHTS (LAUNCH SCRUBS/DELAYS, TAL WEATHER, ASCENT I-LOADS, FIRSTS, SIGNIFICANT ANOMALIES, ETC.)
STS-52 SEQ FLT #51 KSC-51 PAD 39B-17 MLP-3	OV-102 (Flight 13) Columbia OMS PODS LPO5 - 2 RPO5 - 2 FRC2 - 13	CDR: James D. Wetherbee (Flt 2 - STS-32) P277/R108/V80/M97 PLT: Michael A. Baker (Flt 2 - STS-43) P278/R133/V81/M118 M/S 1: Charles (Lacy) Veach (Flt 2 - STS-39) P279/R127/V82/M114 M/S 2: William M. Shepherd (Flt 3 - STS-27, STS-41) P280/R96/V56/M87 M/S 3: Tamara E. Jernigan (Flt 2 - STS-40) P281/R130/V83/F14 P/S 1: Steven MacLean (Canada) P282/R156/M138 MCC FCR-1 (31) FLIGHT DIRECTORS: Asc/Ent - J. W. Bantle Ld/O 1 - R. E. Castle O 2 - R. D. Jackson Planning - C. W. Shaw MOD - A. L. Briscoe	KSC 39, PAD B 296:17:09:38.97Z 11:16:00 AM EDT (P) 01:09:39 PM EDT (A) Thursday 13 10/22/92 (6) LAUNCH WINDOW 2H 30M CTOB EOM PLS: KSC TAL: BYD TAL WX: MOR, BEN SELECTED: RTLS: KSC 15/N/N TAL: BYD 32/N/SF AOA: EDW 22/N/N PLS: EDW 04/CI/N TDEL: - 0.16 - 0.438/0.4 MAX Q NAV: 717 PSF 708 PSF SRB STG: 2:03.8 2:05 PERF: NOMINAL 2 ENG TAL (BYD): 2:23 2:26 NEG RETURN: 4:05 4:09 PTA (U/S 235): 4:22 4:25 PTM (U/S 235): 5:08 5:09 MECO CMD: 8:29.82 8:32 VI: 25875 25874 OMS-2: 39:56 39:56 215 FPS	KSC 33 (KSC 13) 306:14:05:53Z 9:05:53 AM EST Sunday 9 11/1/92 (7) DEORBIT BURN: 306:13:11:59Z XRANGE: 223 NM ORBIT DIR: DL 25 AIM PT: NOMINAL MLGTD: 1080 FT 306:14:05:53Z VEL: 219 KGS 211 KEAS HDOT: -0.3 FPS TD NORM 195: 2819 FT DRAG CHUTE DEPLOY: 169 KEAS 306:14:06:06Z NLGTD: 6949 FT 306:14:06:11Z VEL: 151 KGS HDOT: - 3.5 FPS BRK INIT: 101 KGS DRAG CHUTE JETTISON: 51 KGS 306:14:06:36Z AVE BRK DECEL: 5.7 FPS/S WHEELS STOP: 306:14:06:55Z 11788 FT ROLLOUT: 10708 FT 63 SECS WINDS: T-4, R 5 KTS OFFICIAL: H3, L8 DENS ALT: 1643 FT FLT DURATION: 9:20:56:13 236:56:13 S/T: 333:10:41:06 OV-102: 98:17:40:41 DISTANCE: 4,129,028 sm	104/104/ 109% PREDICTED 100/100/ 100/67/104 ACTUAL 100/100/ 95/67/104 1 = 2030 (8) 2 = 2015 (9) 3 = 2034 (1) M 3 EOM WEIGHT: 216043 LBS X CG: 1082.6 LANDING WEIGHT: 215935 LBS X CG: 1084.	BI-055 RSRM 27K ET-55 LWT-48 ET RPT ET BR/UP ET IMPACT LAT: 12.9 S LONG: 163.4 W	28.46 (30)	DIRECT INSERTION POST OMS-2 162.7 X 160.2 NM LAGEOS DEPLOY: 169.5 X 161.1 NM 0/20:47:45 OMS-6: 154.2 X 114 NM 7/19:59:55 OMS-7: 114.1 X 113.9 NM 7/20:46:26 DEORBIT 113 X 110 NM VELOCITY 25666 FPS ENTRY RANGE 4454 NM	OI-21 (5)	CARGO: 26862 LBS PAYLOAD CHARGEABLE: 20132 LBS DEPLOYED: 5577 LBS NON-DEPLOYED: 12475 LBS MIDDECK: 2080 LBS SHUTTLE ACCUMULATED WEIGHTS: DEPLOYED: 678815 LBS NON-DEPLOYED: 641371 LBS CARGO TOTAL: 1509962 LBS PERFORMANCE MARGINS (LBS): FPR: 4671 FUEL BIAS: 983 FINAL TDDP:11107 RECON: 9801 PAYLOADS: PLB: LASER GEODYNAMICS SATELLITE (LAGEOS-II) (DEPLOYED) CTA DEPLOYED (CANADIAN TARGET ASSY) CANEX-2/TPCE, USMP-01 ASP MIDDECK: PSE HPP CPCB BLOCK II SPIE CMIX CVTE CANEX 5 CRYO TK SETS RMS 29 (S.N. 301) USED FOR CTA DEPLOY	KSC W/D: OPF 72, VAB 5, PAD 27=104 days LAUNCH POSTPONEMENTS: - Launch date was 9/24/92 on 8/21/91. - Launch postponed to 10/15/92 on 6/10/92. - Launch postponed to 10/22/92 on 10/10/92 due to engine 3 steerhorn weld anomaly. LAUNCH SCRUB: None. LAUNCH DELAYS: - Delayed for 1H53M39S because of RTLS crosswind exceedance (15-knot limit). A range safety warning (BLAST) existed for part of launch hold. MMT waived crosswind exceedance (0613G21 on center tower). TAL WX: - Prime TAL Banjul had reduced short range visibility but was forecast and observed GO and selected. Moron was forecast and observed NO GO because of low ceiling. Ben Guerir was NO GO during most of prelaunch period because of ceilings and threat of rain, but was observed GO when rain moved away from runway. DOLILU/I-LOADS: - Both nominal and DOLILU (Q-Alpha-4000) for aero DTO. Alternate (Q-Alpha-3250) to backout DTO. Selected DOLILU, DOLILU uplink #4, total uplink (#11). FLIGHT DURATION CHANGE: None. LANDING SITE CHANGE: None. DRAG CHUTE STRATEGY: - Deploy nose in air at 175 kgs/derotation if crosswinds ≤ 5 kts steady state and nose within ± 10 of center line. Dis-reef would occur at touchdown. Drag chute was deployed at 170 KGS (chute deploy #4), chute pulled left and nose went to right. SIGNIFICANT ANOMALIES: - WCS fan separator 1 failed to operate FD 10. - Fuel cell 1 cell performance monitor hangup. - F3L failed off (oxidizer leak). - PRSD O_2 tank 2 heater A2 erratic. - TAGS hard jam, no developer motor motion. - Intermittent surface position indicator (SPI) power. - S-band PM low frequency forward link loss of lock. - S-band FM transmitter RF power output erratic. - Window 3 internal "void" or "bruise" (R&R).

STS052-25-005 1992-11-01 In orbit crew portrait. Caption unavailable, see names above.

STS052-80-024 1992-11-01 Italian Research Interim Stage (IRIS), a spinning solid fuel rocket, lifts the Laser Geodynamic Satellite II (LAGEOS II) out of its support cradle for deployment.

FLT NO.	ORBITER	CREW (5) TITLE, NAMES & EVA'S	LAUNCH SITE, LIFTOFF TIME, LANDING SITES, ABORT TIMES	LANDING SITE/ RUNWAY, CROSSRANGE LANDING TIMES FLT DURATION, WINDS	SSME-TL NOM-ABORT EMERG THROTTLE PROFILE ENG. S.N.	SRB RSRM AND ET	ORBIT INC (12)	HA/HP	FSW	PAYLOAD WEIGHTS, PAYLOADS/ EXPERIMENTS	MISSION HIGHLIGHTS (LAUNCH SCRUBS/DELAYS, TAL WEATHER, ASCENT I-LOADS, FIRSTS, SIGNIFICANT ANOMALIES, ETC.)

STS-53

SEQ FLT #52

KSC-52

PAD 39A-35 MLP-1

OV-103 (Flight 15) Discovery

OMS PODS LPO4-12 RPO3-16 FRC3-15

CDR: David M. Walker (Flt 3 - STS 51-A & STS-30) P283/R48/V40/M45

PLT: Robert D. Cabana (Flt 2 - STS-41) P284/R113/V84/M101

M/S 1: Guion S. Bluford (Flt 4 - STS-8, STS 61-A & STS-39) P285/R22/V25/M21

M/S 2: James S. Voss (Flt 2 - STS-44) P286/R136/V85/M121

M/S 3: Michael R. Clifford P287/R157/M139

MCC FCR-2 (21)

FLIGHT DIRECTORS: Asc/Ent - N. W. Hale Ld/O 2 - R. M. Kelso O 1 - J. M. Heflin Planning - L. J. Ham MOD - B. R. Stone

KSC 39, PAD A 337:13:24:00Z 6:59:00 AM EST (P) 8:24:00 AM EST (A) Wednesday 7 12/2/92 (3)

LAUNCH WINDOW 2H 30M CTOB

EOM PLS: KSC **TAL:** ZZA TAL WX: MRN, BEN

SELECTED: RTLS: KSC 33/CI/N TAL: BEN 36/N/N AOA: NOR 17/N/N PLS: NOR 17/CI/N

TDEL: 0.32   0.722/0.766

MAX Q NAV: 692 PSF   705 PSF

SRB STG: 2:05.6   2:06

PERF: NOMINAL

2 ENG TAL (MRN): 2:32   2:33

NEG RETURN: 4:04   4:06

PTA (U/S 350): 4:56   4:52

PTM (U/S 350): 5:48   5:41

MECO CMD: 8:33.48   8:34

VI: 25885   25885

OMS-2: 37:03   36:53.6 337.3 FPS337.5 FPS

EDW 22, CONC (EDW 37,CONC 18) 344:20:43:47Z 12:43:47 PM PST Wednesday 5 12/9/92 (6)

DEORBIT BURN: 344:19:43:20Z

XRANGE: 791 NM

ORBIT DIR: DR 8

AIM PT: CLOSE IN

MLGTD: 1108 FT 344:20:43:47Z VEL: 209 KGS 212 KEAS HDOT: -2.5 FPS

TD NORM 195: 2682 FT

DRAG CHUTE DEPLOY: 167 KEAS 344:20:44:00Z

NLGTD: 6329 FT 344:20:44:03.6Z VEL: 145 KGS HDOT: -2.2 FPS

BRK INIT: 106 KGS

DRAG CHUTE JETTISON: 60 KGS 344:20:44:25Z

AVE BRK DECEL: 3.5 FPS/S

WHEELS STOP: 344:20:44:59Z 11273 FT

ROLLOUT: 10165 FT 82 SECS

WINDS: H9, R11 2614P19 **OFFICIAL:** H15, R8

DENS ALT: 2961 FT

FLT DURATION: 7:07:19:47 175:19:47

S/T: 340:18:00:53

OV-103: 90:10:42:50

DISTANCE: 3,034,680 sm

104/104/ 109%

PREDICTED: 100/100/ 100/70/ 104/67

ACTUAL: 100/100/ 100/73/ 104/67

1 = 2024 (5) 2 = 2012 (14) 3 = 2017 (7)

M 3 EOM

WEIGHT: 194028 LBS

X CG: 1089.5

LANDING WEIGHT: 193851 LBS X CG: 1091.3

DEORBIT 174 X 169 NM

VELOCITY 25813 FPS

ENTRY RANGE 4237 NM

BI-055

RSRM 28W

ET-49 LWT-42

ET RPT

ET BR/UP

ET IMPACT LAT: 40.95 S **LONG:** 152.6 W

57 (12)

DIRECT INSERTION

POST OMS-2 200 X 199 NM

DOD-1 DEPLOY: 00/05:54 MET 200 X 199 NM

SEP BURN: 00/06:14MET 204 X 200 NM

OMS-3: 01/06:19:12 202 X 175 NM

OMS-4: 01/07:02:03 176 X 175 NM (ODERACS DEPLOY ALT)

OMS-5: 05/05:51 174.9 X 170.3 NM (2ND KSC LANDING EOM +1)

OI-21 (6)

CARGO: 28316 LBS

PAYLOAD CHARGEABLE: 26118 LBS

DEPLOYED: 20789 LBS (NO ODERACS DEPLOY)

NON-DEPLOYED: 4299 LBS (INCLUDES ODERACS)

MIDDECK: 1030 LBS

SHUTTLE ACCUMULATED WEIGHTS: DEPLOYED: 699604 LBS NON-DEPLOYED: 646700 LBS **CARGO TOTAL:** 1538278 LBS

PERFORMANCE MARGINS (LBS): FPR: 3934 FUEL BIAS: 1055 FINAL TDDP:1368 RECON: 2844

PAYLOADS: PLB: DOD-1 (DPLY) GCP ODERACS (FAILED TO DEPLOY)

MIDDECK: HERCULES, STL, BLAST, RME III, CLOUDS-1A, CREAM, FARE

4 CRYO TK SETS

NO RMS

KSC W/D: OPF 247, VAB 5, PAD 24 = 276 days

LAUNCH POSTPONEMENTS:
- Launch date was 10/9/92 on 3/15/91.
- Launch postponed to 11/5/92 on 6/10/92 when decision made to fly STS-52 before STS-53.
- Postponed launch to 12/2/92 due to LP04 replacing LP01, engine steerhorn Xrays, and NWS anomaly.

LAUNCH SCRUB: None.

LAUNCH DELAYS:
- Delayed 1H25M at T-9 minutes because of acreage ice on ET which ice team confirmed melted approx. 35 minutes after sunrise. Addi-tional delay caused by wing LA16 exceedance of 102% based on L-70 minutes and DOLILU I-loads.

TAL WX:
- Zaragoza was prime but forecast intermittent GO (ceiling and rain) but observed GO. Moron forecast NO GO - ceiling, observed marginal GO. Ben Guerir forecast and observed GO (selected).

DOLILU/I-LOADS:
- Nominal and DOLILU I-loads were GO on L-4.25 balloon. DOLILU was selected and uplinked. DOLILU uplink #5, total 12.

FLIGHT DURATION CHANGES:
- Planned extension of flight from 6 to 7 days, if launch was delayed, to provide night passes for GLO experiment.
- Extended one rev because forecast 3.5K broken on first KSC landing opportunity.

LANDING SITE CHANGES:
- Changed landing site to EDW after waving off first opportunity at KSC and forecast NO GO (ceiling on second landing opportunity at KSC).

FIRSTS/LASTS:
- First flight of OV-103 after OMDP-1 with drag chute, INWS, etc.
- Last flight from FCR-2.

SIGNIFICANT ANOMALIES:
- HPOT secondary seal transducer failure.
- Humidity separator B water deposits.
- Supply water dump valve water leaks.
- Couldn't deploy ODERACS space spheres because logic battery was discharged (160 lbs).
- Speedbrake FCS channel 3 position feedback anomaly.
- F1L jet fail leak post FRCS dump (O$_2$ leak).
- PPO$_2$ C transducer shift.
- Water spray boiler 1 steam vent heater anomalous cycles.

EVENTS:
- DOD-1 deployed at 00/05:54 MET.
- Lowered orbit to 176 nm for ODERACS deploy.

STS053-13-021 1992-12-09 In orbit crew group portrait in the aft flight deck (Caption unavailable, see names above).

STS053-09-021   Fluid Acquisition & Resupply Equipment (FARE) middeck experiment. Photo shows the fluid mixture and transfer process in transparent sphere.

SPACE SHUTTLE MISSIONS SUMMARY

FLT NO.	ORBITER	CREW (5) TITLE, NAMES & EVA'S	LAUNCH SITE, LIFTOFF TIME, LANDING SITES, ABORT TIMES	LANDING SITE/ RUNWAY, CROSSRANGE LANDING TIMES FLT DURATION, WINDS	SSME-TL NOM-ABORT EMERG THROTTLE PROFILE ENG. S.N.	SRB RSRM AND ET	INC	ORBIT HA/HP	FSW	PAYLOAD WEIGHTS, PAYLOADS/ EXPERIMENTS	MISSION HIGHLIGHTS (LAUNCH SCRUBS/DELAYS, TAL WEATHER, ASCENT I-LOADS, FIRSTS, SIGNIFICANT ANOMALIES, ETC.)
STS-54 SEQ FLT #53 KSC-53 PAD 39B-18 MLP-2	OV-105 (Flight 3) Endeavour OMS PODS LPO3-14 RPO4-10 FRC5-3	CDR: John H. Casper (Flt 2 - STS-36) P288/R111/V86/M99 PLT: Donald McMonagle (Flt 2 - STS-39) P289/R126/V87/M113 M/S 1: Gregory J. Harbaugh (Flt 2 - STS-39) P290/R125/V88/M112 M/S 2: Mario Runco (Flt 2 - STS-44) P291/R137/V89/M122 M/S 3: Susan J. Helms P292/R158/F19 EMU/TETHERED EVA: EV1 - Greg Harbaugh EV2 - Mario Runco 1/17/93 4:27:50 Duration SS EVA #20 REFINE TRAINING METHODS FOR SPACE STATION EVA'S	KSC 39, PAD B 13:13:59:29.95Z 19:13:37:47Z 8:52:00 AM EST (P) 8:59:30 AM EST (A) Wednesday 8 1/13/93 (6) LAUNCH WINDOW 2H30M, CTOB EOM PLS: KSC TAL: BEN TAL ALT: BYD, MRN SELECTED: RTLS: KSC 33/N/N TAL: BEN 36/N/N AOA: NOR 17/N/N PLS: NOR 17/N/N TDEL: -0.32 0.322/0.36 MAX Q NAV: 709 PSF 715 PSF SRB STG: 2:05.1 2:06 PERF: NOMINAL 2 ENG TAL (BEN): 3:00 3:06 NEG RETURN: 3:57 4:00 PTA (U/S 235): 5:12 5:14 PTM (U/S 235): 5:54 5:56 MECO CMD: 8:28.66 8:30.6 VI: 25876 25872 OMS-2: 39:53 39:53	KSC 33 (KSC-14) 19:13:37:47Z 8:37:47 AM EST Tuesday 9 1/19/93 (5) DEORBIT BURN: 19:12:38:10Z XRANGE: 320 NM ORBIT DIR: DL 26 AIM PT: CLOSE IN MLGTD: 1536 FT VEL: 205 KGS 212 KEAS HDOT: -1 FPS TD NORM 195: 2710 FT DRAG CHUTE DEPLOY: 166 KEAS 19:13:38:00Z NLGTD: 6247 FT 19:13:38:02Z VEL: 150 KGS HDOT: -3.1 FPS BRK INIT: 107 KGS DRAG CHUTE JETTISON: 52 KGS 19:13:38:23Z AVE BRK DECEL: 7.3 FPS/S WHEELS STOP: 19:13:38:36Z PTA (U/S 235): 10259 FT ROLLOUT: 8723 FT 49 SECS WINDS: 4H, R2 OFFICIAL: H3, R2 DENS ALT: -151 FT FLT DURATION: 5:23:38:17 143:38:17 S/T: 346:17:39:10 OV-105: 22:19:26:18 DISTANCE: 2,501,277 sm	104/104/ 109% PREDICTED: 100/104/ 99/70/ 104/67 ACTUAL: 100/104/ 104/72/ 104/67 1 = 2019(11) 2 = 2033 (2) 3 = 2018 (9) (2018 WAS REBUILT) M 3 EOM WEIGHT: 197481 LBS X CG: 1091.6 LANDING WEIGHT: 197353 LBS X CG: 1093.4	BI-056 RSRM 29W ET-51 LWT-44 ET RPT ET BR/UP ET IMPACT LAT: 12.92 N LONG 163.3 W	28.45 (31)	DIRECT INSERTION POST OMS-2 164 X 160 NM SEP BURN: OMS-3: 173 X 160 NM OMS-4: 14:16:08:42Z 164 X 163 NM MET 1:02:08:42 DEORBIT 165 X 159 NM VELOCITY 25780 FPS ENTRY RANGE 4213 NM	OI-21 (7)	CARGO: 49039 LBS PAYLOAD CHARGEABLE: 46540 LBS DEPLOYED: 37497 LBS NON-DEPLOYED: 7991 LBS MIDDECK: 1052 LBS SHUTTLE ACCUMULATED WEIGHTS: DEPLOYED: 737101 LBS NON-DEPLOYED: 655743 LBS CARGO TOTAL: 1587317 LBS PERFORMANCE MARGINS (LBS): FPR: 3934 FUEL BIAS: 1055 FINAL TDDP:2659 RECON: 3421 PAYLOADS: PLB: TDRS-F/IUS (DEPLOYED) DXS MIDDECK: CHROMEX CGBA PARE SSCE 4 CRYO TK SETS NO RMS	KSC W/D: OPF 55, VAB 6, PAD 27 = 88 days LAUNCH POSTPONEMENTS: - Baselined launch date of 11/19/92 on 4/4/91. - Postponed launch date to 12/15/92 on 5/8/92. - Postponed launch date to 1/13/93, after holi-days, to allow the required OPF processing time. LAUNCH SCRUB: None. LAUNCH DELAYS: - Delayed 7M30S while holding at T-9 minutes while discussing load indicator A16 Q-plane exceedance (101%) at M=1.55. Approved a waiver. TAL WX: - Ben Guerir and Moron forecast and observed GO. Banjul forecast and observe NO GO - VIS (haze). DOLILU/I-LOADS: - DOLILU selected and uplinked. DOLILU #6, total uplink #13. FLIGHT DURATION CHANGES: None. FIRSTS: - First flight with a planned fuel cell shut-down/restart. FC2 shut down for 10 hours per DTO 412 at 04/20:00 - First flight of EDO Waste Collection System (WCS). - First Military Woman in Space - Susan J. Helms SIGNIFICANT ANOMALIES: - EDO WCS commode, urinal, and compactor microswitch problem. - PLB floodlights problems: Both mids and fwd starboard. - R1R jet failed off during RCS hot fire. - Rudder speedbrake secondary hydraulic switching valve indication. - Hydraulic sys 3 residual pressure post APU shutdown. - APU 3 overheat during ascent (WSB 3 not cooling). - DOLILU GPC dump display format error. - EVA - No hitch pin in PFR pip-pin. - R RSRM had 18 psi chamber pressure spike at 67 seconds. EVENTS: - TDRS-F deployed at 06:12:57 MET. - OMS4 to bring in additional ldg opportunities. - EVA started at 03:20:50:25 MET. - Deorbit burn on rev 95, landing rev 96. NOTE: SSME 2018 was rebuilt to new engine status.

MCC FCR-1 (32)

FLIGHT DIRECTORS:
Ascent - J. W. Bantle
Entry - R. D. Jackson
Ld/O2 - P. L. Engelauf
O 1 - C. W. Shaw
Plan - J. W. Muratore
MOD - A. L. Briscoe

STS054-02-008 - In orbit crew portrait (caption not available) Susam Helms, 1st Military Woman in space, at top.

Top: STS054-80-000U DTO 1210 EVA : Harbaugh carries Runco
Bottom: STS054-71-025 TDRS/IUS Deploy

FLT NO.	ORBITER	CREW (5) TITLE, NAMES & EVA'S	LAUNCH SITE, LIFTOFF TIME, LANDING SITES, ABORT TIMES	LANDING SITE/ RUNWAY, CROSSRANGE LANDING TIMES FLT DURATION, WINDS	SSME-TL NOM-ABORT EMERG THROTTLE PROFILE ENG. S.N.	SRB RSRM AND ET	ORBIT		FSW	PAYLOAD WEIGHTS, PAYLOADS/ EXPERIMENTS	MISSION HIGHLIGHTS (LAUNCH SCRUBS/DELAYS, TAL WEATHER, ASCENT I-LOADS, FIRSTS, SIGNIFICANT ANOMALIES, ETC.)
							INC	HA/HP			
STS-56 SEQ FLT #54 KSC-54 PAD 39B-19 MLP-1	OV-103 (Flight 16) Discovery Eleventh Spacelab Flight Igloo (4) OMS PODS LPO1-19 RPO3-17 FRC3-16	CDR: Kenneth D. Cameron (Flt 2 - STS-37) P293/R121/V90/M109 PLT: Stephen S. Oswald (Flt 2 - STS-42) P294/R139/V91/M124 M/S 1: C. Michael Foale (Flt 2 - STS-45) P295/R143/V92/M127 M/S 2: Kenneth D. Cockrell P296/R159/M140 M/S 3: Ellen Ochoa P297/R160/F20 MCC FCR-1 (33) FLIGHT DIRECTORS: Ascent - J. W. Bantle Entry - R. D. Jackson Ld/O1 - C. W. Shaw O 2 - J. W. Muratore O 3 - R. E. Castle MOD - A. L. Briscoe	KSC 39, PAD B 98:05:28:59.95Z 1:29:00 AM EDT (P) 1:29:00 AM EDT (A) Thursday 14 4/8/93 (9) LAUNCH WINDOW Closes on ATMOS Tangent Ray Constraint - 2H28M EOM PLS: KSC TAL: ZZA TAL ALT: MRN, BEN SELECTED: RTLS: KSC 33/N/N TAL: ZZA 30/CI/N AOA: NOR 17/N/N PLS: EDW 22/N/N (ORBIT 7) EDW 04/CI/N (ORBIT 3) TDEL: 0.00 0.24 MAX Q NAV: 675 PSF 676 PSF SRB STG: 2:05.3 2:06 PERF: NOMINAL 2 ENG TAL (MRN): 2:24 2:26 NEG RETURN: 4:10 4:13 PTA (U/S 280): 4:22 4:23 PTM (U/S 280): 5:09 5:12 MECO CMD: 8:28.8 8:35 VI: 25829 25825 OMS-2: 37:08 37:07 252 FPS 254 FPS	KSC 33 (KSC-15) 107:11:37:19Z 7:37:19 AM EST Saturday 10 4/17/93 (8) DEORBIT BURN: 107:10:34:25Z XRANGE: 6 NM ORBIT DIR: DL 27 AIM PT: CLOSE IN MLGTD: 1074 FT 107:11:37:19Z VEL: 196 KGS 206 KEAS HDOT: -2.5 FPS TD NORM 195: 1948 FT DRAG CHUTE DEPLOY: 169 KEAS 107:11:37:30Z NLGTD: 5587 FT 107:11:37:34Z VEL: 144 KGS HDOT: -3.4 FPS BRK INIT: 92 KGS DRAG CHUTE JETTISON: 55 KGS 107:11:37:59Z AVE BRK DECEL: 4.9 FPS/S WHEELS STOP: 107:11:38:22Z 10603 FT ROLLOUT: 9529 FT 63 SECS WINDS: H6, 1L OFFICIAL: H6, 1L DENS ALT: -74 FT FLT DURATION: 9:06:08:19 222:08:19 S/T: 355:23:47:29 OV-103: 99:16:51:09 DISTANCE: 3,853,997 sm	104/104/ 109% PREDICTED: 100/100/ 89/67/104 ACTUAL: 100/100/ 89/69/104 1 = 2024 (6) 2 = 2033 (3) 3 = 2018 (10) M 3 EOM WEIGHT: 208052 LBS X CG: 1084.6 LANDING WEIGHT: 207946 LBS X CG: 1086.3	BI-058 RSRM 31KM ET-54 LWT-47 ET RPT ET BR/UP ET IMPACT LAT: 42.4 N LONG: 154.36 W	57 (13)	DIRECT INSERTION POST OMS-2 159.8 X 159.1 NM DEPLOY: 161.1 X 158.2 NM RNDZ: 160.5 X 156.9 NM DEORBIT 160 X 150 NM VELOCITY 25797 FPS ENTRY RANGE 4375 NM	OI-21 (8)	CARGO: 21000 LBS PAYLOAD CHARGEABLE: 16439 LBS DEPLOYED: 0 LBS NON-DEPLOYED: 12568 LBS MIDDECK: 1031 LBS SHUTTLE ACCUMULATED WEIGHTS: DEPLOYED: 737101 LBS NON-DEPLOYED: 669342 LBS CARGO TOTAL: 1608317 LBS PERFORMANCE MARGINS (LBS): FPR: 3934 FUEL BIAS: 1055 FINAL TDDP: 9521 RECON: 10718 PAYLOADS: PLB: ATMOSPHERE LABORATORY FOR APPLICATIONS AND SCIENCE (ATLAS-2) SSBUV/A SPARTAN 201 (DEPLOYED & RETRIEVED) GBP SUVE MIDDECK: CMIX STL PARE SAREX-II HERCULES RME-III AMOS CREAM 4 CRYO TK SETS RMS 30 (S.N. 301) USED FOR SPARTAN DEPLOY, CAPTURE & BERTH	KSC W/D: OPF 63, VAB 10, PAD 22 = 95 days LAUNCH POSTPONEMENTS: - Launch date of 3/23/93 was postponed to 4/6/93 because of STS-55 launch delays which were caused by SSME HPOTP tip seal retainer problems, hydraulic flex hoses, and range conflicts with Delta and Atlas launches. LAUNCH SCRUB: - Launch on 4/6/93 was scrubbed after an RSLS breakout at T-11 seconds caused by failure to get "close" indication when LH$_2$ high point bleed valve closed. LAUNCH DELAYS: None. TAL WX: All three TAL sites (ZZA, MOR, and BEN) were forecast and observed GO. ZZA selected. DOLILU/I-LOADS: - Nominal I-loads were selected (were uplinked because DOLILU I-loads had been uplinked for 4/6/93 launch attempt). NIGHT LAUNCH: Shuttle night launch #8. FLIGHT DURATION CHANGES: - Waved off two landing opportunities at KSC because of forecast low ceiling at KSC. - Extended 1 day because WX forecast NO GO at KSC. FIRSTS: - First flight with 90% reefed drag chute (same deploy strategy). 90% more stable than baseline. - First TV uplink to American Spacecraft via SAREX-II (UHF fast scan TV). SIGNIFICANT ANOMALIES: - RSRM 7 to 8 psi pressure spike at 74 seconds. - Loose thermal blanket on aft (1307) bulkhead. - FC1 O$_2$ reactant valve falsely indicated closed. - FC1 substack 3 delta voltage increased during purges. - ATVC Channel 4 power failure. - Ku-band singed processor problem - Spacelab data exceeding 2 MPS was degraded. - S-band low frequency interference problem. - TAGS jam. - TIPS on first flight worked OK on S-band, bad on Ku-band (TAGS master switch was turned off). - L5D injector temps high indicated htr failed on. RNDZ: Rendezvous #13 with SPARTAN for retrieval and return. EVENTS: - SAREX contact with Russian Space Station, MIR, at 2:17:55 MET. - SPARTAN was deployed at 3:00:42 MET on orbit 49, grapple was at 05:01:51 MET, and berthed at 05:02:32 MET.

Crew inflight portrait: In front are CDR Cameron (left) and Foal/MS1. In back are (left to right) Ochoa/MS3, PLT Oswald and Cockrell/MS2.

Top: STS056-91-050 ATLAS-2 pallet in PLB
Bottom: STS056-90-034 freeflying SPARTAN-2

FLT NO.	ORBITER	CREW (7) TITLE, NAMES & EVA'S	LAUNCH SITE, LIFTOFF TIME, LANDING SITES, ABORT TIMES	LANDING SITE/ RUNWAY, CROSSRANGE LANDING TIMES FLT DURATION, WINDS	SSME-TL NOM-ABORT EMERG THROTTLE PROFILE ENG. S.N.	SRB RSRM AND ET	ORBIT		FSW	PAYLOAD WEIGHTS, PAYLOADS/ EXPERIMENTS	MISSION HIGHLIGHTS (LAUNCH SCRUBS/DELAYS, TAL WEATHER, ASCENT I-LOADS, FIRSTS, SIGNIFICANT ANOMALIES, ETC.)
							INC (32)	HA/HP			
STS-55 SEQ FLT #55 KSC-55 PAD 39A-36 MLP-3	OV-102 (Flight 14) Columbia Twelfth Spacelab Flight Long Module (8) OMS PODS LPO5-3 RPO5-3 FRC2-14	CDR: Steven R. Nagel (Flt 4 - STS 51-G, STS 61-A, & STS-37) P298/R64/V23/M59 PLT: Terence T. Henricks (Flt 2 - STS-44) P299/R135/V93/M120 M/S 1 (PYLD CDR): Jerry L. Ross (Flt 4 - STS 61-B, STS-27, & STS-37) P300/R86/V38/M78 M/S 2: Charles J. Precourt P301/R161/M141 M/S 3: Bernard A. Harris, Jr. P302/R162/M142 P/S 1: Ulrich Walter (Germany) P303/R163/M143 P/S 2: Hans W. Schlegel (Germany) P304/R164/M144 MCC FCR-1 (34) FLIGHT DIRECTORS: A/E/O1 - N. W. Hale Ld/O 2 - G. E. Coen O 3 - J. M. Heflin MOD - B. R. Stone	KSC 39, PAD A 116:14:49:59.98Z 10:50:00 AM EDT (P) 10:50:00 AM EDT (A) Monday 8 4/26/93 (10) LAUNCH WINDOW 2H 30M - CTOB EOM PLS: KSC TAL: BYD TAL ALT: BEN, MRN SELECTED: RTLS: KSC 15/N/N TAL: BYD 32/N/SF AOA: EDW 22/CI/N PLS: EDW 22/CI/N TDEL: -0.16 0.322/0.36 MAX Q NAV: 714 PSF 715 PSF SRB STG: 2:04.6 2:06 PERF: NOMINAL 2 ENG TAL (BYD): 2:30 2:31 NEG RETURN: 3:58 4:03 PTA (U/S 235): 4:52 4:56 PTM (U/S 235): 5:28 5:33 MECO CMD: 8:28.18 8:30.9 VI: 25877 25870 OMS-2: 39:54 39:54	EDW 22 CONC (EDW 38,CONC 19) 126:14:29:59Z 7:29:59 AM PDT Thursday 6 5/6/93 (5) DEORBIT BURN: 126:13:29:20Z XRANGE: 640 NM ORBIT DIR: DL 28 AIM PT: CLOSE IN MLGTD: 1819 FT VEL: 210 KGS 217 KEAS HDOT: -1.5 FPS TD NORM 205: 2589 FT DRAG CHUTE DEPLOY: 165 KEAS 126:14:30:15Z NLGTD: 7283 FT 126:30:17Z VEL: 149 KGS HDOT: -4.6 FPS BRK INIT: 131 KGS DRAG CHUTE JETTISON: 54 KGS 126:30:41Z AVE BRK DECEL: 4.8 FPS/S WHEELS STOP: 126:14:31:00Z 11944 FT ROLLOUT: 10125 FT 61 SECS WINDS: H13, L5 OFFICIAL: H15, L12 DENS ALT: 3166 FT FLT DURATION: 9:23:39:59 239:39:59 S/T: 365:23:27:28 OV-102: 108:17:20:40 DISTANCE: 4,164,183 sm	104/104/ 109% PREDICTED: 100/100/ 100/70/104 ACTUAL: 100/100/ 100/72/104 1 = 2031 (8) 2 = 2109 (10) 3 = 2029 (7) 2109 was rebuilt M 3 EOM WEIGHT: 227484 LBS X CG: 1078.4 LANDING WEIGHT: 227209 LBS X CG: 1079.7	BI-057 RSRM 30W ET-56 LWT-49 ET RPT ET BR/UP ET IMPACT LAT: 12.75 N LONG: 163.68 W	DIRECT INSERTION POST OMS-2 162 X 160 NM TRIM BURN #1: 0:10:33:00 MET 160.9 X 160.7 NM TRIM BURN #1: 2:21:34 :30 MET 162 X 158 NM DEORBIT 163 X 153 NM VELOCITY 25779 FPS ENTRY RANGE 4299 NM	OI-21 (9)	CARGO: 33416 LBS PAYLOAD CHARGEABLE: 26881 LBS DEPLOYED: 0 LBS NON-DEPLOYED: 24599 LBS MIDDECK: 2282 LBS SHUTTLE ACCUMULATED WEIGHTS: DEPLOYED: 737101 LBS NON-DEPLOYED: 696223 LBS CARGO TOTAL: 1641733 LBS PERFORMANCE MARGINS (LBS): FPR: 3934 FUEL BIAS: 1055 FINAL TDDP: 6248 RECON: 7559 PAYLOADS: PLB: SPACELAB SL-D2/LM (Germany) + USS + GAS GAS MIDDECK: SAREX-II 5 CRYO TK SETS NO RMS	KSC W/D: OPF 77, VAB 5, PAD 73 = 155 days LAUNCH POSTPONEMENTS: - 2/25/93 launch date was postponed to 3/21/93 because of SSME HPOT tip seal retainer pro-blem, SSME 3 LH₂ umbilical hydraulic supply flex hose break and range conflicts with Delta and Atlas launches. LAUNCH SCRUBS AND PAD ABORT #3: - 3/21/93 launch date was scrubbed on 3/18/93 shortly after countdown started because of Delta launch scrub due to high winds. - 3/22/93 launch was scrubbed with a pad abort at T-3 seconds when SSME 3 (S.N. 2011) oxidizer preburner shutdown pressure exceeded 50 psi limit. Oxidizer preburner ASI purge checkvalve (N9) failed to close due to contamination. Decision was made to replace all three SSME's and moved STS-56 ahead of STS-55 (PAD abort #3). - Replaced all 3 engines at pad. - 4/24/93 launch scrubbed after tanking at L-6.5 hours due to an IMU-2 failed BITE test. LAUNCH DELAYS: None. DOLILU/I-LOADS: - Both nominal and DOLILU were go. DOLILU selected because of increased Q-plane margin at Mach 1.55. DOLILU uplink #7, total I-load uplink #14. FLIGHT DURATION CHANGES: - Extended 1 day for additional science. - Extended one rev because of forecast variable broken ceiling and changed landing site to EDW concrete. LANDING SITE CHANGE: KSC to EDW. FIRSTS: First flight of operational TIPS. DRAG CHUTE: - Baseline chute used with strategy to deploy at derotation similar to STS-56. SIGNIFICANT ANOMALIES: - RSRM 6 PSI pressure spike at 69 seconds MET. - LSRM 10-12 PSI pressure spike at 71 seconds. - S/L DDS 1 and 2 problems. - MMU 1 SM checkpoint fail transient. - CRT-4 I/O error (lost aft CRT), CRT-1 dim. - Waste water tank outer shell punctured. Used CWC for wastewater. - FES primary A shut down (ice in core). - ARD Sys parameter incorrect during first launch attempt. - TV camera and WCCU anomalies. - L4D RCS jet heater fail on. - Right OMS GN₂ accumulator leak. - Prime OR/F (refrigerator/freezer) failed to operate. - Enhanced OR/F had thermal problems. - 48 total payload anomalies written. RADIATOR DEPLOYED #14

D-2 mission patch (PRECOURT NAGEL HENRICKS / SCHLEGEL HARRIS ROSS WALTER)

STS055-106-056 - German payload specialists Walter and Schlegel at work in SL-D2.

STS055-203-009 1993-05-06 Inflight crew portrait in SL- Deutsche 2 science module. Front (lt to rt) PLT Henricks, CDR Nagel, Walter/PS1 (Germany) & Precourt/MS2. Rear (lt to rt) Harris/MS3, Schlegel/PS2 (Germany), & Ross/MS1/ PLC.

FLT NO.	ORBITER	CREW (6) TITLE, NAMES & EVA'S	LAUNCH SITE, LIFTOFF TIME, LANDING SITES, ABORT TIMES	LANDING SITE/ RUNWAY, CROSSRANGE LANDING TIMES FLT DURATION, WINDS	SSME-TL NOM-ABORT EMERG THROTTLE PROFILE ENG. S.N.	SRB RSRM AND ET	INC	ORBIT HA/HP	FSW	PAYLOAD WEIGHTS, PAYLOADS/ EXPERIMENTS	MISSION HIGHLIGHTS (LAUNCH SCRUBS/DELAYS, TAL WEATHER, ASCENT I-LOADS, FIRSTS, SIGNIFICANT ANOMALIES, ETC.)
STS-57 SEQ FLT #56 KSC-56 PAD 39B-20 MLP-2	OV-105 (Flight 4) Endeavour Spacehab 1 OMS PODS LPO3-15 RPO4-11 FRC5-4	CDR: Ronald J. Grabe (Flt 4 - STS 51-J, STS-30 & STS-42) P305/R76/V41/M70 PLT: Brian Duffy (Flt 2 - STS-45) P306/R142/V94/M126 M/S 1 (PAYLOAD CDR): G. David Low (Flt 3 - STS-32 & STS-43) - P307/R110/V64/M98 M/S 2: Nancy J. Sherlock P308/R165/F21 M/S 3: Peter J. K. (Jeff) Wisoff P309/R166/M145 M/S 4: Janice E. Voss P310/R167/F22 EMU/TETHERED EVA: EV 1: G. David Low EV 2: Jeff Wisoff EVA 1 - 6/25/93 5:50 Duration Continued. . .	KSC 39B 172:13:07:21.95Z 9:07:00 AM EDT (P) 9:07:22 AM EDT (A) Monday 9 6/21/93 (6) LAUNCH WINDOW: 71M48S PLANAR/ PHASE WINDOW EOM PLS: KSC TAL: BYD TAL WX: BEN, MRN SELECTED: RTLS: KSC15/CI/N TAL: BEN36/N/N AOA: EDW22/CI/N PLS: EDW22/CI/N TDEL: 0.00 0.722/0.76 MAX Q NAV: 695 PSF 722 PSF SRB STG: 2:04 2:06 PERF: NOMINAL 2 ENG TAL (BEN): 2:33 2:37 NEG RETURN: 3:45 4:07 PTA (U/S 395): 4:10 4:12 PTM (U/S 427): 5:32 5:31 MECO CMD: 8:32.47 8:33 VI: 26028 26025 OMS-2: 42:11.7 42:13 318 FPS 316 FPS	KSC 33 (KSC 16) 182:12:52:16Z 8:52:16 AM EDT Thursday 7 7/1/93 (6) DEORBIT BURN: 182:11:41:42Z XRANGE: 587 NM ORBIT DIR: DL 29 AIM PT: CLOSE IN MLGTD: 2296 FT 182:12:52:16Z VEL: 202 KGS 207 KEAS HDOT: -1.0 FPS TD NORM 205: 2461 FT DRAG CHUTE DEPLOY: 175 KEAS 182:12:52:25Z NLGTD: 7498 FT 182:12:52:34Z VEL: 135 KGS HDOT: -3.4 FPS BRK INIT: 101 KGS DRAG CHUTE JETTISON: 56 KGS 182:12:52:57Z AVE BRK DECEL: 4.4 FPS/S WHEELS STOP: 182:12:53:21Z 12251 FT ROLLOUT: 9955 FT 65 SEC Continued. . .	104/104/ 109% PREDICTED: 100/100/100/ 67/104 ACTUAL: 100/100/100/ 72/104 1 = 2019 (12) 2 = 2034 (2) 3 = 2017 (8) M 3 EOM: WEIGHT: 224752 LBS X CG: 1081.1 LANDING WEIGHT: 224468 LBS X CG: 1082.5	BI-059 RSRM 32 KM ET-58 LWT 51 ET RPT ET BR/UP ET IMPACT LAT: 16.09 N LONG: 142.90 W	28.45 (33)	DIRECT INSERTION POST OMS-2: 252 X 212 NM NC3 BURN: 56 FPS 2:04:00:35 MET 257/251 NM TI BURN: 258 X 255 NM ORB ADJ 3: 3:19:01 MET 256 X 209 NM DEORBIT: 256 X 208 NM VELOCITY: 25988 FPS ENTRY RANGE: 4210 NM	OI-22 (1)	CARGO: 29119 LBS PAYLOAD CHARGEABLE: 19630 LBS DEPLOYED: 132 LBS NON-DEPLOYED: 18244 LBS MIDDECK: 1254 LBS SHUTTLE ACCUMULATED WEIGHTS: DEPLOYED: 737233 LBS NON-DEPLOYED: 715721 LBS CARGO TOTAL: 1670852 LBS PERFORMANCE MARGINS (LBS): FPR: 3934 FUEL BIAS: 1055 FINAL TDDP: 2030 RECON: 2162 PAYLOADS: PLB: SPACEHAB-1 EURECA CAPTURE AND RETURN SHOOT, GBA, CONCAP-IV MIDDECK: FARE AMOS SAREX-II 4 CRYO TK SETS RMS 31 (S.N. 303) RMS used to grapple and berth EURECA and EVA DTO	KSC W/D: OPF 52, VAB 16, PAD 51 = 119 days total. LAUNCH POSTPONEMENTS: - Launch date was 5/10/93 then postponed to 5/18/93. - Launch date was postponed from 5/18/93 to 6/3/93 because of STS-55 and STS-56 launch delays. - Launch date was postponed from 6/3/93 to 6/20/93 because SSME 3 HPOTP required changeout (QA electrochemical etch marking found in a high stress area of HPOTP turbine bearing preload spring). LAUNCH SCRUBS: - 6/20/93 launch was scrubbed during hold at T-5 minutes when 71 minute 48 second launch window expired. All three TAL sites were NO-GO (Banjul for thunderstorms and Ben Guerir and Moron for crosswind exceedences.) LAUNCH DELAYS: - Launch delayed 22 seconds because of an intruder aircraft. Countdown was at T-5 minutes awaiting a GO for RTLS weather when the aircraft entered KSC airspace (Launch danger area). TAL WX: - Banjul was forecast and observed NO GO for ceiling and rain. Ben Guerir (selected) was forecast and observed GO. Moron was forecast NO GO for ceiling, rain, and crosswinds but was observed GO. DOLILU/I-LOADS: - Nominal I-loads were GO and selected because of better Q-plane than DOLILU. No uplink required. FLIGHT DURATION CHANGES: 3 days extension - Extended 1 day for additional science. - Extended 1 day because of forecast low ceiling on rev 124 and convective development and potential thunderstorms on rev 125. - Extended 1 day because of forecast thunderstorms on revs 139 and 140. FIRSTS/LASTS: - Last flight of TAGS, next to last flight of teleprinter. - First flight of the improved APU controller (APU #2). - Last flight of drag chute without ribbons removed. (Was second flight with 90 percent reefed.) EVENTS: - Started EVA at 3:23:59:51 MET (planned 4 hours). David Low pushed on EURECA antenna and ESOC commanded latches. David had to move antennas in "z" to get them latched. Both antennas confirmed latched at EVA time of 2:25, when they started the scheduled EVA DTO 1210. (EURECA deployed on STS-46) Continued. . .

STS057-94-017 1993-07-01 Front row left to right: Wisoff/MS3, PLT Duffy, Voss/MS4. In rear (left to right): CDR Grabe, Sherlock/MS2 and Low/MS1/PLC.

FLT NO.	ORBITER	CREW (6) TITLE, NAMES & EVA'S	LAUNCH SITE, LIFTOFF TIME, LANDING SITES, ABORT TIMES	LANDING SITE/ RUNWAY, CROSSRANGE LANDING TIMES FLT DURATION, WINDS	SSME-TL NOM-ABORT EMERG THROTTLE PROFILE ENG. S.N.	SRB RSRM AND ET	INC	ORBIT HA/HP	FSW	PAYLOAD WEIGHTS, PAYLOADS/ EXPERIMENTS	MISSION HIGHLIGHTS (LAUNCH SCRUBS/DELAYS, TAL WEATHER, ASCENT I-LOADS, FIRSTS, SIGNIFICANT ANOMALIES, ETC.)
STS-57 Continued		Continued. . . SPACE SHUTTLE EVA #21 SCHEDULED EVA #17 REFINE EVA TRAINING CONCEPTS AND DEMON-STRATE EVA TECHNIQUES FOR FUTURE EVA'S. ADDED UNSCHEDULED MANUAL LATCHING OF EURECA ANTENNAS MCC FCR-1 (35) FLIGHT DIRECTORS: A/E - J. W. Bantle LD/O 1 - G. A. Pennington O 2 - P. L. Engelauf PLNG - R. M. Kelso MOD - G. E. Coen		Continued. . . WINDS: H6, L2 KTS OFFICIAL: H10, L2 DENS ALT: 1571 FT FLT DURATION: 9:23:44:54 239:44:54 S/T: 375:23:12:22 OV-105: 32:19:11:12 DISTANCE: 4,118,037 sm							Continued. . . RENDEZVOUS #14: - Rendezvous with EURECA for capture, retrieval, and return. SIGNIFICANT ANOMALIES: - O_2 manifold valve tank 1 failed to close. - Fuel cell 3 H_2 reactant valve failed to close. - PPO2 sensor B is biased low. - MCA logic MCA power AC3 3-phase mid 4 CB anomaly. - AC3 phase-to-phase short/Spacehab PDU fuses blown and replaced (command error). - Mid starboard and aft port floodlights failure. - EVA waist tether small tether hook failure. - Leaking EMU 1200-series battery. - RMS grapple fixture/EURECA thermal control unit switch problem (installed reversed). - Jet R5D heater failed on. - EURECA antennas failed to latch (crew manually latched them during planned EVA). - S-band intermittent forward and return links on lower left quad antenna. - Ammonia boilers failed to cool post landing.

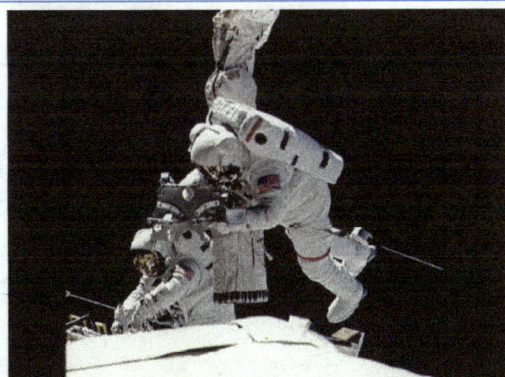

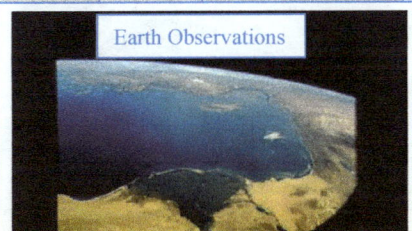

Above: STS057-97-056 1993-07-01 -- Low and Wisoff perform DTO 1210 EVA in OV-105's payload bay .

Earth Observations

ABOVE: STS057-80-09 --- Agriculural development in Rio Bermejo, Argentina.
BELOW: STS057-73-075 --- Eastern Mediterranean, Nile River, Asia Minor - looking north over the Nile.

STS057-93-052 1993-07-01 EURECA is retrieved by RMS to be stowed in PLB for return to earth.

sts057-s-089 -- Post mission in the MCC are Greg Smith/FAO (Flight Activities Officer), holding mission plaque, and CAPCOM Curt Brown (right).

FLT NO.	ORBITER	CREW (5) TITLE, NAMES & EVA'S	LAUNCH SITE, LIFTOFF TIME, LANDING SITES, ABORT TIMES	LANDING SITE/ RUNWAY, CROSSRANGE LANDING TIMES FLT DURATION, WINDS	SSME-TL NOM-ABORT EMERG THROTTLE PROFILE ENG. S.N.	SRB RSRM AND ET	INC	ORBIT HA/HP	FSW	PAYLOAD WEIGHTS, PAYLOADS/ EXPERIMENTS	MISSION HIGHLIGHTS (LAUNCH SCRUBS/DELAYS, TAL WEATHER, ASCENT I-LOADS, FIRSTS, SIGNIFICANT ANOMALIES, ETC.)
STS-51 SEQ FLT #57 KSC-57 PAD 39B-21 MLP-3	OV-103 (Flight 17) Discovery OMS PODS LPO1-20 RPO3-18 FRC3-17	CDR: Frank L. Culbertson (Flt 2 - STS-38) P311/R116/V95/M104 PLT: William F. Readdy (Flt 2 - STS-42) P312/R140/V96/M125 M/S 1: James H. Newman P313/R168/M146 M/S 2: Daniel W. Bursch P314/R169/M147 M/S 3: Carl E. Walz P315/R170/M148 SPACE SHUTTLE EVA #22 SCHEDULED EVA #18 DTO 1210 EVA OPERATIONS/ PROCEDURES/TRAINING FOR FUTURE EVA'S EMU/TETHERED EVA: EV 1: Carl Walz EV 2: Jim Newman 9/16/93 7:05:28 Duration MCC FCR-1 (36) FLIGHT DIRECTORS: A/E - R. D. Jackson LD/O 1 - R. E. Castle O 2 - R. M. Kelso PLNG - N. W. Hale MOD - B. R. Stone	KSC 39B 255:11:44:59.97Z 7:45:00 AM EDT (P) 7:45:00 AM EDT (A) Sunday 7 9/12/93 (4) LAUNCH WINDOW: 1H55M ACTS/TOS RAAN ORBIT 23A EOM PLS: KSC TAL: BYD TAL WX: BEN SELECTED: RTLS: KSC15/CI/N TAL: BEN36/N/N AOA: EDW22/CI/N PLS: EDW22/CI/N TDEL: 0.16 0.322 MAX Q NAV: 700 PSF 707 PSF SRB STG: 2:04.6 2:05.0 PERF: NOMINAL 2 ENG TAL (BEN): 3:15 3:12 NEG RETURN: 3:56 3:59 PTA (U/S 245): 5:15 5:07 PTM (U/S 245): 6:12 6:06 MECO CMD: 8:28.15 8:29.8 VI: 25873 25874 OMS-2: 39:53.7 39:53.7 222 FPS 222 FPS	KSC 15 (KSC 17) 265:07:56:06Z 3:56:06 AM EDT Wednesday 6 9/22/93 (6) DEORBIT BURN: 265:06:55:30Z XRANGE: 89 NM ORBIT DIR: DL 30 AIM PT: CLOSE IN MLGTD: 2099 FT VEL: 198 KGS 194 KEAS HDOT: -1.0 FPS TD NORM 195: 2080 FT DRAG CHUTE DEPLOY: 165 KEAS 265:07:56:16Z NLGTD: 6539 FT 265:07:56:21Z VEL: 144 KGS HDOT: -3.9 FPS BRK INIT: 113 KGS DRAG CHUTE JETTISON: 47 KGS 265:07:56:43Z AVE BRK DECEL: 6.9 FPS/S WHEELS STOP: 265:07:56:56Z 10370 FT ROLLOUT: 8271 FT 50 SEC WINDS: T2, L1 KTS OFFICIAL: H2, L1 DENS ALT: 1049 FT FLT DURATION: 9:20:11:06 236:11:06 S/T: 385:19:23:28 OV-105: 109:13:02:15 DISTANCE: 4,106,411 sm	104/104/ 109% PREDICTED: 100/104/104/ 67/104 ACTUAL: 100/104/104/ 69/104 M 3 EOM WEIGHT: 207043 LBS X CG: 1084.8 LANDING: WEIGHT: 206932 LBS X CG: 1086.5	BI-060 RSRM 33 ET-59 LWT 52 ET RPT ET BR/UP ET IMPACT LAT: 12.89 N LONG: 163.4 W	28.45 (34)	DIRECT INSERTION POST OMS-2: 161.1 X 160.3 NM ACTS/TOS DEPLOY: 0/7:58:09 MET (P) 0/9:28:28 MET (A) 173.5 X 160.9 NM ORFEUS-SPAS DEPLOY: 1/03:21:00 MET 164.6 X 147.2 NM ORFEUS-SPAS GRAPPLE: 7/00:05 MET DEORBIT: 166 X 141 NM VELOCITY: 25794 FPS ENTRY RANGE: 4250 NM	OI-22 (2)	CARGO: 46685 LBS PAYLOAD CHARGEABLE: 42637 LBS DEPLOYED: 26889 LBS NON-DEPLOYED: 7305 LBS MIDDECK: 1122 LBS SHUTTLE ACCUMULATED WEIGHTS: DEPLOYED: 764122 LBS NON-DEPLOYED: 724148 LBS CARGO TOTAL: 1717537 LBS PERFORMANCE MARGINS (LBS): FPR: 3934 FUEL BIAS: 1055 FINAL TDDP: 1358 RECON: 1273 PAYLOADS: PLB: ACTS/TOS (DEPLOYED) ORFEUS-SPAS (DEPLOYED AND RETRIEVED) LDCE (2 CANS) MIDDECK: IMAX CPCG - BLOCK-II CHROMEX, HRSGS-A, APE-B, IPMP, RME-III, AMOS 4 CRYO TK SETS RMS 32 (S.N. 201) RMS USED FOR SPAS DEPLOY, GRAPPLE AND REBERTH	KSC W/D: OPF 57, VAB 8, PAD 69 = 134 days total. LAUNCH POSTPONEMENTS: - Launch date was 2/22/93 as of 6/28/91 but was postponed to 6/30/93 on 7/32/92 to reflect changes in manifest. - 6/30/93 launch was postponed to 7/13/93 on 3/31/93 based on STS-55, STS-56, and STS-57 launch delays. - 7/13/93 launch was postponed to 7/17/93 because of STS-57 launch delays. - 8/4/93 launch date was postponed on 7/30/93 to avoid Perseid Meteoroid (Comet Swift-Tuttle) event on 8/11/93. Launch rescheduled for 8/12/93. (See 8/12/93 scrub below.) - 9/10/93 launch postponed to 9/12/93 on 9/3/93 to allow ACTS/TOS to complete a review/analysis of transistor alert (suspected as potential cause of NOAA-I and MARS Observer failures). LAUNCH SCRUBS/PAD ABORT #4: - 7/17/93 launch was scrubbed at L-31 minutes. At approximately L-2 hours, nine "B" systems PIC's indicated they were charged (four on each SRB holddown post and one on ET vent arm). - 7/24/93 launch was scrubbed at T-19 seconds with an RSLS breakout caused by right SRB tilt HPU underspeed. - 8/12/93 launch aborted at T-3 seconds when SSME #2 (S.N. 2033) fuel flow sensor A2 miscompared with sensor A1. (Pad abort #4.) Launch reset to 9/10/93. Replaced all 3 engines at pad. TAL WX: Banjul (prime) was forecast and observed NO-GO - ceiling. Ben Guerir (selected)was forecast and observed GO. DOLILU/I-LOADS: Both nominal and DOLILU I-loads were GO but DOLILU was selected and uplinked to provide a slight increase in performance and drainback time. DOLILU uplink #8, I-load uplink #15. FLIGHT DURATION CHANGES: - Waved off rev 142 landing at KSC because of rain within 30 nm. Extended flight 1 day minus 1 rev. (Total extension 15 revs.) FIRSTS: - First flight of drag chute with five ribbons removed. - First flight with night landing at KSC. - First flight with wake up music (used Heartbreak Hotel by Carl Walz) sung by a crewmember. - First flight with two U.S. and two Russian EVA's at same time. EVENTS: Fuel cell 1 shut down for 24 hours for DTO 412. RENDEZVOUS #15: - Rendezvous with ORFEUS-SPAS for grapple, berth, and return. NIGHT LANDING: Space Shuttle #6, first night landing at KSC. SIGNIFICANT ANOMALIES: - Right SRB tilt HPU underspeed problem. (Scrub #2.) - SSME #2 fuel flow sensor A2 failed low. (Scrub #3.) - FA2 MDM BITE. - EECOM-01 - Loose thermal blanket on aft bulkhead. - PSA slider door stuck open. - Thruster L3L failed off. - Thruster R1R chamber pressure transducer failure (post-flight found fuel/oxidizer reaction products (FORP) in tube.) - TOS SuperZip damage, both detonation cords fired simultaneously damaging 1307 bulkhead and PLB blankets. - Humidity separator B water carryover.

STS051-44-005 In-flight crew portrait (lt to rt): PLT Readdy, Bursch/MS, CDR Culbertson, Walz/MS & Newman/MS.

Top: STS051-06- 037 Newman & Walz evaluate tools for HST servicing mission. Bottom: STS051(S)158 First night landing at KSC.

FLT NO.	ORBITER	CREW (7) TITLE, NAMES & EVA'S	LAUNCH SITE, LIFTOFF TIME, LANDING SITES, ABORT TIMES	LANDING SITE/ RUNWAY, CROSSRANGE LANDING TIMES FLT DURATION, WINDS	SSME-TL NOM-ABORT EMERG THROTTLE PROFILE ENG. S.N.	SRB RSRM AND ET	ORBIT INC	ORBIT HA/HP	FSW	PAYLOAD WEIGHTS, PAYLOADS/ EXPERIMENTS	MISSION HIGHLIGHTS (LAUNCH SCRUBS/DELAYS, TAL WEATHER, ASCENT I-LOADS, FIRSTS, SIGNIFICANT ANOMALIES, ETC.)
STS-58 SEQ FLT #58 KSC-58 PAD 39B-22 MLP-1	OV-102 (Flight 15) Columbia SLS-2/LM 13th Spacelab Flight Long Module 9 EDO 2 OMS PODS LPO5-4 RPO5-4 FRC2-15	CDR: John E. Blaha (Flt 4 - STS-29, STS-33 & STS-43) P316/R97/V48/M88 PLT: Richard A. Searfoss P317/R171/M149 M/S 1 (PAYLOAD CDR): M. Rhea Seddon (Flt 3 - STS 51-D & STS-40) P318/R55/V63/F5 M/S 2: William S. McArthur P319/R172/M150 M/S 3: David A. Wolf P320/R173/M151 M/S 4: Shannon W. Lucid (Flt 4 - STS 51-G, STS-34 & STS-43) P321/R65/V45/F6 P/S 1: Martin J. Fettman P322/R174/M152 Colorado State University MCC FCR-1 (37) FLIGHT DIRECTORS: A/E - N. W Hale LD/O 1 - L. J. Ham O 2 - P. L. Engelauf O 3 - G. E. Coen O 4 - J. F. Muratore MOD - A. L. Briscoe	KSC 39, PAD B 291:14:53:09.97Z 10:53:00 AM EDT (P) 10:53:10 AM EDT (A) Monday 10 10/18/93 (7) LAUNCH WINDOW: 2H 30M, CTOB EOM PLS: EDW TAL: BEN TAL WX: MRN, ZZA SELECTED: RTLS: KSC33/N/N TAL: BEN36/N/N AOA: EDW22/N/N PLS: EDW22/N/N TDEL: 0.00 0.82/0.12 MAX Q NAV: 687 PSF 684 PSF SRB STG: 1:58.9 1:59 PERF: NOMINAL 2 ENG TAL (BEN): 2:50 2:53 NEG RETURN: 4:02 4:06 PTA (U/S 218): 5:30 5:30 PTM (U/S 218): 6:19 6:18 MECO CMD: 8:33.5 8:36 VI: 25867 25862 OMS-2: 41:41 41:55 200 FPS 198 FPS	EDW 22 CONC (EDW 39, CONC 20) 305:15:05:42Z 7:05:42 AM PST Monday 11 11/1/93 (8) DEORBIT BURN: 305:14:05:30Z XRANGE: 144 NM ORBIT DIR: DR 9 AIM PT: NOMINAL MLGTD: 3380 FT VEL: 205 KGS 198 KEAS HDOT: -2.2 FPS TD NORM 205: 2800 FT DRAG CHUTE DEPLOY: 173 KEAS 305:15:05:51Z NLGTD: 6948 FT 305:15:05:53Z VEL: 167 KGS HDOT: -3.7 FPS BRK INIT: 138 KGS DRAG CHUTE JETTISON: 47 KGS 305:15:06:25Z AVE BRK DECEL: 5.5 FPS/S WHEELS STOP: 305:15:06:44Z 13020 FT ROLLOUT: 9640 FT 62 SEC WINDS: T2, R1 KTS OFFICIAL: T2, R2 DENS ALT: 1827 FT FLT DURATION: 14:00:12:32 336:12:32 S/T: 399:19:36:00 OV-102: 122:17:33:12 DISTANCE: 5,840,450 sm	104/104/ 109% PREDICTED: 100/100/100/ 67/104 ACTUAL: 100/100/100/ 69/104 1 = 2024 (7) 2 = 2109 (11) 3 = 2018 (11) M 3 EOM WEIGHT: 229481 LBS X CG: 1078.8 LANDING: WEIGHT: 229369 LBS X CG: 1080.4	BI-061 RSRM 34 ET-57 LWT 50 ET PRED RPT LH2 TK RPT AT 285.2K LO2 TK RPT AT 283K ET IMPACT 1:25:22 MET LAT: 3.9 N LONG: 173.8 W	39.0 (2)	DIRECT INSERTION POST OMS-2: 155 X 154 NM DEORBIT: 151 X 136 NM VELOCITY: 25755 FPS ENTRY RANGE: 4378 NM	OI-22 (3)	CARGO: 32011 LBS PAYLOAD CHARGEABLE: 23127 LBS DEPLOYED: 0 LBS NON-DEPLOYED: 23127 LBS MIDDECK: 1373 LBS SHUTTLE ACCUMULATED WEIGHTS: DEPLOYED: 764122 LBS NON-DEPLOYED: 747275 LBS CARGO TOTAL: 1749548 LBS PERFORMANCE MARGINS (LBS): FPR: 3934 FUEL BIAS: 1055 FINAL TDDP: 767 RECON: 1114 PAYLOADS: PLB: SPACELAB LIFE SCIENCES (SLS-2/LM) Cardiovascular/ Cardiopulmonary, Neurovascular, and Regulatory Physiology Experiments MIDDECK: SAREX-II 4 CRYO TK SETS + 4 EDO SETS NO RMS	KSC W/D: OPF 82, VAB 17, PAD 28 = 127 days total. LAUNCH POSTPONEMENTS: - Launch date was 8/25/93 as of 7/31/92. - Postponed launch to 9/10/93, then 10/7/93 because of STS-55, STS-56, STS-57, and STS-51 launch delays. Postponed launch to 10/14/93 to replace two APUs. LAUNCH SCRUBS: - Scrubbed 10/14/93 launch at 16:57:20Z while holding at T-31 seconds when drainback time expired with 25M40S left in launch window. Scrub caused by range safety command system problem, and KSC weather caused lengthy hold. - Scrubbed 10/15/93 launch caused by S-Band PM transponder 2 problem. Rescheduled launch for 10/18/93 to change out transponder. LAUNCH DELAYS: - 10/18/93 launch delayed 10 seconds at T-5 minutes because of intruder aircraft in launch area. TAL WX: - Ben Guerir - prime and selected, Moron forecast and observed GO, Zaragoza forecast and observed NO-GO - rain. DOLILU/I-LOADS: - Nominal I-loads were selected. FLIGHT DURATION CHANGES: None. EVENTS: Special attitude flown for OARE data on FD 12. RECORDS: - Longest Shuttle flight - 14:00:12:32 - exceeds STS-50 by 4H 42M 28S (only exceeded by SKYLAB flights). - Shannon Lucid set Shuttle flight time record - 34:22:52:09. SIGNIFICANT ANOMALIES: - S-band transponder 2 uplink failure on second launch attempt (changed out for flight). - S-Band FM transmitter power output degraded. - Engine 1 and 2 dome-mounted heat shield blanket damage. - External tank intertank acreage loss of TPS. - Water leak at WCS/odor/bacteria filter, switched to WCS fan sep 2 (low torque), performed IFM using wand to remove water. - False low battery beep from AIU. - Payload recorder tape broke during track change. - Spacelab overhead container OH5 jammed. - LOMS PC failed off scale low. - RAHF-7 quad temps high - FCL FPV to P/L.

STS058-16-008 Clockwise from top: Seddon/PLC, Lucid/MS, McArthur/MS, Fettman/PS, Wolf/MS, PLT Searfoss, & CDR Blaha.

STS058-92-064 1993-10-30 SPACELAB-2 in PLB flys over northeast Egypt.

FLT NO.	ORBITER	CREW (7) TITLE, NAMES & EVA'S	LAUNCH SITE, LIFTOFF TIME, LANDING SITES, ABORT TIMES	LANDING SITE/ RUNWAY, CROSSRANGE LANDING TIMES FLT DURATION, WINDS	SSME-TL NOM-ABORT EMERG THROTTLE PROFILE ENG. S.N.	SRB RSRM AND ET	INC	ORBIT HA/HP	FSW	PAYLOAD WEIGHTS, PAYLOADS/ EXPERIMENTS	MISSION HIGHLIGHTS (LAUNCH SCRUBS/DELAYS, TAL WEATHER, ASCENT I-LOADS, FIRSTS, SIGNIFICANT ANOMALIES, ETC.)
STS-61 SEQ FLT #59 KSC-59 PAD 39B-23 MLP-2	OV-105 (Flight 5) Endeavour OMS PODS LPO3-16 RPO4-12 FRC5-5	CDR: Richard O. Covey (Flt 4 - STS 51-I, STS-26 & STS-38) P323/R73/V30/M67 PLT: Kenneth D. Bowersox (Flt 2 - STS-50) P324/R146/V97/M130 M/S 1 AND EV3: Kathryn C. Thornton (Flt 3 - STS-33 & STS-49) P325/R107/V73/F11 M/S 2: Claude Nicollier (Flt 2 - STS-46) P326/R150/V98/M134 Switzerland M/S 3 AND EV 1: Jeffrey A Hoffman (Flt 4 - STS 51-D, STS-35 & STS-46) P327/R57/V59/M52 M/S 4, P/L CDR & EV 2: F. Story Musgrave (Flt 5 - STS-6, STS 51-F, STS-33 & STS-44) (P328/R15/V19/M15 M/S 5 AND EV 4: Thomas D. Akers (Flt 3 - STS-41 & STS-49) P329/R115/V74/M103 MCC FCR-1 (38) FLIGHT DIRECTORS: A/E - R. D. Jackson LD/O 2-EVA - J. M. Heflin O 2-SYS - J. W. Bantle O 1 - R. E. Castle PLNG - J. F. Muratore MOD - B. R. Stone Continued . . .	KSC 39, PAD B 336:09:26:59.95Z 4:27:00 AM EST (P) 4:27:00 AM EST (A) Thursday 15 12/2/93 (4) LAUNCH WINDOW: 67 MINUTES, PLANAR WINDOW EOM PLS: KSC TAL: BYD TAL WX: BEN,MRN SELECTED: RTLS: KSC15/N/N TAL: BEN32/N/SF AOA: EDW04/N/N PLS: EDW04/N/N TDEL: 0.32 0.402/.44 MAX Q NAV: 701 PSF 705 PSF SRB STG: 2:05.6 2:07 PERF: NOMINAL 2 ENG TAL (BYD): 2:08 2:07 NEG RETURN: 4:04 4:07 PTA (U/S 500): 4:02 4:07 PTM (U/S 500): 5:24 5:18 MECO CMD: 8:32.8 8:31.9 VI: 26123 26115 OMS-2: 42:39 43:30 322 FPS 324 FPS TGO: 3:18 3:20	KSC 33 (KSC 18) 347:05:25:33Z 00:25:33 AM EST Monday 12 12/13/93 (7) DEORBIT BURN: 347:04:14:45Z XRANGE: 3 NM ORBIT DIR: AR 6 AIM PT: NOMINAL MLGTD: 2903 FT 347:05:25:33Z VEL: 192 KGS 201 KEAS HDOT: -1.7 FPS TD NORM 195: 3415 FT DRAG CHUTE DEPLOY: 170 KEAS 347:05:25:41Z NLGTD: 6635 FT 347:05:25:45Z VEL: 148 KGS HDOT: -3.5 FPS BRK INIT: 118 KGS DRAG CHUTE JETTISON: 49 KTS 347:05:26:08Z AVE BRK DECEL: 6.6 FPS/S WHEELS STOP: 347:05:26:26Z 10825 FT ROLLOUT: 7922 FT 53 SEC WINDS: 6H, 0X KTS OFFICIAL: H7, L1 Continued . . .	104/104/ 109% PREDICTED: 100/100/100/ 74/104 ACTUAL: 100/100/100/ 73/104 1 = 2019 (13) 2 = 2033 (5) 3 = 2017 (9) M 3 EOM: WEIGHT: 212947 LBS X CG: 1078.9 LANDING: WEIGHT: 212836 LBS X CG: 1080.6	BI-063 RSRM 23 ET-60 LWT 53 ET PRED RPT 285 K ET BR/UP 214 K ET IMPACT 1:29:01 MET LAT: 16.4 N LONG: 142.1 W	28.45 (35)	DIRECT INSERTION POST OMS-2: 308.4 X 214.4 NM RNDZ BRAKING: 1:22:34:49 MET 319.6 X 313.4 NM ARRAY JETTISON: 3:19:26:00 MET 320.5 X 313.2 NM HST REBOOST: 6:16:59:23 MET 321.7 X 320.8 NM DEORBIT: 320.4 X 319.3 NM VELOCITY: 26096 FPS ENTRY RANGE: 4220 NM	OI-22 (4)	CARGO: 24363 LBS PAYLOAD CHARGEABLE: 17401 LBS DEPLOYED: 2308 LBS NON-DEPLOYED: 14428 LBS MIDDECK: 665 LBS SHUTTLE ACCUMULATED WEIGHTS: DEPLOYED: 766430 LBS NON-DEPLOYED: 762368 LBS CARGO TOTAL: 1773911 LBS PERFORMANCE MARGINS (LBS): FPR: 3981 FUEL BIAS: 987 FINAL TDDP: 927 RECON: 554 PAYLOADS: PLB: HUBBLE SPACE TELESCOPE (HST) SERVICING MISSION (SM-1) (REPLACEMENT HARDWARE) ICBC MIDDECK: IMAX AMOS 5 CRYO TK SETS RMS 33 (S.N. 303) RMS USED FOR HST GRAPPLE, SERVICE, AND DEPLOY, AND EVA WORK PLATFORM	KSC W/D: OPF 103, VAB 6, PAD 33 = 142 days total. LAUNCH POSTPONEMENTS: - Launch date was 12/2/93 as of 7/17/92. - Launch date was changed to 12/7/93, then 12/2/94, then 12/1/93 on 10/25/93. - Moved from Pad A to Pad B to protect payload from contamination caused by Pad A sandblasting. LAUNCH SCRUBS: - 12/1/93 launch was scrubbed while holding at T-5 minutes when 67-minute window expired. Primary causes of delay were RTLS crosswind exceedence and rain within 20 nm. Other factors were BLAST, COLA, ceiling violation (6.5K broken), and intruder ship in SRB recovery area. LAUNCH DELAYS: None. TAL WX: - Banjul, Ben Guerir, and Moron all forecast and observed GO. DOLILU/I-LOADS: - DOLILU uplink #9, I-load uplink #15. NIGHT LAUNCH: Shuttle night launch #9. FLIGHT DURATION CHANGES: - Shortened flight one rev because cloud cover forecast to move in at nominal landing time. FIRSTS: - First flight with four EVA crewmembers. - First flight with five EVA's (alternating crew on alternating days). - Minimum shuttle crossrange (3 nm). RENDEZVOUS #16: - Rendezvous with HST for grapple, berth, repair, and deploy. NIGHT LANDING: Space Shuttle #7, second night landing at KSC. SIGNIFICANT ANOMALIES: - Aft mission timer circuit breaker popped. - In-suit drink bags leaked. - Large in-suit drink bags not stowed. - EMU 3 intermittent loss of 298.6 receive and all hardline comm. - HST power tool S.N. 1001 failed. - EMU 2 failed 0.5 psi leak check. - -Y star tracker temporary loss. - APU 2 gas generator/fuel pump heater failure. - Right OMS helium tank pressure transducer P2 bias - Jet L2U failed off. - Loss of biomed data on EMU 2 during EVA #5. - +V2 solar array outer bi-stem bowed, hence jettisoned old array. - Missing TPS on forward edge of RSRM RH forward center segment.

STS061-05-031   Crew: Lt to Rt, Musgrave/MS, CDR Covey, Nicollier/MS, Hoffman/MS, PLT Bowersox, Thornton/MS, and Akers/MS.

FLT NO.	ORBITER	CREW (7) TITLE, NAMES & EVA'S	LAUNCH SITE, LIFTOFF TIME, LANDING SITES, ABORT TIMES	LANDING SITE/ RUNWAY, CROSSRANGE LANDING TIMES FLT DURATION, WINDS	SSME-TL NOM-ABORT EMERG THROTTLE PROFILE ENG. S.N.	SRB RSRM AND ET	ORBIT INC	HA/HP	FSW	PAYLOAD WEIGHTS, PAYLOADS/ EXPERIMENTS	MISSION HIGHLIGHTS (LAUNCH SCRUBS/DELAYS, TAL WEATHER, ASCENT I-LOADS, FIRSTS, SIGNIFICANT ANOMALIES, ETC.)
STS-61 Continued		Continued . . . EMU/TETHERED EVA'S: EVA #1 - 12/4/93 SPACE SHUTTLE EVA #23 SCHEDULED EVA #19 BY EV 1 & EV 2 REPLACED RSU'S 2 & 3, ESU'S 1 & 3 AND RELATED GYRO FUSE PLUGS. 7H53M57S EVA #2 - 12/5/93 SPACE SHUTTLE EVA #24 SCHEDULED EVA #20 BY EV 3 & EV 4 REPLACED BOTH SOLAR ARRAYS, OLD +V2 ARRAY JETTISONED 6H35M3S EVA #3 - 12/6/93 SPACE SHUTTLE EVA #25 BY EV 1 & EV 2 SCHEDULED EVA #21 REPLACED WIDE FIELD/PLANETARY CAMERA AND INSTALLED TWO MSS'S 6H47M28S EVA #4 - 12/7/93 SPACE SHUTTLE EVA #26 BY EV 3 & EV 4 SCHEDULED EVA #22 REPLACED HIGH SPEED PHOTOMETER WITH COSTAR AND INSTALLED NEW COPROCESSOR 6H50M55S EVA #5 - 12/8/93 SPACE SHUTTLE EVA #27 BY EV 1 & EV 2 SCHEDULED EVA #23 REPLACED SOLAR ARRAY DRIVE ELECTRONICS, GHRS REDUNDANCY KIT, MLI CONTAMINATION KITS FOR MSS'S, AND MANUALLY OPERATED BOTH SOLAR ARRAY PRIMARY DEPLOYMENT MECHANISMS 7H20M4S		Continued . . . DENS ALT: -1039 FT FLT DURATION: 10:19:58:33 259:58:33 S/T: 410:15:34:33 OV-105: 43:15:09:45 DISTANCE: 4,433,772 sm						

STS061-86-030 1993-12-04 Hubble Space Telescope is berthed in Endeavour's payload bay after capture.

STS061-94-050 Thornton on end of RMS (foreground) and Akers install COSTAR during EVA for HST repair.

At right: STS061-90-028 1993-12-09 After servicing, HST flys away on new "Solar Wings".

Bottom left: m100_wfpcHSTBefore, HST Galaxy photo before repairs.
Bottom right: m100_smalHSTAfter, HST Galaxy photo after repairs.

STS061-74-046 Hoffman on RMS and Musgrave installing Wide Field/Planetary Camera (WFPC II).

FLT NO.	ORBITER	CREW (6) TITLE, NAMES & EVA'S	LAUNCH SITE, LIFTOFF TIME, LANDING SITES, ABORT TIMES	LANDING SITE/ RUNWAY, CROSSRANGE LANDING TIMES FLT DURATION, WINDS	SSME-TL NOM-ABORT EMERG THROTTLE PROFILE ENG. S.N.	SRB RSRM AND ET	INC	ORBIT HA/HP	FSW	PAYLOAD WEIGHTS, PAYLOADS/ EXPERIMENTS	MISSION HIGHLIGHTS (LAUNCH SCRUBS/DELAYS, TAL WEATHER, ASCENT I-LOADS, FIRSTS, SIGNIFICANT ANOMALIES, ETC.)
STS-60 SEQ FLT #60 KSC-60 PAD 39A-37 MLP-3	OV-103 (Flight 18) Discovery Spacehab 2 OMS PODS LPO1-21 RPO3-19 FRC3-18	CDR: Charles F. Bolden (Flt 4 - STS 61-C STS-31 & STS-45)) P330/R88/V52/M80 PLT: Kenneth S. Reightler (Flt 2 - STS-48) P331/R134/V99/M119 M/S 1: N. Jan Davis (Flt 2 - STS-47) P332/R153/V100/F17 M/S 2: Ronald M. Sega P333/R175/M153 M/S 3: Franklin R. Chang-Diaz (Flt 4 - STS 61-C, STS-34 & STS-46) P334/R89/V46/M81 M/S 4: Sergei Krikalev (Flt 3 SOYUZ TM-7, MIR SOYUZ TM-12/MIR) Russian Cosmonaut (P335/R176/M154) MCC FCR-1 (39) FLIGHT DIRECTORS: A/E - J. W. Bantle LD/O 2/C. W. Shaw O 1 - G. A. Pennington PLNG - R. E Castle MOD - G. E. Coen	KSC 39, PAD B 34:12:09:59.965Z 7:10:00 AM EST (P) 7:10:00 AM EST (A) Thursday 16 2/3/94 (3) LAUNCH WINDOW: 2H30M CTOB EOM PLS: KSC TAL: ZARAGOZA TAL ALT: MORON, BEN GUERIR SELECTED: RTLS: KSC33/CI/ N TAL: BEN36/N/N AOA: NOR17/N/N PLS: EDW04/N/N TDEL: 0.00 0.081/0.12 MAX Q NAV: 708 PSF 717 PSF SRB STG: 2:05.3 2:06 PERF: NOMINAL 2 ENG TAL (BEN): 2:49 2:49 NEG RETURN: 4:03 4:06 PTA (U/S 350): 5:06 5:12 PTM : N/A MECO CMD: 8:33.1 8:32.7 VI: 25924 25916 OMS-2: 42:17 42:17 268 FPS 268 FPS	KSC 15 (KSC 19) 42:19:19:22Z 2:19:22 PM EST Friday 6 2/11/94 (2) DEORBIT BURN: 42:18:18:45Z XRANGE: 376 NM ORBIT DIR: DL 31 AIM PT: NOMINAL MLGTD: 2324 FT 42:19:19:22Z VEL: 192 KGS 205 KEAS HDOT: -2.3FPS TD NORM 195: 3016 FT DRAG CHUTE DEPLOY: 172 KEAS 42:19:19:32Z NLGTD: 7522 FT 42:19:19:41Z VEL: 118 KGS HDOT: -4.1 FPS BRK INIT: 97 KGS DRAG CHUTE JETTISON: 52 KGS 42:19:19:55Z AVE BRK DECEL: 6.2 FPS/S WHEELS STOP: 42:19:20:13Z 10144 FT ROLLOUT: 7820FT 51 SEC WINDS: H11, R1 OFFICIAL: H20, R0 DENS ALT: 1377FT FLT DURATION: 8:07:09:22 199:09:22 S/T: 418:07:43:55 OV-103: 117:20:12:37 DISTANCE: 3,439,704 sm	104/104/ 109% PREDICTED: 100/104/104/ 70/104 ACTUAL: 100/104/104/ 70/104 1 = 2012 (15) 2 = 2034 (4) 3 = 2032 (2) M 3 EOM: WEIGHT: 216663 LBS X CG: 1079.6 LANDING: WEIGHT: 216503 LBS X CG: 1081.3	BI-062 RSRM 35 ET-61 LWT 54 ET PRED RPT 285 K ET BR/UP 214 K ET IMPACT 1:27:21 MET LAT: 2.69 N LONG: 123.2 W	57 (14)	DIRECT INSERTION POST OMS-2: 191 X 189 NM ODERACS DEPLOY: 6:02:43:24 MET BREMSAT DEPLOY: 06:07:13:40 MET DEORBIT: 194.4 X 189.1 NM VELOCITY: 25858 FPS ENTRY RANGE: 4349 NM	OI-22 (5)	CARGO: 28957 LBS PAYLOAD CHARGEABLE: 22296 LBS DEPLOYED: 171 LBS NON-DEPLOYED: 21015 LBS MIDDECK: 1110 LBS SHUTTLE ACCUMULATED WEIGHTS: DEPLOYED: 766601 LBS NON-DEPLOYED: 784493 LBS CARGO TOTAL: 1802868 LBS PERFORMANCE MARGINS (LBS): FPR: 3981 FUEL BIAS: 987 FINAL TDDP: 110 RECON: 306 PAYLOADS: PLB: WSF-1 SPACEHAB-2 CAPL-1 ODERACS/ BREMSAT GBA (WITH 4 GAS CANS) MIDDECK: SAREX-II APE-B 4 CRYO TK SETS RMS 34 (S. N. 201) RMS used for WSF deberth but did not deploy because of WSF problems	KSC W/D: OPF 81 VAB 5, PAD 22 = 108 days total. LAUNCH POSTPONEMENTS: - 10/31/93 launch date baselined on 7/31/92, later changed to 10/21/93 and 11/10/93. - Postponed STS-60 to 1/20/94 and moved STS-61 ahead on 9/2/93 (KSC work flows would not allow two flights before holidays). LAUNCH SCRUBS: None. LAUNCH DELAYS: None. TAL WX: - Zaragoza was prime but forecast NO GO for visibility (rain/fog) and 4K ceiling; hence, Ben Guerir was selected. ZZA was observed GO. Moron forecast NO GO (headwinds and ceiling), observed NO GO (headwinds). DOLILU/I-LOADS: - Both DOLILU and Nominal I-loads were GO. DOLILU was selected because they provided approx. 300 lbs performance and 1.1-minute additional hold time. DOLILU uplink #10, total I-load uplink #16. FLIGHT DURATION CHANGES: - Extended flight one orbit because KSC was forecast NO GO for ceiling and crosswinds FIRSTS: - First flight of Russian Cosmonaut on U.S. spacecraft (Krikalev's previous flights were Soyuz TM-7 and Soyuz TM-12 with more than 1 year 3 months aboard Mir.) SIGNIFICANT ANOMALIES: - Supply H20 dump valve leak (several burps after water dumps). - Unable to place diffuser cap into tunnel adapter. - O2 tank 2 quantity transducer erratic. - ARD nominal margin showed major thrust/mass difference with on-board data. - Pilot HIU failed. - Both MCC DVIS CPU's (A and B) went down). - Tunnel adapter stowage net, not stowed. - Hasselblad shutter failed. - Payload retention latch SW 2 position indicated release instead of off. - Air/ground crosstalk from ICOM to A/G loop. - Wakeshield horizon sensor signals bad, hence, did not deploy WSF resulting in limited scientific data. - WOW WONG anomaly.

STS060-15-003 SPACEHAB-2 in Payload Bay

STS060-31-028 Crew squeezes through tunnel to SPACEHAB in PLB. CDR Bolden is at upper right. Others, clockwise from him are: Sega/MS, Davis/MS, Chang-Diaz/PLC, Krikalev/MS & first Russian on U.S. spacecraft, and PLT

FLT NO.	ORBITER	CREW (5) / TITLE, NAMES & EVA'S	LAUNCH SITE, LIFTOFF TIME, LANDING SITES, ABORT TIMES	LANDING SITE/ RUNWAY, CROSSRANGE / LANDING TIMES FLT DURATION, WINDS	SSME-TL NOM-ABORT EMERG / THROTTLE PROFILE ENG. S.N.	SRB RSRM AND ET	ORBIT INC	HA/HP	FSW	PAYLOAD WEIGHTS, / PAYLOADS/ EXPERIMENTS	MISSION HIGHLIGHTS (LAUNCH SCRUBS/DELAYS, TAL WEATHER, ASCENT I-LOADS, FIRSTS, SIGNIFICANT ANOMALIES, ETC.)
STS-62 SEQ FLT #61 KSC-61 PAD 39B-24 MLP-1	OV-102 (Flight 16) Columbia EDO 3 OMS PODS: LPO5-5 RPO5-5 FRC2-16	CDR: John H. Casper (Flt 3 - STS-36 & STS-54) P336/R111/V86/M99 PLT: Andrew M. Allen (Flt 2 - STS-46) P337/R149/V101/M133 M/S 1 (PAYLOAD CDR): Pierre J. Thuot (Flt 3 - STS-36 & STS-49) P338/R112/V72/M100 M/S 2: Charles D. (Sam) Gemar (Flt 3 - STS-38 & STS-48) P339/R118/V67/M106 M/S 3: Marsha S. Ivins (Flt 3 - STS-32 & STS-46) P340/R109/V77/F12 MCC FCR-1 (40) FLIGHT DIRECTORS: A/E/T 1 - N. W. Hale LD/T 2 - P. L. Engelauf T 3 - C. W. Shaw T 4 - J. M. Heflin MOD - A. L. Briscoe	KSC 39, PAD B 63:13:52:59.97Z 8:53:00 AM EDT (P) 8:53:00 AM EDT (A) Friday 9 3/4/94 (4) LAUNCH WINDOW: 2H30M, CTOB EOM PLS: KSC TAL: BEN TAL WX: MRN, ZZA SELECTED: RTLS: KSC33/CI/N TAL: BEN36/N/N AOA: KSC33/CI/N PLS: EDW04/N/N TDEL: 0:00 0.162/0.20 MAX Q NAV: 709 ~708 SRB STG: 2:05.4 2:05 PERF: NOMINAL 2 ENG TAL (BEN): 2:41 2:44 NEG RETURN: 4:00 4:02 PTA (U/S 250): 5:09 5:07 PTM (U/S 250): 6:03 6:02 MECO CMD: 8:30.3 8:30.8 VI: 25886 25877 OMS-2: 42:19.7 42:19.7 208 FPS 208 FPS	KSC 33 (KSC 20) 77:13:09:41Z 08:09:41 AM EST Friday 7 3/18/94 (4) DEORBIT BURN: 77:12:16:50Z XRANGE: 116 NM ORBIT DIR: DR 10 AIM PT: NOMINAL MLGTD: 2905 FT 77:13:09:41Z VEL: 210 KGS 207 KEAS HDOT: -3.4 FPS TD NORM 205: 2974 FT DRAG CHUTE DEPLOY: 166 KEAS 77:13:09:55Z NLGTD: 8764 FT 77:13:10:00Z VEL: 148 KGS HDOT: -3.7 FPS BRK INIT: 123 KGS DRAG CHUTE JETTISON: 57 KGS 77:13:10:22Z AVE BRK DECEL: 7 FPS/S WHEELS STOP: 77:13:10:35Z 13071 FT ROLLOUT: 10166 FT 54 SEC WINDS: T4, L3 KTS OFFICIAL: 1905P08 T4, L3 DENS ALT: 333 FT FLT DURATION: 13:23:16:41 335:16:41 S/T: 432:22:00:36 OV-102: 136:16:49:53 DISTANCE: 5,820,146 sm	104/104/ 109% PREDICTED: 100/104/104/ 67/104 ACTUAL: 100/104/104/ 67/104 1 = 2031 (9) 2 = 2109 (12) 3 = 2029 (8) M 3 EOM: WEIGHT: 228360 LBS X CG: 1082.6 LANDING: WEIGHT: 228250 LBS X CG: 1084.1	BI-064 RSRM 36 KM ET-62 LWT 55 ET PRED RPT 271K ET BKUP 214K ET IMPACT 1:27:04 MET LAT: 8.1 N LONG: 132.9 W	39 (3)	DIRECT INSERTION POST OMS-2: 163 X 161 NM OMS-3: 9:17:09:39 MET 33.4 FPS 161 X 180 NM OMS-4: 9:17:50:30 MET 37.6 FPS 140 X 140 NM OMS-5: 11:18:15:34 MET 37.6 FPS 140 X 105 NM DEORBIT: 138 X 105 NM VELOCITY: 25708 FPS ENTRY RANGE: 4391 NM	OI-22 (6)	CARGO: 30016 LBS PAYLOAD CHARGEABLE: 19792 LBS DEPLOYED: 0 LBS NON-DEPLOYED: 18512 LBS MIDDECK: 1280 LBS SHUTTLE ACCUMULATED WEIGHTS: DEPLOYED: 766601 LBS NON-DEPLOYED: 804285 LBS CARGO TOTAL: 1832884 LBS PERFORMANCE MARGINS (LBS): FPR: 3981 FUEL BIAS: 987 FINAL TDDP: 871 RECON: 1795 PAYLOADS: PLB: U. S. Microgravity Payload (USMP-2) Solidification of metals and semiconductors dendritic growth OAST-2 Technology experiments DEE SSBUV/A LDCE MIDDECK: APCG, PSE, CPCG, CGBA, MODE, AMOS, APE-B 4 CRYO TK SETS + 4 EDO RMS 35 (S.N. 301) RMS used for DEE tests	KSC W/D: OPF 62, VAB 5, PAD 19 = 86 days total. LAUNCH POSTPONEMENTS: - 2/8/94 launch date baselined on 10/2/92. - Postponed launch to 2/24/94 on 9/2/93. - Postponed launch to 3/3/94 on 10/20/93. LAUNCH SCRUBS: - Scrubbed 3/3/94 launch at L-16 hours because excessive RTLS winds were forecast. LAUNCH DELAYS: None. TAL WX: - Ben Guerir, Moron, and Zaragoza were forecast and observed GO, Ben Guerir was prime and selected. NOMINAL/DOLILU/I-LOADS: - Nominal I-loads were NO-GO with PLB torque box indicator at 102 percent. DOLILU was selected and uplinked. DOLILU #11, total I-load uplink #17. FLIGHT DURATION CHANGES: None. FIRSTS: - First flight of DC vacuum cleaner. - First flight of Ku-Band Comm Adapter (KCA) uplink video. SIGNIFICANT ANOMALIES: - Galley overdispensed hot water. - Excessive gas bubbles in food containers. - WCS Fan Sep 1 stalled and popped all three circuit breakers. - Water Coolant Loop 1 accumulator quantity transducer drift. - Supply Water Tank B transducer dropout. - Cryo H_2 Tank A heater failure. - Mid-port and Mid-starboard PLB floodlight failures. - O_2 Tank 7 quantity measurement failure. - TV Cameras A, D, and end effector problems. - Ops Recorder poor quality data on several tracks. - APU 3 high fuel pump inlet pressure (line froze). - LBNP fuse blew when vacuum cleaner operated., caused by a 20-volt peak-to-peak ripple - PDIP power failure. - KCA comm link anomaly. RADIATOR DEPLOYED #15 (PORT RADIATOR ONLY).

STS062-81-024 Features activity with Dexterous End Effector (DEE) on RMS. Also seen are U.S. Microgravity Payload 2 (USMP) and OAST-2.

STS062-17-025 Crew in aft flight deck: Front: CDR Casper (left), & Thuot/MS. Rear: (left to right) are PLT Allen, Ivins/MS (and hair) & Gemar/MS.

FLT NO.	ORBITER	CREW (6) TITLE, NAMES & EVA'S	LAUNCH SITE, LIFTOFF TIME, LANDING SITES, ABORT TIMES	LANDING SITE/ RUNWAY, CROSSRANGE LANDING TIMES FLT DURATION, WINDS	SSME-TL NOM-ABORT EMERG THROTTLE PROFILE ENG. S.N.	SRB RSRM AND ET	INC	ORBIT HA/HP	FSW	PAYLOAD WEIGHTS, PAYLOADS/ EXPERIMENTS	MISSION HIGHLIGHTS (LAUNCH SCRUBS/DELAYS, TAL WEATHER, ASCENT I-LOADS, FIRSTS, SIGNIFICANT ANOMALIES, ETC.)
STS-59 SEQ FLT #62 KSC-62 PAD 39A-38 MLP-2	OV-105 (Flight 6) Endeavour OMS PODS: LPO4-5 RPO1-19 FRC5-6	CDR: Sidney M. Gutierrez (Flt 2 - STS-40) P341/R129/V102/M116 PLT: Kevin P. Chilton (Flt 2 - STS-49) P342/R145/V103/M129 M/S 1: Jerome (Jay) Apt (Flt 3 - STS-37 & STS-47) P343/R123/V79/M110 M/S 2: Michael R. Clifford (Flt 2 - STS-53) P344/R157/V104/M139 M/S 3 (PAYLOAD CDR): Linda M. Godwin (Flt 2 - STS-37) P345/R122/V105/F13 M/S 4: Thomas D. Jones P346/R177/M155 MCC FCR-1 (41) FLIGHT DIRECTORS: A/E/O 1 - R. D. Jackson LD/O 2 - G. A. Pennington O 3 - R. E. Castle MOD - B. R. Stone	KSC 39, PAD A 99:11:04:59.99Z 7:05:00 AM EDT (P) 7:05:00 AM EDT (A) Saturday 4 4/9/94 (11) LAUNCH WINDOW: 2H30M (CTOB) EOM PLS: KSC TAL: ZZA TAL WX: BEN, MRN SELECTED: RTLS: KSC15/CI/N TAL: ZZA30/CI/N AOA: NOR23/N/N PLS: NOR23/N/N TDEL: .16 .042/.08 MAX Q NAV: 701 >694 SRB STG: 2:04 2:05 PERF: NOMINAL 2 ENG TAL (MRN): 2:57 2:56 NEG RETURN: 4:04 4:04 PTA (U/S 190): 5:47 5:38 DROOP (ZZA) 5:28 5:42 PTM (U/S 190): 6:08 5:56 MECO CMD: 8:34:3 8:33 VI: 25778 25774 OMS-2: 35:09.2 35:10.3 163.5 FPS 163.7 FPS	EDW 22, CONC EDW 40, CONC 21 110:16:54:30Z 9:54:30 AM PDT Wednesday 7 4/20/94 (9) DEORBIT BURN: 110:16:00:35Z XRANGE: 721 NM ORBIT DIR: DR 11 AIM PT: NOMINAL MLGTD: 1619 FT 110:16:54:30Z VEL: 228 KGS 215 KEAS HDOT: -3.7 FPS TD NORM 205: 2636 FT DRAG CHUTE DEPLOY: 180 KEAS 110:16:54:41Z NLGTD: 7115 FT 110:16:54:45Z VEL: 171 KGS HDOT: -4.4 FPS BRK INIT: 118 KGS DRAG CHUTE JETTISON: 49 KGS 110:16:55:12Z AVE BRK DECEL: 7.6 FPS/S WHEELS STOP: 110:16:55:23Z 12255 FT ROLLOUT: 10636 FT 53 SEC WINDS: T1, R2 KTS OFFICIAL: 0204 T4, R2 DENS ALT: 3764 FT FLT DURATION: 11:05:49:30 269:49:30 S/T: 444:03:50:06 OV-105: 54:20:59:15 DISTANCE: 4,704,835 sm	104/104/ 109% PREDICTED: 100/100/100/ 67/104 ACTUAL: 100/100/100/ 67/104 1 = 2028 (8) 2 = 2033 (6) 3 = 2018 (12) M 3 EOM: WEIGHT: 221981 LBS X CG: 1079.4 LANDING: WEIGHT: 221865 LBS X CG: 1081.2	BI-065 RSRM 37 ET-63 LWT- 56 ET PRED RPT 271.3K ET BKUP 214K ET IMPACT 1:13:00 MET 45.0 N 158.06 E	57 (15)	DIRECT INSERTION POST OMS-2: 121 X 121 NM DEORBIT: 112 X 110 NM VELOCITY: 25660 FPS ENTRY RANGE: 4468 NM	OI-22 (7)	CARGO: 33758 LBS PAYLOAD CHARGEABLE: 27447 LBS DEPLOYED: 0 LBS NON-DEPLOYED: 27447 LBS MIDDECK: 1445 LBS SHUTTLE ACCUMULATED WEIGHTS: DEPLOYED: 766601 LBS NON-DEPLOYED: 831732 LBS CARGO TOTAL: 1866642 LBS PERFORMANCE MARGINS (LBS): FPR: 3981 FUEL BIAS: 987 FINAL TDDP: 2856 RECON: 1731 PAYLOADS: PLB: SPACE RADAR LABORATORY (SRL-1) SIR-C/X-SAR IMAGING OF EARTH'S SURFACE CONCAP IV GAS (4) MIDDECK: STL (2) VFT-4 SAREX ii 5 CRYO TK SETS6 RMS 36 (S.N. 303) RMS NOT USED PER PLAN	KSC W/D: OPF 67, VAB 5, PAD 21 = 93 days total. LAUNCH POSTPONEMENTS: - Baselined 9/30/93 launch date on 3/11/92. - Postponed launch date to 4/14/94 on 12/21/92. - Advanced launch date to 3/31/94 on 4/2/93. - Postponed launch date to 4/7/94 on 11/5/93. LAUNCH SCRUBS: - Scrubbed 4/7/94 launch approximately 6 hours into count on 4/4/94 to borescope HPOTP preburner volute diffuser vane fillet for undersized radii. - Scrubbed 4/8/94 launch while holding at t-5 minutes. RTLS crosswinds exceeded limits. Decision made to count down to launch 1 hour earlier than nominal launch time on 4/9/94 to improve launch probability (11:05Z vs 12:05Z). LAUNCH DELAYS: None. - Launched 1 hour early as planned. TAL WX: - Zaragoza, Ben Guerir, and Moron forecast and observed GO. DOLILU/I-LOADS: - DOLILU selected because WINGAR18 10 percent more margin than nominal. DOLILU uplink #12, I-load uplink #18. FLIGHT DURATION CHANGES: - Changed from 9 to 10 days to acquire more science. - Waved off landing at KSC on orbits 166 and 167 for fore- cast and observed ceiling violations and rain within 30 nm. Extended flight a second day. - Waved off landing on orbit 182 due to observed ceiling violations and forecast rain within 30 nm. Waved off landing at KSC due to observed and forecast rain. Landed at EDW on orbit 183. - Flight extended 2 days plus one orbit. SIGNIFICANT ANOMALIES: - Right SSME HPOTP turbine discharge temp A biased low (200 degree delta to CH B). - Bubbles in water from SORG ((caused by venturi effect). - Defective (split) LiOH can casing, no LiOH spilled. - FES Feedline A Heater 1 thermostat failure. - H_2 Tank 5 check valve failed to seat. - Sticky cryo H_2 Tank 2 check valve. - GPS DTO status bit static. - MADS recorder tape broke. - Ku-band Channel 3 interferes with Channel 2. - Ku-band range/elevation unit digit inoperative. - Side hatch window impact crew reported. - GO_2 vent arm on pad damaged, caused by shuttle plume effect.

STS059-S-076 Three dimensional image of Isla Isabela in western Galapagos Islands (Earth Surface Imaging).

STS059-44-004 Crew in middeck: CDR Gutierrez (front center) is flanked by Apt/MS & Jones/MS On back row are (left to right) PLT Chilton, Godwin/PLC, & Clifford/MS

FLT NO.	ORBITER	CREW (7) TITLE, NAMES & EVA'S	LAUNCH SITE, LIFTOFF TIME, LANDING SITES, ABORT TIMES	LANDING SITE/ RUNWAY, CROSSRANGE LANDING TIMES FLT DURATION, WINDS	SSME-TL NOM-ABORT EMERG THROTTLE PROFILE ENG. S.N.	SRB RSRM AND ET	INC	ORBIT HA/HP	FSW	PAYLOAD WEIGHTS, PAYLOADS/ EXPERIMENTS	MISSION HIGHLIGHTS (LAUNCH SCRUBS/DELAYS, TAL WEATHER, ASCENT I-LOADS, FIRSTS, SIGNIFICANT ANOMALIES, ETC.)
STS-65 SEQ FLT #63 KSC-63 PAD 39A-39 MLP-3	OV-102 (Flight 17) Columbia 14th Spacelab Flight Long Module 10 EDO 4 OMS PODS: LPO5-6 RPO5-6 FRC2-17	CDR: Robert D. Cabana (Flt 3 - STS-41, STS-53) P347/R113/V84/M101 PLT: James D. Halsell, Jr. P348/R178/M156 M/S 1 (PAYLOAD CDR): Richard J. Hieb (Flt 3 - STS-39, STS-49) P349/R128/V70/M115 M/S 2: Carl E. Walz (Flt 2 - STS-51) P350/R170/V106/M148 M/S 3: Leroy Chiao P351/R179/M157 M/S 4: Donald A. Thomas P352/R180/M158 P/S 1: Chiaki Naito-Mukai P353/R181/F23 (Japan - NASDA) MCC FCR-1 (42) FLIGHT DIRECTORS: A/E/O1 - J. W. Bantle LD/O 2 - J. M. Heflin O 3 - R. E. Castle O4 - P. L. Engelauf MOD - A. L. Briscoe	KSC 39, PAD A 189:16:42:59.977Z 12:43:00 AM EDT (P) 12:43:00 AM EDT (A) Friday 10 7/8/94 (3) LAUNCH WINDOW: 2H30M CTOB EOM PLS: KSC TAL: BYD TAL WX: BEN SELECTED: RTLS: KSC 15/N/N TAL: BYD 32/N/SF AOA: EDW 22/N/N PLS: EDW 22/N/N TDEL: 0.19 -0.048/-0.01 MAX Q NAV: 673 677 SRB STG: 2:03.8 2:05 PERF: NOMINAL 2 ENG TAL (BYD): 2:47 2:43 NEG RETURN: 4:00 4:01 PTA (U/S 244): 5:12 5:01 DROOP (BYD): 5:31 5:27 PTM: 6:03 5:50 MECO CMD: 8:32 8:31 VI: 25877 25870 OMS-2: 39:55 39:55 221 FPS 221 FPS	KSC 33 (KSC 21) 204:10:38:00Z 6:38:00 AM EDT Saturday 11 7/23/94 (4) DEORBIT BURN: 204:09:40:38Z XRANGE: 180 NM ORBIT DIR: DL 32 AIM PT: NOMINAL MLGTD: 2996 FT 204:10:38:00Z VEL: 207 KGS 199 KEAS HDOT: -2.5 FPS TD NORM 205: 2501 FT DRAG CHUTE DEPLOY: 174 KEAS 204:10:38:09Z NLGTD: 8313 FT 204:10:38:18Z VEL: 138 KGS HDOT: -5.7 FPS BRK INIT: 115 KGS DRAG CHUTE JETTISON: 52 KGS 204:10:38:43Z AVE BRK DECEL: 5.7 FPS/S WHEELS STOP: 204:10:39:08Z 13207 FT ROLLOUT: 10211 FT 68 SEC WINDS: T3,0X KTS OFFICIAL: 1503P04 T3,0X KTS DENS ALT: 840 FT FLT DURATION: 14:17:55:00 353:55:00 S/T: 458:21:45:06 OV-102:151:10:44:53 DISTANCE: 6,143,846 sm	104/104/ 109% PREDICTED: 100/104/104/ 67/104 ACTUAL: 100/104/104/ 67/104 SSME S/N: 1 = 2019 (14) 2 = 2030 (9) 3 = 2017 (10) M 3 EOM: WEIGHT: 229368 LBS X CG: 1078.6 LANDING: WEIGHT: 229261 LBS X CG: 1080.1	BI-066 RSRM 39 KM ET-64 LWT 57 ET PRED RPT ET BKUP ET IMPACT 1:21:08 MET LAT: 13.6 S LONG: 163.3 W	28.45 (36)	DIRECT INSERTION POST OMS-2: 163 X 160 NM DEORBIT: 137 X 127 NM VELOCITY: 25720 FPS ENTRY RANGE: 4381 NM	OI-23 (1)	CARGO: 32880 LBS PAYLOAD CHARGEABLE: 24282 LBS DEPLOYED: 0 LBS NON-DEPLOYED: 22521 LBS MIDDECK: 1761 LBS SHUTTLE ACCUMULATED WEIGHTS: DEPLOYED: 766601 LBS NON-DEPLOYED: 856014 LBS CARGO TOTAL: 1899522 LBS PERFORMANCE MARGINS (LBS): FPR: 3981 FUEL BIAS: 987 FINAL TDDP: 2169 RECON: 3531 PAYLOADS: PLB: INTERNATIONAL MICROGRAVITY LABORATORY LIFE SCIENCES AND MATERIAL SCIENCES EXPERIMENTS (IML-2/LM) OARE MIDDECK: CPCG MAST AMOS SAREX-II 4 + 4 EDO CRYO TANK SETS NO RMS	KSC W/D: OPF 62, VAB 5, PAD 20 = 87 days total. LAUNCH POSTPONEMENTS: - Baselined launch date of 6/23/94 on 4/2/93. - Postponed launch date to 7/8/94 on 4/15/93. LAUNCH SCRUBS: None LAUNCH DELAYS: None TAL WX: - Banjul (prime & selected) forecast and observed GO. - Ben Guerir forecast NO GO (rain) but observed GO. DOLILU/I-LOADS: - Both DOLILU and NOMINAL I-loads were GO, NOMINAL I-loads were selected, no uplink required. FLIGHT DURATION CHANGES: - Waved off landing at KSC on orbits 220 and 221 due to forecast and observed rain and potential lightening. Extended flight 1 day. SIGNIFICANT ANOMALIES: - Supply water dump nozzle icing occurred on third dump on FD3. FES was used to dump water for the rest of flight. - WCS problems included commode fault during compaction, commode filter fit and odor problems, and fan sep 1 stall and liquid backflow. - IMU redundant rate BITE messages. - RCS vernier thruster R5D failed off, then nominal ops. - Low wastewater dump flow. Second dump in three cycles. Third dump required seven cycles. - Ops recorder 2 track 2 poor dump quality. - Galley rehydration station did not dispense cold water. - Arriflex magazine jams, Hasselblad jam and lens stuck.

STS065-42-017 Spacelab (IML-2) in payload bay

STS065-20-019 Crew pose in SL: Front row: CDR Cabana flanked by PLT Halsell & Mukai/PS (NASDA). Back row: (left to right) Hieb/PLC, Thomas/MS, Walz/MS, & Chiao/MS

STS065-214-037 -- DR.Chiaki Naito-Mukai enters IML-2 science module in cargo bay to conduct microgravity experiments.

SPACE SHUTTLE MISSIONS SUMMARY

FLT NO.	ORBITER	CREW (6) — TITLE, NAMES & EVA'S	LAUNCH SITE, LIFTOFF TIME, LANDING SITES, ABORT TIMES	LANDING SITE/ RUNWAY, CROSSRANGE — LANDING TIMES, FLT DURATION, WINDS	SSME-TL NOM-ABORT EMERG — THROTTLE PROFILE ENG. S.N.	SRB RSRM AND ET	INC	ORBIT — HA/HP	FSW	PAYLOAD WEIGHTS, PAYLOADS/ EXPERIMENTS	MISSION HIGHLIGHTS (LAUNCH SCRUBS/DELAYS, TAL WEATHER, ASCENT I-LOADS, FIRSTS, SIGNIFICANT ANOMALIES, ETC.)
STS-64 SEQ FLT #64 KSC-64 PAD 39B-25 MLP-2 — MCC FCR-1 (43) — FLIGHT DIRECTORS: A/E/O1 - N. W. Hale LD/O 2 - G. A. Pennington PLNG - W. D. Reeves MOD - B. R. Stone — STS064-114-027 --- Meade & Lee) test the new Simplified Aid for EVA Rescue (SAFER).	OV-103 (Flight 19) Discovery — OMS PODS: LPO1-22 RPO3-20 FRC3-19	CDR: Richard N. Richards (Flt 4 - STS-28, STS-41, STS-50) P354/R101/V55/M92 — PLT: L. Blaine Hammond (Flt 2 - STS-39) P355/R124/V107/M111 — M/S 1: Jerry M. Linenger P356/R182/M159 — M/S 2: Susan J. Helms (Flt 2 - STS-54) P357/R158/V108/F19 — M/S 3/EV2: Carl J. Meade (Flt 3 - STS-38, STS-50) P358/R117/V76/M105 — M/S 4/EV1: Mark C. Lee (Flt 3 - STS-30, STS-47) P359/R100/V78/M91 — SS EVA #28 SAFER FF #1 SCHEDULED EVA #24 9/16/94 EV1 - MARK LEE EV2 - CARL MEADE 6H51M35S DURATION EVALUATED SAFER PERFORMANCE	KSC 39, PAD 39B 252:22:22:54.947Z 4:30:00 PM EDT (P) 6:22:05 PM EDT (A) Friday 11 9/9/94 (5) — LAUNCH WINDOW: 2H30M CTOB — EOM PLS: KSC TAL: ZZA TAL WX: MRN, BEN — SELECTED: RTLS: KSC 15/CI/N TAL: ZZA AOA: NOR 17/N/N PLS: EDW 22/N/N — TDEL: 0.19 -0.088/-0.05 — MAX Q NAV: 688 691 — SRB STG: 2:04.3 2:03 — PERF: NOMINAL — 2 ENG TAL (MRN): 2:38 2:37 — NEG RETURN: 4:08 4:10 — PTA (U/S 250): 4:45 4:43 — DROOP (ZZA): 5:28 5:31 — PTM: 5:31 5:28 — MECO CMD: 8:34.4 8:35.3 — VI: 25805 25800 — OMS-2: 36:09 36:09	EDW 04 CONC (EDW 41, CONC 22) 263:21:12:52Z 2:12:52 PM PDT 9/20/94 (7) — DEORBIT BURN: 263:20:17:00Z — XRANGE: 110 NM — ORBIT DIR: AL 15 — AIM PT: NOMINAL — MLGTD: 2386 FT 263:21:12:52Z VEL: 208 KGS 198 KEAS HDOT: -1 FPS — TD NORM 195: 2627 FT — DRAG CHUTE DEPLOY: 184 KEAS 263:21:12:59Z — NLGTD: 6192 FT 263:21:13:03Z VEL: 163 KGS HDOT: -6.7 FPS — BRK INIT: 133 KGS — DRAG CHUTE JETTISON: 56 KGS 263:21:13:31Z — AVE BRK DECEL: 4.6 FPS/S — WHEELS STOP: 263:21:13:53Z 12042 FT — ROLLOUT: 12045 FT 61 SEC — WINDS: 10H, 3L KTS OFFICIAL: 0204P09 H4, L2 KTS — DENS ALT: 4927 FT — FLT DURATION: 10:22:49:57 262:49:57 — S/T: 469:20:35:03 — OV-103: 128:19:01:34 — DISTANCE: 4,576,174 sm	104/104/ 109% — PREDICTED: 100/100/100/ 67/104 — ACTUAL: 100/100/100/ 67/104 — 1 = 2031 (11) 2 = 2109 (13) 3 = 2029 (10) — M 3 EOM: WEIGHT: 212294 LBS X CG: 1082.3 — LANDING: WEIGHT: 212180 LBS X CG: 1083.9	BI-068 RSRM 41 ET-66 LWT 59 — ET PRED RPT 271K — ET BKUP 214K — ET IMPACT 1:13:57 MET LAT: 43.3 S LONG: 155.5 W	57 (16)	DIRECT INSERTION — POST OMS-2: 141 X 140 NM — DEORBIT: 132.4 X 127.8 NM — VELOCITY: 25727 FPS — ENTRY RANGE: 4433 NM	OI-23 (2)	CARGO: 25621 LBS — PAYLOAD CHARGEABLE: 20417 LBS — DEPLOYED: 0 LBS — NON-DEPLOYED: 16212 LBS — MIDDECK: 1363 LBS — SHUTTLE ACCUMULATED WEIGHTS: DEPLOYED: 766601 LBS NON-DEPLOYED: 873589 LBS CARGO TOTAL: 1925143 LBS — PERFORMANCE MARGINS (LBS): FPR: 3981 FUEL BIAS: 987 FINAL TDDP: 6409 RECON: 9639 — PAYLOADS: PLB: LIDAR In-Space Technology Experiment Atmospheric Research using Laser (LITE) SPARTAN-201 Astronomy (Deploy & retrieve) GBA ROMPS — MIDDECK: SSCE, BRIC, RME-III, MAST, SAREX-II, AMOS — 4 CRYO TK SETS — RMS 37 (S.N. 201) RMS used for SPARTAN deploy, retrieve, and berth, and for SPIFEX and SAFER ops	KSC W/D: OPF 125, VAB 8, PAD 20 = 153 days total. — LAUNCH POSTPONEMENTS: - Launch date was 6/16/94 on 2/19/93. - Launch date postponed to 9/15/94 on 4/2/93. - Launch date advanced to 9/9/94 on 11/19/93. — LAUNCH SCRUBS: None — LAUNCH DELAYS: - Launch delayed 1H52M55S. Held at T-9 minutes for 1H34M18S because of detached opaque thunderstorm anvil and thunderstorms within 20 nm. Picked up count and held at T-5 minutes for 13M37S until KSC weather was GO. — TAL WX: - Zaragosa (prime and selected) Moron and Ben Guerir were all three forecast and observed GO. — DOLILU/I-LOADS: - Both NOMINAL and DOLILU were GO. NOMINAL I-loads were selected, no uplink required. — FLIGHT DURATION/LANDING SITE CHANGES: - Flight was 9+1+1 and was extended 1 day for science. - Waved off landing at KSC on orbits 159 and 160 due to forecast of lightening and thunderstorms with 30 nm and ceiling violations. Extended another day for weather. - Waved off landing at KSC on orbits 175 and 176 due to ceiling and rain within 30 nm. Decision made to change landing site to EDW. — FLIGHT EXTENSION: 2 days plus 2 orbits. — LANDING SITE CHANGE: KSC to EDW due to KSC weather. — RENDEZVOUS #17: To retrieve, berth, and return SPARTAN-201, which was deployed earlier in flight. — SIGNIFICANT ANOMALIES: - FES feedline A accumulator temperature decreased below thermostat spec. - Torn AFRSI blanket on left OMS pod. - Supply H_2O dump valve leakage (burp). - FES outlet temperature oscillations during radiator bypass. - AFT MCA 1 OP STAT 4 indication. - Articulating portable foot restraint simulator fit interference. - Electronic cuff checklist #1 touch screen operation degraded during EVA. - PGSC PL3 hard disk error message and unexplained lockups on flight deck PGSC. - TACAN RM fails. - PROX OPS camera ALC logic lockup. - Side hatch locking device obstruction. - RCS jet L1A fail off.

STS064-24-02 Crew: CDR Richards (upper right), found stability with his back against the overhead in upper right corner. Others, clockwise from him are Meade/MS & Helms/MS PLT Hammond, Lee/MS & Linenger/ MS.

SPACE SHUTTLE MISSIONS SUMMARY

FLT NO.	ORBITER	CREW (6) / TITLE, NAMES & EVA'S	LAUNCH SITE, LIFTOFF TIME, LANDING SITES, ABORT TIMES	LANDING SITE/ RUNWAY, CROSSRANGE / LANDING TIMES, FLT DURATION, WINDS	SSME-TL NOM-ABORT EMERG THROTTLE PROFILE ENG. S.N.	SRB RSRM AND ET	INC	ORBIT HA/HP	FSW	PAYLOAD WEIGHTS, PAYLOADS/ EXPERIMENTS	MISSION HIGHLIGHTS (LAUNCH SCRUBS/DELAYS, TAL WEATHER, ASCENT I-LOADS, FIRSTS, SIGNIFICANT ANOMALIES, ETC.)
STS-68 SEQ FLT #65 KSC-65 PAD 39A-40 MLP-1	OV-105 (Flight 7) Endeavour OMS PODS: LPO4-14 RPO1-20 FRC5-7	CDR: Michael A. Baker (Flt 3 - STS-43 & STS-52) P360/R133/V81/M/118 PLT: Terrence W. Wilcutt P361/R183/M160 M/S 1: Steven V. Smith P362/R184/M161 M/S 2: Daniel W. Bursch (Flt 2 - STS-51) P363/R169/V109/M147 M/S 3: Peter J. K. (Jeff) Wisoff (Flt 2 - STS-57) P364/R166/V110/M145 M/S 4 (PAYLOAD CDR): Thomas D. Jones (Flt 2 - STS-59) P365/R177/V111/M155 MCC FCR-1 (44) FLIGHT DIRECTORS: A/E/O1 - R. D. Jackson LD/O 2 - C. W.Shaw O 3 - R. E. Castle MOD - A. L. Briscoe	KSC 39 PAD A 273:11:15:59.98Z 7:16:00 AM EDT (P) 7:16:00 AM EDT (A) Friday 12 9/30/94 (6) LAUNCH WINDOW: 2H30M CTOB EOM PLS: KSC TAL: ZZA TAL WX: MRN, BEN SELECTED: RTLS: KSC33/N/N TAL: MRN20/N/N AOA: NOR17/N/N PLS: EDW22/N/N TDEL: -0.16 -0.038/0.0 MAX Q NAV: 688 690 SRB STG: 2:03.8 2:03 PERF: NOMINAL 2 ENG TAL (MRN): 2:58 2:59 NEG RETURN: 4:03 4:04 PTA (U/S 180): 5:56 5:49 PTM: 6:18 6:05 MECO CMD: 8:34.8 8:33.9 VI: 25780 25775 OMS-2: 35:09.7 159 FPS	EDW 22, CONC (EDW 42, CONC 23) 284:17:02:08Z 10:02:89 AM PDT Tuesday 11 10/11/94 (6) DEORBIT BURN: 284:16:07:19Z XRANGE: 746 NM ORBIT DIR: DR 12 AIM PT: NOMINAL MLGTD: 3522 FT 284:17:02:08Z VEL: 196 KGS 193 KEAS HDOT: -2.3 FPS TD NORM 205: 2589 FT DRAG CHUTE DEPLOY: 188 KEAS 284:17:02:11Z NLGTD: 7299 FT 284:17:02:21Z VEL: 133 KGS HDOT: -5.1 FPS BRK INIT: 82 KGS DRAG CHUTE JETTISON: 55 KGS 284:17:02:45Z AVE BRK DECEL: 4.0 FPS/S WHEELS STOP: 284:17:03:10Z 12017 FT ROLLOUT: 8495 FT 62 SEC WINDS: H7, L3 KTS OFFICIAL: 2208P10 H8, L1 KTS DENS ALT: 3912 FT FLT DURATION: 11:05:46:08 273:46:08 S/T: 481:02:21:11 OV-105: 66:02:45:23 DISTANCE: 4,703,000 sm	104/104/ 109% PREDICTED: 100/100/100/ 67/104 ACTUAL: 100/100/100/ 67/104 1 = 2028 (9) 2 = 2033 (6) 3 = 2026 (4) M 3 EOM: WEIGHT: 221784 LBS X CG: 1078.7 LANDING: WEIGHT: 221673 LBS X CG: 1080.4	BI-067 RSRM 40 ET-65 LWT 58 ET PRED RPT: 271K ET BKUP: 214K ET IMPACT 1:13:26 MET LAT: 43.9 S LONG: 156.3 W	57 (17)	DIRECT INSERTION POST OMS-2: 120 X 119 NM DEORBIT: 111 X 110 NM VELOCITY: 25658 FPS ENTRY RANGE: 4480 NM	OI-22 (8)	CARGO: 34252 LBS PAYLOAD CHARGEABLE: 27640 LBS DEPLOYED: 0 LBS NON-DEPLOYED: 25997 LBS MIDDECK: 1643 LBS SHUTTLE ACCUMULATED WEIGHTS: DEPLOYED: 766601 LBS NON-DEPLOYED: 901229 LBS CARGO TOTAL: 1959395 LBS PERFORMANCE MARGINS (LBS): FPR: 3981 FUEL BIAS: 987 FINAL TDDP: 1721 RECON: 2071 PAYLOADS: PLB: SPACE RADAR LABORATORY (SRL-2) SIR-C/X-SAR MAPS GAS (5) MIDDECK: CPCG CHROMEX BRIC CREAM MAST 5 CRYO TK SETS RMS 38 (S.N. 303) RMS NOT USED PER PLAN	KSC W/D: OPF 59, VAB 20 (2), PAD 41 (2) = 120 days total. LAUNCH POSTPONEMENTS: - Launch date baselined as 10/27/94 on 7/9/93. - Launch date advanced to 8/18/94 on 9/2/93. - Launch date postponed to 10/2/94 after pad abort #5 on 8/18/94, moving STS-68 after STS-64. - Rolled back on 8/24/94 to VAB to replace all three engines. Returned to pad on 9/13/94. - Advanced launch date to 9/30/94 when range became available. LAUNCH SCRUBS/PAD ABORT #5: - 8/18/94 launch scrubbed with pad abort #5 at -1.86 seconds when HPOTP turbine discharge temp A exceeded 1560 degrees R start redline limit. Rolled back to VAB and replaced all three engines. Rescheduled launch to 10/2/94 and moved STS-64 ahead of STS-68. LAUNCH DELAYS: None TAL WX: - Zaragoza was prime but was forecast and observed NO GO for ceilings. - Moron (selected) and Ben Guerir were forecast and observed GO. DOLILU/I-LOADS: - NOMINAL and DOLILU I-loads were GO, selected NOMINAL, no uplink required. FLIGHT DURATION CHANGES: - Flight extended from 10 to 11 days for additional science. - Waved off landing at KSC on orbit 182 due to late convection activity and forecast (and observed) 3000 ft ceiling variable broken. Waved off landing at KSC on orbit 183 due to continuing convective activity and forecast ceiling violations and chance of rain within 30 nm. Total flight extensions - 1 day plus one orbit. LANDING SITE CHANGE: - Changed landing site to EDW due to forecast of worsening weather at KSC on Wednesday; hence, landed at EDW on orbit 183. SIGNIFICANT ANOMALIES: - MTU accumulator 3 lost. - FES feedline A hi load line temp read off-scale-high. - Rudder channel 3 slow to bypass during FCS checkout. - Simulation termination during DOLILU I-load verification. - Ku-Band CH3 (PL MAX) interference on channels 2 and 1. - CCTV cameras B, C, and D problems. - Linhof, Hasselblad, and Nikon camera problems. - Degraded tracks on payload recorder. - WSB 2 reg pressure increase. - WSB 1 and WSB 3 pressure decay. - RCS jet L3D fail off, low chamber pressure indication. - RCS jet L5D oxidizer injector temp sensor erratic, implemented GMEM and vernier control.

STS068-070-023 --- The Space Radar Laboratory-2 (SRL-2) in the Space Shuttle Endeavour's cargo bay.

STS068-002-016 Crew in middeck: (clockwise from bottom right) Jones/PLC, CDR Baker, Bursch/MS, PLT Wilcutt, Smith/MS, & Wisoff/MS.

FLT NO.	ORBITER	CREW (6) TITLE, NAMES & EVA'S	LAUNCH SITE, LIFTOFF TIME, LANDING SITES, ABORT TIMES	LANDING SITE/ RUNWAY, CROSSRANGE, LANDING TIMES, FLT DURATION, WINDS	SSME-TL NOM-ABORT EMERG THROTTLE PROFILE ENG. S.N.	SRB RSRM AND ET	INC	ORBIT HA/HP	FSW	PAYLOAD WEIGHTS, PAYLOADS/ EXPERIMENTS	MISSION HIGHLIGHTS (LAUNCH SCRUBS/DELAYS, TAL WEATHER, ASCENT I-LOADS, FIRSTS, SIGNIFICANT ANOMALIES, ETC.)
STS-66 SEQ FLT #66 KSC-66 PAD 39B-26 MLP-3	OV-104 (Flight 13) Atlantis 15th Spacelab Flight OMS PODS: LPO3-17 RPO4-13 FRC4-13	CDR: Donald R. McMonagle (Flt 3 - STS-39, STS-54) P366/R126/V87/M113 PLT: Curtis L. Brown (Flt 2 - STS-47) P367/R152/V112/M136 M/S 1 (PAYLOAD CDR): Ellen Ochoa (Flt 2 - STS-56) P368/R160/V113/F20 M/S 2: Joseph R. Tanner P369/R185/M162 M/S 3: Jean-Francois Clervoy P370/R186/M163 (ESA - France) M/S 4: Scott E. Parazynski P371/R187/M164 MCC FCR-1 (45) FLIGHT DIRECTORS: A/E - J. W. Bantle LD/O 2 - R. E. Castle O 1 - J. M. Heflin O3 - P. L. Engelauf O4 - N. W. Hale MOD - A. L. Briscoe	KSC 39 PAD B 307:16:59:42.97Z 11:56:00 AM EST (P) 11:59:43 AM EST (A) Thursday 17 11/3/94 (9) LAUNCH WINDOW: 1H02M, Crista-SPAS Beta Req ≥ 20 deg EOM PLS: KSC TAL: ZZA TAL WX: MRN, BEN SELECTED: RTLS: KSC 33/N/N TAL: BEN 36/N/N AOA: NONE PLS: EDW 04/N/N TDEL: 0.19 0.552/0.59 MAX Q NAV: 688 691 SRB STG: 2:04 2:05 PERF: NOMINAL 2 ENG TAL (BEN): 2:44 2:44 NEG RETURN: 4:07 4:09 PTA (U/S 300): 4:48 4:41 PTM (U/S 215): 5:30 5:32 MECO CMD: 8:35.9 8:34.4 VI: 25832 25826 OMS-2: 36:12 36:13 265 FPS 262 FPS	EDW 22, CONC (EDW 43, CONC 24) 318:15:33:45Z 7:33:45 AM PST Monday 13 11/14/94 (9) DEORBIT BURN: 318:14:31:05Z XRANGE: 745 NM ORBIT DIR: AL 16 AIM PT: NOMINAL MLGTD: 3219 FT VEL: 195 KGS 193 KEAS HDOT: -1.3 FPS TD NORM 195: 3032 FT DRAG CHUTE DEPLOY: 183 KEAS 318:15:33:49Z NLGTD: 6390 FT 318:15:33:56Z VEL: 150 KGS HDOT: -4.4 FPS BRK INIT: 108 KGS DRAG CHUTE JETTISON: 62 KGS 318:15:34:16Z AVE BRK DECEL: 6.0 FPS/S WHEELS STOP: 318:15:34:35Z 10866 FT ROLLOUT: 7647 FT 50 SEC WINDS: T3, R3 KTS 3064 T3, R3 KTS DENS ALT: 645 FT FLT DURATION: 10:22:34:02 262:34:02 S/T: 492:00:55:13 OV-105: 83:08:27:02 DISTANCE: 4,554,791 sm	104/104/ 109% PREDICTED: 100/100/100/ 67/104 ACTUAL: 100/100/100/ 68/104 1 = 2030 (10) 2 = 2034 (5) 3 = 2017 (11) M 3 EOM: WEIGHT: 211562 LBS X CG: 1084.4 LANDING: WEIGHT: 211611 LBS X CG: 1086.1	BI-069 RSRM 38 ET-67 LWT 60 ET RPT 271K ET BR/UP 214K ET IMPACT 1:14:01 MET LAT: 42.2 S LONG: 156.9 W	57 (18)	DIRECT INSERTION POST OMS-2: 164.8 X 164.2 NM DEPLOY (SPAS): 00/19:50:06 MET 164 X 163 NM SPAS GRAPPLE: 08/20:05:35 MET 160 x 157 NM SPAS BERTH: 08/23:50:19 MET DEORBIT: 162 X 156 NM VELOCITY: 25798 FPS ENTRY RANGE: 4387 NM	OI-23 (3)	CARGO: 23560 LBS PAYLOAD CHARGEABLE: 18135 LBS DEPLOYED: 0 LBS NON-DEPLOYED: 9901 LBS MIDDECK: 1080 LBS SHUTTLE ACCUMULATED WEIGHTS: DEPLOYED: 766601 LBS NON-DEPLOYED: 912210 LBS CARGO TOTAL: 1982955 LBS PERFORMANCE MARGINS (LBS): FPR: 3775 FUEL BIAS: 1136 FINAL TDDP: 3284 RECON: 3158 PAYLOADS: PLB: CRISTA/SPAS (Deploy & retrieve)) Atmospheric Science Experiments ATLAS-3 SSBUV-A ESCAPE-II MIDDECK: PARE/NIH-R PCG-TES PCG-STES SAMS, HPP STL/NIH-C 5 CRYO TK SETS RMS 39 (S.N. 202) RMS used for CRISTA/SPAS deploy, grapple and berth, and monitor supply and waste water dump (saw icicle form)	KSC W/D: OPF 110, VAB 6, PAD 24 = 140 days total. LAUNCH POSTPONEMENTS: - Launch baselined as 8/18/94 on 4/22/93. - Postponed launch to 10/27/94 on 9/2/93. - Postponed launch to 11/3/94 on 9/30/94 after STS-68 pad abort. LAUNCH SCRUBS: None. LAUNCH DELAYS: - Launch delayed for 3M43S while holding at T-5 min to discuss TAL weather. ZZA and MRN were NO GO due to forecast ceiling and rain. BEN was forecast NO GO for crosswinds. Decision made to select BEN for launch because observed crosswind trend was downward (last observed at 15 knots). Waiver to flight rule 4-64 was written.) TAL WX: - ZZA (prime) was forecast NO GO for ceiling, tailwind, and light rain within 5 nm. MRN was forecast NO GO for ceiling and light rain with 5 nm. BEN (selected) was forecast NO GO for crosswinds but downward trend. DOLILU/I-LOADS: - Both DOLILU and NOMINAL I-loads were GO, NOMINAL was selected with maximum load indicator at 88 percent. No uplink required. FLIGHT DURATION CHANGES: - Decision made to not try landing at KSC on orbits 174 and 175 due to forecast of gale winds, rain, and ceiling violations caused by Tropical Storm Gordon. Landed at EDW on orbit 176. Extended flight two orbits. LANDING SITE CHANGE: KSC to EDW FIRSTS: - First use of "R-BAR" approach for rendezvous which is required to protect Mir solar arrays on Mir rendezvous flights. RENDEZVOUS #18: To retrieve and return CHRISTA-SPAS, which was deployed earlier in flight. SIGNIFICANT ANOMALIES: - Spacelab ERAU 20 skipped triplet. - GPS 4 MMU1 BCE 18 failure. - Damaged tile at overhead window (W8). - FES oscillations at low heat loads. - FES outlet temp sensor lag. - Av Bay 2 Smoke Detector A concentration transients. - Ice formation on PLBD during simultaneous supply and waste water dump on FD8 (1.5" D X 5-6' long). Canceled icicle removal with RMS when RMS wrist camera failed. At landing, ice (approx 3"x5"x3") was seen on PLBD. - FES B undertemp shutdown. - Fuel Cell 2 H2O through alternate path. - Spacelab subsystem inverter shutdown. - NSP 2 to Ku-Band Channel 1 interface failure. - WSB 3 regulator pressure decay.

STS066-129-005 ATLAS-3 payload in the payload bay.

STS066-56-015 Crew on Flight Deck: left to right in lower row, Tanner/MS, CDR McMonagle, Parazynski/MS, PLT Brown. Floating at top, Ochoa/PLC & Clervoy/MS(ESA).

FLT NO.	ORBITER	CREW (6) TITLE, NAMES & EVA'S	LAUNCH SITE, LIFTOFF TIME, LANDING SITES, ABORT TIMES	LANDING SITE/ RUNWAY, CROSSRANGE LANDING TIMES FLT DURATION, WINDS	SSME-TL NOM-ABORT EMERG THROTTLE PROFILE ENG. S.N.	SRB RSRM AND ET	ORBIT INC	HA/HP	FSW	PAYLOAD WEIGHTS, PAYLOADS/ EXPERIMENTS	MISSION HIGHLIGHTS (LAUNCH SCRUBS/DELAYS, TAL WEATHER, ASCENT I-LOADS, FIRSTS, SIGNIFICANT ANOMALIES, ETC.)
STS-63 SEQ FLT #67 KSC-67 PAD 39B-27 MLP-2	OV-103 (Flight 20) Discovery Spacehab-3 OMS PODS: LPO1-23 RPO3-21 FRC3-20	CDR: James D. Wetherbee (Flt 3 - STS-32, STS-52) P372/R108/V80/M97 PLT: Eileen M. Collins P373/R188/F24 M/S 1/EV2 (PAYLOAD CDR): Bernard A. Harris (Flt 2 - STS-55) P374/R162/V114/M142 M/S 2/EV1: C. Michael Foale (Flt 3 - STS-45, STS-56) P375/R143/V92/M127 M/S 3: Janice E. Voss (Flt 2 - STS-57) P376/R167/V115/F22 M/S 4: Vladimir Titov (SS Flt #1) (Flt 4 - SOYUZ T-8, SOYUZ T-10, MIR SOYUZ TM-4) P377/R189/M165 RUSSIAN COSMONAUT SS EVA #29 EMU/TETHERED EVA SCHEDULED EVA #25 EVA DEVELOPMENT FLIGHT TEST (EDFT) #1 TO DEMONSTRATE EVA PROCEDURES AND ABILITY TO MOVE LARGE OBJECTS. COLD ENVIRONMENT TESTS. 2/9/95 4H38M10S DURATION	KSC 39 PAD B 34:05:22:03.96Z 00:22:04 AM EST (P) 00:22:04 AM EST (A) Friday 13 2/3/95 (4) LAUNCH WINDOW: 5 min Planar/Phase Window for Mir Rendezvous EOM PLS: KSC TAL: ZZA TAL WX: MRN, BEN SELECTED: RTLS: KSC33/CI/N TAL: ZZA30/N/N AOA: KSC33/CI/N PLS: EDW04/N/N TDEL: -0.32 -0.478/0.28 MAX Q NAV: 716 723 SRB STG: 2:05.6 2:05 PERF: NOMINAL 2 ENG TAL (BEN): 2:25 2:22 NEG RETURN: 4:04 4:06 PTA (U/S 293): 4:28 4:24 PTM (U/S 295): 5:54 5:44 SE TAL (ZZA): 5:53 5:59 SE PTM (U/S 810): 6:57 6:57 MECO CMD: 8:30.6 8:31.9 VI: 25885 25892 OMS-2: 42:10.3 252.6 FPS	KSC 15 (KSC 22) 42:11:50:19Z 6:50:19 AM EST Saturday 12 2/11/95 (3) DEORBIT BURN: 42:10:44:04 Z XRANGE: 469 NM ORBIT DIR: DR 13 AIM PT: CLOSE IN MLGTD: 1261 FT 42:11:50:19Z VEL: 206 KGS 212 KEAS HDOT: -2.8 FPS TD NORM 195: 2583 FT DRAG CHUTE DEPLOY: 185 KEAS 42:11:50:27Z NLGTD: 5460 FT 42:11:50:33Z VEL: 148 KGS HDOT: -4.8 FPS BRK INIT: 57 KGS DRAG CHUTE JETTISON: 58 KGS 42:11:51:05Z AVE BRK DECEL: 2.9 FPS/S WHEELS STOP: 42:11:51:40Z 12269 FT ROLLOUT: 11008 FT 70 SEC WINDS: H5, R2 KTS OFFICIAL: 1705P07 H5, R1 KTS DENS ALT: -443 FT FLT DURATION: 8:06:28:15 202:28:15 S/T: 500:07:23:28 OV-103: 137:01:29:49 DISTANCE: 2,922,000 sm	104/104/ 109% PREDICTED: 100/104/97/ 69/104 ACTUAL: 100/104/94/ 69/104 1 = 2035 (1) 2 = 2109 (14) 3 = 2029 (11) M 3 EOM: WEIGHT: 212775 LBS X CG: 1079.5 LANDING: WEIGHT: 212693 LBS X CG: 1081.2	BI-070 RSRM 42 ET-68 LWT 61 ET RPT 271K ET BR/UP 214K ET IMPACT 1:27:07 MET LAT: 0.036 S LONG: 125.6 W	51.66 (1)	DIRECT INSERTION POST OMS-2: 183.9 X 168.9 NM MIR RNDZ: Mir CPA of 37 feet at 3/13:58 MET 37/19:20Z 213.5 X 206 NM Backaway: 3/14:10 MET Flyaround Initiated: 3/14:53 MET Sep Burn: 3/15:50 MET DEORBIT: 212 X 204 NM VELOCITY: 26903 FPS ENTRY RANGE: 4329 NM	OI-23 (4)	CARGO: 24903 LBS PAYLOAD CHARGEABLE: 19051 LBS DEPLOYED: 23 LBS NON-DEPLOYED: 15249 LBS MIDDECK: 1128 LBS SHUTTLE ACCUMULATED WEIGHTS: DEPLOYED: 766624 LBS NON-DEPLOYED: 928587 LBS CARGO TOTAL: 2007858 LBS PERFORMANCE MARGINS (LBS): FPR: 3775 FUEL BIAS: 1136 FINAL TDDP: 1830 RECON: 3476 PAYLOADS: PLB: SPACEHAB-3 CGP/ODERACS-2 (deployed) SPARTAN-204 (deployed and retrieved) MIDDECK: SSCE AMOS 4 CRYO TK SETS RMS 40 (S.N. 201) RMS used for SPARTAN deploy, retrieve, and berth and TCS maneuvers, water dumps and EVA objectives	KSC W/D: OPF 71, VAB 5, PAD 25 = 101 days total. LAUNCH POSTPONEMENTS: - Launch date baselined as 5/19/94 on 1/19/93. - Launch date postponed to 1/26/95 on 11/18/93. - Launch date postponed to 2/2/95 on 3/25/94. LAUNCH SCRUBS: - 2/2/95 launch scrubbed at L-9 hours caused by IMU2 (HAINS) platform fail BITE during transition from STBY to OPERATE. Replaced IMU and rescheduled launch for 2/3/95. LAUNCH DELAYS: None TAL WX: - ZZA (prime and selected) and BEN were forecast and observed GO. MRN was forecast and observed NO GO for visibility (fog). DOLILU/NOMINAL I-LOADS: - Both DOLILU and NOMINAL I-loads were NO GO for Q-plane exceedance with boundary violation for engine knockdown. NOMINAL I-loads were selected because exceedance point on alpha beta envelope was bounded by a wing strut indicator which had adequate margin of safety. Waiver was written. NIGHT LAUNCH: Space Shuttle Night Launch #10. FLIGHT DURATION CHANGES: None FIRSTS: - First flight with a female pilot.- Eileen Collins - First African-American to walk in space - Bernard Harris RENDEZVOUS #19: - Rendezvous with Mir, prox ops and flyaround with closest approach of 37 feet. RENDEZVOUS #20: - Rendezvous with SPARTAN, retrieve and berth. SPARTAN was deployed earlier in flight. EVENTS: - ODERACS deployed at 00/23:35 MET. - SPARTAN deployed at 4/07:05:33 MET, grapple at 6/06:11:16 MET, and berth at 6/06:48:23 MET RADIATOR DEPLOY #16: - Port radiator deployed for approx 7 hours on FD2 for SPARTAN ops (FES INHIBIT period). - Bistable HPOTP on engine 2035 limited throttle bucket to 69 percent. Continued. . .
Continued. . .									STS063-06-018 Crew on aft flight deck: Front row (lt to rt), Harris/PLC & Foale/MS. Back row (lt to rt), Voss/MS, Titov/MS (Russia), CDR Wetherbee, & PLT Collins (first female pilot).	

FLT NO.	ORBITER	CREW (6) / TITLE, NAMES & EVA'S	LAUNCH SITE, LIFTOFF TIME, / LANDING SITES, ABORT TIMES	LANDING SITE/ RUNWAY, CROSSRANGE / LANDING TIMES FLT DURATION, WINDS	SSME-TL NOM-ABORT EMERG / THROTTLE PROFILE ENG. S.N.	SRB RSRM AND ET	ORBIT / INC	HA/HP	FSW	PAYLOAD WEIGHTS, / PAYLOADS/ EXPERIMENTS	MISSION HIGHLIGHTS (LAUNCH SCRUBS/DELAYS, TAL WEATHER, ASCENT I-LOADS, FIRSTS, SIGNIFICANT ANOMALIES, ETC.)
STS-63 Continued		Continued. . . MCC FCR-1 (46) FLIGHT DIRECTORS: A/E - N. W. Hale LD/O 2 - P. L. Engelauf O 1 - R. M. Kelso PLNG - P. F. Dye MOD - B. R. Stone								

STS063-86-028 Collins and Titov get TIPS mail from MCC.

STS063-716-064 Freeflying SPARTAN

Continued. . .

SIGNIFICANT ANOMALIES:
- Cabin pressure transducer shifted low by 0.23 PSI.
- Fuel Cell 2 H_2 motor status increased between 0.6 volts and 0.83 volts.
- EV2 crewman experienced burning sensation in his eyes during repressurization at 5 PSI. Funny odor inside suit was reported.
- During EVA, both EV1 and EV2 electronic cuffs were partially unresponsive.
- THC hotstick event when aft flight controller power was turned on (ref. STS-66), several thrusters fixed.
- TCZ Z-axis system failure during MIR backaway at 322 feet.
- Erratic TCS data sporadically throughout TCS ops on SPARTAN rendezvous day.
- Port radiator latch 1-6 "A" latched indication intermittent.
- Spacehab module pressure decay (air leak into airlock).
- RCS jet R1U failed off (oxidizer temp dropped below RM limit of 30 degree F), oxidizer leak.
- RCS jet L2D failed off. Jet had good driver output with low (< 13 PSI) chamber pressure.
- RCS jet F1F fail leak, indicated oxidizer leak.

STS063-21-011---Harris on RMS foot restraint carries Foale during shared EVA. Harris was first African-American to walk in space.

STS063-712-057 As seen from Discovery: MIR Space Station with docked Soyuz (at bottom of MIR) and Progress at opposite end.

S95-12534 -- Pat Patnesky (left) & unidentified Russian Scientist) with Shuttle mockup in background. Pat was NASA JSC PAO photographer responsible for many, many JSC MCC mission photos. He supported all NASA manned programs from Mercury through Shuttle, retirng in 1997.

FLT NO.	ORBITER	CREW (7) TITLE, NAMES & EVA'S	LAUNCH SITE, LIFTOFF TIME, LANDING SITES, ABORT TIMES	LANDING SITE/ RUNWAY, CROSSRANGE LANDING TIMES FLT DURATION, WINDS	SSME-TL NOM-ABORT EMERG THROTTLE PROFILE ENG. S.N.	SRB RSRM AND ET	ORBIT INC / HA/HP	FSW	PAYLOAD WEIGHTS, PAYLOADS/ EXPERIMENTS	MISSION HIGHLIGHTS (LAUNCH SCRUBS/DELAYS, TAL WEATHER, ASCENT I-LOADS, FIRSTS, SIGNIFICANT ANOMALIES, ETC.)
STS-67 SEQ FLT #68 KSC-68 PAD 39A-41 MLP-1	OV-105 (Flight 8) Endeavour Spacelab Pallet 16th Spacelab Flight EDO 5 OMS PODS: LPO4-15 RPO1-21 FRC5-8	CDR: Stephen S. Oswald (Flt 3 - STS-42, STS-56) P378/R139/V91/M124 PLT: William G. Gregory P379/R190/M166 M/S 1: John M. Grunsfeld P380/R191/M167 M/S 2: Wendy B. Lawrence P381/R192/F25 M/S 3 (PAYLOAD CDR): Tamara E. Jernigan (Flt 3 - STS-40, STS-52) P382/R130/V83/F14 P/S 1: Samuel T. Durrance (Flt 2 - STS-35) P383/R120/V116/M108 P/S 2: Ronald A. Parise (Flt 2 - STS-35) P384/R119/V117/M107 MCC FCR-1 (47) FLIGHT DIRECTORS: A/E - R. E. Jackson O 1 - B. P. Austin O 2 - A. L. Pennington O 3 - J. P. Shannon L/O 4 - C. W. Shaw MOD - A. L. Briscoe MOD - J. W. Bantle	KSC 39A 61:06:38:12.95Z 01:37:00 AM EST (P) 01:38:13 AM EST (A) Thursday 18 3/2/95 (5) LAUNCH WINDOW: 2H30M CTOB EOM PLS: KSC TAL: BEN TAL WX: MRN SELECTED: RTLS: KSC 33/CI/N TAL: BEN 36/CI/N AOA: EDW 22/CI/N PLS: EDW 22/CI/N TDEL: 0.48 0.202/0.24 MAX Q NAV: 728 PSF 739 PSF SRB STG: 2:06.9 2:05 PERF: NOMINAL 2 ENG TAL: 2:38 2:35 NEG RETURN: 3:59 4:01 PTA (U/S 297): 4:22 4:15 PTM (U/S 427): 5:30 5:17 SE T/M (BYD): 5:49 5:49 SE PTM (U/S-897): 6:33 6:33 MECO CMD: 8:27.65 8:27.3 MECO VI: 25922 25914 OMS-2: 40:19.8 40:19.8 279 FPS 279 FPS	EDW 22, CONC (EDW 44, CONC 25) 77:21:47:14Z 1:47:14 PM PST Saturday 13 3/18/95 (5) DEORBIT BURN: 77:20:39:13Z XRANGE: 268 NM ORBIT DIR: AL17 AIM PT: NOMINAL MLGTD: 1672 FT 77:21:47:01Z VEL: 201 KGS 209 KEAS HDOT: -1.4 FPS TD NORM 195: 2980 FT NLGTD: 6240 FT 77:21:47:14Z VEL: 151 KGS HDOT: -6.3 FPS DRAGCHUTE DEPLOY: 147 KEAS 77:21:47:16Z BRK INIT: 142 KGS DRAGCHUTE JETTISON: 54 KGS 77:21:47:43Z AVE BRK DECEL: 5.5 FPS/S WHEELS STOP: 77:21:48Z 11647 FT ROLLOUT: 9935 FT 47 SEC WINDS: H14, R5 KTS OFFICIAL: 2515P22 H14, R4 KTS DENS ALT: 3481 FT Continued. . .	104/104/ 109% PREDICTED: 100/104/104/ 70/104 ACTUAL: 100/104/104/ 67/104 SSME S/N: 1 = 2012 (16) 2 = 2033 (7) 3 = 2031 (12) M 3 EOM: WEIGHT: 217646 LBS X CG: 1083.5 LANDING: WEIGHT: 217437 LBS X CG: 1085.0	BI-071 RSRM 43 ET-69 LWT 62 ET RPT 271K ET BR/UP 214K ET IMPACT 1:22:37 MET LAT: 15.5 S LONG: 159.45 W	28.45 (37) DIRECT INSERTION POST OMS-2: 190.4 X 187.3 NM DEORBIT: 193 X 182 NM VELOCITY: 25852 FPS ENTRY RANGE: 4216 NM	OI-23 (5)	CARGO: 28528 LBS PAYLOAD CHARGEABLE: 20067 LBS DEPLOYED: 0 LBS NON-DEPLOYED: 18303 LBS MIDDECK: 1764 LBS SHUTTLE ACCUMULATED WEIGHTS: DEPLOYED: 766624 LBS NON-DEPLOYED: 948654 LBS CARGO TOTAL: 2036386 LBS PERFORMANCE MARGINS (LBS): FPR: 3775 FUEL BIAS: 1136 FINAL TDDP: 4099 RECON: 6754 PAYLOADS: PLB: ASTRO-2 GAS-2 MIDDECK: CMIX, PGS-TCS PGS-STES SAREX-2, MACE 5 + 4 EDO CRYO TK SETS EDO PALLET RMS 41 (S.N. 303) RMS NOT USED	KSC W/D: OPF 81, VAB 5, PAD 19 = 105 days total. LAUNCH POSTPONEMENTS: - Launch date baselined as 11/3/94 on 6/24/93 - Postponed launch to 12/1/94 on 11/5/93 - Postponed launch to 1/12/95 on 3/25/94 - Postponed launch to 2/23/95 on 9/26/94 - Postponed launch to 3/2/95 on 11/30/94 LAUNCH SCRUBS: None LAUNCH DELAYS: - Delayed coming out of T-9 min hold awaiting confirmation that FES feedline B heater 1 was operating after switching from heater 2 at T-18 mins. Launch delay of 1M13S. TAL WX: - Ben Guerir (prime & selected) and Moron were forecast and observed GO. Banjul was not available because of local instability. DOLILU/NOMINAL I-LOADS: - Both DOLILU and nominal were NO GO for ET load indicator ES-73 using L-1 data base. Using M data base, both were GO, DOLILU was selected because we had a better data base at MACH 1.4. An LSEAT waiver was written. NIGHT LAUNCH: Space Shuttle night launch #11. FLIGHT DURATION CHANGES/LANDING SITE CHANGE: - Waved off landing at KSC on orbits 246, 247, and 248 because of forecast ceiling violations and thunderstorms within 30 nm. Extended flight 1 day. - Waved off landing at KSC on orbits 262 and 263. Forecast of low ceiling and 0.2 cloud cover under 12K. Decision made to change landing site to EDW. - Total flight duration extension 1 day plus 1 orbit. LANDING SITE CHANGE: KSC to EDW EVENTS: - Most persons in orbit at one time, total eleven (11). Mir 18 was launched at 9:11 a.m. Moscow time (12:11 a.m. CST) on March 14 from Baikonur cosmodrome with Norm Thagard, Vladimir Dezhurov and Gennady Strekalov on board (planned return on Atlantis on STS-71). Three Russians went on Mir plus 7 Americans on Endeavor). Continued. . .

STS067-317-002 Crew in aft flight deck: Front (lt to rt): Jernigan/PLC, CDR Oswald, and PLT Gregory. Back (lt to rt) Lawrence/MS, Parise/PS, Durrance/PS; and Grunsfeld/MS.

FLT NO.	ORBITER	CREW (7) TITLE, NAMES & EVA'S	LAUNCH SITE, LIFTOFF TIME, LANDING SITES, ABORT TIMES	LANDING SITE/ RUNWAY, CROSSRANGE LANDING TIMES FLT DURATION, WINDS	SSME-TL NOM-ABORT EMERG THROTTLE PROFILE ENG. S.N.	SRB RSRM AND ET	ORBIT		FSW	PAYLOAD WEIGHTS, PAYLOADS/ EXPERIMENTS	MISSION HIGHLIGHTS (LAUNCH SCRUBS/DELAYS, TAL WEATHER, ASCENT I-LOADS, FIRSTS, SIGNIFICANT ANOMALIES, ETC.)
							INC	HA/HP			
STS-67 Continued			Continued. . . FLT DURATION: 16:15:08:48 S/T: 516:22:32:16 OV-105: 82:17:54:11 DISTANCE: 6,892,836 sm								Continued . . . SIGNIFICANT ANOMALIES: - Spacelab SCOS cache addressing error. - FES primary A failed to come out of standby. - Noisy supply water tank D quantity transducer. - High N_2 flow on PCS system 2, 14.7 cabin regulator. - Middeck audio terminal unit failure (main bus current spike). - CCPI failure to power portable light or camcorder. - Handheld mike was inoperative on both middeck and airlock ATU's. Possible short. - TEAC 8 mm VCR anomaly (degraded picture quality). - Unexplained external IPS disturbances. Pointing performance degraded. - Water spray boiler 2 excessive water usage (most of water was accidentally off-loaded prelaunch.) - L5D oxidizer injector temperature erratic (GMEM uplinked). - R4R jet fail leak, jet stopped leaking at 21:53 MET.

STS067-713-072 ASTRO-2 cluster of telescopes and Instrument Pointing System in payload bay.

STS067-368-008 Oswald (center), Grunsfeld (back), and Gregory (Right) involved in Middeck Experiments.

STS067-721A-087 Flying over the "Roof of the World", the Plateau of China. Himalalyan (foreground) & Gangdise Mountains.

Sts067-s-046-- Space Shuttle Program Manager (and former Flight Director), Tommy Holloway, presents STS-67 Wall Plaque to Flight Control Team for "Mission Well Done".

sts067-s-041 -- Glynn Lunney (left), VP & Program Manager USA (and former NASA Flight Director & Shuttle Porgram MGR) and Flight Director Randy Stone in MCC.

SPACE SHUTTLE MISSIONS SUMMARY

FLT NO.	ORBITER	CREW (10) 7 UP, 8 DOWN / TITLE, NAMES & EVA'S	LAUNCH SITE, LIFTOFF TIME, LANDING SITES, ABORT TIMES	LANDING SITE/ RUNWAY, CROSSRANGE / LANDING TIMES FLT DURATION, WINDS	SSME-TL NOM-ABORT EMERG / THROTTLE PROFILE ENG. S.N.	SRB RSRM AND ET	INC	ORBIT HA/HP	FSW	PAYLOAD WEIGHTS, PAYLOADS/ EXPERIMENTS	MISSION HIGHLIGHTS (LAUNCH SCRUBS/DELAYS, TAL WEATHER, ASCENT I-LOADS, FIRSTS, SIGNIFICANT ANOMALIES, ETC.)
STS-71 SEQ FLT #69 KSC-69 PAD 39A-42 MLP-3	OV-104 (Flight 14) Atlantis Spacelab-Mir LM-11 17th Spacelab Flight OMS PODS: LPO3-18 RPO4-14 FRC4-14	CDR: Robert L. (Hoot) Gibson (Flt 5 - STS-41-B, STS 61-C, STS-27, STS-47) P385/R30/V27/M29 PLT: Charles J. Precourt (Flt 2 - STS-55) P386/R161/V118/M141 M/S 1 (PAYLOAD CDR): Ellen S. Baker (Flt 3 - STS-34, STS-50) P387/R105/V75/F10 M/S 2: Gregory T. Harbaugh (Flt 3 - STS-39, STS-54) P388/R125/V88/M112 M/S 3: Bonnie J. Dunbar (Flt 4 - STS 61-A, STS-32, STS-50) P389/R79/V49/F7 MIR 19 CREW UP: MIR-19 CDR: Anatoly Y. Solovyev P390/R193/M168 MIR-19 FLIGHT ENGINEER: Nikolai Budarin P391/R194/M169 MIR-18 CREW DOWN: MIR-18 CDR: Vladimir Dezhurov P392/R195/M170 ...Continued ...	KSC 39A 178:19:32:18.95Z 3:32:19 PM EDT (P) 3:32:19 PM EDT (A) Tuesday 9 6/27/95 (7) LAUNCH WINDOW: 10M19S Mir Planar/ Phase Window EOM PLS: KSC TAL: ZZA TAL WX: MRN, BEN SELECTED: RTLS: KSC 33/CI/N TAL: MRN 20/N/N AOA: NOR 23/N/N PLS: EDW 22/N/N TDEL: -0.13 0.192/0.23 MAX Q NAV: 708 716 SRB SEP: 2:03.7 1:59:10 PERF: NOMINAL 2 ENG TAL (MRN): 2:25 2:31 NEG RETURN: 4:04 4:05 PTA (U/S 267): 4:39 4:32 DROOP (ZZA): 5:21 5:23 PTM: 6:02 5:56 SE TAL (ZZA): 5:58 6:07 SE PTM (U/S 784): 7:01 6:59 Continued . . .	KSC 15 (KSC 23) 188:14:54:35Z 10:54:35 AM EDT Friday 8 7/7/95 (6) DEORBIT BURN: 188:13:45:19Z XRANGE: 645 NM ORBIT DIR: AL 18 AIM PT: NOMINAL MLGTD: 2243 FT 188:14:54:35Z VEL: 206 KGS 201 KEAS HDOT: -1.8 FPS TD NORM 195: 2575 FT DRAG CHUTE DEPLOY: 184 KEAS 108:14:54:39Z NLGTD: 5471 FT 188:14:54:44Z VEL: 166 KGS HDOT: -6.0 FPS BRK INIT: 144 KGS DRAG CHUTE JETTISON: 52 KGS 188:14:55:09Z AVE BRK DECEL: 5.6 FPS/S WHEELS STOP: 188:14:55:28Z 10607 FT ROLLOUT: 8364 FT 53 SEC WINDS: T3, L5 KTS OFFICIAL: 0307 T4, L6 DENS ALT: 1376 FT FLT DURATION: 9:19:22:15 S/T: 526:12:54:31	104/104/ 109% PREDICTED: 100/104/104/ 68/104 ACTUAL: 100/104/104/ 68/104 SSME S/N: 1 = 2028 (10) 2 = 2034 (6) 3 = 2032 (3) M 3 EOM: WEIGHT: 216527 LBS X CG: 1079.7 LANDING: WEIGHT: 216352 LBS X CG: 1081.3	BI-072 RSRM 45 ET-70 LWT 63 ET RPT 271.3K ET BR/UP 214K ET IMPACT 1:26:57 MET LAT: 0.08 S LONG: 125.4 W	51.63 (2)	DIRECT INSERTION POST OMS-2 159.5 x 85.2 NM DOCKING CAPTURE: 1/17:27:57 MET HARD MATE: 1/17:35:54 MET SHUTTLE HATCH OPEN: 1/19:28:56 MET HAND SHAKE: 1/19:28:56 MET SOYOZ UNDOCKING: 6/15:32:34 MET DEORBIT 215 X 209 NM VELOCITY: 25913 FPS ENTRY RANGE: 4321 NM	OI-24 (1)	CARGO: 26577 LBS PAYLOAD CHARGEABLE: 17941 LBS DEPLOYED: 0 LBS NON-DEPLOYED: 17251 LBS MIDDECK: 690 LBS SHUTTLE ACCUMULATED WEIGHTS: DEPLOYED: 766624 LBS NON-DEPLOYED: 966595 LBS CARGO TOTAL: 2062963 LBS PERFORMANCE MARGINS (LBS): FPR: 3775 FUEL BIAS: 1136 FINAL TDDP: 1040 RECON: 1398 PAYLOADS: PLB: SHUTTLE-MIR MISSION 1 SL-M/LM ODS MIDDECK: IMAX, SAREX-II 5 CRYO TK SETS NO RMS	KSC W/D: OPF 115, VAB 6, PAD 44 = 165 days total. LAUNCH POSTPONEMENTS: - Baselined 5/30/95 as launch date on 10/21/93. - Changed launch date to 5/24/95 on 9/1/94. - Postponed launch date to NET 6/19/95 due to delays in SPECKTR launch. STS-70 was moved ahead of STS-71. - Postponed launch date to NET 6/22/95 due to Mir EVA's to allow time to configure Mir docking ports and solar arrays. - Postponed launch date to NET 6/23/95 (docking on FD4 would be same date as 6/24/95 launch with docking on FD3). LAUNCH SCRUBS: - Scrubbed 6/23/95 launch at T-6.25 hours when tanking window ran out. Tanking violation of lightning within 5 miles. - Scrubbed 6/24/95 launch at L-44 mins while holding at T-9 minutes due to ceiling violations, rain, and thunderstorms in KSC area. LAUNCH DELAYS: None TAL WX: - ZZA (prime) was forecast NO GO for ceiling and thunderstorms within 20 nm. MRN (selected) and BEN were both forecast and observed GO. DOLILU/I-LOADS: - Selected and uplinked, DOLILU uplink #14, I-load uplink #20, last use of DOLILU I-load. FLIGHT DURATION CHANGES: None FIRSTS/SPECIAL EVENTS: - Lowest perigee of all space shuttle flights of 85 nm (phasing maneuver) achieved during initial orbit. - Smallest OMS-2 Delta V of 75.5 FPS. - First permanent transfer of Russian/American crews (Mir-19 up and Mir-18 crew down on Atlantis - 7 up, 8 down. - Carried up orbiter docking system and attached to Mir. - First docking of U.S. & Russian spacecraft since Apollo-Soyuz in 1975. EVENTS: - Thagard lifted off from Baikonur Cosmodrome in Kazakhstan on March 14, 1995, at 9:11:00 AM local time (73:06:11:00Z). - Total Soyuz/Mir time for Thagard 107:09:57:18, total flight time 115:08:43:35. - Mir/Shuttle capture at 180:13:00:14Z, docking complete at 180:13:08:18Z. - Crews transfer time at 180:16:08:18Z (Mir 19 from Atlantis to Mir, and Mir 18 to Atlantis, when seat liners transferred to Atlantis). - Transferred equipment, experiments, 1067 lbm H_2O, 48 lbm O_2, and 87 lbm N_2 to Mir. - Undocking completed at 185:11:09:42Z. Continued. . .

STS-71 KSC-95EC-0913 Liftoff of 100th U.S. human space flight. It featured the 1st docking between the U.S. Space Shuttle and the Russian Space Station Mir.

FLT NO.	ORBITER	CREW (10) 7 UP - 8 DOWN / TITLE, NAMES & EVA'S	LAUNCH SITE, LIFTOFF TIME, LANDING SITES, ABORT TIMES	LANDING SITE/ RUNWAY, CROSSRANGE / LANDING TIMES FLT DURATION, WINDS	SSME-TL NOM-ABORT EMERG THROTTLE PROFILE ENG. S.N.	SRB RSRM AND ET	INC	ORBIT HA/HP	FSW	PAYLOAD WEIGHTS, PAYLOADS/ EXPERIMENTS	MISSION HIGHLIGHTS (LAUNCH SCRUBS/DELAYS, TAL WEATHER, ASCENT I-LOADS, FIRSTS, SIGNIFICANT ANOMALIES, ETC.)

STS-71 Continued...

MCC FCR-1 (48)
FLIGHT DIRECTORS:
A/E - N. W. Hale
LD/O 1 - R. E. Castle
O 2 - P. L. Engelauf
PLNG - P. F. Dye
FD Moscow - W. D. Reeves
MOD - A. L. Briscoe

Continued . . .

MIR-18 FLIGHT ENGINEER:
  Gennady Strekalov
  P393/R196/M171

MIR-18 COSMONAUT RESEARCHER:
  Norman E. Thagard
  (Flt 5 - STS-7, STS 51-B, STS-30, STS-42)
  P394/R20/V14/M19

Continued . . .

MECO CMD:
8:30.72   8:31.1
VI:
25876.5   25871

OMS-2:
42:57.2   42:57.2
Delta V   75.5 FPS
TGO   =   00:47

Continued . . .

OV-104:
93:03:49:17

DISTANCE:
4,100,000 sm

Continued . . .

RENDEZVOUS #21:
- Rendezvous and dock with Russian Mir Space Station (first docking).

SIGNIFICANT ANOMALIES:
- Postflight disassembly of RSRM nozzle joint 3 revealed RTV gas paths with slight heat effect and erosion to primary O-rings of STS-71 LH RSRM and STS-70 RH RSRM. Technique developed to remove RTV from joint and do a vacuum backfill for STS-69 and STS-73 RSRM's.
- GPC 4 annunciated GPC BITE fault message followed by GPC 4 fail. Determined to be single event upset, GPC 4 was assigned string 4 and used successfully during entry.
- Slow docking module vestibule depress rate.
- H_2 manifold valve tank 1 failed open.
- Cryo O_2 tank 1 leak through flight cap of fill/drain line QD.
- H_2 manifold valve 1 microswitch failure.
- Erratic O_2 tank 5 heater temperature.
- VHF system transmit failure.
- PDIP power fail.
- S-band comm string 2 uplink problem.
- RCS jettison R2U fail off (low chamber pressure).

KSC-95EC-0544 Spacelab-Mir module and transfer tunnel at KSC. In foreground is Obiter Docking system (ODS) topped with red Russian Androgynous Peripheral Docking System (APDS).

ABOVE RIGHT: NM18-309-028 -- As Atlantis approaches Mir docking node, MCC/CSR Rep James Nise reported that MIR Cosmonaut Strekalov happily yelled, "The banana truck is here!" (A reference to the days when Russia imported bannas from Cuba.)

BELOW: Soyuz photo of Shuttle docked to MIR from link:
http://io.jsc.nasa.gov/photos/10280/hires/sts071-s-072.jpg
Provided by Gregory A. Lange JSC-/DA8

STS071-013 1995 First permanent transfer of Russian/American crews (Mir-19 up and Mir-18 crew down) on STS-71. Clockwise from Anatoly Y. Solovyev (at bottom center, arms folded) are Gregory J. Harbaugh, Robert L. Gibson, Charles J. Precourt, Nikolai M. Budarin, Ellen S. Baker, Bonnie J. Dunbar, Norman E. Thagard, Gennadiy M. Strekalov (angle) and Vladimir N. Dezhurov.

s95-16417.jpg -- MOD FD, Alan Briscoe (left) leads Post-Mission toast in CSR to success of first Shuttle-MIR docking and first permanent transfer of Russian/American crews (Mir-19 up and Mir-18 crew down).

FLT NO.	ORBITER	CREW (5) / TITLE, NAMES & EVA'S	LAUNCH SITE, LIFTOFF TIME, LANDING SITES, ABORT TIMES	LANDING SITE/ RUNWAY, CROSSRANGE / LANDING TIMES FLT DURATION, WINDS	SSME-TL NOM-ABORT EMERG / THROTTLE PROFILE ENG. S.N.	SRB RSRM AND ET	INC	ORBIT HA/HP	FSW	PAYLOAD WEIGHTS, PAYLOADS/ EXPERIMENTS	MISSION HIGHLIGHTS (LAUNCH SCRUBS/DELAYS, TAL WEATHER, ASCENT I-LOADS, FIRSTS, SIGNIFICANT ANOMALIES, ETC.)
STS-70 SEQ FLT #70 KSC-70 PAD 39B-28 MLP-2	OV-103 (Flight 21) Discovery OMS PODS: LPO1-24 RPO3-22 FRC3-21	**CDR:** Terence T. (Tom) Henricks (Flt 3 - STS-44, STS-55) P395/R135/V93/M120 **PLT:** Kevin R. Kregel P396/R197/ M172 **M/S 1:** Donald A. Thomas (Flt 2 - STS-65) P397/R180/V119/M158 **M/S 2:** Nancy J. (Sherlock) Currie (Flt 2 - STS-57) P398/R165/V120/F21 **M/S 3:** Mary Ellen Weber P399/R198/F26 MCC FCR-1 (49) (A/E & TDRS DEPLOY) WHITE FCR (1) (ON ORBIT OPS) FLIGHT DIRECTORS: A/E - R. D. Jackson LD/O 2 - R. M. Kelso O 1 - J. P. Shannon PLNG - B. P. Austin MDR 1 - C. W. Shaw MDR 2 - J. M. Heflin MOD - B. R. Stone	KSC PAD 39B 194:13:41:55Z 9:41:00 AM EDT (P) 9:41:55 AM EDT (A) Thursday 19 7/13/95 (4) LAUNCH WINDOW: 2H30M CTOB EOM PLS: KSC TAL: BEN TAL WX: MRN SELECTED: RTLS: KSC 15/N/N TAL: BEN 36/N/N AOA: EDW 22/N/N PLS: EDW 22/N/N TDEL: 0.0 0.12/.05 MAX Q NAV: 692 686 SRB STG: 2:02.7 2:05 PERF: NOMINAL 2 ENG TAL (BEN): NEG RETURN: 3:59 4:03 PTA (U/S 244): 5:03 5:01 DROOP (BYD): 5:00 5:31 PTM (U/S): 5:46 5:47 SE TAL (BYD): 5:59 6:06 SE PTM (U/S 537): 7:01 7:01 MECO CMD: 8:30.75 8:30.7 VI: 25876 25874 OMS-2: 39:54.9 39:55 DELTA V 222 FPS	KSC 33 (KSC 24) 203:12:02:00Z 8:02:00 AM EDT Saturday 14 7/22/95 (6) DEORBIT BURN: 203:11:00:13Z XRANGE: 430 NM ORBIT DIR: DL 33 AIM PT: NOMINAL MLGTD: 2601 FT 203:12:02:00Z VEL: 198 KGS 194 KEAS HDOT: -1.4 FPS TD NORM 195: 2400 FT DRAG CHUTE DEPLOY: 189 KEAS 203:12:02:03Z NLGTD: 5460 FT 203:12:02:09Z VEL: 164 KGS HDOT: -6.1 FPS BRK INIT: 89 KGS DRAG CHUTE JETTISON: 59 KGS 203:12:02:35Z AVE BRK DECEL: 4.6 FPS/S WHEELS STOP: 203:12:02:58Z 11066 FT ROLLOUT: 8465 FT 58 SEC WINDS: T2, L2 KTS OFFICIAL: 2005 P8 T3, L4 KTS DENS ALT: 1117 FT FLT DURATION: 8:22:20:05 214:20:05 S/T: 535:16:14:36 OV-103: 145:23:49:54 DISTANCE: 3,700,000 sm	104/104/ 109% PREDICTED: 100/104/104/ 67/104 ACTUAL: 100/104/104/ 67/104 SSME S/N: 1 = 2036 (1) 2 = 2019 (15) 3 = 2017 (12) M 3 EOM: WEIGHT: 194267 LBS X CG: 1097.2 LANDING: WEIGHT: 194190 LBS X CG: 1099.1	BI-073 RSRM 44 ET-71 LWT 64 ET RPT 271K ET BR/UP 214K ET IMPACT 1:20:13 MET LAT: 13.75 S LONG: 163 W	28.45 (38)	DIRECT INSERTION POST OMS-2: 160.9 X 160.7 NM DEORBIT: 166 X 155 NM VELOCITY: 25789 FPS ENTRY RANGE: 4265 NM	OI-24 (2)	CARGO: 46799 LBS PAYLOAD CHARGEABLE: 44445 LBS DEPLOYED: 37714 LBS NON-DEPLOYED: 5585 LBS MIDDECK: 1086 LBS SHUTTLE ACCUMULATED WEIGHTS: DEPLOYED: 804398 LBS NON-DEPLOYED: 973266 LBS CARGO TOTAL: 2109762 LBS PERFORMANCE MARGINS (LBS): FPR: 3775 FUEL BIAS: 1136 FINAL TDDP: 3789 RECON: 5299 PAYLOADS: PLB: TDRS-G/IUS (DEPLOYED) MIDDECK: PARE/NIH-R, BDS, CPCG, STL/NIH-C, BRIC(2), SAREX-II, VFT-4, HERCULES, MIS-B, MSX, MAST, WINDEX, RME-III 4 CRYO TK SETS NO RMS	KSC W/D: OPF 63, VAB 14 (2) PAD 43 (2) = 120 days total. LAUNCH POSTPONEMENTS: - Baselined launch date 6/29/95 on 3/18/94. - Advanced launch date to 6/22/95 on 9/26/94. - Advanced launch date to 6/8/95 on 5/2/95, moving STS-70 ahead of STS-71. Delays on SPEKTR launch & docking with Mir caused STS-71 launch to be postponed. - Postponed 6/8/95 launch to 7/13/95 on 6/2/95 based on decision to rollback to VAB and repair holes (>200) in ET caused by a pair of woodpeckers (Northern Flickers). Moved STS-70 after STS-71. LAUNCH SCRUBS: None LAUNCH DELAYS: - Launch delayed 55 seconds while holding at T-31 seconds due to Range Safety ET destruct package receiver fluctuating AGC (possible multipath). TAL WX: - BEN was prime and selected. MRN was forecast and observed NO GO due to crosswinds. Banjul in plane site was down for runway repair. DOLILU/NOMINAL I-LOADS: - First planned use of DOLILU II I-loads. DOLILU II was selected and uplinked. DOLILU II uplink #1, I-load uplink #21. FLIGHT DURATION CHANGES: - Waved off landing at KSC on orbits 127 and 128 because of forecast and observed low ceiling and ground fog. - Waved off landing at KSC on orbit 142. Weather was observed GO but marginal with potential for ground fog but observed GO at landing time. - Total flight extensions 1 day plus 1 orbit. FIRSTS: - First flight to be controlled by White FCR in new MCC (Bldg 30S) for most of orbit operations. Ascent and entry plus early and late orbit ops being controlled from old MCC FCR-1. - First flight with Block I SSME (2036). SIGNIFICANT ANOMALIES: - Postflight disassembly of RSRM nozzle joint 3 revealed gas paths with slight heat effect and corrosion to primary o-ring of STS-70 RH RSRM. - Erratic supply water tank C transducer. - Ops recorder 2 track 3 degradation. - Vacuum cleaner power cable pinched (IFM fixed). - Crew reported W6 impact crater. - Lost MOC capability when MOC went to 100% CPU.

STS-70 mission patch (HENRICKS, KREGEL, CURRIE, THOMAS, WEBER)

1st flight of one Block I SSME, shown in test, courtesy, Dan Hausman,/P&W/ Rocketdyne/ KSC

STS070-368-003 -- Inflight crew portrait With Ohio flag as backdrop: Left to right, Thomas/MS, Currie/MS, CDR Henricks, Weber/MS, PLT Kregel.

| FLT NO. | ORBITER | CREW (5) TITLE, NAMES & EVA'S | LAUNCH SITE, LIFTOFF TIME, LANDING SITES, ABORT TIMES | LANDING SITE/ RUNWAY, CROSSRANGE LANDING TIMES FLT DURATION, WINDS | SSME-TL NOM-ABORT EMERG THROTTLE PROFILE ENG. S.N. | SRB RSRM AND ET | ORBIT | | FSW | PAYLOAD WEIGHTS, PAYLOADS/ EXPERIMENTS | MISSION HIGHLIGHTS (LAUNCH SCRUBS/DELAYS, TAL WEATHER, ASCENT I-LOADS, FIRSTS, SIGNIFICANT ANOMALIES, ETC.) |
							INC	HA/HP			
STS-69 SEQ FLT #71 KSC-71 PAD 39A-43 MLP-1	OV-105 (Flight 9) Endeavour OMS PODS: LPO4 - 16 RPO5 - 7 FRC5 - 9	CDR: David M. Walker (Flt 4 - STS 51-A, STS-30, STS-53) P400/R48/V40/M45 PLT: Kenneth D. Cockrell (Flt 2 - STS-56) P401/R159/V121/M140 M/S 1 (PAYLOAD CDR): James S. Voss (Flt 3 - STS-44, STS-53) P402/R136/V85/M121 M/S 2/EV-1: James H. Newman (Flt 2 - STS-51) P403/R168/V122/M146 M/S 3/EV-2: Michael L. Gernhardt P404/R199/M173 SS EVA #30 EMU/Tethered EVA Scheduled EVA #26 EVA flight test (EDFT) #2 to evaluate space suit mods to protect space walkers from the cold of space, including heated gloves & LCVG leg bypass) PET 6H46M11S .	KSC 39A 250:15:08:59.96Z 11:09:00 AM EDT (P) 11:09:00 AM EDT (A) Thursday 20 9/7/95 (7) LAUNCH WINDOW: 2H30M CTOB EOM PLS: KSC TAL: BEN TAL WX: MRN SELECTED: RTLS: KSC 15/CI/N TAL: BEN 36/N/N AOA: EDW 22/N/N PLS: EDW 22/N/N TDEL: 0.0 0.032/-0.09 MAX Q NAV: 705 PSF 715 PSF SRB SEP: 2:03.7 1:59.1 PERF: NOMINAL 2 ENG TAL (BEN): 2:40 2:49 NEG RETURN: 4:01 4:02 PTA (U/S 328): 4:18 4:14 DROOP (BYD): 5:28 5:30 PTM (U/S 328): 5:24 5:24 SE TAL (BYD): 5:51 5:52 LAST TAL (BEN): 6:28 MECO CMD: 8:30.2 8:30.2 MECO VI: 25946 25940 OMS-2: 41:43 41:43 293.4 FPS 293.4 FPS	KSC 33 (KSC 25) 261:11:37:55Z 7:37:55 AM EDT Monday 14 9/18/95 (8) DEORBIT BURN: 261:10:35:13Z XRANGE: 202 NM ORBIT DIR: DL 34 AIM PT: CLOSE IN MLGTD: 1912 FT 261:11:37:55Z VEL: 218 KGS 212 KEAS HDOT: -4 FPS TD NORM 205: 2468 FT DRAG CHUTE DEPLOY: 187 KEAS 261:11:38:03Z NLGTD: 6325 FT 261:11:38:08Z VEL: 167 KGS HDOT: -6.5 FPS BRK INIT: 97 KGS DRAG CHUTE JETTISON: 62 KGS 261:11:38:36Z AVE BRK DECEL: 5.6 FPS/S WHEELS STOP: 261:11:38:55Z 12142 FT ROLLOUT: 10230 FT 60 SEC WINDS: T2, L4 KTS OFFICIAL: 2205P06, T2, L5 KTS DENS ALT: 1315 FT FLT DURATION: 10:20:28:55 S/T: 546:12:43:31 OV-105: 93:14:23:06 DISTANCE: 4,500,000 sm	104/104/ 109% PREDICTED: 100/104/104/ 67/104 ACTUAL: 100/104/104/ 67/104 1 = 2035 (2) 2 = 2109 (16) 3 = 2029 (12) M 3 EOM: WEIGHT: 219395 LBS X CG: 1080.7 LANDING: WEIGHT: 219298 LBS X CG: 1082.3	BI-074 RSRM 48 KM ET-72 LWT 65 ET RPT 271K ET BR/UP 214K ET IMPACT 1:24:54 MET LAT: 18.8 S LONG: 151.9 W	28.45 (39)	DIRECT INSERTION POST OMS-2: 201 x 199 NM DEORBIT: 186 x 181 NM VELOCITY: 25839 FPS ENTRY RANGE: 4332 NM	OI-24 (3)	CARGO: 31549 LBS PAYLOAD CHARGEABLE: 25346 LBS DEPLOYED: 0 LBS NON-DEPLOYED: 16739 LBS MIDDECK: 1301 LBS SHUTTLE ACCUMULATED WEIGHTS: DEPLOYED: 804398 LBS NON-DEPLOYED: 991306 LBS CARGO TOTAL: 2141311 LBS PERFORMANCE MARGINS (LBS): FPR: 3775 FUEL BIAS: 1136 FINAL TDDP: 5409 RECON: 7966 PAYLOADS: PLB: WSF (Wakeshield Facility), IEH, Spartan-201-03 CAPL-II/GBA MIDDECK: STL/NIH-C CGBA, BRIC, EPICS CMIX 5 CRYO TK SETS RMS 42 (S.N. 303) RMS USED TO DEPLOY AND RETRIEVE SPARTAN AND WSF. SUPPORT FOR EVA AND CLAWS.	KSC W/D: OPF 81, VAB 7 PAD 47 (2) = 135 days total. LAUNCH POSTPONEMENTS: - Baselined launch date of 3/16/95 on 11/18/93. - Postponed launch date to 5/4/95 on 3/24/94. - Postponed launch date to 7/20/95 on 10/6/94. - Postponed launch date to 8/5/95 caused by delays in STS-71 and STS-70. - Postponed launch date to 8/31/95 while program analyzed RTV gas paths in nozzle joint #3 on STS-71 and STS-70, then developed a fix for STS-69. - Rolled back to VAB on 8/1/95 under threat of Hurricane Erin. - Returned to pad on 8/8/95. LAUNCH SCRUBS: - Scrubbed 8/31/95 launch at approx. L-7.5 hours when fuel cell 2 condenser exit temperature exceeded LCC limit of 160 deg F. - Rescheduled launch for 9/7/95. LAUNCH DELAYS: None TAL WX: - BEN (prime and selected), MRN forecast NO GO for ceiling and rain but observed GO 10 mins prior to landing time. DOLILU II/NOMINAL I-LOADS: - Nominal I-loads were not certified for September. DOLILU-II I-loads uplinked. DOLILU-II uplink #2, total DOLILU uplink #16 I-load uplink #22. FLIGHT DURATION CHANGES: None EVENTS: - SPARTAN released 1:00:38:59, grapple 2:23:53, latched 3:00:03 MET. - WSF released 3:20:16:15, grapple 6:22:50:11 MET. RENDEZVOUS #22: - Rendezvous, grapple & berth WSF. RENDEZVOUS #23: - Rendezvous, grapple & berth SPARTAN 201-03. SIGNIFICANT ANOMALIES: - CRT 1 dim display. - Fuel cell 2 condenser exit temp high (scrubbed launch attempt). - Waste dumpline blockage. IFM to bypass dump filter was unsuccessful, so off loaded waste tank into CWC. - EVA power tool failed. - Portable foot restraint fit problem. - S-band preamp 2 degraded causing intermittent forward link. - Middeck speaker ATU failure. - Camcorder tape eject failure. - Camera D downlink lost. - Loss of Ku-band forward link. - Random ops recorder commands issued when panel brightness control adjusted in new MCC. - Hydraulics pump 3 stuck in norm press (cycled switch twice to get response then started APU - WSB 3 lub oil overcooling during entry.

MCC FCR-1 (50) (A/E) WHITE FCR (2) (ORBIT) FLIGHT DIRECTORS: A/E - N. W. Hale LD/O 1 - J. W. Bantle O 2 - P. F. Dye PLNG - G. A. Pennington MOD - A. L. Briscoe

S95-07799 -- FD's team in MCC. FD Al Pennington (left front) & CAPCOM David Wolf shaking hands.

sts069- 714-042-Voss (top) & Gernhardt EVA

STS069-715-050 Crew in middeck: Front (lt to rt) PLT Cockrell and CDR Walker. Backrow: (lt to rt) Voss/MS/PLC, Gearhardt/MS, and Newman/MS.

FLT NO.	ORBITER	CREW (7) TITLE, NAMES & EVA'S	LAUNCH SITE, LIFTOFF TIME, LANDING SITES, ABORT TIMES	LANDING SITE/ RUNWAY, CROSSRANGE LANDING TIMES FLT DURATION, WINDS	SSME-TL NOM-ABORT EMERG THROTTLE PROFILE ENG. S.N.	SRB RSRM AND ET	INC	ORBIT HA/HP	FSW	PAYLOAD WEIGHTS, PAYLOADS/ EXPERIMENTS	MISSION HIGHLIGHTS (LAUNCH SCRUBS/DELAYS, TAL WEATHER, ASCENT I-LOADS, FIRSTS, SIGNIFICANT ANOMALIES, ETC.)
STS-73 SEQ FLT #72 KSC-72 PAD - 39B - 29 MLP-3	OV-102 (Flight 18) Columbia 18th Spacelab Flight LM-12 EDO 6 OMS PODS: LPO5-7 RPO1-22 FRC2-18	CDR: Kenneth D. Bowersox (Flt 3 - STS-50, STS-61) P405/R146/V97/M130 PLT: Kent V. Rominger P406/R200/M174 M/S 1: Catherine G. Coleman P407/R201/F27 M/S 2: Michael E. Lopez-Alegria P408/R202/M175 M/S 3/Payload CDR: Kathryn C. Thornton (Flt 4 - STS-33, STS-49, STS-61) P409/R107/V73/F11 P/S 1: Fred Leslie P410/R203/M176 P/S 2: Al Sacco, Jr. P411/R204/M177 MCC FCR-1 (51) (ASCENT/ENTRY) WHITE FCR (3) (ORBIT OPS) FLIGHT DIRECTORS: A/E - R. D. Jackson O 1 - B. P. Austin LD/O 2 - G. A. Pennington O 3 - J. P. Shannon O 4 - R. M. Kelso MOD - A. L. Briscoe	KSC 39 PAD B 293:13:52:59.98Z 9:53:00 AM EDT (P) 9:53:00 AM EDT (A) Friday 14 10/20/95 (8) LAUNCH WINDOW: 2H30M CTOB Extended to 3H45M (BEN Darkness) EOM PLS: KSC TAL: BEN TAL WX: MRN, ZZA SELECTED: RTLS: KSC 33/N/N TAL: BEN 36/N/N AOA: EDW 22/N/N PLS: EDW 04/N/N TDEL: 0.00 -0.078/-0.04 MAX Q NAV: 708 713 SRB STG: 2:04.5 2:04 PERF: NOMINAL 2 ENG TAL (BEN): 2:48 2:47 NEG RETURN: 3:59 4:02 PTA (U/S): 5:29 5:19 DROOP (109): 5:28 5:19 PTM (U/S-220): 6:00 5:48 SE TAL (BEN): 6:02 6:08 MECO CMD: 8:29.5 8:29.7 VI: 25866 25860 OMS-2: 41:29 41:29 186.1 FPS 186.0 FPS	KSC 33, (KSC 26) 309:11:45:21Z 7:45:21 AM EDT Sunday 10 11/5/95 (10) DEORBIT BURN: 309:10:46:40Z XRANGE: 231 NM ORBIT DIR: DR 14 AIM PT: CLOSE IN MLGTD: 2500 FT 309:11:45:21Z VEL: 214 KGS 212 KEAS HDOT: -1.7 FPS TD NORM 205: 3079 FT DRAG CHUTE DEPLOY: 187 KEAS 309:11:45:29Z NLGTD: 7098 FT 309:11:45:29Z VEL: 157 KGS HDOT: -5.7 FPS BRK INIT: 125 KGS DRAG CHUTE JETTISON: 50 KGS 309:11:45:58Z AVE BRK DECEL: 6.0 FPS/S WHEELS STOP: 309:11:106:17Z 11532 FT ROLLOUT: 9032 FT 71 SEC WINDS: H3, R4 KTS OFFICIAL: 0305P07, H2, R4 KTS DENS ALT: 206 FT FLT DURATION: 15:21:52:21 381:52:21 S/T: 562:10:35:52 OV-102: 167:08:37:14 DISTANCE: 6,600,000 sm	104/104/ 109% PREDICTED: 100/104/104/ 67/104 ACTUAL: 100/104/104/ 67/104 SSM3 S/N: 1 = 2037 (1) 2 = 2031 (3) 3 = 2038 (1) M 3 EOM: WEIGHT: 230603 LBS X CG: 1080.7 LANDING: WEIGHT: 230479 LBS X CG: 1082.3	BI-075 RSRM 50 ET-73 LWT 67 ET RPT 271K ET BR/UP 214K ET IMPACT 1:24:50 MET LAT: 2.8 S LONG: 138.97 W	39.0 (4)	DIRECT INSERTION POST OMS-2: 151 X 147 NM DEORBIT: 140 x 136 NM VELOCITY: 25744 FPS ENTRY RANGE: 4519 NM	OI-24	CARGO: 33705 LBS PAYLOAD CHARGEABLE: 25310 LBS DEPLOYED: 0 LBS NON-DEPLOYED: 23302 LBS MIDDECK: 2008 LBS SHUTTLE ACCUMULATED WEIGHTS: DEPLOYED: 804398 LBS NON-DEPLOYED: 1016616 LBS CARGO TOTAL: 2175016 LBS PERFORMANCE MARGINS (LBS): FPR: 3775 FUEL BIAS: 1136 FINAL TDDP: 1906 RECON: 4902 PAYLOADS: PLB: U.S. MICROGRAVITY LABORATORY (USML-2) FLUIDS PHYSICS, MATERIALS SCIENCE, BIOTECHNOLOGY, AND COMBUSTION SCIENCE OARE MIDDECK: 5 + 4 EDO CRYO TANK SETS EDO PALLET NO RMS	KSC W/D: OPF 100, VAB 7, PAD 48 = 155 days total. LAUNCH POSTPONEMENTS: - Baselined 9/24/95 as launch date on 6/30/94. - Postponed launch to 9/28/95 on 9/8/95 caused by delay to STS-69 launch (RSRM nozzle joint #3 repairs). LAUNCH SCRUBS: - Scrubbed 9/28/95 launch at L-5:40 hrs when engine #1 main fuel valve leaked hydrogen. Rescheduled launch for 10/5/95. - Scrubbed 10/5/95 launch prior to L-1 day MMT due to forecast of high winds and rain under influence of Hurricane Opal, rescheduled launch for 10/6/95. - Scrubbed 10/6/95 launch at L-6:35 hrs while holding up tanking due to failure to service hydraulic sys 1 NLG section when MFV was replaced. Rescheduled launch for 10/7/95. - Scrubbed 10/7/95 launch while holding at T-20 minutes due to MEC 1, CORE B failure. Rescheduled launch for 10/14/95. - Scrubbed 10/14/95 launch at L-1 day MMT to measure high pressure oxidizer duct weld after test stand duct failure caused an oxidizer leak. Rescheduled launch for 10/15/95. - Scrubbed 10/15/95 launch while holding at T-5 mins. due to forecast and observed range and RTLS NO GO for ceiling (launch window extended to 3H49M (BEN dark). LAUNCH DELAYS: - Launch delayed 3M0S while holding at T-5 mins. due to R/S command problem. TAL WX: - BEN (prime & selected) with MRN and ZZA forecast and observed GO. DOLILU-II/NOMINAL I-LOADS: Both GO - DOLILU-II selected and uplinked . DOLILU uplink #3, DOLILU uplink #17, total uplink #23. FLIGHT DURATION CHANGES: None FIRSTS: - First flight with 2 block I SSME's (S/N 2037 & 2038). SIGNIFICANT ANOMALIES: - CRT-2 display flickered (IFM to replace with ORT-4). - FES feedline A mid 2 thermostat/heater failure. - FCL 1 P/L head exchanger flow degraded. - FC 3 cell performance monitor failed. - H₂ manifold valve tank 1 failed open. - S-band lower right quad antenna degraded. - Spacelab high rate dump data bad. - APU 1 fuel pump inlet pressure decrease. - F1F jet failed off, chamber pressure deceased. - R5D and R5R transient fail off. - TDRSS STGT failure.

STS073-736-018 - -Crew worked in this science module in PLB, shown here flying over Africa.

STS073-303-015 Crew portrait in science module: Front (arms folded), Lopez-Alegria/MS. Others, counter clockwise from him, Thornton/PLC, Coleman/MS, Sacco/PS, PLT Rominger, Leslie PS, & CDR Bowersox.

SPACE SHUTTLE MISSIONS SUMMARY

FLT NO.	ORBITER	CREW (5) TITLE, NAMES & EVA'S	LAUNCH SITE, LIFTOFF TIME, LANDING SITES, ABORT TIMES	LANDING SITE/ RUNWAY, CROSSRANGE LANDING TIMES FLT DURATION, WINDS	SSME-TL NOM-ABORT EMERG THROTTLE PROFILE ENG. S.N.	SRB RSRM AND ET	INC	ORBIT HA/HP	FSW	PAYLOAD WEIGHTS, PAYLOADS/ EXPERIMENTS	MISSION HIGHLIGHTS (LAUNCH SCRUBS/DELAYS, TAL WEATHER, ASCENT I-LOADS, FIRSTS, SIGNIFICANT ANOMALIES, ETC.)
STS-74 SEQ FLT #73 KSC-73 PAD- 39A-44 MLP-2	OV-104 (Flight 15) Atlantis OMS PODS: LPO3-19 RPO4-15 FRC4-15	CDR: Kenneth D. Cameron (Flt 3 - STS-37, STS-56) P412/R121/V90/M109 PLT: James D. Halsell (Flt 2 - STS-65) P413/R178/V123/M156 M/S 1: Chris Hadfield (Canada) P414/R205/M178 M/S 2: Jerry L. Ross (Flt 5 - STS 61-B, STS-27, STS-37, STS-55) P415/R89/V38/M80 M/S 3: William McArthur (Flt 2 - STS-58) P416/R172/V124/M150 MCC FCR-1 (52) Ascent/Entry WHITE FCR (3) (Orbit Ops) FLIGHT DIRECTORS: A/E - N. W. Hale LD/O 1 - W. D. Reeves O 2 - P. F. Dye PLNG - P. E. Engelauf MOD - R. E. Castle sts074-716-021 -- Mir as seen from Atlantis.	KSC 39A 316:12:30:42.98Z 7:30:43 AM EST (P) 7:30:43 AM EST (A) Sunday 8 11/12/95 (10) LAUNCH WINDOW: 7 minutes MIR PLANAR/ PHASE WINDOW EOM PLS: KSC TAL: ZZA TAL WX: MRN, BEN SELECTED: RTLS: KSC 33/CI/N TAL: ZZA 30/N/N AOA: KSC 33/CI/N PLS: EDW 22/N/N TDEL: 0.04 0.122/0.16 MAX Q NAV: 711 PSF 711 PSF SRB STG: PERF: NOMINAL 2 ENG TAL (MRN): 2:22 2:22 NEG RETURN: 4:06 4:08 PTA (U/S 255): 4:27 4:22 DROOP (ZZA): 5:24 5:26 PTM (U/S 255): 6:04 6:03 SE TAG (ZZA): 5:56 5:56 SE PTM (U/S 842): 7:00 6:54 MECO CMD: 8:33.7 8:33.2 VI: 25878 25870 OMS-2: 41:50 41:51.9 212 FPS 212 FPS	KSC 33 (KSC 27) 324:17:01:27Z 12:01:29 PM EST Monday 15 11/20/95 (11) DEORBIT BURN: 324:15:53:49Z XRANGE: 612 NM ORBIT DIR: DR 15 AIM PT: NOMINAL MLGTD: 2471 FT 324:17:01:27Z VEL: 196 KGS 201 KEAS HDOT: -1.4 FPS TD NORM 195: 2955 FT DRAG CHUTE DEPLOY: 180 KEAS 324:17:01:33Z NLGTD: 5565 FT 324:17:01:37Z VEL: 156 KGS HDOT: -6.7 FPS BRK INIT: 72 KGS 324:17:02:00Z DRAG CHUTE JETTISON: 55 KGS 324:17:02:07Z AVE BRK DECEL: 5.0 FPS/S WHEELS STOP: 324:17:02:25Z 11078 FT ROLLOUT: 8607 FT 58 SEC WINDS: H6, R4 KTS OFFICIAL: 0107P10 H5, R4 DENS ALT: 670 FT FLT DURATION: 8:04:30:44 196:30:44 S/T: 570:15:06:36 OV-104 TOTAL: 101:08:20:01 DISTANCE: 3,400,000 sm	104/104/109% PREDICTED: 100/104/104/ 67/104 ACTUAL: 100/104/104/ 67/104 M 3 EOM: WEIGHT: 202767 LBS X CG: 1078.7 LANDING: WEIGHT: 202718 LBS X CG: 1080.6	BI-076 RSRM 51 ET-74 LWT-67 SSME S/N: 1 = 2012 (17) 2 = 2026 (5) 3 = 2032 (4) ET RPT 273.1K ET BR/UP 214K ET IMPACT 1:26:05 MET LAT: 0.31 S LONG: 125.6 W	51.65 (3)	DIRECT INSERTION POST OMS-2: 162 X 162 NM DEORBIT: 185 X 184 NM VELOCITY: 25840 FPS ENTRY RANGE: 4346 NM	OI-24 (4)	CARGO: 23687 LBS PAYLOAD CHARGEABLE: 14064 LBS DEPLOYED: 10015 LBS NON-DEPLOYED: 3135 LBS MIDDECK: 914 LBS SHUTTLE ACCUMULATED WEIGHTS: DEPLOYED: 814413 LBS NON-DEPLOYED: 1020665 LBS CARGO TOTAL: 2198703 LBS PERFORMANCE MARGINS (LBS): FPR: 3775 FUEL BIAS: 1136 FINAL TDDP: 1823 RECON: 3689 PAYLOADS: PLB: SHUTTLE/MIR MISSION 2 ICBC, GPP ORBITER DOCKING SYSTEM DOCKING MODULE MIDDECK: SAREX-II 5 CRYO TK SETS RMS 43 (S.N. 301) RMS used for docking module installation on Mir and monitor plume impingement.	KSC W/D: OPF 76, VAB 8 PAD 23 = 107 days total. LAUNCH POSTPONEMENTS: - Baselined launch date of 10/26/95 on 5/5/94. - Postponed launch date to 11/2/95 on 9/8/95, caused by SRB nozzle joints #3 and #4 repairs to STS-69, STS-73, and STS-74. - Advanced launch date to 11/1/95 on 10/4/95. - Postponed date to 11/16/95 on 10/27/95 caused by STS-73 launch scrubs. LAUNCH SCRUBS: - Scrubbed 11/11/95 launch at T-4 minutes while holding at T-5 mins, when all 3 TAL sites (BEN, MRN, ZZA) were forecast and observed NO GO for weather. LAUNCH DELAYS: None TAL WX: - ZZA (prime & selected) was forecast GO but observed NO GO for 7000' broken ceiling. MRN forecast and observed TO. BEN forecast observed NO GO for ceilings and crosswinds. DOLILU-II I-LOADS: - Selected and uplinked DOLILU-II I-loads, DOLILU-II uplink #4, DOLILU uplink #18, I-load uplink #23. (Last flight with nominal I-load availability). FLIGHT DURATION CHANGES: None RENDEZVOUS #24: - Rendezvous and dock with Russian Mir space station (second docking). EVENTS: - Docking module unberth 1/18:01, capture 1/18:46:12, hardmate 1/18:53:41. - Docking module APDS-1 to Mir docking at 2/17:56:57 MET, hardmate at 2/18/05:05 MET. - Transferred 993 lbm H_2O, 59 lbm O_2, and 44 lbm N_2 to Mir. - Undocking from Mir at 5/19:45:01 MET. RADIATOR DEPLOY #17: - Deployed radiator to make water available for transfer to Mir. - Port RAD deployed to make water 83:23:14 GMT. SIGNIFICANT ANOMALIES: - Fuel cell 3 cell performance monitor delta volt measurements for all 3 substacks shifted approximately 5 millivolts. - Cryo O_2 manifold tank 1 valve failed open. - PLB aft port and aft starboard lights failed. - H_2 manifold valve 1 microswitch failure. - TCS 1 lost calibration, TCS 2 self-test failures. - ODS stowage bag adapter plate jammed. - OPS-1 recorder track 8 data degradation. - Mir camcorder battery low capacity. - WSB 2 regulator pressure erratic postlanding.

STS074-318-005 -- Crew in Docking Module delivered to Mir: Holding camera at bottom center, McArthur/MS, Clockwise from him: PLT Halsell, Hadfield/MS, Ross/MS, and CDR Cameron.

FLT NO.	ORBITER	CREW (6) TITLE, NAMES & EVA'S	LAUNCH SITE, LIFTOFF TIME, LANDING SITES, ABORT TIMES	LANDING SITE/ RUNWAY, CROSSRANGE LANDING TIMES FLT DURATION, WINDS	SSME-TL NOM-ABORT EMERG THROTTLE PROFILE ENG. S.N.	SRB RSRM AND ET	ORBIT INC	HA/HP	FSW	PAYLOAD WEIGHTS, PAYLOADS/ EXPERIMENTS	MISSION HIGHLIGHTS (LAUNCH SCRUBS/DELAYS, TAL WEATHER, ASCENT I-LOADS, FIRSTS, SIGNIFICANT ANOMALIES, ETC.)
STS-72 SEQ FLT #74 KSC-74 PAD-39B-30 MLP-1	OV-105 (Flight 10) Endeavor OMS PODS: LPO4 - 17 RPO5 - 8 FRC5 - 10	CDR: Brian Duffy (Flt 3 - STS-45, STS-57) P417/R142/V94/M126 PLT: Brent W. Jett, Jr. P418/R206/M179 M/S 1/EV 1: Leroy Chiao (Flt 2 - STS-65) P419/R179/V125/M157 M/S 2/EV 3: Winston E. Scott P420/R207/M180 M/S 3: Koichi Wakata (Japan) P421/R208/M181 M/S 4/EV 2: Daniel T. Barry P422/R209/M182 SS EVA #31: EMU/Tethered EVA EVA1 - 1/14/96 to 1/15/96 Scheduled EVA #27 by EV 1 and EV 2 6H09M19S Duration SS EVA #32: EVA 2 - 1/16/96 to 1/7/96 Scheduled EVA #28 EMU/Tethered EVA by EV 1 and EV 3 6H53M41S Duration. To test and evaluate EVA hardware for Space Station use.	KSC 39B 11:09:40:59:98Z 4:18:00 AM EST (P) 4:41:00 AM EST (A) Thursday 21 1/11/96 (7) LAUNCH WINDOW: 49M33S SFU PLANAR/ PHASE WINDOW EOM PLS: KSC TAL: BEN TAL WX: NONE SELECTED: RTLS: KSC 15/N/N TAL: BEN 36/N/N AOA: EDW 04/CI/N PLS:EDW 04/CI/N TDEL: 0.00 0.002/0.10 MAX Q NAV: 710 PSF 713 PSF SRB STG: 2:05.1 2:05 PERF: NOMINAL 2 ENG TAL (BEN): 2:05 NO CALL NEG RETURN: 4:03 4:07 PTA (U/S 411): 3:34 3:33 DROOP: 5:23 5:24 PTM (U/S 411): 4:42 4:34 SE PTM (U/S-1073): 6:23 6:20 MECO CMD: 8:27.3 8:27.1 VI: 26025.7 26025 OMS-2: 43:30 43:30 115.7FPS 115.7 FPS	KSC 15 (KSC 28) 20:07:41:40Z 2:41:40 AM EST Saturday 15 1/20/96 (6) DEORBIT BURN: 20:06:41:23Z XRANGE: 220 NM ORBIT DIR: DL 35 AIM PT: NOMINAL MLGTD: 3386 FT 20:07:41:40Z VEL: 193 KGS 185 KEAS HDOT: -1.7 FPS TD NORM 195: 2768 FT DRAG CHUTE DEPLOY: 179 KEAS 20:07:41:43Z NLGTD: 6574 FT 20:07:41:51Z VEL: 146 KGS HDOT: -6.7 FPS BRK INIT: 86 KGS DRAG CHUTE JETTISON: 58 KGS 20:07:42:17Z AVE BRK DECEL: 4.7 FPS/S WHEELS STOP: 20:07:42:46Z 12155 FT ROLLOUT: 8767 FT 66 SEC WINDS: T6, R2 KTS OFFICIAL: 3206P08 T6, R1 DENS ALT: -1007 FT FLT DURATION: 8:22:00:40 S/T: 579:13:07:16 OV-105: 102:12:23:46 DISTANCE: 3,700,00 sm	104/104/ 109% PREDICTED: 100/104/104/ 67/104 ACTUAL: 100/104/104/ 67/104 1 = 2028 (11) 2 = 2039 (1) 3 = 2036 (2) M 3 EOM: WEIGHT: 218496 LBS X CG: 1081.7 LANDING: WEIGHT: 218345 LBS X CG: 1083.3	BI-077 RSRM 52 ET-75 LWT-68 ET RPT 271.3K ET BR/UP 214K ET IMPACT 1:27:10 MET LAT: 18.4 S LONG: 145.5 W	28.45 (40)	DIRECT INSERTION POST OMS-2: 248 x 94.9 NM SFU GRAPPLE 2:01:16:19 MET 256.8 x 251 NM ORBIT ADJ: 2:04:56:13 MET 254.7 x 164.9 NM CIRC MNVR: 2:05:43:29 MET 165.2 X 164.7 NM OAST REL: 3:01:51:53 MET 166 X 164 NM DEORBIT: 167 x 161 NM VELOCITY: 25799 FPS ENTRY RANGE: 4340 NM	OI-24 (5)	CARGO: 21018 LBS PAYLOAD CHARGEABLE: 14087 LBS DEPLOYED: 0 LBS NON-DEPLOYED: 10546 LBS MIDDECK: 898 LBS SHUTTLE ACCUMULATED WEIGHTS: DEPLOYED: 814413 LBS NON-DEPLOYED: 1032109 LBS CARGO TOTAL: 2219721 LBS PERFORMANCE MARGINS (LBS): FPR: 3775 FUEL BIAS: 1136 FINAL TDDP: 11447 RECON: 13346 PAYLOADS: PLB: SPACE FLYER UNIT (SFU) RETRIEVED (JAPAN) OAST FLYER (DEPLOYED/ RETRIEVED) SSBUV/A SLA-01/GAS (5) MIDDECK: PARE/NIH-R STL/NIH-C PCG-STES CPCG 5 CRYO TK SETS RMS 44 (S.N. 303) RMS used for SFU grapple & berth, OAST deploy & retrieve & EVA support	KSC W/D: OPF 64, VAB 5, PAD 21 = 90 days total. LAUNCH POSTPONEMENTS: - Baselined launch date of 8/24/95 on 6/6/94. - Postponed launch date to 11/30/95 on 10/6/94. - Postponed launch date to 1/11/96 on 9/8/95. LAUNCH SCRUBS: None LAUNCH DELAYS: - 23 minute launch delay while holding at T-5 minutes due to MCC old front end processor and associated problems. 100% CPU caused by not loading a necessary S/W patch. TAL WX: - No TAL site available but no TAL site required (29 seconds overlap between RTLS and AOA). BEN was manned but NO GO for ceiling. NIGHT LAUNCH: #12 NIGHT LANDING: #8 DOLILU-II I-LOADS: - First flight with only DOLILU-II I-Loads. DOLILU-II uplink #5. Total I-load uplink #24. FLIGHT DURATION CHANGES: None EVENTS: - Japanese SFU grapple at 2:01:16:19 MET, latch at 2:01:58:30 MET. Launched from Tanagashima, Japan. - OAST release 3/01:51:33 MET, grapple 5:00:06:15 MET, latch 5:00:31:40 MET. - EVA 1 started at 3:19:52:51 MET. - EVA 2 started at 5:19:59:06 MET. RENDEZVOUS #25: - Rendezvous, grapple, berth, and return of SFU. RENDEZVOUS #26: - Deploy, rendezvous, grapple, and return of OAST Flyer. SIGNIFICANT ANOMALIES: - FCS shutdowns and topping FES case icing. - EMU helmet light damage. - EMU glove cut damage. - Loss of reception in left ear piece of EV 1. - Several EDFT-03 anomalies. - OAST-FLYER unexpected trajectory dispersions. - MOC front end processors operating at 100%. - RCS jet L1A fail off with maximum chamber pressure of 16 PSI. - RCS jet R2U fail leak. Jet had oxidizer leak. - Failure of SFU solar array panels to retract for capture and berthing, jettisoned solar arrays. - SFU AHIU thermal discrepancies. Flight SFU not wired same as training SFU. - RMS wrist roll joint rate degradation. - LO₂ ET umbilical frangible nut detonator did not fire (pyro wiring problem).

MCC FCR-1 (53) ASCENT/ENTRY WHITE FCR (4) FOR ORBIT OPS

FLIGHT DIRECTORS: A/E - J. W. Bantle LD/O 1 - B. P. Austin O 2 - R. M. Kelso PLNG - J. P. Shannon MOD - J. W. Bantle & A. L. Briscoe

STS072-344-019 ---- Crew: Front, lt to rt right, Barry/MS, CDR Duffy, & Chiao/MS. Rear: Wakata/MS, PLT Jett. & Scott/MS.

TOP: EVA 2 -- Barry, lower left, & Chaio, upper right

BOTTOM: EVA 1 -- Scott in P/L bay, Chiao is out of frame.

Both EVA's used to demonstrate ISS assembly techniques .

FLT NO.	ORBITER	CREW (7) TITLE, NAMES & EVA'S	LAUNCH SITE, LIFTOFF TIME, LANDING SITES, ABORT TIMES	LANDING SITE/ RUNWAY, CROSSRANGE LANDING TIMES FLT DURATION, WINDS	SSME-TL NOM-ABORT EMERG THROTTLE PROFILE ENG. S.N.	SRB RSRM AND ET	ORBIT INC	HA/HP	FSW	PAYLOAD WEIGHTS, PAYLOADS/ EXPERIMENTS	MISSION HIGHLIGHTS (LAUNCH SCRUBS/DELAYS, TAL WEATHER, ASCENT I-LOADS, FIRSTS, SIGNIFICANT ANOMALIES, ETC.)
STS-75 SEQ FLT #75 KSC-75 PAD 39B-31 MLP-3	OV-102 (Flight 19) Columbia OMS PODS: LPO5-8 RPO1-23 FRC2-19	CDR: Andrew M. Allen (Flt 3 - STS-46, STS-62) P423/R149/V101/M133 PLT: Scott J. Horowitz P424/R210/M183 M/S 1: Jeffrey A. Hoffman (Flt 5 - STS 51-D, STS-35, STS-46, STS-61) P425/R57/V59/M52 M/S 2: Maurizio Cheli (Italy-ESA) P426/R211/M184 M/S 3: Claude Nicollier (Switzerland - ESA) (Flt 3 - STS-46, STS-61) P427/R150/V98/M134 M/S 4/PAYLOAD CDR: Franklin R. Chang-Diaz (Flt 5 - STS 61-C, STS-34, STS-46, STS-60) P428/R89/V46/M81 P/S1: Humberto Guidoni (Italy) P429/R212/M185	KSC 39B 53:20:17:59:97Z 3:18:00 PM EST (P) 3:18:00 PM EST (A) Thursday 22 2/22/96 (5) LAUNCH WINDOW: 2H30M CTOB EOM PLS: KSC TAL: BEN TAL WX: MRN SELECTED: RTLS: KSC 15/N/N TAL: BEN 36/N/N AOA: KSC 15/N/N PLS: KSC 15/N/N TDEL: 0.0 0.182/0.22 MAX Q NAV: 690 697 SRB STG: 2:06.9 2:09 PERF: NOMINAL 2 ENG TAL (BEN): 3:06 3:07 NEG RETURN: 3:57 3:59 PTA (U/S 242): 4:59 5:00 DROOP: PTM: 6:02 5:58 MECO CMD: 8:27.4 8:28.3 VI: 25877 25869 OMS-2: 39:56 39:52 223 FPS 222 FPS	KSC 33 (KSC-29) 69:13:58:20Z 8:58:20 AM EST Saturday 16 3/9/96 (6) DEORBIT BURN: 69:12:55:43Z XRANGE: 234 NM ORBIT DIR: DL 36 AIM PT: CLOSE IN MLGTD: 2175 FT 69:13:58:20Z VEL: 189 KGS 211 KEAS HDOT: -1.0 FPS TD NORM 205: 2706 FT DRAG CHUTE DEPLOY: 193 KEAS 69:13:58:28Z NLGTD: 6451 FT 69:13:58:36Z VEL: 130 KGS HDOT: -5.2 FPS BRK INIT: 100 KGS DRAG CHUTE JETTISON: 62 KGS AVE BRK DECEL: 3.8 FPS/S WHEELS STOP: 69:13:59:25Z 10635 FT ROLLOUT: 8460 FT 65 SEC WINDS: H13. 0X KTS OFFICIAL: 3312P20 H12, L2 DENS ALT: -1645 FT FLT DURATION: 15:17:40:21 S/T: 595:06:47:37 OV-102: 183:02:17:35 DISTANCE: 6,500,000 sm	104/104/ 109% PREDICTED: 100/104/104/ 67/104 ACTUAL: 100/104/104/ 67/104 1 = 2029 (13) 2 = 2034 (7) 3 = 2017 (13) M 3 EOM WEIGHT: 226443 LBS X CG: 1079.40 LANDING WEIGHT: 226287 LBS X CG: 1080.94	BI-078 RSRM 53 ET-76 LWT-69 ET RPT 271K ET BR/UP 214K ET IMPACT 1:20:58 MET LAT: 13.6 S LONG: 163.3 W	28.46 (41)	DIRECT INSERTION POST OMS-2: 161.9 x 160.2 NM USMP PRCS 1 5/21:45:00 160.1 x 153.5 MEPHESTO: 10:12:25:00 MET 158.4 X 149.4 NM DEORBIT: 173 x 146 NM VELOCITY: 25816 FPS ENTRY RANGE: 4375 NM	OI-24 (6)	CARGO: 32006 LBS PAYLOAD CHARGEABLE: 23353 LBS DEPLOYED: 1494 LBS NON-DEPLOYED: 20490 LBS MIDDECK: 1369 LBS SHUTTLE ACCUMULATED WEIGHTS: DEPLOYED: 815907 LBS NON-DEPLOYED: 1053968 LBS CARGO TOTAL: 2251727 LBS PERFORMANCE MARGINS (LBS): FPR: 3775 FUEL BIAS: 1136 FINAL TDDP: 1594 RECON: 638 PAYLOADS: PLB: TETHERED SATELLITE SYSTEM REFLIGHT (TSS-1R) U.S. MICROGRAVITY PAYLOAD SEMICONDUCTOR EXPERIMENTS (USMP-3) OARE MIDDECK: TSS SUPPORT EQUIPMENT MGBX CPCG 5 CRYO TK SETS PLUS 4 EDO EDO PALLET NO RMS	KSC W/D: OPF 64, VAB 5, PAD 25 = 94 days total. LAUNCH POSTPONEMENTS: - Baselined launch date of 2/15/96 on 10/13/94. - Postponed launch date to 2/22/96 on 12/1/95. LAUNCH SCRUBS: NONE LAUNCH DELAYS: NONE TAL WX: - Both BEN (prime & selected) and MRN were forecast and observed GO. BYD was not available as an intact abort site due to local situation. DOLILU-II I-LOADS: - DOLILU II uplink #6, I-load uplink #25. FLIGHT DURATION CHANGES: - Extended flight 1 day for additional USMP science. - Decision to not try to land on orbit 235 due to forecast of low ceiling. Waved off landing on orbits 236 and 237 due to forecast of low ceiling. Extended flight second day for weather. - Waved off landing at KSC on orbit 251 due to forecast of low ceiling. - Total flight duration extension of 2 days plus one orbit. FIRSTS/LASTS: - First flight with thermocouple transducers on all 3 engines. EVENTS: - TSS deployed at 03:00:27:00 MET, tether broke at 03/05:11:35, tether length of 19,695 meters, and TSS separated rapidly from orbiter. Tether was rewound starting at 03:21:49:00 MET and boon retraction completed at 03:02:41 MET. SIGNIFICANT ANOMALIES: - Left main engine chamber pressure read 40% in lieu of 104%. - FA1 MDM card 0 failure during FCS C/O, aerosurfaces not receiving commands from FA1 (waiver written to F/R 2-30A.2a, MDF or next PLS). - Topping FES core icing used, ice flush procedure. - Fuel cell 3 CPM not doing self-test. - H₂ tank 4 heater A failure. - AC 1 phase B short caused loss of utility outlets J31 and J7. - IMU 3 X and Y axis drift, compensations up to 8 sigma. Powered off to preserve lifetime. Used for entry but continued high drift rates. - MLS 2 did not lock on in range. - S-band transponder 2 failed to acquire TDRS (forward link). - MOC processing problems. - APU 1 fuel pump inlet pressure decay. - TSS was lost when tether parted when being deployed (at 19.7 kilometers). - Uncommanded SFMDM warm starts. - LH aft structure attach (to ET) blade valve not fully closed (debris catcher).

MCC FCR-1 (54) ASCENT/ENTRY

WHITE FCR (5) FOR ORBIT OPS

FLIGHT DIRECTORS: A/E - R. D. Jackson LD/O 2 - C. W. Shaw O 1 - G. A. Pennington O 3 - R. E. Castle O 4 - J. P. Shannon MOD - A. L. Briscoe

Tethered Satellite System (TSS)

sts075-772-020 ---- Inflight crew portrait: (bottom center) CDR Allen. Clockwise from him: Chang-Diaz/PLC, Cheli/MS, Nicollier/MS- ESA, PLT Horowitz, Guidoni/PS-ASI, Hoffman/MS.

FLT NO.	ORBITER	CREW (6) TITLE, NAMES & EVA'S	LAUNCH SITE, LIFTOFF TIME, LANDING SITES, ABORT TIMES	LANDING SITE/ RUNWAY, CROSSRANGE LANDING TIMES FLT DURATION, WINDS	SSME-TL NOM-ABORT EMERG THROTTLE PROFILE ENG. S.N.	SRB RSRM AND ET	ORBIT INC	HA/HP	FSW	PAYLOAD WEIGHTS, PAYLOADS/ EXPERIMENTS	MISSION HIGHLIGHTS (LAUNCH SCRUBS/DELAYS, TAL WEATHER, ASCENT I-LOADS, FIRSTS, SIGNIFIC'T ANOMALIES, ETC.)
STS-76 SEQ FLT #76 KSC-76 PAD 39B-32 MLP-2	OV-104 (Flight 16) Atlantis Spacehab 4 OMS PODS: LPO3-20 RPO4-16 FRC4-16	CDR: Kevin P. Chilton (Flt 3 - STS-49, STS-59) P430/R145/V103/M129 PLT: Richard A. Searfoss (Flt 2 - STS-58) P431/R171/V126/M149 M/S 1 (PAYLOAD CDR): Ronald M. Sega (Flt 2 - STS-60) P432/R175/V127M153 M/S 2/EV 2: M. Richard Clifford (Flt 3 - STS-53, STS-59) P433/R157/V104/M139 M/S 3/EV 1: Linda M. Godwin (Flt 3 - STS-37, STS-59) P434/R122/V105/F13 M/S 4: Shannon W. Lucid (Flt 5 - STS 51-G, STS-34, STS-43, STS-58, to return on STS-79) P435/R65/V45/F6 SS EVA #33 Tethered with SAFER CTGY EV 1 - Linda Godwin EV 2 - Rich Clifford Scheduled EVA #29 To install MEEP on Mir DM, evaluate EVA H/W, aids & tools. 3/27/96 - 6:02:28 Duration MCC FCR-1 (55) ASCENT ONLY Continued...	KSC PAD 39B 82:08:13:03.9Z 3:13:04 AM EST (P) 3:13:04 AM EST (A) FRIDAY 15 3/22/96 (6) LAUNCH WINDOW: 6M59S MIR PLANAR/ PHASE WINDOW EOM PLS: KSC TAL: ZZA TAL WX: MRN, BEN SELECTED: RTLS: KSC 33/CI/N TAL: ZZA 30/N/N AOA: KSC 33/CI/N PLS: EDW 22/N/N TDEL: 0.09 0.492/0.49 MAX Q NAV: 720 PSF 724 PSF 52 SECS MET SRB STG: 2:05.5 2:09 PERF: NOMINAL 2 ENG TAL (BEN): 2:25 2:28 NEG RETURN: 4:06 4:09 PTA (U/S 242): 4:23 4:24 DROOP (ZZA): 5:24 5:23 PTM: 5:54 5:58 SE TAL (ZZA): 5:54 6:09 MECO CMD: 8:32.6 8:33.2 Continued . . .	EDW 22, CONC (EDW 45, CONC 26) 91:13:28:57Z 5:28:57 AM PST SUNDAY 11 3/31/96 (7) DEORBIT BURN: 91:12:23:08Z XRANGE: 763 NM ORBIT DIR: DR 16 AIM PT: NOMINAL MLGTD: 2185 FT 91:13:28:57Z VEL: 204 KGS 198 KEAS HDOT: -1.6 FPS TD NORM 195: 2433 FT DRAG CHUTE DEPLOY: 188 KEAS 91:13:29:00Z NLGTD: 5747 FT 91:13:29:08Z VEL: 154 KGS HDOT: -5.0 FPS BRK INIT: 116 KGS DRAG CHUTE JETTISON: 54 KGS 91:13:29:31Z AVE BRK DECEL: 5.4FPS/S WHEELS STOP: 91:13:29:52Z 10579 FT ROLLOUT: 8394 FT 55 SEC WINDS: H0, L1 KTS OFFICIAL: 1301P04 T0, L1 Continued . . .	104/104/ 109% PREDICTED: 100/104/104/ 67/104 ACTUAL: 100/104/104/ 69/104 1 = 2035 (3) 2 = 2109 (16) 3 = 2019 (16) M 3 EOM: WEIGHT: 211913 LBS X CG: 1082.76 LANDING: WEIGHT: 211805 LBS X CG: 1084.46	BI-079 RSRM 46 ET-77 LWT-70 ET RPT 271K ET BR/UP 269K ET IMPACT 1:25:49 MET LAT: 0.1 N LONG: 125.4 W	51.65 (4)	DIRECT INSERTION POST OMS-2: 158.5 x 85.1 NM MIR-RNDZ MNVR AT 1/01:11 MET 210 x 127 NM TI: 1:15:28:01 MET 215.8 x 206.3 NM DEORBIT: 216 X 206 NM VELOCITY: 25898 FPS ENTRY RANGE: 4243 NM	OI-24 (7)	CARGO: 24605 LBS PAYLOAD CHARGEABLE: 14152 LBS DEPLOYED: 2814 LBS NON-DEPLOYED: 10578 LBS MIDDECK: 760 LBS SHUTTLE ACCUMULATED WEIGHTS: DEPLOYED: 818721 LBS NON-DEPLOYED: 1065306 LBS CARGO TOTAL: 2276332 LBS PERFORMANCE MARGINS (LBS): FPR: 3775 FUEL BIAS: 1136 FINAL TDDP: 3140 RECON: 3563 PAYLOADS: PLB: SHUTTLE/MIR MISSION 3 SPACEHAB 4 ORBITER DOCKING SYSTEM (ODS) MIDDECK: KIDSAT SAREX-II 5 CRYO TK SETS NO RMS	KSC W/D: OPF 68, VAB 6, PAD 22 = 96 days total. LAUNCH POSTPONEMENTS: - Baselined launch date of 3/21/96 on 12/14/94. LAUNCH SCRUBS: - Scrubbed 3/21/96 launch at ET tanking MMT on 3/20/96 at approx. L-8 hours due to weather forecast of excessive RTLS crosswinds, chance of 5000' broken ceiling at KSC, and high seas in SRB recovery area. LAUNCH DELAYS: None TAL WX: - ZZA (prime and selected) and MRN were forecast and observed GO. BEN forecast and observed NO GO for ceiling and visibility. DOLILU-II I-LOADS: - DOLILU-II I-Loads uplinked (#8), I-Load uplink #27. SPACE SHUTTLE NIGHT LAUNCH: #13 FLIGHT DURATION CHANGES/LANDING SITE CHANGE: - MMT decision on 3/28/96 to land 1 day early on 3/30 (forecast of low ceiling & fog). - Loss of APU 3 imposed weather placards, flight rule 10-4A. - Waved off landing at KSC on orbit 129 due to overcast ceiling. - Waved off landing at KSC on orbit 130. Extended flight 1 day to original duration. - Waved off landing at KSC on orbit 144 due to ground fog. Changed landing site to EDW. - Total flight duration extension: one orbit. FIRSTS/LASTS: - Mir docking at 01:18:39:26, hatch opening at 01:20:18:00 MET. - Shannon Lucid transferred to Mir 21 crew at 02:04:29:00 MET (84:12:42:04Z) and will return on STS-79. - Fifteen CWC's, total of 1506 lbm water, 42 lbm N₂, 62 lbm O₂, 614 lbm food transferred to Mir. - First EVA during orbiter/Mir docked operations at 04:22:23 MET. - Mir undocking at 06:16:54:59 MET. - Last flight from old MCC (FCR-1). First flight controlled from old MCC was Gemini 4. RADIATOR DEPLOY #18: - Port radiator deployed for 47 hours to conserve water for transfer to Mir. RENDEZVOUS #27: - Rendezvous and third docking with Mir Space Station (third docking flight). Continued . . .

NM21-727-030 (23 March 1996) --- Atlantis as seen from Mir during rendezvous.

FLT NO.	ORBITER	CREW (6) TITLE, NAMES & EVA'S	LAUNCH SITE, LIFTOFF TIME, LANDING SITES, ABORT TIMES	LANDING SITE/ RUNWAY, CROSSRANGE LANDING TIMES FLT DURATION, WINDS	SSME-TL NOM-ABORT EMERG THROTTLE PROFILE ENG. S.N.	SRB RSRM AND ET	ORBIT INC	HA/HP	FSW	PAYLOAD WEIGHTS, PAYLOADS/ EXPERIMENTS	MISSION HIGHLIGHTS (LAUNCH SCRUBS/DELAYS, TAL WEATHER, ASCENT I-LOADS, FIRSTS, SIGNIFIC:T ANOMALIES, ETC.)
STS-76 Continued		Continued... WHITE FCR (6) ORBIT OPS & ENTRY FLIGHT DIRECTORS: A/E - J. W. Bantle LD/O 1 - P. L. Engelauf O 2 - W. D. Reeves PLNG - P. F. Dye MOD - R. E. Castle	Continued . . . VI: 25878 25871 OMS-2: 42:18.5 42:21.9 77.1 FPS 76.8 FPS	Continued . . . DENS ALT: 1536 FT FLT DURATION: 9:05:15:53 S/T: 604:12:03:30 OV-105: 110:13:35:54 DISTANCE: 3,800,000 sm						Continued . . .	Continued . . . SIGNIFICANT ANOMALIES: - Hydraulic System 3 leak during ascent (approximately 20% fluid lost), kept in low pressure for entry, F/R waiver S063689CU. - WSB 3A failed to cool during ascent. - WSB 2 overcooked post-MECO. - Loss of PLBD centerline 9-12 release microswitch inclinations postlanding wave-off. - WSB 3B steam vent heater transient failure. - R4R fail off (low chamber pressure). - L2L fail leak (oxidizer leak). - L2U fail off (low chamber pressure). - EVA camera bracket not onboard. - EV 2 biomed (ECG) signal conditioner failed. - EMU 2 battery power discrete fail on. - MCC loss of forward link during countdown. - Loss of KCA forward link. - Water transfer mineral syringe failed to inject.

STS076-371-002 (25 March 1996) --- Inflight crew portrait on mid deck. From left on front row: Godwin/MS, CDR Chilton, and PLT Searfoss. Left to right on back row: Clifford/MS, Lucid/MS and payload commander Sega/PLC. Lucid later joined Mir-21 crew for first leg of her five-month stay.

Above: STS076-724-016 -- Clifford works at restraining bar on Mir Docking Module. Clifford and Godwin mark first EVA while MIR & Shuttle are docked.

Below: NM21-399-001 --- Aboard Mir Base Block Module Lucid works out on treadmill.

NO.	ORBITER	CREW (6) TITLE, NAMES & EVA'S	LAUNCH SITE, LIFTOFF TIME, LANDING SITES, ABORT TIMES	LANDING SITE/ RUNWAY, CROSSRANGE LANDING TIMES FLT DURATION, WINDS	SSME-TL NOM-ABORT EMERG THROTTLE PROFILE ENG. S.N.	SRB RSRM AND ET	ORBIT INC	HA/HP	FSW	PAYLOAD WEIGHTS, PAYLOADS/ EXPERIMENTS	MISSION HIGHLIGHTS (LAUNCH SCRUBS/DELAYS, TAL WEATHER, ASCENT I-LOADS, FIRSTS, SIGNIFICANT ANOMALIES, ETC.)
STS-77 SEQ FLT #77 KSC-77 PAD 39B-33 MLP-1	OV-105 (Flight 11) Endeavour OMS PODS LPO4-18 RPO5-9 FRC5-11	CDR: John H. Casper (Flt 4 - STS-36, STS-54, STS-62) P436/R111/V86/M99 PLT: Curtis L. Brown (Flt 3 - STS-47, STS-66) P437/R152/V112/M136 M/S 1: Andrew S. W. Thomas P438/R213/M186 M/S 2: Daniel W. Bursch (Flt 3 - STS-51, STS-68) P439/R169/V109/M147 M/S 3: Mario Runco, Jr. (Flt 3 - STS-44, STS-54) P440/R137/V89/M122 M/S 4: Marc Garneau (Canada) (Flt 2 - STS 41-G) P441/R47/V128/M44 MCC WHITE FCR (7) (ALL OPS) FLIGHT DIRECTORS: A/E - R. D. Jackson LD/O 2 - N. W. Hale O 1 - B. P. Austin PLNG - L. J. Ham MOD - J. W. Bantle s77e5053-- RMS holds Spartan 207 free flyer above PLB. SPACEHAB is in foreground.	KSC PAD 39B 140:10:29:59.973Z 6:30:00 AM EDT (P) 6:30:00 AM EDT (A) Sunday 9 5/19/96 (3) LAUNCH WINDOW: 2H30M CTOB EOM PLS: KSC TAL: BEN TAL WX: MRN, ZZA SELECTED: RTLS: KSC 33/N/N TAL: MRN 20/N/N AOA: KSC 33/N/N PLS: EDW 22/N/N TDEL: 0.1 -0.108/0.09 MAX Q NAV: 693 701 SRB STG: 2:05.4 2:05 PERF: NOMINAL 2 ENG TAL (MRN): 2:40 2:36 NEG RETURN: 3:59 4:00 PTA (U/S 249): 4:45 4:36 DROOP (BYD): 5:23 5:23 PTM: 5:41 5:32 MECO CMD: 8:27.66 8:28.1 VI: 25865 25856 OMS-2: 41:47 41:47 2:06 2:07 198.5 FPS 198.6 FPS	KSC 33 (KSC 30) 150:11:09:20Z 7:09:20 AM EDT Wednesday 8 5/29/96 (6) DEORBIT BURN: 150:10:09:30Z XRANGE: 314 NM ORBIT DIR: DR 17 AIM PT: CLOSE IN MLGTD: 1687 FT 150:11:09:20Z VEL: 216 KGS 216 KEAS HDOT: -4.6 FPS TD NORM 205: 2536 FT DRAG CHUTE DEPLOY: 191 KEAS 150:11:09:27Z NLGTD: 6612 FT 150:11:09:35Z VEL: 150 KGS HDOT: -4.8 FPS BRK INIT: 99 KGS DRAG CHUTE JETTISON: 59 KGS 150:11:09:56Z AVE BRK DECEL: 6.8 FPS/S WHEELS STOP: 150:11:10:11Z 10978 FT ROLLOUT: 9291 FT 51 SEC WINDS: H0, L6 KTS OFFICIAL: 2607P9 H2, L7 DENS ALT: 1012 FT FLT DURATION: 10:00:39:20 S/T: 614:12:42:50 OV-105: 112:13:03:06 DISTANCE: 4,100,000 sm	104/104/ 109% PREDICTED: 100/104/104/ 67/104 ACTUAL: 100/104/104/ 67/104 1 = 2037 (2) 2 = 2040 (1) 3 = 2038 (3) M 3 EOM WEIGHT: 222399 LBS X CG: 1080.45 LANDING WEIGHT: 222276 LBS X CG: 1082.04	BI-080 RSRM 47 ET-78 LWT 71 ET RPT 271K ET BR/UP 214K ET IMPACT 1:24:57 MET LAT: 2.97 N LONG: 138.89 W	39.03 (5)	DIRECT INSERTION POST OMS-2: 152.9 x 152.8 NM SPARTAN DEPLOY: 153.6 x 150.4 NM SPARTAN GRAPPLE: 153.1 x 152.0 NM PAMS/STU DEPLOY: 152.6 x 152.0 NM DEORBIT: 154 x 147 NM VELOCITY: 25763 FPS ENTRY RANGE: 4378 NM	OI-24 (8)	CARGO: 35205 LBS PAYLOAD CHARGEABLE: 27393 LBS DEPLOYED: 1104 LBS NON-DEPLOYED: 23586 LBS MIDDECK: 866 LBS SHUTTLE ACCUMULATED WEIGHTS: DEPLOYED: 819825 LBS NON-DEPLOYED: 1089758 LBS CARGO TOTAL: 2311537 LBS PERFORMANCE MARGINS (LBS): FPR: 3080 FUEL BIAS: 900 FINAL TDDP: 5381 RECON: 8528 PAYLOADS: PLB: SPACEHAB-4 SPARTAN 207/IAE TEAMS (GANE, LMTE, VTRE, PAMS/STU (deployed)) GBA (12 BETSCE MIDDECK: ARF-01 BRIC-07 5 CRYO TK SETS RMS 45 (S.N. 301) RMS used for SPARTAN 207 deploy, retrieve, and berth (IAE deployed from SPARTAN).	KSC W/D: OPF 69, VAB 5, PAD 27 = 101 days total. LAUNCH POSTPONEMENTS: - Baselined launch date of 4/25/96 on 6/19/95. - Postponed launch date to 5/16/96 on 9/11/95. - Postponed launch date to 5/19/96 on 5/14/96 (Atlas launch had range priority). LAUNCH SCRUBS: None. LAUNCH DELAYS: None. TAL WX: - BEN (prime) was forecast NO-GO for broken ceiling but observed GO at TAL landing time. MRN was forecast GO, selected, and observed GO. ZZA was forecast GO but observed NO-GO for broken ceiling at TAL landing time. DOLILU-II I-LOADS: - DOLILU-II uplink #8, I-load uplink #27. FLIGHT DURATION CHANGES: None. Flight was planned to be 10 days assuming 5/19/96 liftoff; hence, this does not count as a flight duration change. FIRSTS/LASTS: - First flight with all 3 Block I engines. - First flight to be controlled completely from the new MCC (White FCR). EVENTS: - SPARTAN deployed at 1:01:59:12 MET. - SPARTAN grappled at 2:04:22:34 MET and berthed at 2:05:25:41 MET. - PAMS/STU deployed at 2:22:50:00 MET. RENDEZVOUS #28: Rendezvous, capture, and berth (return) of SPARTAN-207). RENDEZVOUS #29, #30, & #31: Rendezvous & PROXIVOUS OPS with PAMS/STU payload. "STS-77 still holds the record for most number of rendezvous operations of any space flight" - From Wayne Hale's blog: http://blogs.nasa.gov/cm/newui/blog/viewpostlist.jsp?blogname=waynehalesblog - "My Favorite Shuttle Flight" posted May 26, 2010. SIGNIFICANT ANOMALIES: - IPS file server (MPSR1) disk crash prelaunch. - FES failure to come out of standby. - PCS 1 O2 supply transducer failed. - WSB 2 failed to cool during ascent. - APU 2 fuel pump seal cavity drain line pressure decay. - WSB 3 overcool during entry. - RCS jet F2F fail leak (oxidizer leak). - RCS jet R3A heater failed off.

STS077-314-011 Inflight crew portrait. Left to right, front: Thomas/MS, CDR Casper and Runco/MS. Back row: PLT Brown, Garneau /MS/CSA & Bursc/MS.

SPACE SHUTTLE MISSIONS SUMMARY

FLT NO.	ORBITER	CREW (7) TITLE, NAMES & EVA'S	LAUNCH SITE, LIFTOFF TIME, LANDING SITES, ABORT TIMES	LANDING SITE/ RUNWAY, CROSSRANGE LANDING TIMES FLT DURATION, WINDS	SSME-TL NOM-ABORT EMERG THROTTLE PROFILE ENG. S.N.	SRB RSRM AND ET	INC	ORBIT HA/HP	FSW	PAYLOAD WEIGHTS, PAYLOADS/ EXPERIMENTS	MISSION HIGHLIGHTS (LAUNCH SCRUBS/DELAYS, TAL WEATHER, ASCENT I-LOADS, FIRSTS, SIGNIFICANT ANOMALIES, ETC.)
STS-78 SEQ FLT #78 KSC-78 PAD 39B-34 MLP-3	OV-102 (Flight 20) Columbia 19th Spacelab Flight LM-13 EDO 8 OMS PODS: LP05-9 RP01-24 FRC2-20	CDR: Terence T. (Tom) Henricks (Flt4 - STS-44, STS-55 STS-70) P442/R135/V93/M120 PLT: Kevin Kregel (Flt 2 - STS-70) P443/R197/V129/M172 M/S 1: Richard M. Linnehan P444/R214/M187 M/S 2 (PAYLOAD CDR): Susan J. Helms (Flt 3 - STS-54, STS-64) P445/R158/V108/F19 M/S 3: Charles E. Brady, Jr. P446/R215/M188 P/S 1: Jean-Jacques Favier (France) P447/R216/M189 P/S 2: Robert B. Thirsk (Canada) P448/R217/M190 MCC WHITE FCR (8) FLIGHT DIRECTORS: A/E - J. W. Bantle LD/O 2 - J. P. Shannon O 1 - P. L. Engelauf O 3 - B. P. Austin O 4 - C. W. Shaw MOD - A. L. Briscoe	KSC PAD 39B 172:14:48:59.98Z 10:49:00 AM EDT (P) 10:49:00 AM EDT (A) Thursday 23 6/20/96 (8) LAUNCH WINDOW: 2H30M CTOB EOM PLS: KSC TAL: BEN TAL WX: MRN, ZZA SELECTED: RTLS: KSC 33/N/N TAL: BEN 36/N/N AOA: EDW 22/N/N PLS: EDW 22/N/N TDEL: 0 -0.178/0.02 MAX Q NAV: 705 714 SRB STG: 2:04.6 204 PERF: NOMINAL 2 ENG TAL (BEN): 2:43 2:41 NEG RETURN: 3:57 3:59 PTA (U/S 240): 5:15 5:15 DROOP: 5:25 5:24 PTM (U/S 240): 5:47 5:45 MECO CMD: 8:27.9 8:29.6 VI: 25865.4 25856 OMS-2: 41:28.7 41:28.6 185.6 FPS 185.7 FPS 1:59 1:59	KSC (KSC 31) 189:12:36:36Z 8:39:36 AM EDT Sunday 12 7/7/96 (7) DEORBIT BURN: 189:11:36:36Z XRANGE: 91 NM ORBIT DIR: DR 18 AIM PT: NOMINAL MLGTD: 2300 FT 189:12:36:36Z VEL: 214 KGS 208 KEAS HDOT: -1.3 FPS TD NORM 205: 2515 FT DRAG CHUTE DEPLOY: 191 KEAS 189:12:36:40Z NLGTD: 6537 FT 189:12:36:48Z VEL: 158 KGS HDOT: -5.2 FPS BRK INIT: 124 KGS DRAG CHUTE JETTISON: 189:12:37:12 Z 59KGS AVE BRK DECEL: 5.6FPS/S WHEELS STOP: 189:12:37:31Z 11639 FT ROLLOUT: 9339 FT 55 SEC WINDS: T3, L1 KTS OFFICIAL: 1803P5 T3, L2 DENS ALT: 854 FT FLT DURATION: 16:21:47:35 S/T: 631:10:30:25 OV-102: 200:00:05:10 DISTANCE: 7,046,000 sm	104/104/ 109% PREDICTED: 100/104/104/ 67/104 ACTUAL: 100/104/104/ 67/104 1 = 2041 (1) 2 = 2039 (2) 3 = 2036 (3) M 3 EOM: WEIGHT: 229134 LBS X CG: 1081.88 LANDING: WEIGHT: 228986 LBS X CG: 1083.40	BI-081 RSRM 55 ET-79 LWT 72 ET RPT 271.3K ET BR/UP 214K ET IMPACT 1:24:50 MET 2.86 N LAT: LONG: 138.9 W	39.03 (6)	DIRECT INSERTION POST OMS-2: 153.6 X 146.7 NM TRIM 1 BURN: 4:30:00 MET 146.6 X 146.4 NM TRIM 2 BURN: 15:23:29:00Z 142.3 X 129.6 NM DEORBIT: 142 X 130 NM VELOCITY: 25749 FPS ENTRY RANGE: 4466 NM	OI-24 (9)	CARGO: 31854 LBS PAYLOAD CHARGEABLE: 23666 LBS DEPLOYED: 0 LBS NON-DEPLOYED: 21598 LBS MIDDECK: 2066 LBS SHUTTLE ACCUMULATED WEIGHTS: DEPLOYED: 819825 LBS NON-DEPLOYED: 1113422 LBS CARGO TOTAL: 2343391 LBS PERFORMANCE MARGINS (LBS): FPR: 3080 FUEL BIAS: 900 FINAL TDDP: 3683 RECON: 4245 PAYLOADS: PLB: LIFE AND MICROGRAVITY SCIENCES (LMS) Musculoskeletal Physiology, Fluid Physics, Advanced Semiconductory and Metal Alloys Processing (SPACELAB LM) OARE MIDDECK: BRIC SAREX II 5 CRYO TK SETS + 4 EDO, 5 GN2 TANKS EDO PALLET NO RMS	KSC W/D: OPF 63, VAB 7, PAD 19 = 89 days total. LAUNCH POSTPONEMENTS/ADVANCEMENTS: - Baselined launch date of 6/27/96 on 3/30/95. - Advanced launch date to 6/20/96 on 3/21/96. LAUNCH SCRUBS: None LAUNCH DELAYS: None TAL WX: - BEN (prime and selected) and MRN were forecast and observed GO. ZZA was forecast and observed NO-GO (thunderstorms within 20 NM). DOLILU-II I-LOADS: DOLILU-II uplink #9, I-load uplink #28 FLIGHT DURATION CHANGES: - Extended flight 1 day to 17 days for additional science (planned 16 + 1). EVENTS: - Longest space shuttle flight to date. RADIATOR DEPLOY #19: Full deploy for cooling. SIGNIFICANT ANOMALIES: - Main engine 2036 violated thrust build up rate at engine start (>14,000 lbs thrust change for any two consecutive 20 millisec time intervals). - MPS LH2 low level cutoff sensors indicated dry (flashed) 2.3 seconds after MECO during shutdown transient flow (changed mixture ratio for STS-79 to 6.020). - Heavy sooting and heat effect (discoloration and charring) observed on insulation interfaces within STS-78 field joints. No heat effects to metal interface or capture feature o-ring, no gas past CF O-rings. (Environment process change this flight to J-leg adhesive and joint cleaning process.) Postponed STS-79 to use STS-80 stack with old processing. - Center MPS LH2 inlet pressure failed OSH. - BFS I/O TERMINATE B discrete toggling low. BFS moved to GPC 2 for entry. - FES high-load duct temps low during ascent and high-load core freeze-up during deorbit prep. High-load core was flushed. - FES topping core freezeup at 2 days 1 hour MET and during deorbit prep. Core flush procedure performed. - Cryo N2 tank 4B heater failed. - Spacelab EPDB 2 AC phase A amps and EPDB 3 AC phase C amps transducer failures. - Loss of MCC read/write (aka HA) servers. - APU 1 fuel pump seal leakage more severe that seen on STS-75. - APU 1 turbine speed transducer erratic. - WSB 1 ready indication intermittent (or bypass valve indication).

sts078-730-033 -- Life & Microgravity Sciences (LMS) in PLB.

sts078-397-030 Crew poses in LMS-1 with Helms/MS at bottom center. Others, clockwise: Favier/PS (France), Thirsk/MS (Canada), PLT Kregel, Brady/MS, Linnehan/MS, and CDR Henricks.

FLT NO.	ORBITER	CREW (7) 6 UP / 6 DOWN — TITLE, NAMES & EVA'S	LAUNCH SITE, LIFTOFF TIME, — LANDING SITES, ABORT TIMES	LANDING SITE/ RUNWAY, CROSSRANGE — LANDING TIMES FLT DURATION, WINDS	SSME-TL NOM-ABORT EMERG — THROTTLE PROFILE ENG. S.N.	SRB RSRM — AND ET	INC	ORBIT — HA/HP	FSW	PAYLOAD WEIGHTS, — PAYLOADS/ EXPERIMENTS	MISSION HIGHLIGHTS (LAUNCH SCRUBS/DELAYS, — TAL WEATHER, ASCENT I-LOADS, FIRSTS, SIGNIFICANT ANOMALIES, ETC.)

STS-79
SEQ FLT #79
KSC-79
PAD 39A-45
MLP-1

ORBITER:
OV-104 (Flight 17)
Atlantis
Spacehab 5
OMS PODS:
LPO3-21
RPO4-17
FRC4-17

CREW:
CDR:
William F. Ready (Flt 3 - STS-42, STS-51)
P449/R140/V96/M125
PLT:
Terrence W. Wilcutt (Flt 2 - STS-68)
P450/R183/V130/M160
M/S 1:
Jay Apt (Flt 4 - STS-37, STS-47, STS-59)
P451/R123/V79/M110
M/S 2:
Thomas D. Akers (Flt 4 - STS-41, STS-49, STS-61)
P452/R115/V74/M103
M/S 3:
Carl E. Walz (Flt 3 - STS-51, STS-65)
P453/R170/V106/M148
M/S 4:
Ascent
John E. Blaha (Flt 5 - STS-29, STS-33, STS-43, STS- 58, stay on Mir 22, and return on STS-81)
P454/R97/V48/M88
M/S 4:
Descent
Shannon Lucid (Flt 5 - STS-51-G, STS-34, STS-43, STS-58, Ascent on STS-76, on-orbit stay on Mir 21 and Mir 22)
P455/R65/V45/F6
MCC WHITE FCR (9)
FLIGHT DIRECTORS:
ASC - R. D. Jackson
ENT- L. J. Ham
LD/O 1 - P. F. Dye
O 2 - R. E. Castle
PLNG - W. D. Reeves
MOD - A. L. Briscoe

LAUNCH SITE, LIFTOFF TIME:
KSC PAD 39A
260:08:54:48.96Z
4:54:49 AM EDT (P)
4:54:49 AM EDT (A)
Monday 11
9/16/96 (8)
LAUNCH WINDOW:
5:47M
MIR PLANAR/ PHASE WINDOW
EOM PLS: KSC
TAL: ZZA
TAL WX: MRN, BEN
SELECTED:
RTLS: KSC 33/N/SF
TAL: ZZA 30/N/SF
AOA: KSC 15/CI/N
PLS: EDW 22/N/N
TDEL:
0.06   -0.018/0.02
MAX Q NAV:
697   705
SRB STG:
2:02.4   2:05
PERF: NOMINAL
2 ENG TAL (ZZA):
2:38   2:35
NEG RETURN:
4:06   4:03
PTA (U/S 260):
4:46   4:48
PTM (U/S 260):
5:20   5:24
DROOP (BYD):
5:28   5:28
MECO CMD:
8:33   8:34.6
VI:
25878   25880
OMS-2:
42:50.9   42:50.9
75.9 FPS   75.9 FPS
00:47   00:47

LANDING SITE/RUNWAY, CROSSRANGE:
KSC 15 (KSC 32)
270:12:13:13Z
8:13:13 AM EDT
Thursday 8
9/26/96 (9)
DEORBIT BURN:
270:11:06:14Z
XRANGE: 777 NM
ORBIT DIR: DR 19
AIM PT: CLOSE IN
MLGTD: 807 FT
270:12:13:13Z
VEL: 217 KGS
217 KEAS
HDOT: -4.3 FPS
TD NORM 195:
2496 FT
DRAG CHUTE DEPLOY: 192 KEAS
270:12:13:22Z
NLGTD: 5760 FT
270:12:13:29Z
VEL: 150 KGS
HDOT: -4.2 FPS
BRK INIT: 89 KGS
DRAG CHUTE JETTISON: 55 KGS
270:12:13:57Z
AVE BRK DECEL:
3.1FPS/S
WHEELS STOP:
270:12:14:34Z
11788 FT
ROLLOUT:
10981 FT
81 SEC
WINDS:
H4, L3 KTS
OFFICIAL: 1206P09
H5, L3

SSME-TL NOM-ABORT EMERG:
104/104/ 109%
PREDICTED:
100/104/104/ 67/104
ACTUAL:
100/104/104 67/104
1 = 2012 (18)
2 = 2031 (14)
3 = 2033 (8)
M 3 EOM:
WEIGHT:
215990 LBS
X CG:
1081.31
LANDING:
WEIGHT:
215904 LBS
X CG:
1083.02

SRB RSRM AND ET:
BI-083
RSRM 56
ET-82
LWT 75
ET PRED RPT:
271.3K
ET BRKUP:
214K
ET IMPACT
1:26:47 MET
LAT:
0.65 S
LONG:
125.96 W

INC:
51.67 (5)

ORBIT HA/HP:
DIRECT INSERTION
POST OMS-2:
158.6 X 85.3 NM
NC6:
2:14:05:33 MET
203.7 X 201 NM
NC2:
2:15:38:10 MET
208.8 X 201.9 NM
SEP BURN:
7:16:49:15 MET
211 X 201.3 NM
DEORBIT:
209.1 X 197.7 NM
VELOCITY:
25892 FPS
ENTRY RANGE:
4276 NM

FSW:
OI-25 (1)

PAYLOAD WEIGHTS, PAYLOADS/EXPERIMENTS:
CARGO:
27812 LBS
PAYLOAD CHARGEABLE:
19039 LBS
DEPLOYED:
3170 LBS
NON-DEPLOYED:
15151 LBS
MIDDECK:
718 LBS
SHUTTLE ACCUMULATED WEIGHTS:
DEPLOYED:
822995 LBS
NON-DEPLOYED:
1129291 LBS
CARGO TOTAL:
2371203 LBS
PERFORMANCE MARGINS (LBS):
FPR: 4456
FUEL BIAS: 432
FINAL TDDP: 462
RECON: 716
PAYLOADS:
PLB:
SHUTTLE/MIR MISSION 4
SPACEHAB 5 (DOUBLE MODULE)
ODS
MIDDECK:
SAREX
IMAX
MSX
CPCG
MGM
SAMS
CGBA
MGBX
5 CRYO TK SETS
4 GN₂ TANKS
NO RMS

MISSION HIGHLIGHTS:
KSC W/D: OPF 73 (2), VAB 17 (3), PAD 25 (3) = 115 days total
LAUNCH POSTPONEMENTS:
- Baselined 8/1/96 launch date on 5/4/95.
- 5/4/95 launch date was postponed when the shuttle was rolled back from pad A to VAB on 7/10/96 under threat from Hurricane Bertha.
- Due to STS-78 booster sooting and heat effects in field joints, decision was made to restack using STS-80 SRB's (and ET) which used old process. Set launch date to 9/12/96.
- Rolled out to pad A on 8/30/96.
- Rolled back to VAB 9/4/96 under threat of Hurricane Fran. Postponed launch to 9/16/96. Rolled to pad on 9/6/96.
LAUNCH SCRUBS: None
LAUNCH DELAYS: None
LAUNCH WINDOW:
- Mir rendezvous planar/phase window was 7M00S; however, it was limited to 5M47S due to a negative performance margin (-523 lbs) at window opening. Liftoff was delayed (per plan) for 36 seconds for zero performance margin plus an additional 10 seconds (total delay 46 seconds) which allowed approx + 200 lbs APM (wind, loads allowance).
SHUTTLE NIGHT LAUNCH #14
DOLILU-II I-LOADS: DOLILU-II uplink #10, I-load uplink #29.
FLIGHT DURATION CHANGES: Extended 1 day for additional science.
FIRSTS:
- First U.S. spaceflight with female flight director for entry/ landing (Linda Ham).
RENDEZVOUS #32: Rendezvous and dock with Mir (fourth docking).
EVENTS:
- Shannon Lucid was carried to Mir 21 on STS-76 and was replaced on Mir 22 by John Blaha on this flight.
- Shannon Lucid's total flight time: 188:04:00:09 and total Mir time: 178:22:23:45.
- Docking complete at 263:03:21:20Z, 2:18:26:31 MET.
- Transferred 2025 lbm H₂O, 69 lbm O₂, and 43 lbm N₂ to Mir.
- At 3:02:11 MET, Shannon Lucid transferred to STS-79 and John Blaha transferred to Mir-22 crew. (263:11:05:49Z)
- Undocking at 268:01:31:29Z, 07:16:36:40 MET.
RADIATOR DEPLOY #19:
- Both port and starboard radiators were deployed for cooling and to conserve water for transfer to Mir.
- Transferred 20 CWC's with 2025 lbs water.
Continued . . .

NM22-427-023 --- STS-79 Atlantis as seen on approach to MIR.

FLT NO.	ORBITER	CREW (6) TITLE, NAMES & EVA'S	LAUNCH SITE, LIFTOFF TIME, LANDING SITES, ABORT TIMES	LANDING SITE/ RUNWAY, CROSSRANGE LANDING TIMES FLT DURATION, WINDS	SSME-TL NOM-ABORT EMERG THROTTLE PROFILE ENG. S.N.	SRB RSRM AND ET	ORBIT		FSW	PAYLOAD WEIGHTS, PAYLOADS/ EXPERIMENTS	MISSION HIGHLIGHTS (LAUNCH SCRUBS/DELAYS, TAL WEATHER, ASCENT I-LOADS, FIRSTS, SIGNIFIC:T ANOMALIES, ETC.)
							INC	HA/HP			
STS-79 Continued				Continued . . . DENS ALT: 1084 FT FLT DURATION: 10:03:18:24 S/T: 641:13:48:49 OV-104: 120:16:54:18 DISTANCE: 3,900,000 sm							Continued . . . SIGNIFICANT ANOMALIES: - RH RSRM nozzle erosion beginning in throat ring and extending aft into forward exit cone (approx 60 longitudinal erosion areas up to 0.4 inch diameter). - Supply water tank B quantity transducer dropouts. - Fuel cell O$_2$ flow transducer degraded. - Cryo H$_2$ tank 3 B heater failure. - Single string GPS erroneous time reference, loss of lock and runaway. (Firmware problem.) - TCS range discrepancy. - APU 2 underspeed shutdown at 13:14 MET. Two-APU entry/landing. - APU 2 fuel pump seal cavity drain line pressure decay to vacuum.

S79e5131 --- Mir Changeout: Lucid (left) comes down after 6 mos visit, Blaha stays up.

STS079-349-022 --- Inflight crew portrait, in Mir: Front row, left to right, Aleksandr Y. Kaleri/MIR, Apt, Blaha, Readdy, & Lucid. Back row, left to right, Akers, Walz, Valeri G. Korzun/MIR, Wilcutt.

STS079-S-097-- Left to right, PLT Wilcutt, Lucid/MS, & CDR Readdy on aft flight deck for undocking. Lucid looking to come home.

STS079-810-028 --- Russia's Mir Space Station as seen after undocking.

FLT NO.	ORBITER	CREW (5) TITLE, NAMES & EVA'S	LAUNCH SITE, LIFTOFF TIME, LANDING SITES, ABORT TIMES	LANDING SITE/ RUNWAY, CROSSRANGE LANDING TIMES FLT DURATION, WINDS	SSME-TL NOM-ABORT EMERG THROTTLE PROFILE ENG. S.N.	SRB RSRM AND ET	INC	ORBIT HA/HP	FSW	PAYLOAD WEIGHTS, PAYLOADS/ EXPERIMENTS	MISSION HIGHLIGHTS (LAUNCH SCRUBS/DELAYS, TAL WEATHER, ASCENT I-LOADS, FIRSTS, SIGNIFICANT ANOMALIES, ETC.)
STS-80 SEQ FLT #80 KSC-80 PAD 39B-35 MLP-3	OV-102 (Flight 21 Columbia EDO 9 OMS PODS: LPO5-10 RPO1-25 FRC2-21	CDR: Kenneth D. Cockrell (Flt 3 - STS-56, STS-69) P456/R159/V121/M140 PLT: Kent V. Rominger (Flt 2 - STS-73) P457/R200/V131/M174 M/S 1 (EV1): Tamara E. Jernigan (Flt 4 - STS-40, STS-52, STS-67) P458/R130/V83/F14 M/S 2 (EV2): Thomas D. Jones (Flt 3 - STS-59, STS-68) P459/R177/V111/M155 M/S 3: F. Story Musgrave (Flt 6 - STS-6, STS 51-F, STS-33, STS-44, STS-61) P460/R15/V19/M15 Two 6-hour EVA's planned by Jernigan (EV1) and Jones (EV2) for EDFT. EVA's were canceled when crew could not get "B" hatch open. MCC WHITE FCR (10) FLIGHT DIRECTORS: A/E - N. W. Hale LD/O 2 - G. A. Pennington O 1 - R. M. Kelso O 3 - J. P. Shannon O 4 - B. P. Austin MOD - J. W. Bantle	KSC PAD 39B 324:19:55:46.95Z 2:53:00 PM EST (P) 2:55:47 PM EST (A) Tuesday 10 11/19/96 (11) LAUNCH WINDOW: 2H30M CTOB EOM PLS: KSC TAL: BEN TAL WX: MRN SELECTED: RTLS: KSC15/N/N TAL: BEN36/N/N AOA: EDW22/N/N PLS: EDW22/N/N TDEL: -0.04 -0.238/-0.2 MAX Q NAV: 717 719 SRB STG: 2:04.3 ~2:05 PERF: NOMINAL 2 ENG TAL (BEN): 3:03 3:03 NEG RETURN: 3:58 3:59 PTA (U/S 304): 4:55 4:51 DROOP (BYD): 5:28 5:28 PTM (U/S 304): 5:57 5:55 MECO CMD: 8:29.9 8:30.4 VI: 25922 25915 OMS-2: 40:24 40:24 279 FPS 279 FPS	KSC 33 (KSC 33) 342:11:49:04Z 6:49:04 AM EST Saturday 17 12/7/96 (8) DEORBIT BURN: 342:10:48:02Z XRANGE: 72 NM ORBIT DIR: DL 37 AIM PT: NOMINAL MLGTD: 3068 FT 342:11:49:04Z VEL: 210 KGS 203 KEAS HDOT: -1.0 FPS TD NORM 205: 3063 FT DRAG CHUTE DEPLOY: 193 KEAS 342:11:49:08Z NLGTD: 7100 FT 342:11:49:17Z VEL: 149 KGS HDOT: -5.5 FPS BRK INIT: 121 KGS DRAG CHUTE JETTISON: 54 KGS 342:11:49:40Z BRK DECEL FPS2: AVE 5.1 PK 7.6 WHEELS STOP: 342:11:50:13Z 11789 FT ROLLOUT: 8721 FT 69 SEC WINDS: 2T, 4L KTS OFFICIAL: 2006P9 4T, 4L DENS ALT: 522 FT FLT DURATION: 17:15:53:17 S/T: 659:05:42:06 OV-102: 217:16:58:27 DISTANCE: 7,043,950 sm	104/104/ 109% PREDICTED: 100/104/104/ 67/104 ACTUAL: 100/104/104/ 67/104 1 = 2032 (5) 2 = 2026 (6) 3 = 2029 (14) M 3 EOM: WEIGHT: 227815 LBS X CG: 1079.10 LANDING: WEIGHT: 227670 LBS X CG: 1080.62	BI-084 RSRM 49 ET-80 LWT 73 ET PRED RPT: 271.3K ET BRKUP: 214K ET IMPACT 1:22:40 MET LAT: 15.5 N LONG: 159.6 W	28.45 (42)	DIRECT INSERTION POST OMS-2 190 X 188 NM DEORBIT: 203 X 169 NM VELOCITY: 25877 FPS ENTRY RANGE: 4346 NM	OI-25 (2)	CARGO: 31111 LBS PAYLOAD CHARGEABLE: 21208 LBS DPLY/RETRIEVE: 12524 / 12427 LBS NON-DEPLOYED: 7575 LBS MIDDECK: 1109 LBS SHUTTLE ACCUMULATED WEIGHTS: DEPLOYED: 822995 LBS NON-DEPLOYED: 1137975 LBS CARGO TOTAL: 2402314 LBS PERFORMANCE MARGINS (LBS): FPR: 3100 FUEL BIAS: 884 FINAL TDDP: 486 RECON: 1102 PAYLOADS: PLB: ORFEUS-SPAS (Astronomical observations) WSF-3 (Epitaxial semiconductor) SEM MIDDECK: PARE/NIH-R CMIX VIEW-CPL CCM-A, BRIC, MSX 5 CRYO TK SETS + 4 EDO & 5 N2 TANKS EDO PALLET RMS 46 (S.N. 202) RMS used for ORFEUS-SPAS deploy, grapple & berth and WSF deploy, grapple & berth and EDFT-05	KSC W/D: OPF 80, VAB 6, PAD 33 = 119 days total. LAUNCH POSTPONEMENTS: - Baselined launch date of 11/7/96 on 7/14/95. - Advanced launch date to 10/31/96 on 4/23/96. - Postponed launch date to 11/8/96 on 9/20/96 to analyze implications of STS-79 RH SRM nozzle erosion. - Postponed launch date to 11/15/96 to allow Thiokol time to complete SRM analysis. LAUNCH SCRUBS: - Scrubbed 11/15/96 launch date after L-2 MMT on 11/13/96 due to forecast of high surface winds at KSC from 11/15/96 through 11/18/96. New launch date of 11/19/96. LAUNCH DELAYS: - Launch delayed 2M47S at T-31 secs while measuring H2 gas in aft compartment per preplanned procedure to confirm <600 ppm. TAL WX: - Ben Guerir (prime and selected) was forecast and observed GO. Moron was forecast and observed NO-GO for 300 ft overcast. Banjul was not available. DOLILU-II I-LOADS: - DOLILU-II uplink #12, I-Load uplink #30. FLIGHT DURATION CHANGES: - Extended a day for science, then changed to original landing day due to weather at KSC. Waved off landing at KSC on orbits 248 and 249 (broken ceiling). Waved off landing on orbits 264 and 265 due to forecast and observed ground fog. Total extension of 2 days. RENDEZVOUS #33: Rendezvous, deploy, grapple, berth and return ORFEUS-SPAS. RENDEZVOUS #34: Rendezvous, deploy, grapple, berth and return WSF-3. FIRSTS/LASTS: - First flight with two free-flyers (ORFEUS-SPAS and WSF) and orbiter in constrained motion. EVENTS: - ORFEUS-SPAS deployed by RMS at 325:04:10:50Z, 08:15:03 MET. - SEP 1 maneuver at 325:04:11:48Z, SEP 2 at 325:04:44:11Z. - WSF-3 deployed by RMS at 328:01:37:40Z, 03:05:41:53 MET. - WSF-3 grappled, berthed at 331:02:33:51Z, 06:06:38:04 MET. - Crew attempted opening "B" hatch at 334:02:30Z, 09:06:34 MET. Being unsuccessful, the two EVA's were canceled. - ORFEUS-SPAS grappled at 339:08:25:47Z; berthed at 339:13:03:41Z. SIGNIFICANT ANOMALIES: - Loss of LMG down indications. - Crew unable to unlatch and open "B" hatch (outer airlock). Crew able to turn handle only 30 degrees. Resulted in cancellation of two EVA's. Found screw backed out and in latch actuator planetary gears. - Window W8 impact damage. - IMU 1 BITE annunciations (deselected from selection filter for entry.) - EV2 helmet difficult to latch.

STS080-310-028 -- Musgrave photographs Wake Shield Facility during free flight mode with 600mm camera.

STS080-701-004 -- Middeck inflight crew portrait. Back row, left to right, CDR Cockrell, Jernigan/MS, & PLT Rominger. Front row, Jones/MS (left) & Musgrave/MS.

SPACE SHUTTLE MISSIONS SUMMARY

FLT NO.	ORBITER	CREW (7) 6 UP / 6 DOWN — TITLE, NAMES & EVA'S	LAUNCH SITE, LIFTOFF TIME, — LANDING SITES, ABORT TIMES	LANDING SITE/ RUNWAY, CROSSRANGE — LANDING TIMES, FLT DURATION, WINDS	SSME-TL NOM-ABORT EMERG — THROTTLE PROFILE ENG. S.N.	SRB RSRM AND ET	INC	ORBIT HA/HP	FSW	PAYLOAD WEIGHTS, — PAYLOADS/ EXPERIMENTS	MISSION HIGHLIGHTS (LAUNCH SCRUBS/DELAYS, TAL WEATHER, ASCENT I-LOADS, FIRSTS, SIGNIFICANT ANOMALIES, ETC.)
STS-81 SEQ FLT #81 KSC-81 PAD 39B-36 MLP-2	OV-104 (Flight 18) Atlantis Spacehab 6 OMS PODS: LPO3-22 RPO4-18 FRC4-18	CDR: Michael A. Baker (Flt 4 - STS-43, STS-52, STS-68) P461/R133/V81/M118 PLT: Brent W. Jett, Jr. (Flt 2 - STS-72) P462/R206/V132/M179 M/S 1: Peter J. K. (Jeff) Wisoff (Flt 3 - STS-57, STS-68) P463/R166/V110/M145 M/S 2: John M. Grunsfeld (Flt 2 - STS-67) P464/R191/V133/M167 M/S 3: Marsha S. Ivins (Flt 4 - STS-32, STS-46, STS-62) P465/R109/V77/F12 M/S 4: Ascent Jerry M. Linenger (Flt 2 - STS-64, stay on Mir 22, and return on STS-84) P466/R182/V134/M159 M/S 4: Descent John E. Blaha (Flt 5 - STS-29, STS-33, STS-43, STS-58, Ascent on STS-79, and stay on Mir 22) P467/R97/V48/M88	KSC PAD 39B 12:09:27:23Z 4:27:23 AM EST (P) 4:27:23 AM EST (A) Sunday 10 1/12/97 (9) LAUNCH WINDOW: 6M59S. Mir planar/ phase window and ET heating constraint EOM PLS: KSC TAL: ZZA TAL WX: MRN SELECTED: RTLS: KSC15/N/N TAL: BEN36/N/N AOA: EDW22/N/N PLS: EDW22/N/N TDEL: -0.04 -0.238/-0.2 MAX Q NAV: 717 PSF 719 PSF SRB STG: 2:04.3 2:05 PERF: NOMINAL 2 ENG TAL (BEN): 3:03 3:03 NEG RETURN: 3:58 3:59 PTA (U/S 304): 4:55 4:51 DROOP (ZZA): 5:23 5:24 PTM (U/S 304): 5:57 5:55 MECO CMD: 8:29.9 8:30.4 VI: 25922 25915 OMS-2: 40:25 40:24 279 FPS 279 FPS	KSC 33 (KSC 34) 22:14:22:44Z 9:22:44 AM EST Wednesday 9 1/22/97 (7) DEORBIT BURN: 22:13:17:33Z XRANGE: 34 NM ORBIT DIR: DL 38 AIM PT: NOMINAL MLGTD: 2926 FT 22:14:22:44Z VEL: 199 KGS 195 KEAS HDOT: -1.4 FPS TD NORM 195: 2961 FT DRAG CHUTE DEPLOY: 187 KEAS 22:14:22:55Z NLGTD: 6377 FT 22:14:22:55Z VEL: 144 KGS 136 KEAS HDOT: -6.5 FPS BRK INIT: 79 KGS DRAG CHUTE JETTISON: 56 KGS 22:14:23:26Z BRK DECELFPS² AVE 4.0 PK 7.7 WHEELS STOP: 22:14:23:51Z 12276 FT ROLLOUT: 9350 FT 67 SEC WINDS: 4T, 1R KTS OFFICIAL: 1404P6 4T, 1R DENS ALT: 86 FT FLT DURATION: 10:04:55:21 S/T: 669:10:37:27 OV-104: 130:21:49:39 DISTANCE: 3,900,000 sm	104/104/ 109% PREDICTED: 100/104/104/ 67/104 ACTUAL: 100/104/104/ 70/104 1 = 2041 (2) 2 = 2034 (8) 3 = 2042 (3) M 3 EOM: WEIGHT: 215403 LBS X CG: 1081.41 LANDING: WEIGHT: 215337 LBS X CG: 1083.11	BI-082 RSRM 54 ET-83 LWT 76 ET PRED RPT: 271.3K ET BRKUP: 214K ET IMPACT 1:26:53 MET LAT: 0.38 S LONG: 125.6 W	51.67 (6)	DIRECT INSERTION POST OMS-2: 159.9 X 84.9 NM NC5: 14:09:10:43Z 209.9 X 142.4 NM NC6: 14:23:41:15Z 209.9 X 201.6 NM BRAKING: 15:02:38:46Z 209.5 X 208.9 NM SEP: 20:04:01:40Z 212.7 X 203.2 NM DEORBIT: 207.5 X 181.9 NM VELOCITY: 25891 FPS ENTRY RANGE: 4428 NM	OI-25 (3)	CARGO: 28149 LBS PAYLOAD CHARGEABLE: 19321 LBS DEPLOYED: 4019 LBS NON-DEPLOYED: 14492 LBS MIDDECK: 810 LBS SHUTTLE ACCUMULATED WEIGHTS: DEPLOYED: 827014 LBS NON-DEPLOYED: 1153277 LBS CARGO TOTAL: 2430463 LBS PERFORMANCE MARGINS (LBS): FPR: 3100 FUEL BIAS: 884 FINAL TDDP: 1285 RECON: 2117 PAYLOADS: PLB: ODS SHUTTLE-MIR MISSION 5 SPACEHAB DOUBLE MODULE MIDDECK: CREAM KIDSAT SAMS MSX 5 CRYO TK SETS 4 N2 TANKS NO RMS	KSC W/D: OPF 62, VAB 5, PAD 24 = 91 days total. LAUNCH POSTPONEMENTS: - Baselined 12/5/96 as launch date on 9/1/95. - Postponed launch date to 1/16/97 on 8/1/96 (SRM heat effects and nozzle erosion on STS-79 and STS-80). - Advanced launch date to 1/12/97 on 9/5/96. LAUNCH SCRUBS: None LAUNCH DELAYS: None TAL WX: - Zaragoza was prime but NO-GO due to forecast overcast 500 feet and observed broken 300 feet. Moron was selected. Moron and Ben Guerir were forecast and observed GO. DOLILU-II I-LOADS: - DOLILU-II uplink #12, I-Load uplink #31 SHUTTLE NIGHT LAUNCH #15 FLIGHT DURATION CHANGES: - Waved off landing at KSC on orbit 161 due to forecast of broken 4000 foot ceiling. - Flight duration extended one orbit. EVENTS: - Mir capture at 15:03:54:49Z, 2:18:27:26 MET. - Docking at 15:04:02:28Z, 2:18:35:05 MET. - Blaha transferred to STS-81/Atlantis and Linenger transferred to Mir 22 at 3:00:17:00 MET. - Blaha total flight time 127:05:27:55 and Mir time 116:22:38:34. - Hatch closure at 07:03:19 MET and undocking at 20:02:15:23Z, 07:16:48:00 MET. RENDEZVOUS #35: Rendezvous and dock with Mir (fifth docking). SIGNIFICANT ANOMALIES: - Fuel Cell 1 voltage erratic below MNA voltage. - Fuel Cell 2 cell performance monitor self test anomaly. - OCA video conference VLHS cable adapter failure. - LiOH door latch jammed closed. - EVA protect mode command fails when used in TEC (capability not in software). - VIU S.N. 1025 failure. - IMU3 exhibited large X and Y gyro drift rates. Took to standby.

MCC WHITE FCR (11) FLIGHT DIRECTORS:
ASC - R. D. Jackson
ENT - L. J. Ham
LD/O 1 - W. D. Reeves
O 2 - P. F. Dye
PLNG - P. L. Engelauf
MOD - R. E. Castle

STS081-328-013
Mir as seen from Atlantis.

STS081-369-003 --- Inflight crew portrait of Mir-22 & STS-81 crews. Front: lt to rt, STS-81 CDR Baker, Grunsfeld/MS, Aleksandr Y. Kaleri/FE/Mir-22. Middle row: Mir-22 CDR Valeri G. Korzun, Ivins/MS, & Blaha/Mir-22 now MS. Back: Linenger/MS & current guest researcher, Wisoff/MS, & PLT Jett.

FLT NO.	ORBITER	CREW (7) TITLE, NAMES & EVA'S	LAUNCH SITE, LIFTOFF TIME, LANDING SITES, ABORT TIMES	LANDING SITE/ RUNWAY, CROSSRANGE LANDING TIMES FLT DURATION, WINDS	SSME-TL NOM-ABORT EMERG THROTTLE PROFILE ENG. S.N.	SRB RSRM AND ET	ORBIT	INC	HA/HP	FSW	PAYLOAD WEIGHTS, PAYLOADS/ EXPERIMENTS	MISSION HIGHLIGHTS (LAUNCH SCRUBS/DELAYS, TAL WEATHER, ASCENT I-LOADS, FIRSTS, SIGNIFICANT ANOMALIES, ETC.)

STS-82

SEQ FLT #82

KSC-82

PAD 39A-46

MLP-1

ORBITER: OV-103 (Flight 22) Discovery

OMS PODS: LPO1-25 RPO3-23 FRC3-22

CREW:
CDR: Kenneth D. Bowersox (Flt 4 - STS-50, STS-61, STS-73) P468/R146/V97/M130

PLT: Scott J. (Doc) Horowitz (Flt 2 - STS-75) P469/R210/V135/M183

M/S 1/EV-4: Joseph R. Tanner (Flt 2 - STS-66) P470/R185/V136/M162

M/S 2: Steven A. Hawley (Flt 4 - STS 41-DR, STS 61-C, STS-31) P471/R39/V29/M38

M/S 3/EV-3: Gregory J. Harbaugh (Flt 4 - STS-39, STS-54, STS-71) P472/R125/V88/M112

M/S 4/EV-1: Mark C. Lee (Flt 4 - STS-30, STS-47, STS-64) P473/R100/V78/M91

M/S 5/EV-2: Steven L. Smith (Flt 2 - STS-68) P474/R184/V137/M161

Continued . . .

LAUNCH SITE: KSC 39A 42:08:55:16.98Z 3:55:17 AM EST (P) 3:55:17 AM EST (A) Tuesday 11 2/11/97 (6)

LAUNCH WINDOW: 1H6M30S HST PLANAR/ PHASE WINDOW

EOM PLS: KSC
TAL: BEN
TAL WX: NONE

SELECTED: RTLS: KSC 15/N/N TAL: NONE AOA: KSC 15/N/N PLS: KSC 22/N/N

TDEL: -0.01   0.312/0.35

MAX Q NAV: 745 PSF   754 PSF

SRB STG: 2:04.3   2:05

PERF: NOMINAL

2 ENG TAL (BEN) NO CALL

NEG RETURN: 4:04   4:05

PTA (U/S 500): 3:56   3:51

DROOP: 5:27   5:25

PTM (U/S 500): 5:14   5:04

MECO CMD: 8:30.1   8:29.8

VI: 26129   26119

OMS-2: 44:29.6   44:33.6 273.8 FPS   276 FPS

LANDING SITE: KSC 15 (KSC 35) 52:08:32:24Z 3:32:24 AM EST

Friday 9 2/21/97 (4)

DEORBIT BURN: 52:07:21:55Z
XRANGE: 484 NM
ORBIT DIR: DL 39
AIM PT: CLOSE IN

MLGTD: 2522 FT 52:08:32:24Z VEL: 184 KGS 191 KEAS HDOT: -1.5 FPS

TD NORM 195: 2394 FT

DRAG CHUTE DEPLOY: 184 KEAS 52:08:32:27Z

NLGTD: 5581 FT 52:08:32:34Z VEL: 136 KGS 140 KEAS HDOT: -6.7 FPS

BRK INIT: 94 KGS

DRAG CHUTE JETTISON: 52 KGS 52:08:32:56Z

BRK DECEL FPS²: AVE 5.2 PK 7.7

WHEELS STOP: 52:08:33:16Z 9588 FT

ROLLOUT: 7066 FT 52 SEC

WINDS: 5H, 1L KTS OFFICIAL:1407P13 7H, 1L

Continued . . .

SSME-TL: 104/104/ 109%

PREDICTED: 100/100/100/ 67/104

ACTUAL: 100/100/100/ 68/104

1 =2037 (3) 2 =2040 (2) 3 =2038 (3)

M 3 EOM: AVE WEIGHT: 213949 LBS X CG: 1077.83

LANDING WEIGHT: 213869 LBS X CG: 1079.57

SRB/RSRM: BI-085

RSRM 58

ET-81

LWT-74

ET PRED RPT: 271.3K

ET BRKUP: 214K

ET IMPACT 1:29:22 MET LAT: 17.4 N LONG: 141.1 W

ORBIT/HA/HP: DIRECT INSERTION

POST OMS-2: 312.9 X 186.3 NM

FINAL BRAKES: 322.3 X 316.4 NM

REBOOST 1: 323.7 X 319.2 NM

REBOOST 1A: 325.4 X 320.0 NM

REBOOST 2: 328.9 X 320.5 NM

REBOOST 3: 335.1 X 321.0 NM

DEORBIT: 334.1 X 312.2 NM

DEORBIT BURN: 504 FPS

VELOCITY: 26120 FPS

ENTRY RANGE: 4238 NM

INC: 28.46 (43)

FSW: O1-25 (4)

PAYLOAD WEIGHTS:
CARGO: 24891 LBS

PAYLOAD CHARGEABLE: 17374 LBS

DEPLOYED: 6941 LBS

NON-DEPLOYED: 9921 LBS

MIDDECK: 512 LBS

SHUTTLE ACCUMULATED WEIGHTS: DEPLOYED: 833955 LBS NON-DEPLOYED: 1163710 LBS CARGO TOTAL: 2455354 LBS

PERFORMANCE MARGINS (LBS): FPR: 3100 FUEL BIAS: 884 FINAL TDDP:3503 RECON:4235

PAYLOADS: PLB: Hubble Space Telescope Service Mission 2 (HST SM-02)

MIDDECK: MSX

5 CRYO TK SETS + 5 N2 TANKS

RMS 47 (S.N. 301)

RMS USED FOR HST CAPTURE, BERTH, & DEPLOY

STS082-E-5147 Captured HST

MISSION HIGHLIGHTS:
KSC W/D: OPF 147, VAB 5, PAD 26 = 178 days total.

LAUNCH ADVANCEMENTS:
- Baselined 2/13/96 launch date on 10/27/96.
- Advanced launch date to 2/11/97 on 1/15/97.

LAUNCH SCRUBS: None

LAUNCH DELAYS: None

TAL WX:
- Only Ben Guerir was manned; however, Ben Guerir was NO-GO for ceiling and visibility (overcast 500 feet and ground fog). There was no requirement for a TAL site due to a planned 8-second overlap between RTLS and PTA (actual overlap 14 seconds).

DOLILU-II I-LOADS:
- DOLILU-II uplink #13, I-Load uplink #32

SHUTTLE NIGHT LAUNCH #16

FLIGHT DURATION CHANGES:
- Waved off landing at KSC on orbit 149 due to clouds forming over runway with chance of 3000 feet broken. Landed on orbit 150.
- Extended flight duration 1 rev.

SHUTTLE NIGHT LANDING #9

FIRSTS/LASTS:
- First night landing at KSC with centerline lights.

EVENTS:
- HST grapple at 1:23:38 MET.
- Space Shuttle **altitude record** 335.1 NM X 321.0 NM after Reboost 3 maneuver.

RENDEZVOUS #36:
- Rendezvous, grapple, service, reboost, and release of HST.

HST REBOOST MANEUVERS:
- Reboost 1 was 20M43S at 04:01:09:28 MET.
- Reboost 1A was 10M13S at 04:06:07:02 MET with delta V 33 FPS. Maneuver was to avoid a conjunction with Pegasus debris.
- Reboost 2 was 19M47S at 05:01:15:00 MET.
- Reboost 3 was 31M54S at 07:01:32:58 MET.

SIGNIFICANT ANOMALIES:
- HST + V2 solar array rapid slew during airlock depress. For subsequent airlock depresses, one equalization valve on each hatch was duct-taped to limit air flow.
- EMU gloves had yellow smudges from HST handrails.
- FES feedline A accumulator heater failure.
- Erratic supply water tank D transducer.

FLT NO.	ORBITER	CREW (7) TITLE, NAMES & EVA'S	LAUNCH SITE, LIFTOFF TIME, LANDING SITES, ABORT TIMES	LANDING SITE/ RUNWAY, CROSSRANGE LANDING TIMES FLT DURATION, WINDS	SSME-TL NOM-ABORT EMERG THROTTLE PROFILE ENG. S.N.	SRB RSRM AND ET	INC	HA/HP	FSW	PAYLOAD WEIGHTS, PAYLOADS/ EXPERIMENTS	MISSION HIGHLIGHTS (LAUNCH SCRUBS/DELAYS, TAL WEATHER, ASCENT I-LOADS, FIRSTS, SIGNIFIC:T ANOMALIES, ETC.)
STS-82 Continued		Continued . . . SS EVA #34 EMU/tethered EVA 1 by EV1 and EV2 on 2/13/97 Scheduled EVA #30 6H42M21S duration SS EVA #35 EMU/tethered EVA 2 by EV3 and EV4 on 2/14/97 Scheduled EVA #31 7H27M31S duration SS EVA #36 EMU/tethered EVA 3 by EV1 and EV2 on 2/15/97 Scheduled EVA #32 7H11M00S duration SS EVA #37 EMU/tethered EVA 4 by EV3 and EV4 on 2/16/97 Scheduled EVA #33 6H34M30S duration SS EVA #38 EMU/tethered EVA 5 by EV1 and EV2 on 2/17/97 Unscheduled EVA #5 5H17M21Sduration MCC WHITE FCR (12) FLIGHT DIRECTORS: A/E - N. W. Hale LD/O 1 - J. W. Bantle O 2 - B. P. Austin PLNG - C. W. Shaw MOD - A. L. Briscoe		Continued . . . DENS ALT: 926 FT FLT DURATION: 9:23:37:07 S/T: 679:10:14:34 OV-103: 155:23:27:01 DISTANCE: 3,800,000 sm							Continued . . . SIGNIFICANT ANOMALIES (CONTINUED: - Fuel cell 3 water flow through alternate path causing concern that H_2 gas would get into EMU's during recharge from tank C. - Bent pins on SADE-2R P2 harness. - Three PGSC problems. - No RSRM erosion found.

STS081-E-5948 --- Crew portrait: Both the sign and shirts pay tribute to the HST and ground support team. Front (lt to rt),Tanner/MS, Lee/MS & Harbaugh/MS. Behind them (lt to rt), Hawley/MS, CDR Bowersox & PLT Horowitz. In back is Smith/MS

S82e5307 - Lee/PLC inside HST & Smith/MS on RMS during removal of Goddard High Resolution Spectrometer.

S82e5407 - Harbaugh/MS (left) & Tanner/MS on RMS accessing Fine Guidance Sensor (FGS) in the F site.

STS081-E-5937 HST begins its separation from Discovery following release.

FLT NO.	ORBITER	CREW (7) TITLE, NAMES & EVA'S	LAUNCH SITE, LIFTOFF TIME, LANDING SITES, ABORT TIMES	LANDING SITE/ RUNWAY, CROSSRANGE LANDING TIMES FLT DURATION, WINDS	SSME-TL NOM-ABORT EMERG THROTTLE PROFILE ENG. S.N.	SRB RSRM AND ET	ORBIT INC	ORBIT HA/HP	FSW	PAYLOAD WEIGHTS, PAYLOADS/ EXPERIMENTS	MISSION HIGHLIGHTS (LAUNCH SCRUBS/DELAYS, TAL WEATHER, ASCENT I-LOADS, FIRSTS, SIGNIFICANT ANOMALIES, ETC.)
STS-83 SEQ FLT #83 KSC - 83 PAD 39A - 47 MLP - 3	OV-102 (Flight 22) Columbia 20th Spacelab Flight LM-14 EDO 10 OMS PODS: LPO5-11 RPO5-10 FRC2-22	CDR: James D. Halsell, Jr. (Flt 3 - STS-65, STS-74) P475R178/V123/M156 PLT: Susan L. Still P476/R218/F28 M/S 1 (PAYLOAD CDR): Janice E. Voss (Flt 3 - STS-57, STS-63) P477/R167/V115/F22 M/S 2: Michael L. Gernhardt (Flt 2 - STS-69) P478/R199/V138/M173 M/S 3: Donald A. Thomas (Flt 3 - STS-65, STS-70) P479/R180/V119/M158 P/S 1: Roger Crouch P480/R219/M191 P/S 2: Gregory T. Linteris P481/R220/M192 MCC WHITE FCR (13) FLIGHT DIRECTORS: A/E - L. J. Ham LD/O 3 - R. M. Kelso O 1 - W. D. Reeves O 2 - G. A. Pennington O 4 - J. P. Shannon MOD - J. W. Bantle	KSC, PAD 39A 94:19:20:31.98Z 2:00:00 PM EST (P) 2:20:32 PM EST (A) Friday 16 4/4/97 (12) LAUNCH WINDOW: 2H30M CTOB EOM PLS: KSC TAL: BYD TAL WX: BEN, MRN SELECTED: RTLS: KSC 15/N/N TAL: BYD 32/N/N AOA: KSC 15/N/N PLS: FD1 NONE FD2 DELAY PRESS 12 SECONDS TDEL: 0.01 0.012/0.05 MAX Q NAV: 709 708 SRB STG: 2:03.5 2:03 PERF: NOMINAL 2 ENG TAL (BYD): 2:40 2:41 NEG RETURN: 3:57 4:00 PTA (U/S 154): 5:21 5:16 DROOP (BYD): 5:29 5:30 PTM (U/S 243): 5:45 5:45 MECO CMD: 8:29.7 8:30.7 VI: 25877 25871 OMS-2: 39:53 39:54.7 221.6FPS 222 FPS	KSC 33 (KSC 36) 98:18:33:11Z 2:33:11 PM EDT Tuesday 12 4/8/97 (10) DEORBIT BURN: 98:17:31:18Z XRANGE: 56 NM ORBIT DIR: DL 40 AIM PT: NOMINAL MLGTD: 3127 FT 98:18:33:11Z VEL: 193 KGS 197 KEAS HDOT: -1.3 FPS TD NORM 205: 2553 FT DRAG CHUTE DEPLOY: 186 KEAS 98:18:33:15Z NLGTD: 6654 FT 98:18:33:23Z VEL: 145 KGS 151 KEAS HDOT: -5.8 FPS BRK INIT: 85 KGS DRAG CHUTE JETTISON: 57 KGS 98:18:33:48Z BRK DECELFPS2: AVE 4.8 PK 6.9 WHEELS STOP: 98:18:34:11Z 11729 FT ROLLOUT: 8602 FT 60 SEC WINDS: H10, R2 OFFICIAL: 0209P18 H6, R6 DENS ALT: 963 FT FLT DURATION: 3:23:12:39 S/T: 683:09:27:13 OV-102: 221:15:11:06 DISTANCE: 1,500,000 sm	104/104/ 109% PREDICTED: 100/104/104/ 67/104 ACTUAL: 100/104/104/ 67/104 1 = 2012 (19) 2 = 2109 (17) 3 = 2019 (17) M 3 EOM: WEIGHT: 235510 LBS X CG: 1078.45 LANDING: WEIGHT: 235421 LBS X CG: 1079.99	BI-086 RSRM 59 ET-84 LWT-77 ET PRED RPT: 271.3K ET BRKUP: 214K ET IMPACT 1:21:10 MET LAT: 13.68 N LONG: 163.15 W	28.46 (44)	DIRECT INSERTION POST OMS-2: 163.5 X 160.1 NM DEORBIT: 162.7 X 158.3 NM VELOCITY: 25791 FPS ENTRY RANGE: 4402 NM	OI-25 (5)	CARGO: 34373 LBS PAYLOAD CHARGEABLE: 25556 LBS DEPLOYED: NONE NON-DEPLOYED: 23536 LBS MIDDECK: 2020 LBS SHUTTLE ACCUMULATED WEIGHTS: DEPLOYED: 833955 LBS NON-DEPLOYED: 1189266 LBS CARGO TOTAL: 2489727 LBS PERFORMANCE MARGINS (LBS): FPR: 3100 FUEL BIAS: 884 FINAL TDDP: 4820 RECON: 3741 PAYLOADS: PLB: Microgravity Science Laboratory. Protein Crystallography, Combustion Science, and Materials Sciences (MSL-1/LM) OARE CRYOFD MIDDECK: SAREX-II MSX 5 CRYO TK SETS + 4 EDO 5 N2 TANKS EDO PALLET NO RMS	KSC W/D: OPF 73, VAB 6, PAD = 24, 103 days total. LAUNCH POSTPONEMENTS: - Baselined 3/27/97 as launch date on 12/14/95. - Postponed launch date to 4/3/97 on 1/16/97 LAUNCH SCRUBS: - Scrubbed 4/3/97 launch on 4/1/97 at approximately L-42 hours based on decision to add missing insulation blankets to water coolant lines on 576 bulkhead. LAUNCH DELAYS: - Launch delayed 20M32S during T-9 minute hold because the cabin pressurization probe nose seal was found damaged and was replaced. Followed by high O_2 reading in mid-body caused by cabin vent into PLB. TAL WX: - Banjul (prime and selected) and Moron were forecast and observed GO. Ben Guerir was forecast NO-GO for crosswinds but observed GO. DOLILU-II I-LOADS: - DOLILU-II uplink #16, I-Load uplink #33. FLIGHT DURATION CHANGES: - Planned NEOM was on orbit 251. A Minimum Duration Flight (MDF) was declared due to concern about fuel cell 2 substack 3 increasing delta volts. Landing occurred on orbit 64 (11 days and 11 orbits early). FIRSTS/LASTS: - First U.S. spaceflight with female flight director for ascent (Linda Ham). SIGNIFICANT ANOMALIES: - FC2 substack 3 delta volts unusual start up and continuing on-orbit trend toward 300 mvolts caused a Minimum Duration Flight (MDF) to be declared. Postflight analysis indicated trend in multiple cells, not a single cell. - FC2 H_2 reactant valve failed to close by switch action when shutting down FC2 (regulator vented reactants). Valve closed 6 hours later. - -Y star tracker bypassed by PASS. - -Z star tracker pressure fail. - F3F failed off (low PC). - Subsystem RAU E transient - Multiple ECOS "hang" occurrences.

S98-16095-- In JSC MCC: Linda Ham, first female Ascent Flight Director. (Photo is from STS-095)

STS083-303-002 --- PLT Still floats into the Spacelab Module during activation.

STS083-325-004--- Crew portrait in Spacelab. Front row (lt to rt): Voss/MS/PLC, CDR Halsell, & Thomas/MS. Rear (lt to rt) Crouch/PS, Gernhardt/MS, PLT Still & Linteris/PS

SPACE SHUTTLE MISSIONS SUMMARY

FLT NO.	ORBITER	CREW (8) 7 UP & 7 DOWN / TITLE, NAMES & EVA'S	LAUNCH SITE, LIFTOFF TIME, LANDING SITES, ABORT TIMES	LANDING SITE/ RUNWAY, CROSSRANGE / LANDING TIMES FLT DURATION, WINDS	SSME-TL NOM-ABORT EMERG / THROTTLE PROFILE ENG. S.N.	SRB RSRM AND ET	ORBIT — INC	ORBIT — HA/HP	FSW	PAYLOAD WEIGHTS, PAYLOADS/ EXPERIMENTS	MISSION HIGHLIGHTS (LAUNCH SCRUBS/DELAYS, TAL WEATHER, ASCENT I-LOADS, FIRSTS, SIGNIFICANT ANOMALIES, ETC.)
STS-84 SEQ FLT #84 KSC-84 PAD 39A-48 MLP-2	OV-104 (Flight 19) Atlantis Spacehab 7 OMS PODS: LPO3-23 RPO4-19 FRC4-19	CDR: Charles J. Precourt (Flt 3 - STS-55, STS-71) P482/R161/V118/M141 PLT: Eileen M. Collins (Flt 2 - STS-63) P483/R188/V139/F24 M/S 1 (PAYLOAD CDR): Jean-Francois Clervoy (Flt 2 - STS-66) ESA Astronaut (France) P484/R186/V140/M163 M/S 2: Carlos I. Noriega P485/R221/M193 M/S 3: Edward T. Lu P486/R222/M194 M/S 4: Elena V. Kondakova (Russia) P487/R223/F29 M/S 5: Ascent C. Michael Foale (Flt 4 - STS-45, STS-56 & STS-63, stay on MIR 23, and return on STS-86) P488/R143/V92/M127 M/S 6: Descent Jerry M. Linenger (Flt 2 - STS-64, ascent on STS-81, and stay on Mir 22 and 23) P489/R182/V134/M159	KSC, PAD A 135:08:07:47.9Z 4:07:48 AM EDT (P) 4:07:48 AM EDT (A) Thursday 24 5/15/97 (4) LAUNCH WINDOW: 7M00S MIR PLANAR/ PHASE WINDOW EOM PLS: KSC TAL: ZZA TAL WX: MRN, BEN SELECTED: RTLS: KSC 33/N/N TAL: ZZA 30 AOA: KSC 15/N/N PLS: EDW 22/N/SF TDEL: 0.06 0.142/0.18 MAX Q NAV: 725 728 SRB STG: 2:04.2 2:04 PERF: NOMINAL 2 ENG TAL (BEN): 2:32 2:37 NEG RETURN: 4:03 4:05 PTA (U/S 263): 4:37 4:35 DROOP (ZZA): 5:20 5:25 PTM (U/S 263): 6:07 6:07 MECO CMD: 8:32.1 8:33.4 VI: 25873 25870 OMS-2: 44:01.6 43:04 75.6 FPS 76 FPS	KSC 33 (KSC 37) 144:13:27:43Z 9:27:43 AM EDT Saturday 18 5/24/97 (7) DEORBIT BURN: 144:12:23:33Z XRANGE: 34 NM ORBIT DIR: DL 41 AIM PT: NOMINAL MLGTD: 2882 FT 144:13:27:43Z VEL: 208 KGS 196 KEAS HDOT: -1.0 FPS TD NORM 195: 2989 FT DRAG CHUTE DEPLOY: 183 KEAS 144:13:27:47Z NLGTD: 5720 FT 144:13:27:52Z VEL: 175 KGS 156 KEAS HDOT: -6.9 FPS BRK INIT: 134 KGS DRAG CHUTE JETTISON: 53 KGS 144:13:28:17Z BRK DECEL FPS[2]: AVE 6.2 PK 9.6 WHEELS STOP: 144:13:28:36Z 11266 FT ROLLOUT: 8384 FT 53 SEC WINDS: 6T, R6 KTS OFFICIAL: 1109P13 T7, R6 DENS ALT: 1316 FT FLT DURATION: 9:05:19:55 S/T: 692:14:47:10 OV-104: 140:03:09:34 DISTANCE: 3,600,000 sm	104/104/ 109% PREDICTED: 100/104/104/ 67/104 ACTUAL: 100/104/104/ 67/104 1 = 2032 (6) 2 = 2031 (15) 3 = 2029 (15) M 3 EOM: WEIGHT: 216168 LBS X CG: 1080.95 LANDING: WEIGHT: 216021 LBS X CG: 1082.57	BI-087 RSRM 60 ET-85 LWT-78 ET PRED RPT: 271.3K ET BRKUP: 214K ET IMPACT: 1:26:42 MET LAT: 0.95 S LONG: 128.0 W	51.65 (7)	DIRECT INSERTION POST OMS-2: 160.6 X 85.5 NM TI 1:17:11:52 MET 215.6 X 203.4 NM 7:03:48 214.3 X 199.7 NM 07:08:10:39 214.3 X 199.7 NM DEORBIT: 214.1 X 199.7 NM VELOCITY: 25906 FPS ENTRY RANGE: 4397 NM	OI-25 (6)	CARGO: 28497 LBS PAYLOAD CHARGEABLE: 19643 LBS DEPLOYED: 3902 LBS NON-DEPLOYED: 14605 LBS MIDDECK: 1136 LBS SHUTTLE ACCUMULATED WEIGHTS: DEPLOYED: 1205007 LBS NON-DEPLOYED: 2042864 LBS CARGO TOTAL: 2518224 LBS PERFORMANCE MARGINS (LBS): FPR: 3100 FUEL BIAS: 884 FINAL TDDP: 938 RECON: 868 PAYLOADS: PLB: SHUTTLE/MIR MISSION 6 SPACEHAB DOUBLE MODULE MIDDECK: CREAM MSX SIMPLEX RME-III EPICS PCG-STES LME 5 CRYO TK SETS 4 N2 TANKS NO RMS	KSC W/D: OPF 76, VAB 4, PAD 21 = 101 days total. LAUNCH POSTPONEMENTS: - Baselined 5/1/97 launch date on 1/12/96. - Postponed launch date to 5/15/97 on 2/1/96 due to STS-78 SRB sooting and heat effects in field joints. LAUNCH SCRUBS: - None LAUNCH DELAYS: - None TAL WX: - Zaragoza (prime and selected), Moron, and Ben Guerir All forecast GO and observed GO. DOLILU-II I-LOADS: - DOLILU-II uplink #15, I-Load uplink #34 SHUTTLE NIGHT LAUNCH #17 FLIGHT DURATION CHANGES: - Waved off landing on orbit 144 due to forecast of 5000 feet variable broken and too dynamic. - Extended flight one orbit and landed on orbit 145. EVENTS: - Elena Kondakova's first flight was on Soyuz TM-17. - Mir 23 crew is Commander Vasily Tsibiliyev and Flight Engineer Alexander Lazutkin. - Mir capture at MET 1:18:25:36. Hooks closed at MET 1:18:33. - Hatch open at MET 1:20:16. - Crew transfer time: Foale to Mir 23 and Linenger to STS-84 was 2D6H13M. Linenger stay time on Mir was 122:04:36:25 and total flight time was132:04:00:20. - Transferred equipment, 1038 lbm H₂O, 82 lbm O₂, and 21 lbm N₂ to Mir. - Hatch closing at MET 6:04:32; undocking at MET 6:15:56. FIRSTS: - First EVA by a U.S. astronaut from Mir Space Station to deploy optical properties monitor by Linenger and Tsibiliyev. EVA was on 4/29/97. Exit from KVANT-2 airlock in Orlan M suit. Duration 4:57:30. RENDEZVOUS #37: - Rendezvous and dock with Mir (sixth docking). SIGNIFICANT ANOMALIES: - GPC Transient Mode Switch - dump indicated it was procedural problem. - Aft PL MNC amps measurement failed. - GPS/INS and GPS DTO problems. - Primary VHF and radio interface unit failure. - Window 1 impact reported by crew. - MS4 lightweight seat entry position/"A" hatch interference.

MCC WHITE FCR (14)

FLIGHT DIRECTORS:
A/E - N. W. Hale
LD/O 1 - P. L. Engelauf
O 2 - R. E. Castle
PLNG - P. F. Dye
MOD - A. L. Briscoe

Russia's Mir-post Atlantis sep.

STS084-366-015 --- Crews from STS-84 & Mir-23 onboard Spacehab Double Module tie record (ten) for number of persons in orbitingg spacecraft. Front from left: Linegar, Vasil V. Tsibliyev, Precourt, Aleksandr L. Lazutkin & Foale. Back, from left: Lu, Collins, Clervoy, Kondakova & Noriega.

FLT NO.	ORBITER	CREW (7) TITLE, NAMES & EVA'S	LAUNCH SITE, LIFTOFF TIME, LANDING SITES, ABORT TIMES	LANDING SITE/ RUNWAY, CROSSRANGE LANDING TIMES FLT DURATION, WINDS	SSME-TL NOM-ABORT EMERG THROTTLE PROFILE ENG. S.N.	SRB RSRM AND ET	ORBIT		FSW	PAYLOAD WEIGHTS, PAYLOADS/ EXPERIMENTS	MISSION HIGHLIGHTS (LAUNCH SCRUBS/DELAYS, TAL WEATHER, ASCENT I-LOADS, FIRSTS, SIGNIFICANT ANOMALIES, ETC.)
							INC	HA/HP			
STS-94 (STS-83R) SEQ FLT #85 KSC - 85 PAD 39A-49 MLP-1	OV-102 (Flight 23) Columbia 21st Spacelab Flight LM-15 EDO 11 OMS PODS: LPO5-12 RPO5-11 FRC2-23	CDR: James D. Halsell, Jr. (Flt 4 - STS-65, STS-74, & STS-83) P490/R178/V123/M156 PLT: Susan L. Still (Flt 2 - STS-83) P491/R218/V141/F28 M/S 1 (PAYLOAD CDR): Janice E. Voss (Flt 4 - STS-57, STS-63, & STS-83) P482/R167/V115/F22 M/S 2: Michael L. Gernhardt (Flt 3 - STS-69 & STS-83) P493/R199/V138/M173 M/S 3: Donald A. Thomas (Flt 4 STS-65, STS-70, STS-83) P494/R180/V119/M158 P/S 1: Roger Crouch (Flt 2 - STS-83) P495/R219/V142/M191 P/S 2: Gregory T. Linteris (Flt 2 - STS-83) P496/R220/V143/M192 MCC WHITE FCR (15) FLIGHT DIRECTORS: A/E - L. J. Ham LD/O 3 - R. M. Kelso O 1 - W. D. Reeves O 2 - G. A. Pennington O 3 - J. P. Shannon MOD - A. L. Briscoe	KSC PAD 39A 182:18:01:59.96Z 1:50:00 PM EDT (P) 2:02:00 PM EDT (A) Tuesday 12 7/1/97 (5) LAUNCH WINDOW: 2H30M CTOB EOM PLS: KSC TAL: BYD TAL WX: BEN SELECTED: RTLS: KSC 15/N/N TAL: BYD 32 AOA: EDW 22/N/N PLS: EDW 22/N/N TDEL: 0.01 0.382/0.42 MAX Q NAV: 701 PSF 703 PSF SRB STG: 2:03.5 2:04 PERF: NOMINAL 2 ENG TAL (BYD): 2:41 2:41 NEG RETURN: 3:56 3:58 PTA (U/S): 5:11 5:08 DROOP (BYD): 5:27 5:30 PTM (U/S): 7:03 7:05 MECO CMD: 8:28.6 8:29 VI: 25877 25871 OMS-2: 39:53 39:53 222 FPS 221.7 FPS BURN TIME: 2:23 2:23	KSC 33 (KSC 38) 198:10:46:33Z 6:46:33 AM EDT Thursday 9 7/17/97 (8) DEORBIT BURN: 198:09:43:45Z XRANGE: 81.7 NM ORBIT DIR: DL 42 AIM PT: NOMINAL MLGTD: 3056 FT 198:10:46:33Z VEL: 208 KGS 202 KEAS HDOT: -1.1 FPS TD NORM 205: 2774 FT DRAG CHUTE DEPLOY: 194 KEAS 198:10:46:37Z NLGTD: 6583 FT 198:10:46:44Z VEL: 158 KGS 152 KEAS HDOT: -5.9 FPS BRK INIT: 100 KGS DRAG CHUTE JETTISON: 52 KGS 198:10:47:12Z BRK DECEL FPS²: AVE 5.8 PK 7.2 WHEELS STOP: 198:10:47:31Z 11948 FT ROLLOUT: 8892 FT 58 SEC WINDS: T1, 0X KTS OFFICIAL: 1502P02 T2, 0X KTS DENS ALT: 1113 FT FLT DURATION: 15:16:44:33 S/T: 708:07:31:41 OV-102: 237:07:55:39 DISTANCE: 6,200,000 sm	104/104/ 109% PREDICTED: 100/104/104/ 67/104 ACTUAL: 100/104/104/ 69/104 1 = 2037 (4) 2 = 2034 (9) 3 = 2033 (9) M 3 EOM: WEIGHT: 230818 LBS X CG: 1078.40 LANDING: WEIGHT: 230773 LBS X CG: 1080.10	BI-088 RSRM 62 ET-86 LWT-79 ET PRED RPT: 271.3K ET BRKUP: 214K ET IMPACT 1:21:04 MET LAT: 13.5 N LONG 163.46 W	28.45 (45)	DIRECT INSERTION POST OMS-2: 163.4 X 160.1 NM DEORBIT: 162 X 156.4 NM VELOCITY: 25793 FPS ENTRY RANGE: 4396 NM	OI-25 (7)	CARGO: 34359 LBS PAYLOAD CHARGEABLE: 25568 LBS DEPLOYED: 0 LBS NON-DEPLOYED: 23536 LBS MIDDECK: 2032 LBS SHUTTLE ACCUMULATED WEIGHTS: DEPLOYED: 837857 LBS NON-DEPLOYED: 1230575 LBS CARGO TOTAL: 2552583 LBS PERFORMANCE MARGINS (LBS): FPR: 3200 FUEL BIAS: 809 FINAL TDDP: 2845 RECON: 4193 PAYLOADS: PLB: Microgravity Science Laboratory. Protein Crystallography, Combustion Science, and Materials Sciences (MSL-1/LM) OARE CRYOFD MIDDECK: SAREX-II MSX 5 CRYO TK SETS + 4 EDO 5 N2 TANKS EDO PALLET NO RMS	KSC W/D: OPF 53, VAB 7, PAD 21 = 81 days total. LAUNCH POSTPONEMENTS: None - Reflight of MSL-01/STS-83 was baselined as STS-83R on 4/10/97 with a launch date of 7/1/97. - On 4/25/97, STS-83R was renumbered STS-94. LAUNCH SCRUBS: None - LAUNCH DELAYS/EARLY LAUNCH TIMES: At the L-1 MMT, the weather forecast at KSC for 7/1/97 launch at 1837Z was thunderstorms/rain with 90% probability of NO-GO. The decision was made to move the launch time 47 minutes early to improve the probability of launch, which changed the EDW landing opportunities from 2-2-2 to 1-1-1. New launch time was 1750Z. Counted down to T-9 minutes and held due to thunderstorm forecast for RTLS landing time. Thunderstorms at RTLS time was removed from the forecast. Launch delay was 12M00S. TAL WX: Banjul was prime and selected. Banjul was NO GO for most of the count for 3000 feet broken but became GO late in count. Ben Guerir forecast and observed GO. DOLILU-II I-LOADS: DOLILU-II uplink #16, I-load uplink #35. KSC LANDING WEATHER: - Forecast for landing time was technically NO-GO for rain within 30 NM; however, rain was offshore, moving NE, and approach path was clear. Observed GO at deorbit burn minus 2 minutes. At landing time, rain was 29 ESE. Flight rule waiver written. FLIGHT DURATION CHANGES: None. FIRSTS/LASTS: - First reflight of same payloads (MSL-01 with same crew after STS-83 minimum duration flight declared due to FC2, substack 3 delta volts change). - First flight of Wraparound DAP (called part 5) used for complete entry. RCS usage 500 lbs vs baseline 700 lbs and redline 1430 lbs (28.45 inclination). EVENTS: - Entry was observed at approx 16 degrees elevation in Houston. - Deorbit burn was 298.5 FPS. SIGNIFICANT ANOMALIES: - Fuel cell 3, substack 2, cell performance monitor output increased approximately 32 mv in 20 minutes. - TDRSS Ku-band channel lock dropouts (worse with 48 MBPS on TDRS-E). - Loss of aero surface actuator (ASA) 4 redundant power. - Lower port fastener retainer housing separated from locker L6G (transfer from Spacelab to MF28K & M as DTO). - Ku-band channel 2 frequency shifts. - Ku-band roll/alpha gimbal anomaly. - Window #7 debris impact reported by crew. - APU 3 fuel isolation valves on heated string B cycling low. - Tempus top video camera failure.

STS094-344-001 Crouch (front) & Gernhardt at the NASDA Large Isothermal Furnace (LIF) facility.

STS094-307-001 --- Inflight crew portrait. Front (lt to rt): PLT Still & Voss/PLC. Middle row (lt to rt): Gernhardt/MS, CDR Halsell, & Linteris/PS. Back row: Thomas/MS (left) & Crouch/PS.

FLT NO.	ORBITER	CREW (7) TITLE, NAMES & EVA'S	LAUNCH SITE, LIFTOFF TIME, LANDING SITES, ABORT TIMES	LANDING SITE/ RUNWAY, CROSSRANGE LANDING TIMES FLT DURATION, WINDS	SSME-TL NOM-ABORT EMERG THROTTLE PROFILE ENG. S.N.	SRB RSRM AND ET	ORBIT		FSW	PAYLOAD WEIGHTS, PAYLOADS/ EXPERIMENTS	MISSION HIGHLIGHTS (LAUNCH SCRUBS/DELAYS, TAL WEATHER, ASCENT I-LOADS, FIRSTS, SIGNIFICANT ANOMALIES, ETC.)
							INC	HA/HP			
STS-85 SEQ FLT #86 KSC - 86 PAD 39A-50 MLP-3	OV-103 (Flight 23) Discovery OMS PODS: LPO1-26 RPO3-23 FRC3-23	CDR: Curtis L. Brown, Jr. (Flt 4 - STS-47, STS-66 & STS-77) P497/R152/V112/M136 PLT: Kent V. Rominger (Flt 3 - STS-73, STS-80) P498/R200/V131/M174 M/S 1 (PAYLOAD CDR): N. Jan Davis (Flt 3 - STS-47, STS-60) P499/R153/V100/F17 M/S 2: Robert L. Curbeam, Jr. P500/R224/M195 M/S 3: Stephen K. Robinson P501/R225M196 P/S 1: Bjarni V. Tryggvason (Canada) P502/R226/M197 MCC WHITE FCR (16) FLIGHT DIRECTORS: A/E/O1 - N. W. Hale LD/O 2 - B. P. Austin PLNG - G. A. Pennington MOD - A. L. Briscoe & J. W. Bantle	KSC PAD 39A 219:14:40:59.98Z 10:41:00 AM EDT (P) 10:41:00 AM EDT (A) Thursday 25 8/7/97 (6) LAUNCH WINDOW: 1H39M CHRISTA-SPAS BETA REQUIREMENTS EOM PLS: KSC TAL: ZZA TAL WX: MRN, BEN SELECTED: RTLS: KSC 33/N/N TAL: MRN 20/N/N AOA: NOR 35/N/SF PLS: EDW 22/N/N TDEL: 0.0 -0.198/-0.16 MAX Q NAV: 699 PSF 703 PSF SRB STG: 2:03.8 2:04 PERF: NOMINAL 2 ENG TAL (MRN): 2:53 2:50 NEG RETURN: 4:01 4:02 PTA (U/S 298): 5:11 5:12 DROOP (ZZA): 5:28 5:34 PTM (U/S 579): 7:05 7:10 MECO CMD: 8:30.7 8:32.7 VI: 25831 25823 OMS-2: 33:06 33:06 254 FPS 254 FPS	KSC 33 (KSC 39) 231:11:07:58Z 7:07:58 AM EDT Tuesday 13 8/19/97 (5) DEORBIT BURN: 231:10:07:30Z XRANGE: 346 NM ORBIT DIR: AR 7 AIM PT: NOMINAL MLGTD: 2917 FT 231:11:07:58Z VEL: 199 KGS 192 KEAS HDOT: -1.5 FPS TD NORM 195: 2550 FT DRAG CHUTE DEPLOY: 185 KEAS 231:11:08:01Z NLGTD: 6065 FT 231:11:08:09Z VEL: 153 KGS 144 KEAS HDOT: -6.1 FPS BRK INIT: 84 KGS DRAG CHUTE JETTISON: 55 KGS 231:11:08:37Z BRK DECEL FPS[2]: AVE 5.7 PK 7.2 WHEELS STOP: 231:11:09:07Z 11709 FT ROLLOUT: 8792 FT WINDS: T5, L3 KTS OFFICIAL: 2006P09, T4, L5 KTS DENS ALT: 1565 FT FLT DURATION: 11:20:26:58 S/T: 720:03:58:39 OV-103: 167:19:53:59 DISTANCE: 4,725,000 sm	104/104/ 109% PREDICTED: 100/104/104/ 67/104 ACTUAL: 100/104/104/ 67/104 M 3 EOM: WEIGHT: 221335 LBS X CG: 1081.95 LANDING: WEIGHT: 221264 LBS X CG: 1083.63	BI-089 RSRM 57 ET-87 LWT-80 ET PRED RPT: 271.3K ET BRKUP: 214K ET IMPACT 1:14:30 MET LAT: 42.77 S LONG 154.86 W	57 (19)	DIRECT INSERTION POST OMS-2: 161 X 160 NM SEP-1: 219:22:28:00 160.0 X 158.9 NM TI: 228:12:50:47 157.7 X 154.3 NM DEORBIT: 4492 NM VELOCITY: 25755 FPS ENTRY RANGE: 4492 NM ENTRY ATTITUDE: 139.2 X 138.4 NM	OI-26 (1)	CARGO: 31959 LBS PAYLOAD CHARGEABLE: 24982 LBS DEPLOYED: 0 LBS NON-DEPLOYED: 24982 LBS MIDDECK: 1590 LBS SHUTTLE ACCUMULATED WEIGHTS: DEPLOYED: 837857 LBS NON-DEPLOYED: 1247831 LBS CARGO TOTAL: 2584542 LBS PERFORMANCE MARGINS (LBS): FPR: 3200 FUEL BIAS: 809 FINAL TDDP: 1446 RECON: 3065 PAYLOADS: PLB: CRISTA-SPAS-02 (Atmospheric physics, dynamics, and chemistry by MAHRSI, SESAM, MIDES, GAPS, and IPEX) MFD (Robot Arm) TAS-01 (8 technology and science experiments) IEH-2 (UV exp) MIDDECK: SWUIS, BDS-03, BRIC-10, PCG-STES, SSCE, ACIS, MSX, SIMPLEX 5 CRYO TK SETS 5 N2 TANKS (S.N. 301) RMS 48 RMS Used For CRISTA-SPAS deploy, grapple, and berth	KSC W/D: OPF 102 , VAB 5, PAD 23 = 130 days total. LAUNCH POSTPONEMENTS: - Baselined launch date of 7/17/97 on 3/28/96. - Postponed launch date to 8/7/97 on 4/17/97 caused by remanifest to refly MSL-1 due to STS-83 early termination. LAUNCH SCRUBS: None LAUNCH DELAYS: None TAL WX: - ZZA was prime but forecast NO GO with thunderstorms within 20 nm. MRN (selected) and BEN were forecast and observed GO. DOLILU-II I-LOADS: DOLILU-II uplink #17, I-load uplink #36. PERFORMANCE ENHANCEMENTS (FIRST FLIGHT): - Flight control filter updates. - Yaw gain enhancement. - Constant pitch rate at SRB separation. FLIGHT DURATION CHANGES: - Planned landing time was 230:11:14 on 8/16/97, orbit 174. Waved off this only landing opportunity to land at KSC due to forecast of probability of fog. SLF was observed GO at landing time. Landed on orbit 190. - Flight duration extended 1 day. FIRSTS/LASTS: - First flight of OI-26. - First flight at 57 degrees inclination since STS-66. - First flight of complete Wraparound DAP (DTO 255). Used approx 330 lbm RCS from EI to M=1 (vs. redline of 1630 lbm). FOURTH SHUTTLE CREWMEMBER REPLACEMENT - Jeff Ashby was replaced by Rominger in March 1997. (Third shuttle crewmember replacement occurred on STS-46.). EVENTS: - Launched on Kent Rominger's birthday. - CHRISTA-SPAS deployed at 00:07:46:04 MET, 219:22:27:04Z. - CHRISTA-SPAS captured at 228:15:13Z, 09:00:32 MET. - Berthed and latched at 228:16:30:12Z, 09:01:49:32 MET. RENDEZVOUS #38: Deployed, rendezvoused, grappled, and berthed CHRISTA-SPAS. SIGNIFICANT ANOMALIES: - CRT 1 transient BITE message. - Supply H2O tank A quantity erratic. - APU 1 seal cavity drain line pressure decay. - APU 1 fuel pump thermostat cyclic in narrow band. - Payload commanding problems with MCC input set to 3/sec.

STS085-706-051--- Release of CRISTA-SPAS-2

STS085-326-016 Impromptu in-flight crew portrait: (Left to right) PLT Rominger, Curbeam/MS, Robinson/MS, CDR Brown, Davis/MS/PLC, & Tryggvason/PS Canada).

FLT NO.	ORBITER	CREW (8) 7UP, 7DOWN TITLE, NAMES & EVA'S	LAUNCH SITE, LIFTOFF TIME, LANDING SITES, ABORT TIMES	LANDING SITE/ RUNWAY, CROSSRANGE LANDING TIMES FLT DURATION, WINDS	SSME-TL NOM-ABORT EMERG THROTTLE PROFILE ENG. S.N.	SRB RSRM AND ET	INC	ORBIT HA/HP	FSW	PAYLOAD WEIGHTS, PAYLOADS/ EXPERIMENTS	MISSION HIGHLIGHTS (LAUNCH SCRUBS/DELAYS, TAL WEATHER, ASCENT I-LOADS, FIRSTS, SIGNIFICANT ANOMALIES, ETC.)

STS-86
SEQ FLT #87
KSC - 87
PAD 39A-51
MLP- 2

ORBITER:
OV-104
(Flight 20)
Atlantis
Spacehab 8
OMS PODS:
LPO3-24
RPO4-20
FRC4-20

CREW:
CDR:
James D. Wetherbee
(Flt 4 - STS-32, STS-52, STS-63)
P503/R108/V/80/M97

PLT:
Michael J.Bloomfield
P504/R227/M198

M/S 1 EV2:
Vladimir Titov
(Russia)
(Flt 2 - STS-63)
P505/R189/V144/M165

M/S 2 EV1:
Scott E. Parazynski
(Flt 2 - STS-66)
P506/R187/V145/M164

M/S 3:
Jean-Loup Chretien
(France)
P507/R228/M199

M/S 4:
Wendy B. Lawrence
(Flt 2 - STS-67)
P508//R192/V146/F25

M/S 5:
Ascent
David A. Wolf
(Flt 2 - STS-58, stay on Mir 24 and return on STS-89)
P509/R173/V147/M151

M/S 6:
Descent
Michael C. Foale
(Flt 4 - STS-45, STS-56, STS-63, ascent on STS-84, on-orbit stay on Mir 23 and Mir 24, and descent on STS-86)
P510/R143/V92/M127

Continued...

LAUNCH SITE:
KSC PAD 39A
269:02:34:18.96   Z
10:34:19 PM EDT (P)
10:34:19 PM EDT (A)
Thursday 26
9/25/97 (9)

LAUNCH WINDOW:
6M38S
USING PLT MIR PLANAR /PHASING WINDOW

EOM PLS: KSC
TAL: ZZA
TAL WX: MRN, BEN

SELECTED:
RTLS: KSC 33/N/SF
TAL: MRN 20
AOA: KSC 15/CI/N
PLS: EDW 04/CI/N

TDEL:
0.05   0.162/0.20

MAX Q NAV:
723   721

SRB STG:
2:03.05   2:03

PERF: NOMINAL

2 ENG TAL (MRN):
2:32   2:33

NEG RETURN:
4:02   4:04

PTA (U/S 269):
4:48   4:49

DROOP (MRN):
5:22   5:24

PTM (U/S 760):
6:53   6:55

MECO CMD:
8:30.5   8:31

VI:
25876   25872

OMS-2:
41:50   41:50
171.8 FPS171.8 FPS

LANDING SITE:
KSC 15 (KSC 40)
279:21:55:10Z
5:55:10 PM EDT
Monday 16
10/6/97 (7)

DEORBIT BURN:
279:20:47:45Z

XRANGE: 376 NM

ORBIT DIR: AL 19

AIM PT: NOMINAL

MLGTD: 2420 FT
279:21:55:10Z
VEL: 198 KGS
        194 KEAS
HDOT: -2.2 FPS

TD NORM 195:
2592 FT

DRAG CHUTE DEPLOY: 152KEAS
279:21:55:22Z

NLGTD: 5522 FT
279:21:55:19
VEL: 163 KGS
        159 KEAS
HDOT: -6.1 FPS

BRK INIT: 60 KGS

DRAG CHUTE JETTISON: 67 KGS
279:21:55:57Z

BRK DECEL FPS^2:
AVE 3.7 PK 5.0

WHEELS STOP:
279:21:56:31Z
14367 FT

ROLLOUT:
11947 FT
81 SEC

WINDS:
H2, L9 KTS
OFFICIAL:
0707P14
H2, L9 KTS

Continued...

SSME-TL:
104/104/
109%

PREDICTED:
100/104/104/
67/104

ACTUAL:
100/104/104/
67/104

1 = 2012 (20)
2 = 2040 (3)
3 = 2019 (18)

M 3 EOM:
WEIGHT:
215387 LBS
X CG:
1081.33

LANDING:
WEIGHT:
215303 LBS
X CG:
1083.03

STS086-720-091 --
View of damaged solar panel & radiator on Mir Spektr caused by Progress re-supply ship that collided with Mir June 25,1997, causing Spektr to repressurize. (Atlantis photo during docking.)

SRB RSRM:
BI-090
RSRM 61
ET-88
LWT-81

ET PRED RPT:
271.3K

ET BRKUP:
269.1K

ET IMPACT
1:26:44
MET
LAT:
0.52 S
LONG:
126.53
W

INC:
51.65
(8)

ORBIT:
DIRECT INSERTION

POST OMS-2:
161 X 138.5 NM

NC1:
269:05:59:10Z
201 X 150.9 NM

TI:
270:17:31:56Z
211.2 X 203.5 NM

MCC4:
270:18:52:13Z
211.8 X 204.3 NM

UNDOCK:
276:17:28:34Z
212 X 204.4 NM

SEP: 276

DEORBIT:
207 X 190 NM

VELOCITY:
25898 FPS

ENTRY RANGE:
4380 NM

ENTRY ATTITUDE:
205.9 X 190.8 NM

FSW:
OI-26
(2)

PAYLOAD:
CARGO:
29728 LBS

PAYLOAD CHARGEABLE:
21039 LBS

DEPLOYED:
6058 LBS

NON-DEPLOYED:
14379 LBS

MIDDECK:
602 LBS

SHUTTLE ACCUMULATED WEIGHTS:
DEPLOYED:
843515 LBS
NON-DEPLOYED:
1262812 LBS
CARGO TOTAL:
2614270 LBS

PERFORMANCE MARGINS (LBS):
FPR: 3200
FUEL BIAS: 809
FINAL TDDP: 1446
RECON: 3065

PAYLOADS:
PLB:
SHUTTLE/MIR MISSION 7

SPACEHAB DOUBLE MODULE

ODS, SEEDS - II

MIDDECK:
CREAM
SIMPLEX
KIDSAT
CPCG
CCM-A

5 CRYO TK SETS
4 N2 TANKS

NO RMS

MISSION HIGHLIGHTS:
KSC W/D: OPF 60, VAB 5, PAD 29 = 94 days total.

LAUNCH POSTPONEMENTS:
- Baselined launch date of 9/11/97 on 6/21/96; orbiter OV-104.
- Postponed launch date to 9/18/97 on 8/1/96; multi-flight changes.
- Changed from orbiter OV-104 to OV-105 on 3/27/97.
- Postponed launch date to 9/25/97 on 4/17/97; multi-flight changes for re-flight of MSL-01 (on STS-94).
- Advanced launch date to 9/18/97 on 4/25/97 and moved back to orbiter OV-104 (from OV-105).
- Postponed launch date to 9/26/97 (GMT), 9/25/97 EDT, on 8/21/97.

LAUNCH SCRUBS: None

LAUNCH WINDOWS:
- The total launch window for the two panes was 10:57. However, using the preferred liftoff time of 269:02:34:19 (4m19s into window) the window was only 6m38s.

LAUNCH DELAYS: None

TAL WX:
- ZZA was prime but was forecast NO GO (ceiling) at L-15 minutes, MRN was forecast GO and was selected. Both ZZA and MRN were observed GO at TAL time. BEN was forecast NO GO (ceiling) until L-8 minutes and was observed GO at TAL time.

SHUTTLE NIGHT LAUNCH: #18

DOLILU II I-LOADS:
- DOLILU II uplink #18, total uplink #37.

PERFORMANCE ENHANCEMENTS:
- Flight control filter updates.
- Yaw gain enhancement.
- Constant pitch rate at SRB separation.
- Auto delta psi.

FLIGHT DURATION CHANGES:
- Waved off landing on orbit 155 due to observed broken 4000 feet, but forecast GO.
- Waved off landing on orbit 156 (observed GO), but forecast NO GO 5000 feet broken.
- Landed on orbit 170.
- Flight duration extended 1 day.

FIRSTS/LASTS:
- First flight of auto delta psi.
- First flight using a Preferred Liftoff Time (PLT), which was not at window opening.
- First shuttle EVA with an International Partner (V. Titov, Russia).

Continued...

FLT NO.	ORBITER	CREW (8) 7UP, 7DOWN TITLE, NAMES & EVA'S	LAUNCH SITE, LIFTOFF TIME, LANDING SITES, ABORT TIMES	LANDING SITE/ RUNWAY, CROSSRANGE LANDING TIMES FLT DURATION, WINDS	SSME-TL NOM-ABORT EMERG THROTTLE PROFILE ENG. S.N.	SRB RSRM AND ET	ORBIT INC	HA/HP	FSW	PAYLOAD WEIGHTS, PAYLOADS/ EXPERIMENTS	MISSION HIGHLIGHTS (LAUNCH SCRUBS/DELAYS, TAL WEATHER, ASCENT I-LOADS, FIRSTS, SIGNIFICANT ANOMALIES, ETC.)
STS-86 Continued		Continued… SS EVA #39 EMU/Tethered EVA #32 Scheduled EVA #34 10/1/97 5H01M26S Duration MCC WHITE FCR (17) FLIGHT DIRECTORS: A/E - L. J. Ham LD/O 1 - P. D. Dye O 2 - C. W. Shaw PLNG - P. L. Engelauf MOD - R. E. Castle		Continued… DENS ALT: 1506 FT FLT DURATION: 10:19:20:51 S/T: 730:28:19:30 OV-104: 150:22:30:25 DISTANCE: 4,225,000 sm						Continued . . . EVENTS: - Mir capture at 270:19:57:46Z, 01:17:23:27 MET - Docking complete at 270:20:06:15Z, 01:17:31:56 MET - Foale transfer to STS-86 and David Wolf transfer to Mir 24 at 2D14H00M, 271:16:34:19Z. Foale Mir stay time 134:02:13:31, total flight time 144:13:47:22. - Foale completed a Mir EVA with Anatoliy Solovyev with exit from KVANT-2 airlock in Orlan M suits (5.7 psia). Both were double tethered using U.S. tether reel and waist tethers. EVA duration was 5H59M to inspect Specktr module leak, slew solar arrays, and put out dosimeter. - Scott Parazynski and Vladimir Titov made a Shuttle EVA to retrieve MEEP experiments left on Mir DM on STS-76. - Jean-Loup Chretien flew on Soyuz T-6/Salyut 7 and Soyuz TM-7/Mir11. - Hooks open 276:17:25:59Z, 07:14:51:40 MET - Undock 276:17:28:15Z, 07:14:53:56 MET (one rev late to check Mir computer interface box). - Total consumables transferred to Mir: 1717.2 lbm H_2O (17 CWC's), 75.7 lbm O_2, 130.7 lbm N_2. - Wendy was to replace Foale; however, concerns of inadequate reach in Orlan EVA spacesuit, Wolf moved to STS-86 from STS-89. RENDEZVOUS #39: Rendezvous and dock with Mir Space Station. SIGNIFICANT ANOMALIES: - Fuel Cell 2 substack 1 differential volts transient. - Primary RCS thruster L3D failed off. - EVA Safety Tether Reel failure. - WSB 3 vent heater failure on B controller.

STS086-371-004 -- Seven STS-86 crew members are joined by the three-member Mir-24 crew in the Spacehab Module for in-flight portrait. New Mir-24 crew member Wolf holds a cap (right). Clockwise from him are: Titov/MS/RSA, Mir CDR Anatoliy Y. Solovyev, Parazynski/MS, Pavel V. Vinogradov/Mir/FE, CDR Wetherbee, Lawerence/MS, Foale/MS, PLT Bloomfield, & Chretien/MS.

STS086-332-021--Parazynski tethered to cargo bay handrail during EVA shared Titov (RSA) out of photo.

sts086-720-056 -- Mir as seen by departing Atlantis.

FLT NO.	ORBITER	CREW (6) TITLE, NAMES & EVA'S	LAUNCH SITE, LIFTOFF TIME, LANDING SITES, ABORT TIMES	LANDING SITE/ RUNWAY, CROSSRANGE LANDING TIMES FLT DURATION, WINDS	SSME-TL NOM-ABORT EMERG THROTTLE PROFILE ENG. S.N.	SRB RSRM AND ET	ORBIT		FSW	PAYLOAD WEIGHTS, PAYLOADS/ EXPERIMENTS	MISSION HIGHLIGHTS (LAUNCH SCRUBS/DELAYS, TAL WEATHER, ASCENT I-LOADS, FIRSTS, SIGNIFICANT ANOMALIES, ETC.)
							INC	HA/HP			

STS-87
 	
SEQ
FLT #88
 	
KSC - 88
 	
PAD
39B-37
 	
MLP-1
	OV-102
	(Flight 24)
	Columbia
	
	22nd
	Spacelab
	Flight
	
	EDO 12
	
	OMS PODS:
	LPO5-13
	RPO5-12
	FRC2-24
		CDR:
		Kevin R. Kregel
		(Flt 3 - STS-70, STS-78)
		P511/R197/V129/M172
		PLT:
		Steven W. Lindsey
		P512/R229/M200
		M/S 1:
		Kalpana Chawla
		P513/R230/F30
		M/S 2:
		Winston E. Scott
		(Flt 2 - STS-72)
		P514/R207/V148/M180
		M/S 3:
		Takao Doi
		(Japan)
		P515/R231/M201
		P/S 1:
		Leonid Kadenyuk
		(Ukraine)
		P516/R232/M202
			KSC PAD 39B
			323:19:45:95.6Z
			2:46:00 PM EST (P)
			2:46:00 PM EST (A)
			Wednesday 9
			11/19/97 (12)
			
			LAUNCH WINDOW:
			2H30M CTOB
			
			EOM PLS: KSC
			TAL: BYD
			TAL WX: BEN, MRN
			
			SELECTED:
			RTLS: KSC 33/CI/N
			TAL: BYD 32/N/N
			AOA: EDW 22/N/N
			PLS: NOR 17/N/SF
			
			TDEL:
			-0.03 0.132/0.17
			
			MAX Q NAV:
			731 741
			
			SRB STG:
			2:03.8 2:04
			
			PERF: NOMINAL
			
			2 ENG TAL (BYD):
			2:38 2:41
			
			NEG RETURN:
			3:58 3:59
			
			PTA (U/S 219):
			4:59 4:58
			
			DROOP (BYD):
			5:25 5:30
			
			PTM (U/S 567):
			6:58 7:00
			
			MECO CMD:
			8:28.5 8:29.9
			
			VI:
			25872 25873
			
			OMS-2:
			41:04 41:08.9
			192.9 FPS 193.8 FPS
			2:05 2:08
				KSC 33 (KSC 41)
				339:12:20:04Z
				7:20:04 AM EST
				Friday 10
				12/5/97 (9)
				
				DEORBIT BURN:
				339:11:21:28Z
				
				XRANGE: 66 NM
				
				ORBIT DIR: DL 43
				AIM PT: CLOSE IN
				
				MLGTD: 2549 FT
				339:12:20:04Z
				VEL: 189 KGS
				            196 KEAS
				HDOT: -1.1 FPS
				
				TD NORM 205:
				1821 FT
				
				DRAG CHUTE
				DEPLOY: 188 KEAS
				339:12:20:08Z
				
				NLGTD: 5612 FT
				339:12:20:14Z
				VEL: 147 KGS
				            151 KEAS
				HDOT: -4.6 FPS
				
				BRK INIT: 107 KGS
				
				DRAG CHUTE
				JETTISON: 61 KGS
				339:12:20:38Z
				
				BRK DECEL FPS2:
				AVG 4.7 PK 7.7
				
				WHEELS STOP:
				339:12:21:02Z
				10553 FT
				
				ROLLOUT:
				8004 FT
				58 SEC
				
				WINDS:
				6H, 0X KTS
				OFFICIAL:
				3306P10
				6H, 0X KTS
				DENS ALT: -195 FT
				
				FLT DURATION:
				15:16:34:04
				
				S/T: 746:15:53:34
				
				OV-102:
				253:00:29:43
				
				DISTANCE:
				6,544,000 sm
					104/104/
					109%
					
					PREDICTED:
					100/104/104/
					67/104
					
					ACTUAL:
					100/104/104/
					67/104
					
					1 = 2031 (16)
					2 = 2039 (4)
					3 = 2037 (5)
					
					M 3 EOM:
					WEIGHT:
					232930 LBS
					X CG:
					1080.99
					
					LANDING:
					WEIGHT:
					232849 LBS
					X CG:
					1082.58
						BI-092
						
						RSRM
						63
						
						ET-89
						
						LWT-82
						
						ET
						PRED
						RPT:
						271.3K
						
						ET
						BRKUP:
						269.1K
						
						ET
						IMPACT
						1:25:02
						MET
						LAT:
						20.28 N
						LONG:
						147.99
						W
							28.45
							(46)
								DIRECT
								INSERTION
								
								POST OMS-2:
								155 X 150 NM
								
								SEP BURN:
								02:03:25:30
								MET
								
								NC5
								MANEUVER:
								05:01:33:33
								MET
								
								TI:
								05:03:04:38
								MET
								
								DEORBIT:
								149.7 X 145.5
								NM
								
								VELOCITY:
								25760 FPS
								
								ENTRY
								RANGE:
								4424 NM
									OI-26
									(3)
										CARGO:
										34395 LBS
										
										PAYLOAD
										CHARGEABLE:
										21946 LBS
										
										DEPLOYED:
										0 LBS
										
										NON-DEPLOYED:
										17496 LBS
										
										MIDDECK:
										1452 LBS
										
										SHUTTLE
										ACCUMULATED
										WEIGHTS:
										DEPLOYED:
										843915 LBS
										NON-DEPLOYED:
										1281760 LBS
										CARGO TOTAL:
										2648665 LBS
										
										PERFORMANCE
										MARGINS (LBS):
										FPR: 3085
										FUEL BIAS: 853
										FINAL TDDP: 4384
										RECON: 6115
										
										PAYLOADS:
										PLB:
										SPARTAN-201
										USMP-04
										EDFT-05
										SOLSE
										GAS (1)
										NASBE
										LHP
										TGDF
										AERCAM SPRINT
										
										MIDDECK:
										USMP-04/MGBX
										CUE, MSX,
										SIMPLEX
										
										5 CRYO TK SETS
										+ 4 EDO
										5 N2 TANKS
										RMS 49
										(S.N. 301)
										
										RMS used for
										Spartan deploy,
										capture attempt, and
										assist berthing.
										Also EVA EDFT-5
										ORU activities.
											KSC W/D: OPF 94, VAB 5, PAD 22 = 121 days total.
											
											LAUNCH POSTPONEMENTS:
											- Baselined 10/9/97 launch date on 7/11/97.
											- Postponed launch date to 11/13/97 on 4/17/97.
											- Postponed launch date to 11/19/97 on 5/22/97.
											
											LAUNCH SCRUBS: None.
											
											LAUNCH DELAYS: None.
											
											TAL WX:
											- Banjul (prime and selected), Ben Guerir, and Moron were all forecast and observed GO.
											
											DOLILU/I-LOADS:
											- DOLILU-II uplink #19, total I-load uplink #38.
											
											PERFORMANCE ENHANCEMENTS:
											- Flight control filter updates.
											- Yaw gain enhancement.
											- Constant pitch rate at SRB separation.
											- First stage trim, second stage trim, and roll to headsup.
											
											FLIGHT DURATION CHANGES:
											- None. Landed on orbit 252.
											
											FIRSTS/LASTS:
											- First flight with the following performance enhancements:
											- Roll-to-heads-up at approximately 6:10 MET, APM loss of 70 lbs.
											- Ascent DAP trim (APM gain of approximately 270 lbs).
											- Extended pitch parallel to MECO (APM gain of approximately 125 lbs).
											- Second stage pitch gimbal relief (no APM change).
											
											EVENTS:
											- Spartan deploy was delayed 1 day to allow recovery of SOHO satellite.
											- Spartan deploy at 325:21:04:00Z, 02:01:18 MET. Spartan failed to perform pirouette maneuver indicating a problem. Attempt to grapple Spartan at 02:01:24 MET failed, and a tip-off rate of 2 deg/sec was introduced.
											- Separation burn was made at, 02:03:25:30 MET.
											- Decision to hand capture Spartan by two EVA crew, done at 05:05:18:00 MET (rates were very low). RMS berth assist was required with Spartan grapple at 05:06:53 and berth at 05:07:37:22 MET.
											- EDFT-05 tasks were performed on EVA 1 and evaluated crane.
											- An unscheduled EVA 2 was performed to deploy, maneuver, and retrieve a free flying video camera (AERCam Sprint) and to perform EDFT-05 tasks which were planned for EVA 1.
											
											RADIATOR DEPLOY #20:
											- Starboard and port radiators deployed twice for thermal control and water production.
											
											RENDEZVOUS #40: Deploy Spartan, separate, rendezvous and retrieve Spartan.
											
											SIGNIFICANT ANOMALIES:
											- Sticky supply water A/B check valve.
											- H2 tank 4 quantity measurement failure.
											- EV 2 helmet light intermittent.
											- Left outboard tire pressure measurement lost.
											- Spartan MPESS EVA ingress aid extend/stow difficulty during retrieval.
											- RCS jet R5D heater fail off.
											- Excessive tile damage by ET foam loss.

MCC WHITE FCR (18)

FLIGHT DIRECTORS:
ASC - N. W. Hale
ENT - J. P. Shannon
LD/O 4 - W. D. Reeves
O 2 - J. W. Bantle
O 2 - P. S. Hill (1 Shift)
O 1 - B. P. Austin
O 3 - A. F. Algate
MOD - A. L. Briscoe

SS EVA #40
EMU/Tethered
EVA #33
Scheduled EVA #35
on 11/24/97
7H42M55S Duration
EVA start at 05:04:16:05
MET

SS EVA #41
EMU/Tethered
EVA #34
Unscheduled EVA #6
on 12/3/97
4H59M40S Duration
EVA start at 13:13:24 MET

STS087-E-5048--Scott (left) & Doi grab free-flying Spartan 201 satellite and berth it in shuttle P/L bay.

STS087-307-002 On middeck: In front (lt to rt), PLT Lindsey, Doi/MS (NASDA) & Scott/MS, In back (lt to rt) CDR Kregel, Chawla/MS, Kadenyuk/PS.

STS087-706-020 ---

Spartan 201 satellite in grasp of RMS

FLT NO.	ORBITER	CREW 7 UP, 7 DOWN / TITLE, NAMES & EVA'S	LAUNCH SITE, LIFTOFF TIME, LANDING SITES, ABORT TIMES	LANDING SITE/ RUNWAY, CROSSRANGE / LANDING TIMES, FLT DURATION, WINDS	SSME-TL NOM-ABORT EMERG THROTTLE PROFILE ENG. S.N.	SRB RSRM AND ET	ORBIT INC / HA/HP	FSW	PAYLOAD WEIGHTS, PAYLOADS/ EXPERIMENTS	MISSION HIGHLIGHTS (LAUNCH SCRUBS/DELAYS, TAL WEATHER, ASCENT I-LOADS, FIRSTS, SIGNIFICANT ANOMALIES, ETC.)

STS-89
SEQ
FLT #89
KSC - 89
PAD 39A-52
MLP-3

ORBITER: OV-105 (Flight 12) Endeavour / Spacehab 9 / OMS PODS: LP04-19 RP01-26 FRC5-12

CREW:
CDR: Terrence W. Wilcutt (Flt 3 - STS-68, STS-79) P517/R183/V130/M160
PLT: Joseph F. Edwards Jr. P518/R233/M203
M/S 1: James F. Reilly, II P519/R234/M204
M/S 2: Michael Anderson P520/R235/M205
M/S 3: (PAYLOAD CDR): Bonnie J. Dunbar (Flt 5 - STS-61-A, STS-32, STS-50, STS-71) P521/R79/V49/F7
M/S 4: Salizhan Shakirvich Sharipov (Russia) P522/R236/M206
M/S 5: Ascent Andrew S. W. Thomas (Flt 2 - STS-77) Stay on Mir 24 and Mir 25, return on STS-91. P523/R213/V149/M186
M/S 6: Descent David A. Wolf (Flt 2- STS-58) Ascent on STS-86, stay on Mir 24. P524/R173/V147/M151

MCC WHITE FCR (19)
FLIGHT DIRECTORS:
ASCENT- L.J. Ham
LD/O1- P.L. Engelhauf
O2- R.E. Castle
PLNG- P.S.Hill
ENTRY- J.P. Shannon
MOD - A. L. Briscoe

sts089-742-024-- Atlantis with SPACEHAB on approach to Mir.

LAUNCH SITE, LIFTOFF TIME:
KSC PAD 39A
23:02:48:14.98Z
9:48:15 PM EST (P)
9:48:15 PM EST (A)
Thursday 27
1/22/98 EST (9)

LAUNCH WINDOW:
7M 56S Using PLT MIR PLANAR/ PHASE WINDOW

EOM PLS: KSC
TAL: ZZA
TAL WX: MRN, BEN

SELECTED:
RTLS: KSC 15/CI/N
TAL: ZZA 30/CI/N
AOA: NOR 17/N/SF
PLS: EDW 22/N/SF

TDEL: 0.14   -0.098/0.1

MAX Q NAV: 702 PSF   710 PSF

SRB STG: 2:03.8   2:06

PERF: NOMINAL

2 ENG TAL (ZZA): 2:26   2:25

NEG RETURN: 4:02   4:05

PTA (U/S 265): 4:42   4:35

DROOP (ZZA): 5:20   5:22

PTM (U/S 265): 5:50   5:48

MECO CMD: 8:28.9   8:29

VI: 25876   25873

OMS-2: 41:46   41:48
213 FPS   213 FPS

LANDING SITE/RUNWAY:
KSC 15 (KSC 42)
31:22:35:09Z
5:35:09 PM EST

Saturday 19
1/31/98 (8)

DEORBIT BURN:
31:21:27:55Z

XRANGE: 600NM

ORBIT DIR: AL 20

AIM PT: NOMINAL

MLGTD: 2702 FT
31:22:35:09Z
VEL: 202 KGS
198 KEAS
HDOT: -2.3 FPS

TD NORM 195: 2776 FT

DRAG CHUTE DEPLOY: 190 KEAS
31:22:35:13Z

NLGTD: 6112 FT
31:22:35:20Z
VEL: 152 KGS
149 KEAS
HDOT: -5.9 FPS

BRK INIT: 94 KGS

DRAG CHUTE JETTISON: 56 KGS
31:22:35:53Z

BRK DECEL FPS[2]: AVE 3.6 PK 5.0

WHEELS STOP: 31:22:36:21Z
12492 FT

ROLLOUT: 9790 FT
72 SEC

WINDS: 4T, 4L KTS
OFFICIAL: 0205P11 7T,8L

DENS ALT: -103 FT

FLT DURATION: 8:19:46:54

S/T: 755:11:40:28

OV-105: 121:08:50:00

DISTANCE: 3,610,000 sm

SSME-TL:
104/104/ 109%

PREDICTED:
100/104/104
67/104

ACTUAL:
100/104/104
67/104

1 = 2043 (1)
2 = 2044 (1)
3 = 2045 (1)

M 3 EOM:
WEIGHT: 217475 LBS
X CG: 1086.45

LANDING:
WEIGHT: 217422 LBS
X CG: 1088.16

SRB RSRM AND ET:
BI-093
RSRM 64
ET-90
LWT-83

ET RPT: 271.3K

ET BRKUP: 269.1K

ET IMPACT
1:27:09 MET
LAT: 0.69 N
LONG: 120.7 W

ORBIT:
DIRECT INSERTION

POST OMS-2:
162.4 X 161.1 NM

TI: 1:15:03:04 MET
215.6 X 203.4 NM

SEP1: 6:15:28:26 MET
206.6 X 203.2 NM

DEORBIT: 207.1 X 193.7 NM

VELOCITY: 25900 FPS

ENTRY RANGE: 4341 NM

INC: 51.65 (9)

FSW: OI-26 (4)

PAYLOAD WEIGHTS:
CARGO: 28040 LBS

PAYLOAD CHARGEABLE: 22163 LBS

DEPLOYED: 4596 LBS

NON-DEPLOYED: 16699 LBS

MIDDECK: 868 LBS

SHUTTLE ACCUMULATED WEIGHTS:
DEPLOYED: 848511 LBS
NON-DEPLOYED: 1299327 LBS
CARGO TOTAL: 2676765 LBS

PERFORMANCE MARGINS (LBS):
FPR: 3272
FUEL BIAS: 854
FINAL TDDP: 2309
RECON: 3594

PAYLOADS:
PLB:
SHUTTLE/MIR MISSION 8
SPACEHAB (Double Module)
GAS (4)
ODS

MIDDECK:
HP, MPNE, AST, CREAM, SIMPLEX, SAMS, MGM (2), CEBAS, EARTHCAM

5 CRYO TK SETS
6 GN2 TANKS

NO RMS

MISSION HIGHLIGHTS:
KSC W/D: OPF 202, VAB 7, PAD 26 = 235 days total.

LAUNCH POSTPONEMENTS:
- Baselined 1/15/98 launch date on 10/1/96.
- Moved STS-89 from OV-103 to OV-105 on 5/22/97.
- Postponed launch date to 1/22/98 EST (1/23/98 GMT) on 12/18/97.

LAUNCH SCRUBS: None

LAUNCH DELAYS: None

TAL WX:
- Zaragoza (prime and selected) and Moron forecast and observed GO. Ben Guerir was forecast NO GO for ceiling and visibility (very dense fog).

SHUTTLE NIGHT LAUNCH: #19

DOLILU/I-LOADS:
- DOLILU II uplink #20, total uplink #39.

PERFORMANCE ENHANCEMENTS:
- Standard set plus Block IIA SSME's.

FLIGHT DURATION CHANGES:
- None. Landed on orbit 139.

FIRSTS/LASTS:
- First flight using Block IIA SSME's. (Rocketdyne HPFTP)
- First flight with external airlock.
- Record number of people in orbit: Mir 3 - 2 Russians, 1 American; Soyuz 3 - 2 Russians, 1 French; Endeavour 7 - 6 Americans, 1 Russian.

EVENTS:
- Mir capture at 24:20:14:21Z, 1:17:26:06 MET.
- Docking complete at 24:20:23Z, 1:17:35 MET.
- Andrew Thomas transferred to Mir 24 and David Wolf to STS-89 Endeavour at 26:05:51:15Z, 3D1H3M3M. David Wolf total Mir time 119:23:16:56 and total flight time 127:20:00:50.
- Undocking at 29:16:56:56Z, 6:14:08:41 MET.
- Inert weight adjustment of -200 lbs included in STS OPR chargeable.

RENDEZVOUS #41:
- Rendezvous and dock with Mir.

RADIATOR DEPLOY #21:

SIGNIFICANT ANOMALIES:
- GPC 3 mode switch no apparent detent at standby. Went to halt from run.
- Payload bay floodlights FWB STBD and MID PORT failed (new design).
- TIPS and OCA problems.
- -Z Star Tracker pressure fail BITE.
- S-Band antenna electronics 2 failed to select the best antenna.
- Vestibule vent valves were misconfigured (3 of 4 open).
- Vernier thruster L5D oxidizer temp failed erratic, attitude control passed to Mir jets, then to orbiter PRCS.
- Right RCS fuel helium isolation valve B failed to open.
- Vernier driver F5 RPC 2 failed off.

STS089-391-004 Onboard Mir Base Block: In conventional position (from left) are Wolf/MS(former Mir guest), Pavel V. Vinogradov/ Mir-24/FE, CDR Wilcutt, Mir-24 CDR Anatoly Y. Solovyev, & Dunbar/MS/PLC. Above, head-to-head with bottom row are (from left) Sharipov/MS (RSA), Reilly/MS & PLT Edwards. At 90 deg angle are: Thomas/MS/MirGuest (top) & Anderson/MS.

FLT NO.	ORBITER	CREW (7) TITLE, NAMES & EVA'S	LAUNCH SITE, LIFTOFF TIME, LANDING SITES, ABORT TIMES	LANDING SITE/ RUNWAY, CROSSRANGE LANDING TIMES, FLT DURATION, WINDS	SSME-TL NOM-ABORT EMERG THROTTLE PROFILE ENG. S.N.	SRB RSRM AND ET	ORBIT		FSW	PAYLOAD WEIGHTS, PAYLOADS/ EXPERIMENTS	MISSION HIGHLIGHTS (LAUNCH SCRUBS/DELAYS, TAL WEATHER, ASCENT I-LOADS, FIRSTS, SIGNIFICANT ANOMALIES, ETC.)
							INC	HA/HP			
STS-90 SEQ FLT #90 KSC-90 PAD 39B-38 MLP-2	OV-102 (Flight 25) Columbia 23 RD Spacelab Flight LM-16 EDO 13 OMS PODS: LPO5-14 RPO5-13 FRC2-25	CDR: Richard A. Searfoss (Flt 3 - STS-58, STS-76) P525/R171/V126/M149 PLT: Scott D. Altman P526/R237/M207 M/S 1 (PAYLOAD CDR): Richard M. Linnehan (Flt 2 - STS-78) P527/R214/V150/M187 M/S 2: Kathryn P. Hire P528/R238/F31 M/S 3: Dafydd R. Williams (Canada) P529/R239/M208 P/S 1: Jay C.Buckey, Jr. P530/R240/M209 P/S 2: James A.Pawelczyk P531/R241/M210	KSC 39B 107:18:18:59.99Z 2:19:00 PM EDT (P) 2:19:00 PM EDT (A) Friday 17 4/17/98 (13) LAUNCH WINDOW: 2H30M Neurolab Crew Circadian Constraint EOM PLS: KSC TAL: BEN TAL WX: MRN, ZZA SELECTED: RTLS: KSC 15/Cl/N TAL: BEN 36/N/N AOA: EDW 22/N/N PLS: EDW 22/N/N TDEL: 0.08 0.322/0.36 MAX Q NAV: 694 697 SRB STG: 2:05.1 2:05 PERF: NOMINAL 2 ENG TAL (BEN): 2:50 2:49 NEG RETURN: 3:56 3:58 PTA (U/S 248): 5:31 5:29 DROOP (ALL): 5:24 5:25 PTM (U/S 390): 7:08 7:11 MECO CMD: 8:27.3 8:28.4 VI: 25864 25860 OMS-2: 41:23 41:23 171 FPS 171 FPS	KSC 33 (KSC 43) 123:16:08:59Z 12:08:59 PM EDT Sunday 13 5/3/98 (13) DEORBIT BURN: 123:15:10:10Z XRANGE: 245.9 NM ORBIT DIR: DR 20 AIM PT: NOMINAL MLGTD: 1561 FT 123:16:08:59Z VEL: 224 KGS 218 KEAS HDOT: -6.0 FPS TD NORM 195: 2451 FT DRAG CHUTE DEPLOY: 194 KEAS 123:16:09:06Z NLGTD: 6288 FT 123:16:09:12Z VEL: 167 KGS 161 KEAS HDOT: -4.6 FPS BRK INIT: 122 KGS DRAG CHUTE JETTISON: 56 KGS 123:16:09:37Z BRK DECEL FPS2: AVE 5.7 PK 9.3 WHEELS STOP: 123:16:09:57Z 11559 FT ROLLOUT: 9998 FT 58 SEC WINDS: T1, L4 KTS OFFICIAL: 2204P11 T1, L4 KTS DENS ALT: 1560 FT FLT DURATION: 15:21:49:59 S/T: 771:09:30:27 OV-102: 268:22:19:42 DISTANCE: 6,375,000 sm	104/104/ 109% PREDICTED: 100/104/104 67/104 ACTUAL: 100/104/104 69/104 1 = 2041 (4) 2 = 2032 (7) 3 = 2012 (21) M 3 EOM: WEIGHT: 233031 LBS X CG: 1080.33 LANDING: WEIGHT: 232979 LBS X CG: 1081.94	BI-094 RSRM 65 ET-91 LWT-84 ET RPT: 283K ET BRKUP: 215K ET IMPACT 1:24:30 MET LAT: 1.88 N LONG: 139.9 W	39 (7)	DIRECT INSERTION POST OMS-2: 154 X 138 NM DEORBIT: 149 X 131 NM VELOCITY: 25758 FPS ENTRY RANGE: 4422 NM	OI-26B (1)	CARGO: 35549 LBS PAYLOAD CHARGEABLE: 25625 LBS DEPLOYED: 0 LBS NON-DEPLOYED: 9944 LBS MIDDECK: 2340 LBS SHUTTLE ACCUMULATED WEIGHTS: DEPLOYED: 848511 LBS NON-DEPLOYED: 1325532 LBS CARGO TOTAL: 2712754 LBS PERFORMANCE MARGINS (LBS): FPR: 3085 FUEL BIAS: 853 FINAL TDDP: 3162 RECON: 1999 PAYLOADS: PLB: NEUROLAB SVF GAS (3) MIDDECK: BIOREACTOR DEMO. SYSTEM 5 CRYO TK SETS + 4 EDO & 5 N2 TANKS	KSC W/D: OPF 80, VAB 5, PAD 24 = 109 days total. LAUNCH POSTPONEMENTS: - Baselined launch date of 3/18/98 on 1/10/97. - Postponed launch date to 4/2/98 on 4/17/97. - Postponed launch date to 4/16/98 on 2/13/98. LAUNCH SCRUBS: - Scrubbed 4/16/98 launch attempt at approximately L-6 hours due to an NSP 2 problem (did not tank). Replaced NSP 2. LAUNCH DELAYS: None TAL WX: - Ben Guerir (prime and selected) was forecast and observed GO. Moron was forecast GO late after ceiling violation, Zaragoza was forecast NO GO for crosswinds and low ceiling, but observed GO at TAL time. DOLILU/I-LOADS: - DOLILU II uplink #21, I-Load uplink #40. PERFORMANCE ENHANCEMENTS: - Standard set plus OMS assist is 4000 lbs. FLIGHT DURATION CHANGES: None. Landed on orbit 256. FIRSTS/LASTS: - First use of OMS assist during ascent (102 seconds) 4000 lbs. - Final flight of Spacelab. - Total size of the seven crewmembers was the largest. - Largest number of animals (over 2000 animals on board). EVENTS: - SSME 1 Block IIA and SSME 2 & 3 Phase 2 engines. RADIATOR DEPLOY #22: Port radiator only. SIGNIFICANT ANOMALIES: - Water spray boiler 3 failed to cool, APU3 shutdown at 13:05 MET. Also failed to cool during FCS C/O, so was not started until TAEM for entry. - Icing in topping FES core (did FES core flush). - CO2 removal system failure. RCRS recovered with IFM. - Waste water dump clogged filter. IFM preformed but urine filter clogged. - APU 2 Gas Gen/Fuel Pump B heaters failed. - DOLILU processor integrity rule violation at L-6.5 hours.

STS090-378-022 (17 April - 3 May 1998) --- Crew floats as a unit in Spacelab. From left are: Hire/MS, Buckey/PS (top), CDR Searfoss, Pawelczyk/PS, PLT Altman, Williams/CSA/MS (top); and Linnehan/PL CDR.

MCC WHITE FCR (20)

FLIGHT DIRECTORS:
A/E - J. P. Shannon
LD/O 2 - G. A. Pennington
O 1 - B. P. Austin
O 3 - R. M. Kelso
MOD - J. W. Bantle

STS090-383-018 (17 April - 3 May 1998): Last Spacelab Science Module (center), hosted 16-days of Neurolab research. Tunnel from cabin to Spacelab is at bottom. Sunrise and airglow are seen in background.

FLT NO.	ORBITER	CREW 6 UP, 7 DOWN / TITLE, NAMES & EVA'S	LAUNCH SITE, LIFTOFF TIME, LANDING SITES, ABORT TIMES	LANDING SITE/ RUNWAY, CROSSRANGE / LANDING TIMES FLT DURATION, WINDS	SSME-TL NOM-ABORT EMERG THROTTLE PROFILE ENG. S.N.	SRB RSRM AND ET	ORBIT INC	HA/HP	FSW	PAYLOAD WEIGHTS, PAYLOADS/ EXPERIMENTS	MISSION HIGHLIGHTS (LAUNCH SCRUBS/DELAYS, TAL WEATHER, ASCENT I-LOADS, FIRSTS, SIGNIFICANT ANOMALIES, ETC.)
STS-91 SEQ FLT #91 KSC-91 PAD 39A-53 MLP-1	OV-103 (Flight 24) Discovery Spacehab 10 OMS PODS: LPO1-27 RPO3-24 FRC3-24	CDR: Charles J. Precourt (Flt 4 - STS-55, STS-71, STS-84) P532/R161/V118/M141 PLT: Dominic L. Gorie P533/R242/M211 M/S 1 (PAYLOAD CDR): Franklin R. Chang-Diaz (Flt 6 - STS 61-C, STS-34, STS-46, STS-60, STS-75) P534/R89/V46/M81 M/S 2: Wendy B. Lawrence (Flt 3 - STS-67, STS-86) P535/R192/V146/F25 M/S 3: Janet L. Kavandi P536/R243/F32 M/S 4: Valery V. Ryumin (Russia) P537/R244/M212 M/S 5: Andrew Thomas (Flt 2 - STS-77) P538/R213/V149/M186 Launch on STS-89, stay on Mir 24 and 25, return on STS-91. MCC WHITE FCR (21) FLIGHT DIRECTORS: A/E – N. W. Hale LD/O 1 – P. F. Dye O 2 – A. F. Algate PLNG – P. L. Engelauf MOD – A. L.Briscoe	KSC PAD 39A 153:22:06:24Z 6:06:24 EDT (P) 6:06:24 EDT (A) Tuesday 13 6/2/98 (9) LAUNCH WINDOW: 7M42S USING MIR PLANAR/ PHASING WINDOW IN LIEU OF PLT. EOM PLS: KSC TAL: ZZA TAL WX: MRN, BEN SELECTED: RTLS: KSC 15/N/N/S TAL: ZZA 30/N/N AOA: KSC 15/N/N PLS: EDW 22/N/N TDEL: 0.04 0.082/0.12 MAX Q NAV: 692 663 SRB STG: 2:03.4 2:03 PERF: NOMINAL 2 ENG TAL (AAZ): 2:34 2:29 NEG RETURN: 4:00 4:02 PTA (U/S 274): 4:45 4:42 DROOP (ZZA): 5:30 5:30 PTM (U/S 780): 6:16 6:16 MECO CMD: 8:29.4 8:30.2 VI: 25931 25924 OMS-2: 44:11 44:11 161 FPS 161 FPS	KSC 15 (KSC 44) 163:18:00:17Z 2:00:17 PM EDT Friday 11 6/12/98 (4) DEORBIT BURN: 163:16:52:26Z XRANGE: 317 NM ORBIT DIR: AL 21 AIM PT: NOMINAL MLGTD: 1218 FT 163:18:00:17Z VEL: 215 KGS 207 KEAS HDOT: -3.4 FPS TD NORM 195: 2366 FT DRAG CHUTE DEPLOY: 162 KEAS 163:18:00:29Z NLGTD: 4518 FT 163:18:00:27Z VEL: 176 KGS 167 KEAS HDOT: -6.6 FPS BRK INIT: 147 KGS DRAG CHUTE JETTISON: 57 KGS 163:18:00:58Z BRK DECEL FPS[2]: AVE 4.7 PK 11.2 WHEELS STOP: 163:18:01:28Z 11935 FT ROLLOUT: 10717 FT 71 SEC WINDS: T3, L6 KTS OFFICIAL: 0407P11 T2, L7 KTS DENS ALT: 2260 FT FLT DURATION: 9:19:53:53 S/T: 781:05:24:20 OV-103: 177:16:47:42 DISTANCE: 3,800,000 sm	104/104/ 109% PREDICTED: 100/104/104/ 67/104 ACTUAL: 100/104/104/ 67/104 1 = 2047 (1) 2 = 2040 (4) 3 = 2042 (3) Block IIA 2047 Throttled to 104.5 M 3 EOM: WEIGHT: 226968 LBS X CG: 1079.49 LANDING: WEIGHT: 226872 LBS X CG: 1081.09	BI-091 RSRM 66 ET-96 SLWT-1 ET PRED RPT: 283K ET BRKUP: 215K ET IMPACT: 1:26:24 MET LAT: 2.68 S LONG: 127.2 W	51.65 (10)	DIRECT INSERTION POST OMS-2: 177 X 129 NM DEORBIT: 204 X 187 NM VELOCITY: 25889 FPS ENTRY RANGE: 4418 NM	OI-26B (2)	CARGO: 35549 LBS PAYLOAD CHARGEABLE: 25625 LBS DEPLOYED: 2419 LBS NON-DEPLOYED: 2 LBS MIDDECK: 891 LBS SHUTTLE ACCUMULATED WEIGHTS: DEPLOYED: 850930 LBS NON-DEPLOYED: 1348738 LBS CARGO TOTAL: 2748303 LBS PERFORMANCE MARGINS (LBS): FPR: 3783 FUEL BIAS: 720 FINAL TDDP: 631 RECON: 403 PAYLOADS: PLB: ODS SHUTTLE/MIR MISSION 9 SPACEHAB (Single Module) AMS, SEM (2), GAS (6) MIDDECK: SSCE SIMPLEX CPCG 5 CRYO TK SETS 5 N2 TANKS RMS 50 Used to check out S.N. 201 With new digital SPA H/W.	KSC W/D: OPF 168, VAB 4, PAD 29 = 201 days total. LAUNCH POSTPONEMENTS: - Baselined launch date of 5/28/98 EDT (5/29/98 GMT) on 2/20/97. - Changed launch date to 5/28/98 EDT (5/29/98 GMT) on 8/25/97. - Postponed launch date to 6/2/98 to allow AMS additional time. LAUNCH SCRUBS: None LAUNCH WINDOW: - 7M42S based on Mir planar/phase window (not PLT) to increase LO₂ drainback time. LAUNCH DELAYS: None TAL WX: - ZZA prime and selected. - ZZA, MRN, and BEN were forecast and observed GO. DOLILU/I-LOADS: - DOLILU II uplink #22, I-Load uplink #41. PERFORMANCE ENHANCEMENTS: - Standard set plus MECO altitude is 52 NM, plus Delta psi. - First use of MECO is 52 NM. FLIGHT DURATION CHANGES: - None. Landed on orbit 155. FIRSTS/LASTS: - First flight of Super Light Weight tank - First flight of Block IIA SSME 2047 - Last Shuttle flight to Mir (ninth docking). EVENTS: - Valery Ryumin's previous flights were Soyuz-25, Soyuz/Salyut-6 (2 flights). - WRAP DAP entry. - Andrew Thomas, last American to visit Mir. Andy transferred to STS-91 from Mir at 155:18:33:24Z. Mir time is 129:02:42:09 and total flight time is 140:15:11:45. RENDEZVOUS #42: - Rendezvous and docking with Mir. SIGNIFICANT ANOMALIES: - Center SSME PC sensor failure. - Fuel cell 3 overboard relief. - Fuel cell monitoring time word problem. - MAGR-S3S GPS ascent performance anomaly. - Cyclic GNC GPC errors caused by bad GPS SV caused by handshaking problem between GPS and the GNC. - Failure of Ku-Band to radiate (no Ku-Band return link). - Camera C pan and tilt failure. - Thrusters R2U and F2U failed off at first command firing of both jets (low chamber pressures). - LOMS ball valve I failed open. - RPOP error during approach.

STS091-707-060 (2-12 June 1998): MIR as seen during final flyaround by NASA Shuttles.

STS091-718-010 98) ---- Crew portrait: Bottom, from left, CDR Precourt, Kavandi/MS, & Chang-Diaz/PLC. At top, from left, PLT Gorie, Lawrence/MS, Thomas/MS & Ryumin/MS(RSA). After 4 months stay Thomas was last American to visit MIR.

FLT NO.	ORBITER	CREW (7) TITLE, NAMES & EVA'S	LAUNCH SITE, LIFTOFF TIME, LANDING SITES, ABORT TIMES	LANDING SITE/ RUNWAY, CROSSRANGE LANDING TIMES, FLT DURATION, WINDS	SSME-TL NOM-ABORT EMERG THROTTLE PROFILE ENG. S.N.	SRB RSRM AND ET	INC	ORBIT HA/HP	FSW	PAYLOAD WEIGHTS, PAYLOADS/ EXPERIMENTS	MISSION HIGHLIGHTS (LAUNCH SCRUBS/DELAYS, TAL WEATHER, ASCENT I-LOADS, FIRSTS, SIGNIFICANT ANOMALIES, ETC.)
STS-95 SEQ FLT #92 KSC-92 PAD 39B-39 MLP-2	OV-103 (Flight 25) Discovery OMS PODS: LPO1 - 28 RPO3 - 26 FRC3 - 25	CDR: Curtis L. Brown (Flt 5 - STS-47, STS-66, STS-77, STS-85) P539/R152/V112/M136 PLT: Steven W. Lindsey (Flt 2 - STS-87) P540/R229/V131/M200 M/S 1: Stephen K. Robinson (Flt 2 – STS-85) P541/R222/V152/M196 M/S 2: Scott E. Parazynski (Flt 3 - STS-66, STS-86) P542/R187/V145/M164 M/S 3: Pedro Duque (ESA-Spain) P543/R245/M213 P/S 1: Chiaki Mukai (Japan) (Flt 2 - STS-65) P544/R181/V153/F23 P/S 2: Senator John H. Glenn (2) P545/R246/M214 MCC WHITE FCR (22) FLIGHT DIRECTORS: A/E - L. J. Ham LD/O1 - P. L. Engelauf O 2 - P. S. Hill O 3 - P. F. Dye MOD - J. W. Bantle	KSC 39B 302:19:19:33:98Z 2:00:00 PM EST (P) 2:19:34 PM EST (A) Thursday 28 10/29/98 (10) LAUNCH WINDOW: 2H30M CTOB EOM PLS: KSC TAL: BYD TAL WX: BEN, MRN SELECTED: RTLS: EDW 15/N/SF TAL: BYD 32/N/SF AOA: EDW 22 PLS: EDW 22/CI/N TDEL: -0.03 -0.108/0.07 MAX Q NAV: 755 765 SRB STG: 2:03.7 2:03 PERF: NOMINAL 2 ENG TAL (BYD): 2:11 2:13 NEG RETURN: 3:45 3:52 PTA (U/S 500): 4:12 4:08 DROOP: 5:21 PTM (U/S 500): 5:13 5:06 MECO CMD: 8:20.7 8:21.6 VI: 26102 26092 OMS-2: 41:57 41:57 5.02 FPS 5.02 FPS	KSC 33 (KSC 45) 311:17:03:30Z 12:03:30 PM EST Saturday 20 11/7/98 (11) DEORBIT BURN: 311:15:52:54Z XRANGE: 172 NM ORBIT DIR: DL 44 AIM PT: NOMINAL MLGTD: 3243 FT 311:17:03:30Z VEL: 199 KGS 196 KEAS HDOT: -1.0 FPS TD NORM 205: 2559 FT DRAG CHUTE DEPLOY: NOT USED NLGTD: 6248 FT 311:17:03:40Z VEL: 164 KGS 164 KEAS HDOT: -6.6 FPS BRK INIT: 138 KGS 8726 FT DRAG CHUTE JETTISON: NOT USED BRK DECEL FPS[2]: AVE 5.8 PK 7.8 WHEELS STOP: 311:17:04:30Z 12751 FT ROLLOUT: 9508 FT 60 SECS WINDS: 0H, 10R KTS OFFICIAL: 0609P14 T0, R9 KTS Continued…	104.5/104.5/ 109% PREDICTED: 100/104.5/ 104.5/67/ 104.5 ACTUAL: 100/104.5/ 104.5/72/ 104.5 1 = 2048 (1) 2 = 2043 (2) 3 = 2045 (2) ALL BLOCK II A ENGINES M 3 EOM WEIGHT: 228455 LBS X CG: 1076.83 LANDING WEIGHT: 228388 LBS X CG: 1078.45	BI-096 RSRM 68 ET-98 SLWT-2 SLWT RPT MAX: 283K MIN: 215K SLWT IMPACT 1:28:02 MET LAT: 20.8 N LONG: 147.2 W	28.45 (47)	DIRECT INSERTION POST OMS-2: 303 X 295 NM SEP 1: 2:23:46:30 MET 302.2 X 294 NM SEP 2: 3:06:16:40 MET TI:5:22:01:37 MET 301.5 X 293.5 NM DEORBIT ALT: 301.5 X 285.9 NM VELOCITY 26063 FPS ENTRY RANGE 4290 NM	OI-26B (3)	CARGO: 38618 LBS PAYLOAD CHARGABLE: 28520 LBS DEPLOYED: 125 LBS NON-DEPLOYED: 24108 LBS MIDDECK: 1314 LBS SHUTTLE ACCUMULATED WEIGHTS: DEPLOYED: 850155 LBS NON-DEPLOYED: 1378355 LBS CARGO TOTAL: 2824652 LBS PERFORMANCE MARGINS (LBS): FPR: 3783 FUEL BIAS: 720 FINAL TDDP: 1587 RECON: 2740 PAYLOADS: PLB: SPACEHAB (Single) HOST SPARTAN–201 (Deploy & Retrieve) (Solar Wind Exp.) GAS (2) IEH-3 (PANSAT) (Deployed) MIDDECK: PCG-STES SAREX-II BRIC 5 CYRO TK SETS 5 GN2 TANKS Continued…	KSC W/D: OPF 76, VAB 5, PAD 29 = 110 days LAUNCH POSTPONMENTS: - Baselined launch date of 10/8/98 on 7/31/97. - Postponed launch date to 10/29/98 on 12/18/97. LAUNCH SCRUBS: None LAUNCH DELAYS: - Held for 9 minutes 36 seconds during T-9 minute hold to understand the cause of the three master alarms (MA) during cabin leak checks. First MA was cabin P reached 15.35 psi during cabin leak check. Two MA's were differential pressure/differential time alarms. It was concluded that the alarms were expected and count was resumed. - Held for 9 minutes 58 seconds at T-5 minutes for range safety hold call for two intruder aircraft in Launch Danger Area. Resumed count but two calls were made to hold at T-31 seconds, one for engine 2 pitch position NO GO and the second for range safety NO GO. These holds were removed before count reached T-31 seconds; hence, no additional delay. TAL WX: - Banjul, Ben Guerir, and Moron were forecast and observed GO. Banjul was prime and selected. DOLILU/I-LOADS: - DOLILU II uplink # 23, I-Load uplink # 42. PERFORMANCE ENHANCEMENTS: - Standard set plus PE High Q. FLIGHT DURATION CHANGES: None FLIGHT RULE WAIVER: - Forecast at deorbit burn time was a maximum crosswind of 16 knots. Flight rule limit is 15 knots. Observed crosswind < 10 knots. Landed on orbit 135. FIRSTS/LASTS/RECORDS: - First flight using High Q flight design. - First flight with three Block IIA SSME's (Rocketdyne HPFTP). - John Glenn's first flight was Mercury-Atlas 6 on 2/20/62. - Glenn's age at first flight 40Y7.5M, second flight 77Y4M, 36Y8.5M between flights. - First flight using space-to-space comm system (as DTO). - Second flight of Super Lightweight Tank (SLWT). Continued…

STS095-328-031 (29 Oct.-7 Nov. 1998): CDR Brown (rt ctr), then clockwise, PLT Lindsey, Robinson/MS, Duque/MS/ESA, Naito-Mukai/PS/NASDA, Parazynski/MS, & Sen. Glenn/PS.

SPACE SHUTTLE MISSIONS SUMMARY

FLT	ORBITER	CREW (7)	LAUNCH SITE, LIFTOFF TIME,	LANDING SITE/ RUNWAY, CROSSRANGE	SSME-TL NOM- ABORT EMERG	SRB RSRM	ORBIT		FSW	PAYLOAD WEIGHTS,	MISSION HIGHLIGHTS (LAUNCH SCRUBS/DELAYS,
NO.		TITLE, NAMES & EVA'S	LANDING SITES, ABORT TIMES	LANDING TIMES FLT DURATION, WINDS	THROTTLE PROFILE ENG. S.N.	AND ET	INC	HA/HP		PAYLOADS/ EXPERIMENTS	TAL WEATHER, ASCENT I-LOADS, FIRSTS, SIGNIFICANT ANOMALIES, ETC.)

STS-95
Continued…

S62-00303 (2-20-62): Lt, John Glenn 1st american to orbit Earth, Friendship7/MA6. STS095-E-5032 (10-29-98): Rt, Sen. John Glenn, at 77, oldest human in space, STS-95.

98-E-03312 (14 April 1998) --- President Bill Clinton (at lectern) addresses JSC employees. Seated behind him (from left): JSC Director George W.S. Abbey, U.S. Rep. Nick Lampson (D.-TX), NASA Administrator Daniel Goldin and Houston Mayor Lee Brown. Standing are STS-95 crew: (from left) Pedro Duque, Chiaki Mukai, U.S. Sen. John H. Glenn Jr. (D.-Ohio), Stephen K. Robinson, Scott E. Parazvnski, PLT Steven W. Lindsey & CDR Curtis L. Brown.

Continued…

DENS ALT: 965 FT

FLT DURATION:
8:21:43:56

S/T: 790:03:08:16

OV-103:
186:13:31:48

DISTANCE:
3,644,459 sm

S98-16165 (10-29-98) --- In MCC: From left, CAPCOMS Susan Still &, Scott J. Horowitz; & Flight Directors Jeffrey W. Bantle, Linda Ham and Wayne Hale following launch of STS-95.

STS095-E-5077 (11-01-98)- Spartan201-05 departs discovery as a free flyer for several days recording solar wind and sun corona data.

Continued…

RMS 51
(S.N. 201)
RMS used for SPARTAN-201 deploy, retrieve and berth, ACVS, OSVS, and VGS OPS.

Continued…

EVENTS:
- SPARTAN-201 release 305:19:00:12Z, 2:23:40:36 MET.
- Due to drag chute anomaly, drag chute was not armed and deployed.
- Inert weight adjustment -200 lbs included in STS OPR chargeable.
- SPARTAN capture 307:20:47:49Z, 5:01:28 MET. Berth 5:01:46 MET.

RENDEZVOUS # 43:
- Deployed, separated, rendezvoused with SPARTAN-201.

RADIATOR DEPLOY # 23:
- Both port and starboard panels deployed.

SIGNIFICANT ANOMALIES:
- Low Iodine Residual System (LIRS) large spraying leak. Used backup galley iodine removal system.
- Unpleasant taste (rubber hose) from LIRS.
- During space-to-space comm tests, no data from EMU 1 in primary.
- Drag chute door fell off during ME throttle up at T-5 seconds; hence, not deployed during landing.
- Decision made to disable chute for STS-88.
- WSB 2 overcooled six times during entry.
- SPARTAN ground command problem.
- RCS jet L3L failed off, then failed leak.

FLT NO.	ORBITER	CREW (6) TITLE, NAMES & EVA'S	LAUNCH SITE, LIFTOFF TIME, LANDING SITES, ABORT TIMES	LANDING SITE/ RUNWAY, CROSSRANGE LANDING TIMES FLT DURATION, WINDS	SSME-TL NOM-ABORT EMERG THROTTLE PROFILE ENG. S.N.	SRB RSRM AND ET	ORBIT		FSW	PAYLOAD WEIGHTS, PAYLOADS/ EXPERIMENTS	MISSION HIGHLIGHTS (LAUNCH SCRUBS/DELAYS, TAL WEATHER, ASCENT I-LOADS, FIRSTS, SIGNIFICANT ANOMALIES, ETC.)
							INC	HA/HP			
STS-88/ ISS-2A First Shuttle Flight to ISS SEQ FLT #93 KSC-93 PAD 39A-54MLP-3	OV-105 (Flight 13) Endeavour OMS PODS LPO4-20 RPO1-27 FRC5-13	CDR: Robert D. Cabana (Flt 4 - STS-41, STS-53, & STS-65) P546/R113/V84/M101 PLT: Frederick W. Sturckow P547/R247/M215 M/S 1/EV 1: Jerry L. Ross (Flt 6 - STS-61-B, STS-27, STS-37, STS-55, &STS-74) P548/R89/V38/M80 M/S 2: Nancy J. Currie (Flt 3 - STS-57, STS-70) P549/R165/V120/F21 M/S 3/EV 2: James H. Newman (Flt 3 - STS-51, STS-69) P550/R168/V122/M146 M/S 4: Sergei Krikalev (Russia) (Flt 2 - STS-60) P551/R176/V154/M154 SS EVA #42: EMU/Tethered EVA #35 Scheduled EVA #36 on 12/7/98 Duration 7H21M EVA start at 3D13H34M MET SS EVA #43: EMU/Tethered EVA #36 Scheduled EVA #37 on 12/9/98 Duration 7H02M EVA start at 5D11H57M30S MET SS EVA #44: EMU/Tethered EVA #37 Scheduled EVA #38 on 12/12/98 Duration 6H59M EVA start at 8D11H57M50S MET Continued…	KSC 39, PAD A 338:08:35:34Z 3:35:34 AM EST (P) 3:35:34 AM EST (A) Friday 18 12/4/98 (5) LAUNCH WINDOW 4M59S Based on Preferred Launch Time and FGB Planar/Phase. EOM PLS: KSC TAL: ZZA TAL WX: MRN, BEN SELECTED: RTLS: KSC 33/N/N TAL: MRN 20/N/N AOA: KSC 33 PLS: KSC 33/N/N TDEL: -0.15 -0.008/-0.03 MAX Q NAV: 707 715 SRB STG: 2:05.3 2:05 PERF: NOMINAL 2 ENG TAL (ZZA): 2:25 2:25 NEG RETURN: 3:55 3:55 PTA (U/S 500): 4:48 4:45 DROOP: N/A 5:24 PTM (U/S 273): 5:57 5:56 MECO CMD: 8:22.8 8:22.6 VI: 25931 25929 OMS-2 TIG: 43:38 43:41 103 FPS 103 FPS	KSC 15 (KSC 46) 350:03:53:30Z 10:53:30 PM EST Tuesday 14 12/15/98 (10) DEORBIT BURN: 350:02:48:04 340 FPS XRANGE: 134 NM ORBIT DIR: AL 22 AIM PT: NOMINAL MLGTD: 3163 FT 350:03:53:30Z VEL: 197 KGS 197 KEAS HDOT: -2.3 FPS TD NORM 195: 3293 FT DRAG CHUTE DEPLOY: CHUTE WAS DISABLED. NLGTD: 6009 FT 350:03:53:38Z VEL: 164 KGS 158 KEAS HDOT: -6.2 FPS BRK INIT: 135 KGS 8153 FT DRAG CHUTE JETTISON: CHUTE WAS DISABLED. BRK DECEL FPS[2]: AVE 7.7 PK 9.3 WHEELS STOP: 350:03:54:16Z 11506 FT ROLLOUT: 8343 FT 44 SEC Continued…	104.5/104.5/ 109% PREDICTED 100/104.5/ 104.5/72/ 104.5 ACTUAL 100/104.5/ 104.5/72/ 104.5 1 = 2050 (1) 2 = 2044 (2) 3 = 2041 (5) ALL BLOCK IIA SSME'S	BI-095 RSRM 67 ET-97 SLWT-3 SLWT RPT 283K SLWT BR/UP 207K	51.60 (1)	DIRECT INSERTION POST OMS-2 175 X 97 NM DEPLOY: SEP BURN: 347:21:49Z 213.1 X 209 NM RCS-2 COLLISION AVOIDANCE	OI-26B (4)	CARGO: 37731 LBS PAYLOAD CHARGABLE: 30986 LBS DEPLOYED: 26791 LBS NON-DEPLOYED: 3073 LBS MIDDECK: 1122 LBS SHUTTLE ACCUMULATED WEIGHTS: DEPLOYED: 877846 LBS NON-DEPLOYED: 1378355 LBS CARGO TOTAL: 2824652 LBS PERFORMANCE MARGINS (LBS): FPR: 3783 FUEL BIAS: 720 FINAL TDDP: 2365 RECON: 1043 PAYLOADS: PLB: ISS - 2A Node 1/PMA 1&2 (Deployed to ISS) ICBC Mighty Sat (Deployed) SAC-A (Deployed) GAS (1), SEM RMS, ODS MIDDECK: SIMPLEX 5 CYRO TK SETS 6 GN2 TANKS RMS 52 RMS used to grapple Node 1 and position on ODS. Grapple FGB and dock with Node 1.	***Brief Mission Summary:*** *The STS-88/2A "Unity" mission was the first manned ISS assembly flight. The primary mission objective was to rendezvous with the already launched Zarya control module and successfully attach the Unity connecting module. This mission provided the foundation for assembly of future ISS components.* KSC W/D: OPF 187, VAB 5, PAD 37 = 229 days LAUNCH POSTPONEMENTS: - Baselined launch date of 12/4/97 on 6/21/96. - Postponed launch date to 7/9/98 on 5/27/97. - Postponed launch date to 12/3/98 on 6/4/98. LAUNCH SCRUBS: - Scrubbed 12/3/98 launch attempt after LO2 drainback hold time of 3M42S expired based on preferred launch time (PLT) 5-minute window (LO2 drainback hold time was 5M19S based on T-0 at PLW opening and 3M42S nominal T-0 at PLT). The Planar Launch Window was 7M48S (opened at 337:08:55:31 and closed at 337:09:03:19). Opted for use of the Preferred Launch Time of 377:08:58:19 which provided a window of 5M00S. An unexpected master alarm (MA), associated with hydraulic system 1 momentary pressure spike, caused an automatic hold at T-4 minutes. After holding at T-4 minutes for 3 minutes, the count was resumed. at T-31 seconds, another hold was called while troubleshooting the MA. Resolution of the MA occurred slightly after the expiration of the 3M42S LO2 drainback hold time. The count was resumed; however, the launch window had expired. Post-flight, it was concluded that the most probable cause of the pressure spike was a "Switch Tease" which momentarily reenergized the systems 1 hydraulic pump pressure solenoid valve. SHUTTLE NIGHT LAUNCH #20 LAUNCH DELAYS: None. Launched on-time at 338:08:35:34Z, 3:35:34 AM EST, on Friday, December 4, 1998. TAL WX: - Zaragoza (prime) forecast and observed NO GO (ceiling and crosswind), Moron (selected) forecast and observed GO. Ben Guerir forecast NO GO (ceiling & rain) but observed GO. DOLILU-II I-LOADS: - Uplink #24, I-Load uplink #43. Continued…
				M 3 EOM WEIGHT: 201538 LBS X CG: 1084.33 LANDING WEIGHT: 201492 LBS X CG: 1086.18		SLWT IMPACT: 1:27:30 MET LAT: 1.72 N LONG: 127.2 W		DEORBIT 213.6 X 208.8 NM VELOCITY 25898 FPS ENTRY RANGE 4343 NM			

s99_03770 ---- In Dec 1998, assembly of the ISS began with the joining of the U.S.-built Unity Node & the Russian-built Zarya module (IMAXR camera view).

FLT NO.	ORBITER	CREW (6) TITLE, NAMES & EVA'S	LAUNCH SITE, LIFTOFF TIME, LANDING SITES, ABORT TIMES	LANDING SITE/ RUNWAY, CROSSRANGE LANDING TIMES FLT DURATION, WINDS	SSME-TL NOM-ABORT EMERG THROTTLE PROFILE ENG. S.N.	SRB RSRM AND ET	ORBIT		FSW	PAYLOAD WEIGHTS, PAYLOADS/ EXPERIMENTS	MISSION HIGHLIGHTS (LAUNCH SCRUBS/DELAYS, TAL WEATHER, ASCENT I-LOADS, FIRSTS, SIGNIFICANT ANOMALIES, ETC.)
							INC	HA/HP			

STS-88/ ISS-2A

Continued

Continued...

MCC WHITE FCR (23)

FLIGHT DIRECTORS:
A/E/O 4 - J. P. Shannon
LD/O 1 - R. E. Castle
O 2 - P. L. Engelauf
Plng/O 3 - A. F. Algate
MOD - J. W. Bantle
ISS LD/O 2 - M. A. Kirasich
ISS/O 1 - S. P. Davis
ISS/Plng/O 3 - J. M. Hanley

Continued...

WINDS:
5T, 1R KTS
OFFICIAL:
3105P09
R2, T5 KTS

DENS ALT: -854 FT

FLT DURATION:
11:19:17:56

S/T: 801:22:26:12

OV-105:
133:04:07:58

DISTANCE:
4,650,000 sm

STS088-370-006 ---Crew in U.S. -built Unity Node: Bottom row (left to right) are PLT Sturckow, CDR Cabana, & Currie/MS. Top row, Krikalev/MS (Russia), Newman/MS, & Ross/MS.

STS088-E-5059 (12-08-98) --- Newman (left) & Ross mated 40 cables & connectors running 76 ft between Zarya & Unity (foreground).

BELOW: 98e09779 In MCC on console: Scott Altman, Dominic Gorie, & Scott Horowitz .

SIGNIFICANT ANOMALIES:
- Galley iodine removal assembly hose QD incompatibility.
- Five PLB floodlights failed.
- Anomalous SAFER S/N 1007 GN2 and tank pressure reading.
- GPS anomalies.
- APU 2 fuel pump drain line pressure decay.
- RCS jet R2D fail leak.
- Right Pad A heater circuit failure.
- Right RCS 1/2 tank isolation valves fail open.
- Right inboard tire pressure indication failed low.
- Failed portable foot restraint attachment device hatch pin came out, then broke.

STS088-703-032 --- Blanketing clouds form the backdrop for the connected Zarya and Unity modules after release from Endeavour's cargo bay.

Continued...

PERFORMANCE ENHANCEMENTS:
- Standard set plus PE High Q WIN/DEC, OMS assist 4000 lbs, 52 NM MECO, and Del Psi.

FLIGHT DURATION CHANGES: None

FIRSTS/LASTS:
- First Shuttle flight to International Space Station (FGB), docked node to PMA/FGB.
- First ISS assembly flight.

SHUTTLE NIGHT LANDING #10
- Landed on orbit 186 on KSC 15.

EVENTS:
- STS-88/2A first International Space Station (ISS) assembly flight carried NODE, Unity.
- First ISS element, the FGB Zarya, was launched from Baikonur Cosmodrome by a PROTON at 324:06:40:006Z into an orbit of 191.4 X 100 NM at inclination of 51.62 degrees.
- STS-88/2A was the first rendezvous and docking of the ISS Program.
- RMS grapple of PMA-1/Node 1/PMA-2 at 339:21:54:19Z, unberth at 339:22:08:10Z, installed on ODS at 339:23:52:40Z, ungrapple at 340:00:09:30Z.
- RMS grapple of FGB at 340:23:47:02Z, FGB ungrapple at 341:02:43:52Z.
- EVA 1 start at 341:22:09:51Z, end at 342:05:30:42Z, duration 7H21M51S.
- ISS reboost burn start at 342:20:35:34Z, duration____.
- EVA 2 start at 343:20:33:04Z, end at 344:03:34:34Z, duration 7H01M30S.
- Node 1 (Unity) ingress at 344:19:54Z, FGB ingress at 344:21:11Z.
- EVA 3 start at 346:20:33:24Z, end at 347:03:32:01, duration 6H58M37S.
- SAC-A deployed at 9:20:15 MET.
- Mighty SAT deployed at 10:17:13 MET.
- Drag Chute was disarmed pending resolution of STS-95 Drag Chute door anomaly. (Mortar was removed.)
- Undock at 347:20:24:34Z.
- ISS Visitor time 6D17H34M20S

RENDEZVOUS #44
- Rendezvous and dock with ISS PMA 2 Node 1 forward port.

FLT NO.	ORBITER	CREW (7) TITLE, NAMES & EVA'S	LAUNCH SITE, LIFTOFF TIME, / LANDING SITES, ABORT TIMES	LANDING SITE/ RUNWAY, CROSSRANGE / LANDING TIMES FLT DURATION, WINDS	SSME-TL NOM-ABORT EMERG / THROTTLE PROFILE ENG. S.N.	SRB RSRM AND ET	ORBIT INC / HA/HP	FSW	PAYLOAD WEIGHTS, PAYLOADS/ EXPERIMENTS	MISSION HIGHLIGHTS (LAUNCH SCRUBS/DELAYS, TAL WEATHER, ASCENT I-LOADS, FIRSTS, SIGNIFICANT ANOMALIES, ETC.)

STS-96 ISS-2A.1

SEQ FLT #94
KSC-94
PAD 39B-40
MLP-2

Second Shuttle Flight to ISS

First Flight With Logistics and Maintenance Spacehab #13

MCC WHITE FCR (24)

FLIGHT DIRECTORS:
A/E - L. J. Ham
LD/O1 - N. W. Hale
O 2 - P. F. Dye
PLNG - W. D. Reeves
MOD - J. W. Bantle
ISS LD/O1 - P. S. Hill
ISS/O2 - M. J. Kirasich
ISS/PLNG - M. J. Ferring

ORBITER
OV-103 (Flight 26) Discovery

OMS PODS:
LPO1 - 29
RPO3 - 27
FRC3 - 26

CREW
CDR: Kent Rominger (Flt 4 - STS-73, STS-80, STS-85) P552/R200/V131/M174

PLT: Rick D. Husband P553/R248/M216

M/S 1/EV 1: Tamara E. Jernigan (Flt 5 - STS-40, STS-52, STS-67, STS-80) P554/R130/V83/F14

M/S 2: Ellen Ochoa (Flt 3 - STS-56, STS-66) P555/R160/V113/F20

M/S 3/EV 2: Daniel T. Barry (Flt 2 – STS-72) P556/R209/V155/M182

M/S 4: Julie Payette (Canada) P557/R249/ F33

M/S 5: Valery Tokarev (Russia) P558/R250/M217

SS EVA #45: EMU/Tethered EVA #38 on 5/29/99 Scheduled EVA #39 ISS EVA #4 7H55M Duration

LAUNCH SITE
KSC 39B
147:10:49:42Z
6:49:42 AM EDT (P)
6:49:42 AM EDT (A)
Thursday 29
5/27/99 (5)

LAUNCH WINDOW:
8M6S USING PREFFERED LAUNCH TIME

EOM PLS: KSC
TAL: ZZA
TAL WX: MRN, BEN

SELECTED:
RTLS: KSC 33/CI/N
TAL: MRN 20/N/N
AOA: KSC 15/N/N
PLS: EDW 22/CI/N

TDEL:
0.1   -0.18/+0.18

MAX Q NAV:
740   740

SRB STG:
2:04.6   2:05

PERF: NOMINAL

2 ENG TAL (MRN):
2:17   2:21

NEG RETURN:
3:54   3:57

PTA (U/S 272):
4:21   4:24

DROOP (ZZA):
5:22   5:24

PTM (U/S 272):
5:30   5:39

SE TAL (ZZA):
5:51   5:52

SE PTM:
6:41   6:48

MECO CMD:
8:22.1   8:22

VI:
25931   25929

Continued…

LANDING SITE
KSC 15 (KSC 47)
157:06:02:43Z
02:02:43 AM EDT
Sunday 14
6/6/99 (5)

DEORBIT BURN:
157:04:54:09Z

XRANGE: 712 NM

ORBIT DIR: AL 23

AIM PT: CLOSE IN

MLGTD: 1963 FT
157:06:02:43Z
VEL: 210 KGS
210 KEAS
HDOT: -1.0 FPS

TD NORM 205:
2290 FT

DRAG CHUTE DEPLOY:184 KEAS
157:06:02:51Z

NLGTD: 6504 FT
157:06:02:57Z
VEL: 156 KGS
149 KEAS
HDOT: -5.8 FPS

BRK INIT: 112 KGS

DRAG CHUTE JETTISON: 53 KGS
157:06:03:18Z

BRK DECEL FPS2:
AVE 7.1 PK 9.0

WHEELS STOP:
157:06:03:35Z
10829 FT

ROLLOUT:
8866 FT
52 SECS

WINDS:
2H, 5L KTS
OFFICIAL: 0904P07
2H, 3L KTS

Continued…

SSME
104/104/ 109%

PREDICTED:
100/104.5/
104.5/72/
104.5

ACTUAL:
100/104.5/
104.5/72/
104.5

1 = 2047 (2)
2 = 2051 (1)
3 = 2049 (1)

ALL BLOCK IIA SSME'S

SRB/ET
BI-098

RSRM 70

ET-100

SLWT-4

ET IMPACT:
1:26:12 MET
LAT: 2.46 S
LONG: 127.99 W

ORBIT
51.60 (2)

DIRECT INSERTION

POST OMS-2:
182.7 X 177.1 NM

TI:
149:01:35:18Z
MET
208.3 X 202.4 NM

MC4:
149:02:55:18Z
MET
209.3 X 208.4 NM

REBOOST:
154:09:36:53Z
213.9 X 208.6 NM

FSW
OI-27 (1)

PAYLOAD
CARGO: 33808 LBS

PAYLOAD CHARGABLE: 22707 LBS

DEPLOYED: 4228 LBS

NON-DEPLOYED: 17994 LBS

MIDDECK: 1034 LBS

SHUTTLE ACCUMULATED WEIGHTS:
DEPLOYED: 882074 LBS
NON-DEPLOYED: 139783 LBS
CARGO TOTAL: 2858460 LBS

PERFORMANCE MARGINS (LBS):
FPR: 3783
FUEL BIAS: 720
FINAL TDDP: 4435
RECON: 4306

PAYLOADS:
PLB:
ISS 2A.1
SPACEHAB (Double Module)
ODS, OTD
STARSHINE (DEPLOYED)
SVF
ICC

MIDDECK:
DTO
EMU H/W
EMU TOOLS

5 CYRO TK SETS
5 GN2 TANKS

RMS 53 (S.N. 303)

Continued…

M 3 EOM
WEIGHT: 222365 LBS
X CG: 1080.18

LANDING:
WEIGHT: 222300 LBS
X CG: 1081.83

DEORBIT
213.9 X 208.6 NM

ENTRY VELOCITY 25915 FPS

ENTRY RANGE 4358 NM

MISSION HIGHLIGHTS

Brief Mission Summary: The major objective of STS-96-2A.1, 2nd ISS mission, was to transfer nearly 2 tons of logistical supplies to the ISS. These supplies would be used to continue the outfitting of the Unity and Zarya modules and for later use to set up the Russian Service Module for occupancy by a three-man crew. In addition, a small educational satellite called STARSHINE was deployed for observation by international students.

KSC W/D: OPF 122, VAB 12 (2), PAD 30 (2) = 164 days total (Rollback to repair ET foam)

LAUNCH POSTPONEMENTS:
- Baselined launch date of 12/9/98 on 10/2/97.
- Postponed launch date to 5/13/99 on 6/4/98 (Multi-flight changes to ISS flights), then to 5/20/99, to 5/24/99, and to 5/20/99 on 4/21/99.
- Postponed launch date to NET 5/27/99 based on 5/13/99 decision to roll back to VAB on 5/16/99 to repair hail damage to ET foam (648 divots, 459 required repair).
- Rolled back to pad on 5/21/99 and confirmed 5/27/99 as the launch date.

LAUNCH SCRUBS: None

LAUNCH WINDOW:
- The launch window was in two panes. Pane 1 opened at 147:10:48:46Z and closed at 147:10:54:42Z. There was a 10-second cutout with pane 2 opening at 147:10:54:52Z and closing at 147:10:57:48Z. The total launch window was 9M2S with a 10-second cutout between panes based on the ISS Planar/Phase window. The decision was made to use the Preferred Launch Time (PLT) of 147:10:49:42Z for a launch window of 8 minutes 6 seconds, in two panes with a 10-second gap.

LAUNCH DELAYS: None
- Launch occurred on time at 147:10:49:42Z, 6:49:42 AM EDT on Friday, May 27, 1999.

TAL WX:
- ZARAGOZA (Prime) was forecast NO GO – tailwind (at landing time observed NO GO, tailwind and thunderstorms). Moron (Selected) and Ben Guerir were both forecast GO and observed GO at landing time.

PERFORMANCE ENHANCEMENTS:
- Standard set plus: (1) PE High Q SUM/MAY, (2) OMS assist is 4000 lbs, (3) 52 nm MECO, and (4) Del Psi.

Continued…

STS096-E-5037 (29 May 1999) --- Rookie Pilot, Rick Husband, signals thumbs up during rendezvous with ISS.

FLT	ORBITER	CREW (7)	LAUNCH SITE, LIFTOFF TIME,	LANDING SITE/ RUNWAY, CROSSRANGE	SSME-TL NOM-ABORT EMERG	SRB RSRM	ORBIT		FSW	PAYLOAD WEIGHTS,	MISSION HIGHLIGHTS (LAUNCH SCRUBS/DELAYS,
NO.		TITLE, NAMES & EVA'S	LANDING SITES, ABORT TIMES	LANDING TIMES FLT DURATION, WINDS	THROTTLE PROFILE ENG. S.N.	AND ET	INC	HA/HP		PAYLOADS/ EXPERIMENTS	TAL WEATHER, ASCENT I-LOADS, FIRSTS, SIGNIFICANT ANOMALIES, ETC.)
STS-96 **ISS-2A.1** Continued			Continued… OMS-2: 43:11 43:10.6 255 FPS 255 FPS 2:43 2:43	Continued… DENS ALT: 1321 FT FLT DURATION: 9:19:13:01 S/T: 811:17:39:13 OV-103: 196:08:44:49 DISTANCE: 4,051,000 sm	SIGNIFICANT ANOMALIES: - Humidity separator B water carryover. - Vestibule leakage during airlock depress. - SSOR anomalies: choppy EVA comm, EVA comm squeal, SSOR noise malfunctions during EVA, EMU TLM from SSOR static. - Difficulty attaching SCU 1 to DCM. - Lost LG/SM retractable tether - came off fish stringer. - Small equipment hook failed open - tether release from D-ring on miniworkstation. - SAFER Pyro Valve Fired/Manual Isolation Valve open. - F4R Thruster declared failed leak by RM.					Continued… RMS USED FOR EVA SUPPORT AND SURVEY SVS (SPACE VISION SYSTEM)	Continued… FLIGHT DURATION CHANGES: None - Landed on orbit 154 as planned. FIRSTS/LASTS/RECORDS: - First flight of Functional Drag Chute with strengthened door pins after problem on STS-95 (Inconel was aluminum). - First logistics/maintenance flight to ISS, Third ISS flight, 2nd Docking Flight to ISS (PMA2) Node 1 forward port. SHUTTLE NIGHT LANDING # 11: KSC runway 15 EVENTS: - OMS assist burn 147:10:51:57Z with a duration of 2M42S. - RCS MC4 at 149:02:55:17/01:16:05:35 MET. - ISS ring capture 149:04:23:51Z, docking 149:04:37:38Z/01:17:47:56 MET at PMA2, Node 1 Forward Port. - STARSHINE deployed at 156:07:21Z/08:20:32 MET. - Crew ingressed ISS PMA2 at 149:07:00Z/01:20:10 MET. - IFM: Replaced FGB Battery MIRT's, and Replaced ECOMM Transceiver and Power Distribution Box. - EVA Start Time 150:16:21:36Z/03:05:31:54 MET. EVA End Time 151:00:16:36Z/03:13:26:54 MET. EVA tasks include Installation of FGB target mask, installed Orbital Transfer Device and IAPFR on PAM 1, installed Strela crane on PMA2, installed trunnion pin cover, and transferred EVA tools to Node 1. - Reboost Start 154:09:36:54Z/06:22:47:11 MET. Reboost End 154:10:11:40Z, Delta V 21.8 fps, altitude increased 6 nm, orbit 212.1 by 206.2 nm. - Undocking complete 154:22:39:17Z/07:11:36 MET. - ISS Visitor time is 5:18:01:39. - Final transfers to ISS: EVA 661 lbs, IVA transfers 2881 lbs, and water transfers 686 lbs (7 CWC's), Total to ISS 4228 lbs. To Shuttle 197 lbs. - Return IVA transfers to Discovery 213 lbs. - Landed on orbit 154, Ascending Left 23, Crossrange 712 NM, range 4370 NM, Runway 15. RENDEZVOUS # 45: - Rendezvous and dock with ISS. RADIATOR DEPLOY # 24:

STS096-E-5168--Iinflight crew portrait: At bottom center: CDR Rominger, flanked by Barry/MS & Ochoa/MS. Above Barry (left) Tokaerev/MS(RSA), Jernigan/MS & Payette/MS (CSA). PLT Husband is between Payette & Ochoa.

STS096-357-003 (30 May 1999) --- MS1 Jernigan totes part of a Russian-built crane, Strela (a Russian word meaning "arrow").

STS096-E-5219 --- ISS as seen from Discovery after separation.

STS096-(S)-010 --- First flight of Functional Drag Chute with strengthened door pins after STS-95 problem (door fell off at SSME throttle-up). Inconel replaced aluminum pins.

FLT NO.	ORBITER	CREW (5) TITLE, NAMES & EVA'S	LAUNCH SITE, LIFTOFF TIME, LANDING SITES, ABORT TIMES	LANDING SITE/ RUNWAY, CROSSRANGE LANDING TIMES FLT DURATION, WINDS	SSME-TL NOM-ABORT EMERG THROTTLE PROFILE ENG. S.N.	SRB RSRM AND ET	ORBIT INC	HA/HP	FSW	PAYLOAD WEIGHTS, PAYLOADS/ EXPERIMENTS	MISSION HIGHLIGHTS (LAUNCH SCRUBS/DELAYS, TAL WEATHER, ASCENT I-LOADS, FIRSTS, SIGNIFICANT ANOMALIES, ETC.)

STS-93

SEQ
FLT # 95

KSC- 95

PAD 39B-41

MP-1L

ORBITER:
OV-102
(Flight 26)
Columbia

OMS PODS:
LPO5-15
RPO5-14
FRC2-26

CREW:
CDR:
Eileen M. Collins
(Flt 3 - STS-63, STS-84)
P559/R188/V139/F24

PLT:
Jeffrey S. Ashby
P560/R251/M218

M/S 1:
Cady G. Coleman
(Flt 2 - STS-73)
P561/R201/V156/F27

M/S 2:
Steven A. Hawley
(Flt 5 - STS 41-D,
STS 61-C,STS-31 & STS-82)
P562/R39/V29/M38

M/S 3:
Michel Tognini
(CNES-France)
P563/R252/M219

MCC WHITE FCR (25)

FLIGHT DIRECTORS:
A/E/O1 - J. P. Shannon
LD/O 2 - B. P. Austin &
P. F. Dye
PLNG - C. W. Shaw
MOD - B. R. Stone &
J. W. Bantle

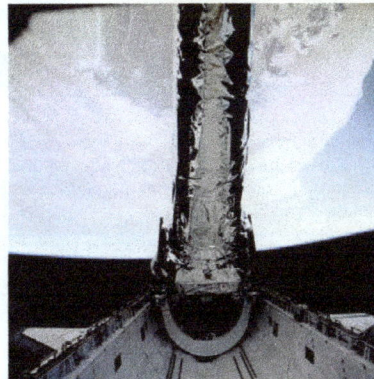

Eileen M. Collins, first female Shuttle CDR

LAUNCH SITE, LIFTOFF TIME:
KSC PAD 39B
204:04:31:00Z
12:24:00 AM EDT (P)
12:31:00 AM EDT (A)
Friday 19
7/23/99 (6)

LAUNCH WINDOW:
46 Minutes

EOM PLS: KSC
TAL: BYD
TAL WX: BEN

SELECTED:
RTLS: KSC 15/N/N
TAL: BEN 36/N/N
AOA: EDW 22/N/N
PLS: EDW 22/CI/N

TDEL:
0.05   0.092/0.13

MAX Q NAV:
673   675

SRB STG:
2:03.5   2:04

PERF: NOMINAL

2 ENG TAL (BEN):
3:20   3:18

NEG RETURN:
3:52   3:59

PTA (U/S 219):
5:25   5:19

DROOP:
5:26   5:25

SE TAL (BYD):
6:02   5:59

PTM (U/S 219):
6:20   6:10

MECO CMD:
8:28   8:28

VI:
25876   25859

Continued…

LANDING SITE/ RUNWAY:
KSC 33 (KSC 48)
209:03:20:35Z
11:03:20:35 PM EDT

Wednesday 10
7/28/99 (9)

DEORBIT BURN:
209:02:19:00Z

XRANGE: 83 NM

ORBIT DIR: DL 45

AIM PT: NOMINAL

MLGTD: 2533 FT
209:03:20:35Z
VEL: 201 KGS
         196 KEAS
HDOT: -1.0 FPS

TD NORM 195:
2628 FT

DRAG CHUTE
DEPLOY: 190 KEAS
209:03:20:37Z

NLGTD: 5470 FT
209:03:20:44Z
VEL: 159 KGS
         149 KEAS
HDOT: -4.1 FPS

BRK INIT: 122 KGS

DRAG CHUTE
JETTISON:
43 KGS
209:03:21:05Z

BRK DECEL FPS[2]:
AVE 9.1 PK 10.4

WHEELS STOP:
209:03:21:19Z
9384 FT

ROLLOUT:
6851 FT
44 SEC

Continued…

SSME-TL:
104/104/
109%

PREDICTED:
100/104/104/
67/104

ACTUAL:
100/104/104/
67/104

1 = 2012 (22)
2 = 2031 (17)
3 = 2019 (19)

M 3 EOM:
WEIGHT:
202872 LBS

X CG:
1097.54

LANDING:
WEIGHT:
202796 LBS

X CG:
1099.36

SRB RSRM AND ET:
BI-097

RSRM
69

ET-99

SLWT-5

ET
RPT:
283K

ET
BR/UP:
          K

ET
IMPACT:

MET:
1:23:16
LAT:
17.54 N
LONG:
154.66 W

ORBIT INC:
28.45
(48)

HA/HP:
DIRECT
INSERTION

POST OMS-2:
154 X 145 NM

DEORBIT:
151 x 139 NM

ENTRY
VELOCITY:
25762 FPS

ENTRY
RANGE:
4332 NM

FSW:
OI-26B
(5)

PAYLOAD WEIGHTS:
CARGO:
52382 LBS

PAYLOAD
CHARGEABLE:
49798 LBS

DEPLOYED:
43080 LBS

NON-DEPLOYED:
5171 LBS

MIDDECK:
1538 LBS

SHUTTLE
ACCUMULATED
WEIGHTS:
DEPLOYED:
925154 LBS
NON-DEPLOYED:
1404092 LBS
CARGO TOTAL:
2910842 LBS

PERFORMANCE
MARGINS (LBS):
FPR: 3553
FUEL BIAS: 720
FINAL TDDP: 2081
RECON: -3981

PAYLOADS:
PLB:
AXAF-I/IUS
(CHANDRA
deployed)

MIDDECK:
MSX, SIMPLEX,
SWUIS, GOSMAR,
STL-B, LFSAH,
CCM, SAREX-II,
EARTHKAM, PGIM,
CGBA, MEMS, BRIC

3 CRYO TK SETS
(Off Load)

4 GN2 TANKS

NO RMS

MISSION HIGHLIGHTS:

Brief Mission Summary: The primary objective of the STS-93 mission was to deploy the $1.5 billion Chandra, the world's most powerful X-Ray Observatory, and third in NASA's series of "Great Observatories". Other objectives included execution of jet firings for Air Force satellite plume study and operation of the Southwest Ultraviolet Imaging System. This was also the first Shuttle mission commanded by a female, CDR Eileen M. Collins.

KSC W/D: OPF 223, VAB 5, PAD 43 = 271 days total.

LAUNCH POSTPONEMENTS:
- Baselined 8/27/98 as launch date on 5/16/97.
- Postponed launch date to 12/3/98 and to 1/21/99 (Multi-flight change to ISS flights).
- Postponed to 3/18/99, to 3/25/99, to 4/8/99, to 4/15/99, to 7/9/99, to 7/22/99, and to 7/20/99 (primarily Chandra AXAF/IUS delays).

LAUNCH SCRUBS:
- 7/20/99 (12:36 AM EDT.) Launch attempt was halted with a manual GLS cutoff at T-7 seconds (approximately 200 milliseconds prior to Main Engine Start) due to a (false) spike indication of 640 ppm H2 concentration in the aft. Insufficient time to wait for the confirmation sample at T-8 seconds and allow time to issue a manual GLS cutoff before Main Engine Start at T-6.33 seconds. The manual cutoff call was made at T-10 seconds. A 48-hour scrub turnaround was required to replace the Hydrogen Long-Throw Igniters. KSC, BYD, and BEN were forecast and observed GO. Launch reset for 7/22/99. Technical Scrub.
- 7/22/99 (12:28 AM EDT.) Launch attempt was scrubbed at T+47:30 due to Range and RTLS weather. During count, rain and lightning hits within 20 NM, and thunderstorms within 20 NM. Counted down to T-5 minutes and held awaiting improved weather. Mission Director gave ok to extend window 36 minutes by giving up first day deploy. Scrubbed launch at 203:05:17:35Z (T+47:30) with no signs of improvement in weather (lightning within 8.6 miles of SLF and thundershowers within 20 NM). Banjul was NO GO for ceiling/rain. Ben Guerir was GO. Launch reset for 7/23/99. Weather Scrub.

LAUNCH WINDOW:
46 minutes planned window. During count, the customer relaxed contingency deploy opportunities and IUS battery eclipse constraints to extend window to 116 minutes; however, launch window was limited to Range availability (60 minutes).

Continued…

STS093-702-048 --- Chandra X-Ray observatory, back-dropped against a desert in Namibia, just before release from Columbia's payload bay.

FLT NO.	ORBITER	CREW (5)	LAUNCH SITE, LIFTOFF TIME,	LANDING SITE/ RUNWAY, CROSSRANGE	SSME-TL NOM-ABORT EMERG	SRB RSRM	ORBIT		FSW	PAYLOAD WEIGHTS,	MISSION HIGHLIGHTS (LAUNCH SCRUBS/DELAYS,
		TITLE, NAMES & EVA'S	LAUNCH SITES, ABORT TIMES	LANDING TIMES FLT DURATION, WINDS	THROTTLE PROFILE ENG. S.N.	AND ET	INC	HA/HP		PAYLOADS/ EXPERIMENTS	TAL WEATHER, ASCENT I-LOADS, FIRSTS, SIGNIFICANT ANOMALIES, ETC.)
STS-93 Continued			Continued… **OMS-2:** 41:04 41:06.9 200 FPS 200 FPS 2:14 2:14	Continued… **WINDS:** 04T, 5L KTS OFFICIAL: 2405P06 SS: OT, 5L PK: OT, 6L **DENS ALT:** 1551 FT **FLT DURATION:** 4:22:49:35 S/T: 816:16:28:48 **OV-102:** 273:21:09:17 **DISTANCE:** 1,796,000 sm							Continued…

STS093-706-039 --- Chandra X-Ray Observatory after release from Columbia's payload bay.

STS093-322-017 --- Collins, first female Shuttle CDR, and crew are shown on-orbit. In front are CDR Collins and Tognini/MS (France). In rear are (from the left) Hawley/MS, Ashby/PLT, and Coleman/MS.

ABOVE: Hawley/MS shown with Micro-Electromechanical Systems (MEMS) experiment. MEMS monitors a suite of sensors under flight conditions. ABOVE RIGHT: Mark Sowa (PAO photographer) recorded the fly-over of Space Shuttle Columbia above the JSC Rocket Park. The Saturn V is below the streak left by the shuttle Columbia re-entering the atmosphere.

SIGNIFICANT ANOMALIES:
- At approximately Liftoff plus 5 seconds, there was a short circuit on AC1 Phase A for approximately 0.5 seconds. The resultant under voltage caused SSME 1 "A" and SSME 3 "B" controllers to be disqualified. Postflight, it was determined the short was on AC1 Phase A to SSME 1 "A" controller.
- At liftoff, the right SRB hydraulic pressure sensor 2 was erratic.
- Four ET LO_2 sensors indicated dry resulting in low-level cutoff of main engines and slightly early MECO.
- Right SSME multiple performance parameters deviations (Post-flight inspection revealed ruptures in three Engine 2019 nozzle tubes caused by an impact of a loose LO_2 post deactivation pin. LH2 leak resulted in controller compensating for fuel loss with additional LOX flow, a 16 fps underspeed, and 8 nm lower altitude.
- CRT 3 Critical BITE.
- High-load FES excessive water carryover.
- Camcorder tape jam.
- Primary thruster F2D low fuel injector temperature.

Continued…

LAUNCH DELAY:
- Launch was delayed 7M0S during T-20 minute hold for MILA to change out A Frame Sync Box to restore the forward link.
- Launched at 204:04:31:00Z, 12:31:00 AM EDT on July 23, 1999.

SHUTTLE NIGHT LAUNCH #21

TAL WX:
- Banjul (prime) was forecast NO GO (thunderstorms and anvil clouds) and observed NO GO (thunderstorms and ceiling). Ben Guerir (selected) was forecast and observed GO.

PERFORMANCE ENHANCEMENTS:
- Standard set.
- PE LO Q SUM/JUL

SHUTTLE NIGHT LANDING # 12: KSC 33 on Wednesday, 7/28/99 at 11:20:35 PM EDT - moonlit landing.

FLIGHT DURATION CHANGES: None
- Landed on orbit 80 as planned.

FIRSTS/LASTS:
- First space flight with female Commander (Eileen Collins).
- First U.S. flight for Michel Tognini (CNES-France). Michel's first space flight was to Mir on Soyuz TM-15S.
- Last flight of phase 2 engines.
- Most aft landing Xcg (1099.36)

FLT NO.	ORBITER	CREW (7) TITLE, NAMES & EVA'S	LAUNCH SITE, LIFTOFF TIME, LANDING SITES, ABORT TIMES	LANDING SITE/ RUNWAY, CROSSRANGE LANDING TIMES FLT DURATION, WINDS	SSME-TL NOM-ABORT EMERG THROTTLE PROFILE ENG. S.N.	SRB RSRM AND ET	INC	ORBIT HA/HP	FSW	PAYLOAD WEIGHTS, PAYLOADS/ EXPERIMENTS	MISSION HIGHLIGHTS (LAUNCH SCRUBS/DELAYS, TAL WEATHER, ASCENT I-LOADS, FIRSTS, SIGNIFICANT ANOMALIES, ETC.)
STS-103 SEQ FLT #96 KSC-96 PAD 39B-42 MLP-2 HST FLT #4 (SM-3A) HST SERVICE FLT #3	OV-103 (Flight 27) Discovery OMS PODS: LPO1-30 RPO3-28 FRC3-27	CDR: Curtis L. Brown (Flt 6 - STS-47, STS-66, STS-77, STS-85, & STS-95) P564/R152/V112/M136 PLT: Scott J. Kelly P565/R253/M220 M/S 1/EV 1: Steven L. Smith (Flt 3 - STS-68, STS-82) P566/R184/V137M161 M/S 2: Jean-Francois Clervoy (ESA-France) (Flt 3 - STS-66, STS-84) P567/R186/V140/F163 M/S 3/EV 2: John M. Grunsfeld (Flt 3 - STS-67, STS-81) P568/R191/V133/M167 M/S 4/EV 3: Michael Foale (Flt 5 - STS-45, STS-56, STS-63, Up STS-84,& Dn STS-86) P569/R143/V92/M127 M/S 5/EV 4: Claude Nicollier ESA-Switzerland) (Flt 4 - STS-46, STS-61, & STS-75) P570/R150/V98/M134 SS EVA #46 EMU/TETHERED EVA #39 ON 12/22/99 SCHEDULED EVA #40 DURATION 8:15:30 SS EVA #47 EMU/TETHERED EVA #40 ON 12/23/99 SCHEDULED EVA #41 DURATION 8:10 Continued…	KSC, PAD 39B 354:00:50:00Z 7:50:00 PM EST (P) 7:50:00 PM EST (A) Sunday 11 12/19/99 (6) LAUNCH WINDOW: 42M16S HST Planar/Phase Window EOM PLS: KSC TAL: BYD TAL WX: BEN SELECTED: RTLS: KSC 15/N/N TAL: BEN 36/N/N AOA: EDW 04/N/N PLS: EDW 22/N/N TDEL: 0.08 -0.158/-0.12 MAX Q NAV: 718 720 SRB STG: 2:05.3 2:05 PERF: NOMINAL 2 ENG TAL (BEN): 2:05 2:05 NEG RETURN: 3:51 3:54 PTA (U/S 500): 3:09 3:08 PTM (U/S 500): 4:16 4:15 SE TAL (BYD): 5:37 5:43 MECO CMD: 8:24.4 8:25.9 VI: 26128 26124 Continued…	KSC 33 (KSC 49) 362:00:00:47Z 7:00:47 PM EST Monday 17 12/27/99 (11) DEORBIT BURN: 361:22:48:26Z XRANGE: 155 NM ORBIT DIR: DL 46 AIM PT: NOMINAL MLGTD: 2804 FT 362:00:00:47Z VEL: 187 KGS 186 KEAS HDOT: -2.9 FPS TD NORM 195: 2237 FT DRAG CHUTE DEPLOY: 176 KEAS 362:00:00:50Z NLGTD: 5955 FT 362:00:00:58Z VEL: 141 KGS 138 KEAS HDOT: -4.6 FPS BRK INIT: 111 KGS DRAG CHUTE JETTISON: 54 KGS 362:00:01:18Z BRK DECEL FPS2: AVE 6.5 PK 10.0 WHEELS STOP: 362:00:01:35Z 9809 FT 48 SECS ROLLOUT: 7005 FT Continued…	104/104/ 109% PREDICTED: 100/104.5/ 104.5/67/ 104.5 ACTUAL: 100/104.5/ 104.5/67/ 104.5 1 = 2053 (1) 2 = 2043 (3) 3 = 2049 (2) ALL IIA ENGINES	BI-099 RSRM 73 ET-101 SLWT-6 ET RPT: 283K ET IMPACT: 1:19:15 MET LAT: 17.4 N LONG: 141.4 W	28.45 (49)	DIRECT INSERTION POST OMS-2 315.4 X 170.2 NM	OI-26B (6)	CARGO: 20276 LBS PAYLOAD CHARGABLE: 13208 LBS DEPLOYED: 5423 LBS NON-DEPLOYED: 6451 LBS MIDDECK: 1334 LBS SHUTTLE ACCUMULATED WEIGHTS DEPLOYED: 930577 LBS NON-DEPLOYED: 1411877 LBS CARGO TOTAL: 2931118 LBS PERFORMANCE MARGINS (LBS): FPR: 3783 FUEL BIAS: 720 FINAL TDDP: 13576 RECON: 13308 PAYLOADS PLB: HST SM-3A (3rd HST Service Flight) 5 CYRO TK SETS 6 GN2 TANKS RMS 54 (S.N. 301) RMS USED FOR HST GRAPPLE, BERTH, AND RELEASE AND EVA SUPPORT	**Brief Mission Summary:** The STS-103 mission was the third Servicing Mission to ensure the health of the Hubble Space Telescope (HST), the first of NASA's "Great Observatories". Included were four spacewalks designed to install new equipment and replace old. The primary objective was to replace the six gyroscopes to restore the three Rate Sensor Units to full power. Other replacements included: an upgraded computer, a set of Fine Guidance Sensors, and a new Solid State Recorder. Deteriorated insulation on the HST's outer surface was also repaired. KSC W/D: OPF 141, VAB 9, PAD 36 = 186 days LAUNCH POSTPONEMENTS: - Baselined 10/14/99 as launch date on 3/18/99. - Postponed launch to 11/19/99 on 9/16/99. OV-103 wire inspections and repair. - Postponed launch to 12/2/99 on 10/22/99. OV-103 wire inspections and repair. - Postponed launch to 12/6/99 on 11/10/99. OV-103 wire inspections and repair. - Postponed launch to 12/11/99 on 12/7/99. Replacement of damaged SSME wiring harness. - Postponed launch to 12/16/99 on 12/9/99. Changeout of dented LH2 4-in Recirc manifold. LAUNCH SCRUBS: - Scrubbed 12/16/99 launch attempt at 9:18 AM EST at ET Tanking MMT while holding at T-6 hours. ET weld wire issue caught by vendor X-ray inspection. ET cleared ET hardware. Orbiter needed 24 hours to review orbiter weld processes and personnel records to evaluate possible impact to orbiter hardware. Review found no issue to orbiter fleet. Reset launch to 12/17/99. Technical Scrub. - Scrubbed 12/17/99 launch attempt at 8:47 PM EST at 4 minutes into window due to KSC range and RTLS weather. Weather concerns were low ceiling (broken 6500 feet), rain, turbulence, thick cloud layer (triggered lightning), and RTLS crosswinds at limit. Had difficulty getting Jimsphere balloons to altitude due to icing conditions. Use of 450 MHz radar profiler as backup confirmation of wind persistence was being worked. EDW runway distance lighting markers power failure. FD switched to NOR for AOA and first day PLS. Launch was scrubbed when it became evident bad weather conditions would continue throughout the remainder of the window. Ben Guerir and Banjul TAL sites were GO. Ben Guerir was selected. Reset launch to 12/18/99. Window was 42M11S first pane, 10 second cutout, and then 4M11S in second pane. Weather Scrub. Continued…

Additional data in lower cells:

Landing Site/Runway column (M 3 EOM):
WEIGHT: 212288 LBS
X CG: 1080.64
LANDING:
WEIGHT: 212217 LBS
X CG: 1082.39
Continued…

Orbit column:
DEORBIT: 330 X 301 NM
ENTRY VELOCITY: 26114 FPS
ENTRY RANGE: 4237 NM

STS103-713-048 (19-27 December 1999) — Smith and Grunsfeld replacing gyroscopes, contained in rate sensor units (RSU), inside HST.

SPACE SHUTTLE MISSIONS SUMMARY

FLT NO.	ORBITER	CREW (7) TITLE, NAMES & EVA'S	LAUNCH SITE, LIFTOFF TIME, LANDING SITES, ABORT TIMES	LANDING SITE/ RUNWAY, CROSSRANGE LANDING TIMES FLT DURATION, WINDS	SSME-TL NOM-ABORT EMERG THROTTLE PROFILE ENG. S.N.	SRB RSRM AND ET	ORBIT INC	HA/HP	FS W	PAYLOAD WEIGHTS, PAYLOADS/ EXPERIMENTS	MISSION HIGHLIGHTS (LAUNCH SCRUBS/DELAYS, TAL WEATHER, ASCENT I-LOADS, FIRSTS, SIGNIFICANT ANOMALIES, ETC.)
STS-103 Continued		Continued… SS EVA #48 EMU/TETHERED EVA #41 ON 12/24/99 SCHEDULED EVA #42 DURATION 8:09 MCC WHITE FCR (26) FLIGHT DIRECTORS: A/E/O 4 - N. W.Hale LD/O 1 - L. J. Ham O 2 - B. P. Austin Plng - J. M. Hanley MOD - J. W. Bantle	Continued… OMS-2: 44:15 44:08 252 FPS 247 FPS 2:34 2:34 DISTANCE:	Continued… WINDS: 1T, 7L KTS OFFICIAL: 2406P12 DENS ALT: -107 FT FLT DURATION: 7:23:10:47 S/T: 824:15:39:35 OV-103: 204:07:55:46 DISTANCE: 3,267,360 sm							

STS103-726-081 (19-27 December 1999) --- Repaired HST after release from RMS.

S99-15923 --View of JSC MCC during Flight Day 3 activity. Lead Orbit 1 FD Linda Ham is at rear right.

STS103-397-035 -- Crew portrait. Front: (lt to rt) Nicollier/MS(ESA), PLT Kelly, & Grunsfeld/MS. Back row: (lt to rt) Smith/MS, Foale/MS, CDR Brown, & Clervoy/MS(ESA).

STS103-731-051 (19-21 December 1999) --- Foale (left) and Nicollier/ESA (on end of RMS) replacing one of HST's Fine Guidance Sensors (FGS).

SIGNIFICANT ANOMALIES:
- Jammed PFR roll joint.
- Loss of power indication on middeck EMU battery charger.
- HST PFR pitch joint would not lock.
- Release hatch Pip Pin on Starboard Airlock hinge.
- EMU 2 Power up failure.
- Bent pin on EMU3 DCM.

Continued… LAUNCH SCRUBS:

- Scrubbed 12/18/99 launch attempt at 8:21 AM EST at ET Tanking MMT while holding at T-6 hours due to observed and forecast bad Range and RTLS weather: Rain, low ceiling, and thick clouds triggered lightning conditions. Decision to evaluate 8 + 2, 3 EVA flight, evaluate landing as late as 12/29/99, and vehicle configuration for holiday standdown. At MMT Meeting at 8:30 AM EST on 12/19/99, decision was made to recommend GO for launch on 12/19/99 at 7:50 PM EST. Weather forecast was good and ET MMT gave a GO to tank. Range and RTLS Weather Scrub.

LAUNCH WINDOW:
Launch window 42M16S in one pane.

LAUNCH DELAYS: None
- Launched at 354:00:50:00Z (GMT date 12/20/99), 7:50:00 PM EST, on Friday, 12/19/99.

TAL WX:
- Banjul (prime) was forecast and observed NO GO with visibility 3 miles (smoke/haze). Ben Guerir (selected) was forecast and observed GO.

PERFORMANCE ENHANCEMENTS:
- Standard set. PE LO Q WIN/DEC

SHUTTLE NIGHT LAUNCH #22

FLIGHT DURATION CHANGES:
- Planned landing at KSC on orbit 119. Extended flight one orbit for weather. Waved off landing at KSC on orbit 119 due to crosswinds of 18 knots, peak 19 knots and STA reported turbulence at 500 feet. Landed on KSC 33 on orbit 120.

SHUTTLE NIGHT LANDING #13
- Landed on KSC 33 on orbit 120 at 362:00:00:47Z, 7:00:47 PM EST on Monday, December 27, 1999.

EVENTS:
- HST grapple at 356:00:34:01Z; HST berth 356:01:42:00Z.
- EVA-1 - Start at 356:18:41:01Z; MET 02:18:04:40 to 03:02:19 MET; duration 8:15:30.
- EVA 2 - Start MET 03:18:16 to 04:02:26; duration 8:10.
- EVA 3 - MET 04:13:27 to 05:02:36; duration 8:09.
- HST unberth at 359:21:18:41Z; HST release 359:23:03:01Z.

RENDEZVOUS # 46:
- Rendezvous, capture, service, and release HST.

FLT NO.	ORBITER	CREW (6) TITLE, NAMES & EVA'S	LAUNCH SITE, LIFTOFF TIME, LANDING SITES, ABORT TIMES	LANDING SITE/ RUNWAY, CROSSRANGE LANDING TIMES FLT DURATION, WINDS	SSME-TL NOM-ABORT EMERG THROTTLE PROFILE ENG. S.N.	SRB RSRM AND ET	INC	ORBIT HA/HP	FSW	PAYLOAD WEIGHTS, PAYLOADS/ EXPERIMENTS	MISSION HIGHLIGHTS (LAUNCH SCRUBS/DELAYS, TAL WEATHER, ASCENT I-LOADS, FIRSTS, SIGNIFICANT ANOMALIES, ETC.)
STS-99 SEQ FLT #97 KSC-97 PAD 39A-55 MLP-3	OV-105 (Flight 14) Endeavour OMS PODS: LPO4-21 RPO1-28 FRC5-14	CDR: Kevin R. Kregel (Flt 4 - STS-70, STS-78, & STS-87) P571/R197/V129/M172 PLT: Dom L. Gorie (Flt 2 - STS-91) P572/R242/V157/M211 M/S 1: Gerhard P. J. Thiele ESA Germany P573/R254/M221 M/S 2: Janet L. Kavandi (Flt 2 - STS-91) P574/R243/V158/F32 M/S 3: Janice Voss (Flt 5 - STS-57, STS-63, STS-83,& STS-94) P575/R167/V115/F22 M/S 4: Mamoru Mohri Japan (Flt 2 - STS-47) P576/R155/V159/M137 MCC WHITE FCR (27) FLIGHT DIRECTORS: A/E - J. P. Shannon LD/O2 - P. F. Dye O-1 - L. E. Cain O3 - B. P. Austin MOD - J. M. Heflin	KSC 39A 42:17:43:40Z 12:30:00 PM EST (P) 12:43:40 PM EST (A) Friday 20 2/11/2000 (7) LAUNCH WINDOW: 2H10M Closed on SRTM BETA ANGLE CONSTRAINT EOM PLS: KSC TAL: ZZA TAL WX: MRN,BEN SELECTED: RTLS: KSC 33/CI/N TAL: ZZA 30/N/N AOA: NOR 23/CI/N PLS: EDW 22/CI/N TDEL: 0.12 -0.38/-0.04 MAX Q NAV: 727 733 SRB STG: 2:05.6 2:06 PERF: NOMINAL 2 ENG TAL (ZZA): 2:48 2:46 NEG RETURN: 3:52 3:55 PTA (U/S 187): 5:26 5:21 DROOP(ZZA): 5:16 N/A PTM (U/S 187): 6:15 6:11 SE TAL (ZZA): 6:03 MECO CMD: 8:22.5 8:23.42 VI: 25776 25769 Continued…	KSC 33 (KSC 50) 53:23:22:24Z 6:22:24: PM EST Tuesday 15 2/22/2000 (5) DEORBIT BURN: 53:22:25:10Z XRANGE: 242 NM ORBIT DIR: DL 47 AIM PT: NOMINAL MLGTD: 2885 FT 53:23:22:24:Z VEL:206 KGS 207 KEAS HDOT: -1.6 FPS TD NORM 205: 3004 FT DRAG CHUTE DEPLOY: 166 KEAS 53:23:22:36Z NLGTD: 6520 FT 53:23:22:34Z VEL:169 KGS 168 KEAS HDOT: -65 FPS BRK INIT: 115 KTS DRAG CHUTE JETTISON: 52 KGS 53:23:23:05Z AVE BRK DECEL: AVE 5.9 PK 7.8 FPS/S WHEELS STOP: 53:23:22:23:Z 12828 FT ROLLOUT: 9943 FT 59 SEC WINDS: 1R, 7R KTS OFFICIAL: 0507P09 SS: 2T, 7R PK: 3T, 12R DENS ALT: 72 FT Continued…	104/104/ 109% PREDICTED: 100/104.5/ 104.5/72/ 104.5 ACTUAL: 100/104.5/ 104.5/72/ 104.5 1 = 2052 (1) 2 = 2044 (3) 3 = 2047 (3) ALL BLOCK IIA SSME'S M 3 EOM: WEIGHT: 225092 LBS X CG: 1078.48 LANDING: WEIGHT: 225030 LBS X CG: 1080.19	BI-100 RSRM 71 ET-92 LWT 85 ET RPT: 283K ET IMPACT 1:12:05 MET LAT: 47.41 S LONG: 162.19 W	57.0 (20)	DIRECT INSERTION POST OMS-2: 129.5 X 126.1 NM RCS OA MANEUVER 4:14:00 MET: 126.5 X 128.7 NM DEORBIT: 127.9 X 124.4 NM ENTRY VELOCITY: 25714 ENTRY RANGE: 4624	OI-27 (2)	CARGO: 35410 LBS PAYLOAD CHARGEABLE: 29069 LBS DEPLOYED: 260 LBS NON-DEPLOYED: 26987 LBS MIDDECK: 1822 LBS SHUTTLE ACCUMULATED WEIGHTS: DEPLOYED: 930837 LBS NON-DEPLOYED: 1440686 LBS CARGO TOTAL: 2966528 LBS PERFORMANCE MARGINS (LBS): FPR: 3272 FUEL BIAS: 854 FINAL TDDP: 1085 RECON: 395 PAYLOADS: PLB: SRTM/SRL-3 with radar antennas on 200 ft boom. MIDDECK: EARTHKAM 5 CRYO TK SETS 5 GN2 TANKS NO RMS	*Brief Mission Summary:* STS-99 was the first shuttle flight of the new century. The primary payload was a space radar, known as Shuttle Radar Topography Mission (SRTM). The SRTM successfully mapped the Earth in 3-D, 30 times more accurately than current global maps. The system used two radar antennas mounted in the shuttle payload bay and two on a 200-foot-long mast extended out of the payload bay. This mast was the longest rigid structure deployed in space at this time. The SRTM is an outgrowth of the Spaceborne Imaging Radar flown on STS-59 and STS-68. KSC W/D: OPF 257, VAB 10, PAD 44 = 311 days total. LAUNCH POSTPONEMENTS: - Baselined launch date of 6/30/99 on 3/5/98 (OV-104); then to 1/22/99 on 6/4/98 (Multi-flight changes ISS SM delay). - Advanced launch date to 9/16/99 on 7/23/98. OV-104, OV-103 on 7/30/98 to achieve additional GPS DTO Flight. Updates to flight dates and baseline STS-101 OV-105 on 10/5/98. - Postponed launch date to NET 11/19/99 on 9/16/99. STS-103 also NET 11/19/99 due to wire inspections and repairs. - Postponed launch date to 1/13/00; additional wire work and STS-103 to fly first. - Postponed launch date to 1/31/00. STS-103 flight delays and Y2K testing. LAUNCH SCRUBS: - Scrubbed 1/31/00 launch attempt at 31:19:08:55Z (T-9M12S) with 40M05S left in 2H02M launch window while counting to T-9 minutes. At T-29 minutes, a preflight BITE test to the MEC's was executed. MEC 2 (an EMEC) first response was anomalous (bad address, bad parity, bad SEV). Scrub at 19:08:55Z (T-9M12S). Decision on a 2/1/00 launch at MMT early Tuesday morning. The Range and RTLS was observed and forecast NO GO for 1/31 launch (low ceiling, rain within 20 NM, field mills in and out, thick cloud layer, and triggered lightning potential). All 3 TAL sites were GO. Technical/ Weather Scrub. New launch date 2/1/00 at 12:44 PM EST. - Scrubbed 2/1/00 launch attempt at approximately 3:00 AM EST with the decision to change out MEC 2. MEC changeout and retest is 5 to 7 days. Tried to get range for 2/9/00. MCC changeout/retest and range availability set next launch to 2/11/00. Technical scrub. LAUNCH WINDOW: - The Launch Window was 2H10M00S. Opened at 42:17:30:00Z and closed at 42:19:40Z. Closed on 0 degrees beta angle constraint for SRTM operations. Continued…

JSC2000E01556 (January 2000) --- Artist's concept of SRTM Earth mapping operation.

SPACE SHUTTLE MISSIONS SUMMARY

FLT NO.	ORBITER	CREW (6) TITLE, NAMES & EVA'S	LAUNCH SITE, LIFTOFF TIME, LANDING SITES, ABORT TIMES	LANDING SITE/ RUNWAY, CROSSRANGE LANDING TIMES FLT DURATION, WINDS	SSME-TL NOM-ABORT EMERG THROTTLE PROFILE ENG. S.N.	SRB RSRM AND ET	INC	ORBIT HA/HP	FSW	PAYLOAD WEIGHTS, PAYLOADS/ EXPERIMENTS	MISSION HIGHLIGHTS (LAUNCH SCRUBS/DELAYS, TAL WEATHER, ASCENT I-LOADS, FIRSTS, SIGNIFICANT ANOMALIES, ETC.)
STS-99 Continued			Continued… OMS-2: 34:59.5 35:03 181 FPS 181 FPS	Continued… FLT DURATION: 11:05:38:44 S/T: 835:21:18:19 OV-105: 144:09:46:40 DISTANCE: 4,708,821 sm						

S99-E-5034 (12 February 2000)--- The 200 ft.-long mast supporting the Shuttle Radar Topography Mission juts into space from Endeavour (out of frame at left).

STS099-318-015 --- A "star-burst" pose. Top Center: Voss/MS, (clockwise from her) PLT Gorie, Kavandi/MS, Thiele/MS (ESA), Mohri/MS (NASDA), and CDR Kregel.

JSC2000-E-02781 PIA02733 (Release Date: 21 February 2000) --- Perspective view of San Andreas Fault near Palmdale, CA. The view was created by draping a Landsat satellite image (showing residential and agricultural development) over an SRTM elevation model. Topography is exaggerated 1.5 times vertically.

ABOVE: JSC2000-01451 -- SRTM personnel support STS-99 in JSC Payload Operations Control Center (POCC). From left are Mike Kobrick, Ian Joughin and Diane Ainsworth. ABOVE RIGHT: JSC2000-01454 ---- Scott D. Vangen "ta ks topography" at the Crew Interface Console (CIC) in JSC POCC.

Continued…

LAUNCH DELAYS:
- Launch delay was 13M40S. Held at T-9 minutes hold to clear the IPR's: (1) MPS LH2 manifold P, (2) cabin pressure leak check at lower pressure, and (3) Hyd Sys 1 Circ Pump pressure low. Launched at 42:17:43:40Z, 12:43:40 PM EST, on Friday, February 11, 2000.

TAL WX:
- Zaragoza (prime and selected); Moron (2-engine TAL Call), and Ben Guerir were all forecast and observed GO.

PERFORMANCE ENHANCEMENTS:
- Standard Set plus: (1) Interim generic High Q WIN/FEB, and (2) OMS Assist is 4000 lbs.

FLIGHT DURATION CHANGES: Extended One Rev due to Crosswind Violations at KSC. Waved off landing on orbit 181.

FIRSTS/LASTS:
- First shuttle flight in the year 2000.
- First flight of Shuttle Radar Topography Mission using dual-antenna imaging radar with antennas mounted on 200 foot extended boom.
- Last flight of Lightweight ET.

EVENTS:
- Landed on KSC runway 33 on orbit 182 at 53:23:22:24Z, 6:22:24 PM EST on Tuesday, 2/22/00.

SIGNIFICANT ANOMALIES:
- GPC I/O Errors and EMEC preflight BITE error.
- LH2 Manifold Pressure Tape Meter Oscillations.
- WSB 2 under cool during ascent.
- CRT 1 BITE.
- ET GH2 Ullage Pressure Low at MECO.
- Forward Mission Timer Display Elements Failed.
- RRCS Fuel Regulator B Primary Stage Leakage.
- Vernier Thruster L5D Oxidizer Temperature Erratic.
- Supply water dump nozzle blockage.
- APU 1 GG Injector tuber temperature failure.

FLT NO.	ORBITER	CREW (7) TITLE, NAMES & EVA'S	LAUNCH SITE, LIFTOFF TIME, LANDING SITES, ABORT TIMES	LANDING SITE/ RUNWAY, CROSSRANGE LANDING TIMES FLT DURATION, WINDS	SSME-TL NOM-ABORT EMERG THROTTLE PROFILE ENG. S.N.	SRB RSRM AND ET	ORBIT INC	ORBIT HA/HP	FSW	PAYLOAD WEIGHTS, PAYLOADS/ EXPERIMENTS	MISSION HIGHLIGHTS (LAUNCH SCRUBS/DELAYS, TAL WEATHER, ASCENT I-LOADS, FIRSTS, SIGNIFICANT ANOMALIES, ETC.)
STS-101/ ISS 2A.2a SEQ FLT #98 KSC-98 PAD 39A-56 MLP-1 THIRD SHUTTLE FLIGHT TO ISS SPACEHAB #14	OV-104 (Flight 21) Atlantis OMS PODS: LPO3-25 RPO4-21 FRC4-21	CDR: James D. Halsell (Flt 5 - STS-65, STS-74, STS-83, STS-94) P577/R178/V123/M156 PLT: Scott J. Horowitz (Flt 3 - STS-75, STS-82) P578/R210/V135/M183 M/S 1: Mary Ellen Weber (Flt 2 - STS-70) P579/R198/V160/F26 M/S 2/EV1: Jeffrey N. Williams P580/R255/M222 M/S 3/EV2: James S. Voss (Flt 4 - STS-44, STS-53, STS-69) P581/R136/V85/M121 M/S 4: Susan J. Helms (Flt 4 - STS-54, STS-64, STS-78) P582/R158/V108/F19 M/S 5: Yuri Usachev (Russia) P583/R256/M223 SSEVA #49 EMU TETHERED EVA #42 SCHEDULED EVA #43 DURATION 6:44 MCC WHITE FCR (28) FLIGHT DIRECTORS: A/E - J. P. Shannon LD/O1 - P. L. Engelauf O2 - K. B. Beck PLNG - C. W. Shaw PLNG/O2 - L. E. Cain (Beck, Shaw, and Cain switched shifts during flight.) ISS LD/O1 - P. S. Hill ISS O2 - A. F. Algate ISS PLNG - J. M. Curry MOD - J. W. Bantle	KSC 39A 140:10:11:10Z 6:11:10 AM EDT (P) 6:11:10 AM EDT (A) Friday 21 5/19/00 (6) LAUNCH WINDOW: 5M4S BASED ON ISS IN-PLANE TIME EOM PLS: KSC TAL: ZZA TAL WX: MRN, BEN SELECTED: RTLS: KSC 15/CI/N TAL: ZZA 30/CI/N AOA: KSC 15/CI/N PLS: EDW 04/N/SF TDEL: 0.09 -0.388/-0.19 MAX Q NAV: 714 709 SRB STG: 2:04.8 2:04 PERF: NOMINAL 2 ENG TAL (ZZA): 2:27 2:27 NEG RETURN: 3:52 3:56 PTA (U/S 269): 4:42 4:47 DROOP (ZZA): 5:26 5:28 PTM (U/S 269): 5:59 6:06 SE TAL (ZZA): 6:02 6:02 MECO CMD: 8:23.8 8:25.3 VI: 25931 25930 OMS-2: 43:04 43:04 81.3 FPS 81.4 FPS	KSC 15 (KSC 51) 150:06:20:19Z 2:20:19 AM EDT Monday 18 5/29/00 (9) DEORBIT BURN: 150:05:12:10Z XRANGE: 95.8 NM ORBIT DIR: AL 24 AIM PT: NOMINAL MLGTD: 3269 FT 150:06:20:19 Z VEL: 202 KGS 199 KEAS HDOT: -2.0 FPS TD NORM 205: 2731 FT DRAG CHUTE DEPLOY: 189 KEAS 150:06:20:22 Z NLGTD: 6752 FT 150:06:20:30 Z VEL: 154 KGS 152 KEAS HDOT: -4.2 FPS BRK INIT: 102 KGS DRAG CHUTE JETTISON: 54 KGS 150:06:20:57 Z BRK DECEL (fps/s): AVE 5.3 PK 6.6 WHEELS STOP: 150:06:21:07 Z 12182 FT ROLLOUT: 8913 FT 48 SEC WINDS: 2407P09 SS:OH 7R PK:IH 9R DENS ALT: 1591 FT Continued...	104/104/ 109% PREDICTED: 100/104.5/ 104.5/72 104.5 ACTUAL: 100/104.5/ 96/72/ 104.5 1 = 2043 (4) 2 = 2054 (1) 3 = 2049 (3) ALL BLOCK IIA SSME'S M 3 EOM: WEIGHT: 226277 LBS X CG: 1081.20 LANDING WEIGHT: 226212 LBS X CG: 1082.85	BI-101 (3) RSRM 74 ET-102 SLWT-7 ET IMPACT 1:26:29 MET LAT: 1.955 LONG: 127.3 W	51.60 (3)	DIRECT INSERTION POST OMS-2 178.9 X 85.2 NM DEORBIT: APOGEE: 207.2 NM PERIGEE: 189.3 NM ENTRY VELOCITY: 25899 FPS ENTRY RANGE: 4449 NM	OI-27 (3)	CARGO: 35604 LBS PAYLOAD CHARGEABLE: 24733 LBS DEPLOYED: 3371 LBS NON-DEPLOYED: 20159 LBS MIDDECK: 1262 LBS SHUTTLE ACCUMULATED WEIGHTS: DEPLOYED: 934208 LBS NON-DEPLOYED: 1462107 LBS CARGO TOTAL: 3002132 LBS PERFORMANCE MARGINS (LBS): FPR: 3783 FUEL BIAS: 720 FINAL TDDP: 733 RECON: 998 PAYLOADS: PLB: ISS 2A.2a Spacehab DM ICC, SEM-06, MARS RMS, ODS MIDDECK: CPCG PCG-BAG BIOTUBE AST 5 CRYO TK SETS 6 GH2 TANKS RMS 55 RMS USED FOR EVA SUPPORT	**Brief Mission Summary:** STS-101, 3rd mission to ISS, was initially designed to outfit the Russian Zvezda crew quarters. However, Zvezda's launch was delayed and the mission was changed to ISS maintenance and logistics support. Outfitting Zvezda would await STS-106 later in the year. A high priority of this flight was the replacement of four of six 800 amp Zarya batteries. Also, this was first flight of Shuttle "Glass Cockpit" upgrade. KSC W/D: OPF 333, VAB 8, PAD 50 = 391 days total. LAUNCH POSTPONEMENTS: - Baselined 8/5/99 as launch date on 10/5/98. Postponed to 10/14/99, then 12/2/99. TACAN scars removed for GPS scar then reinstated TACAN. - Postponed launch to 11/19/99 on 9/16/99. OV-103 wire inspections and repair. - Postponed launch to 12/2/99 on 10/22/99. OV-103 wire inspections and repair. - Postponed launch to 4/14/00 on 4/16/00. CDR training accident (ankle) - Postponed launch to 4/24/00 on 4/16/00. OV-104 Rudder/ Speed Brake PDU R&R from OV-102. LAUNCH SCRUBS: - Scrubbed 3:17:17 PM EDT (115:20:17:17Z) 4/24/00 launch attempt while holding at T-9 minutes due to high RTLS crosswinds. Scrub was declared at approximately L-15 minutes, when RTLS crosswinds observed and forecast to exceed the 15-knot limit. - Scrubbed 2:53:17 PM EDT (116:19:53:17Z) 4/25/00 launch attempt at L-1:35:00 by Launch Director when RTLS crosswinds persisted in 29-30 knots range and were forecast to exceed limit. RTLS Weather Scrub. - Scrubbed 2:34:16 PM EDT (117:19:34:17Z) 4/26/00 launch attempt at 117:19:21Z (L-0H13M) while holding in T-9 min hold due to no TAL site. All three TAL sites were observed and forecast NO GO: ZZA for showers within 20 nm and forecast chance of broken 4000 feet. MRN for showers/thundershowers and forecast chance of broken 3000 feet. BEN was observed and forecast NO GO for crosswind violation. BEN wind swing from around 285 degrees to around 300 degrees after sundown did not materialize - crosswind forecast was steady state R11 and P16. The launch window opened 117:19:24:42Z and closed at 117:1934:16Z and the PLT was 117:19:29:13Z for a launch window of 4M55S. TAL WX Scrub. - Unable to get May 9 launch date due to GOES launch delays. Scheduled a May 18 launch at 6:32:00 AM EDT. At approximately L-36 hours, the Atlas III launch scrub due to high winds caused a slip to May 19. LAUNCH WINDOW: - Window opened at 140:10:09:29Z and closed at 140:10:16:14Z for a total window of 6M45S. Selected Preferred Launch Time (PLT) of 140:10:11:10Z for a launch window of 5M4S. Continued...

S99- 01417-- 1st flight MEDS cockpit

FLT NO.	ORBITER	CREW (7) TITLE, NAMES & EVA'S	LAUNCH SITE, LIFTOFF TIME, LANDING SITES, ABORT TIMES	LANDING SITE/ RUNWAY, CROSSRANGE LANDING TIMES FLT DURATION, WINDS	SSME-TL NOM-ABORT EMERG THROTTLE PROFILE ENG. S.N.	SRB RSRM AND ET	ORBIT INC	HA/HP	FSW	PAYLOAD WEIGHTS, PAYLOADS/ EXPERIMENTS	MISSION HIGHLIGHTS (LAUNCH SCRUBS/DELAYS, TAL WEATHER, ASCENT I-LOADS, FIRSTS, SIGNIFICANT ANOMALIES, ETC.)
STS-101/ ISS 2A.2a Continued…			Continued… FLT DURATION: 9:20:09:09 S/T: 845:17:27:28 OV-104: 160:18:39:34 DISTANCE: 5,076,281 sm							Continued… LAUNCH DELAYS: None - Launched on time at 140:10:11:10Z, 6:11:10 AM EDT on Friday, May 19, 2000. TAL WX: - Zaragoza (Prime and Selected), Moron, and Ben Guerir all forecast and observed GO. PERFORMANCE ENHANCEMENTS: - Standard Set Plus: (1) PE Operational - High Q TRN/APR, (2) OMS Assist is 4000 lbs, (3) 52 NM MECO, and (4) Del psi FLIGHT DURATION CHANGES: - One-day extension. Extended flight one day to accomplish ISS tasks. SHUTTLE NIGHT LAUNCH #23 SHUTTLE NIGHT LANDING #14 - Landed on KSC runway 15 at 150:06:20:19Z, 2:20:19 AM EDT on Monday, May 29, 2000. FIRSTS/LASTS: - First flight of glass cockpit (MEDS) - First flight of OV-104 since STS-86 after OMDP.

S101-E-5048 – Williams/MS attaches Russian Crane (Strela) to ISS. Strela was delivered on STS-96.

STS101-717-094 --- Inflight crew portrait on ISS Unity (Node 1). Rear (from left): Weber/MS, CDR Halsell, Williams/MS, & PLT Horowitz. Front: Helms/MS, Usachev/MS (RSA), & Voss/MS.

EVENTS:
- ISS ring capture at 142:03:56:10Z
- Docked with ISS PMA2 Node 1 Forward Port at 142:04:44:09Z, 1:18:32:59 MET.
- EVA 1 Start at 143:01:52:58Z, 2:15:41:48 MET and End at 143:08:36:58Z, 2:21:25:48 MET, duration 6:44.
- Reboost #1 - Start at 145:00:02:11Z, 4:13:51:01 MET, 29.06 fps, final orbit 190 by 184 nm, increase approximately 9 nm.
- Reboost #2 - Start at 146:02:14:01Z, 5:16:02:51 MET, 29 fps, final orbit 196 by 195 nm, increase approximately 9 nm.
- Reboost #3 - Start at 146:23:32:38Z, 6:13:21:28 MET, 28.2 fps, final orbit 206.7 by 199.5 nm.
- Undocked at 147:23:02:38Z, 7:12:51:18 MET
- STS-101/2A.2a ISS Visitor Time is 5D:18H:18M:29S (Docking to Undocking)
- Total transfers: To ISS, 3371 lbs consisting of 2657 lbs dry cargo (IVA), 4 CWC's with 387 lbs H2O, and External (EVA) 327 lbs. From ISS, 1391 lbs. Net transfer to ISS was 1980 lbs.
- Completed air quality work, R&R FGB failed electrical equipment and FGB lifetime equipment. EVA tasks completed include installation of OTD and Strela cranes and ECOMM antenna R&R.

RENDEZVOUS #47
- Rendezvous and dock with ISS at PMA2, Node 1 Forward Port.

SIGNIFICANT ANOMALIES:
- Left OMS Engine Bipropellant Valve 2 indicates open.
- Left OMS Engine GN2 regulator pressure low during Post-Firing Purges.
- Ku-band radiating within RF Protect Box.
- PRSD Oxygen Tank 4 Heater temporarily failed.
- Collins TACAN BITE faults.
- Slump tile at wing leading edge with internal flow.
- APCU 1 converter B failure.
- MEDS MDU CRT 2 display screen came up blank.
- Speedbrake Ch 3 secondary Delta Pressure delayed response

STS101-390-025 (19-29 May 2000) --- Helms/MS performs battery maintenance below floor of Zarya.

JSC 2000-04279 - In JSC MCC: Flight Controllers huddle over I-load update for Day-of-Launch winds. From left: Larry Bourgeois/Space Ops, Steve Hawley/FCOD, FD Jeff Bantle, and Henry Cordova & Ed Gonzalez/Flight Design & Dynamics.

FLT NO.	ORBITER	CREW (7) TITLE, NAMES & EVA'S	LAUNCH SITE, LIFTOFF TIME, LANDING SITES, ABORT TIMES	LANDING SITE/ RUNWAY, CROSSRANGE LANDING TIMES FLT DURATION, WINDS	SSME-TL NOM-ABORT EMERG THROTTLE PROFILE ENG. S.N.	SRB RSRM AND ET	INC	ORBIT HA/HP	FSW	PAYLOAD WEIGHTS, PAYLOADS/ EXPERIMENTS	MISSION HIGHLIGHTS (LAUNCH SCRUBS/DELAYS, TAL WEATHER, ASCENT I-LOADS, FIRSTS, SIGNIFICANT ANOMALIES, ETC.)
STS-106/ ISS 2A.2b SEQ FLT #99 KSC-99 PAD 39B-43 MLP-2 FOURTH SHUTTLE FLIGHT TO ISS SPACEHAB #15	OV-104 (Flight 22) Atlantis OMS PODS: LPO3-26 RPO4-22 FRC4-22	CDR: Terrence W. Wilcott (Flt 4 - STS-68, STS-79, STS-89) P584/R183/V130/M160 PLT: Scott D. Altman (Flt 2 - STS-90) P585/R237/V161/M207 M/S 1/EV1: Edward T. Lu (Flt 2 - STS-84) P586/R222/V162/M194 M/S 2: Richard A. Mastracchio P587/R257/M224 M/S 3: Daniel C. Burbank P588/R258/M225 M/S 4/EV2: Yuri Malenchenko (Russia) P589/R259/M226 M/S 5: Boris Morukov (Russia) P590/R260/M227 SS EVA #50 EMU/TETHERED EVA #43 SCHEDULED EVA #44 DURATION 6:14 Continued…	KSC PAD 39B 252:12:45:47Z 8:45:47 AM EDT (P) 8:45:47 AM EDT (A) Friday 22 9/8/00 (10) LAUNCH WINDOW: 3:54 USING PLT (IN-PLANE TIME) EOM PLS: KSC TAL: ZZA TAL WX: MRN, BEN SELECTED: RTLS: KSC 33 N/N TAL: ZZA 30 N/N AOA: NOR 17N/SFD PLS: EDW 22 N/N TDEL: 0.09 -0.348/-0.31 MAX Q NAV: 710 712 SRB STG: 2:03.4 2:02 PERF: NOMINAL 2 ENG TAL (ZZA): 2:28 2:23 NEG RETURN: 3:52 3:52 PTA (U/S 267): 4:39 4:38 PTM (U/S 267): 5:47 5:46 SE TAL (ZZA): 5:52 6:05 SE PTM (U/S 827): 6:49 6:48 MECO CMD: 8:24.3 8:25.6 Continued…	KSC 15 (KSC 52) 264:07:56:44Z 3:56:44 AM EDT Wednesday 11 9/20/00 (10) DEORBIT BURN: 264:06:50:07 Z XRANGE: 203 NM ORBIT DIR: AL 25 AIM PT: CLOSE IN MLGTD: 2951 FT 264:07:56:44Z VEL: 187 KGS 186 KEAS HDOT: -2.5 FPS TD NORM 205: 1643 FT DRAG CHUTE DEPLOY: 180 KEAS 264:07:56:46Z NLGTD: 5485 FT 264:07:56:52Z VEL: 153 KGS 153 KEAS HDOT: -6.3 FPS BRK INIT: 71 KGS DRAG CHUTE JETTISON: 56 KGS 264:07:57:23Z BRK DECEL FPS²: AVE 2.7 PK 4.8 WHEELS STOP: 264:07:58:02Z 12078 FT ROLLOUT: 9127 FT 78 SEC WINDS: 1306P09 SS: 5H 2L PK: 8H 4L DENS ALT: 1761 FT Continued…	104/104/ 109% PREDICTED: 100/104.5/ 104.5/72 104.5 ACTUAL: 100/104.5/ 98/72/104.5 ALL BLOCK IIA SSME'S M 3 EOM: WEIGHT: 222835 LBS X CG: 1080.07 LANDING WEIGHT: 222774 LBS X CG: 1081.73	BI-102 RSRM 75 ET-103 SLWT-8 ET IMPACT 1:26:12 MET LAT: 2.46 S LONG: 128.1 W	51.60 (4)	DIRECT INSERTION POST OMS-2: 176.4 X 85.0 NM 1 = 2052 (2) 2 = 2044 (4) 3 = 2047 (4) DEORBIT: APOGEE 206 NM PERIGEE 205 NM ENTRY RANGE: 4390 NM ENTRY VELOCITY: 25892	OI-27 (4)	CARGO: 34991 LBS PAYLOAD CHARGEABLE: 23967 LBS DEPLOYED: 5399 LBS NON-DEPLOYED: 17935 LBS MIDDECK: 1172 LBS SHUTTLE ACCUMULATED WEIGHTS: DEPLOYED: 939607 LBS NON-DEPLOYED: 1481214 LBS CARGO TOTAL: 3037123 LBS PERFORMANCE MARGINS (LBS): FPR: 3274 FUEL BIAS: 818 FINAL TDDP: 1940 RECON: 317 PAYLOADS: PLB: ISS-2A.2b Spacehab/DM ICC (SHOSS Box, SOAR) GAS (2) RMS, ODS MIDDECK: CGBA DTO EMU H/W EVA Tools 5 CRYO TK SETS 6 GN2 TKS RMS 56 RMS USED FOR EVA SUPPORT	Brief Mission Summary: The goal of the STS-106 mission, 4th mission to ISS, was to prepare the Zvezda Service Module for the arrival, later in the year, of the first residents, Expedition 1 crew, to start a permanent human presence on the ISS outpost. KSC W/D: OPF 66, VAB 5, PAD 22 = 93 days total. LAUNCH POSTPONEMENTS: - Baselined launch date of 8/19/00 on 2/17/00. - Postponed launch to 9/8/00 on 5/17/00. LAUNCH SCRUBS: None LAUNCH WINDOW: - Launch window opened at 252:12:42:01Z and closed at 252:12:49:41Z for a total window of 7M40S. Preferred Launch Time (PLT) (In-Plane Time) was 252:12:45:47Z, 8:45:47 AM EDT, resulting in a launch window of 3M54S. LAUNCH DELAYS: None - Launch occurred on time at 252:12:45:47Z, 8:45:47 AM EDT on Friday, September 8, 2000. TAL WX: - Zaragoza (Prime and Selected) and Moron (2-engine TAL) were both forecast and observed GO, Ben Guerir was forecast and observed NO GO for crosswinds. KSC RTLS forecast and observed precipitation within 20 nm; however, was GO based on Flight Rule A2.1.1-6C4e, f, and g, LANDING SITE WEATHER CRITERIA [HC], "2-nm vertical clearance from the top of that shower and a 10-nm lateral clearance must be maintained along the approach paths…" PERFORMANCE ENHANCEMENTS: - Standard Set plus: (1) PE Operational High Q SUM/SEP, (2) OMS assist is 4000 lbs, (3) 52 NM MECO, and (4) Del Psi FLIGHT DURATION CHANGES: - One-day extension. Extended Flight one day to accomplish additional ISS tasks. SHUTTLE NIGHT LANDING #15: - Landed on KSC runway 15, orbit 185 at 264:07:56:44Z, 3:56:44 AM EDT on Wednesday, September 20, 2000. Continued…

STS106-712-028 -- Atlantis crew found a much larger ISS since STS-101 departure with the addition of the Russian Zvezda and a docked Progress resupply ship.

FLT NO.	ORBITER	CREW (7) TITLE, NAMES & EVA'S	LAUNCH SITE, LIFTOFF TIME, LANDING SITES, ABORT TIMES	LANDING SITE/ RUNWAY, CROSSRANGE LANDING TIMES FLT DURATION, WINDS	SSME-TL NOM-ABORT EMERG THROTTLE PROFILE ENG. S.N.	SRB RSRM AND ET	ORBIT INC	HA/HP	FSW	PAYLOAD WEIGHTS, PAYLOADS/ EXPERIMENTS	MISSION HIGHLIGHTS (LAUNCH SCRUBS/DELAYS, TAL WEATHER, ASCENT I-LOADS, FIRSTS, SIGNIFICANT ANOMALIES, ETC.)
STS-106/ ISS 2A.2b Continued...		Continued... MCC WHITE FCR (29) FLIGHT DIRECTORS: A/E - N. W. Hale LD/O1 - P. L. Engelauf O2 - P. F. Dye O3 - K. B. Beck O4 - W. D.Reeves ISS LD/O1 - M. J. Ferring ISS O2 - J. M. Hanley ISS PLNG - R. E. LaBrode MOD - J. W. Bantle	Continued... VI: 25926 25928 OMS-2: 44:00 44:00 81 FPS 81 FPS 00:52 00:54	Continued... FLT DURATION: 11:19:10:57 S/T: 857:12:38:25 OV-102: 172:13:50:31 DISTANCE: 4,919,243 sm							Continued... EVENTS: - OMS Assist Start 2:23 MET - Orbiter/ISS capture at 254:05:51:16Z, 1:17:05:59 MET - Docked to ISS PMA2 Node 1 Forward Port at 254:06:04:53Z, 1:17:19:06 MET. - Shuttle ISS EVA #6. EVA Start at 255:04:46:47Z, 2:16:01:50 MET, EVA End 255:11:00:47Z, duration 6:14. Routed and connected 9 power, data, and comm cables between Zvezda (SM) and Zarya (FGB). Installed magnatometer to ISS for use as compass relative to Earth. - Inert weight adjustment is -200 lbs. - Reboost #1 - Start at 255:12:28:47Z, 2:23:43:00 MET, 11 fps, altitude increase 3.2 nm, orbit 201 by 191 nm. - Reboost #2 - Start at 258:06:13:17Z, 5:17:27:30 MET, 11.4 fps, altitude increase 3.2 nm, orbit 203.4 by 195.3 nm. - Reboost #3 - Start at 259:06:45:47Z, 6:18:00:00 MET, 11.4 fps, altitude increase 3.4 nm, orbit 206.3 by 199.2 nm. - Reboost #4 (Unplanned pre-mission) - Start at 261:03:25:47Z, 8:14:40:00 MET, 11.6 fps, altitude increase 3.3 nm, orbit 208.6 by 203.8 nm. - Undocked at 262:03:46:05Z, 9:15:00:18 MET - STS-106/2A.2b crew ISS Visitor Time is 7:21:41:05 (Docking to Undocking). - Total Transfers - Shuttle to ISS, 5399 lbs (Includes 10 CWC's with 780 lbs of H2O.) ISS to Shuttle, 948 lbs. Net transfer to ISS is 4451 lbs. - Installed magnetometer and three SM battery blocks. Connected FGB/SM cables. R&R'ed and C/O two FGB battery systems. R&R'ed FGB limited life items, delivered exercise devices. Prepared crew quarters for Expedition 1 crew. RENDEZVOUS #48: - Rendezvous and dock with ISS at PMA2, Node 1 Forward Port SIGNIFICANT ANOMALIES: - MNB APC5 60 ampere bus transient, power supply fail BITE - Fuel Cell 1 H2 flowmeter failed OSL - Aft Main Bus B current spike - Loss of crew audio for OCA video conferencing - Ku-band forward link lost - -Z Star Tracker failure - Left OMS Forward Fuel Probe failure - Ops Recorder 1 defective tape segment - ODS C/L Camera Harness Assembly failure - ODS C/L Camera misalignment - Camera C iris failed to fully close - Left Vent 8 and 9 Drive Microswitch failures - MSBLS 2 range failure

STS106-349-002 (8-20 September 2000) --- This unique picture captures the cabin of Atlantis, the RMS arm, and part of the ISS.

STS106-373-019 --- Inflight crew portrait on ISS. Front, from the left, Malenchenko/MS (RSA), CDR Wilcutt, PLT Altman. Back, from left, Burbank/MS, Lu/MS & Mastracchio/MS. & Morukov/MS (RSA).

IN THE JSC MCC --- LEFT: (l to r) FD's Leroy Cain, Wayne Hale & Jeff Bantle await launch for "baton" handoff from Florida to Houston. CENTER: FCT Planning with FD Kelly Beck holding flight insignia. RIGHT: FD Orbit 4 Bill Reeves on console.

FLT NO.	ORBITER	CREW (7) TITLE, NAMES & EVA'S	LAUNCH SITE, LIFTOFF TIME, LANDING SITES, ABORT TIMES	LANDING SITE/ RUNWAY, CROSSRANGE LANDING TIMES FLT DURATION, WINDS	SSME-TL NOM-ABORT EMERG THROTTLE PROFILE ENG. S.N.	SRB RSRM AND ET	INC	ORBIT HA/HP	FSW	PAYLOAD WEIGHTS, PAYLOADS/ EXPERIMENTS	MISSION HIGHLIGHTS (LAUNCH SCRUBS/DELAYS, TAL WEATHER, ASCENT I-LOADS, FIRSTS, SIGNIFICANT ANOMALIES, ETC.)
STS-92/ ISS 3A SEQ FLT # 100 KSC-100 PAD 39A-57 MLP-3 FIFTH SHUTTLE FLIGHT TO ISS	OV-103 (Flight 8 (Discovery) OMS PODS: LP01-31 RP03-29 FRC3-28	CDR: Brian Duffy (Flt 4 - STS-45, STS-57, STS-72) P591/R142/V94/M126 PLT: Pamela A. Melroy P592/R261/F34 M/S 1/EV1: Leroy Chiao (Flt 3 - STS-65, STS-72) P593/R179/V125/M157 M/S 2/EV2: William S. McArthur (Flt 3 - STS-58, STS-74) P594/R172/V124/M150 M/S 3/EV3: Peter J. K. (Jeff) Wisoff (Flt 4 - STS-57, STS-68, STS-81) P595/R166/V110/M145 M/S 4/EV4: Michael E. Lopez-Alegria (Flt 2 - STS-73) P596/R202/V163/M175 M/S 5: Koichi Wakata (Japan) (Flt 2 - STS-72) P597/R208/V164/M181 SS EVA #51 EMU/TETHERED EVA #44 SCHEDULED EVA #45 DURATION 6:28 SS EVA #52 EMU/TETHERED EVA #45 SCHEDULED EVA #46 DURATION 7:08 Continued…	KSC 39A 285:23:17:00 Z 6:17:00 PM EST 6:17:00 PM EST Wednesday 10 10/11/00 (10) LAUNCH WINDOW: 4:12 USING PLT (IN-PLANE TIME) EOM PLS: KSC TAL: ZZA TAL WX: MRN, BEN SELECTED: RTLS: KSC 33 N/N TAL: BEN 36 CI/N AOA: KSC 33 N/N PLS: EDW CI/N TDEL: 0.00 -0.04 MAX Q NAV: 752 748 SRB STG: 2:02.6 2:02 PERF: NOMINAL 2 ENG TAL (BEN): 2:25 2:27 NEG RETURN: 3:57 3:57 PTA (U/S 282): 4:40 4:41 PTM (U/S 282): 5:56 6:05 SE ZZA: 6:02 6:02 SE PTM: 6:48 6:55 WINDS: 2009P16 KTS SS: 8H 4L PK: 15H 7L MECO CMD: 8:25.3 8:25.6 Continued…	EDW 22, CONC EDW 46, CONC 27 298:20:59:42 Z 12:59:42 PM PST Tuesday 16 10/24/00 (8) DEORBIT BURN: 298:19:52:00Z XRANGE: 200 NM ORBIT DIR: AL 26 AIM PT: NOMINAL MLGTD: 2656 FT 298:20:59:42Z VEL: 205 KGS 201 KEAS HDOT: -2.9 FPS TD NORM 195: 3287 FT DRAG CHUTE DEPLOY: 188 KEAS 298:20:59:46Z NLGTD: 6504 FT 298:20:59:54Z VEL: 144 KGS 152 KEAS HDOT: -6.7 FPS BRK INIT: 67 KGS DRAG CHUTE JETTISON: 55 KGS 298:21:00:21Z BRK DECELFPS[2]: AVE 3.5 PK 5.3 WHEELS STOP: 298:21:00:49Z 11746 FT ROLLOUT: 9090 FT 67 SEC DENS ALT: 3743 FT Continued…	104/104/ 109% PREDICTED: 100/104.5/ 104.5/72/ 104.5 ACTUAL: 100/104.5/ 104.5/72/ 104.5 1 = 2045 (3) 2 = 2053 (2) 3 = 2048 (2) ALL BLOCK IIA ENGINES M 3 EOM: WEIGHT: 205188 LBS X CG: 1079.95 LANDING: WEIGHT: 205129 LBS X CG: 1081.77	BI-104 (5) RSRM 76 ET-104 SLWT 9 ET BRKUP: 283 K ET IMPACT 1:26:22 MET	51.60	DIRECT INSERTION POST OMS-2: 175.1 x 85.4 NM TI BURN: 1/14:52 MET ORBIT: 206.2 X 200.1 NM DEORBIT: APOGEE 213 NM PERIGEE 200.9 NM ENTRY RANGE: 4352 NM ENTRY VELOCITY: 25901	OI-27 (5) LAT: 2.00 S LONG: 127.7 W	CARGO: 35250 LBS PAYLOAD CHARGEABLE: 28009 LBS DEPLOYED: 21998 LBS NON-DEPLOYED: 4678 LBS MIDDECK: 1333 LBS SHUTTLE ACCUMULATED WEIGHTS: DEPLOYED: 961605 LBS NON-DEPLOYED: 1487225 LBS CARGO TOTAL: 3072373 LBS PERFORMANCE MARGINS (LBS): FPR: 3274 FUEL BIAS: 818 FINAL TDDP: 1532 RECON: 2330 PAYLOADS: PLB: ISS-3A ISS Z1 TRUSS CMG'S KU/S-BAND PMA-3/SLP ICBC30 RMS, ODS MIDDECK: DTO EMU H/W EVA TOOLS 5 CRYO TK SETS 6 GH2 TKS Continued…	***Brief Mission Summary:*** *STS-92, the 5th mission to ISS, delivered the first framework structure, Z1 truss, to house communications and motion control equipment; and delivered the third Pressurized Mating Adapter docking station. This was the 100th mission of America's Space Shuttle.* KSC W/D: OPF 197, VAB 10, PAD 21 = 238 days total. LAUNCH POSTPONEMENTS: - Baselined launch date of 7/23/98 on 3/13/97 - Postponed launch to 1/14/99 on 5/27/97. ISS Flight Delays - Postponed launch to 6/17/99 on 6/4/98. ISS Flight Delays - Postponed launch to 12/2/99 on 2/4/99. ISS Flight Delays - Postponed launch to 6/14/00, then to 10/28/99, to 9/21/00, to 10/5/00 due to ISS Service Module Delays LAUNCH SCRUBS: - Scrubbed launch on EST date of 10/5/00 at ET Tanking MMT due to Orb/ET Attach Bolt Protrusion. Launch was scheduled for 9:38:46 PM EST (280:01:38:46Z GMT date of 10/6/00). A Review of STS-106 ET 35 mm film revealed RH Orbiter/ET attach bolt protruding several inches causing concern for bolt contact with Orbiter during sep sequence with potential for a tip load and subsequent ET/Orbiter contact. Film review of additional flights and loads analyses needed to clear STS-92 launch. During recycle, POGO valve #2 did not get an open indication when valve was cycled open. Replaced POGO valve with launch date of 10/9/00. Completed film review and analyses which cleared protruding bolt concern (within pogo valve replacement time.). Technical Scrub. Reset launch for 10/9/00 EST, 10/10/00 GMT. - Scrubbed launch on EST date of 10/9/00 at ET Tanking MMT due to wind gusts greater than 42 knots holding up extension of the GO₂ Vent Arm. Ran out of time to complete work in time for launch at 8:05:17 PM EST, 284:00:05:17Z GMT date of 10/10/00 (3.5 hours work after arm extension before tanking could start at L-8.5 hour). Weather Scrub. Reset launch for 10/10/00 at 7:39:36 EST. - Scrubbed 10/10/00 launch at L- 1H07M due to a concern for debris damage by a wayward pip pin and tether seen on the LO₂ feedline foam inboard support bracket. Pip pin was discovered during ice/debris team walkdown. (Launch had been scheduled for 7:39:36 EST. Technical scrub. Reset launch for 10/11/00. LAUNCH WINDOW: - Total launch window was 7M58S. Window opened at 285:23:13:14Z and closed at 285:23:21:12Z. Selected Preferred Launch Time (PLT) of 285:23:17:00Z (in-plane time) giving a launch window of 4M12S. LAUNCH DELAYS: None - Launched on time at 285:23:17:00Z, 6:17:00 PM EST on Wednesday, October 11, 2000. Continued…

STS092-S-022 [EC00-0311-3] (24 OCTOBER 2000) --- Successful landing at EAFB of the 100th Shuttle mission – "Still young at 100", PAO.

FLT NO.	ORBITER	CREW (7) TITLE, NAMES & EVA'S	LAUNCH SITE, LIFTOFF TIME, LANDING SITES, ABORT TIMES	LANDING SITE/ RUNWAY, CROSSRANGE LANDING TIMES FLT DURATION, WINDS	SSME-TL NOM-ABORT EMERG THROTTLE PROFILE ENG. S.N.	SRB RSRM AND ET	ORBIT INC	HA/HP	FSW	PAYLOAD WEIGHTS, PAYLOADS/ EXPERIMENTS	MISSION HIGHLIGHTS (LAUNCH SCRUBS/DELAYS, TAL WEATHER, ASCENT I-LOADS, FIRSTS, SIGNIFICANT ANOMALIES, ETC.)
STS-92/ ISS 3A Continued...		Continued... SS EVA #53 EMU/TETHERED EVA #46 SCHEDULED EVA #47 DURATION 6:48 SS EVA #54 EMU/TETHERED EVA #47 SCHEDULED EVA #48 DURATION 6:56 MCC WHITE FCR (30) FLIGHT DIRECTORS: Asc - N. W. Hale Ent - L. E. Cain LD/O3 - C. W. Shaw O1 - R. E. Castle O2 - J. P. Shannon O4 - B. P. Austin ISS LD/O1 - S. P. Davis ISS O2 - M. A. Kirasich ISS Plng/O3 - R. E. LaBrode MOD - J. M. Heflin	Continued... VI: 25931 25928 OMS-2: 43.30 43.33 82.4 FPS 82.1 FPS 00:54 00:54	Continued... FLT DURATION: 12:21:42:42 S/T: 870:10:21:07 OV-103: 217:05:38:18 DISTANCE: 5,331,301 sm						Continued... RMS 57 (S.N. 301) RMS USED FOR OSVS checkout, Z1 truss grapple and install on ISS and EVA support PMA3/SLP on Z1	Continued... TAL WX: - Zaragoza (prime) forecast and observed NO GO for rain, Moron forecast and observed NO GO for violent storms, Ben Guerir (selected) Qbar 353 vs. 350 limit at 1100 feet cleared by L-10 minute balloon. NOTE: PTA set on AOA FOR KSC even though forecast showed chance of rain and chance 4000 ft broken and peak winds of 13 knots. EDW and NOR down for AOA/PLS, FD2 PLS would have resulted in additional 10 second TAL exposure. PERFORMANCE ENHANCEMENTS: - Standard Set Plus: (1) PE Operational High Q TRN/OCT, (2) OMS assist is 4000 lbs, (3) 52 nm MECO, and (4) Del Psi. - Note: OMS Assist Time reduced from 102 seconds to 41 seconds with DOLILU uplink (2400 lbs more OMS to orbit). - Inert weight adjustment is 199 lbs; was -200 lbs. SHUTTLE NIGHT LAUNCH #24 FLIGHT DURATION CHANGES: - Total Flight duration extension was 2 days plus 3 orbits. - EDW was not called up for NEOM. - Did not close PLBD's. Waved-off landing at KSC on orbits 170 and 171 due to sustained high SLF crosswinds. EOM+1. Waved-off landing at KSC on orbits 186 and 187 (Did not close PLBD's or crew in suits) due to high crosswinds. - Retargeted to EDW on orbit 187, then waved-off due to broken ceiling and showers within 30 nm. - Targeted EDW on orbit 188, closed PLBD's, and put crew in suits. Waved-off landing at EDW on orbit 188 at Tig-16 minutes due to forecast and observed showers and rain within 30 nm. Waved-off landing at EDW on orbit 189 at Tig-1 hour for showers and rain within 30 nm. NOEM+2. Activated NOR for EOM+2. Did not attempt to land at KSC on orbits 201 and 202 due to forecast and observed high crosswinds, low ceiling, and rain within 30 nm. Landed at EDW runway 22 on orbit 203 at 298:20:59:42Z, 12:59:42 PM PST, Tuesday, October 24, 2000. EVENTS: - Ring capture at 287:17:45:10Z, 1:18:28:10 MET - Docked at PMA2 Node 1 Forward Port at 287:17:57:55Z - Z1 Truss grapple at 288:15:57:14Z, Z1 release 288:19:05:30Z - EVA 1 Start at 289:14:26Z, duration 6H28M. - PMA grapple at 290:15:43:30Z, PMA release at 290:17:59:35Z - EVA 2 Start at 290:14:13Z, duration 7H08M. - ISS Reboost maneuver #1 Start at 290:21:03:00Z, 4:21:46:00 MET, Delta-V was 6 fps, 1.5 nm, 208 by 202 nm. - EVA 3 Start at 291:14:29Z, duration 6H48M. - ISS Reboost maneuver #2 Start at 291:22:45:59Z, 5:23:28:59 MET, 5.8 fps, 1.5 nm, 211 by 202 nm. - EVA 4 Start at 292:15:00Z, duration 6H56M. - ISS Reboost maneuver #3 Start at 292:22:23:32Z, 6:23:06:32 MET, 5.6 fps, 1.5 nm, 214 by 202 nm. Continued...

STS092-342-011 --- In-flight crew portrait. Front, from the left: Wisoff/MS, Wakata/MS (NASDA), CDR Duffy, & McArthur/MS. Rear, from the left: PLT Melroy, Chiao/MS, Lopez-Alegria/MS.

ГИДЗЕНКО КРИКАЛЁВ
SHEPHERD

FLT NO.	ORBITER	CREW (7) TITLE, NAMES & EVA'S	LAUNCH SITE, LIFTOFF TIME, LANDING SITES, ABORT TIMES	LANDING SITE/ RUNWAY, CROSSRANGE LANDING TIMES FLT DURATION, WINDS	SSME-TL NOM-ABORT EMERG THROTTLE PROFILE ENG. S.N.	SRB RSRM AND ET	ORBIT INC	HA/HP	FSW	PAYLOAD WEIGHTS, PAYLOADS/ EXPERIMENTS	MISSION HIGHLIGHTS (LAUNCH SCRUBS/DELAYS, TAL WEATHER, ASCENT I-LOADS, FIRSTS, SIGNIFICANT ANOMALIES, ETC.)
STS-92/ ISS 3A Continued…											Continued…

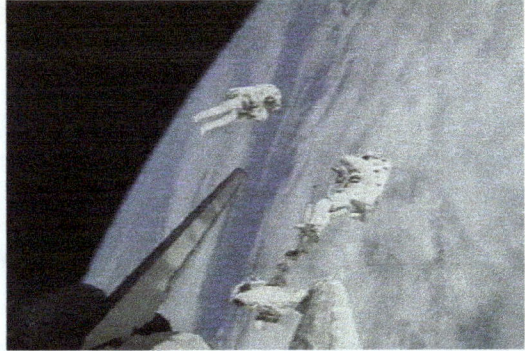

LEFT: JSC2000-E-26675 --- Astronauts Peter J.K. (Jeff) Wisoff and Michael Lopez-Alegria participate in final of four STS-92 space walks, including a run with SAFER backpack."

BELOW: JSC2000-06403 --- Wayne Hale (front center), Ascent Flight Director for the STS-92 mission, poses with the 50-odd flight controllers who supported his shift.

JSC2000-E-26636--- ISS after installation of Z1 Truss. From the top, elements are the Zvezda, the FGB or Zarya, Node 1 or Unity, and Z1.

Continued…

EVENTS: (Continued)
- Undocked at 294:15:08:21Z, 8:15:51:21 MET
- Total transfers to ISS - 21998 lbs (includes Z1=18351 and PMA3=2549 lbs).
- Delivered Z1 Truss. Mated Z1 to Node 1 zenith port. Installed CMG jumper. Z1 umbilicals connected and powered. Delivered PMA3 and berthed to Node 1 Nadir Port, umbilicals connected. SGANT deployed. Relocated IAPFR and Z1 FRGF. Installed two DDCU's and ETSD on Z1.
- STS-92/3A ISS Visitor Time 6:21:10:26.
- ISS Visitor time 6D21H10M26S

RENDEZVOUS #49:
- Rendezvous and dock with ISS at PMA2 Node 1 Forward Port

SIGNIFICANT ANOMALIES:
- Airlock Depress Valve Cap came loose from tether and was lost
- FES Primary B shutdown in Full-Up mode.
- Cabin Payload 3 Bus loss, which powered OIU 1, OSVS, ODS C/L Camera.
- EMU Middeck Battery Charger ready indication failure
- APFR/IAPFR interference with flush side-mounted WIF's
- Modular Mini Workstation anomaly
- Pistol Grip tool chatter
- Difficulty mating PMA 3 P607 to Node J609
- Ku-band lost forward link
- WSB 2 failed to cool
- ODS C/L Camera misalignment
- WSB 2 GN_2 Relief Valve high cracking P and low reseat P.
- DSC OM2 Card 22 failure
- WSB 3 Steam Vent Heater erratic

SPACE SHUTTLE MISSIONS SUMMARY

FLT NO.	ORBITER	CREW (5) TITLE, NAMES & EVA'S	LAUNCH SITE, LIFTOFF TIME, LANDING SITES, ABORT TIMES	LANDING SITE/ RUNWAY, CROSSRANGE LANDING TIMES FLT DURATION, WINDS	SSME-TL NOM-ABORT EMERG THROTTLE PROFILE ENG. S.N.	SRB RSRM AND ET	INC	ORBIT HA/HP	FSW	PAYLOAD WEIGHTS, PAYLOADS/ EXPERIMENTS	MISSION HIGHLIGHTS (LAUNCH SCRUBS/DELAYS, TAL WEATHER, ASCENT I-LOADS, FIRSTS, SIGNIFICANT ANOMALIES, ETC.)
STS-97/ ISS 4A SEQ FLT #101 KSC-101 PAD 39B-44 MLP- 1 SIXTH SHUTTLE FLIGHT TO ISS	OV-105 (Flight 15) Endeavour OMS PODS: LPO4-22 RPO1-29 FRC5-15	CDR: Brent W. Jett (Flt 3 - STS-72, STS-81) P598/R206/V132/M179 PLT: Michael J. Bloomfield (Flt 2 - STS-86) P599/R227/V165/M198 M/S 1/EV1: Joseph R. Tanner (Flt 3 - STS-66, STS-82) P600/R185/V136/M162 M/S 2: Marc Garneau (Canada) (Flt 3 - STS-41-G, STS-77) P601/R47/V128/M44 M/S 3/EV2: Carlos I. Noriega (Flt 2 - STS-84) P602/R221/V166/M193 SS EVA #55 EMU/TETHERED EVA #48 SCHEDULED EVA #49 DURATION 7:33:23 SS EVA #56 EMU/TETHERED EVA #49 SCHEDULED EVA #50 DURATION 6:37:19 SS EVA #57 EMU/TETHERED EVA #50 SCHEDULED EVA #51 DURATION 5:09:49	KSC 39B 336:03:06:01 Z 10:06:01 PM EST (P) 10:06:01 PM EST (A) Thursday 30 11/30/00 (13) LAUNCH WINDOW: 4M01S USING PLT (IN-PLANE TIME) EOM PLS: KSC TAL: ZZA TAL WX: MRN, BEN SELECTED: RTLS: KSC 33 N/N TAL: ZZA 30 SF/N AOA: KSC 33 N/N PLS: EDW 4 N/N TDEL: 0.11 -0.048/-0.01 MAX Q NAV: 758 753 SRB STG: 2:03.5 2:03.0 PERF: NOMINAL 2 ENG TAL (ZZA): 2:43 2:40 NEG RETURN: 3:51 3:54 PTA (U/S 265): 4:54 4:54 PTM (U/S 265): 5:54 5:53 SE TAL (ZZA): 5:55 5:55 SE PTM 6:55 6:58 MECO CMD: 8:24.3 8:25.9 Continued…	KSC 15 (KSC 53) 346:23:03:23Z 6:03:23 PM EST Monday 19 12/11/00 (12) DEORBIT BURN: 346:21:57:31Z XRANGE: 20 NM ORBIT DIR: AR 8 AIM PT: CLOSE IN MLGTD: 2360 FT 346:23:03:23Z VEL: 196 KGS 199 KEAS HDOT: -3.5 FPS TD NORM 195: 2783 FT NLGTD: 5839 FT 346:23:03:35Z VEL: 138 KGS 144 KEAS HDOT: -6.5 FPS DRAG CHUTE DEPLOY: 189 KEAS 346:23:03:27Z DRAG CHUTE JETTISON: 70 KGS 346:23:03:53Z BRK INIT: 88 KGS BRK DECEL FPS/S: AVE 4.6 PK 6.7 WHEELS STOP: 346:23:04:20Z 10340 FT ROLLOUT: 7980 FT 57 SEC WINDS: 6H 2L OFFICIAL: 1406P09 SS: 6H 1L PK: 9H 2L Continued…	104/104/ 109% PREDICTED: 100/104.5/ 104.5/72/ 104.5 ACTUAL: 100/104.5/ 104.5/72/ 104.5 1 = 2054 (2) 2 = 2043 (5) 3 = 2049 (4) ALL BLOCK IIA ENGINES	BI-103 RSRM 72 ET-105 SLWT 10 ET IMPACT 1:26:32 MET LAT: 1.54 S LONG: 127.4 W	51.60 (6)	DIRECT INSERTION POST OMS-2: 175.1 X 106.2 NM TI BURN: 1:14:26:43 MET ORBIT: 199.6 X 204 NM MC-4: 1:15:50:55Z ORBIT: 205.5 X 201.3 NM M 3 EOM: WEIGHT: 197829 X CG: 1085.85 LANDING: WEIGHT: 197781 LBS X CG: 1087.73	OI-27 (6)	CARGO: 42804 LBS PAYLOAD CHARGEABLE: 37486 LBS DEPLOYED: 36213 LBS NON-DEPLOYED: 719 LBS MIDDECK: 1021 LBS SHUTTLE ACCUMULATED WEIGHTS: DEPLOYED: 997818 LBS NON-DEPLOYED: 1488965 LBS CARGO TOTAL: 3115177 LBS PERFORMANCE MARGINS (LBS): FPR: 3274 FUEL BIAS: 818 FINAL TDDP: 1920 RECON: 2032 PAYLOADS: PLB: ISS-4A PV module P6 ICBC3D RMS, ODS MIDDECK: HEDS tech demo EMU H/W, EVA Tools 5 CRYO TK SETS 5 GN2 Tanks RMS 58 RMS USED FOR P6 TRUSS AND EVA SUPPORT DEORBIT: APOGEE 198 NM PERIGEE 188.5 NM ENTRY RANGE: 4338 NM ENTRY VELOCITY: 25877	***Brief Mission Summary:*** *The STS-97/4A mission, 6th mission to ISS, helped "Station spread its wings". The 17-ton P6 Integrated Truss Segment (the 1st of four such sets) was delivered and installed on ISS. With the deployment of its 240-foot solar arrays the ISS could now provide more electrical power than on any spacecraft before it. This was also the 1st Shuttle to visit an inhabited ISS.* KSC W/D: OPF 203, VAB 5, PAD 26 = 234 days total. LAUNCH POSTPONEMENTS: - Baselined launch date of 4/8/99 on 11/6/97 - Postponed launch to 8/5/99, 2/3/00, 3/23/00, 7/20/00, 12/2/00, and then 11/30/00 EST (12/1/00 GMT date). The primary cause for postponements was Service Module late delivery to ISS. LAUNCH SCRUBS: None LAUNCH WINDOW: - Total launch window was 7M45S. Window opened at 336:03:02:17Z and closed at 336:03:10:02Z. Selected Preferred Launch Time (PLT) of 336:03:06;01Z (In-plane time) resulting in a launch window of 4M01S. LAUNCH DELAYS: None - Launched on time at 336:03:06:01 GMT on December 1, 2000 (at 10:06:01 PM EST on Thursday, November 30, 2000). - Note: During the count, a loose Firex line bracket/clamp was discovered on OAA, which was rolled back to allow access and removal using a 180 foot condor crane. No impact to launch. TAL WX: - Zaragoza (prime and selected) was forecast and observed GO, Moron was forecast and observed NO GO due to low ceiling, and Ben Guerir (2-engine TAL call) was forecast and observed GO. PERFORMANCE ENHANCEMENTS: - Standard Set plus: (1) PE Operational High Q WIN/DEC, (2) OMS assist is 4000 lbs, (3) 52 NM MECO, (4) No roll to heads up, and (5) Del Psi FLIGHT DURATION CHANGES: None - Landed at KSC runway 15 on orbit 170. MLGTD at 346:23:03:23Z (10:19:57:22 MET) on Monday, December 11, 2000. SHUTTLE NIGHT LAUNCH #25 SHUTTLE NIGHT LANDING #16 - Landed on KSC runway 15 on orbit 170 at 346:23:03:23Z, 6:03:23 PM EST on Monday, December 11, 2000. Continued…

FLT NO.	ORBITER	CREW (5) TITLE, NAMES & EVA'S	LAUNCH SITE, LIFTOFF TIME, LANDING SITES, ABORT TIMES	LANDING SITE/ RUNWAY, CROSSRANGE LANDING TIMES FLT DURATION, WINDS	SSME-TL NOM-ABORT EMERG THROTTLE PROFILE ENG. S.N.	SRB RSRM AND ET	ORBIT INC	HA/HP	FSW	PAYLOAD WEIGHTS, PAYLOADS/ EXPERIMENTS	MISSION HIGHLIGHTS (LAUNCH SCRUBS/DELAYS, TAL WEATHER, ASCENT I-LOADS, FIRSTS, SIGNIFICANT ANOMALIES, ETC.)
STS-97/ ISS 4A Continued...		Continued... MCC WHITE FCR (31) FLIGHT DIRECTORS: Asc - N. W. Hale Ent - L. E. Cain LD/O1 - W. D. Reeves O2 - P. L. Engelauf PLNG - K. B. Beck ISS LD/O2 - J. M. Hanley ISS O1 - J. M. Curry ISS PLNG - P. S. Hill MOD - J. W. Bantle	Continued... VI: 25930 25928 OMS-2: 43:10.6 43:14.6 121 FPS 119 FPS	Continued... DENS ALT: 1068 FT FLT DURATION: 10:19:57:22 S/T: 881:06:18:29 OV-105: 155:05:44:02 DISTANCE: 4,476,164 sm							Continued... EVENTS: - Ring capture at 337:19:59:35Z - Docked with ISS PMA3 Node 1 Nadir Port at 337:20:11:47Z (1:17:03:59 MET) - RMS grapple of P6 Truss from PLB at 337:22:16:57Z, 1:19:19:59 MET. P6 moved to overnight park position and grapple released at 338:20:17:25Z, 2:17:11 MET. - Hatch between orbiter and PMA3 was opened at 338:00:22:01Z, 1:21:16 MET - EVA 1 Start at 338:18:34:46Z, 2:15:29:45 MET and End at 2:23:02:06 MET, duration 7:33:23. 2B Solar Array wing deployed, but had tensioning problem. - RMS used to deploy P6 Truss to Z1 Truss. P6 Truss 4B SAW deployed. - EVA 2 Start at 340:17:20:52Z, 4:14:14:51 MET and End 4:20:52:10 MET, duration 6:37:19 - EVA 3 Start at 342:16:12:13Z, 6:13:06:12 MET and End 6:18:16:01 MET, duration 5:09:49. EVA crew successfully tensioned SAW 2B. - Undocked at 344:19:13:00Z (8:16:06:59 MET) - Total Transfers from orbiter to ISS 1457 lbs, includes 773 lbs hardware and 7 CWC's with 684 lbs H_2O. Transfers from ISS to orbiter 227 lbs. - ISS Visitor time 6:23:01:13 (docking to undocking). - Delivered and mated P6 Truss to Z1. Deployed and activated 2B and 4B Solar Array wings. Deployed and activated PMV radiator, EETCS aft radiator. Relocated S-band Antenna Support assembly. ISS EPS reconfigured to power U.S. and Russian Segments. FPP assembled and tested. RENDEZVOUS #50: - Rendezvous and dock with ISS at PMA2 Node 1 Nadir Port. SIGNIFICANT ANOMALIES: - Waste water quantity sensor dropouts - Crew could not remove Cabin Temp Controller Actuator Pip Pin - APCU 1 converters shutdown and APCU 2 tripped off. - During EVA 1, EV2 reported equipment hook inadvertently opened. - EV1's WVS EMU TV not received - EV2 reported during helmet light battery charging, battery overheated (bad battery). - IPS workstation crashed, delaying execute package - CPS application on IPS crashed - Sequential Still Video processing anomaly - ICBC3D Camera stopped filming - Erratic RCS jet L5D oxidizer injector temp transducer - F5R Fuel Injector temp sensor failure - OCA/Audio malfunctions

STS097-326-031 (8 December 2000) --- The STS-97 and Expedition 1 crews pose for an historic portrait (1st Shuttle visit to inhabited ISS): Front row are (left to right) STS-97 CDR Jett, EXP 1 CDR William M. Shepherd, & STS-97 MS/Tanner. 2nd row (from the left) EXP 1 FE/Sergei K. Krikalev, STS-97 MS/Noriega, EXP 1 Soyuz CDR/Yuri P. Gidzenko, & STS-97 PLT/Bloomfield. In the rear is STS-97 MS/Garneau representing the Canadian Space Agency (CSA). Krikalev and Gidzenko represent the Russian Aviation and Space Agency.

JSC2000-E-29413 --- Flight Directors: Front row: Lead FD Bill Reeves (left), and Jeff Hanley. Back row, from the left: John Curry, Wayne Hale, LeRoy Cain, Paul Hill and Kelly Beck.

STS097-704-074 (9 December 2000) --- New ISS configuration following Endeavour undocking.

FLT NO.	ORBITER	CREW (5) — TITLE, NAMES & EVA'S	LAUNCH SITE, LIFTOFF TIME, LANDING SITES, ABORT TIMES	LANDING SITE/ RUNWAY, CROSSRANGE — LANDING TIMES, FLT DURATION, WINDS	SSME-TL NOM-ABORT EMERG — THROTTLE PROFILE ENG. S.N.	SRB RSRM AND ET	INC	ORBIT — HA/HP	FSW	PAYLOAD WEIGHTS, PAYLOADS/ EXPERIMENTS	MISSION HIGHLIGHTS (LAUNCH SCRUBS/DELAYS, TAL WEATHER, ASCENT I-LOADS, FIRSTS, SIGNIFICANT ANOMALIES, ETC.)
STS-98/ ISS 5A SEQ FLT # 102 KSC-102 PAD: 39A-58 MLP-2 SEVENTH SHUTTLE FLIGHT TO ISS	OV-104 (Flight 23) Atlantis OMS PODS: LPO3-27 RPO4-23 FRC4-23	CDR: Kenneth D. Cockrell (Flt 4 - STS 56, STS-69, STS-80) P603/R159/V121/M140 PLT: Mark L. Polansky P604/R262/M228 M/S 1/EV2: Robert L. Curbeam (Flt 2 - STS-85) P605/R225/V167/M195 M/S 2: Marsha S. Ivins (Flt 5 - STS-32, STS-46, STS-62, STS-81) P606/R108/V77/F12 M/S 3/EV1: Thomas D. Jones (Flt 4 - STS-59, STS-68, STS-80) P607/R177/V111/M155 SS EVA #58 EMU/TETHERED EVA #51 SCHEDULED EVA #52 DURATION 7:33:58 SS EVA #59 EMU/TETHERED EVA #52 SCHEDULED EVA #53 DURATION 6:50 SS EVA #60 EMU/TETHERED EVA #53 SCHEDULED EVA #54 DURATION 5:25 Continued…	KSC 39A 38:23:11:16Z 6:11:16 PM EST 6:13:02 PM EST P603/R159/V121/M140 Wednesday 11 2/7/01 (8) LAUNCH WINDOW: 4M42S USING PLT (IN-PLANE TIME) EOM PLS: KSC TAL: ZZA TAL WX: MRN, BEN SELECTED: RTLS: KSC 33 N/N TAL: ZZA 30 AOA: KSC 33 N/N PLS: EDW 22 N/N TDEL: 0.00 0.22/0.06 MAX Q NAV: 727 735 SRB STG: 2:05.6 2:06 PERF: NOMINAL 2 ENG TAL (BEN): 2:34 2:37 NEG RETURN: 3:53 3:55 PTA (U/S): 4:48 4:46 PTM: 5:50 5:46 SE ZZA: 6:02 5:58 SE PTM: 6:51 6:51 Continued…	EDW 22, CONC EDW 47, CONC 28 51:20:33:06Z 12:33:06 PM PST Tuesday 17 2/20/01 (6) DEORBIT BURN: 51:19:27:20Z XRANGE: 381 NM ORBIT DIR: AL 27 AIM PT: CLOSE IN MLGTD: 1994 FT 51:20:33:06Z VEL: 199 KGS 209 KEAS HDOT: -2.5 FPS TD NORM 195: 3540 FT NLGTD: 5635 FT 51:20:33:18Z VEL: 133 KGS 144 KEAS HDOT: -5.9 FPS DRAG CHUTE DEPLOY: 206 KEAS 51:20:33:08Z BRK INIT: 58 KGS DRAG CHUTE JETTISON: 64 KGS 51:20:33:36Z BRK DECEL FPS^2: AVE 4.7 PK 6.7 WHEELS STOP: 51:20:34:02Z 9964 FT ROLLOUT: 7970 FT 56 SEC WINDS: 20H 1L OFFICIAL: 23020P27 SS: 20H 2R PK: 27H 3R Continued…	104/104/ 109% PREDICTED: 100/104.5/ 104.5/72/ 104.5 ACTUAL: 100/104.5/ 104.5/67/ 104.5 1 = 2052 (3) 2 = 2044 (5) 3 = 2047 (5) ALL 3 BLOCK IIA ENGINES	BI-105 RSRM 77 ET-106 SLWT-11 ET IMPACT 1:26:23 MET	51.60 (7)	DIRECT INSERTION POST OMS-2: 175.1 X 110.3 NM LAT: 1.73 S LONG: 127.9 W M 3 EOM: WEIGHT: 197909 LBS X CG: 1080.06 LANDING: WEIGHT: 197854 LBS X CG: 1081.98	OI-28 (1)	CARGO: 39162 LBS PAYLOAD CHARGEABLE: 33286 LBS DEPLOYED: 32270 LBS NON-DEPLOYED: 583 LBS MIDDECK: 983 LBS SHUTTLE ACCUMULATED WEIGHTS: DEPLOYED: 1030088 LBS NON-DEPLOYED: 1490535 LBS CARGO TOTAL: 3154339 LBS PERFORMANCE MARGINS (LBS): FPR: 3274 FUEL BIAS: 818 FINAL TDDP: 2138 RECON: 1538 PAYLOADS: PLB: ISS-5A (DESTINY) U.S. LABORATORY RMS, ODS, SPDU MIDDECK: SIMPLEX BMRRM (LON) 5 CRYO TK SETS 6 GH2 TANKS RMS 59 RMS USED FOR U.S. LAB TO NODE 1, PMA-2 TO LAB, AND EVA SUPPORT	*Brief Mission Summary:* The STS-98/5A mission, 7th mission to ISS, delivered and installed the U.S. Destiny Laboratory onto the forward port of the Unity Node. Destiny is the centerpiece for research on the ISS. The lab is 28 feet long by 14 feet wide. Atlantis landed at EAFB, CA after two consecutive days of wave offs at KSC, due to high winds, then clouds and rain on the third day. KSC W/D: OPF 70, VAB 30 (2), PAD 28 (2) = 128 days total (Rollback to inspect SRB cables). LAUNCH POSTPONEMENTS: - Baseline launch date of 5/20/99 on 11/20/97 - Postponed to 10/28/99, 2/3/00, 3/2/00, 4/20/00, 8/29/00, and 1/18/01 - Postponed launch date to NET 2/6/01 when decision made to roll back to VAB and inspect/x-ray SRB cables (Replaced damaged cables). - Set 2/7/01 launch date at FRR. LAUNCH SCRUBS: None LAUNCH WINDOW: - The total launch window was 9M02S, which opened at 38:23:06:56Z and closed at 38:23:15:58Z. The decision was made to use the Preferred Launch Time (PLT) of 38:23:11:16Z (In-plane time) with a 4M42S launch window. LAUNCH DELAYS: - During T-9 hold, a step function was seen on APU 1 Turbine Speed (OA1 card 6). This proved to be a ground-processing problem; however, coming out of T-9 minute hold was 1m46s late, resulting in a launch delay of 1m46s. Launch occurred at 38:23:13:02Z, 6:13:02 PM EST on Wednesday, February 7, 2001. TAL WX: - Zaragoza (prime and selected) and Ben Guerir (2-engine TAL call) were forecast and observed GO. Moron was forecast and observed NO GO for ceiling and showers within 20 nm. PERFORMANCE ENHANCEMENTS: - Standard Set Plus: (1) PE Operational High Q WIN/JAN, (2) OMS assist is 4000 lbs, (3) 52 NM MECO, (4) Del Psi FLIGHT DURATION CHANGES: - Total extension 2 days plus two orbits and changed landing site to EDW. - EDW was not called up for NEOM. Closed PLBD's, but waved-off landing at KSC on NEOM orbits 170 (Tig-24 mins) and 171 (Tig-36 mins) due to observed and forecast crosswind violations. Activated EDW for EOM+1. Closed PLBD's for EOM+1 but waved-off landing at KSC on orbit 186 for crosswind violations and orbit 187 due to observed and forecast crosswind violations and precipitation. Waved-off landing at EDW on orbits 188 and 189 due to forecast ceiling, crosswind, and precipitation violations. EOM+2. Waved-off landing at KSC on orbits 201 and 202 due to forecast of low ceiling and precipitation. Landed at EDW runway 22 on orbit 203 at 12:33:06 PST on Tuesday, February 20, 2001. Continued…

STS098-331-0017 (7-20 February 2001) --- RMS lifts Destiny from Atlantis payload bay for installation on ISS.

DEORBIT: APOGEE 210.8 NM PERIGEE 196.2 NM

ENTRY VELOCITY: 25893

ENTRY RANGE: 4350 NM

FLT NO.	ORBITER	CREW (5) TITLE, NAMES & EVA'S	LAUNCH SITE, LIFTOFF TIME, LANDING SITES, ABORT TIMES	LANDING SITE/ RUNWAY, CROSSRANGE LANDING TIMES FLT DURATION, WINDS	SSME-TL NOM-ABORT EMERG THROTTLE PROFILE ENG. S.N.	SRB RSRM AND ET	ORBIT		FSW	PAYLOAD WEIGHTS, PAYLOADS/ EXPERIMENTS	MISSION HIGHLIGHTS (LAUNCH SCRUBS/DELAYS, TAL WEATHER, ASCENT I-LOADS, FIRSTS, SIGNIFICANT ANOMALIES, ETC.)
							INC	HA/HP			

STS-98/ ISS 5A

Continued…

Continued…

MCC WHITE FCR (32)

FLIGHT DIRECTORS:
A/E - L. E. Cain
LD/O1 - R. E. Castle
O2 - K. B. Beck
PLNG/O3 - B. P. Austin

ISS LD/O2 - A. F. Algate
ISS O1 - M. A. Kirasich
ISS O3 - M. J. Ferring
MOD - J. W. Bantle

Continued…

MECO CMD:
8:25.1   8:24.7

VI:
25928   25928

OMS-2:
43:46   43:45
127.1 FPS   127.1 FPS

Continued…

DENS ALT:
2334 FT

FLT DURATION:
12:21:20:04
S/T:
894:03:38:33
OV-102:
185:11:10:35

DISTANCE:
5,369,576 sm

STS98-E-5276 --- Group portrait of Shuttle & ISS crews on board ISS. Front, from the left: CDR Cockrell, EXP 1 CDR William M. (Bill) Shepherd, & Curbeam/MS. Rear, from the left: Sergei K. Krkalev/FE EXP 1, Ivins/MS, PLT Polansky, Yuri P. Gidzenko/ Soyuz CDR EXP 1, & Jones/MS.

Continued…

FIFTH SHUTTLE CREWMEMBER REPLACEMENT
- Mark Lee was replaced by Curbeam in February 2001. (Fourth Shuttle crewmember replacement occurred on STS-85.)

EVENTS:
- OMS assist at 2:16 MET, duration 102.2 seconds
- MC-4 at 40:15:41:20Z, 1:16:28:18 MET.
- Docked with ISS PMA3 Node 1 Nadir Port at 40:16:50:49Z, 01:17:37:47 MET
- Collision avoidance maneuver for ISS at 41:11:48:02Z, 02:12:35:00 MET Delta V +2.5 ft/sec, 186.5 by 199.4 nm
- RMS grappled PMA2 on Node 1 at 41:14:12Z, 2:14:59 MET. PMA2 installed on Z1 Truss at 41:17:00Z, 2:17:47 MET.
- U.S. Laboratory grappled in PLB at 41:17:22Z, 2:18:00 MET. U.S. Lab (Destiny) was attached to Node at 41:19:00Z, 2:19:47 MET.
- EVA 1 Start at 41:15:51Z, 2:16:36 MET. EVA duration 7H33M56S.
- First ISS Reboost maneuver Started at 42:17:13Z, 3:18:00 MET.
- Second Reboost maneuver Started at 42:18:18Z, 3:19:05 MET. Altitude increase of 3.6 nm, 203.0 by 188.9.
- EVA 2 Start at 43:15:58Z, 4:16:45 MET, duration 6H50M.
- Third Reboost maneuver Started at 44:15:53:02Z, 5:16:40:00 MET lasted 4 hours.
- Fourth Reboost Started at 44:20:06:02Z, 5:20:53:00 MET. 5 nm altitude increase, orbit 206.5 by 193.7 nm
- EVA 3 Start at 45:14:30Z, 6:15:16:58 MET, duration 5H25M.
- Fifth Reboost at 45:23:08Z, 6:23:54:58 MET, 1.4 nm altitude increase, orbit 209 by 195 nm.
- Sixth Reboost at 46:15:23Z, Delta V of 4.4 fps, orbit 209.4 by 195.5 nm.
- Seventh Reboost at 46:16:56Z, duration 3h41m, Delta V 11.9 fps, orbit 212.5 by 199.2 nm.
- Hatch closed at 47:13:22Z, 8:14:08:58 MET.
- Undocked at 47:14:06Z, 8:14:53 MET.
- Relocated PMA2 from Node 1 to fwd CBM. Delivered and installed U.S. Lab on Node 1 fwd CBM and connected umbilicals, activated U.S. Lab core systems. Activated and C/O CMG's, then handed over attitude control to U.S. GN&C system.
- ISS Visitor Time is 6:21:15:11.

TRANSFERS:
- To ISS: Dry cargo IVA 3036 lbs, U.S. Lab 29866 lbs, external EVA 368 lbs = total 33270 lbs. (Included H2O transfer to ISS: 10 CWC's = 993 lbs)
- Transfers from ISS to shuttle 872 lbs.

RENDEZVOUS #51:
- Rendezvous and dock with ISS at PMA3, Node 1 Nadir Port.

ABOVE: STS98-E-5143 --- Inside newly opened Destiny (lt to rt): Ivins/MS, CDR Cockrell & CDR EXP 1 William Shepherd

BELOW: STS098-713a-016 --- New ISS configuration as viewed from departing Atlantis.

------- CREW GREETINGS AT ELLINGTON ON RETURN HOME -------
ABOVE: PLT Polansky (left), CDR Cockrell (center), gretted by Steve Hawley/Flt Crew Ops Dir.
RIGHT: JSC Center Director George W.S. Abbey also greets crew .

IN MCC: Orbit 1 FCT in Shuttle FCR . FD Robert Castle ,near center, holds crew insignia.

SIGNIFICANT ANOMALIES:
- CDR and PLT HUD runway misalignment. PLT saw about 600 foot offset to the right of the runway, CDR was about half of this offset.
- PCA vent cover bolts did not fit 5/16-in socket. PCA vent bolts were difficult to start with power tool.
- EV2 EMU boot pressure point during EVA #1 and EVA #2.
- Broken connector bail linkage, one of rivets on connector bail broke.
- Sticky mini-workstations end effectors, occasionally stuck open.
- SASA P4 connector O-ring loose.
- Bad video for proshare video conferencing.
- STS-98 Vent Command error for Reboost 5.
- Ku-band radar Alpha gimbal angle error

FLT NO.	ORBITER	CREW (10) 7 UP/7 DOWN — TITLE, NAMES & EVA'S	LAUNCH SITE, LIFTOFF TIME, LANDING SITES, ABORT TIMES	LANDING SITE/ RUNWAY, CROSSRANGE — LANDING TIMES FLT DURATION, WINDS	SSME-TL NOM-ABORT EMERG THROTTLE PROFILE ENG. S.N.	SRB RSRM AND ET	INC	ORBIT HA/HP	FSW	PAYLOAD WEIGHTS, PAYLOADS/ EXPERIMENTS	MISSION HIGHLIGHTS (LAUNCH SCRUBS/DELAYS, TAL WEATHER, ASCENT I-LOADS, FIRSTS, SIGNIFICANT ANOMALIES, ETC.)
STS-102/ ISS 5A.1 SEQ FLT # 103 KSC-103 ISS-5A.1 PAD 39B-45 MLP-3 EIGHTH SHUTTLE FLIGHT TO ISS	OV-103 (Flight 29) Discovery OMS PODS: LPO1-32 RPO3-30 FRC3-29	CDR: James D. Wetherbee (Flt 5 - STS-32, STS-52, STS-63, STS-86) P608/R108/V80/M198 PLT: James M. Kelly P609/R263/M229 M/S 1 UP/EV3: Andrew S. W. Thomas (Flt 3 - STS-77, Up to Mir on STS-89, Down STS-91) P610/R213/V149/M186 M/S 2/EV4: Paul Richards P611/R264/M230 M/S 3 UP/EV1/EXP2 Flt Eng 1: James S. Voss (Flt 5 - STS-44, STS-53, STS-69, STS-101) P612/R136/V85/M121 M/S 4 UP/EV2/EXP2 Flt Eng 2: Susan Helms (Flt 5 - STS-54, STS-64, STS-78, STS-101) P613/R158/V108/F19 M/S 5 UP/EXP2 CDR: Yury Usachev (Russia) (Flt 2 - STS-101) P614/R256/M223 M/S 3 DN/EXP1 Flt Eng: Sergei Krikalev (Russia) (Soyuz UP, STS-102 DN) (Flt 3 - STS-60, STS-88) P615/R177/V154/M154 M/S 4 DN/EXP1 CDR: William M. Shepard (Flt 4 - STS-27, STS-41, STS-52, Soyuz TM UP to ISS, STS-102 DN) P616/R96/V56/M87 Continued...	KSC 39B 67:11:42:09Z 6:42:09 AM EST (P) 6:42:09 AM EST (A) Thursday 31 3/8/01 (7) LAUNCH WINDOW: 4:59 USING PLT (IN-PLANE TIME) EOM PLS: KSC TAL: ZZA TAL WX: MRN, BEN SELECTED: RTLS: KSC 33 CI/N TAL: BEN 36 AOA: KSC 33 CI/N PLS: EDW 22 N/N TDEL: 0.03 -0.118/-0.08 MAX Q NAV: 740 748 SRB STG: 2:05.6 2:04 PERF: NOMINAL 2 ENG TAL (BEN): 2:24 2:24 NEG RETURN: 3:51 3:55 PTA (U/S 152): 4:48 4:48 DROOP: 4:43 PTM (U/S 152): 6:02 6:01 MECO CMD: 8:21.9 8:23.1 VI: 25823 25824 Continued...	KSC 15 (KSC 54) 80:07:31:41Z 2:31:41 AM EST Wednesday 12 3/21/01 (7) DEORBIT BURN: 80:06:26:06Z XRANGE: 373 NM ORBIT DIR: AR 9 AIM PT: NOMINAL MLGTD: 2839 FT 80:07:31:41Z VEL: 199 KGS 203 KEAS HDOT: -1.0 FPS TD NORM 205: 2529 FT NLGTD: 6190 FT 80:07:31:52Z VEL: 165 KGS 159 KEAS HDOT: -6.3 FPS DRAG CHUTE DEPLOY: 153 KEAS 80:07:31:55Z BRK INIT: 98 KGS DRAG CHUTE JETTISON: 57 KGS 80:07:32:31Z BRK DECEL FPS^2: AVE 3.5 PK 5.4 WHEELS STOP: 80:07:33:06Z ROLLOUT: 11244 FT 85 SEC WINDS: 2H 9R OFFICIAL: 2309P16 KTS SS: 2H 9R PK: 4H 16R Continued...	104/104/ 109% PREDICTED: 100/104.5/ 104.5/72/ 104.5 ACTUAL: 100/104.5/ 104.5/72/ 104.5 1 = 2048 (3) 2 = 2053 (3) 3 = 2045 (4) ALL BLOCK IIA ENGINES M 3 EOM: WEIGHT: 218094 LBS X CG: 1083.19 LANDING: WEIGHT: 218031 LBS X CG: 184.92	BI-106 RSRM 78 ET-107 SLWT-12 ET RPT: 283 K ET IMPACT 1:12:24 MET LAT: 36.5 S LONG: 158.1 W	51.60 (8)	DIRECT INSERTION POST OMS-2 126/86.2 NM DEORBIT: APOGEE: 206.5 NM PERIGEE: 206 NM ENTRY VELOCITY: 25899 FPS ENTRY RANGE: 4391 NM	OI-28 (2)	CARGO: 37328 LBS PAYLOAD CHARGEABLE: 28739 LBS DEPLOYED: 9649 LBS NON-DEPLOYED: 3517 LBS MIDDECK: 472 LBS SHUTTLE ACCUMULATED WEIGHTS: DEPLOYED: 1039900 LBS NON-DEPLOYED: 1494524 LBS CARGO TOTAL: 3191667 LBS PERFORMANCE MARGINS (LBS): FPR: 3274 LBS FUEL BIAS: 818 LBS FINAL TDDP: 2847 RECON: 3031 PAYLOADS: PLB: ISS-5A.1 MPLM PMA3 Logistics GAS (2) WSVFM ICC RMS, ODS MIDDECK: NONE 5 CRYO TK SETS 6 GN2 TANKS RMS 68 RMS used for PMA3 install on lab, MPLM grapple, deploy, retrieve, and berth, and EVA Support	*Brief Mission Summary:* STS-102, 8th mission to ISS, provided the first ISS crew changeout and, the first flight of the Italian-built Multipurpose Logistics Module (MPLM) named Leonardo. Among the MPLM cargo was the first scientific rack for U.S. Lab, Destiny, delivered on STS-98. With the ISS crew changeout, three crews participated in the STS-102 mission. KSC W/D: OPF 84, VAB 8, PAD 24 = 113 days total. LAUNCH POSTPONEMENTS: - Baselined launch date of 3/16/00 on 1/28/99. - Postponed launch to 4/13/00, 6/29/00, 10/19/00, 2/15/01, then 3/8/01. (Postponements caused by replacement of 9 damaged RCS thrusters, STS-98 launch postponements, and SRB x-rays/inspections and replacement of damaged cables. LAUNCH SCRUBS: None LAUNCH WINDOW: - Launch window opened at 67:11:37:10Z and closed at 67:11:47:08Z for a total window of 9M58S. - Selected the Preferred Launch Time (In-plane time) of 67:11:42:09Z, 6:42:09 AM EST, giving a launch window of 4M59S. Note: Sunrise was 2 minutes before launch. This was a daylight launch. LAUNCH DELAYS: None - Launch occurred on time at 67:11:42:09Z, 6:42:09 AM EST on Thursday, March 8, 2001. TAL WX: - Zaragoza (prime) was forecast NO GO for crosswinds (observed GO at launch and TAL landing times), Moron was NO GO for ceiling and showers within 20 nm. Ben Guerir (2-engine TAL call) was GO and selected. PERFORMANCE ENHANCEMENTS: - Standard Set Plus: (1) PE OPS High Q WIN/MAR, (2) OMS assist is 3717 lbs, (3) 52 nm MECO, (4) Del Psi FLIGHT DURATION CHANGES: - Total flight duration extensions 1 day plus 1 orbit. - Extended 1 day for MPLM stowage exceeding planned time and 1 orbit for showers and low clouds at KSC. Plan was to land at KSC on orbit 201; however, KSC was forecast NO GO for the next 3 days. Waved-off the planned landing at KSC for orbit 201 due to weather forecast NO GO for showers and low clouds. Plan was to land at KSC on orbit 202; if not, then land at EDW on orbit 203. Minutes before Tig, the weather forecast was GO and forecast GO to land at KSC on orbit 202. (Observed crosswinds at landing time were 16 knots, a 4-knot violation.) Low ceiling at 4200 feet became scattered minutes before landing. SHUTTLE NIGHT LANDING #17: - Landed at KSC runway 15 on orbit 202 at 80:07:31:41Z, 2:31:41 AM EST Wednesday, March 21, 2001. Flight duration 12:19:49:32. Landed at KSC Orbit 101. Continued...

STS102-326-034 --- First Shuttle flight to transport EXP crews. ISS is lined up for rendezvous with Shuttle Discovery.

FLT NO.	ORBITER	CREW (10) 7 UP/7 DOWN — TITLE, NAMES & EVA'S	LAUNCH SITE, LIFTOFF TIME, LANDING SITES, ABORT TIMES	LANDING SITE/ RUNWAY, CROSSRANGE — LANDING TIMES FLT DURATION, WINDS	SSME-TL NOM-ABORT EMERG THROTTLE PROFILE ENG. S.N.	SRB RSRM AND ET	ORBIT — INC	HA/HP	FSW	PAYLOAD WEIGHTS, PAYLOADS/ EXPERIMENTS	MISSION HIGHLIGHTS (LAUNCH SCRUBS/DELAYS, TAL WEATHER, ASCENT I-LOADS, FIRSTS, SIGNIFICANT ANOMALIES, ETC.)

STS-102/ ISS 5A.1

Continued…

JSC2000-E-06202 --- At their MOCR console, Flight Directors Wayne Hale (left) and John Shannon discuss a mission detail.

Crew column:

Continued…

M/S 5 DN/EXP1 Soyuz PLT:
Yuri Gidzenko
(Russia)
(Soyuz Up, STS-102 DN)
P617/R265/M231

SS EVA #61
EMU/TETHERED
EVA #54
SCHEDULED EVA #55
DURATION 8:56

SS EVA #62
EMU/TETHERED
EVA #55
SCHEDULED EVA #56
DURATION 8:21

MCC WHITE FCR (33)

FLIGHT DIRECTORS:
A/E - N. W. Hale
LD/O1 - J. P. Shannon
O2 - P. S. Hill
PLNG/O3 - P. F. Dye
MOD - J. W. Bantle

STATION:
LD/O1 - R. E. LaBrode
O2 - S. P. Davis
PLNG/O3 - R. E. Castle

Launch Site column:

Continued…

OMS-2:
38:35   38:37
95.6 FPS   97.2 FPS
1:02   1:03.8

Landing Site column:

Continued…

DENS ALT:
264 FT

FLT DURATION:
12:19:49:32

S/T:
906:23:28:05

OV-103:
230:01:27:50

DISTANCE:
5,357,432 sm

STS102-319-028 --- STS-102, EXP 1, & EXP 2 crews in Destiny. Front (l to r): Gidzenko/RSA, Krikalev/RSA, Shepherd, Helms, Usachev/RSA & Voss. Rear (l to r): Kelly, Richards, Wetherbee & Thomas.

STS102-712-005 --- Backdropped against the blackness of space, the ISS as viewed after Shuttle separation.

STS102-312-004 --- During EVA 1 Voss (and Helms – out of frame) prepared for MPLM docking to ISS Unity Node.

Mission Highlights column:

Continued…

FIRSTS/LASTS:
- First shuttle flight transporting an Expedition crew - Expedition 2 up, Expedition 1 down. Expedition 1 Crew launched on Flight 2R, Russian Soyuz rocket from Baikonur Cosmodrome, Kazakhstan on October 31, 2000 at 2:53 AM EST (305:07:53Z). Soyuz docked with ISS on 11/2/2000 at 4:21 AM EST (307:09:21Z). Expedition 1 Crew: CDR - William Shepherd, Soyuz pilot - Yuri Gidzenko, Flight Engineer - Sergei Krikalev.
- Sheperd flight time 80:07:31:41

EVENTS:
- TI maneuver at 69:03:12:39Z, 1:15:30:30 MET, orbit 199.2 by 205.3. MC-4 at 69:04:33:21Z, 1:16:51:12 MET, orbit 199.1 by 206.1 nm.
- ISS capture at 69:06:38:26Z, 1:18:56:17 MET; Docked at PMA2 Lab Forward Port at 69:06:58:23Z; hatch opened at _____.
- EVA 1 Start at 2:17:29 MET and End at 3:02:25 MET, duration 8:56.
- PMA3 grappled, unberthed, and installed on Node 1 Port ACBM at 70:13:50Z.
- MPLM grapple at 71:03:36Z, 3:15:54 MET, and installed on Node 1 Nadir ACBM at 71:06:08Z, 3:18:46 MET.
- EVA 2 Start at 4:17:45 MET and End 5:00:06 MET, duration 6:21.
- Collision avoidance maneuver/ISS Reboost #1 at 73:12:12:09Z, 6:02:30:00 MET, duration 47M22S, orbit 200.1 by 210.8 nm, Delta V 11.8 fps.
- ISS Reboost #2 at 75:11:32:23Z, 7:23:50:14 MET, 7.2 fps, orbit 203 by 212 nm.
- ISS Reboost #3 at 76:09:17:45Z, 8:22:33:52 MET, 7.4 fps, orbit 204.5 by 213.7 nm.
- MPLM grappled at 9:20:22 MET, reberthed in orbiter, and ungrappled at 10:00:05 MET
- ODS hatch was closed at 78:02:48Z, 10:15:06 MET.
- Undocked at 78:04:31:53Z, 10:16:50 MET.
- Transfers: Shuttle to ISS: 9649 lbs cargo plus 980 lbs water in 10 CWC's. ISS to Shuttle: 1647 lbs cargo.
- Crew rotation (Expedition 1 to Expedition 2). Relocated PMA3 from Node 1 Nadir to Node 1 Port. Berthed MPLM to Node 1 Nadir. Transferred RSP's, RSR's, HRF, ISPR, etc. to ISS.
- Krikalev flew two long-duration missions to Mir.
- ISS Visitor Time is 8:21:33:30

RENDEZVOUS #52:
- Rendezvous and dock with ISS at PMA2 Lab Forward Port.

SIGNIFICANT ANOMALIES:
- Flash evaporator left topping Evaporator Duct Heater String A failure
- WCS Fan Sep Rotary Switch 2 position failure
- Freon® loop flow degradation
- EV1 burning sensation in eyes during Airlock depress
- PMA3 J603 loose O-ring EVA
- Unable to remove PMA3 P608 connector cap
- TCS failure during rendezvous termination operation
- OCAC fan failure (running slow at all speed settings)
- Right OMS Vapor Isolation Valve #2 anomaly
- C&W limits set volts pushbutton rotary switch down position not working on panel R13U

FLT NO.	ORBITER	CREW (7) — TITLE, NAMES & EVA'S	LAUNCH SITE, LIFTOFF TIME, LANDING SITES, ABORT TIMES	LANDING SITE/ RUNWAY, CROSSRANGE — LANDING TIMES FLT DURATION, WINDS	SSME-TL NOM-ABORT EMERG — THROTTLE PROFILE ENG. S.N.	SRB RSRM AND ET	INC	ORBIT HA/HP	FSW	PAYLOAD WEIGHTS, PAYLOADS/ EXPERIMENTS	MISSION HIGHLIGHTS (LAUNCH SCRUBS/DELAYS, TAL WEATHER, ASCENT I-LOADS, FIRSTS, SIGNIFICANT ANOMALIES, ETC.)
STS-100/ ISS 6A SEQ FLT # 104 KSC-104 PAD: 39A-59 ISS-6A MLP-1 NINTH SHUTTLE FLIGHT TO ISS	OV-105 (Flight 16) Endeavor OMS PODS: LPO4-23 RPO1-30 FRC5-16	CDR: Kent V. Rominger (Flt 5 - STS-73, STS-80, STS-85, STS-96) P618/R200/V131/M174 PLT: Jeffrey S. Ashby (Flt 2 - STS-93) P619/R251/V169/M218 M/S 1/EV1: Chris A. Hadfield (Flt 2 - STS-74) P620/R202/V170/M178 M/S 2: John L. Phillips CSA/Canada P621/R266/M232 M/S 3/EV2: Scott E. Parazynski (Flt 4 - STS-66, STS-86, STS-95) P622/R187/V144/M165 M/S 4: Umberto Guidoni (Flt 2 - STS-75) (ESA-Italy) P623/R212/V171/M185 M/S 5: Yuri V. Lonchokov (Russia) P624/R267/M233 Continued…	KSC PAD 39A 109:18:40:41:99Z 2:40:42 PM EDT (P) 2:40:42 PM EDT (A) Thursday 32 4/19/01 (14) LAUNCH WINDOW: 4M49S BASED ON IN-PLANE TIME (PLT) EOM PLS: KSC TAL: ZZA TAL WX: MRN, BEN SELECTED: RTLS: KSC 33/N/N TAL: MRN 20/N/N AOA: KSC 33/N/N PLS: KSC 15/N/N TDEL: 0.10 -0.018/0.02 MAX Q NAV: 725 728 SRB STG: 2:03.7 2:04 PERF: NOMINAL 2 ENG TAL (MRN): 2:33 2:33 NEG RETURN: 3:54 3:55 PTA (U/S 243): 4:47 4:46 PTM (U/S 243): 5:56 5:50 SE TAL (ZZA): 6:04 6:03 SE PTM (U/S 701): 6:53 6:53 MECO CMD: 8:24.2 8:25.4 Continued…	EDW 22, CONC EDW 48, CONC 29 121:16:10:43Z 9:10:43 AM PDT Tuesday 18 5/1/01 (10) DEORBIT BURN: 121:15:02:47Z XRANGE: 527 NM ORBIT DIR: AL 28 AIM PT: NOMINAL MLGTD: 2159 FT 121:16:10:43Z VEL: 207 KGS 195 KEAS HDOT: -3.6 FPS TD NORM 195: 2148 FT NLGTD: 5410 FT 121:16:10:53Z VEL: 157 KGS 149 KEAS HDOT: -5.2 FPS DRAG CHUTE DEPLOY: 191 KEAS 121:16:10:45Z BRK INIT: 106 KGS DRAG CHUTE JETTISON: 53 KGS 121:16:11:16Z BRK DECEL FPS^2: AVE 6.5 PK 10.6 WHEELS STOP: 121:16:11:34Z 10123 FT ROLLOUT: 7964 FT 51 SEC WINDS: 2H 3R OFFICIAL: 28006P10 SS: 5H 4R PK: 8H 7R Continued…	104/104/ 109% PREDICTED: 100/104.5/ 104.5/72/ 104.5 ACTUAL: 100/104.5/ 104.5/72/ 104.5 1 = 2054 (3) 2 = 2043 (6) 3 = 2049 (5) ALL BLOCK IIA ENGINES M 3 EOM: WEIGHT: 220693 LBS X CG: 1083.79 LANDING: WEIGHT: 220556 LBS X CG: 1085.49	BI-107 RSRM 79 ET-108 SLWT-13 BRKUP: 283 K ET IMPACT: 1:26:38 MET LAT: 1.23 S LONG: 127.14 W	51.60 (9)	DIRECT INSERTION POST OMS-2: 178.7 X 85.7 NM DEORBIT: APOGEE 219 NM PERIGEE 204 NM ENTRY VELOCITY: 25919 FPS ENTRY RANGE: 4387 NM	OI-28 (3)	CARGO: 38330 LBS PAYLOAD CHARGEABLE: 29472 LBS DEPLOYED: 6346 LBS NON-DEPLOYED: 4282 LBS MIDDECK: 781 LBS SHUTTLE ACCUMULATED WEIGHTS: DEPLOYED: 1046246 LBS NON-DEPLOYED: 1499587 LBS CARGO TOTAL: 3229997 LBS PERFORMANCE MARGINS (LBS): FPR: 3274 FUEL BIAS: 818 FINAL TDDP: 2670 RECON: 2296 PAYLOADS: PLB: ISS-6A ICBC3D MPLM SLP-06A RMS, ODS MIDDECK: DTO EMU H/W EVA Tools 5 CRYO TK SETS 7 GN2 TANKS RMS 61 RMS used to grapple, deploy, retrieve, and berth Spacelab Pallet and MPLM, and for EVA Support	***Brief Mission Summary:*** STS-100/6A, 9th mission to ISS, delivered and installed the ISS Canadarm2 robotic arm. The first job for the arm was to attach a new airlock on ISS, to be delivered on the next flight, STS-104. In addition, the second MLPM, Raffaelo, flown on this flight, transferred needed cargo to ISS and returned items from ISS to Earth. KSC W/D: OPF 82, VAB 5, PAD 23 = 110 days total. LAUNCH POSTPONEMENTS: - Baselined launch date of 12/2/99 - Postponed launch to 4/20/00, then 7/13/00, 7/27/00, 11/30/00. - Postponed launch to 4/19/01 on 2/24/00. LAUNCH SCRUBS: None LAUNCH WINDOW: - Launch window opened at 109:18:36:12Z and closed at 109:45:31Z, giving a total window of 9M29S. The Preferred Launch Time (PLT) was 109:18:40:42 (In-plane time) 2:40:42 PM EDT, giving a launch window of 4M49S. LAUNCH DELAYS: None - Launch occurred on time at 109:18:40:42Z, 2:40:42 PM EDT on Thursday, April 19, 2001. TAL WX: - Zaragoza (prime) was NO GO for head wind violations until approximately L-3 minutes when head winds dropped to 25 knots. Moron (selected early) was GO and decision made to stay with a solid Moron. Ben Guerir was NO GO for forecast and observed showers/virga. PERFORMANCE ENHANCEMENTS: - Standard Set Plus: (1) PE Operational High Q TRN/APR, (2) OMS assist is 4000 lbs, (3) 52 nm MECO, (4) Del Psi FLIGHT DURATION CHANGES: - Total ext 1 day + 2 orbits. Planned landing was on orbit 170. - Extended 1 docked day due to ISS C&C MDM (computer) problems resulting in a planned landing on orbit 185. Did not close PLBD's and waved-off landing at KSC on orbits 185 and 186 due to forecast of showers, crosswinds, and low ceiling weather violations. Similar weather violations were forecast for KSC for the next 2 days. EDW had been called up for EOM because KSC WX violations were forecast to continue through the majority of the week. Decision was made to land at EDW on orbit 187. KSC WX was observed NO GO on the two extension days. Weather observations forecast KSC was NO GO for all 3 days. EDW was GO on EOM+1. Landed on EDW runway 22 on orbit 187 at 121:16:10:43Z, 8:10:43 AM PST on May 1, 2001, 11:21:30:01 MET. Continued…

ISS02-E-5829 (21 April 2001) --- Endeavour, with MPLM Raffaello & Canadarm2 on board, approaching ISS for docking.

SPACE SHUTTLE MISSIONS SUMMARY

<table>
<tr><th>FLT
NO.</th><th>ORBITER</th><th>CREW
(7)
TITLE, NAMES
& EVA'S</th><th>LAUNCH SITE,
LIFTOFF TIME,
LANDING SITES,
ABORT TIMES</th><th>LANDING SITE/
RUNWAY,
CROSSRANGE
LANDING TIMES
FLT DURATION,
WINDS</th><th>SSME-TL
NOM-ABORT
EMERG
THROTTLE
PROFILE
ENG. S.N.</th><th>SRB
RSRM
AND
ET</th><th>ORBIT
INC</th><th>ORBIT
HA/HP</th><th>FSW</th><th>PAYLOAD
WEIGHTS,
PAYLOADS/
EXPERIMENTS</th><th>MISSION HIGHLIGHTS
(LAUNCH SCRUBS/DELAYS,
TAL WEATHER, ASCENT I-LOADS,
FIRSTS, SIGNIFICANT ANOMALIES, ETC.)</th></tr>
<tr><td>STS-100/
ISS 6A

Continued…</td><td></td><td>Continued…

SS EVA #63
EMU/TETHERED
EVA #56
SCHEDULED EVA #57
DURATION 7:09:51

SS EVA #64
EMU/TETHEREDEVA #57
SCHEDULED EVA #58
DURATION 7:39:23

MCC WHITE FCR (34)

<u>FLIGHT DIRECTORS</u>:
A/E - L. E. Cain
LD/O 1 - P. L. Engelauf
O 2 - K. B. Beck
PLNG/O3 - B. P. Austin
MOD - J. M. Heflin

<u>STATION</u>:
LD/O 1 - J. M. Curry
O 2 - M. J. Ferring
PLNG/O 3- R. E. Castle</td><td>Continued…

<u>VI</u>:
25930 25920

<u>OMS-2</u>:
43:40 43:42</td><td>Continued…

<u>DENS ALT</u>:
3925 FT

<u>FLT DURATION</u>:
11:21:30:01
<u>S/T</u>:
918:20:58:06

<u>OV-105</u>:
167:03:14:03

<u>DISTANCE</u>:
4,910,188 sm</td><td></td><td></td><td></td><td></td><td></td><td></td><td>Continued…

<u>EVENTS</u>:
- MC-4 (RCS) at 1:18:00:36 MET, orbit 199.1 by 206.1 nm
- Docked at ISS PMA2 Lab Forward Port at 111:14:10:42Z
- EVA 1 Start at 2:17:04:41 MET, duration 7:09:51
- RMS grappled the Spacelab Pallet, unberthed from orbiter, and installed on Lab Cradle Assembly at 2:16:07:18 MET
- ISS hatch opening and crew ingress into ISS at approximately 3:14:40 MET.
- MPLM in PLB at 3:19:45 MET grappled and positioned over Node 1 Nadir CBM and installed at 3:21:04 MET.
- First ISS Reboost maneuver Started at 4:01:09:54 MET, duration 59M36S, Delta V 7.41 fps, orbit 205.5 by 212.2, raised orbit 2.1 nm.
- EVA 2 Start at 4:17:53:12 MET, duration 7h39M22S
- Second ISS Reboost maneuver Started at 7:16:40:00 MET (RCS), ended at 1 hour, Delta V was 15.9 fps, orbit 210 by 206.
- RMS berthed MPLM in PLB and powered down at 8:02:43 MET. SSRMS to RMS handoff of SLP berthed at 9:02:02 MET.
- Delivered and installed SSRMS and connected cables to U.S. Lab. UHF antenna on U.S. Lab, removed starboard ECOMM antenna. Delivered and installed express racks with payloads. Replaced failed CMC MDM #1.
- Undocked at 119:17:34:04Z (Extended flight 1 docked day due to ISS C&C MDM and Node MDM problems).
- Transferred 6346 lbs cargo to ISS and 1608 lbs from ISS to Shuttle. Transferred 1380 lbs water in 14 CWC's.
- ISS Visitor time is 8:03:23:22.

<u>RENDEZVOUS #53</u>:
- Rendezvous and dock with ISS at PMA2 Lab Forward Port

<u>SIGNIFICANT ANOMALIES</u>:
- FES Feedline B Mid 2 Htr 1 failed off
- RMS End Effector Capture Switch sticky
- WSB 3 anomalous temperature response when operating on WSB 3B controller
- Humidity Separator B water carryover
- RCS Jet R5D low chamber pressure
- EV1 eye irritation during EVA 1 and EVA 2 (Disposable in-suit drink bag leaked)
- ISS Early Comm Antenna connector fell apart
- Video Signal Converter failed to release from SLP during EVA 2
- SIGI data check bad status indications
- SRB - Unburned propellant (3 percent) in RH Forward Booster Separation Motor (BSM). Conclusion is water intrusion.
- LOMS POD inboard Y-web dithering/erratic System A Heater
- In video of launch, the lower left hand OMS Pod TPS appeared to be flexing during SSME startup. Similar but smaller motion has been seen on the pods in the past.</td></tr>
</table>

STS100-341-003 --- STS-100 and EXP 2 crews in-flight portrait in Destiny. Bottom, from left: Hadfield/CSA
Guidoni/ESA
row: James S
Lonchakov/R

JSC2001-E-12120 -- Ascent Flight Director LeRoy Cain (left) discusses mission with FD Jeffrey Bantle in the MOCR.

STS100-E-5238 (22 April 2001) --- Hadfield/MS representing CSA, stands on one Canadian-built robot arm (RMS) to work with another one, called Canadarm2, for ISS.

STS100-E-5958 -- ISS, sporting a readily visible new addition in the form of the Canadarm2 robotic arm, as seen from Shuttle post separation.

FLT NO.	ORBITER	CREW (5) TITLE, NAMES & EVA'S	LAUNCH SITE, LIFTOFF TIME, LANDING SITES, ABORT TIMES	LANDING SITE/ RUNWAY, CROSSRANGE LANDING TIMES FLT DURATION, WINDS	SSME-TL NOM-ABORT EMERG THROTTLE PROFILE ENG. S.N.	SRB RSRM AND ET	INC	ORBIT HA/HP	FSW	PAYLOAD WEIGHTS, PAYLOADS/ EXPERIMENTS	MISSION HIGHLIGHTS (LAUNCH SCRUBS/DELAYS, TAL WEATHER, ASCENT I-LOADS, FIRSTS, SIGNIFICANT ANOMALIES, ETC.)
STS-104/ ISS 7A

SEQ FLT #105

KSC-105

PAD 39B-46

MLP-2

TENTH SHUTTLE FLIGHT TO ISS | OV-104 (Flight 24) Atlantis

OMS PODS: LPO3-28 RPO4-24 FRC4-24 | CDR: Steven W. Lindsey (Flt 3 - STS-87, STS-95) P625/R229/V131/M200

PLT: Charles O. Hobaugh P626/R268/M234

M/S 1/EV1: Michael L. Gernhardt (Flt 4 - STS-69, STS 83, STS-94) P627/R198/V138/M173

M/S 2: Janet L. Kavandi (Flt 3 - STS-91, STS-99) P628/R243/V158/F32

M/S 3/EV2: James F. Reilly (Flt 2 - STS-89) P629/R234/V172/M204

SS EVA #65 EMU/TETHERED EVA #58 SCHEDULED EVA #59 DURATION 5:59

SS EVA #66 EMU/TETHERED EVA #59 SCHEDULED EVA #60 DURATION 6:29:20

SS EVA #67 DOCKED EVA 1 FROM QUEST A/L #1 EMU/TETHERED EVA #60 SCHEDULED EVA #61 DURATION 4:01:30

MCC WHITE FCR (35)

FLIGHT DIRECTORS: A/E/O 2 - N. W. Hale LD/O 1 - P. S.Hill PLNG/O3 - J. P. Shannon

ISS LD/O 2 - M. A. Kirsich ISS O 1 - S. P. Davis ISS PLNG/O3 - J. M. Hanley MOD - R. E. Castle | KSC 39B 193:09:03:59Z 5:03:59 AM EDT (P) 5:03:59 AM EDT (A) Thursday 33 7/12/01 (7)

LAUNCH WINDOW: 7M57S USING PLT (IN-PLANE TIME)

EOM PLS: KSC TAL: ZZA TAL WX: MRN

SELECTED: RTLS: KSC 33 N/N 199 KEAS TAL: ZZA 30 N/SF AOA: KSC 15 N/N PLS: EDW 22 N/N

TDEL: 0.01   0.012/0.05

MAX Q NAV: 732   732

SRB STG: 2:02.1   2:02

PERF: NOMINAL

2 ENG TAL (MRN): 2:23   2:26

NEG RETURN: 3:54   3:57

PTA (U/S 159): 4:39   4:36

SE OPS 3: 5:20   NC

PTM (U/S 159): 6:02   6:02

SE TAL (ZZA): 6:03   6:06

SE PTM (U/S 755): 6:49   6:52

Continued... | KSC 15 (KSC 55) 206:03:38:55Z 11:38:55 PM EDT

Tuesday 19 7/24/01 (10)

DEORBIT BURN: 206:02:31:35Z

XRANGE: 391 NM

ORBIT DIR: AL 29

AIM PT: NOMINAL

MLGTD: 2183 FT 206:03:38:55Z VEL: 198 KGS 199 KEAS HDOT: -1.4 FPS

TD NORM 195: 2499 FT

NLGTD: 5442 FT 206:03:39:06Z VEL: 148 KGS 148 KEAS HDOT: -5.7 FPS

DRAG CHUTE DEPLOY: 191 KEAS 206:03:38:58Z

BRK INIT: 56 KGS

DRAG CHUTE JETTISON: 57 KGS 206:03:39:39Z

BRK DECEL (FPS): AVE 1.6 PK 5.1

WHEELS STOP: 206:03:40:06Z 13041 FT

ROLLOUT: 10858 FT 68 SEC

WINDS: 4H 1L OFFICIAL: 13005P07 SS: 5H 2L PK: 6H 3L

Continued... | 104/104/ 109%

PREDICTED: 100/104.5/ 104.5/72/ 104.5

ACTUAL: 100/104.5/ 104.5/72/ 104.5

1 = 2056 (1) 2 = 2051 (2) 3 = 2047 (6)

ENG 1 & 3 BLOCK IIA ENG 2 BLK II

M 3 EOM: WEIGHT: 209142 LBS X CG: 1083.81

LANDING: WEIGHT: 209097 LBS X CG: 1085.59 | BI-108

RSRM 80

ET-109

SLWT 14

ET RPT: 283 K

ET IMPACT 1:14:17 MET

LAT: 36.32 S LONG: 158.55 W | 51.60 (10) | DIRECT INSERTION

POST OMS-2: 127 X 85 NM

DEORBIT: APOGEE: 211.0 NM PERIGEE: 207.5 NM

VELOCITY: 25905 FPS

ENTRY RANGE: 4405 NM | OI-28 (4) | CARGO: 35135 LBS

PAYLOAD CHARGEABLE: 26424 LBS

DEPLOYED: 19792 LBS

NON-DEPLOYED: 6060 LBS

MIDDECK: 582 LBS

SHUTTLE ACCUMULATED WEIGHTS: DEPLOYED: 1066028 LBS NON-DEPLOYED: 1506229 LBS CARGO TOTAL: 3265132 LBS

PERFORMANCE MARGINS (LBS): FPR: 3274 FUEL BIAS: 818 FINAL TDDP: 2884 RECON: 2990

PAYLOADS: PLB: ISS-7A ISS Airlock Spacehab Double Pallet (O2 and N2 TKS) ICBC3D RMS, ODS

MIDDECK: ICBC SPT EQUIP, EMU H/W, EVA TOOLS

5 CRYO TK SETS 7 GH2 TKS RMS 62

RMS used to view A/L Installation, OSVS, and EVA Support | *Brief Mission Summary:* STS-104, 10th mission to ISS, delivered, installed, and operated the first ISS airlock, Quest – "Giving ISS a Doorway to Space". Quest provided the capability for conducting EVA's without the presence of Shuttle, for EVA's using either Russian Orlan or U.S spacesuits, and for a new pre-breathing protocol to prevent "the bends". Also, this was first mission support from Houston's ISS Flight Control Room (BFCR).

KSC W/D: OPF 82, VAB 11, PAD 21 = 114 days total.

LAUNCH POSTPONEMENTS: - Baselined launch date of 8/24/00 on 7/29/99 - Postponed launch date to 2/8/01 on 11/10/99 - Postponed launch date to 5/15/01 on 2/24/01 - Postponed launch date to 7/12/01

LAUNCH SCRUBS: NONE

LAUNCH WINDOW: - Launch window opened at 193:08:59:00Z and closed at 193:09:11:56Z in two panes with a 10 second cutout between panes, resulting in a total window of 12M56S. The Preferred Launch Time was 193:09:03:59Z (Pane 1 In-Plane Time) resulting in a launch window of 7M57S.

LAUNCH DELAYS: NONE - Launch occurred On-Time at 193:09:03:59Z (5:03:59 AM EDT) on Thursday, July 12, 2001.

TAL WX: - Zaragoza (Prime and Selected) forecast and observed GO, Moron (2-Eng TAL Call) was forecast and observed GO. Ben Guerir was not available due to security concerns (BEN was forecast and observed GO).

PERFORMANCE ENHANCEMENTS: - Standard Set Plus: PE Operational High Q SUM/JUL, 52 nm MECO, and Del Psi

SHUTTLE NIGHT LAUNCH #26

SHUTTLE NIGHT LANDING #18 - Landed on orbit 201 on KSC runway 15 at 206:03:38:55Z, 11:38:55 PM EDT on 7/24/2001.

FLIGHT DURATION CHANGES: - Total extension 2 days. One day for ISS Ops and one day for weather at KSC. - Extended Flight 1 day due to delays in completing ISS activities primarily caused by airlock leaks. - Closed PLBD's and fluid loaded crew for planned landing on orbit 186 at KSC at 11:19:32:47 MET. At Tig -10 mins, waved-off when small cluster of showers formed SW of SLF and forecast to be within 30 nm at landing. At Tig -11 mins, waved-off landing on orbit 187 at KSC with observed precipitation and low ceiling within 30 nm and forecast precipitation within 30 nm at landing time.

Continued... |

STS104-E-5178 --- STS-104 & EXP2 crews pose in new Quest airlock: Front: PLT Hobaugh. 2nd row, from left: Reilly/MS, CDR Lindsey, CDR/EXP2 Yury V. Usachev & Gernhardt/MS. In rear: Kavandi/MS, James S. Voss/EXP2/FE and Susan J. Helms EXP2/FE.

FLT NO.	ORBITER	CREW (5) TITLE, NAMES & EVA'S	LAUNCH SITE, LIFTOFF TIME, LANDING SITES, ABORT TIMES	LANDING SITE/ RUNWAY, CROSSRANGE LANDING TIMES FLT DURATION, WINDS	SSME-TL NOM-ABORT EMERG THROTTLE PROFILE ENG. S.N.	SRB RSRM AND ET	ORBIT		FSW	PAYLOAD WEIGHTS, PAYLOADS/ EXPERIMENTS	MISSION HIGHLIGHTS (LAUNCH SCRUBS/DELAYS, TAL WEATHER, ASCENT I-LOADS, FIRSTS, SIGNIFICANT ANOMALIES, ETC.)
							INC	HA/HP			

STS-104/ ISS 7A

Continued...

1st Flight Blk II SSME (P&W HPFTP) Courtesy: Dan Hausman/P&W/Rocketdyne/ KSC

JSC2001-01944 (June 2001) --- First mission from ISS MCC: Members of Orbit 2 team pose for group portrait in the ISS flight control room (BFCR) in Houston's MCC. Orbit 2 Flight Director Mark Kirasich (blue shirt) stands near front at frame center. Lisa Holmesly, lead operations planner for ISS, is standing in front of Kirasich between the two logos.

Continued...

MECO CMD:
8:23.8   8:26

VI:
25824   25823

OMS-2:
38:29   38:33
96.7 FPS   96.6 FPS

Photo at right: JSC2001-E-21323 -- During Pre-launch in MCC (lt to rt) Robert Gest /USA ; Steven Hawley, Dep. Dir. FCOD; Lee Briscoe, Ch. Eng. MOD; & Milt Heflin, Ch. Flt Director's Office.

Continued...

DENS ALT:
1346 FT

FLT DURATION:
12:18:34:56

S/T: 931:15:33:02

OV-104:
198:05:45:31

DISTANCE:
5,309,429 sm

Photo at Right: STS104-E-5237 --- Astronaut James F. Reilly participates in a bit of space history as he joins astronaut Michael L. Gernhardt (out of frame) in utilizing the new Quest airlock for the first ever space walk to egress from ISS.

SIGNIFICANT ANOMALIES:

- Water Loop 1 floodlight coldplate low temperature
- FES Feedline A heater failure
- EMU 3 battery electrolyte leakage
- EV1 right foot discomfort
- Airlock Handhold 0535 installation failure
- Non-tending retractable tether
- Proshare video conferencing anomaly
- Failed hand held microphone
- Sequential Still Video (SSV) not operating
- Ku-Band failed to detect and track Ku forward signal.
- ODS C/L Camera misalignment
- Left Vent doors 8 and 9 Open 2 sticky microswitch

Continued...

FLIGHT DURATION CHANGES:
- Second Extension Day. Called up EDW for EOM+1. Landed on first KSC opportunity on orbit 201 on runway KSC 15 at 206:03:38:55Z, 12:18:34:56 MET, 11:38:55 PM EDT (Tuesday, July 24, 2001 EDT).

FIRSTS/LASTS:
- First flight of SSME with alternate Pratt & Whitney HPFTP (S/N 2051) Block II engine.
- First operational use of SSRMS since delivery on STS-100/6A. Used to grapple Airlock and install on Node 1 Starboard Port.
- First use of exercise pre-breathe of pure oxygen to purge nitrogen from EVA crew for EVA 3 (12 minute pre-breathe).
- First use of ISS Joint Airlock for EVA (by Shuttle Crew on EVA 3).

EVENTS:
- Docked at ISS PMA2 Lab Fwd Port. ISS contact at 1:18:04:02 MET, 195:03:08:01Z; Docking complete at 1:18:19:16 MET, 195:03:23:15Z.
- ISS Hatch open (first) 1:20:24 MET, 195:05:28Z.
- Airlock grapple.
- EVA 1 started at 2:18:07 MET, 196:03:12Z; ended at 3:00:06 MET, 196:09:11Z, duration 5H59M.
- ISS Reboost 1 maneuver started at 196:01:18:06Z, 3:16:14:07 MET, Delta V=6.8 ft/sec, altitude increase 2.3 nm, altitude 206 by 201 nm.
- EVA 2 started at 199:03:05Z; ended at 199:09:34Z, duration 6H29M20S.
- ISS Reboost 2 maneuver started at 199:09:59:12Z, 6:00:55:13 MET, delta V=6.9 ft/sec, altitude increase 2.0 nm, altitude 207.8 by 203.7 nm.
- ISS Reboost 3 maneuver started at 200:07:35:04Z, 6:22:31:05 MET, delta V=14.9 ft/sec, altitude increase 4.3 nm, altitude 211.1 by 208.6 nm.
- EVA 3 started by 202:08:35Z, and ended at 202:08:37Z, duration 4H01M30S. EVA from Joint Airlock.
- Delivered and installed ISS Joint Airlock on Node 1 Stbd port using SSRMS. Delivered and installed four HPGT's (two O2 and two H2) on Airlock. End of ISS Phase 2.
- ISS Hatch close (Final) at 9:17:51 MET, 203:02:55Z.
- Undocked at 9:19:50:00 MET, 203:04:53:59 Z.
- Transfers: Shuttle to ISS: 19782 lbs cargo (includes Airlock, 13299 lbs) plus 897 lbm water in 9 CWC's. ISS to Shuttle: 626 lbs.
- ISS Visitor Time is 8:01:45:58.

RENDEZVOUS #54:
- Rendezvous and dock with PMA2 Lab Forward Port

FLT NO.	ORBITER	CREW (7 UP/7 DOWN) TITLE, NAMES & EVA'S	LAUNCH SITE, LIFTOFF TIME, LANDING SITES, ABORT TIMES	LANDING SITE/ RUNWAY, CROSSRANGE LANDING TIMES FLT DURATION, WINDS	SSME-TL NOM-ABORT EMERG THROTTLE PROFILE ENG. S.N.	SRB RSRM AND ET	INC	ORBIT HA/HP	FSW	PAYLOAD WEIGHTS, PAYLOADS/ EXPERIMENTS	MISSION HIGHLIGHTS (LAUNCH SCRUBS/DELAYS, TAL WEATHER, ASCENT I-LOADS, FIRSTS, SIGNIFICANT ANOMALIES, ETC.)
STS-105/ ISS 7A.1 SEQ FLT #106 KSC-106 PAD 39A-60 MLP-3 ELEVENTH SHUTTLE FLIGHT TO ISS	OV-103 (Flight 30) (Discovery) OMS PODS: LPO1-33 RPO3-31 FRC3-30	CDR: Scott J. Horowitz (Flt 4 - STS-75, STS-82, STS-101) P630/R210/V135/M183 PLT: Frederick W. Sturckow (Flt 2 - STS-88) P631/R247/V173/M215 M/S 1/EV2: Patrick G. Forrester P632/R269/M235 M/S 2/EV1: Daniel T. Barry (Flt 3 - STS-72, STS-96) P633/R209/V155/M182 M/S 3 UP/EXP 3 CDR: Frank L. Culbertson, Jr. (Flt 3 - STS-38, STS-51) P634/R116/V95/M105 M/S 4 UP/EXP 3 SPLT: Vladimir N. Dezhurov (Russia) (Flt 2 - STS-71) P635/R195/V174/M170 M/S 5 UP/EXP 3 Flt Eng: Mikail Tyurin (Russia) P636/R270/M236 M/S 3 DN/EXP 2 Flt Eng 1: James S. Voss (Flt 5 - STS-44, STS-53, STS-69, STS-101, STS-102 UP) P637/R136/V85/M121 Continued…	KSC 39A 222:21:10:14Z 5:10:14 PM EDT (P) 5:10:14 PM EDT (A) Friday 23 8/10/01 (7) LAUNCH WINDOW: 9M58S ISS WINDOW OPEN EOM PLS: KSC TAL: ZZA TAL WX: MRN, BEN SELECTED: RTLS: KSC 15 N/N TAL: MRN 20 N/N AOA: KSC 15 N/N PLS: EDW 22 N/N TDEL: 0.05 -0.148/-0.11 MAX Q NAV: 723 715 SRB STG: 2:02.2 2:07 PERF: NOMINAL 2 ENG TAL (BEN): 2:27 2:21 NEG RETURN: 3:55 3:58 PTA (U/S 163): 4:35 4:36 SE OPS³: 5:25 PTM (U/S 163): 6:36 6:44 Continued…	KSC 15 (KSC 56) 234:18:22:59Z 2:22:59 PM EDT Wednesday 13 8/22/01 (6) DEORBIT BURN: 234:17:15:23Z XRANGE: 793 NM ORBIT DIR: AR 10 AIM PT: NOMINAL MLGTD: 1508 FT 234:18:22:59Z VEL: 210 KGS 202 KEAS HDOT: -3.2 FPS TD NORM 195: 2256 FT NLGTD: 4971 FT 234:18:23:10Z VEL: 157 KGS 149 KEAS HDOT: -6.9 FPS DRAG CHUTE DEPLOY: KEAS 234:18:23:01Z BRK INIT: 78 KGS DRAG CHUTE JETTISON: 56 KGS 234:18:23:43Z BRK DECEL FPS²: AVE 3.8 PK 4.9 WHEELS STOP: 234:18:24:05Z 11544 FT ROLLOUT: 10036 FT 66 SEC Continued…	104/104/ 109% PREDICTED: 100/104.5/ 104.5/72/ 104.5 ACTUAL: 100/104.5/ 104.5/72/ 104.5 1 = 2052 (4) 2 = 2044 (6) 3 = 2045 (6) ALL BLOCK IIA SSME'S	BI-109 (11) RSRM 81 ET-110 SLWT 15	51.60 (11)	DIRECT INSERTION POST OMS-2: 125.9 X 84.8 NM	OI-28 (5)	CARGO: 33107 LBS PAYLOAD CHARGEABLE: 29305 LBS DEPLOYED: 9657 LBS NON-DEPLOYED: 4654 LBS MIDDECK: 475 LBS SHUTTLE ACCUMULATED WEIGHTS: DEPLOYED: 1075685 LBS NON-DEPLOYED: 1511356 LBS CARGO TOTAL: 3298239 LBS PERFORMANCE MARGINS (LBS): FPR: 3065 FUEL BIAS: 937 FINAL TDDP: 705 RECON: 631 PAYLOADS: PLB: ISS-7A.1 (MPLM, ICC crew rotation) Heat, GAS (2) RMS, ODS MIDDECK: None 5 CRYO TK SETS 6 GN2 Tanks RMS 63 RMS used to install MPLM on Node 1 and berth in PLB, to install EAS on P6 truss, and EVA Support	**_Brief Mission Summary:_** _The STS-105/7A. 1 (11th ISS mission) provided a new crew to the ISS, transfer of supplies and equipment via the second flight of the Leonardo MPLM. This flight completed the first round trip for Expedition rotation crews (EXP 2)._ KSC W/D: OPF 79, VAB 8, PAD 31 = 118 days total. LAUNCH POSTPONEMENTS: - Baselined launch date of 6/21/01 on 6/22/00 - Postponed launch date to 7/12/01 - Postponed launch date to NET 8/5/01 on 6/7/01 - Postponed launch date to NET 8/9/01 on 7/11/01 LAUNCH SCRUBS: - Scrubbed the 8/9/01 launch attempt. The launch window was in two planes; however, at the L-2 day MMT, it was decided not to use Plane 2 for the first launch attempt on Thursday, August 9, 2001. Window opened at 221:21:32:47Z and closed at 221:21:42:46Z or 9M59S total window. With a Preferred Launch Time (PLT) of 221:21:37:46Z, the launch window was 5M00S. Launch attempt was scrubbed at L-25 minutes due to thunderstorms within 20nm, lightening strikes at 12 nm, and detached anvils over the Pad and SLF. All three TAL sites were GO. Weather Scrub. Launch set for Friday, August 10. LAUNCH WINDOW: - Launch window opened at 222:21:10:14Z and closed at 222:21:20:12Z, giving a total launch window of 9M58S. The PLT (Preferred Launch Time) of 222:21:15:13Z (In Plane Time) was selected, which gave a planned window of 4M59S. During the late count, thunderstorms were moving toward the launch site from the Southwest and forecast to be within 30 nm of the Pad and SLF at launch time. At L-27 minutes, the Ops Manager made the decision to increase the probability of launching by moving the Launch Time to the opening of the launch window (222:21:10:14Z), giving the ultimate launch window of 9M58S. Weather was observed GO at RTLS landing time for PLT and Window Open Time. LAUNCH DELAYS: NONE - Launch occurred On-Time at 222:21:10:14Z, Friday, August 10, 2001 at 5:10:14 PM EDT. TAL WX: - All three TAL sites were forecast and observed GO (Zaragoza (prime), Moron, and Ben Guerir). Moron was selected because it had the best weather (ZZA had potential for winds and rain). PERFORMANCE ENHANCEMENTS: - Standard Set plus PE Operational High Q SUM/AUG, 52 nm MECO, and Del Psi. FIRSTS/LASTS: - First Shuttle round trip with Expedition rotation crews (Expedition 3 crew up, Expedition 2 crew down). RENDEZVOUS #55: Rendezvous and dock with ISS-PMA 2 Lab Forward Port Continued…

(extra landing data rows)

M 3 EOM: WEIGHT: 220682 LBS X CG: 1083.96 LANDING: WEIGHT: 222620 LBS X CG: 1085.62

ET RPT: 283K ET IMPACT 1:14:21 MET LAT: 36.7 S LONG: 157.75 W

DEORBIT: APOGEE 218.8 NM PERIGEE 199.2 NM ENTRY VELOCITY: 25909 FPS ENTRY RANGE: 4286 NM

STS105-E-5067 (12 August 2001) --- Close-up view of Shuttle/ISS docking.

FLT NO.	ORBITER	CREW (7 UP/7 DOWN) / TITLE, NAMES & EVA'S	LAUNCH SITE, LIFTOFF TIME, / LANDING SITES, ABORT TIMES	LANDING SITE/ RUNWAY, CROSSRANGE / LANDING TIMES FLT DURATION, WINDS	SSME-TL NOM-ABORT EMERG / THROTTLE PROFILE ENG. S.N.	SRB RSRM AND ET	ORBIT / INC	HA/HP	FSW	PAYLOAD WEIGHTS, / PAYLOADS/ EXPERIMENTS	MISSION HIGHLIGHTS (LAUNCH SCRUBS/DELAYS, TAL WEATHER, ASCENT I-LOADS, FIRSTS, SIGNIFICANT ANOMALIES, ETC.)
STS-105/ ISS 7A.1 Continued…		Continued…	Continued…	Continued…						Continued…

Crew (column 3):

Continued…

M/S 4 DN/EXP 2 Flt Eng 2:
Susan J. Helms
(Flt 5 - STS-54, STS-64, STS-78, STS-101, STS-102 UP)
P638/R158/V108/F19

M/S 5 DN/EXP 2 CDR:
Yuri V. Usachev
(Flt 2 - STS-101)
(Russia)
(STS-102 UP)
P639/R256/V168/M223

SS EVA #68
EMU/TETHERED
EVA #61
SCHEDULED EVA #62
DURATION 6:16

SS EVA #69
EMU/TETHERED
EVA #62
SCHEDULED EVA #63
DURATION 5:29

MCC WHITE FCR (36)

FLIGHT DIRECTORS:
A/E/ O1 - J. P. Shannon
LD/O1 - P. F. Dye
O 2 - K. B. Beck
PLNG/O3 - B. P. Austin

ISS LD/O1 - M. J. Ferring
ISS O2 - R. E. La Brode
ISS P/O3 - J. M. Curry
MOD - N. W. Hale

Launch Site/Abort Times (column 4):

Continued…

SE TAL (ZZA):
6:04   5:59

MECO CMD:
8:24.4   8:27

OMS-2:
38:34   38:34
96.4   96.2

Landing Site/Flt Duration/Winds (column 5):

Continued…

WINDS:
3T 6L
OFFICIAL:
04007P11
SS: 6L 3T
PK: 10L 4T

DENS ALT:
1816 FT

FLT DURATION:
11:21:12:45

S/T:
943:12:45:47

OV-103:
241:22:40:35

DISTANCE:
4,912,390 sm

Center image caption:

STS105-E-5326 (17 August 2001) --- The STS-105 mission involved three crews, shown in U.S. Lab. EXP 3 crew (white shirts) front to back, Culbertson/RSA, Dezhurov/RSA, & Tyurin/RSA; STS-105 crew (stripped shirts) front row, Forrester & Barry, and back row, Horowitz and Sturckow. EXP 2 crew (red shirts) front to back, Usachev/RSA, Voss, & Helms.

SIGNIFICANT ANOMALIES:
- Loss of AC2 phase A during MPM stow
- Zero-G connector loose O-rings
- Safety tether hook lock guard inadvertently released on EV2's safety tether
- GPS ADL-CC-15 anomaly (MAGR tracking difficulty)
- Ku-Band Power Output low
- OPS Recorder 1 degraded tracks
- Nose Wheel Steering switch anomaly
- Left OMS Crossfeed low point drain line heater failure
- TCS power supply under-voltage annunciations

STS105-E-5265 --- Barry (left) and Forrester surround Early Ammonia Servicer (EAS), to be installed on P6 during EVA 1.

Mission Highlights (column 11):

Continued…

FLIGHT DURATION CHANGES:
- Total changes-one orbit weather extension. NEOM was to land at KSC on orbit 186 at approximately 12:46 PM EDT. EDW was not called up. At Tig-25 minutes, waved-off landing due to observed and forecast thunderstorms and rain showers within 20 nm of SLF. STA reported there was not-a-cloud-in-the-sky over Florida except for the rain cell that persisted at 1 or 2 miles south of the SLF, which caused the wave-off. Landed at KSC 15 on orbit 187 at 234:18:22:59Z, 2:2:59 PM EDT, on Wednesday, August 22, 2001.

EVENTS:
- ISS capture was at 1:21:31:27 MET, 224:18:41:41Z.
- ISS hard dock at PMA2 Lab Forward Port at 1:21:53:39Z, 224:19:03:53Z.
- First ISS hatch opening at 1:23:30 MET, 224:20:41:14Z.
- RMS grapple of the MPLM at 2:15:41:46 MET, 225:12:52:00Z.
- MPLM installed on Node 1 at 2:18:35:37 MET, 225:15:45:51Z.
- IELK time and Command Handover Time (ISS transfer from Exp 2 crew to Exp 3 crew and Cmd from Usachev to Culbertson) at 225:19:15Z.
- Exp 2 habitant time (Usachev=156:08:35, Voss=154:14:17, Helms=152:10:34). OV-105 crew ISS Visitor Time=7:19:47:44.
- EVA 1 Start time 228:13:58:14Z, 5:16:48:00, duration 6H16M.
- EAS installed on P6 Truss and Pip Pin in at 228:15:40:02Z, 5:18:29:47 MET.
- First Reboost maneuver started at 226:17:56:26Z, 3:20:48:12 MET, delta V 6.0 ft/sec, altitude increase 1.7 nm, orbit 218 by 208 nm.
- Second Reboost maneuver started at 229:12:12:27Z, 6:15:02:13 MET, delta V 6.4 ft/sec, altitude increase 1.8 nm, orbit 218.8 by 209.5 nm.
- EVA 2 started at 230:14:32Z, 7:16:32 MET, and ended at 230:20:01Z, duration 5M29S.
- SimpleSat deployed from Gas Can at 232:18:29:14Z, 9:21:19:00 MET.
- Total transferred to ISS 10651 lbs; 9657 lbs cargo (MPLM 6314, ICC 1549, MD 1794, H2O 10 CWC's with 993.8 lbs). Total transferred from ISS 3802 lbs (MPLM 2564, ICC 0, MD 1238). Net transfer from Shuttle to ISS=6849 lbs.
- Crew rotation, Exp 3 up and Exp 2 down. Delivered and installed EAS on P6 Truss and attached cables. Clamped MISSE to ISS Airlock handrails. Installed 11 handrails on U.S. Lab.
- Undocked at 232:14:51:37Z.
- ISS Visitor Time is 7:19:47:44. Exp 2 Crew ISS Flight Time 167:06:40:50 (New U.S. record). Exp 2 Crew ISS Habitant Times: Usachev 156:08:35:00 (ISS record), Voss 154:14:17:00, Helms 152:10:34:00 (Times based on Exp 2 to Exp 3 IELK transfer times).

FLT NO.	ORBITER	CREW 7 UP/7DOWN / TITLE, NAMES & EVA'S	LAUNCH SITE, LIFTOFF TIME, LANDING SITES, ABORT TIMES	LANDING SITE/ RUNWAY, CROSSRANGE / LANDING TIMES FLT DURATION, WINDS	SSME-TL NOM-ABORT EMERG THROTTLE PROFILE ENG. S.N.	SRB RSRM AND ET	ORBIT INC	ORBIT HA/HP	FSW	PAYLOAD WEIGHTS, PAYLOADS/ EXPERIMENTS	MISSION HIGHLIGHTS (LAUNCH SCRUBS/DELAYS, TAL WEATHER, ASCENT I-LOADS, FIRSTS, SIGNIFICANT ANOMALIES, ETC.)
STS-108/ ISS UF-1 SEQ FLT #107 KSC-107 PAD: 39B-47 MLP-1 TWELFTH SHUTTLE FLIGHT TO ISS	OV-105 (Flight 17) Endeavor OMS PODS: LPO4-24 RPO1-31 FRC5-17	CDR: Dominic L. Gorie (Flt 3 - STS-91, STS-99) P640/R242/V157/M211 PLT: Mark E. Kelly P641/R271/M237 M/S 1: Linda M. Godwin (Flt 4 - STS-37, STS-59, STS-76) P642/R122/V105/F13 M/S 2: Daniel M. Tani P643/R272/M238 M/S 3/EXP 4 Flt Eng: Carl E. Walz (Flt 4 - STS-51, STS-65, STS-79) P644/R170/V106/M148 M/S 4 UP/EXP 4 Flt Eng: Daniel W. Bursch (Flt 4 - STS-51, STS-68, STS-77) P645/R169/V109/M147 M/S 5 UP/EXP 4 CDR: Yuri I. Onufrienko (Russia) P646/R273/M239 M/S 3 DN/EXP 3 CDR: Frank L. Culbertson, Jr. (Flt 3 - STS-38, STS-51, STS-105 UP) P647/R116/V95/M105 M/S 4 DN/EXP 3 SPLT: Vladimir N. Dezhurov (Russia) (Flt 2 - STS-71, STS-105 UP) P648/R195/V174/M170 Continued...	KSC 39B 339:22:19:28Z 5:19:28 PM EST (P) 5:19:28 PM EST (A) Wednesday 12 12/5/01 (7) LAUNCH WINDOW: 7M34S USING PLT (IN-PLANE TIME) EOM PLS: KSC TAL: ZZA TAL WX: MRN, BEN SELECTED: RTLS: KSC 33/N/N TAL: ZZA 30/N/N AOA: NOR 17/N/SF PLS: EDW 22/N/N TDEL: 0.03 -0.1568 MAX Q NAV: 714 708 SRB STG: 2:05 2:04 PERF: NOMINAL 2 ENG TAL (MRN): 2:19 2:26 NEG RETURN: 3:48 3:53 PTA (U/S 154): 4:51 4:58 SE TAL (ZZA 104): 6:03 6:06 PTM (U/S 154): 6:20 6:20 SE PTM (U/S 736): 6:52 6:57 MECO CMD: 8:23.8 8:25.7 Continued...	KSC 15 (KSC 57) 351:17:55:12Z 11:55:12 AM EST Monday 20 12/17/01 (13) DEORBIT BURN: 351:16:48:13Z XRANGE: 26 NM ORBIT DIR: AR 11 AIM PT: NOMINAL MLGTD: 3024 FT 351:17:55:12Z VEL: 198 KGS 201 KEAS HDOT: -1.6 FPS TD NORM 205: 2734 FT NLGTD: 6901 FT 351:17:55:24Z VEL: 143 KGS 146 KEAS HDOT: -6.3 FPS DRAG CHUTE DEPLOY: 191 KEAS 351:17:55:16Z BRK INIT: 92 KGS DRAG CHUTE JETTISON: 57 KGS 351:17:56:18Z BRK DECEL FPS[2]: AVE 4.2 PK 6.9 WHEELS STOP: 351:17:56:18Z 11965 FT ROLLOUT: 8941 FT 66 SEC WINDS: 6H, 2L OFFICIAL: 14006P13 SS: 6H, 2L PK: 13H, 2L Continued...	104/104/ 109% PREDICTED: 100/104.5/ 104.5/72/ 104.5 ACTUAL: 100/104.5/ 93/72/ 104.5 1 = 2049 (6) 2 = 2043 (7) 3 = 2050 (2) ENGINE 2050 IS BLOCK II ENGINE. OTHER TWO BLOCK IIA ENGINES M 3 EOM: WEIGHT: 220623 LBS X CG: 1083.79 LANDING: WEIGHT: 220556 LBS X CG: 1085.49	BI-110 RSRM 82 ET-111 SLWT 16 ET IMPACT 1:14:20 MET LAT: 36.3 S LONG: ET	51.60 (12)	DIRECT INSERTION POST OMS-2: 124.2 X 121.6 NM DEORBIT: 204 X 191 NM VELOCITY: 25888 FPS ENTRY RANGE: 4416 NM	OI-28 (6)	CARGO: 38177 LBS PAYLOAD CHARGEABLE: 31393 LBS DEPLOYED: 6454 LBS NON-DEPLOYED: 8635 LBS MIDDECK: 690 LBS SHUTTLE ACCUMULATED WEIGHTS: DEPLOYED: 1082139 LBS NON-DEPLOYED: 1520683 LBS CARGO TOTAL: 3336416 LBS PERFORMANCE MARGINS (LBS): FPR: 3065 FUEL BIAS: 937 FINAL TDDP: 2881 RECON: 1182 PAYLOADS: PLB: ISS UF-1 (MPLM, LMC) MACH-1, SEM (1), GAS (5), RMS, ODS, Crew Transfer MIDDECK: ADF CBTM SIMPLEX ISS UF-1 5 CRYO TK SETS 6 GN₂ TANKS RMS 64 RMS used for ISS MPLM deploy and retrieve and EVA support	*Brief Mission Summary:* The STS-108/UF 1 (12th ISS mission) provided a new crew to the ISS, transfer of supplies and equipment via the Raffaello MPLM, and an EVA to install thermal blankets at the bases of the solar panels. Launch was scrubbed twice; first due to debris in ISS docking port from Progress 6 soft dock, and second due to RTLS and Range weather. KSC W/D: OPF 142, VAB 6, PAD 34 = 182 days total. LAUNCH POSTPONEMENTS: - Baselined launch date of 10/4/01 on 9/21/00 - Postponed launch date to NET 11/1/01 - Postponed launch date to 11/29/01 LAUNCH SCRUBS: - Scrubbed Thursday 11/29/01 EDT (11/30/01 GMT) Launch at ET Tanking MMT at L-9.5 Hours due to an ISS problem. Progress 6 had Soft Docked with SM Aft Port; however, did not achieve Hard Dock. Suspect debris within the docking interface. U.S. ISS Mgmt wanted to work problem and it was decided to go into a 24-hour scrub turnaround, then 48-hr scrub turnaround. Initially IP Russia was GO. U.S. ISS management wanted to scrub to work problem. Then IP Russia announced at ISS MMT on 11/30/01 that they planned an EVA on 12/3/01 to clear debris in docking mechanism. SSP MMT on 11/30/01 set launch for 12/4/01 to allow review of results of EVA. IP Russia EVA crew removed damaged material from previous Progress enabling Progress 6 to Hard Dock. ISS Technical Scrub (new category of scrub). - Scrubbed Tuesday 12/4/01 launch due to RTLS and Range weather (light precipitation and low ceiling). Low clouds moved into launch area from the Northeast bringing dynamic weather conditions particularly in last hour before launch. RTLS runway selection alternated between 33 and 15. Light rain was reported only by the STA as it was not visible on radar or by SLF Observer. Counted down to T-5 minutes and held while evaluating the observed and forecast weather. Scrubbed at 338:22:44:43Z (Preferred Launch Time was 22:45:08Z) while holding at T-5 minutes based on STA observations of precipitation and cloud cover and a late update SMG forecast of broken clouds over SLF runway. RTLS and Range WX Scrub. Went into a 24 hour scrub turnaround. All 3 TAL sites were GO. LAUNCH WINDOW: - Window opened at 339:22:15:35Z and closed at 339:22:27:02Z giving a total window of 11:37 in two panes with a 19-second gap between panes. Preferred Launch Time (PLT) in-plane time for pane 1 was 339:22:19:28Z giving a window of 7M34S. LAUNCH DELAYS: None - Launch occurred On-Time at 339:22:19:27.951Z, 5:19:28 PM EST, on Wednesday, 12/5/01. TAL WX: - All three TAL sites (ZZA, MRN, and BEN) were GO. Zaragoza was prime but it was a low energy day there, so Moron was selected. - MRN was 2-Eng TAL Call PERFORMANCE ENHANCEMENTS: - Standard Set plus PE Operational High Q, OMS Assist is 4000 lbs, 52 nm MECO, and Del Psi. Continued...

ISS003-E-8272 (7 December 2001) --- Endeavour approaches ISS with ISS P/L ISS UF-1 (MPLM, LMC)

FLT NO.	ORBITER	CREW 7 UP/7 DOWN / TITLE, NAMES & EVA'S	LAUNCH SITE, LIFTOFF TIME, / LANDING SITES, ABORT TIMES	LANDING SITE/ RUNWAY, CROSSRANGE / LANDING TIMES, FLT DURATION, WINDS	SSME-TL NOM-ABORT EMERG / THROTTLE PROFILE ENG. S.N.	SRB RSRM AND ET	INC	ORBIT HA/HP	FSW	PAYLOAD WEIGHTS, / PAYLOADS/ EXPERIMENTS	MISSION HIGHLIGHTS (LAUNCH SCRUBS/DELAYS, / TAL WEATHER, ASCENT I-LOADS, FIRSTS, SIGNIFICANT ANOMALIES, ETC.)

| STS-108/ ISS UF-1

Continued... | | Continued...

M/S 5 DN/EXP 3 Flt Eng:
Mikail Tyurin
(Russia)
(STS-105 UP)
P649/R270/M236

SS EVA #70
EMU/TETHERED
EVA #63
SCHEDULED EVA #64
DURATION 4:11

MCC WHITE FCR (37)

FLIGHT DIRECTORS:
A/E - L. E. Cain
Shuttle LD/O 1 - N. W. Hale
Shuttle O 2 - P. S. Hill
Shuttle Plng - C. A. Koerner

ISS LD/O 1 - S. P. Davis
ISS O 2 - R. E. Castle
ISS PLNG - J. A. McCullough
MOD - J. M. Heflin | Continued...

VI:
25822   25823

OMS-2:
37:42   37:47
164 FPS   164 FPS
1:48   1:48 | Continued...

DENS ALT:
1607 FT

FLT DURATION:
11:19:35:44

S/T: 955:08:21:31

OV-105
178:22:49:47

DISTANCE:
4,817,649 sm | | | | | | | |

SIGNIFICANT ANOMALIES:
- GSE Gaseous Hydrogen (GH2) Vent Arm did not latch-back and the GUCP rebounded beyond FSS. GH2 Vent Arm contacted side of support structure (Constraint to next flight)
- RCS Thruster R4U Failed-Off and was auto deselected
- RCS Thruster F3F Failed-Off and was auto deselected
- Loud white noise was heard on A/G 2 after SSOR 1 was tied to Orbiter Audio Bus
- IMU 2 Platform fail and redundant rate BITE
- Left RCS Oxidizer B Regulator Low Flow-Pressure
- FES Secondary Hi-Load Not Controlling
- Tear or hole on drag chute main canopy during dis-reef, 5 ribbons torn and 2 stretched
- Failed Ties Between Sabot and Pilot Chute Bag

ABOVE:STS108-328-007 (16 December 2001) --- A small satellite called STARSHINE 2 is deployed for 30,000 students studying density of Earth's upper atmosphere

BELOW: STS108-E-5359 (10 December 2001) --- Godwin & Tani install insulation blankets on ISS solar array rotation mechanisms.

STS108-E-5390 --- Crews: Exp 4 (green shirts), STS-108 (blue shirts), & Exp 3 (white shirts) in ISS Destiny Lab. Exp 4 from front to back, CDR Onufrienko, Bursch/FE, & Walz/FE. STS-108 back row, Godwin/MS, PLT Kelly, CDR Gorie, & Tani/MS. Exp 3 crew from front to back, CDR Culbertson, Dezhurov/FE & Tyurin/FE.

Continued...

FLIGHT DURATION CHANGES:
- Extended flight one docked day to allow time for additional ISS tasks. Initially planned (before extension) to land at KSC on orbit 170. After one day extension, planned landing at KSC on orbit 186. Endeavour landed at KSC on Runway 15 on orbit 186 at 351:17:55:11Z, 122:55:11 PM EST on Monday, December 17, 2002.

FLIGHT DURATION CHANGES:
- SMG Weather forecast for KSC on Tig orbit 185/Landing orbit 186 was forecast NO GO due to ceiling (3000 broken and 6500 broken). However, STA was reporting an observed GO and several positive factors provided the FD confidence to give a GO for landing on orbit 186. A Flight Rule waiver was approved post flight.

FIRSTS/LASTS:
- First flight of Block II SSME (S/N 2050) in position 3.

EVENTS:
- MC4 maneuver at 341:16:52Z, 01:18:32 MET, orbit 195.8 by 209.7 nm
- ODS captured ISS at 341:20:03:25Z, 1:21:43:58 MET
- MPLM grappled by RMS at 342:16:14Z, 2:17:54 MET, unberthed at MPLM at 342:17:00Z, 2:18:40 MET and installed on NODE, RMS ungrappled MPLM at 342:18:09:20Z, 2:19:49 MET.
- Reboost #1 Start at 343:15:11:40Z, 3:16:52:12 MET, Delta V = 6.3 FPS, altitude increase 1.9 nm, resulting orbit 210.6 by 199.0 nm.
- EVA 1 Start at 344:19:34Z, 4:21:14 MET, duration of 4 hours 11 minutes. Installed MLI blankets on Beta Gimbal Assembly on solar arrays 4B and 2B. Removed SASA blanket and pre-positioned Circuit Interrupt Devices (CID's).
- Reboost #2 Start at 345:16:19:40Z, 5:18:00:12 MET, Delta V = 6.5 FPS, altitude increase 1.8 nm, resulting orbit 211.3 by 201.2 nm.
- Reboost #3 Start at 346:15:22:32Z, 6:17:03:04 MET, Delta V = 14.1 FPS, altitude increase of 4.0 nm, resulting orbit 213.4 by 206.9 nm.
- Reboost #4 was performed for collision avoidance. Started at 349:14:55:40Z, 9:16:36:13 MET, Delta V = 2.1 FPS, altitude increase of 0.6 nm, resulting orbit 213.8 by 206.3 nm.
- Undocking: 349:17:28:35Z, 9:19:08 MET
- ISS Separation burn at 349:17:28:35Z, 9:19:09:08 MET
- Total water transferred to ISS was 299 lbm (210.3 lbm in 3 CWC's plus 88.7 lbm in 4 PWR's).
- Total transfers from Shuttle to ISS was 6244 lbs (from MPLM 5249 lbs and Middeck 995 lbs), total transfer from ISS was 4156 lbs (in MPLM 3007 lbs and to Middeck 1149 lbs).
- Endeavour/ISS Visitor Time is 7:21:25:11.
- Expedition 4 Crew Up, Expedition 3 Crew Down.
- Expedition 3 Crew ISS Habitant Time - 117:02:57:00.
- Expedition 3 Crew Flight Time - 128:20:44:58
- Culbertson Total Flight Time - 143:14:50:31
- Official transfer time from Expedition 3 to Expedition 4 crew was 342:22:12:00Z.

RENDEZVOUS #56:
- Rendezvous and dock with ISS to PMA2 Lab Fwd Port. Expedition 4 Crew Up, Expedition 3 Crew Down.

FLT NO.	ORBITER	CREW (7) TITLE, NAMES & EVA'S	LAUNCH SITE, LIFTOFF TIME, LANDING SITES, ABORT TIMES	LANDING SITE/ RUNWAY, CROSSRANGE LANDING TIMES FLT DURATION, WINDS	SSME-TL NOM-ABORT EMERG THROTTLE PROFILE ENG. S.N.	SRB RSRM AND ET	INC	ORBIT HA/HP	FSW	PAYLOAD WEIGHTS, PAYLOADS/ EXPERIMENTS	MISSION HIGHLIGHTS (LAUNCH SCRUBS/DELAYS, TAL WEATHER, ASCENT I-LOADS, FIRSTS, SIGNIFICANT ANOMALIES, ETC.)
STS-109 SEQ FLT # 108 KSC-108 PAD 39A-61 MLP-2 Fourth HST Service Flight	OV-102 (Flight 27) Columbia OMS PODS: LPO5-16 RPO5-15 FRC2-27	CDR: Scott D. Altman (Flt 3 - STS-90, STS-106) P650/R237/V161/M207 PLT: Duane Carey P651/R274/M240 M/S 1/EV1: John Grunsfeld (Flt 4 - STS-67, STS-81, STS-103) P652/R191/V133/M167 M/S 2: Nancy Currie (Flt 4 - STS-57, STS-81, STS-103) P653/R165/V120/F21 M/S 3/EV2: Richard Linnehan (Flt 3 - STS-78, STS-90) P654/R214/V150/M187 M/S 4/EV3: James Newman (Flt 4 - STS-51, STS-69, STS-88) P655/R168/V122/M146 M/S 5/EV4: Michael Massimino P656/R275/M241 SS EVA #71 EMU/TETHERED EVA #64 SCHEDULED EVA #65 DURATION 7:01 SS EVA #72 EMU/TETHERED EVA #65 SCHEDULED EVA #66 DURATION 7:16 Continued…	KSC 39A 60:11:22:01.99Z 6:22:02 AM EST (P) 6:22:02 AM EST (A) Friday 24 3/1/02 (8) LAUNCH WINDOW: HST Planar/Phase Window 61M51S EOM PLS: KSC TAL: BEN TAL WX: NONE SELECTED: RTLS: KSC 15/CI/N TAL: BEN 36/N/N AOA: EDW 22/N/N PLS: EDW 04/CI/N TDEL: -0.03 -0.26/-0.023 MAX Q NAV: 693 ??? 754 SRB STG: 2:06 2:07 PERF: NOMINAL 2 ENG TAL (BEN): 2:17 2:16 NEG RETURN: 3:55 3:59 PTA (U/S 530): 3:50 3:55 PTM (U/S 500): 5:06 5:08 SE TAL (BYD): 5:50 5:50 MECO CMD: 8:21.5 8:23.9 Continued…	KSC 33 (KSC 58) 71:09:31:53Z 4:31:53 AM EST Tuesday 20 3/12/02 (9) DEORBIT BURN: 71:08:22:39Z XRANGE: 268 NM ORBIT DIR: DL 48 AIM PT: NOMINAL MLGTD: 3433 FT 71:09:31:53Z VEL: 196 KGS 186 KEAS HDOT: -2.7 FPS TD NORM 195: 2993 FT NLGTD: 6286 FT 71:09:32:01Z VEL: 156 KGS 149 KEAS HDOT: -5.6 FPS DRAG CHUTE DEPLOY: 181 KEAS 71:09:31:55Z BRK INIT: 66 KGS DRAG CHUTE JETTISON: 63 KGS 71:09:32:37Z BRK DECEL (FPS^2): AVE 3.7 PK 7.2 WHEELS STOP: 71:09:33:05Z 13552 FT ROLLOUT: 10119 FT 72 SEC WINDS:T5, R2 OFFICIAL: 13005P08 SS: T5, R2 PK: T8, R3 Continued…	104/104/ 109% PREDICTED: 100/104.5/ 104.5/72/ 104.5 ACTUAL: 100/104.5/ 101/72/ 104.5 1 = 2056 (2) 2 = 2053 (4) 3 = 2047 (7) ALL SSME's BLOCK IIA	BI-111 RSRM 83 ET-112 SLWT-17 ET IMPACT 1:28:35 MET LAT: 16.3 N LONG: 143.6 W	28.45 (50)	DIRECT INSERTION POST OMS-2 310.5 x 105.0 NM DEORBIT: 312.6 x 259 NM VELOCITY: 26082 FPS ENTRY RANGE: 4274 NM	OI 28 (7)	CARGO: 27564 LBS PAYLOAD CHARGEABLE: 20144 LBS DEPLOYED: 8256 LBS NON-DEPLOYED: 10672 LBS MIDDECK: 1216 LBS SHUTTLE ACCUMULATED WEIGHTS: DEPLOYED: 1090395 LBS NON-DEPLOYED: 1532571 LBS CARGO TOTAL: 3363980 LBS PERFORMANCE MARGINS (LBS): FPR: 3065 FUEL BIAS: 937 FINAL TDDP: 3309 RECON: 4170 PAYLOADS: PLB: HST Service Mission 3B RMS MIDDECK: NONE 5 CRYO TK SETS 5 GN2 TANKS RMS 65 RMS USED FOR: HST GRAPPLE, BERTH, SERVICE, AND RELEASE.	*Brief Mission Summary:* The STS-109 mission was the 4th Servicing Mission to the Hubble Space Telescope to rejuvenate the World's Greatest Observatory. During five EVA's the crew replaced the Reaction Wheel Assembly, the solar arrays, the Power Control Unit (down since 1999) and installed a new scientific instrument, the Advanced Camera for Surveys (ACS). The ACS is able to survey a field of the cosmos twice as large as previous instruments, with ten times the resolution and four times the speed. KSC W/D: OPF 253, VAB 8, PAD 32 = 293 days total. LAUNCH POSTPONEMENTS: - Baselined launch date of 11/1/01 on 9/21/00 - Postponed launch date to NET 11/19/01 on 5/4/01 - Postponed launch date to 1/17/02 on 5/10/01 - Postponed launch date to 2/14/02 on 10/4/01 - On 12/21/01, postponed launch date to NET 2/21/02 to allow manifest of new RWA (new HST problem) and train EVA crew. - On 1/10/02, postponed launch date to 2/28/02, had to prepare and ship another RWA to KSC. First RWA was faulty. LAUNCH SCRUBS: - 2/28/02 Launch was scrubbed at approximately L-16 hours due to forecast of cold weather at pad at LCC limits. Forecast was for 38 deg, 73 percent humidity, winds 7 to 10 knots. This forecast is one degree above the minimum temperature, and MMT decided to scrub and reschedule launch for 3/1/02. Observation S at launch time were 28 deg, RH 71 percent, winds 7 to 10 knots. Wx scrub #36. LAUNCH WINDOW: - Window was in 2 panes: Pane 1 opened at 60:11:22:02Z and closed at 60:11:27:23Z (5M21S window), pane 2 opened at 60:11:27:33Z and closed at 60:12:23:53Z (56M20S window), and combined panes 1 & 2 yielded a window of 61M51S with a cutout from 11:23:20 to 11:24:20. LAUNCH DELAYS: NONE - Launched On-Time at 60:11:22:02Z, 6:22:02 AM EST, on March 1, 2002. TAL WX: - Ben Guerir was the only TAL site available. Ben Guerir was forecast and observed GO. SHUTTLE NIGHT LAUNCH #27 RENDEZVOUS #57: Rendezvous and berth HST, performed service operations, and released HST. Continued…

M 3 EOM:
WEIGHT:
222447 LBS
X CG:
1082.87
LANDING:
WEIGHT:
222366 LBS
X CG:
1084.57

Intro_im2_smACS.jpg --- in the Clean Room at GSFC two men in "bunny suits" stand near the new ACS to be installed on HST.

FLT NO.	ORBITER	CREW (7) TITLE, NAMES & EVA'S	LAUNCH SITE, LIFTOFF TIME, LANDING SITES, ABORT TIMES	LANDING SITE/ RUNWAY, CROSSRANGE LANDING TIMES FLT DURATION, WINDS	SSME-TL NOM-ABORT EMERG THROTTLE PROFILE ENG. S.N.	SRB RSRM AND ET	ORBIT INC	HA/HP	FSW	PAYLOAD WEIGHTS, PAYLOADS/ EXPERIMENTS	MISSION HIGHLIGHTS (LAUNCH SCRUBS/DELAYS, TAL WEATHER, ASCENT I-LOADS, FIRSTS, SIGNIFICANT ANOMALIES, ETC.)
STS-109 Continued...		Continued... SS EVA #73 EMU/TETHERED EVA #66 SCHEDULED EVA #67 DURATION 6:48 SS EVA #74 EMU/TETHERED EVA #67 SCHEDULED EVA #68 DURATION 7:30 SS EVA #75 EMU/TETHERED EVA #68 SCHEDULED EVA #69 DURATION 7:20	Continued... VI: 26114 26113 OMS-2: 44:00 43:57 134 FPS 134 FPS 1:27 1:27	Continued... DENS ALT: 326 FT FLT DURATION: 10:22:09:51 S/T: 966:06:31:22 OV-102: 284:19:19:08 DISTANCE: 3,941,705 sm							Continued...

MCC WHITE FCR (38)

FLIGHT DIRECTORS:
LD/O 1 - B. P. Austin
O 2 - A. J. Ceccacci
PLNG - J. M. Hanley
A/E - J. P. Shannon
MOD - N. W. Hale

STS109-E-6032 --- Crew on middeck. From left (front row): Currie/MS, CDR Altman, & PLT Carey. From the left (back row): Grunsfeld/PLC, Linnehan/MS, Newman/MS, & Massimino/MS.

PERFORMANCE ENHANCEMENTS:
- Standard Set Plus PE Operational High Q, WIN/FEB

SHUTTLE NIGHT LANDING #19

KSC NIGHT LANDING #14

FLIGHT DURATION CHANGES: NONE
- Planned landing at KSC on orbit 166. Landed at KSC Runway 33 on orbit 166, MLGTD at 71:09:31:53Z on Tuesday, March 12, 2002.

EVENTS:
- OMS-2 Start at 60:16:43:49Z, 13.8 duration, Delta V 10.3 ft/sec, resultant orbit 105.0 by 310.5 nm.
- NH maneuver (OMS-4) at 62:04:07:30Z, 207 seconds duration, Delta V 326.6 ft/sec, resultant orbit 302.2 by 309.2 nm. MC-4 at 62:08:23:29Z, resultant orbit 303.4 by 314.9 nm.
- HST capture by RMS at 62:09:31:21Z, and HST berth on FSS in PLB at 62:10:31:Z. 1:22:09:19 MET.
- EVA 1 Start at 63:06:37Z, 2:19:15 MET, End at 63:13:38Z, duration 7H01M. Replaced old SA with -V2 Solar Array 3 and diode box.
- EVA 2 Start at 64:06:41Z, 3:19:19 MET, End at 64:13:57Z, duration 7H16M. Replaced old SA with +V2 Solar Array 3 and diode box. Preplaced Reaction Wheel Assembly. Installed NOBL in Bay 6 and two doorstop extensions (one on -V2 side and one on +V2 side.)
- EVA 3 Start 2 hrs late at 65:08:28Z, 04:21:06 MET (EMU 1 got water in suit), hence had to resize EMU 3 for use by EV1. EVA duration 6H48M. Powered down HST and replaced PCU (Power Control Unit).
- EVA 4 Start at 66:09:00Z, 5:21:38 MET, duration 7H30M. Replaced FOC (Faint Object Camera) with new ACS (Advanced Camera for Surveys), installed Electronics Support Module and PCU clean up tasks.
- EVA 5 Start at 67:08:46Z, 6:21:24 MET. Installed NICMOS Camera and cryogenic cooler, duration 7:20.
- HST Reboost started at 67:17:18:04Z, 7:05:56:02 MET, Delta V 11.8 fps, altitude increase 3.6 nm, orbit of 314.7 by 310.6 nm.
- HST unberthed from Orbiter at 68:08:34Z, 7:21:12 MET and released at 68:10:04Z, 7:22:42 MET.
- Orbit Adjust maneuver at 70:10:07:32Z, 48.3 seconds, Delta V 11.6 fps, orbit 259 by 312.5 nm.
- Last flight of Block IIA Engines.

STS109-713-014 (8 March 2002) --- Grunsfeld/MS (right) and Linnehan/MS during 5th EVA completing HST upgrades.

STS109-331-005 (9 March 2002) --- Rejuvenated HST flies away.

SIGNIFICANT ANOMALIES:
- Freon® Loop 1 Aft Coldplate Flow Blockage
- Loss of EV1 Suit data during EVA
- Starboard Slidewire Slider Anomaly
- Inner Airlock "A" Hatch locking device difficult to actuate
- APU 3 Drain Line Pressure Decay
- MPS LH2 4-Inch Recirculation Disconnect Slow to Close
- Forward THC -X Contact Lost During One Burn
- FES Accumulator/Hi-Load Feedline B Heater System 2 Failure
- Primary RCS Thruster R3R Failed Off
- Water leaking from EMU 1 PLSS

SPACE SHUTTLE MISSIONS SUMMARY

FLT NO.	ORBITER	CREW (7) TITLE, NAMES & EVA'S	LAUNCH SITE, LIFTOFF TIME, LANDING SITES, ABORT TIMES	LANDING SITE/ RUNWAY, CROSSRANGE LANDING TIMES FLT DURATION, WINDS	SSME-TL NOM-ABORT EMERG THROTTLE PROFILE ENG. S.N.	SRB RSRM AND ET	INC	ORBIT HA/HP	FSW	PAYLOAD WEIGHTS, PAYLOADS/ EXPERIMENTS	MISSION HIGHLIGHTS (LAUNCH SCRUBS/DELAYS, TAL WEATHER, ASCENT I-LOADS, FIRSTS, SIGNIFICANT ANOMALIES, ETC.)
STS-110/ ISS 8A SEQ FLT #109 KSC-109 PAD 39B-48 MLP-3 THIRTEENTH SHUTTLE FLIGHT TO ISS	OV-104 (Flight 25) Atlantis OMS PODS: LPO3-29 RPO4-25 FRC4-25	CDR: Michael J. Bloomfield (Flt 3 - STS-86, STS-97) P657/R227/V165/M198 PLT: Stephen N. Frick P658/R276/M242 MS1/EV2: Rex J. Walheim P659/R277/M243 M/S 2: Ellen Ochoa (Flt 4 - STS-56, STS-66, STS-96) P660/R180/V113/F20 M/S/EV4: Lee M. E. Morin P661/R278/M244 MS 4/EV3: Jerry L. Ross (Flt 7 - STS 61-B, STS-27, STS-37, STS-55, STS-74 STS-88) P662/R89/V38/M80 MS5/EV1: Steven L. Smith (Flt 4 - STS-68, STS-82, STS-103) P663/R184/V137/M161 Continued..	KSC 39B 98:20:44:19Z 4:39:31 PM EDT (P) 4:44:19 PM EDT (A) Monday (12) 4/8/02 (15) LAUNCH WINDOW: 4M59S PLT (In-Plane Time) with ISS EOM PLS: KSC TAL: ZZA TAL WX: MRN, BEN SELECTED: RTLS: KSC 15/CI/N TAL: ZZA 30/CI/N AOA: KSC 15/CI/N PLS: EDW 22/N/N TDEL: 0.02 -0.58/-0.2 MAX Q NAV: 737 742 SRB STG: 1:59:58 PERF: NOMINAL 2 ENG TAL (BEN): 2:29 2:37 NEG RETURN: 3:53 4:02 PTA (U/S 160): 4:46 5:02 PTM (U/S 160): 6:02 6:20 SE TAL (ZZA) 104: 6:00 6:02 SE PTM (U/S 675): 6:51 6:53 Continued…	KSC 33 (KSC 59) 109:16:26:58Z 12:26:58 PM EDT Friday 12 4/19/02 (11) DEORBIT BURN: 109:15:18:59Z XRANGE: 73 NM ORBIT DIR: AL 30 AIM PT: NOMINAL MLGTD: 3058 FT 109:16:26:58Z VEL: 197 KGS 193 KEAS HDOT: -2.2 FPS TD NORM 195: 3070 FT NLGTD: 6353 FT 109:16:27:08Z VEL: 146 KGS 137 KEAS HDOT: -5.9 FPS DRAG CHUTE DEPLOY: 186 KEAS 109:16:27:00Z BRK INIT: 75 KGS DRAG CHUTE JETTISON: 54 KGS 109:16:27:42Z BRK DECEL FPS^2: AVE 4.4 PK 5.5 WHEELS STOP: 109:16:28:08Z 12677 FT ROLLOUT: 9619 FT 70 SEC WINDS: 0T, 8R OFFICIAL: 08008P11 SS: 3T, 8R PK: 4T, 10R Continued…	104/104/ 109% PREDICTED: 100/100/100/ 67/104 ACTUAL: 100/100/100/ 72/104 1 = 2048 (4) 2 = 2051 (3) 3 = 2045 (6) ALL THREE SSME'S BLOCK II M 3 EOM: WEIGHT: 201513 LBS X CG: 1085.32 LANDING: WEIGHT: 201463 LBS X CG: 1087.17	BI-112 RSRM 85 ET-114 SLWT-18 ET IMPACT 1:14:19 MET LAT: 35.8 S LONG: 158.8 W	51.60 (13)	DIRECT INSERTION POST OMS-2: 124.1 X 84.8 NM ENTRY: HA/HP 218.7 X 166 NM ENTRY VELOCITY: 25917 FPS ENTRY RANGE: 4354 NM	OI-29 (1)	CARGO: 35849 LBS PAYLOAD CHARGEABLE: 28379 LBS DEPLOYED: 30600 LBS NON-DEPLOYED: 0 LBS MIDDECK: 757 LBS SHUTTLE ACCUMULATED WEIGHTS: DEPLOYED: 1122264 LBS NON-DEPLOYED: 1533328 LBS CARGO TOTAL: 3399829 LBS PERFORMANCE MARGINS (LBS): FPR: 3065 FUEL BIAS: 937 FINAL TDDP: 1256 RECON: 2670 PAYLOADS: PLB: ISS 8A S0 Truss and ITS RMS, ODS MIDDECK: ISS 8A Simplex RAMBO 5 CRYO TK SETS 6 GN2 TANKS RMS 66 RMS USED TO MATE S0 TRUSS AND EVA SUPPORT	*Brief Mission Summary:* The STS-110/8A (13th mission to ISS) was the most complex ISS assembly flight to date with four EVA's and extensive use of Shuttle and ISS robotic arms. The EVA included successful beam assemblies, bolting of girders, and installing work lights and electrical connections. The ISS Canadarm2 transferred the 13.5 ton, 43-foot long S0 Truss (ISS backbone) from Shuttle payload bay for installation on U.S. Lab, Destiny. Also, the first railcar was operated on the new truss, paving the way for eventual transportation for the Canadarm2 along the length of the ISS. KSC W/D: OPF 132, VAB 6, PAD 28 = 166 days total. LAUNCH POSTPONEMENTS: -Baselined launch date of 1/17/02 on 11/15/00. -Postponed launch date to 2/28/02 on 5/4/01 and Postponed launch date to 3/21/02 on 10/4/01. -Postponed launch date to 4/4/02 on 1/10/02 due to ground processing delays requiring OMS Pod removal. LAUNCH SCRUBS: - Scrubbed 4/4/02 Launch at approximately L-8 hours, during ET Fill operations, due to a Hydrogen leak in the MLP 3 Hydrogen Vent Line which is fed by Orbiter Hi-Point Bleed line. The leak was found to be from a 1/8 wide crack in a weld location in the 16-inch double walled aluminum line. Weld is more than 20 years old. Decision was made to repair using a clam-shell technique. New launch date was set for Monday, 4/8/02. Line was repaired using a two-piece clam-shell that was welded to the 16-inch outer line. LAUNCH WINDOW: - The Launch Window opened at 98:20:34:32Z and closed at 98:20:44:30Z for a total window of 9M58S. Using a Preferred Launch Time (In-Plane Time) of 98:20:39:31Z, the Launch Window was 4M59S. LAUNCH DELAYS: - Day-of-Launch Delay was 4M48S. LPS system detected consecutive sync errors in all three Stand-by PCM FEP'S (OI, GPC, PLD). The count was held at T-5 Min for 4M48S to execute Front End Processor resynchronization procedure which was successfully completed. Came out of the T- 5 Min hold, and picked up the count at 98:20:39:19Z (4:39:19 PM EDT) with 5M11S remaining to Launch Window closure. Launch occurred at 98:20:44:19Z, 4:44:19 PM EDT, on Monday, April 8, 2002. Only 11 seconds remained in the Launch Window at Liftoff. TAL WX: - Zaragoza (Prime and Selected) was Forecast and Observed GO. Moron was Forecast and Observed NO GO for Showers within 20 nm. Ben Guerir was Forecast GO but Observed NO GO for precipitation within 20 nm. PERFORMANCE ENHANCEMENTS: - Standard Set plus: (1) PE Operational High Q TRN/APR, (2). OMS Assist, (3) 52 NM MECO, (4) Del Psi Continued…

STS110-341-002 (11 April 2002) --- Canadarm2, operated by Ochoa & Bursch, moves S0 truss from Atlantis to temp location on ISS Destiny Lab.

FLT NO.	ORBITER	CREW (7)	LAUNCH SITE, LIFTOFF TIME,	LANDING SITE/ RUNWAY, CROSSRANGE	SSME-TL NOM-ABORT EMERG	SRB RSRM	ORBIT		FSW	PAYLOAD WEIGHTS,	MISSION HIGHLIGHTS (LAUNCH SCRUBS/DELAYS,
		TITLE, NAMES & EVA'S	LANDING SITES, ABORT TIMES	LANDING TIMES FLT DURATION, WINDS	THROTTLE PROFILE ENG. S.N.	AND ET	INC	HA/HP		PAYLOADS/ EXPERIMENTS	TAL WEATHER, ASCENT I-LOADS, FIRSTS, SIGNIFICANT ANOMALIES, ETC.)

STS-110/ ISS 8A

Continued…

Continued…

SS EVA 76
DOCKED QUEST EVA 2
SCHEDULED EVA 70
EMU/TETHERED EVA 69
DURATION 7:48

SS EVA 77
DOCKED QUEST EVA 3
SCHEDULED EVA 71
EMU/TETHERED EVA 70
DURATION 7:30

SS EVA 78
DOCKED QUEST EVA 4
SCHEDULED EVA 72
EMU/TETHERED EVA 71
DURATION 6:27

SS EVA 79
DOCKED QUEST EVA 5
SCHEDULED EVA 73
EMU/TETHERED EVA 72
DURATION 6:37

MCC WHITE FCR (39)

FLIGHT DIRECTORS:
Flt & ISS Ld/O2 - R. E. Castle
ISS O 1 - A. F. Algate
ISS PLNG - N. D. Knight
STS LD/O 1 - J. M. Hanley
O 2 - P. F. Dye
O 3/PLNG - J. S. Stich
A/E - L. E. Cain
MOD - J. M. Heflin

Continued…

VI:
25821   25822

OMS-2:
96 FPS   96 FPS
38:38   38:45

Continued…

DENS ALT:
1260 FT

FLT DURATION:
10:19:42:39

S/T:
977:02:14:01

OV-104:
209:01:28:10

DISTANCE:
4,525,299 sm

ABOVE: STS110-E-5732 --- STS-110 & Exp 4 crews in ISS Destiny Lab. From the left (front row): Ellen Ochoa/MS, CDR Bloomfield, & Exp 4 CDR Yury I. Onufrienko. From the left (middle row): Daniel W. Bursch Exp 4/FE, Walheim/MS, & Carl E. Walz, Exp 4/FE. From the left (back row): PLT Frick, Ross/MS, Morin/MS, & Smith/MS.

ABOVE: STS110-718-013 (13 April 2002) --- Morin anchored on Canadarm2 (& Ross, not shown) worked in tandem on S0 Truss during EVA 2.

LEFT: STS110-E-5926 (17 April 2002) --- New ISS configuration as viewed from departing Atlantis.

SIGNIFICANT ANOMALIES:
- Pre-Launch Scrub of 4/4/02 Launch due to Hydrogen Leak in MLP-3 16-inch Hydrogen Vent Line.
- Sync errors on LPS RF TLM FEP reload required at L-5M11S (Launched occurred with 11 seconds in window.)
- MED'S IDP-2 MSU BITE and FCW Buffer Overflow Error
- Primary RCS Thruster L1A Failed Off and was auto-deselected (Chamber P Max 20 psia)
- Low Chamber Pressure on Primary RCS Thruster F1D (Pc = 63-65 psia)
- Low Chamber Pressure on Primary RCS Thruster F3L (Pc = 63-65 psia)
- Lack of Digital Video from PD100 Camcoder to DTV MUX
- ICOM Problem with BPSMU
- ODS Upper Hatch Delta Pressure Gauge Bias
- Loss of Biomed Data during EVA 2
- Payload Bay Flood Light Failure
- Problems with Proshare Audio and Video during PMC
- Window 2 impact

Continued…

RENDEZVOUS #58:
- Rendezvous and Dock with ISS to PMA 2 Lab Fwd Port.

FIRSTS:
- First flight with all three Block II SSME's.
- First flight of FSW OI-29.
- First operation availability of delayed TAL.

EVENTS:
- MC-4 Maneuver at 100:15:04:09, 1:18:19:50, Delta V 2 fps, resultant altitude 204.0 by 211.3 nm.
- ISS Capture at 01/19:20:09 MET, 100:16:04:28Z.
- ISS Hard Dock at 1/19:34:46 MET, 100:16:19:05Z.

EVENTS (Continued):
- EVA 1 Start at 2:17:52 MET, 101:14:36Z, duration 7H48M. Installed Port & Stbd Fwd Struts to S0 truss and Port & Stbd avionics trays, deployed aft Umbilical tray, and installed TUS-1 cable.
- EVA 2 Start at 4:17:25 MET, 103:14:09Z, duration 7H30M. Installed Aft Port & Stbd Struts, installed TUS-2 cables, installed A/L handrail, Mated MT/MBS feed through cable.
- Reboost 1 at 5:01:59 MET, 103:22:44Z, Delta V 3.2 fps, alt. increase 0.95 nm, orbit 212 x 205 nm.
- EVA 3 Start at 5:17:04 MET, 104:13:48Z, duration 6H27M. Installed J300/400 panels, released capture claw, installed CID's 7 & 8, removed MT Launch restraints. Removed MT RPCM Thermal cover.
- Reboost 2 at 6:01:00 MET, 104:21:44Z, Delta V 3.4 fps, alt. increase 1.0 nm, orbit 212 x 206 nm.
- EVA 4 Start at 7:17:45:17 MET, 106:14:29:36Z, duration 6H37M. Installed Node & U.S. Lab EVA lights, released LCA guides, S0 handrails, MT energy absorbers, and deployed A/L spur & EV-CPDS.
- Reboost 3 at 8:14:35:01 MET, 107:11:19:20Z, Delta V 12.8 fps, alt. increase orbit to 213.8 by206.3 nm.
- Cargo transferred to ISS = 28944 lbs (S0 ITS 26716, middeck 2228); ISS to Atlantis middeck 2607 lbs.
- Transfers to ISS: O_2 146 lb, N_2 45 lb, and water 1465 lb (1397 lb in 14 CWC's +68 lbs in three PWR's)
- Total transfers to ISS = 30600 lbs, net transfer 27993 lbs (30600 minus 2607)
- Hatch close between ISS and Atlantis at 107:16:04Z, 11:04 AM CDT, Wednesday, 4/17/02
- Undocked at 107:18:31Z, 8:21:47 MET, 1:31 AM CDT, 4/17/02
- ISS Visitor Time is 7:02:12:30.
- Jerry Ross total EVA time is U.S. record of 58H18m.

FLIGHT DURATION CHANGES: NONE
- Planned Landing at KSC on orbit 171. MLGTD on orbit 171 at KSC runway 33 at 109:16:26:58Z, 4:26:58 PM EDT, 10:19:42:39 MET.

FLT NO.	ORBITER	CREW 7 UP/7 DOWN — TITLE, NAMES & EVA'S	LAUNCH SITE, LIFTOFF TIME, LANDING SITES, ABORT TIMES	LANDING SITE/ RUNWAY, CROSSRANGE — LANDING TIMES FLT DURATION, WINDS	SSME-TL NOM-ABORT EMERG — THROTTLE PROFILE ENG. S.N.	SRB RSRM AND ET	INC	ORBIT — HA/HP	FSW	PAYLOAD WEIGHTS, PAYLOADS/ EXPERIMENTS	MISSION HIGHLIGHTS (LAUNCH SCRUBS/DELAYS, TAL WEATHER, ASCENT I-LOADS, FIRSTS, SIGNIFICANT ANOMALIES, ETC.)
STS-111/ ISS UF-2 SEQ FLT #110 KSC-110 PAD 39A-62 MLP-1 14TH SHUTTLE FLIGHT TO ISS	OV-105 (Flight 18) Endeavour OMS PODS: LPO4-25 RPO1-32 FRC5-18	**CDR:** Kenneth D. Cockrell (Flt 5 - STS-56, STS-69, STS-80, STS-98) P664/R159/V121/M140 **PLT:** Paul S. Lockhart P665/R279/M245 **M/S 1/EV2:** Philippe Perrin (France - CNES) P666/R280/M246 **M/S 2/EV1:** Franklin R. Chang-Diaz (Flt 7 - STS 61-C, STS-34, STS-46, STS-60, STS-75, STS-91) P667/R89/V46/M81 **M/S 3 UP/EXP 5 Flt Eng:** Peggy A. Whitson P668/R281/F35 **M/S 4 UP/EXP 5 CDR:** Valery C. Korzun (Russia) P669/R282/M247 **M/S 5 UP/EXP 5 Flt Eng:** Sergei Y. Treschev (Russia) P670/R283/M248 **M/S 3 DN/EXP 4 Flt Eng** Carl Walz (Flt 4 - STS-51, STS-65, STS-79, STS-108 Up) P671/R170/V106/M148 **M/S 4 DN/EXP 4 Flt Eng** Daniel Bursch (Flt 4 - STS-51, STS-68, STS-77, STS-108 Up) P672/R169/V109/M147 Continued…	KSC 39A 156:21:22:49Z 5:22:49 PM EDT (P) 5:22:49 PM EDT (A) Wednesday 13 6/5/02 (10) **LAUNCH WINDOW:** 4M39S PLT (In-Plane Time) ISS Planar/Phase **EOM PLS:** KSC **TAL:** ZZA **TAL WX:** MRN, BEN **SELECTED:** RTLS: KSC 33/N/SFD TAL: MRN 20/N/N AOA: KSC 15/CI/N PLS: EDW 22/N/N **TDEL:** 0.12 -0.058/-0.20 **MAX Q NAV:** 748 722 **SRB STG:** 2:04 2:05 **PERF:** NOMINAL **2 ENG TAL (MRN):** 2:24 2:29 **NEG RETURN:** 3:52 3:57 **PTA (U/S 182):** 4:49 4:45 **DROOP (ZZA 109):** 5:23 5:24 **PTM (U/S 182):** 6:11 6:06 **SE TAL (ZZA 104):** 6:03 6:06 **VI:** 25821 25815 Continued…	EDW 22, CONC EDW 49, CONC 30 170:17:57:42Z 10:57:42 AM PDT Wednesday 14 6/19/02 (6) **DEORBIT BURN:** 170:16:50:26Z **XRANGE:** 603 NM **ORBIT DIR:** AL 31 **AIM PT:** NOMINAL **MLGTD:** 3058 FT 170:17:57:42Z VEL: 197 KGS 193 KEAS HDOT: -2.2 FPS **TD NORM 195:** 3070 FT **NLGTD:** 6353 FT 170:17:57:53Z VEL: 146 KGS 137 KEAS HDOT: -5.9 FPS **DRAG CHUTE DEPLOY:** 186 KEAS 170:17:57:45Z **BRK INIT:** 75 KGS **DRAG CHUTE JETTISON:** 54 KGS 170:17:58:23Z **BRK DECEL FPS²:** AVE 4.4 PK 5.5 **WHEELS STOP:** 170:17:58:46Z 12677 FT **ROLLOUT:** 9619 FT 64 SEC **WINDS:** 3T, 4R OFFICIAL: 35005p08 SS: H3, R4 PK: H5, R6 **DENS ALT:** 1260 FT Continued…	104/104/ 109% **PREDICTED:** 100/104.5/ 104.5/72/ 104.5 **ACTUAL:** 100/104.5/ 98/72/104.5 1 = 2050 (3) 2 = 2044 (7) 3 = 2054 (4) ALL BLOCK II SSME's **M 3 EOM:** WEIGHT: 220334 LBS X CG: 1083.62 **LANDING:** WEIGHT: 220279 LBS X CG: 1085.30	BI-113 RSRM 84 ET-113 SLWT-19 **ET IMPACT** 1:13:47 MET **LAT:** 37.3 S **LONG:** 160.1 W	51.60 (14)	DIRECT INSERTION **POST OMS-2:** 126.7 X 84.8 NM **DEORBIT:** HA 210.5 HP 187.1 **ENTRY VELOCITY:** 25902 FPS **ENTRY RANGE:** 4360 NM	OI-29 (2)	**CARGO:** 36082 LBS **PAYLOAD CHARGEABLE:** 29712 LBS **DEPLOYED:** 9512 LBS **NON-DEPLOYED:** 906 LBS **MIDDECK:** 288 LBS **SHUTTLE ACCUMULATED WEIGHTS:** DEPLOYED: 1130507 LBS NON-DEPLOYED: 1534522 LBS CARGO TOTAL: 3435911 LBS **PERFORMANCE MARGINS (LBS):** FPR: 3065 FUEL BIAS: 937 FINAL TDDP: 2484 RECON: 1870 **PAYLOADS:** PLB: ISS UF-2 (MPLM, MBS, PDGF, SMOP, SSRMS, WRJ, RMS, ODS) MIDDECK: ISS UF-2 RAMBO 5 CRYO TK SETS 6 GN2 TANKS RMS 67 RMS USED FOR ISS MPLM DEPLOY AND RETRIEVE AND EVA SUPPORT	*Brief Mission Summary: The STS-111/UF 2 (14th ISS mission) provided a new crew to the ISS, transfer of supplies and equipment via the Leonardo MPLM, and three EVA's for ISS assembly. The Shuttle RMS was used to successfully install the Mobile Remote Service Base System to the Mobile Transporter on the Destiny Lab. This allows the Canadarm2 to travel the length of the ISS for future construction tasks.* KSC W/D: OPF 92, VAB 7, PAD 33 = 132 days total. **LAUNCH POSTPONEMENTS:** - Launch was scheduled for 5/2/02. - Postponed launch to 5/31/02 to the end of a Beta Cutout and allow time to train EVA crew to R&R SSRMS failed Wrist Roll Joint. - Advanced launch to 5/30/02 after analysis indicated adequate power generation using an ISS Pitch attitude bias. **LAUNCH SCRUBS:** - Scrubbed Thursday 5/30/02 Launch at L-24M53S due to opaque anvils within 30 nm circle while holding at T-9 minutes. PLT was 7:44:26 PM EDT with a window of 4M9S. Lightning was present throughout a wide area in Florida with occasional strike within 30 nm circle and thunderstorms were forecast. Weather forecast 70 percent chance NO GO for launch due to continuing anvil clouds, lightning, and thunderstorms through Monday, June 3. An upper Low is bringing in moist air from the tropics. Decision was made to hold a tanking MMT on Friday, May 31, where it was decided not to tank. Forecast included thunderstorms, anvil clouds, and chance of hail.. During the count, the L OME GN2 Regulator leaked and increased the accumulator pressure. Regulator locked up after a test. Went into a 24-hour Scrub turnaround. RTLS and Range Weather Scrub. - A Tanking MMT was held on Friday, 5/31/02 and a decision was made not to tank due to inclement observed and forecast weather. There was a tanking weather violation with observed lightning within 5 nm. Launch forecast was for attached anvil clouds, thunderstorms, lightning, and precipitation. Tanking, RTLS, and Range Weather Scrub. - A tentative decision was made to try for a Monday, 6/3 launch but keep an eye on the weather and hold a special MMT at 6:30 PM CDT (Later changed to 1:00 PM CDT) to decide whether to hold a tanking MMT on Saturday, 6/1. - At the 1:00 PM CDT MMT, it was decided to top-off the cryos and reload the GN₂ (and at the same time to run another GN2 regulator test) with a target of a Monday evening launch. This would allow three launch opportunities based on Range schedule on Monday, Tuesday, and Wednesday. Tentative plans were made for a tanking MMT on Monday. On Friday, the GN₂ was reloaded and the regulator failed the leak test. At a Saturday morning management meeting, it was decided to replace the L OME GN₂ Regulator, and with success oriented schedule, it would lead to a launch date of NET Tuesday 6/4/02. On Sunday morning, management decided to re-target the launch date to Wednesday, 6/5 due to delays in completing GSE work. Wednesday launch was confirmed later. Technical Scrub. Continued…

ISS004-E-13246 (7 June 2002) — Endeavour approaches ISS with Leonardo (MPLM) supplies.

FLT NO.	ORBITER	CREW 7 UP/7 DOWN / TITLE, NAMES & EVA'S	LAUNCH SITE, LIFTOFF TIME, / LANDING SITES, ABORT TIMES	LANDING SITE/ RUNWAY, CROSSRANGE / LANDING TIMES FLT DURATION, WINDS	SSME-TL NOM-ABORT EMERG / THROTTLE PROFILE ENG. S.N.	SRB RSRM AND ET	ORBIT INC	HA/HP	FSW	PAYLOAD WEIGHTS, / PAYLOADS/ EXPERIMENTS	MISSION HIGHLIGHTS (LAUNCH SCRUBS/DELAYS, TAL WEATHER, ASCENT I-LOADS, FIRSTS, SIGNIFICANT ANOMALIES, ETC.)
STS-111/ ISS UF-2 Continued...		Continued... M/S 5 DN/EXP 4 CDR: Yury I. Onufrienko (Russia) (Flt 1 - STS-108 Up) P673/R273/R239 SS EVA 80 DOCKED QUEST EVA 6 EMU/TETHERED EVA 73 SCHEDULED EVA 74 DURATION 7:14 SS EVA 81 DOCKED QUEST EVA 7 EMU/TETHERED EVA 74 SCHEDULED EVA 75 DURATION 5:00 SS EVA 82 DOCKED QUEST EVA 8 EMU/TETHERED EVA 75 SCHEDULED EVA 76 DURATION 7:17 MCC WHITE FCR (40) FLIGHT DIRECTORS: ISS Ld/O1-R. E. LaBrode ISS O 2 - J. M. Curry ISS PLNG - B. C. Lunney STS LD/O 1 - P. S. Hill STS O 2 - A. J. Ceccacci STS O 3/PLNG - K. B. Beck A/E - J. P. Shannon MOD - R. E. Castle	Continued... OMS-2: 38:42 38:45 98 FPS 95 FPS	Continued... FLT DURATION: 13:20:34:53 S/T: 990:22:48:54 OV-105: 192:19:24:40 DISTANCE: 5,781,115 sm						

ISS004-E-13426 --- Exp 4 (dark blue shirts), STS-111 (green shirts), and Exp 5 (medium blue shirts) crews in ISS Destiny Lab. Exp 4 crew, from front to back, CDR Onufrienko (RSA), Bursch/FE, & Walz/FE. STS-111 crew, from front to back, CDR Cockrell, Chang-Diaz/MS, PLT Lockhart, & Perrin/MS (CNES). Exp 5 crew, from front to back, CDR Korzun (RSA), Whitson/FE, & Treschev/FE (RSA).

STS111-E-5095 (7 June 2002) --- EXP 4 CDR Onufrienko (Russia) greets EXP 5 CDR Korzun (Russia, back to camera) with STS-111 CDR Cockrell partially visible at right.

Continued...

LAUNCH WINDOW:
- The June 4, 2002 launch window opened at 156:21:18:19Z and closed at 156:21:27:28Z giving a total window of 9M09S. Using a Preferred Launch Time of 156:21:22:49Z (5:22:49 PM EDT), the window was 4M39S.

LAUNCH DELAYS: NONE
- Launch occurred On-Time at 156:21:22:49Z (5:22:49 PM EDT) on Wednesday, June 5, 2002.

TAL WX:
- Zaragoza (Prime) was forecast and observed NO GO for precipitation. Ben Guerir was forecast and observed NO GO for Head Winds of 27 Knots. Moron (Selected) was forecast and observed GO.

PERFORMANCE ENHANCEMENTS:
- Standard Set plus: (1) PE Operational High Q TRN/MAY, (2). OMS Assist, (3) 52 NM MECO, (4) Del Psi

FLIGHT DURATION CHANGES:
- Total Extensions: 2 Days Plus 2 Revs. Planned landing at KSC on Orbit 186 at 12:59 PM EDT on June 17, 2002. Did not call up EDW.   Closed PLBD's but did not fluid load crew. Waved off Orbit 186 due to forecast ceiling, precipitation, crosswinds, and thunderstorms and observed precipitation, thunderstorms within 20 nm, ceiling 2600 broken and visibility violations. Waved off landing at KSC on Orbit 187 with similar forecast and observed at landing time. Extended one day. Brought up EDW for EOM+1. Waved off landing at KSC on Orbit 201 due to forecast ceiling, precipitation, and thunderstorms. Observed ceiling, precipitation, thunderstorms, and visibility violations. Waved off Orbit 202 due to similar forecasts and observations. Extended the second day.
- EOM+2 was "pick the landing site" day. EOM-2 PLBD's were closed for Planned landing at KSC on Orbit 216 at 170:14:52Z. Crew not in suits and no fluid load. Waved off landing at KSC on Orbit 216 at approximately Tig -40 minutes due to forecast and observed thunderstorms, attached anvil clouds, and low ceiling within 30 nm. Waved off landing at KSC on Orbit 217 at approximately Tig -20 minutes due to thunderstorms, attached anvils, and low clouds. (Two orbits wave-off).
- Decision made to land at EDW 22 on Orbit 218. MLGTD at 170:17:57:42Z, 10:57:42 AM PDT (MET 13:20:34:57) on Wednesday, June 19, 2002.
- NLGTD at 170:17:57:53Z.
- Total Flight Duration Extensions: Two Days plus two orbits.

FIRSTS:
- First use of orbiter oxygen for EVA pre-breathe for astronauts in ISS Joint Airlock.

Continued...

FLT NO.	ORBITER	CREW 7 UP/7 DOWN TITLE, NAMES & EVA'S	LAUNCH SITE, LIFTOFF TIME, LANDING SITES, ABORT TIMES	LANDING SITE/ RUNWAY, CROSSRANGE LANDING TIMES FLT DURATION, WINDS	SSME-TL NOM-ABORT EMERG THROTTLE PROFILE ENG. S.N.	SRB RSRM AND ET	ORBIT		FSW	PAYLOAD WEIGHTS, PAYLOADS/ EXPERIMENTS	MISSION HIGHLIGHTS (LAUNCH SCRUBS/DELAYS, TAL WEATHER, ASCENT I-LOADS, FIRSTS, SIGNIFICANT ANOMALIES, ETC.)
							INC	HA/HP			

STS-111/ ISS UF-2

Continued...

STS111-E-5238 (11 June 2002) --- Perrin/MS1 (France) installs the Mobile Remote Servicer Base System (MBS) on the ISS railcar.

JSC2002-E-23106 --- J. Milton (Milt) Heflin (standing), Chief, Flight Director's Office, along with Dan Carpenter (background), Director, Public Affairs Office, and Rob Navias, lead STS-111 PAO commentator, discuss mission in JSC MCC WFCR

JSC2002-E-23100 --- Flight Directors Steve Stich (right foreground) and John Shannon; along with astronauts William A. Oefelein and Kenneth T. Ham, spacecraft communicators (CAPCOM), watch the large MOCR screens.

SIGNIFICANT ANOMALIES:
- Right Main Engine High Pressure Fuel Pump Speed Sensor Failure
- Flash Evaporator Controller Primary B failure
- WIF Adapter Hitch Pin Anomaly
- EV2 Boot Fit Problems during EVA 1
- EVA Communications Anomaly on STS-111 EVA 3
- AVIU-Camcorder Failed
- BPSMU XMIT/ICOM Dey causes Video to Flicker
- LL QUAD Reflected Power Spikes
- Loss of BIOMED Data on EVA 1

Continued...

EVENTS:
- MC4 Maneuver Start at 158:15:16:16Z, 1:127:53:27 MET, 1.2 ft/sec, altitude 203.3 by 211.9 nm.
- ISS Capture at 158:16:24Z, 1:19:01 MET.
- ISS Hard-Docked at 158:17:26:32Z, 1:20:03:43 MET.
- Official Transfer Time (IELK time) from Expedition 4 Crew to Expedition 5 Crew = 158:22:55Z, 5:55 PM CDT, June 7, 2001.
- Expedition 4 ISS Habitant Time is 181:00:43.
- MPLM installed on Node 1 by RMS at 159:14:28Z, 2:17:05 MET.
- EVA 1 Start at 160:15:26Z, 3:18:03 MET and End 160:22:40Z, 04:01:17 MET, duration 7:14. Installed PDGF on P6 Truss, mated heater cables from MBS to MT, and installed SM debris protectors on PMA1 for future installation on SM.
- Photographed failed ISS CMG-1.
- Reboost Maneuver 1 Start at 161:20:53:24Z, 4:23:30:35 MET, Delta V 3.0 fps, 0.8 nm altitude increase, altitude 212 by 205 nm.
- EVA 2 Start at 162:15:19Z, 5:17:58 MET and End 162:20:19Z, 5:22:58 MET, duration 5:00, final installation of MBS to MT (Connected video and data cables), attached bag with contingency extension cable to MBS.
- Reboost Maneuver 2 Start at 163:12:08:02Z, 6:15:45:13 MET, Delta V 3.0 fps, altitude increase .81 nm, Orbit 212.8 by 206.2 nm
- EVA 3 Start at 164:15:16Z, 7:17:53 MET, duration 7:17. R&R SSRMS Wrist Roll Joint (WRJ).
- Reboost Maneuver 3 Start at 165:11:51:26Z, 6:14:28:37 MET, Delta V 12.5 fps, altitude increase 3.6 nm, orbit 214.4 by 211.1 nm.
- Transfers from shuttle to ISS = 9512 lbs (from MPLM = 8062 lbs and from middeck = 1450 lbs). Transfers from ISS to Shuttle = 6342 lbs (to MPLM = 4668 lbs and to middeck = 1675 lbs). Consumables transfer: Total water = 884.9 lbm (8 CWC's with 798.9 and 4 PWR's with 86.0 lbm). Total shuttle O2 transferred = 34 lbm for the 3 EVA prebreathes in JAL, N2 tank transfer of 18.9 lbm.
- Undocked at 166:14:31Z, 9:17:08 MET
- STS-111/ISS Visitor Time is 7:31:04:28 (Docking to Undocking)
- Expedition 4 ISS Habitant Time is 181:00:43:00 (IELK S/L Xfer to IELK S/L Xfer), Expedition 4 broke U.S. Flight Time record, flight time is 195:19:38:14 (STS-108 L/O to STS-111 MLGTD).
- Carl Walz record total flight time is 230:13:02:44. Dan Bursch Total Flight Time is 226:22:14:48.
- Sep Burn 166:16:14:27Z, 6:18:51:38 MET.
- Orbit Adjust Maneuver at 166:17:57:48Z, 9:20:34:59 MET, Delta V 45.6 fps, orbit was 186.1 by 211.9 nm.

RENDEZVOUS # 59: Rendezvous and Dock with ISS (Dock to PMA2 Lab Fwd Port)

FLT NO.	ORBITER	CREW (6) TITLE, NAMES & EVA'S	LAUNCH SITE, LIFTOFF TIME, LANDING TIMES ABORT TIMES	LANDING SITE/ RUNWAY, CROSSRANGE LANDING TIMES FLT DURATION, WINDS	SSME-TL NOM-ABORT EMERG THROTTLE PROFILE ENG. S.N.	SRB RSRM AND ET	INC	ORBIT HA/HP	FSW	PAYLOAD WEIGHTS, PAYLOADS/ EXPERIMENTS	MISSION HIGHLIGHTS (LAUNCH SCRUBS/DELAYS, TAL WEATHER, ASCENT I-LOADS, FIRSTS, SIGNIFICANT ANOMALIES, ETC.)

STS-112/ ISS 9A
 	
SEQ FLT #111
 	
KSC-111
 	
PAD 39B-49
 	
MLP-3
 	
15TH SHUTTLE FLIGHT TO ISS
 	
MCC WHITE FCR (41)
 	
FLIGHT DIRECTORS:
ISS Ld/O1 - A. F. Algate
ISS O 2 - M. A. Kirasich
ISS PLNG - A. P. Hasbrook
STS LD/O 1 - P. L. Engelauf
STS O 2 - C. A. Koerner
STS O 3/PLNG - J. M. Curry
A/E - J. P. Shannon
MOD - R. E. Castle
	OV-104 (Flight 26)
	
	Atlantis
	
	OMS PODS:
	LPO3-30
	RPO4-26
	FRC4-26
		CDR:
		Jeffrey S. Ashby
		(Flt 3 - STS-93, STS-100)
		P674/R251/V169/M218
		
		PLT:
		Pamela A. Melroy
		(Flt 2 - STS-92)
		P675/R261/V175/F34
		
		M/S 1/EV1:
		David A. Wolf
		(Flt 3 - STS-58, Up to Mir on STS-86, Dn on STS-89)
		P676/R173/V147/M151
		
		M/S 2:
		Sandra H. Magnus
		P677/R284/F36
		
		M/S 3/EV2:
		Piers J. Sellers
		P678/R285/M249
		
		M/S 4:
		Fyodor N. Yurchikhin (Russia)
		P679/R286/M250
		
		SS EVA 83
		DOCKED QUEST EVA 9
		EMU/TETHERED EVA 76
		SCHEDULED EVA 77
		DURATION 7:01
		
		SS EVA 84
		DOCKED QUEST EVA 10
		EMU/TETHERED EVA 77
		SCHEDULED EVA 78
		DURATION 6:04
		
		SS EVA 85
		DOCKED QUEST EVA 11
		EMU/TETHERED EVA 78
		SCHEDULED EVA 79
		DURATION 6:36
			KSC 39B
			280:19:45:51Z
			3:45:51 PM EDT (P)
			3:45:51 PM EDT (A)
			Monday (13)
			10/7/02 (11)
			
			LAUNCH WINDOW:
			4M59S USING PLT (ISS IN-PLANE TIME)
			
			EOM PLS: KSC
			TAL: ZZA
			TAL WX: MRN
			
			SELECTED:
			RTLS: KSC 33/N/N
			TAL: ZZA 30/N/SFD
			AOA: KSC 33/N/N
			PLS: EDW 04/N/N
			
			TDEL:
			-0.11 -0.368/-0.490
			
			MAX Q NAV:
			726 725
			
			SRB STG:
			2:04 2:02
			
			PERF: NOMINAL
			
			2 ENG TAL (MRN):
			2:33 2:30
			
			NEG RETURN:
			3:54 3:54
			
			PTA (U/S 182):
			4:57 4:55
			
			PTM (U/S 182):
			6:14 6:10
			
			SE TAL (ZZA):
			6:04 6:08
			
			MECO CMD:
			8:21.5 8:24.5
			
			VI:
			25822 25815
			
			OMS-2:
			38:40 38:42
			96.1 FPS 95.9 FPS Continued...
				KSC 33 (KSC 60)
				291:15:43:41Z
				11:43:41 AM EDT
				
				Friday 13
				10/18/02 (9)
				
				DEORBIT BURN:
				291:14:36:14Z
				
				XRANGE: 21 NM
				
				ORBIT DIR: AR 12
				
				AIM PT: NOMINAL
				
				MLGTD: 3072 FT
				291:15:43:41Z
				VEL: 186 KGS
				        187 KEAS
				HDOT: -1.0 FPS
				
				TD NORM 195:
				2851 FT
				
				NLGTD: 5475 FT
				291:15:43:48Z
				VEL: 161 KGS
				        160 KEAS
				HDOT: -6.2 FPS
				
				DRAG CHUTE DEPLOY: 157 KEAS
				291:15:43:51Z
				
				BRK INIT: 86 KGS
				
				DRAG CHUTE JETTISON:
				51 KGS
				291:15:44:18Z
				
				BRK DECEL FPS2:
				AVE 6.9 PK 9.1
				
				WHEELS STOP:
				291:15:44:33Z
				11377 FT
				
				ROLLOUT:
				8305 FT
				52 SEC
				
				WINDS:
				11H, 5R KTS
				OFFICIAL:
				01011P17
				AVE: 8H 11R
				PK: 13H 11R
				
				DENS ALT: 1019 FT
					104/104/ 109%
					
					PREDICTED:
					100/104.5/104.5/ 72/104.5
					
					ACTUAL:
					100/104.5/97/ 72/104.5
					
					1 = 2048 (5)
					2 = 2051 (4)
					3 = 2047 (8)
					
					M 3 EOM
					
					WEIGHT:
					202688 LBS
					
					X CG:
					1087.08
					
					LANDING:
					
					WEIGHT:
					202621 LBS
					
					X CG:
					1088.94
						BI-115
						
						RSRM 87
						
						ET-115
						
						SLWT-20
						
						M EOM
						
						ET IMPACT
						
						1:14:01 MET
						
						LAT:
						36.97 S
						
						LONG:
						159.3 W
							51.60 (15)	DIRECT INSERTION
								
								POST OMS-2
								126.4 x 85.0 NM
								
								DEORBIT:
								HA 220.0 NM
								HP 146.0 NM
								
								VELOCITY:
								25917 FPS
								
								ENTRY RANGE:
								4342 NM
									OI-29 (3)	CARGO:
										37441 LBS
										
										PAYLOAD CHARGEABLE:
										29502 LBS
										
										DEPLOYED:
										29543 LBS
										
										NON-DEPLOYED:
										0 LBS
										
										MIDDECK:
										382 LBS
										
										SHUTTLE ACCUMULATED WEIGHTS:
										DEPLOYED:
										1160050 LBS
										NON-DEPLOYED:
										1534904 LBS
										CARGO TOTAL:
										3473352 LBS
										
										PERFORMANCE MARGINS (LBS):
										FPR: 3065
										FUEL BIAS: 937
										FINAL TDDP: 2744
										RECON: 3860
										
										PAYLOADS:
										PLB:
										ISS 9A
										(ITS S1 TRUSS)
										CETA CART A
										RMS, ODS
										
										MIDDECK:
										ISS 9A
										(SHIMMER, RAMBO)
										
										5 CRYO TK SETS
										6 GN2 TANKS
										RMS 69
										
										RMS USED FOR TV SUPPORT DURING S1 INSTALL (SSRMS INSTALL)
											Brief Mission Summary: The STS-112/9A (15th ISS mission) delivered the 45-foot long, 15 ton S1 Truss for further assembly of ISS. The S1 Truss was attached to the starboard side of the Center S0 Truss allowing for the outboard expansion of the rail system to prepare for future ISS growth. This truss also contains a new cooling system, S-band Comm, and the first Thermal Radiator Rotary Joint (TRRJ).
											
											KSC W/D: OPF 106, VAB 6, PAD 25 = 139 days total.
											
											LAUNCH POSTPONEMENTS:
											- Launch was postponed from June after Post-STS-110 visual inspections of OV-104 Inconel 12" MPS LH2 Flowliners revealed three cracks to SSME 2. Subsequent inspections found cracks in other Orbiter LH2 Flowliners:
											  - OV-103 - three cracks (SSME 1)
											  - OV-105 - one crack (SSME 1) and one crack (SSME 2)
											  - MPTA - one crack (SSME 1)
											  - OV-102 three cracks (SSME 2). OV-102 flowliners are CRES. After analyses, tests, etc., including consideration of other repair techniques, the decision was made to use weld-repair technique and polishing of Flowliner holes.
											- Severe cracks were found in Mobile Launch Platform Crawler-Transporter (CT-2) jacking cylinder bearings. CT-2 was repaired using undamaged spare and new bearings. CT-2 bearings will be replaced incrementally.
											- These postponements resulted in rescheduling STS-112 and STS-113 ahead of STS-107. STS-112 launch date was set to October 2, 2002.
											
											LAUNCH SCRUBS:
											- Scrubbed October 2 Launch at approximately L-27 hours at an MMT due to the threat to JSC/MCC posed by Hurricane Lili in the Gulf of Mexico. Launch delayed for at least 24 hours. At approximately L-21 hours, the Space Shuttle and ISS Programs decided there was less risk to the MCC by implementing an orderly powerdown of the MCC with a launch in the Sunday/Monday timeframe. Weather Scrub.
											- Early Wednesday morning, October 2, MCC-H transitioned USOS operations support to BCC HSG Moscow.
											- At the October 2, 6:45 AM CST MMT, the decision was made not to launch earlier than Monday, October 7. This presumes a GO to begin Restoration of the MCC late Wednesday or early Thursday.
											- MCC powerup/restoration began early Thursday morning, October 3. ISS operations in MCC will be resumed Thursday night. Launch scheduled for Monday, October 7.
											
											Continued...

STS112_ETCAM_typical ---- Typical view during ascent from first ET Shuttle Observation Camera . (Courtesy MSFC ET Project Office)

SPACE SHUTTLE MISSIONS SUMMARY

FLT NO.	ORBITER	CREW (6) TITLE, NAMES & EVA'S	LAUNCH SITE, LIFTOFF TIME, LANDING SITES, ABORT TIMES	LANDING SITE/ RUNWAY, CROSSRANGE LANDING TIMES FLT DURATION, WINDS	SSME-TL NOM-ABORT EMERG THROTTLE PROFILE ENG. S.N.	SRB RSRM AND ET	ORBIT INC	HA/HP	FSW	PAYLOAD WEIGHTS, PAYLOADS / EXPERIMENTS	MISSION HIGHLIGHTS (LAUNCH SCRUBS/DELAYS, TAL WEATHER, ASCENT I-LOADS, FIRSTS, SIGNIFICANT ANOMALIES, ETC.)

STS-112/ ISS 9A

Continued…

STS112-709-033 (12 October 2002) --- Newly installed Starboard S1 Truss and Canadarm2.

STS112-326-033 --- Wolf (left) & Sellers during 2nd EVA. Wolf is anchored to a foot restraint on ISS's Canadarm2 while Sellers traverses along the airlock spur.

Continued…

FLT DURATION: 10:19:57:50

S/T: 1001:18:46:44

OV-104: 219:21:26:00

DISTANCE: 4,513,015 sm

ISS005-E-16524 ---- Atlantis on approach to ISS for rendezvous and docking operations to deliver the 15 ton S1 Truss.

BELOW: STS112-331-031 -- The EXP 5 & STS-112 crews in Destiny Lab on ISS. From left, front row EXP 5 crew: Peggy A. Whitson/FE, Valery G. Korzun/CDR(RSA), & Sergei Y. Treschev/FE(RSA). From left, back row STS-112 crew: Wolf/MS, Magnus/MS, Melroy/PLT, Ashby/CDR, Sellers/MS, and Yurchikhin/MS(RSA).

Continued…

LAUNCH WINDOW:
- Launch window opened at 280:19:40:51Z and closed at 280:19:50:50Z for a total launch window of 9m59s. In-plane time was 280:19:45:51Z for a launch window of 4m59s.

LAUNCH DELAYS: NONE
- Launch occurred On-Time at 280:19:45:51Z, 3:45:51 PM EDT on Monday, October 7, 2002.

TAL WX:
- Zaragoza (prime and selected) and Moron (2-Eng TAL Call) were forecast and observed GO. Moron earlier forecast was NO GO for showers and anvils. Ben Guerir was not available.

PERFORMANCE ENHANCEMENTS:
- Standard Set plus: (1) PE Operational High Q TRN/OCT, (2) OMS Assist, (3) 52 NM MECO, (4) Del Psi

FLIGHT DURATION CHANGES: NONE
- Planned landing at KSC on Orbit 171. MLGTD at KSC Runway 33 on Orbit 171 at 291:15:43:41Z, 11:43:41 AM EDT, 10:19:57:50 MET. NLGTD at 291:15:43:48Z, 11:43:48 AM EDT. STS-112 was the 75th planned landing at KSC, but the 60th actual landing at KSC, and the 36th landing on Runway 33.

FIRSTS/LASTS:
- First use of ET Shuttle Observation Camera during ascent.

EVENTS:
- MC4 Start at 282:14:18:46Z, 3.2 fps, orbit 200.4 by 213.6 nm.
- ISS Capture at MET 1:19:30:19, 282:15:16:10Z.
- Hard dock to PMA2 Lab Fwd Port complete at 1:19:44:06 MET, 282:15:29:57Z.
- PMA/APAS Hatch Open at 282:16:40Z, 1:20:55:09 MET. ODS Hatch open at 282:16:50Z, 1:21:05:09 MET.
- EVA 1 (JAL) Start at 283:15:21Z, 2:19:35 MET End at 283:22:22Z, 3:02:36 MET, duration 7h01m (Attached S1 to S0 Truss using SSRMS. Released CETA cart launch locks. Connected Zenith side power umbilicals and deployed S-Band Antenna. Installed S1 nadir ETVCG).
- First Reboost maneuver start at 285:10:52:48Z, 4:15:06:57 MET, delta V of 11.9 fps, altitude increase of 3.4 nm, orbit 216 by 204 nm.
- EVA 2 (JAL) Start at 285:14:30Z, 4:18:44 MET, End 285:20:34Z, 05:00:48 MET, duration 6h04m. (Installed Z1/P6, Z1/Lab and RBVM SPD's. Connected ATA Umbilicals. Installed Lab ETVCG. ZCG Activation).

Continued…

FLT NO.	ORBITER	CREW (6) TITLE, NAMES & EVA'S	LAUNCH SITE, LIFTOFF TIME, LANDING SITES, ABORT TIMES	LANDING SITE/ RUNWAY, CROSSRANGE LANDING TIMES FLT DURATION, WINDS	SSME-TL NOM-ABORT EMERG THROTTLE PROFILE ENG. S.N.	SRB RSRM AND ET	ORBIT		FSW	PAYLOAD WEIGHTS, PAYLOADS/ EXPERIMENTS	MISSION HIGHLIGHTS (LAUNCH SCRUBS/DELAYS, TAL WEATHER, ASCENT I-LOADS, FIRSTS, SIGNIFICANT ANOMALIES, ETC.)
							INC	HA/HP			
STS-112/ ISS 9A Continued...											

STS112-382-003 (16 October 2002) --- New ISS configuration as viewed from departing Atlantis.

LEFT: JSC2002-E-41249--STS Lead FD Phil Engelauf in MCC WFCR reviewing Flight Day 2 activities.

JSC2002-01809 -- Members of MOD Planning Team in JSC MCC shuttle flight control room (WFCR). CAPCOM Stephanie D. Wilson holds the STS-112 mission logo. Flight Director John Curry stands to right of Wilson.

JSC2002-01806 -- STS-112/ISS-9A Orbit 1 Team in the ISS Flight Control Room (BFCR) in JSC MCC. Flight Director Mark Kirasich stands near center on front row. Left of center, ISS SPAN Team Lead Dan Bahadorani holds ISS logo.

Continued...

EVENTS (Continued):
- Second Reboost maneuver (c3) start at 287:11:20:50Z, 6:15:34:59Z MET, delta V = 6.9 fps, altitude increase 1.96 nm, orbit 219.4 by 203.3 nm.
- EVA 3 (JAL) Start at 287:14:11:25Z, 6:18:25:34 MET, End 287:20:47Z, 07:01:01 MET, EVA duration 6h36m. (IUA on MT R&R. S1 to S0 fluid (ammonia) jumper connections, removal of port and starboard keel pins, last of TRRJ SPD's, TRRJ bolts).
- Total cargo transfers from Orbiter to ISS = 29120 lbm (S1 Segment = 27676 lbm), Total cargo transfers from ISS to Orbiter = 1351 lbm Consumables Transfer: H_2O Total = 1658.1 lbm (16 CWC's with 1603.7 lbm and 3 PWR's with 54.4 lbm). Total N2 (Tank) = 68.2 lbm.
- Total O_2 = 60 lbm (Pre-Breathe: EVA 1 = 10 lbm, EVA 2 = 10 lbm, EVA 3 = 10 lbm, Tank Transfer= 28 lbm).
- Undocking at 289:13:13:25Z, 8:17:27:34 MET.
- Total ISS Visitor Time = 6:21:33:28.
- Post-undocking initial separation maneuver began at 289:13:13Z. ISS flyaround terminated at 289:14:30Z, 8:18:44 MET.
- Final Separation at 289:15:00Z, 8:19:14 MET, delta V= 5.5 fps, resulting Orbit = 200.8 nm by 219.9 nm.
- Orbit Adjust Maneuver at 290:20:26:51Z, 10:00:41:00 MET, delta V = 93.9 fps, Orbit 146.6 nm by 219.9 nm
- Note: At 291:08:35Z, using Progress engines, raised the ISS 6.9 miles.

RENDEZVOUS # 60:
- Rendezvous and Dock with ISS (Dock to PMA2 Lab Fwd Port)

SIGNIFICANT ANOMALIES:
- Piece of debris impacted ETA ring near IEA box on LH SRB at 33 seconds.
- Insulating foam was lost on ET-115 left bipod ramp (approx 4" X 5" X12") exposing bipod housing SLA closeout.
- Primary Thruster L4D failed off due to low chamber pressure (IFA STS-112-V-01).
- Panel F7 SM Alert Light Brightness
- Supply Water Crossover Valve Circuit Breaker did not indicate Open
- System A Pyros for SRB Holddown Posts and ET Vent Arm Systems did not fire at T-0 (IFA STS-112-K-01).
- EVA Glove Wrist Tether Point Torn
- RPOP PGSC (STS-5) Network Problem
- Emergency Egress Net Daisy Wheel Knob broke
- PCS 1 O2 Supply Pressure Indication failed OSH
- MADS recorder "stuck" at beginning of tape (tape came off reel)
- Forward RCS Primary Thruster F3F Failed On Heater
- ICOM A from Shuttle to Station not operating
- Handheld Microphone failed

FLT NO.	ORBITER	CREW 7 UP/7 DOWN TITLE, NAMES & EVA'S	LAUNCH SITE, LIFTOFF TIME, LANDING TIMES ABORT TIMES	LANDING SITE/ RUNWAY, CROSSRANGE LANDING TIMES FLT DURATION, WINDS	SSME-TL NOM-ABORT EMERG THROTTLE PROFILE ENG. S.N.	SRB RSRM AND ET	INC	ORBIT HA/HP	FSW	PAYLOAD WEIGHTS, PAYLOADS/ EXPERIMENTS	MISSION HIGHLIGHTS (LAUNCH SCRUBS/DELAYS, TAL WEATHER, ASCENT I-LOADS, FIRSTS, SIGNIFICANT ANOMALIES, ETC.)

STS-113/ ISS 11A
 	
SEQ FLT #112
 	
KSC-112
 	
PAD 39A-63
 	
MLP-2
 	
16TH SHUTTLE FLIGHT TO ISS
	OV-105 (Flight 19)
	
	ENDEAVOUR
	
	OMS PODS:
	LPO4-26
	RPO1-33
	FRC5-19
		CDR:
		James D. Wetherbee (Flt 6 - STS-32, STS-52, STS-63, STS-86, STS-102) P680/R108/V80/M198
		
		PLT:
		Paul S. Lockhart (Flt 2 - STS-111) P681/R279/V176/M245
		
		M/S 1/EV1:
		Michael E. Lopez-Alegria (Flt 3 - STS-73, STS-92) P682/R202/V163/M175
		
		M/S 2/EV2:
		John B. Herrington P683/R287/M251
		
		M/S 3 UP/EXP 6 CDR:
		Kenneth D. Bowersox (Flt 5 - STS-50, STS-61, STS-73, STS-82) P684/R146/V97/M130
		
		M/S 4 UP/EXP 6 Flt Eng 1:
		Nikolai Budarin (Russia) P685/R288/M252
		
		M/S 5 UP/EXP 6 Flt Eng 2:
		Donald R. Pettit P686/R289/M253
		
		M/S 3 DN/EXP 5 Flt Eng 2:
		Sergei Y. Treschev (Russia) (STS-111 Up) P687/R283/M248
		
		M/S 4 DN/EXP 5 CDR:
		Valery C. Korzun (Russia) (STS-111 Up) P688/R282/M247
		
		M/S 5 DN/EXP 5 Flt Eng 1:
		Peggy A. Whitson (STS-111 Up) P689/R281/F35
		
		Continued…
			KSC 39A
			328:00:49:47Z
			7:49:47 PM EST (P)
			7:49:47 PM EST (A)
			Saturday 5
			11/23/02 (EST) (14)
			
			LAUNCH WINDOW:
			7M08S IN 2 PANES
			ISS PLANAR/PHASE
			
			EOM PLS: KSC
			TAL: ZZA
			TAL WX: MRN
			
			SELECTED:
			RTLS: KSC 33/N/N
			TAL: ZZA 30/N/SF
			AOA: KSC 33/N/N
			PLS: EDW 22/N/N
			
			TDEL:
			0.04 -0.278/-0.24
			
			MAX Q NAV:
			763 765
			
			SRB STG:
			2:04.8 2:04
			
			PERF: NOMINAL
			
			2 ENG TAL (BEN):
			2:33 2:35
			
			NEG RETURN:
			3:52 3:55
			
			PTA (U/S 183):
			5:01 5:01
			
			PTM (U/S 183):
			6:05 6:10
			
			SE TAL (ZZA):
			6:00 6:01
			
			SE PTM (U/S 646):
			6:53 6:59
			
			MECO CMD:
			8:22.4 8:22.9
			
			VI:
			25821 25823
			
			Continued…
				KSC 33 (KSC 61)
				341:19:37:13Z
				2:37:13 PM EST
				
				Saturday 21
				12/7/02 (14)
				
				DEORBIT BURN:
				341:18:31:33Z
				
				XRANGE: 2.1 NM
				
				ORBIT DIR: AL 32
				
				AIM PT: NOMINAL
				
				MLGTD: 2846 FT
				341:19:37:13Z
				VEL: 194 KGS
				         197 KEAS
				HDOT: -2.8 FPS
				
				TD NORM 195:
				3009 FT
				
				NLGTD: 5814 FT
				341:19:37:23Z
				VEL: 163 KGS
				         159 KEAS
				HDOT: -5.8 FPS
				
				DRAG CHUTE DEPLOY: 155 KEAS
				341:19:37:25Z
				
				BRK INIT: 65 KGS
				
				DRAG CHUTE JETTISON: 57 KGS
				341:19:38:00Z
				
				BRK DECEL FPS[2]:
				AVE 3.9 PK 5.1
				
				WHEELS STOP:
				341:19:38:28Z
				13420 FT
				
				ROLLOUT:
				10574 FT
				75 SEC
				
				WINDS:
				H3 R7 KTS
				OFFICIAL:
				0308P13
				H4 R7
				
				DENS ALT: 580 FT
				
				FLT DURATION:
				13:18:47:26
				
				Continued…
					104/104/ 109%
					
					PREDICTED:
					100/104.5/104.5/ 72/104.5
					
					ACTUAL:
					100/104.5/99/ 72/104.5
					
					1 = 2050 (4)
					2 = 2044 (8)
					3 = 2045 (7)
					
					M 3 EOM:
					WEIGHT:
					200993 LBS
					X CG:
					1087.63
					
					LANDING:
					WEIGHT:
					200939 LBS
					X CG:
					1089.52
						BI-114
						
						RSRM 86
						
						ET-116
						
						SLWT-21
						
						ET IMPACT
						
						1:14:10 MET
						
						LAT:
						36.54 S
						
						LONG:
						158.67 W
							51.60 (16)	DIRECT INSERTION
								
								POST OMS-2:
								169.9 x 125.7 NM
								
								DEORBIT:
								APOGEE:
								214 NM
								PERIGEE:
								212 NM
								
								VELOCITY:
								25907 FPS
								
								ENTRY RANGE:
								4351 NM
									OI-29 (4)	CARGO:
										38393 LBS
										
										PAYLOAD CHARGEABLE:
										30217 LBS
										
										DEPLOYED:
										29672 LBS
										
										NON-DEPLOYED:
										46 LBS
										
										MIDDECK:
										288 LBS
										
										SHUTTLE ACCUMULATED WEIGHTS:
										DEPLOYED:
										1189722 LBS
										NON-DEPLOYED:
										1559554 LBS
										CARGO TOTAL:
										3547208 LBS
										
										PERFORMANCE MARGINS (LBS):
										FPR: 3065
										FUEL BIAS: 937
										FINAL TDDP: 1736
										RECON: 2486
										
										PAYLOADS:
										PLB:
										ISS 11A (ITS P1 TRUSS) CETA CART B SRMS, ODS
										
										MIDDECK:
										ISS 11A
										
										5 CRYO TK SETS 6 GN2 TANKS RMS 70
										
										RMS USED TO UNBERTH P1 ITS AND HAND-OFF TO SSRMS FOR MATE TO S0 TRUSS.
											Brief Mission Summary: STS-113 was the 16th American assembly mission to the ISS. The primary goals achieved on this mission were to transport the EXP 6 crew to the ISS and return the EXP 5 crew to earth after 5 months in space and to install the Port (P1) Integrated Truss Assembly. The 45-ft long 14-ton P1 truss is the opposite side mate to the Starboard S1 truss delivered on STS-112. It is the 4th of 11 truss structures that ultimately will extend the ISS length to that of a football field. The P1 truss contains the Active Thermal Control System (to be activated later), a second UHF comm system, a second CETA cart, and a Thermal Radiator Rotary Joint (TRRJ).
											
											KSC W/D: OPF 79, VAB 9, PAD 35 = 123 days total.
											
											LAUNCH POSTPONEMENTS:
											- Launch was postponed from July after Post-STS-110 visual inspections of OV-104 Inconel 12" MPS LH2 Flowliners revealed three cracks to SSME 2. Subsequent inspections found cracks in other orbiter LH2 Flowliners:
											- OV-103 - three cracks (SSME 1)
											- OV-105 - one crack (SSME 1) and one crack (SSME 2)
											- MPTA - one crack (SSME 1)
											- OV-102 three cracks (SSME 2). OV-102 flowliners are CRES. After analyses, tests, etc., including consideration of other repair techniques, the decision was made to use weld-repair technique and polishing of Flowliner holes.
											- As a result, STS-113 and STS-112 moved ahead of STS-107. STS-113 launch date was set to November 6, 2002 EST.
											- At FRR, STS-113 Launch was postponed 1 day to November 7, 2002 EST at 11:56 PM (311:04:56Z).
											
											LAUNCH SCRUBS:
											- Scrubbed Monday, November 7 Launch at approximately L-3 hours due to an O2 leak in PCS 2 between ECLSS Supply Valve and 576 Bulkhead. Leak was first noticed when Haz Gas Detection System indicated an O_2 concentration of approximately 150 ppm in the Mid-Body. Troubleshooting procedures isolated the leak to PCS 2 outside the cabin between ECLSS O2 Supply valve and Crew Module 576 bulkhead. Launch date set to NET Monday, November 18. Inspection/troubleshooting found a blowing leak in PCS 2 O_2 flex hose near the 576 bulkhead. Replaced PCS 2 O_2 and N2 flex hoses. During preparation to get access to PCS 2 O_2 line under PLB liner, an Access Platform came in contact with the RMS damaging the TPS, Kevlar honeycomb with minor delamination to composite boom. Tests and analyses proved it is OK to fly-as-is. On November 20, set launch date to 11/22/02. Technical Scrub.
											- Scrubbed 11/22/02 launch planned for 8:15:30 PM EST at L-8 minutes due to unstable weather at ZZA and MRN. Early forecasts were showers within 20nm at Zaragoza and occasional overcast 1500 feet and showers at MRN. At L-1 hour, Moron weather had improved and FD updated TAL to Moron. However, both TAL sites were forecast and observed NO GO at the L-8 minute scrub time and at TAL landing times. TAL weather Scrub. Ben Guerir was not available as a TAL site; however, Ben Guerir was observed NO GO for ceiling and showers.
											
											Continued…

ISS005-E-21546 (25 November 2002) --- Endeavour approaches the ISS with the Port One (P1) truss in the cargo bay.

FLT NO.	ORBITER	CREW 7 UP/7 DOWN / TITLE, NAMES & EVA'S	LAUNCH SITE, LIFTOFF TIME, LANDING SITES, ABORT TIMES	LANDING SITE/ RUNWAY, CROSSRANGE / LANDING TIMES FLT DURATION, WINDS	SSME-TL NOM-ABORT EMERG / THROTTLE PROFILE ENG. S.N.	SRB RSRM AND ET	ORBIT INC	HA/HP	FSW	PAYLOAD WEIGHTS, PAYLOADS/ EXP	MISSION HIGHLIGHTS (LAUNCH SCRUBS/DELAYS, TAL WEATHER, ASCENT I-LOADS, FIRSTS, SIGNIFICANT ANOMALIES, ETC.)

STS-113/ ISS 11A

Continued…

SS EVA 86
DOCKED QUEST EVA 12
EMU/TETHERED EVA 79
SCHEDULED EVA 80
DURATION 6:45

SS EVA 87
DOCKED QUEST EVA 13
EMU/TETHERED EVA 80
SCHEDULED EVA 81
DURATION 6:10

SS EVA 88
DOCKED QUEST EVA 14
EMU/TETHERED EVA 81
SCHEDULED EVA 82
DURATION 7:00

MCC WHITE FCR (42)

FLIGHT DIRECTORS:
ISS LD/O1 - A. F. Algate
ISS O 2 - M. A. Kirasich
ISS PLNG - A. P. Hasbrook
STS LD/O 1 - P. L. Engelauf
STS O 2 - C. A. Koerner
STS O 3/PLNG - J. M. Curry
A/E - J. P. Shannon
MOD - R. E. Castle

Continued…

OMS-2:
38.12   37:49.2
250 FPS   256 FPS
2:42   5:31

Continued…

S/T: 1015:13:34:10

OV-105: 206:14:12:06

DISTANCE: 5,735,600 sm

S113E05230

Continued…

LAUNCH WINDOW:
- ISS first Planar window opened at 328:00:44:48Z and closed at 328:0054:46Z with PLT at 328:00:49:47Z (7:49:47 PM EST) for a 7M08S launch window. Second Planar window opened at 328:00:47:56Z and closed at 328:00:57:55Z.

LAUNCH DELAYS: NONE
- Launch occurred on time at 328:00:49:47Z, 7:49:47 PM EST Sat. 11/ 23/ 2002.

TAL WX:
- Zaragoza (prime and selected) was forecast and observed GO. Moron was forecast NO GO for ceiling (BKN 2500 ft and showers within 20 nm) but verified GO at landing time. 2-Eng TAL call ZZA. Ben Guerir was N/A, but was NO GO.

PERFORMANCE ENHANCEMENTS:
- Standard Set plus: (1) PE Operational High Q (WIN/DEC), (2) OMS Assist, (3) 52 NM MECO, (4) Del Psi

FIRSTS/LASTS:
- First flight with 3 Days Extension due to weather wave-offs.
- Record Minimum Crossrange of 2.1 nautical miles.
- **John Herrington/MS2 is the first & as of 2010 the only Native American to fly in space. He is an enrolled member of the Chickasaw Nation.**

6th & 7th SHUTTLE CREWMEMBER REPLACEMENTS
- Gus Loria was replaced by Lockhart in Aug. 2002 and Don Thomas (to join EXP 6) by Pettit in Jul. 2002 - both due to medical issues. (Fifth Shuttle crewmember replacement occurred on STS-98.)

FLIGHT DURATION CHANGES: Extended flight 3 days total.
- EOM - Planned landing at KSC on orbit 170 (Tig orbit 169) at 338:20:49Z, 3:49 PM EST on Wednesday, December 4, 2002. Waved-off landing on orbit 170 (Tig orbit 169) at Tig-21 minutes due to NO GO forecast for ceiling (broken 6000 feet). Weather reported that at landing time ceiling was 8000 feet and showers at 30 nm (GO Observation).
- Waved-off landing on orbit 171 (Tig orbit 170) at Tig –24 minutes due to NO GO Forecast of ceiling 6500 feet. (One day extension) waveoff 1 day. Landing observations verified NO GO (BKN 6500 feet).
- EOM+1 - Waved-off landing at KSC on orbit 185 (Tig orbit 184) at 339:19:54Z, 2:54 PM EST on Thursday, December 5, 2002 at approximately Tig-3H15M due to observed 18 knot crosswinds, moisture within 30 nm and broken 7000 feet.
- Waved-off landing at KSC on orbit 186 (Tig orbit 185) a few minutes later for crosswind, moisture, and ceiling violations. (Second day Extension) waveoff 2 days.
- EOM+2 - Waved-off landing at KSC on orbit 200 at 340:18:57Z, 1:57 PM EST on Friday, December 6, 2002 at Tig-3H03M due to NO GO forecast and observed drizzle at SLF and overcast 900 ft.
- Decided to proceed with Deorbit Prep for orbit 201 landing but not fluid load. Closed the PLBD's and gave GO for OPS 3 transition. Weather violations continued. Waved-off landing at Tig-1H12M due to continued NO GO observed and forecast drizzle/fog, visibility 3 miles and overcast 600 feet. (Third Day Extension) waveoff 3 days.
- EOM+3 - Landed at KSC Runway 33 on orbit 216 at 341:19:37:13Z, 2:37:13 PM EST, Saturday, December 7, 2002 (MET 13:18:47:26). Total extensions 3 Days (Record for three days extension due to weather, landed on EOM+4). STS-57 was extended 3 days; however, the first day extension was for science and the last 2 days were weather extensions. Record minimum crossrange of 2.1 miles

Continued…

FLT NO.	ORBITER	CREW 7 UP/7 DOWN / TITLE, NAMES & EVA'S	LAUNCH SITE, LIFTOFF TIME, / LANDING SITES, ABORT TIMES	LANDING SITE/ RUNWAY, CROSSRANGE / LANDING TIMES FLT DURATION, WINDS	SSME-TL NOM-ABORT EMERG / THROTTLE PROFILE ENG. S.N.	SRB RSRM AND ET	ORBIT INC	HA/HP	FSW	PAYLOAD WEIGHTS, / PAYLOADS/ EXP	MISSION HIGHLIGHTS (LAUNCH SCRUBS/DELAYS, / TAL WEATHER, ASCENT I-LOADS, FIRSTS, SIGNIFICANT ANOMALIES, ETC.)

STS-113/ ISS 11A

Continued...

STS113-714-039 --- John B. Herrington (left) and Michael E. Lopez-Alegrias, work on the newly installed Port One (P1) truss.

SIGNIFICANT ANOMALIES:
- O_2 concentration in Mid Body above expected baseline. Replaced secondary O2 line and secondary GN_2 flex hoses (IFA STS-113-V-01).
- Right OMS Engine Bi-Propellant Valve 2 position indicator indicated 96 percent Open at start of OMS Assist Burn and continued to indicate 96 percent Open after burn (IFA STS-113-V-02).
- S-Band Power Amplifier 2 power output low (IFA STS-113-V-04).
- Hardware C&W pushbutton failures
- APU 2 GG Bed Heater Cycles Abnormal
- Wireless Video System video problems
- FES Primary B Shutdown - Ice in Topping Core (IFA STS-113-V-03)
- RMS Wrist Roll Sluggish Joint Response
- OCA failure during private medical and private family conferences
- PGSC for RPOP RS 422 cable bad
- Film review indicates very small engine 1 coldwall nozzle fuel leak, no performance impact.

Continued...

RENDEZVOUS #61:
- Rendezvous and Dock with ISS (PMA2 Lab Fwd Port).

SHUTTLE NIGHT LAUNCH #28:

EVENTS:
- NC1 maneuver at 328:03:42:05Z (02:52:28 MET) resultant altitude of 170.2 by 186.7 nm.
- MC4 maneuver at 329:20:27Z (01:19:37 MET) resultant altitude 203.3 by 215.5 nm.
- ISS Capture (PMA 2 Lab Fwd Port) at 329:21:20:27Z (01:21:08:53 MET)
- ISS Hard dock at 329:22:10:49Z (01:21:21:02 MET).
- ODS Upper Hatch Open (all hatches open) at 329:23:29:47Z (01:22:40 MET)
- IELK S/L Transfer (Official transfer of ISS from Expedition 5 Crew to Expedition 6 Crew) at 330:02:28Z (02:01:39:13 MET)
- SRMS unberth of P1 ITS at 330:15:19:51Z (02:14:30 MET) and positioned P1 over orbiter Port Wing for handoff to SSRMS. (Thereafter SRMS camera was used only for video support of EVA activities.)
- SSRMS used to mate P1 ITS to S0 truss at 330:18:50:14Z (02:18:00:27 MET)
- EVA 1 Start at 330:19:48Z (02:18:57 MET), EVA 1 End at 331:02:33Z (03:01:43 MET) on November 26, 2002, duration 6H45M. All three EVA's used Pre-Breathe Protocol while exercising on Shuttle Ergometer located in mid-deck. Crew had to use Shuttle Ergometer as the CEVAS had a problem. Made connections between P1 and S0 Trusses. Released launch restraints on CETA Cart, DLA, and TARJ Stinger, installed Node 1 WETA.
- Reboost 1 at 331:17:10:47Z (03:16:21 MET) delta V + 2.4 fps, altitude increase 2.4 nm, altitude 216 by 207 nm
- EVA 2 Start at 332:18:36Z (04:17:46 MET), EVA 2 End at 333:00:47Z (04:23:57 MET) on November 28, 2002, duration 6H10M. Installed fluid jumpers between P1 & S0. Removed P1 Port & Stbd keel pins. Installed WVS TX Assy on P1. Relocated CETA Cart from P1 to S1. Released P1/P3 line clamps. Removed & stowed Radiator beam launch locks.
- Reboost 2 at 333:16:50:59Z (05:16:01:12 MET), delta V = 2.56 fps, altitude increase 0.7 nm, altitude 216 by 209 nm.
- EVA 3 Start at 334:19:24Z (06:18:34 MET) and End at 335:02:24Z (07:01:34 MET) on November 30, 2002, duration 7H00M. Installed Z1/P6/Lab, Lab HX, and P1 RBVM SPD's. Reconfigured electrical harnesses, route power through Main Bus switching units.
- Reboost 3 at 335:16:36:47Z (07:15:49 MET), delta V = 8.6 fps, altitude increase 2.4 nm, final orbit 216.6 by 211.4 nm.
- Farewell 336:17:18Z (08:16:28 MET)
- ODS Upper Hatch closed at 336:17:47:47Z (08:16:58 MET), Lab Fwd Hatch (all hatches closed) closed at 326:18:15:47Z (08:17:26 MET)
- Undocking complete at 336:20:04:50Z (08:19:15:03 MET)
- Transfers: Shuttle to ISS 2160 lbs plus P1 ITS of 27514 lbs, 690 lbs H_2O (672 lbs in 7 CWC's and 18 lbs in one PWR), 32 lbs O_2 used during prebreathe for 3 EVA's. Plus 6 LiOH cans. Transfer ISS to Shuttle 2250 lbs.
- MEPSI deploy at approx. 336:22:25Z (08:21:36 MET)

STS113-E-05433 (2 December 2002) --- The ISS post undocking of Endeavour as the two spacecraft flew over northwestern Australia. The newly installed Port One (P1) truss now complements the Starboard One (S1) truss in center frame.

JSC2002-01994 --- The Ascent/Entry FCT pose for group portrait in the shuttle flight control room (WFCR) in Houston's MCC. Ascent/Entry Flight Director Wayne Hale is in center front row.

FLT NO.	ORBITER	CREW (7) TITLE, NAMES & EVA'S	LAUNCH SITE, LIFTOFF TIME, LANDING SITES, ABORT TIMES	LANDING SITE/ RUNWAY, CROSSRANGE LANDING TIMES FLT DURATION, WINDS	SSME-TL NOM-ABORT EMERG THROTTLE PROFILE ENG. S.N.	SRB RSRM AND ET	INC	ORBIT HA/HP	FSW	PAYLOAD WEIGHTS, PAYLOADS/ EXPERIMENTS	MISSION HIGHLIGHTS (LAUNCH SCRUBS/DELAYS, TAL WEATHER, ASCENT I-LOADS, FIRSTS, SIGNIFICANT ANOMALIES, ETC.)
STS-107 SEQ FLT #113 KSC-113 PAD 39A-40 MLP-1	OV-102 (Flight 28) Columbia OMS PODS: LPO5-17 RPO5-16 FRC2-28 EDO FLT 15 S/H RDM 1	CDR: Rick D. Husband (Flt 2 - STS-96) P690/R248/V177/M216 PLT: William C. McCool P691/R290/M254 M/S 1: David M. Brown P692/R291/M255 M/S 2: Kalpana Chawla (Flt 2 - STS-87) P693/R230/V178/F30 M/S 3 (PAYLOAD CDR): Michael P. Anderson (Flt 2 - STS-89) P694/R235/V179/M205 M/S 4: Laurel Blair Salton Clark P695/R292/F37 P/S 1: Ilan Ramon (ISRAEL) P696/R293/M256  MCC WHITE FCR (43) FLIGHT DIRECTORS: LD/O 2 - K. B. Beck O 1 - J. S. Stich O 3 - B. P. Austin O 4 - J. M. Hanley A/E - L. E. Cain MOD - P. L. Engelauf	KSC 39A 16:15:39:00Z 10:39:00 AM EST (P) 10:39:00 AM EST (A) Thursday (34) 1/16/03 (10) LAUNCH WINDOW: 2H30M CTOB LAUNCH WINDOW: 2H30M CTOB DEORBIT BURN: 32:13:15:18Z Sunday, February 1, 2003 EOM PLS: KSC TAL: MRN TAL WX: ZZA SELECTED: RTLS: KSC 15 CI/N TAL: MRN 20 N/N AOA: EDW 04 CI/N PLS: EDW 04 N/N TDEL: 0.11 0.032/0.070 MAX Q NAV: 756 749 SRB STG: 2:05.4 2:07 PERF: NOMINAL 2 ENG TAL (MRN): 2:39 2:50 NEG RETURN: 3:50 3:52 PTA (U/S 242): 5:15 5:14 SE OPS 3: 5:25 PTM (U/S 242): 5:54 6:05 SE TAL (ZZA): 5:56 6:05 SE PTM (U/S 459): 7:00 7:05 MECO CMD: 8:20.9 8:23 VI: 25863 25860 OMS-2: 41.18 41:24 186 FPS 186 FPS	DEORBIT BURN: 32:13:15:18Z Sunday, February 1, 2003 PLANNED LANDING: On KSC 33 at 9:15:50 AM EST ORBIT DIRECTION: DL 49 **IN MEMORIAM -- See next page.** FLT DURATION: 15:22:20:32 Lost contact with Columbia at 8:59:32 AM EST S/T: 1031:11:54:42 OV-102: 300:17:39:40 DISTANCE: 6,649,757 sm	104/104/ 109% PREDICTED: 100/104.5/72/ 72/104.5 ACTUAL: 100/104.5/72/ 72/104.5 1 = 2055 (1) 2 = 2053 (5) 3 = 2049 (7) EI: WEIGHT: 234495 LBS X CG: 1078.53 EI + 15 MIN: WEIGHT: 234167 LBS X CG: 1077.87	BI-116 RSRM 88 ET-93 LWT-86 ET IMPACT 1:24:35 MET LAT: 2.28 N LONG: 139.42 W	39.0 (8)	DIRECT INSERTION POST OMS-2: 156 x 147 NM DEORBIT: Ha 151.6 NM Hp 135.0 NM VELOCITY: 25762 FPS ENTRY RANGE: 4439 NM	OI-29 (5)	CARGO: 35463 LBS PAYLOAD CHARGEABLE: 24316 LBS DEPLOYED: 0 LBS NON-DEPLOYED: 23515 LBS MIDDECK: 801 LBS SHUTTLE ACCUMULATED WEIGHTS: DEPLOYED: 1189722 LBS NON-DEPLOYED: 1559554 LBS CARGO TOTAL: 3547208 LBS PERFORMANCE MARGINS (LBS): FPR: 3047 FUEL BIAS: 1112 FINAL TDDP: 1335 RECON: 1348 PAYLOADS: PLB: SPACEHAB RDM FREESTAR OARE (MORE THAN 80 EXPERIMENTS) MIDDECK: FREESTAR - MIDDECK H/W RAMBO S/H SUPPORT EQUIPMENT 9 CRYO TK SETS (EDO PALLET) 5 GN2 TANKS NO RMS	*Brief Mission Summary:* The STS-107 crew carried out a 16-day mission dedicated to a mix of life and physical sciences on board the first SPACEHAB Research Double Module (RDM). The crew of seven included the first Israeli astronaut. During descent for landing at KSC at an altitude of 203,000 feet over north central Texas, a breach in the TPS on Columbia's left wing resulted in loss of vehicle and crew. Communications with the crew were lost at 9 AM EST, Saturday, Feb. 1, 2001. Second loss of vehicle and crew in Shuttle program. KSC W/D: OPF 79, VAB 9, PAD 35 = 123 days total. LAUNCH POSTPONEMENTS: - Baselined launch date of 1/11/01 on 11/10/99. - Postponed launch date to 2/22/01 on 3/3/00. - Postponed launch date to 4/15/01, then 6/14/01, others(?), then to 9/2/03, moved after STS-112 and STS-113 (Priority flights to HST and ISS flights that had been ppd. due to flow-liner cracks.) - Postponed launch date to 1/16/03. LAUNCH SCRUBS: None LAUNCH WINDOW: - Launch Window was 2H30M (Crew Time On Back). LAUNCH DELAYS: NONE - KSC weather was excellent, perhaps the best launch weather experienced in Shuttle Program. - Launch occurred On-Time at 16:15:39:00Z, 10:39:00 AM EST, on Thursday, January 16, 2003. TAL WX: - Moron was prime and selected. Both Moron and Zaragoza were forecast and observed GO. Ben Guerir was not available. PERFORMANCE ENHANCEMENTS: - Standard Set plus: PE Operational High Q (WIN/JAN) and OMS Assist. FIRSTS/LASTS: - First flight of Space Shuttle in CY 2003. - First flight of Spacehab RDM (Research Double Module) with more than 80 Experiments. Science: Biological, Physiological & Countermeasures, Physical Sciences, Earth and Space Science, Space & Technology Development. - First EDO Pallet Flight since STS-90 (April 17, 1998) - First flight of Israeli Astronaut - Ilan Ramon FLIGHT DURATION CHANGES: - Planned landing at KSC on orbit 256 (TIG orbit 255) on Saturday, February 1, 2003. Deorbit maneuver was initiated at 32:13:15:18Z, 8:15:18 AM EST on Saturday, February 1, 2003 (TIG orbit 255, landing orbit 256). Planned landing time was 32:14:15:50Z, 9:15:50 AM EST. - Orbiter weight and Xcg at entry interface was 234,495 lbm, Xcg was 1078.53. - Orbiter weight and Xcg at entry interface plus 15 minutes 234,167 lbm, Xcg was 1077.87. - Flight controllers reported increased temperatures on some sensors and some failed sensors in left wing area. Off-nominal indications started at approximately 32:13:52:17Z. Columbia contact loss (Loss-of-Signal) occurred at 32:13:59:32Z, 8:59:32 AM EST (15:22:20:32 MET), 16 minutes prior to planned landing Continued…

FLT NO.	ORBITER	CREW (7) TITLE, NAMES & EVA'S	LAUNCH SITE, LIFTOFF TIME, LANDING SITES, ABORT TIMES	LANDING SITE/ RUNWAY, CROSSRANGE LANDING TIMES FLT DURATION, WINDS	SSME-TL NOM-ABORT EMERG THROTTLE PROFILE ENG. S.N.	SRB RSRM AND ET	ORBIT		FSW	PAYLOAD WEIGHTS, PAYLOADS/ EXPERIMENTS	MISSION HIGHLIGHTS (LAUNCH SCRUBS/DELAYS, TAL WEATHER, ASCENT I-LOADS, FIRSTS, SIGNIFICANT ANOMALIES, ETC.)
							INC	HA/HP			

STS-107

Continued...

CAIB REPORT:
Accident Analysis indicated that the physical cause of the loss of Columbia and its crew was a breach in the Thermal Protection System on the leading edge of the left wing. The breach was initiated by a piece of insulating foam that separated from the left bipod ramp area of the External Tank and struck the wing in the vicinity of the lower half of Reinforced Carbon-Carbon panel 8 at 81.9 seconds after launch. During re-entry, this breach in the Thermal Protection System allowed superheated air to penetrate the leading-edge insulation and progressively melt the aluminum structure of the left wing, resulting in a weakening of the structure until increasing aerodynamic forces caused loss of control, failure of the left wing, and breakup of the Orbiter.

Shuttle Legacy Mural - Hanging in LCC Firing Room at KSC

COLUMBIA TRIBUTE
By Mike Leinbach/Launch Director & Amy Simpson/KSC PH-2, May 2010

IN MEMORIAM

The STS 107 crew is shown on-orbit in SPACEHAB research module aboard Columbia. From left (bottom row) wearing red shirts to signify their work shift color, are Kalpana Chawla/MS2, CDR Rick D. Husband, Laurel B. Clark/MS4, and Ilan Ramon/PS1(Israel). From left (top row), wearing blue shirts, are David C. Brown/MS1, PLT William C. McCool, and Michael P. Anderson/PL-CDR.

Continued...

FLIGHT DURATION CHANGES: (continued)
time. Communications and tracking were lost at an altitude of approximately 203,000 feet while Columbia was traveling at approximately 12,500 miles per hour at Mach 18.
- Columbia and 7 astronauts were lost over Texas.

RED SHIFT: Rick Husband, Kalpana Chawla, Laurel Clark, Ilan Ramon.

BLUE SHIFT: William McCool, David Brown, Michael Anderson (PL CDR)

STS-107 EVENTS:
Orbital Altitude was 150 nm.

STS-107 FLIGHT OBJECTIVES/EXPERIMENTS:
- Flight was a dedicated and successful science/research mission.
- Primary payload is SPACEHAB Research Double Module (SHRDM) with International, NASA and SPACEHAB commercial payloads including Life Sciences, Materials, and Microgravity Science Research Experiments.
- Fast Reacting Experiments Enabling Science, Technology, Applications and Research (FREESTAR) is a complex Secondary Payload which is a cross bay carrier with following payloads: MEIDEX (Mediterranean Israeli Dust Experiment), Solar Constant-3 (SOLCON-3), Shuttle Ozone Limb Sounding Experiment-2 (SOLSE-2), Critical Viscosity of Xenon-2 (CVX-2), Low Power Transceiver (LPT), and Space Experiment Module-14 (SEM-14)
- Ram Burn Observation (RAMBO)

SIGNIFICANT ANOMALIES:
- ET Foam loss during ascent at approximately 81 seconds (likely from Bi-pod area) (IFA). Re-design constraint to flight.
- RSRM Nozzle Flex Boot Separation (IFA). Constraint to flight.
- O$_2$ Tank 7 Heater failed off in Manual Mode (IFA STS-107-V-02)
- Suspected Fuel Cell Monitoring System Data Cable problem. FCMS is suspect after same problem with backup cable.
- SM I/O Errors on IP Bus
- DSR 20 Error Message 32 (Loss of tape recording and playback)
- 70 mm Hasselblad Intermittent Motor Drive (Binds or jams)
- 2nd 70 mm Hasselblad Motor Jam
- STGT site outage
- Payload No I-COM B Transmission in Spacehab (Not being heard in Spacehab)
- Spacehab water loop Degradation (Flow rates decreasing)
- Payload Ku Channel 2 Data Dropouts (Ku-Band and S-Band)
- AC2 Phase B "Sluggish" Current Signature on Orbiter (IFA STS-107-V-01)
- Forward DAP Auto A Contact Deselected by RM
- Spacehab Rotary Separator flooding short
- Loss of Columbia and crew during Entry - IFA STS-107-V-03

KSC-2010-4452 (http://mediaarchive.ksc.nasa.gov/index.cfm). This Tribute Display features Columbia, the "first of the fleet", rising above earth at the dawn of the Space Shuttle Program. Crew-designed patches for each of Columbia's missions lead from earth toward our remembrance of the STS-107 crew. In the background are images from the Chandra X-Ray Observatory (launched aboard STS-93) representing Columbia's contributions toward scientific discovery. Other significant accomplishments include the first space shuttle landing at White Sands with STS-3, first deployment of commercial satellites during STS-5, first four-member crew on STS-5, first Spacelab mission and first six-member crew on STS-9, first female mission commander (Eileen Collins) on STS-93, as well as multiple laboratory missions—many with international partnership. (May 2010)

FLT NO.	ORBITER	CREW (7) TITLE, NAMES & EVA'S	LAUNCH SITE, LIFTOFF TIME, LANDING SITES, ABORT TIMES	LANDING SITE/ RUNWAY, CROSSRANGE LANDING TIMES FLT DURATION, WINDS	SSME-TL NOM-ABORT EMERG THROTTLE PROFILE ENG. S.N.	SRB RSRM AND ET	INC	ORBIT HA/HP	FSW	PAYLOAD WEIGHTS, PAYLOADS/ EXPERIMENTS	MISSION HIGHLIGHTS (LAUNCH SCRUBS/DELAYS, TAL WEATHER, ASCENT I-LOADS, FIRSTS, SIGNIFICANT ANOMALIES, ETC.)
STS-114/ LF-1 SEQ FLT #114 KSC-114 PAD 39B-50 MLP-3 17TH SHUTTLE FLIGHT TO ISS ISS LOGISTICS FLIGHT 1	OV-103 (Flight 31) Discovery OMS PODS: LPO1-34 RPO3-32 FRC3-31	CDR: Eileen Collins (Flt 4 - STS-63, STS-84, STS-93) P697/R188/V139/F24 PLT: James M. Kelly (Flt 2 - STS-102) P698/R263/V180/M229 M/S 1/EV-1: Soichi Noguchi (Japan JAXA) P699/R294/M257 M/S 2/EV-2: Stephen K. Robinson (Flt 3 - STS-85, STS-95) P700/R222/V152/M196 M/S 3: Andrew S.W. Thomas (Flt 4 - STS-77, Up to Mir on STS-89, Down on STS-91, STS-102) P701/R213/V149/M186 M/S 4: Wendy B. Lawrence (Flt 4 - STS-67, STS-86, STS-91) P702/R192/V146/F25 M/S 5: Charles Camarda P703/R295/M258 SS EVA 89 EMU/TETHERED EVA 82 SCHEDULED EVA 83 DURATION 6:50 SS EVA 90 EMU/TETHERED EVA 83 SCHEDULED EVA 84 DURATION 7:14 SS EVA 91 EMU/TETHERED EVA 84 SCHEDULED EVA 85 DURATION 6:01 Continued…	KSC 39B 207:14:39:00Z 10:14:39 AM EDT (P) 10:39:00 AM EDT (A) Tuesday 14 7/26/05 (8) LAUNCH WINDOW: 4M52S (In-Plane Time) with ISS EOM/PLS: KSC TAL: ZZA TAL WX: MRN, FMI SELECTED: RTLS: KSC 33/N/N TAL: ZZA 30/N/SFD AOA: KSC 33/N/N PLS: EDW 22/N/SFD TDEL: 0.02 -0.178 MAX Q NAV: 775 709 SRB STG: 122.4 126.76 PERF: NOMINAL: 2 ENG TAL (ZZA): 2:43 2:44 NEG RETURN: 3:52 3:57 PTA (U/S 182): 5:10 5:14 SE TAL (ZZA 104): 6:09 6:14 PTM (U/S 614): 6:10 6:14 SE PRESS 104: 6:57 7:02 MECO CMD: 8:24.2 8:24.9 VI: 25819 25819.6 OMS-2: 37:40 38:00 100.7 FPS 99 FPS	EDW 22, CONC EDW 50, CONC 31 221:12:11:23Z 5:11:23 AM PDT Tuesday 21 8/9/05 (7) DEORBIT BURN: 221:11:06:18Z XRANGE: 46 NM ORBIT DIR: AL 33 AIM PT: NOM MLGTD: 1311 FT 221:12:11:23Z VEL: 226 KGS 222 KEAS HDOT: -5.5 FPS TD NORM 205: 2761 FT DRAG CHUTE DEPLOY: 192 KEAS 221:12:11:31.9Z NLGTD: 6573 FT 221:12:11:38Z VEL: 163 KGS 156 KEAS HDOT: -6.4 FPS BRK INIT: 90 KGS DRAG CHUTE JETTISON: 53 KGS 221:12:12:08Z BRK DECEL FPS: AVE 5.1 PK 6.6 WHEELS STOP: 221:12:12:31Z 12657 FT ROLLOUT: 11346 FT 68 SEC NO BLACKOUT DURING ENTRY Continued…	104/104/109% PREDICTED: 100/104.5/104.5/ 72/104.5 ACTUAL: 100/104.5/104.5/ 72/104.5 1 = 2057 (1) 2 = 2054 (5) 3 = 2056 (3) ALL BLOCK II ENGINES M 3 EOM: WEIGHT: 225792 LBS X CG: 1086.58 LANDING: WEIGHT: 225727 LBS X CG: 1088.21	BI-125 RSRM-92 (17) ET-121 SLWT-22 ET IMPACT: 1:14:10 MET LAT: 36.56°S LONG: 158.7°E	51.60 (17)	DIRECT INSERTION POST-OMS-2 123.6 NM X 85.0 NM DEORBIT: Ha 191.0 NM Hp 168.0 NM ENTRY VELOCITY: 25858 FPS ENTRY RANGE: 4416 NM	OI-30 (1)	CARGO: 38652 LBS PAYLOAD CHARGEABLE: 29807 LBS DEPLOYED: 26413 LBS NON-DEPLOYED: 3231 LBS MIDDECK: 163 LBS SHUTTLE ACCUMULATED WEIGHTS: DEPLOYED: 1216135 LBS NON-DEPLOYED: 1562948 LBS CARGO TOTAL: 3585860 LBS PERFORMANCE MARGINS (LBS): FPR: 3098 FUEL BIAS: 1269 FINAL TDDP: 2111 RECON: 3792 PAYLOADS: PLB: ISS LF-1 MPLM RAFFAELLO, ESP2, LMC, RMS, ODS, OBSS MIDDECK: ISS LF-1 RAMBO 5 CRYO TK SETS 6 GN2 TANKS RMS 71 RMS USED FOR TPS SURVEYS AND TWO GAP FILLER REMOVALS	*Brief Mission Summary:* With STS-114/LF-1 (17th ISS mission), NASA initiated Return to Flight 2 years after the Columbia accident. The crew was charged with a busy to-do list that included testing new safety techniques and delivering much-needed supplies to ISS. KSC/WD: OPF 994, VAB 25, PAD 85 = 1104 days total LAUNCH POSTPONEMENTS: - Baselined OV-104 Atlantis as ULF-1 Crew Rotation flight with launch date of 1/16/03 on 12/6/01 - Postponed launch date to NET 3/1/03 on 9/16/02. Postponement caused by Engine Flowliner cracks. - Subsequent postponements after STS-107 Accident to NET 7/21/03, NET 10/1/03, NET 12/18/03, NET 3/11/04, NET 9/12/04. - Postponed launch date to NET 3/6/05 on 3/22/04. Changed flight to ISS Logistics Flight LF-1, canceled crew rotation, and changed orbiters to Discovery OV-103. - Tanking Test 1 on 4/24/05 experienced two intermittent LH2 ECO anomalies. (ECO sensors #3 & #4 failed WET). Replaced MPS Point Sensor Box (PSB) and all Sensor #3 & #4 wiring to LH2 monoball. Subsequent to completion of this work, the Tanking Test #2 LH2 Sensor performance was nominal. - Postponed launch date to NET 5/12/05, 5/15/05, 5/22/05, 7/13/05 - Rolled back from pad 39B to VAB on 5/26/05 to swap stacks with STS-121, due to a late all-flights requirement for a heater on the ET LO2 Feedline upper bellows, to prevent formation of critical ascent ice debris in that area. Installation of the bellows heater was started on ET-121 (STS-114 was ET-120) in the VAB before the STS-114 stack was rolled-back. Removed and replaced an out-of-spec H2 diffuser. - Replaced MPS PSB after a power card failure. - Rolled out to Pad 39B on 06/15/05 and set launch date of 07/13/05 on 05/22/05. LAUNCH SCRUBS: - Scrubbed 07/13/05 launch attempt at 194:17:30Z (L-2:14:51 to Window Opening) when LH2 ECO Sensor #2 failed WET (failed to transition to DRY with Sim Commands). This violated OMRSD and LCC MPS-22 requirements for four functional LH2 sensors. Extensive tests were conducted that identified a degraded PSB ground and some evidence of EMI as potential causes of the false WET problem. At MMT on 07/20/05, decided to set launch for 07/26/05 (without a special tanking test), allowing sufficient time to clean up the ground and EMI. Decision was made to perform ECO Sensor #2 and #4 pin swap that provides additional troubleshoot results. (Note: ECO sensors operated normally on 7/26/05; further analyses and tests have significantly reduced the concerns about PSB grounding and EMI as causes of the STS-114 anomalies, but this remains a UA as of February 2006). - Weather: All three TAL sites were forecast and observed GO. RTLS and AOA1 landing site KSC was forecast NO GO for precipitation and thunderstorms within 20 NM and observed NO GO for thunderstorms within 20 NM (Anvil). 07/13/05 Launch Attempt was a combined Technical/Weather Scrub. LAUNCH WINDOW: Window opened at 207:14:34:33Z and closed at 207:14:43:52Z for a total window of 9M19S. The Preferred Launch Time (In-Plane Time) was 207:14:39:00Z resulting in a Launch Window of 4M52S. Continued…

JSC2005-E-16245 (April 2005) — Art panel for STS-114 Return to Flight - Features Shuttle, ISS Assembly, crew patch, first step for humans return to the Moon, and onward to Mars & beyond.

FLT NO.	ORBITER	CREW (7) TITLE, NAMES & EVA'S	LAUNCH SITE, LIFTOFF TIME, LANDING SITES, ABORT TIMES	LANDING SITE/ RUNWAY, CROSSRANGE LANDING TIMES FLT DURATION, WINDS	SSME-TL NOM-ABORT EMERG THROTTLE PROFILE ENG. S.N.	SRB RSRM AND ET	ORBIT INC	HA/HP	FSW	PAYLOAD WEIGHTS, PAYLOADS/ EXPERIMENTS	MISSION HIGHLIGHTS (LAUNCH SCRUBS/DELAYS, TAL WEATHER, ASCENT I-LOADS, FIRSTS, SIGNIFICANT ANOMALIES, ETC.)
STS-114/ LF-1 Continued...		Continued... MCC WHITE FCR (44) FLIGHT DIRECTORS: SHUTTLE: A/E - LeRoy Cain LD/O 1 - Paul Hill O 2 - Anthony Ceccacci O 3/Plng - Catherine Koerner Team 4 - Kelly Beck WX - Steven Stich MOD - Phil Engelauf ISS: LD/O 2 - Mark Ferring O 1 - Bryan Lunney O 3/Plng - Joel Montalbano Team 4 - Richard LaBrode		Continued... DENS ALT: 3799 FT FLT DURATION: 13:21:32:23 S/T: 1045:09:27:05 OV-103: 255:20:12:58 DISTANCE: 5,796,419 sm						

S114-E-5070 (26 July 2005) --- Photo shows a large piece of foam detached from ET PAL Ramp (light spot centered just below LO₂ feedline). The debris was also seen on ET live video camera, in photo below at left, and indicated no impact to Discovery.

First use of the 50-foot-long robotic arm known as Orbiter Boom Sensor System (OBSS) equipped with laser imager and cameras to inspect for ascent damage of Wing Leading Edges RCC and Shuttle Bottom Tiles during approach and docking with ISS.

STS-114 ET camera view of large debris 6 seconds after SRB separation

Based on an estimated, bounded distance from the camera of 171 to 236 inches:

Length 1 (L1) = 24 to 33 inches   Width 1 (W1) = 2.3 to 3.5 inches
Length 2 (L2) = 10 to 14 inches   Width 2 (W2) = 6.9 to 9.6 inches
                                    Width 3 (W3) = 5.6 to 7.8 inches

Lengths and widths are projected from the image plane.
The actual length of the object may be greater than that stated

From MMT Brief of IFA: "ET TPS Foam Loss During Ascent – Constraint to next flight"

Continued...

LAUNCH DELAYS:
None. Launch occurred at 207:14:39:00Z, 10:39:00 AM EDT on Tuesday, 07/26/05.

TAL WX:
Zaragoza (Primary and Selected) was forecast and observed GO. Moron was forecast and observed NO GO for Crosswind. FMI (Istres) was forecast GO but observed NO GO for Tailwind violation.

PERFORMANCE ENHANCEMENTS:
Standard Set plus: (1) PE Operational High Q SUM/JUL, (2) OMS Assist, (3) 52 NM MECO, (4) Del Psi

FLIGHT DURATION CHANGES:
- On Flight Day 4, decision made to extend flight 1 day to give more time to transfer activities to and from ISS. EOM Day: Deorbit Tig on Orbit 201 was at 220:07:43Z and landing time at KSC on Orbit 202 at 12/18:07 MET 220:08:46Z (4:46 AM EDT). EDW was not called up for support on EOM day.

- Early weather forecast was GO except for a chance of showers. Gave crew a GO for PLBD closure at 220:05:15Z. Light rain was observed at SLF for a few minutes. At 220:06:15Z gave crew a GO for fluid loading. Last forecast changed to NO GO at 220:0643Z with observed broken low clouds at 1000 feet in SLF area. At 220:07:16Z, due to low clouds, decision was made to wave off first opportunity at KSC. KSC was observed GO at landing time. Flight extension 1 day plus one orbit. KSC opportunity 2 Deorbit Tig on Orbit 202 was at 220:09:19Z and landing time at KSC was 220:10:22Z (5:42 AM CDT). Last forecast at 220:08:46Z was GO. However, due to unstable conditions in low clouds, FD made decision to wave off landing at KSC on second opportunity. KSC was observed NO GO due to precipitation in SLF area. Flight extension now 2 days.

- EOM + 1 Day: All three EOM landing sites KSC, EDW, and NOR were called up on pick-em day with Discovery landing at one of the three sites. First opportunity for a KSC landing was on Orbit 218 at 221:09:08Z with Tig at 220:08:05Z on Orbit 217. Gave a GO for PLBD closing at 221:05:05Z but did not give a GO for crew fluid loading. Weather was NO GO with showers, thunderstorms, and confirmed electrified cloud within 30 NM. Showers and thunderstorms were forecast within 30 NM at landing time. At 221:06:55Z, waved off landing at KSC on Orbit 218. Flight extensions 2 days + one orbit.

- Changed Landing site to EDW. Targeted landing at KSC on Orbit 219 at 221:10:43Z. Gave crew a GO to fluid load at 221:08:40Z. At 221:08:43Z, weather forecaster reported two cells developing rapidly northeast of field moving NE with lightning in a northeast cell. At 221:08:57Z, Crew reported APU prestart complete. Current observations at SLF had showers within 30 NM with electrified cirrus (anvil) within 30 NM with forecast of thunderstorms within 30 NM moving NE. At 221:09:00, Flight Director advised crew to stop fluid loading. Waved off landing at KSC on Orbit 219, the last opportunity on FD 13. Decision made to change landing sites to EDW concrete runway 22 on Orbit 220. Flight extensions 2 days + two orbits.

Continued...

FLT NO.	ORBITER	CREW (7) TITLE, NAMES & EVA'S	LAUNCH SITE, LIFTOFF TIME, LANDING SITES, ABORT TIMES	LANDING SITE/ RUNWAY, CROSSRANGE LANDING TIMES FLT DURATION, WINDS	SSME-TL NOM-ABORT EMERG THROTTLE PROFILE ENG. S.N.	SRB RSRM AND ET	ORBIT INC HA/HP	FSW	PAYLOAD WEIGHTS, PAYLOADS/ EXPERIMENTS	MISSION HIGHLIGHTS (LAUNCH SCRUBS/DELAYS, TAL WEATHER, ASCENT I-LOADS, FIRSTS, SIGNIFICANT ANOMALIES, ETC.)
STS-114/ LF-1 Continued…									

JSC2004-E-01407

CDR Collins

Discovery was about 600 ft from ISS when CDR Collins performed the first R-Bar (back-flip) maneuver to allow inspection of the vehicle heat shield. Photos were analyzed on the ground to assess any damage during ascent. (Photos shown top to bottom are: iss011e11255, iss011e11257, Iss011e11260, iss011e11263, Iss011e11270)

S114-E-6751 (2 August 2005) --- Crew portrait in Destiny Lab. From left (front row) are Thomas/MS, CDR Collins, & Noguchi/MS (JAXA). From left (back row) are PLT Kelly, Camarda/MS, Robinson/MS, & Lawrence/MS.

S114-E-6062 --- Noguchi (JAXA) participates in Mission's first EVA demonstrating Shuttle thermal protection repair techniques.

Continued…

Targeted landing at EDW on orbit 220. Discovery landed with MLGTD at EDW 22 at 221:12:11:23Z, 13:22:32:23 MET, 5:11:23 AM PDT on August 9, 2005. NLGTD was at 221:12:11:38Z.

FIRSTS/LASTS:
- First flight in Return-To-Flight after Columbia STS-107.
- First launch in 922 days after STS-107 launch.
- First flight with Istres, France as a TAL site.
- First flight with ET bipod redesign to eliminate large insulating foam ramps as a debris source and replace them with electric heaters.
- First use of the 50-foot-long robotic arm extension known as Orbiter Boom Sensor System (OBSS) equipped with Laser Imager and cameras to inspect Wing Leading Edges RCC and the Shuttle Bottom tiles for damage.
- First use of upgraded Ground Camera Ascent Imagery System, two WB-57 aircraft based video, and ship and ground based radar.
- First use of WLE instrumentation behind RCC panels to gather and downlink acceleration and temperature data during ascent phase.
- First use of orbiter back-flip pirouette (R-bar pitch maneuver) to allow ISS based photography of orbiter bottom TPS.
- First EVA crew to make repairs on shuttle bottom. Removed gap fillers protruding approximately 1 inch from black tiles in two areas of orbiter bottom black tiles, each extended approximately 1 inch. Gap fillers were removed during EVA 3.
- First flight with ET design change to use heater in bipod ramp area to prevent ice/frost buildup (in lieu of insulating foam in that area).
- Mandated day-time launch for STS-114 and STS-121 to provide proper lighting for video and film cameras observation of ET debris shedding during ascent.
- First flight with ET LOX Feedline upper bellows heater to prevent formation of critical ascent ice debris in that area.

EVENTS:
- ET Separation at 207:14:47:00Z, 8:46 GET
- MC-1 maneuver at 01:17:37:53, delta V 0.44 ft/sec Orbit 199.7 by 213.1 NM
- FD2 SRMS/OBSS survey of Wing Leading Edges and nose cap
- FD2 SRMS survey of orbiter upper surfaces
- ISS capture at 209:11:17:20Z (01:20:38:20 MET)
- Hard Dock: 209:11:31:53Z (01:20:52:53 MET)
- Open Lab Fwd Hatch at 209:11:51:00Z (01:21:12 MET)
- Open APAS Hatch at 209:12:35:00Z (01:21:56:00 MET)
- Open ODS Hatch at 209:12:14:00Z (01:22:14 MET) ISS ingress
- FD4 OBSS survey of heat-protection tiles. MPLM docked to Node 1. MPLM and Middeck transfers begin.
- EVA 1 start at 211:09:45:50Z, 3:19:06:50 MET, duration 6H50M, on 07/30/05. Crew members performed EWA & NOAX TPS sample repair DTO 848 in PLB. Crew used OBSS to scan pre-damaged RCC samples on DTO pallet.

Continued…

Continued…

FLT NO.	ORBITER	CREW (7) TITLE, NAMES & EVA'S	LAUNCH SITE, LIFTOFF TIME, LANDING SITES, ABORT TIMES	LANDING SITE/ RUNWAY, CROSSRANGE LANDING TIMES FLT DURATION, WINDS	SSME-TL NOM-ABORT EMERG THROTTLE PROFILE ENG. S.N.	SRB RSRM AND ET	INC	ORBIT HA/HP	FSW	PAYLOAD WEIGHTS, PAYLOADS/ EXPERIMENTS	MISSION HIGHLIGHTS (LAUNCH SCRUBS/DELAYS, TAL WEATHER, ASCENT I-LOADS, FIRSTS, SIGNIFICANT ANOMALIES, ETC.)
STS-114/ LF-1 Continued…											

ISS011-E-11517 (5 August 2005) --- ISS Canadarm2 grasps the MPLM for transfer from ISS Unity Node back to Discovery's cargo bay for return to Earth. James Kelly/Pilot, and Wendy Lawrence/MS controlled the transfer.

S114-E-6642 --- Robinson anchored to a foot restraint on ISS Canadarm2, participates in the mission's third EVA which included removal of two gap fillers protruding from orbiter bottom tiles.

JSC2004-E-45140 ---Lead Flight Director Paul Hill (foreground) and CAPCOM Stephen N. Frick monitor communications in the Shuttle Flight Control Room (WFCR) in JSC MCC with the STS-114 crewmembers during a fully-integrated - simulation - one of many to establish readiness for Return to Flight.

JSC2005-E-32556 (5 August 2005) --- U.S. Senator Kay Bailey Hutchison (R.-Texas) and U.S. Representative Tom DeLay (R.-Texas) talk to CDR Eileen M. Collins aboard Discovery. Looking on are NASA Administrator Mike Griffin (left) and Flight Director Jeff Hanley.

Continued…

EVENTS (Continued):
- EVA 2 start at 213:08:43:00, 5:18:04:00 MET, duration 7H14M, on 08/01/05. EVA crew removed, replaced, and performed checkout of ISS CMG 1. Crew started CMG 1.
- EVA 3 start at 215:08:48:00Z, 7:18:09:00 MET, duration 6H01M, on 08/03/05. Installed External Stowage Platform (ESP-2) on ISS airlock. Removed gap filler material (two) protruding from orbiter bottom tiles.
- Orbiter undocked from ISS at 218:07:23:45Z (10:16:44:45 MET)
- Total Consumables transferred to ISS 1855.2 lbm (18 CWC's & 5 PWR's), N₂ = 29 lbm tank-to-tank; Stack-to-stack O₂ = 60.85 lbm (27.6 lbm atmo & 33.3 metabolic), N₂ to ISS cabin transfer = -7.7 lbm.
- Total MPLM transfers to ISS 3695 lbs (2095 Cargo and 1600 HRF). 6600 lbs transferred to MPLM/Discovery for return to earth
- ISS Visitor Time was 8D19H51M52S (Hard dock to Undock)
- Sep 1 Burn at 218:08:36:26Z Ha 193.5 Hp 189.3, Sep Burn 2 at 218:09:04:26Z Ha 194.1 Hp 168.1 NM
- Orbit Adjust Burn at 221:11:06:18Z H

RENDEZVOUS # 62: Rendezvous and dock with ISS.

SPACE SHUTTLE NIGHT LANDING: # 20 total and sixth night landing at EDW.

SIGNIFICANT ANOMALIES:
- LH₂ ECO sensor #2 stayed wet when commanded dry caused launch scrub.
- ET TPS damages and TPS foam losses during ascent constraint to next flight:
- LH₂ PAL ramp, Ice/Frost ramp, Acreage, Intertank flange foam losses.
- +Y thrust strut flange and -Y Bipod spindle closeout foam losses.
- TPS Blanket damage near window 1
- TPS Gap Filler Protuberances (removed during EVA 3)
- Nose Landing Gear TPS tile damage
- APU 2 momentary loss of Press & Temp Indications
- ODS Capture Latch manual release talkback showed "Open" prior to hooks drive
- Airlock Aft "B" Hatch Closure difficulties
- Airlock Depress Off-Nominal
- TCS repeated loss of Track
- VRCS thruster R5R Low Pc. Heater may have failed on.
- MPS/SSME low pressure helium decay rate exceeded
- WSB GN₂ Regulator outlet pressure low
- High O₂ concentration in aft compartment during ascent
- Loss of several Orbiter tile putty repairs during ascent
- Late release of two FRCS Thruster TYVEK rain covers during ascent
- Orbiter forward ET attach point NSI pyro bolt ejection after nominal NSI firing

FLT NO.	ORBITER	CREW (7 up, 6 down) TITLE, NAMES & EVA'S	LAUNCH SITE, LIFTOFF TIME, LANDING SITES, ABORT TIMES	LANDING SITE/ RUNWAY, CROSSRANGE LANDING TIMES FLT DURATION, WINDS	SSME-TL NOM-ABORT EMERG THROTTLE PROFILE ENG. S.N.	SRB RSRM AND ET	INC	ORBIT HA/HP	FSW	PAYLOAD WEIGHTS, PAYLOADS/ EXPERIMENTS	MISSION HIGHLIGHTS (LAUNCH SCRUBS/DELAYS, TAL WEATHER, ASCENT I-LOADS, FIRSTS, SIGNIFICANT ANOMALIES, ETC.)
STS-121/ ULF1.1 SEQ FLT# 115 KSC 115 PAD 39B-51 MLP-1 18th Shuttle Flight to ISS ISS Logistics Flight 2	OV-103 (Flight 32) Discovery OMS PODS: LPO1-27 RPO3-34 FRC3-32	CDR: Steven W. Lindsey (Flt 4 - STS-87, STS-95, STS-104) P704/R229/V131/M200 PLT: Mark E. Kelly (Flt 2 (STS-108)) P705/R271/V181/M237 EV2/M/S 1 (PAYLOAD CDR): Michael E. Fossum P706/R296/M259 M/S 2: Lisa M. Nowak P707/R297/F38 M/S 3: Stephanie D. Wilson P708/R298/F39 EV1/M/S 4: Piers J. Sellers (Flt 2 (STS-112)) P709/R285/V182/M249 M/S 5 UP, stay as ISS EXP 13 FE: Thomas Reiter P710/R299/M260 (ESA - Germany) SS EVA 92 DOCKED QUEST EVA 15 EMU/TETHERED EVA 85 SCHEDULED EVA 86 DURATION 7:31 SS EVA 93 DOCKED QUEST EVA 16 EMU/TETHERED EVA 86 SCHEDULED EVA 87 DURATION 6:47 SS EVA 94 DOCKED QUEST EVA 17 EMU/TETHERED EVA 87 UNSCHEDULED EVA 7 DURATION 7:11	KSC 39B 185:18:37:55 Z 2:37:55 PM EDT (P) 2:37:55 PM EDT (A) Tuesday 15 7/4/06 (9) LAUNCH WINDOW: 3M43S (In-plane time with ISS) EOM PLS: KSC TAL: MRN TAL WX: ZZA, FMI SELECTED: RTLS: KSC 33/N/N TAL: MRN 20/CI/N AOA: KSC 15/N/N PLS: EDW 22/N/N TDEL: 0.09 .172 MAX Q NAV: 684 660 SRB STG: 2:03 2.02 PERF: NOMINAL 2 ENG TAL: 2:49 2:52 NEG RETURN: 3:58 4.02 PTA (U/S 160): 5:48 5:42 SE TAL (FMI 104): 606 6:17 PTM (U/S 160): 6:34 6:45 SE PRESS 104: 7:04 7:12 MECO CMD: 8:29.8 8:30.1	KSC 15 (KSC 62) 198:13:14:42 Z 9:14:42 AM EDT Monday 21 7/17/06 (11) DEORBIT BURN: 198:12:06.55 Z XRANGE: 258 NM ORBIT DIR: AL 34 AIM PT: NOMINAL MLGTD: 3273 FT 198:13:14:42 Z VEL: 198 KGS 199 KEAS HDOT: -1.8 FPS TD NORM 205: 2662 FT DRAG CHUTE DEPLOY: 189 KEAS 198:13:14:45 Z NLGTD: 6646 FT 198:13:14:53Z VEL: 149 KGS 145 KEAS HDOT: -5.8 FPS BRK INIT: 100 KGS DRAG CHUTE JETTISON: 54 KGS 198:13:15:18 Z BRK DECEL FPS[2]: AVE 5.6 PK 6.7 WHEELSTOP: 198:13:15:56 Z 12238 FT ROLLOUT: 8965 FT 74 SEC	104/104/109% PREDICTED: 100/104.5/ 104.5/67 104.5 ACTUAL: 100/104.5/ 104.5/67 104.5 1 = 2045 (8) 2 = 2051 (5) 3 = 2056 (4) All Block II Engines M 3 EOM: WEIGHT: 226063 LBS X CG: 1084.58 LANDING: WEIGHT: 225972 LBS X CG: 1086.32	BI-126 RSRM 93 ET-119 SLWT 23 ET IMPACT MET 1:14:32 LAT: 35.845S LONG: 157.76 W	51.60 (18)	DIRECT INSERTION POST OMS-2: 123.6 NM BY 85.0 NM DEORBIT: HA 190.7 NM HP 176.7 NM ENTRY VELOCITY: 25862 FPS ENTRY RANGE: 4494 NM	OI-30 (2)	CARGO: 37736 LBS PAYLOAD CHARGEABLE: 29280 LBS DEPLOYED: 23696 LBS NON-DEPLOYED: 5426 LBS MIDDECK: 158 LBS SHUTTLE ACCUMULATED WEIGHTS: DEPLOYED: 1239831 LBS NON-DEPLOYED: 1568532 LBS CARGO TOTAL: 3623596 LBS PERFORMANCE MARGINS (LBS): FPR: 3519 FUEL BIAS: 825 FINAL TDDP: 2290 RECON: N/A (sensor fail) PAYLOADS: PLB: ISS ULF1.1 ICC MPLM LMC RMS, ODS, OBSS MIDDECK: ISS ULF1.1, RAMBO, MAUI 5 CRYO TK SETS 6 GN2 TANKS RMS 72 USED FOR OBSS/LDRI ACTIVITIES	*Brief Mission Summary:* STS-121/ULF1.1 (18th ISS mission) continued the testing of new equipment and procedures for increasing Space Shuttle safety of flight. Specifically, this mission continued the testing of ET design and process changes for minimizing potentially damaging debris during launch, ground and flight camera systems for vehicle observations during launch, and techniques for on-orbit inspection and repair of vehicle TPS. The flight also delivered critical supplies and cargo for the repair and future expansion of the ISS. KSC W/D: OPF 264, VAB 7, PAD 41 = 312 days total. LAUNCH POSTPONEMENTS: - Baselined OV-103 launch date of 11/15/04 on 10/26/03 - Postponed launch date to NET 5/5/05 on 3/26/04. Slip due to Columbia accident - Postponed launch date to NET 7/10/05 on 10/29/04. Slip due to Columbia accident - Postponed launch date to NET 7/12/05 on 2/17/05 to provide on acceptable launch lighting conditions - Postponed launch date to NET 9/9/05 on 5/23/05 to reflect latest planning decisions - Postponed launch date to **TBD** on 11/15/05 - Postponed launch date to 5/10/06 on 3/16/06 - Postponed launch date to 7/1/06 LAUNCH SCRUBS: - Scrubbed Saturday 7/1/2006 launch attempt at 182:19:46Z (at L- 0h2m41s) while holding count at L-9 min. The window opened at 182:19:43:41 and closed at 19:53:41Z. The Preferred Launch Time was 183:19:26:11Z. Last forecast for KSC RTLS was forecast and observed NO-GO for thunderstorm attached anvils within 20 NM. KSC AOA1 and NOR AOA2 were forecast and observed NO-GO for thunderstorms within 20 NM. KSC PLS3 was forecast GO but observed crosswind of 19 knots. Primary TAL Moron and alternates Zaragoza and Istres (France) were forecast and observed GO. Weather scrub for KSC RTLS, AOA1 and PLS3. - Scrubbed Sunday 7/2/2006 launch attempt at 183: 17:14Z (at L-2h12m). The window opened at 183:19:21:09Z and closed at 183:19:31:09Z. The preferred launch time was 183:19:26:09Z. At the time of the scrub, there remained 7m41s to window closure. KSC RTLS was forecast NO-GO thunderstorm anvils within 20 NM and chance of broken 3000 ft and observed thunderstorms within 20 NM. KSC AOA1 was forecast NO-GO for thunderstorm anvils within 30 NM and chance of broken 3000 ft and observed thunderstorms. NOR AOA2 was forecast NO-GO for chance of thunderstorms within 30 NM and observed GO Primary TAL site Moron and alternate Istres (FMI) were forecast and observed GO. Zaragoza was forecast slight chance of thunderstorms within 20 NM but observed GO. All three TAL sites were observed GO. Weather Scrub - KSC RTLS, AOA. Management made the decision to go for a 48-hour turnaround so the fuel cell cryos could be topped off for a possible 1-day extension, power permitting. KSC RTLS/AOA/Launch weather scrub. Continued…

ISS013-E-48774 --- Discovery approaches ISS for docking with Leonardo Multipurpose Logistics Module (MPLM) in the payload bay.

FLT NO.	ORBITER	CREW (7) TITLE, NAMES & EVA'S	LAUNCH SITE, LIFTOFF TIME, LANDING SITES, ABORT TIMES	LANDING SITE/ RUNWAY, CROSSRANGE LANDING TIMES FLT DURATION, WINDS	SSME-TL NOM-ABORT EMERG THROTTLE PROFILE ENG. S.N.	SRB RSRM AND ET	ORBIT INC	HA/HP	FSW	PAYLOAD WEIGHTS, PAYLOADS/ EXPERIMENTS	MISSION HIGHLIGHTS (LAUNCH SCRUBS/DELAYS, TAL WEATHER, ASCENT I-LOADS, FIRSTS, SIGNIFICANT ANOMALIES, ETC.)
STS-121/ ULF1.1 Continued...		Continued... MCC WHITE FCR (45) FLIGHT DIRECTORS: A/E - Steve Stich LD/O 1 - Anthony Ceccacci O 2 - Norman Knight PLNG - Paul Dye MOD - Phil Engelauf ISS: LD/O 2 - R.E. LaBrode O 1 - A.P. Hasbrook O 3/PLNG - P. F. Dye ADO EVA's	Continued... VI: 25819 25821 HaHp: 123.6 x 31.1 OMS-2: 38:00 38:00 98.1 FPS 98.6 FPS	Continued... WINDS: 21008 P10 AVE: 5H, 7R PEAK: 6H, 8R DENS ALT: 1691 FT FLT DURATION: 12:18:36:47 OV-103: 263:14:49:45 S/T: 1058:04:03:42 DISTANCE: 5,293,923 sm							Continued...

STS121-E-05156 (4July 2006)--- ET was photographed by orbiter umbilical well camera for damage studies by ground experts.

S121-E-06239 --- STS-121 (green shirts) & Exp 13 crews in ISS Destiny Lab. From left (front row): Reiter/FE13 (ESA), Exp 13 CDR Pavel V. Vinogradov/RSA, & Jeffrey N. Williams/FE13. From the left (middle row): Wilson/MS, CDR Lindsey, & Nowak/MS. From the left (back row): Sellers/MS. Fossum/MS. & PLT Kelly.

STS-E-06058 (8 July 2006) --- Fossum and Sellers test the Shuttle RMS and the OBSS as a platform for making repairs to a damaged orbiter.

LAUNCH WINDOW:
- The July 4th launch window opened at 185:18:32:55Z and closed at 185:18:42:56Z giving a total window of 10 minutes plus 1 second. The Preferred Launch Time (In-Plane Time) was 185:18:37:55Z.
- Performance close time was 185:18:41:38Z, giving a launch window of 3m43s.

LAUNCH DELAYS:
- None. Launch occurred on time at 185:18:37:55Z (2:37:55 PM EDT) on Tuesday, July 4, 2006. SLF crosswinds were forecast at 16 knots but STA evaluation raised RTLS crosswind limit to 17 knots. All three TAL sites were forecast GO but Zaragoza was observed NO-GO for showers within 25 NM.

TAL WEATHER:
- MRN (Primary TAL), Istres, and Zaragoza were all three forecast GO. Zaragoza was observed NO-GO for showers within 25 nm.

PERFORMANCE ENHANCEMENTS:
- Standard Set plus (1) PE Low Q SUM/JUL, (2) OMS Assist, (3) 52 NM MECO, (4) Del Psi

FLIGHT DURATION CHANGES/LANDING:
- Total flight extension is 1 day.
- On FD4 , MMT made decision to extend flight 1 day (from 12+1+2 to 13+2) to permit additional EVA to accomplish RCC/tile repair materials DTO's. The plan was to land at one of the two EOM opportunities at KSC: (1) Deorbit 202 with landing on orbit 2023 (2) Deorbit 203 with landing on orbit 204. EDW was not called up. If unable to land at KSC on EOM, EDW would be called up for a "pick 'em?" KSC or EDW. TD 6-hr weather forecast for Deorbit 202 chance of showers within 30 nm. The weather forecast update at 1155Z removed showers within 30 nm and detached anvils were removed from the forecast changing the forecast to GO for deorbit. (Deorbit 203 forecast showers within 30 nm)
- Deorbit burn was at 198:12:06:55Z with KSC runway 33 as the preferred runway. At EI-15, an unexpected rain shower moved toward the SLF that was expected close to HAC for runway 33 by touchdown. Re-designated from runway 33 to runway 15 at M15 (185,000 feet) to avoid the weather buildup south of the SLF. MLG touchdown was at 198:13:14:42Z (9:14:42 AM EDT) on Monday July 17, 2006 for a flight duration of 12:18:36:47. NLG touchdown was at 198:13:14:53Z. There were no further flight duration changes. Total 1 day extension for operations.

EIGHTH SHUTTLE CREWMEMBER REPLACEMENT
- Carlos Noriega (medical issue) was replaced by Sellers in July 2004. (6th & 7th Shuttle crewmembers replacements occurred on STS-113.)

RENDEZVOUS # 63: Rendezvous and dock with ISS

Continued...

FLT NO.	ORBITER	CREW (7) TITLE, NAMES & EVA'S	LAUNCH SITE, LIFTOFF TIME, LANDING SITES, ABORT TIMES	LANDING SITE/ RUNWAY, CROSSRANGE LANDING TIMES FLT DURATION, WINDS	SSME-TL NOM-ABORT EMERG THROTTLE PROFILE ENG. S.N.	SRB RSRM AND ET	ORBIT		FSW	PAYLOAD WEIGHTS, PAYLOADS/ EXPERIMENTS	MISSION HIGHLIGHTS (LAUNCH SCRUBS/DELAYS, TAL WEATHER, ASCENT I-LOADS, FIRSTS, SIGNIFICANT ANOMALIES, ETC.)
							INC	HA/HP			

STS-121/ ULF1.1

Continued...

S121-E-06199 (10 July 2006) --- Fossum and Sellers (partially out of frame) restored ISS Mobile Transporter rail car to full operation and delivered a spare cooling system pump.

In JSC MCC Chris Lessmann/Entry Console Operator/USA (foreground) reviewing abort entry performance predictions and John Davidson/Abort Support/USA updating the Abort Region Determinator for DOL winds & atmosphere.

JSC2006-E-27890 --- Orbit-1 Flight Control Team group portrait in the Shuttle White Flight Control Room of JSC MCC. Flight Director Tony Ceccacci holds the STS-121 mission logo.

SIGNIFICANT ANOMALIES:
- L5L thruster heater fail off (first launch attempt)
- ET LH2 5% fill-point sensor failed wet when commanded to dry state (during loading attempts)
- FES Full up PRI B Shutdown
- Protruding Gap Fillers
- Personal hygiene hose leak
- TPS Blanket Damage
- 85-ft safety tether #24 retraction issue
- Scratch reported on crewlock external hatch sealing surface
- SAFER 5000 (EV1) unlatched during EVA. Relocked by EV2
- APU 1 Fuel Tank Leak
- APU 3 GG/FU Pump Heaters cycling in over temp range
- Two-inch spatula inadvertently released during EVA 3
- Waste Dump Nozzle Temps A&B unusual signature during condensate dump
- Right Air Data Probe initial fail to deploy
- WLEIS Inadvertent Software Shutdown (GFE)
- MCC GNC ISP Server Issue
- DOLILU PLOAD Procedural error (PLOAD LOX estimate high)

Continued...

FIRSTS/LASTS/NEW:
- First flight of an ET without the Protuberance Air Load ramps as a safety improvement to reduce potential for debris.
- First test of 50-ft robotic arm boom extension as a work platform.
- First flight with hardened tiles on NLG doors.
- First use of SRMS/OBSS/Laser Dynamic Range Imager (LDRI) to scan Orbiter WLE and Nose Cap (RCC).
- DTO 848 RCC crack repair tasks using caulk guns to dispense the NOAX (non-oxide adhesive experimental) material.
- First flight of Orbiter MLG with four new larger, smoother tires that can withstand higher loads at landing.
- New procedures developed to ensure gap fillers between heat-shielding tiles stay in place (5000 replaced prior to launch).
- First flight to take GPS to NAV (BFS). Incorporated after processing TACAN approx. 140K. Performed well.
- ISS has three crew members for first time since May 2003.

EVENTS:
- ET Separation at 185:18:46:46Z, 000:00:08:51 MET.
- OMS-2 ignition at 185:19:15:55Z, 98.7 fps, resultant orbit 124.4 by 85.1 nm.
- TI ignition 187:12:04:46Z, 16.8 seconds, resulting orbit 190.1 by 177.9 nm.
- SRMS/OBSS/Laser Dynamic Range Imager (LDRI) scanned both WLE and nose cap, no anomalous conditions identified.
- ISS captured at 187:14:51:45Z (1:20:13:49 MET).
- Hard dock at 187:15:10:28Z (1:20:32:33 MET).
- ISS Hatch Open at 187:16:29Z (1:21:51 MET). Welcomed by Expedition 13 two-person crew (Vinogradov and Williams).
- IELK Seat Liner transfer at 187:19:13Z (002:00:35:05 MET which is Reiter's Shuttle time). This is the official transfer of Thomas Reiter from Space Shuttle STS-121 crew to ISS Expedition 13 crew. ISS crew increased to three persons for first time since May 2003.
- Leonardo MPLM grappled and installed on Unity **Module**.
- EVA 1 Start at approximately 3/18:38 MET (189:13:15:55Z) July 8. Duration 7h 31m. Blade blocker inserted into Zenith IUA of MS, OBSS/SRMS Characterization. Rerouted TUS cable. EVA from ISS Quest A/L.
- EVA 2 Start at approximately 5/17:36 met (191:12:13:55Z) July 10. Duration 6h 47m. Nadir IUA R&R, Pump Module (w/FGB) transferred from ICC to ESP-2, R&R TUS. Piers' SAFER became detached, Mike re-locked it.
- EVA 3 start 193:11:20:30Z (7:16:42:35 MET), July 12. Duration 7h 11m. Completed 5 samples of NOAX DTO & IR imaging. Grapple Bar transferred to ISS.
- STS-121 crew farewell to ISS crew (Commander Pavel Vinogradov, Flight Engineers Jeffrey Williams & Thomas Reiter).
- APAS Hatch Close at 10/13:36 MET, ODS Hatch close 10/13:38 MET (196:08:15:55Z).
- STS-121 Undock from ISS at 10/15:29 MET, 196:10:06:55Z.
- Total consumables transferred from Orbiter to ISS: Water 1545.8 lbm (1454.9 lbm in 15 CWC's and 90.9 lbm in 4 PWR's); N2 74.2 lbm transferred to Joint Air Lock tanks. No oxygen transferred between tanks.
- Cargo transferred from Orbiter to ISS total 10903.35 lbs (7423.99 from MPLM, 1862.93 from Middeck, 1616.43 from ICC).
- Cargo transferred from ISS to Orbiter total 6450.92 lbs (4389.14 plus unplanned 241.52 lbs to MPLM and 1820.26 lbs to Middeck).
- No communications blackout during entry.

FLT NO.	ORBITER	CREW (6) — TITLE, NAMES & EVA'S	LAUNCH SITE, LIFTOFF TIME, LANDING SITES, ABORT TIMES	LANDING SITE/ RUNWAY, CROSSRANGE — LANDING TIMES, FLT DURATION, WINDS	SSME-TL NOM-ABORT EMERG THROTTLE PROFILE ENG. S.N.	SRB RSRM AND ET	INC	ORBIT HA/HP	FSW	PAYLOAD WEIGHTS, PAYLOADS/ EXPERIMENTS	MISSION HIGHLIGHTS (LAUNCH SCRUBS/DELAYS, TAL WEATHER, ASCENT I-LOADS, FIRSTS, SIGNIFICANT ANOMALIES, ETC.)
STS-115/ ISS 12A SEQ FLT# 116 KSC 116 PAD 39B-52 MLP-2 19TH SHUTTLE FLIGHT TO ISS	OV-104 (Flight 27) Atlantis OMS PODS: LPO4-RPO FRC4-27	CDR: Brent W. Jett (Flt 4 - STS-72, STS-81, STS-97) P711/R206/V132/M179 PLT: Christopher J. Ferguson P712/R300/M261 MS1/EV1: Joseph R. Tanner (Flt 4 - STS-66, STS-82, STS-97) P713/R185/V136/M162 MS2/EV2: Daniel C. Burbank (Flt 2 - STS-106) P714/R258/V183/M225 MS3/EV3: Heidimarie M. Stefanyshyn-Piper P715/R301/F40 MS4/EV4: Steven G. MacLean (Flt 2 - STS-52) P716/R156/V184/M138 (CSA-Canada) SS EVA 95 DOCKED QUEST EVA 18 EMU/TETHERED EVA 88 SCHEDULED EVA 88 DURATION 6:26 SS EVA 96 DOCKED QUEST EVA 19 EMU/TETHERED EVA 89 SCHEDULED EVA 89 DURATION 7:11 SS EVA 97 DOCKED QUEST EVA 20 EMU/TETHERED EVA 90 SCHEDULED EVA 90 DURATION 6:42 Continued…	KSC 39B 252:15:14:55 Z 11:14:55 AM EDT (P) 11:14:55 AM EDT (A) Saturday 5 9/9/06 (11) LAUNCH WINDOW: 4M41S (PLT in-plane) EOM PLS: KSC TAL: MRN TAL WX: ZZA, FMI SELECTED: RTLS: KSC 33/N/N TAL: MRN 20/N/N AOA: KSC 33/N/N PLS: EDW 22/N/N TDEL: 0.10 .062 MAX Q NAV: 731.36 723.09 SRB STG: 2:05 2.08 PERF: NOMINAL 2 ENG TAL (MRN): 2:42 2:47 NEG RETURN: 3:52 4.00 PTA (U/S 155): 5:16 5:26 SE TAL (FMI 104): 6:09 PTM (U/S 575): 6:19 6:24 SE PRESS 104: 7:00 7:00 MECO CMD: 8:23.7 8:24.8	KSC 33 (KSC 63) 264:10:21:23 Z 6:21:23 AM EDT Thursday 10 9/21/06 (11) DEORBIT BURN: 264:09:14.23 Z XRANGE: 225 NM ORBIT DIR: AL 35 AIM PT: NOMINAL MLGTD: 3131 FT 264:10:21.23 Z VEL: 191 KGS 189 KEAS HDOT: -1.5 FPS TD NORM 195: 2639 FT DRAG CHUTE DEPLOY: 181 KEAS 264:10:21:26 Z NLGTD: 5775 FT 264:13:21:32Z VEL: 158 KGS 156 KEAS HDOT: -6.4 FPS BRK INIT: 107 KGS DRAG CHUTE JETTISON: 63 KGS 264:10:21:53 Z BRK DECEL FPS^2: AVE 5.8 PK 8.5 WHEELSTOP: 264:10:22:15 Z 10670 FT ROLLOUT: 7539 FT 52 SEC	104/104/109% PREDICTED: 100/104.5/ 104.5/72 104.5 ACTUAL: 100/104.5/ 104.5/72 104.5 1 = 2044 (9) 2 = 2048 (6) 3 = 2047 (9) All 3 Block II Engines M 3 EOM: WEIGHT: 199711 LBS X CG: 1084.99 LANDING: WEIGHT: 199642 LBS X CG: 1086.98	BI-127 RSRM 94 ET-118 SLWT 24 ET IMPACT MET 1:13:36 LAT: 37.58S LONG: 160.16 W	51.60 (19)	DIRECT INSERTION POST OMS-2: 154.0 NM X 123.8 NM DEORBIT: HA 190 NM HP 179 NM ENTRY VELOCITY: 25867 FPS ENTRY RANGE: 4378 NM	OI-30 (3)	CARGO: 41848 LBS PAYLOAD CHARGEABLE: 35758 LBS DEPLOYED: 35552 LBS NON-DEPLOYED: 0 LBS MIDDECK: 206 LBS SHUTTLE ACCUMULATED WEIGHTS: DEPLOYED: 1275483 LBS NON-DEPLOYED: 1568738 LBS CARGO TOTAL: 3665444 LBS PERFORMANCE MARGINS (LBS): FPR: 2886 FUEL BIAS: 921 FINAL TDDP: 1749 RECON: 349 PAYLOADS: PLB: ISS 12A (P3/P4) Segment MIDDECK: RAMBO, MAUI, RMS, ODS, OBSS 5 CRYO TK SETS 5 N2 TANKS RMS 73 RMS USED FOR OBSS/LDRI SURVEYS AND UNBERTH P3/P4	***Brief Mission Summary:*** STS-115/12A (19th ISS mission), for the first time since late 2002, resumed assembly of the ISS. Atlantis left ISS with a new, second pair of 240-foot solar wings attached to a new 17.5-ton truss segment P3/P4 with batteries, electronics, and a giant rotating joint for sun tracking. The new solar arrays would double the ISS on-board power when the electrical systems were brought online during the STS-116 mission to follow. KSC W/D: OPF 264, VAB 7. PAD 41 = 312 days total. LAUNCH POSTPONEMENTS: - Baselined OV-104 launch date of 4/10/03 on 3/7/02 - Postponed launch date to 5/23/03 on 10/8/02; delays due to engine crack repairs - Postponed launch date to NET 8/21/03 on 3/13/03 - Postponed launch date to NET 10/30/03 on 4/17/03 - Postponed launch date to NET 1/22/04 on 5/28/03 - Postponed launch date to NET 7/24/04 on 7/29/03 - Postponed launch date to NET 2/10/05 on 10/3/03 - Postponed launch date to NET 8/28/05 on 3/22/04 - Postponed launch date to NET 12/8/05 on 10/29/04 - Postponed launch date to NET 2/16/06 on 5/23/04 - Postponed launch date to NET 7/1/06 on 10/31/05 - Changed launch date to TBD on 11/15/05 - Changed launch date to NET 8/28/06 on 3/16/06 - Advanced launch to 8/27/06 on 8/3/06 (actual launch date was 9/9/06) LAUNCH SCRUBS: - Scrubbed Sunday, 8/27/06 launch scheduled for 4:30 PM EDT at approximately L-26 hours to allow all Shuttle elements time to evaluate the lightning strike on Pad 39B on 8/26. Technical scrub. Launch rescheduled to NET 8/28/06 at 4:04 PM EDT. The Saturday, 10:00 PM EDT MMT decision was to spend another day analyzing the probability of damage to the SRB pics. The launch countdown was to continue for a NET Tuesday 8/29 launch. - Scrubbed Tuesday, 8/29/06 launch at approximately L-37 hours based on a KSC forecast of 50 knots, gusts to 65 with a potential of reaching the Pad maximum of 70 knots due to Tropical Storm Ernesto. Decision made at 3:45 AM EDT on 8/29/06 morning to roll back to the VAB with option to stop and reverse the rollback if the forecast improved. Rollback to VAB started at 10:04 AM EDT. The 11 AM forecast was in fact improved. KSC would sustain winds of less than 45 knots with gusts to 60 knots that is within the pad limit of 70 knots mph. The STS-115 stack was midway between Pad B and the VAB at 2:45 PM EDT when the decision was made to stop the Rollback and return the stack to Pad B. The launch date is under assessment. Weather Scrub. Rescheduled launch to 11:29 AM EDT on 9/6/06. Continued…

ISS013-E-79714 --Atlantis, carrying a crew of six, approached the orbital outpost with major elements for continuing construction of ISS.

FLT NO.	ORBITER	CREW (7) TITLE, NAMES & EVA'S	LAUNCH SITE, LIFTOFF TIME, LANDING SITES, ABORT TIMES	LANDING SITE/ RUNWAY, CROSSRANGE LANDING TIMES FLT DURATION, WINDS	SSME-TL NOM-ABORT EMERG THROTTLE PROFILE ENG. S.N.	SRB RSRM AND ET	ORBIT INC	HA/HP	FSW	PAYLOAD WEIGHTS, PAYLOADS/ EXPERIMENTS	MISSION HIGHLIGHTS (LAUNCH SCRUBS/DELAYS, TAL WEATHER, ASCENT I-LOADS, FIRSTS, SIGNIFICANT ANOMALIES, ETC.)
STS-115/ ISS 12A Continued…		Continued… MCC WHITE FCR (46) FLIGHT DIRECTORS: SHUTTLE: A/E - J. S. Stich LD/O 1 - P. F. Dye O 2 - C. A. Koerner O 3/PLNG - B. C. Lunney MOD- ISS: LD/O 2 - J. A. Mccullough O 1 - K. B. Beck O 3/PLNG - K. L. Alibaruho	Continued… VI: 25819 25818 OMS-2: 37:21 37:20.7 222 FPS 220.7 FPS	Continued… WINDS: 2H, 3R OFFICIAL: 2H, 3R 0303P04 DENS ALT: 696 FT FLT DURATION: 11:19:06:28 S/T: 1069:23:10:10 OV-104: 231:16:32:28 DISTANCE: 4,910,268 sm							Continued… LAUNCH SCRUBS: (continued) - Scrubbed Wednesday, 9/6/06 launch at approximately L-8.5 hours due to a fuel cell 1 coolant pump phase A short. (Pump operated on two phases.) 24-hour scrub turnaround with MMT at 1 PM 9/6 to decide launch date. The MMT decision was to press for a launch attempt on Friday, 9/8. Plan was to keep Phase A cb open during ascent. Technical scrub. - Scrubbed Friday, 9/8/06 launch attempt at 251:14:53Z while holding at T-9 minutes when ET LH2 ECO Sensor #3 indicated failed wet when actually sensor was dry. 24-hour scrub turnaround. ECO sensor operated normally during drainback and on Saturday launch day. GO for launch. Technical scrub. LAUNCH WINDOW: - The 9/9/06 launch window opened at 252:15:10:39Z and closed at 252:15:19:36Z for a total launch window of 9 minutes 0 seconds. The Preferred Launch Time (In-Plane time) was 252:15:14:55Z giving a launch window of 4m41s. LAUNCH DELAYS: - None. Launch occurred on time at 252:15:14:55Z (11:14:55 AM EDT) on Saturday, September 9, 2006. TAL WEATHER: - Zaragoza and Moron were forecast NO-GO for thunderstorms within 20. FMI was forecast with a 1-knot tailwind violation (average tailwind forecast to be 11 knots and peak tailwind forecast to be 16 knots). Zaragoza was observed NO-GO for thunderstorms and attached anvil. MRN and FMI were both observed GO at TAL landing time. Moron was selected as Prime TAL site. PERFORMANCE ENHANCEMENTS: - Standard set plus (1) PE Operational High Q SUM/AUG, (2) OMS Assist, (3) 52 NM MECO, (4) Del Psi, (5) Non-standard consumables reduction. FLIGHT DURATION CHANGES/LANDING: - EOM landing was planned for 263:13:04Z on 9/20/06 at KSC. However, during INCO survey of the orbiter after FCS checkout, an unidentified piece of debris was observed in Camera A. Tuesday 9/19/06 MMT decided to investigate the significance of the debris. The MMT extended the flight 1 day to allow time to perform RMS and OBSS surveys. The RMS and OBSS surveys of the PLB, both WLE and flight control surfaces using the RMS elbow camera, did not identify the debris. Atlantis was cleared for landing on EOM +1 day. Deorbit burn occurred at 264:09:14:23Z (11/17:59:28 MET) Orbit 185. Main Landing Gear touchdown on KSC Runway 33 was at 264:10:21:23Z (6:21:23 AM EDT) on Thursday, 9/20/06 for a flight duration of 11:19:06:28. Nose Landing Gear touchdown was at 264:10:21:32Z. Landing winds were forecast 03003P05 and observed 0303P04 (2H, 3R). Total flight duration extensions of 1 day (technical extension). Continued…

ISS013-E-81630 --- Crews in ISS Destiny Lab: Exp 13 from the left (front row): Thomas Reiter/FE (ESA), CDR Pavel V. Vinogradov (RSA), & Jeffrey N. Williams/FE. STS-115 from the left (second row): Tanner/MS, Stefanyshyn-Piper/MS, & CDR Jett; and from the left (top row): PLT Ferguson, Burbank/MS, & MacLean/MS (CSA).

S115-E-05623 (12 Sept. 2006) --- Piper, releases the restraints on the forward Solar Array Blanket Box (SABB) during EVA with Tanner, partially visible at top edge of frame.

JSC2006-E-40208 --- Mike Suffredini, ISS Program Manager, responds to a question from media during STS-115 mission update briefing on Sept. 14, 2006, at JSC. Shuttle Flight Director John McCullough is at left.

FLT NO.	ORBITER	CREW (7) TITLE, NAMES & EVA'S	LAUNCH SITE, LIFTOFF TIME, LANDING SITES, ABORT TIMES	LANDING SITE/ RUNWAY, CROSSRANGE LANDING TIMES FLT DURATION, WINDS	SSME-TL NOM-ABORT EMERG THROTTLE PROFILE ENG. S.N.	SRB RSRM AND ET	ORBIT		FSW	PAYLOAD WEIGHTS, PAYLOADS/ EXPERIMENTS	MISSION HIGHLIGHTS (LAUNCH SCRUBS/DELAYS, TAL WEATHER, ASCENT I-LOADS, FIRSTS, SIGNIFICANT ANOMALIES, ETC.)
							INC	HA/HP			
STS-115/ ISS 12A Continued…											Continued…

S115-E-05801 (13Sept. 2006) --- Burbank (red leg stripes) and MacLean/CSA (above & right) complete activation of SARJ.

Continued...

RENDEZVOUS # 64: Rendezvous and dock with ISS

SPACE SHUTTLE NIGHT LANDING: 21 (landed on runway KSC 33)

FIRSTS/LASTS/NEW:
- Used Airlock Campout Prebreathe Protocol for the first time. Crew spent sleep period isolated in the JAL (Quest Airlock) at reduced pressure of 10.2 psia.

EVENTS:
- Max Q at 252:15:15:45Z (00m50s)
- OMS Assist ignition was 252:15:17:08Z with burn duration of 2m52s
- OMS-2 ignition was at 252:15:52:16Z (37:21 MET), burn duration 2m25s
- TI at 254:08:08:08Z
- SRMS/OBSS/LDRI survey of nosecap, port, and starboard wing RCC on FD2
- ISS Docking capture at 254:10:48:27Z, 1:19:33:32 MET
- Docking complete at 254:11:01:01Z, 1:19:46:06 MET
- ISS Hatch Open at 1d21h19m; ISS crew welcoming
- EVA 1 Crew began campout in ISS Airlock at 10.2 psia in prep for EVA 1.
- EVA 1 Start at 255:09:19Z (3/18:01 MET) on 9/12/06, conducted from the ISS JAL (Quest Airlock). The astronauts used a new prebreathe protocol first tested during the handover of Expedition 12. EV1/Joe Tanner and EV2/Heidimarie Piper spent the night isolated in the JAL (Quest Airlock) with a reduced pressure of 10.2 psi while the ISS remains at 14.7 psi. This prebreathe protocol is called Prebreathe Campout Protocol (PBCOP). The Integrated Truss Segment (ITS) P3/P4 was attached to the Port 1 (P1) segment using the SSRMS. EVA crew connected power cables, released SABB and BGA restraints to prepare SARJ for operations. During removal of launch lock cover, a bolt/spring and a washer were accidentally released and lost. The EVA duration was 6:26.
- EVA 2 Start at 256:09:18Z (4/17.51 MET) on 9/13/06,. EV3/Dan Burbank and EV4/Steven MacLean slept in the JAL for Spacewalk Prebreathe Campout Protocol. They completed preparations for the activation of SARJ for operations. EVA 2 duration was 7:11.
- EVA 3 Start at _____. EV1/Tanner and EV2/Piper used PBCOP protocol. They completed P3 and P4 tasks, R&R SASA on Z1 truss, and installed heat shield on Ku-band antenna group interface tube. The EVA duration was 6:42.
- Hatch closed at 7/19:27 MET after saying goodbyes to Expedition 13 crew.

Continued…

FLT NO.	ORBITER	CREW (7) TITLE, NAMES & EVA'S	LAUNCH SITE, LIFTOFF TIME, LANDING SITES, ABORT TIMES	LANDING SITE/ RUNWAY, CROSSRANGE LANDING TIMES FLT DURATION, WINDS	SSME-TL NOM-ABORT EMERG THROTTLE PROFILE ENG. S.N.	SRB RSRM AND ET	INC	ORBIT HA/HP	FSW	PAYLOAD WEIGHTS, PAYLOADS/ EXPERIMENTS	MISSION HIGHLIGHTS (LAUNCH SCRUBS/DELAYS, TAL WEATHER, ASCENT I-LOADS, FIRSTS, SIGNIFICANT ANOMALIES, ETC.)
STS-115/ ISS 12A Continued...											

S115-E-05493 (11 Sept. 2006) --- ISS Configuration prior to docking of STS-115.

JSC2006-E-40599 --- Flight Director Bryan Lunney monitors data at his console in MOCR.

JSC2006-E-40475 --- STS-115/12A ISS Orbit 2 flight control team portrait in the MCC. Flight Director John McCullough (center right) holds the STS-115 mission logo and CAPCOM Pamela A. Melroy holds the STS-115/12A mission logo.

S115-E-06741 (17 Sept. 2006) --- ISS Configuration after undocking of STS-115

Continued...

EVENTS: (continued)
- Atlantis undocking completed at 260:12:49:50Z, 7/19:27 MET
- Total cargo transferred from Atlantis to the ISS was 36678 lbs (included 35552 lbs for P4/P5, but excluding water)
- Total cargo transferred from ISS to Atlantis was 993 lbs
- Total consumables transferred from Atlantis to ISS was 1110.5 lbm of water (11 CWC's with 1043.8 lbm and four PWR's with 66.1 lbm). Total oxygen transferred to ISS was 103 lbm.

SIGNIFICANT ANOMALIES:
- Fuel Cell 1 Coolant Pump AC1 Phase A short caused launch scrub. (See Launch Scrubs.)
- ARD response to erroneous telemetry (ARD NO-GO)
- Elevon Positioning Procedure callout errors
- ASA 3 Speedbrake driver channel # erratic
- Starboard PLBD aft (B) closed indication ON should be OFF
- F4D Tyvek cover late release
- TPS tile and blanket anomalies (cleared for Entry)
- FES shutdown during Ascent
- Water supply dump line heater A abnormal temperature cycling
- Hydraulic System 3 TVC Pitch Actuator indication
- Water supply dump valve leak
- Sequential Stills Video failure
- APU 2 X-axis accelerometer data erratic
- S-band lower right antenna communication problems
- FES topping left duct sensor erratic/OSL
- MADS BITE indication on FDM 2 MUX D
- Nosecap expansion seal RCC damage
- Engine 2 LO2 inlet pressure transducer reading low
- R4R heater failed on
- Aft sample bottles L1 and R2 leaking
- Starboard radiator MMOD strike

FLT NO.	ORBITER	CREW (6+1 UP/6+1 DN) TITLE, NAMES & EVA'S	LAUNCH SITE, LIFTOFF TIME, LANDING SITES, ABORT TIMES	LANDING SITE/ RUNWAY, CROSSRANGE LANDING TIMES FLT DURATION, WINDS	SSME-TL NOM-ABORT EMERG THROTTLE PROFILE ENG. S.N.	SRB RSRM AND ET	INC	ORBIT HA/HP	FSW	PAYLOAD WEIGHTS, PAYLOADS/ EXPERIMENTS	MISSION HIGHLIGHTS (LAUNCH SCRUBS/DELAYS, TAL WEATHER, ASCENT I-LOADS, FIRSTS, SIGNIFICANT ANOMALIES, ETC.)
STS-116/ ISS 12A.1 SEQ FLT# 117 KSC-117 PAD 39B-53 MLP-1 20TH SHUTTLE FLIGHT TO ISS	OV-103 (Flight 33) DISCOVERY OMS PODS: LPO1-36 RPO3-34 FRC3-33	CDR: Mark L. Polansky (Flt 2 - STS-98) P717/R262/V185/M228 PLT: William A. Oefelein P718/R302/M262 MS1: Nicholas J. M. Patrick (Flt 2 - STS-105) P719/R303/V186/M263 MS2/EV1: Robert L. Curbeam, Jr. (Flt 3 - STS-85, STS-98) P720/R225/V167/M195 MS3/EV2: Christer Fuglesang (ESA) P721/R304/M264 MS4: Joan E. Higginbotham P722/R305/F41 MS5 Up/EV3/EXP14: Sunita L. Williams P723/R306/F42 MS5 Down/EXP14: Thomas Reiter (M/S5 Up on STS-121) P724/R299/M260 SS EVA 98 DOCKED QUEST EVA 18 EMU/TETHERED EVA 91 SCHEDULED EVA 91 DURATION 6:36 SS EVA 99 DOCKED QUEST EVA 19 EMU/TETHERED EVA 92 SCHEDULED EVA 92 DURATION 5:00 SS EVA 100 DOCKED QUEST EVA 20 EMU/TETHERED EVA 93 SCHEDULED EVA 93 DURATION 7:31 SS EVA 101 DOCKED QUEST EVA 21 EMU/TETHERED EVA 94 SCHEDULED EVA 94 DURATION 6:38 Continued…	KSC 39B 344:01:47:35Z 8:47:35 PM EST (P) 8:47:35 PM EST (A) Saturday (6) 12/09/06 (8) LAUNCH WINDOW: 5 Minutes (PLT in-plane) EOM PLS: KSC TAL: ZZA TAL WX: MRN, FMI SELECTED: RTLS: KSC 33 N/N TAL: MRN 20 N/N AOA: NOR 17 N/N PLS: EDW 22 CI TDEL: 0:00 0.232 MAX Q NAV: 760 764 SRB STG: 2:04.16 2:04.64 PERF: NOMINAL 2 ENG TAL (MRN): 2:31 2:28 NEG RETURN: 3:55 3:52 PTA (U/S 160): 4:55 4:56 SE TAL (FMI): 6:07 6:03 PTM (U/S 160): 6:07 6:02 SE PRESS 104: 6:54 6:56 MECO CMD: 8:22.5 8:23.8 VI: 25819.0 25819.0 OMS-2: 37:07.4 37:10 187.2 FPS188.5 FPS	KSC 15 (KSC 64) 356:22:31:58Z 5:31:58 PM EST Friday 14 12/22/06(15) DEORBIT BURN: 356:21:30:53Z XRANGE: 813 NM ORBIT DIR: AR 13 AIM PT: CLOSE IN MLGTD: 1825 FT 356:22:31:58Z VEL: 196 KGS 208 KEAS HDOT: -2.9 FPS TD NORM 205: 2015 FT DRAG CHUTE DEPLOY: 191 KEAS 356:22:32:04Z NLGTD: 5594 FT 356:22:32:11Z VEL: 140 KGS 152 KEAS HDOT: -7.0 FPS BRK INIT: 79 KGS DRAG CHUTE JETTISON: 52 KGS 356:22:32:36Z BRK DECEL FPS^2: AVE 5.3 PK 6.1 WHEELSTOP: 356:22:32:51Z 9980 FT ROLLOUT: 8155 FT 53 SEC	104/104/109% PREDICTED: 100/104.5/ 104.5/72 104.5 ACTUAL: 100/104.5/ 104.5/74 104.5 1 = 2050 (5) 2 = 2054 (6) 3 = 2058 (1) ALL 3 SSME'S BLOCK II M 3 EOM: WEIGHT: 226476 LBS X CG: 1077.4 in LANDING: WEIGHT: 224041 LBS X CG: 1079.6 in	BI-128 RSRM 95 ET-123 SLWT 25 ET IMPACT MET 1:14:00 LAT: 36.83S LONG: 159.1W	51.60 (20)	DIRECT INSERTION POST OMS-2: 134.7x122.7NM DEORBIT: HA 184.5 NM HP 168.1 NM ENTRY VELOCITY: 25837 FPS ENTRY RANGE: 4263 NM	OI-30 (4)	CARGO: 35690 LBS PAYLOAD CHARGEABLE: 22502 LBS DEPLOYED: 5748 LBS NON-DEPLOYED: 16572 LBS MIDDECK: 182 LBS SHUTTLE ACCUMULATED WEIGHTS: DEPLOYED: 1281231 LBS NON-DEPLOYED: 1585492 LBS CARGO TOTAL: 3701134 LBS PERFORMANCE MARGINS (LBS): FPR: 2886 FUEL BIAS: 921 FINAL TDDP: 3768 RECON: 4559 PAYLOADS: PLB: ISS 12A.1 - ITS SPACEHAB SM ICC (W/STP-H2 UTILIZATION PAYLOAD) MIDDECK: ISS 12A.1 RAMBO MAUAI Continued…	***BRIEF MISSION SUMMARY:*** STS-116/12A.1 (20th ISS mission) continued ISS construction with the delivery and installation of Integrated Truss Segment P5 and began the process of reconfiguration and redistribution of the power generated by the pair of U.S. solar arrays. P6 truss was relocated to its final assembly position after 6 years atop the Unity Module. KSC W/D: OPF 105, VAB 8, PAD 28 = 141 days total LAUNCH POSTPONEMENTS: - Baselined OV-104 launch date of 06/05/2003 on 05/05/2002 - Postponed launch date to 07/24/2003 on 10/08/2002; delays due to engine flowliner crack repairs - Postponed launch date to NET 12/18/2003 on 03/13/2003. Slip due to Columbia accident. - Postponed launch date to NET 03/01/2004 on 04/17/2003. Slip due to Columbia accident. - Postponed launch date to NET 05/13/2004 on 05/28/2003. Slip due to Columbia accident. - Postponed launch date to NET 09/13/2004 on 07/29/2003. Slip due to Columbia accident. - Postponed launch date to NET 04/14/2005 on 10/03/2003. Slip due to Columbia accident. - Delete flight from FDRD on 03/22/2004 - Re-baselined STS-116 launch date to NET 02/09/2006 on 12/09/2004 - Postponed launch date to NET 04/23/2006 on 05/23/2005. Slip reflected latest planning decisions. - Postponed launch date to NET 10/01/2006 on 10/31/2005. Slip reflected latest planning decisions. - Postponed launch date to NET 11/16/2006 on 03/16/2006. Slip reflected latest planning decisions. - Postponed launch date to NET 12/14/2006 on 04/04/2006. Slip reflected latest planning decisions. - Advanced launch date to NET 12/07/2006 on 09/28/2006. LAUNCH SCRUBS: - Scrubbed Thursday 12/7/06 EST launch (12/8/06 GMT day 242) while holding at T-5 minutes. The window opened at 342:02:30:48Z and closed at 342:02:40:48Z with a Preferred Launch Time of 342:02:35:48Z. TAL1 (ZZA) was forecast and observed GO at TAL landing time and was selected as Prime TAL site. TAL2 (MRN) was forecast NO-GO thunderstorms WI 20 NM and BKN30 and observed NO-GO BKN. TAL3 (FMI) was forecast and observed NO-GO BKN30/BKN35. Launch Director counted down and held at 5 minutes until window closed. Scrubbed launch due to Range Safety violation of clouds below 6000 feet, thicker than 500 feet (verified at 5500 feet). MMT opted for a 48-hour turnaround and top off cryos and weather forecast was NO-GO. Launch date set for 12/09/06 EST (12/10/06 GMT). Weather Scrub.. Continued…

S116-E-05504 --- View from Discovery AFD of payload bay and approaching ISS (background). Shown in PLB are shuttle's docking mechanism (foreground), Spacehab (partially obscured), Canadian-built RMS robotic arm (right), and RMS/Orbiter Boom Sensor System (left, in stowed position).

FLT NO.	ORBITER	CREW (7) TITLE, NAMES & EVA'S	LAUNCH SITE, LIFTOFF TIME, LANDING SITES, ABORT TIMES	LANDING SITE/ RUNWAY, CROSSRANGE LANDING TIMES FLT DURATION, WINDS	SSME-TL NOM-ABORT EMERG THROTTLE PROFILE ENG. S.N.	SRB RSRM AND ET	ORBIT INC	HA/HP	FSW	PAYLOAD WEIGHTS, PAYLOADS/ EXPERIMENTS	MISSION HIGHLIGHTS (LAUNCH SCRUBS/DELAYS, TAL WEATHER, ASCENT I-LOADS, FIRSTS, SIGNIFICANT ANOMALIES, ETC.)
STS-116/ ISS 12A.1 Continued…		Continued... MCC WHITE FCR (47) FLIGHT DIRECTORS: SHUTTLE: A - J. S. Stich E - N. D. Knight LD/O 1 - A. J. Ceccacci O 2 - M. R. Abbott O 3/PLNG - R. E. LaBrode Team 4 - R. S. Jones MOD - P. L. Engelauf ISS: LD/O 2 - J. M. Curry O 1 - J. D. Hassmann O 3 - J. R. Montalbano TEAM 4/PLNG - D. J. Weigel CAPCOMS: SHUTTLE: A/E - K. T. Ham - C. J. Ferguson (Wx) LD/O1 - K. A. Ford O2 - K. M. McArthur O3/PLNG - S. W. Lucid Team 4 PLNG - N/A ISS: LD/O2 - S. K. Robinson O1 - T. W. Virts O3/PLNG - H. D. Getzelman Team 4 PLNG – N/A		Continued... WINDS: 14H/2R Kts OFFICIAL: 159/14 14/2R Kts DENS ALT: 1229 FT FLT DURATION: 12:20:44:23 S/T: 1082:19:54:33 OV-103: 276:11:34:05 DISTANCE: 5,330,398 sm TOTAL SHUTTLE DISTANCE: 438,715,036 sm					Continued... 5 CRYO TK SETS 6 N2 TANKS RMS 74 RMS USED FOR RMS/OBSS SURVEYS AND GRAPPLE/ UNBERTH P5 HANDOFF TO SSRMS	Continued... LAUNCH WINDOW: - Total launch window was 10 minutes with window open at 344:01:42:35Z and close at 344:01:52:35Z. Preferred Launch Time was 344:01:47:35Z (In-Plane Time) for a launch window of 5m00s. NOTE: In October, the self-imposed post-Columbia daylight launch constraint was relaxed, thus clearing STS-116 for a night launch. LAUNCH DELAYS: - None. Launch occurred on time at 344:01:47:35Z, 8:47:35 PM EST on Saturday, 12/09/06. TAL WEATHER: - All three TAL sites were forecast and observed GO. MRN was selected as Prime TAL site. MRN had best TD energy, ZZA had low TD energy, and FMI had balloon problems. PERFORMANCE ENHANCEMENTS: - Include the standard set plus: (1) PE Operational High Q WIN/DEC, (2) OMS Assist, (3) 52 nm MECO, and (4) Del Psi FLIGHT DURATION CHANGES/LANDING: - Early planning had STS-116 as an 11+1+2 flight that was changed a few weeks before the flight to 12+0+2 as consumables proved adequate. Pre-flight EOM TIG was 11/17:47 MET with landing at 11/18:49 MET. Difficulties with P5 retraction resulted in an FD8 MMT decision to add an unscheduled EVA 4 to inspect P5 for feasibility of retraction by EVA crew. This resulted in a loss of a weather wave-off day and a 13+1 flight. Undocking would be delayed 1 day and FD10 would be used for a late inspection. NIGHT LAUNCH #29: RENDEZVOUS #65: Rendezvous and dock with ISS FIRSTS/LASTS/NEW: - First flight of Advanced Health Monitoring System (AHMS). Flew on right engine in monitor mode. - First use of Quest for four EVA's and four Campout Prebreathes on a Shuttle flight - First flight with four EVA's by one astronaut - Curbeam - First on-orbit retraction of an ISS solar array - First ISS crew rotation through Shuttle since STS-113/11A in November 2002 - First entry of a Shuttle on the day of landing opportunity that was both the first and "pick 'em" days of opportunity for weather Continued…

S116-E-06472 --- STS-116 & Exp 14 crews gather in ISS Destiny Lab. From the left (front row): Reiter/Exp 14FE/MS-Dn, Patrick/MS, Higginbotham/MS, & PLT Oefelein. From the left (center row): Curbeam/MS, Fuglesang/MS (ESA), & CDR Polansky. From the left (back row): CDR Exp14 Lopez-Alegria, Mikhail Tyurin/Exp14/FE (RSA), & Williams/MS-Up/ Exp14FE.

S116E06472

S116E05983

LEFT: S116-e-05983 - Curbeam (left) and Fuglesang conduct EVA1 tasks for installation of P5 Truss. New Zealand and Cook Strait are seen in the background.

SPACE SHUTTLE MISSIONS SUMMARY

FLT NO.	ORBITER	CREW (7) TITLE, NAMES & EVA'S	LAUNCH SITE, LIFTOFF TIME, LANDING SITES, ABORT TIMES	LANDING SITE/ RUNWAY, CROSSRANGE LANDING TIMES, FLT DURATION, WINDS	SSME-TL NOM-ABORT EMERG THROTTLE PROFILE ENG. S.N.	SRB RSRM AND ET	INC	ORBIT HA/HP	FSW	PAYLOAD WEIGHTS, PAYLOADS/ EXPERIMENTS	MISSION HIGHLIGHTS (LAUNCH SCRUBS/DELAYS, TAL WEATHER, ASCENT I-LOADS, FIRSTS, SIGNIFICANT ANOMALIES, ETC.)

STS-116/ ISS 12A.1

Continued...

S116-E-05789 - A kink occurred in the port-side P6 solar array during the first attempt to retract that array on Dec. 13, 2006.

JSC2006-E-54706 ---FD Matt Abbott talks to Paul Hill, Mgr Space Shuttle Mission Ops in FCR during the final deployment of some small satellites.

JSC2006-E-53934 (12 Dec. 2006) --- John Shannon, Deputy Shuttle Program Manager and Manager, MMT, emphasizes a point during a MMT meeting in JSC MCC. Behind Shannon are Wayne Hale (left), Shuttle Program Manager; and Robert D. Cabana, JSC Deputy Director.

S116-E-06854 - FD10: EVA 4 Curbeam & Fuglesang (out of frame), working in tandem, used specially-prepared tape insulated tools to guide the P6 overhead SAW neatly inside its blanket box.

Continued...

EVENTS:
- OMS Assist ignition at 344:01:49:50Z (duration 1m38s)
- SRMS OBSS/LDRI survey of nosecap, port and starboard wing RCC (WLE's) completed
- TI maneuver at 345:19:28:22Z (1:17:40:47 MET). Resultant altitude 176.7 by 192.4 nm
- R-Bar pitch maneuver started at 345:21:04:46Z and was completed 7m33s later. Photos of Discovery's tile surfaces by ISS crew
- Docking capture occurred at 345:22:11:05Z (1:20:23:30 MET).
- Hard dock occurred at 345:22:26:33Z (1:20:38:58 MET).
- ISS hatch open 345:23:54Z (1:22:06 MET), ISS Crew Welcoming
- IELK seat liner transfer at 346:01:00:00Z (1:23:12 MET). At that time, Thomas Reiter became a member (MS5) of STS-116 and Sunita Williams joined the ISS Expedition 14 as Flight Engineer 2.
- EVA 1: EV1 and EV2 completed nominal tasks including P5 truss installed to P4 truss and mated P4-P4 umbilicals. 5/8-in socket lost from Pistol Grip Tool. EVA 1 duration 6h36m
- FD5: P6 4B SAW retraction required a series of partial deploy/retract sessions into 19 bays out for P4 SARJ to be free to rotate. P6 4B SAW now 16.5 bays out
- Solar flares raised radiation level. Crew slept in areas with better shielding.
- EVA 2: EV1 and EV2 Ch 2/3 reconfig and transfer to permanent power. CETA cart relocate. EVA 2 duration 5h00m
- FD7: Several IVA tests "wiggling" SAW, then extension/retraction were unsuccessful, 17.5 bays out
- EVA 3: EV1 and EV3 Ch 1/4 reconfig and transfer to permanent power. T/S P6 SAW. In an attempt to free the wires and grommets, oscillations and retractions were attempted. An additional 6 bays retracted, leaving additional 11 bays out. During EVA, a digital camera floated away. EVA 3 duration 7h31m.
- FD8: ISS and Space Shuttle Programs reached a joint decision to extend STS-116/12A.1 to 13+1 days to perform an unscheduled EVA to troubleshoot and complete P6 SAW retraction. Undocking now on FD11
- EVA 4: Curbeam and Fuglesang, unscheduled EVA 4 start at 352:19:00:00Z (8:17:12:25 MET). EVA crew successfully retracted P6 the last 36 feet by repeated actions of pulling on guide wires, shaking, and retract commands. Array was successfully retracted and folded into box. EVA duration 6h38m
- Total cargo transferred to ISS from Discovery was 4877 lbs (middeck 1305 lbs and logistics single module 3572 lbs).

Continued...

FLT NO.	ORBITER	CREW (7) TITLE, NAMES & EVA'S	LAUNCH SITE, LIFTOFF TIME, LANDING SITES, ABORT TIMES	LANDING SITE/ RUNWAY, CROSSRANGE LANDING TIMES FLT DURATION, WINDS	SSME-TL NOM-ABORT EMERG THROTTLE PROFILE ENG. S.N.	SRB RSRM AND ET	INC	ORBIT HA/HP	FSW	PAYLOAD WEIGHTS, PAYLOADS/ EXPERIMENTS	MISSION HIGHLIGHTS (LAUNCH SCRUBS/DELAYS, TAL WEATHER, ASCENT I-LOADS, FIRSTS, SIGNIFICANT ANOMALIES, ETC.)
STS-116/ ISS 12A.1 Continued...											

S118E07113

S116e07113 - ISS Configuration, FD11 view from departing Shuttle.

------ IN THE JSC CONTROL CENTER ------
LEFT: JSC2006-E-53281 --- Steve Stich, STS-116 Ascent Flight Director, monitors data and video at his console.
CENTER: JSC2006-E-53261 --- Karl A. Silverman with the Space Flight Meteorology Group pores through weather data.
RIGHT: JSC2006-E-53290 --- CAPCOM Christopher J. Ferguson follows the latest data (in background Stephen N. Frick).

Continued...

EVENTS (Continued):
- Total cargo transferred to Discovery from ISS was 4911 lbs (to middeck 1345 lbs and to logistics module 3566 lbs).
- Total consumables transferred to ISS: Oxygen tank transfer 69 lbm and total nitrogen tank transfer 47.2 lbm; total water transferred to ISS was 261.6 lbm (201.9 lbm in two CWC's and 59.7 lbm in three PWR's).
- Undocked at 353:22:09:35Z.
- A flyaround (1/2 lap) was initiated at 353:22:35:13Z.
- Sep 1 and Sep 2 maneuvers resulted in orbit 171.1 by 192.5 nm
- Micrometeoroid Orbital Debris late inspection was completed.
- MEPSI payload was deployed at 355:00:19:35Z (10:22:32:00 MET).
- RAFT payload was deployed at 355:01:56:46Z (11:00:09:11 MET).
- ANDE was deployed at 355:18:23Z (11:16:35 MET).
- No communications blackout during Entry.

SIGNIFICANT ANOMALIES:
Orbiter:
- Loss of RMS End Effector Auto Release Capability
- Fuel Cell O_2 Flowmeter Failed
- FES Primary B Failed To Come Out Of Standby
- Port Mid Payload Bay Floodlight Failed
- A6U Aft Event Thumbwheel Failure
- TPS Tile And Blanket Anomalies
- ML94B Bogen Bracket Shoe Debonded
- Kodak DCS 760 Digital Camera Lost During EVA 3
- Waste Water Dump Degraded Flow
- -Z Star Tracker Pressure BITE Fail Indication
- GPS Receiver Failed To Change Satellites
- MADS Signal Dropout
- WLE IDS Sensor Unit Inadvertent Shutdown
SRB:
- SRB Separation Debris Impact On Orbiter Not A Safety Issue
- T-0 Umbilical 1/4-Inch Frangible Bolt Missing
- Delaminated/Missing BTA on Aft BSM Housing
RSRM: No IFA's
SSME: No IFA's
ET: No IFA's
MOD:
- Erroneous Procedure Callout on OBSS LCS Cue Card
- MCC Automation System (MAS) File Server Failure
Integration:
- Ice Balls Noted Hanging From The North GOX Vent Arm Duct Exit Flange
- Debris Release from SRB LH BSM Area Traveled Fwd And Impacted Orbiter
- Delaminated/missing BTA on Aft BSM Housing with Sooting

SPACE SHUTTLE MISSIONS SUMMARY

FLT NO.	ORBITER	CREW (6+1 UP/6+1 DN) TITLE, NAMES & EVA'S	LAUNCH SITE, LIFTOFF TIME, LANDING SITE, ABORT TIMES	LANDING SITE/ RUNWAY, CROSSRANGE LANDING TIMES FLT DURATION, WINDS	SSME-TL NOM-ABORT EMERG THROTTLE PROFILE ENG. S.N.	SRB RSRM AND ET	INC	ORBIT HA/HP	FSW	PAYLOAD WEIGHTS, PAYLOADS/ EXPERIMENTS	MISSION HIGHLIGHTS (LAUNCH SCRUBS/DELAYS, TAL WEATHER, ASCENT I-LOADS, FIRSTS, SIGNIFICANT ANOMALIES, ETC.)
STS-117/ ISS 13A SEQ FLT# 118 KSC-118 PAD 39A-41 MLP-2 21ST SHUTTLE FLIGHT TO ISS	OV-104 (Flight 28) ATLANTIS OMS PODS: LPO4-28 RPO1-35 FRC4-28	CDR: Frederick W. Sturckow (Flt 3 - STS-88, STS-105) P725/R247/V173/M215 PLT/R2/M1: Lee J. Archambault P726/R307/M265 MS 1/EV 3/R1: Patrick G. Forrester (Flt 2 - STS-105) P727/R269/V186/M235 MS 2/EV4/M2: Steven R. Swanson P728/R308/M266 MS 3/EV2/R1: John D. Olivas P729/R309/M267 MS 4/EV1: James F. Reilly II (Flt 3 - STS-89, STS-104) P730/R234/V172/M204 MS 5 UP/EXP 15/16 FLT ENG: Clayton C. Anderson P731/R310/M268 MS 5 DN/EXP 14/15 FLT ENG: UP ON STS-116, STAY ISS Sunita L. Williams P732/R306/F42 SS EVA 102 DOCKED QUEST EVA 25 EMU/TETHERED EVA 95 SCHEDULED EVA 95 DURATION 6:16 SS EVA 103 DOCKED QUEST EVA 26 EMU/TETHERED EVA 96 SCHEDULED EVA 96 DURATION 7:16 SS EVA 104 DOCKED QUEST EVA 27 EMU/TETHERED EVA 97 UNSCHEDULED EVA 8 DURATION 7:58 Continued…	KSC 39A 159:23:38:04Z 7:38:04 PM EDT (P) 7:38:04 PM EDT (A) Friday (25) 6/8/07 (11) LAUNCH WINDOW: 3M 18S (PLT IN-PLANE) EOM PLS: KSC TAL: FMI TAL WX: ZZA (MRN: N/A RWY REPAIRS) SELECTED: RTLS: KSC 15 CI/N TAL: FMI 33 N/SFD AOA: KSC 15 N/N 1ST DAY PLS: EDW 22 N/N TDEL: 0:000(P) 0.112(A) MAX Q NAV: 720.08(P) 719.70(A) SRB STG: 2:03 (P) 2:03 (A) PERF: NOMINAL 2 ENG TAL (ZZA): 2:48 (P) 2:54 (A) NEG RETURN: 3:47 3:55 PTA (U/S 162): 5:19 5:20 SE TAL (ZZA 104): 6:04 6:08 PTM (U/S 180): 6:19 6:23 SE PRESS 104 7:02 7:03 MECO CMD: 8:24.9 8:24.9 Continued…	EDW 22, CONC EDW 51, CONC 32 173:19:49:37Z 12:49:37 PM PDT Friday (14) 06/22/07 (7) DEORBIT BURN: 173:18:43:47Z XRANGE 772 NM ORBIT DIR: AL 36 AIM PT: NOMINAL MLGTD: 1443 FT 173:19:49:37Z VEL: 219 KGS 205 KEAS HDOT: -4.0 FPS TD NORM 195: 2380 FT DRAG CHUTE DEPLOY: 196 KEAS 173:19:49:40Z NLGTD: 5379 FT 173:19:49:49Z VEL: 158 KGS 140 KEAS HDOT: -6.2 FPS BRK INIT: 88 KGS DRAG CHUTE JETTISON: 55 KGS 173:19:50:18Z BRK DECEL FPS²: AVE 4.0 PK 6.0 WHEELS STOP: 173:19:50:51Z 11422 FT ROLLOUT: 9979 FT 1:04 M:S Continued…	104/104/109% PREDICTED: 100/104.5/ 104.5/72 104.5 ACTUAL: 100/104.5/ 104.5/72 104.5 1 = 2059 (1) 2 = 2052 (5) 3 = 2057 (2) M 3 EOM: WEIGHT: 199418 LBS X CG: 1084.62 IN LANDING: WEIGHT: 199305 LBS X CG: 1086.76 IN	BI-129 RSRM 96 ET-124 SLWT 26 ET IMPACT MET 1:14:15 LAT: 36.38S LONG: 158.48W	51.6 (21)	DIRECT INSERTION POST OMS-2: 123.7x84.7 NM DEORBIT: HA 192.8 NM HP 178.8 NM ENTRY VELOCITY: 25868 FPS ENTRY RANGE: 4226 NM	OI-30 (5)	CARGO: 42641 LBS PAYLOAD CHARGEABLE: 36593 LBS DEPLOYED: 36393 LBS NON-DEPLOYED: 0 LBS MIDDECK: 200 LBS SHUTTLE ACCUMULATED WEIGHTS: DEPLOYED: 1317624 LBS NON-DEPLOYED: 1585692 LBS CARGO TOTAL: 3743775 LBS PERFORMANCE MARGINS (LBS): FPR: 2651 FUEL BIAS: 1063 FINAL TDDP: 1306 RECON: 1431 PAYLOADS: PLB: ISS 13A MIDDECK: ISS 13A RAMBO MAUAI 5 CRYO TK SETS 5 GN2 TANKS RMS 75 ODS, OBSS RMS USED FOR RMS/OBSS SURVEYS AND GRAPPLE/ UNBERTH S3/S4, HANDOFF TO SSRMS	BRIEF MISSION SUMMARY: STS-117/13A (21st ISS mission) continued the construction of the International Space Station with the delivery and installation of the second starboard truss segment (S3/S4), the deployment of the third set of solar arrays, and the retraction of the P6 starboard solar array wing, and one radiator. The truss also contained a Solar Alpha Rotary Joint (SARJ) which rotates 360 degrees for S4 & S6 solar arrays tracking of the sun. In addition, performed unscheduled EVA repair to Port OMS Pod thermal blanket for damage incurred during ascent. KSC W/D: OPF 125, VAB 8, PAD 17, Rollback to VAB, then VAB 72, PAD 25 = 247 Total Work Days LAUNCH POSTPONEMENTS: - Baselined OV-104 launch date of 09/05/2003 on 07/18/2002. - Postponed to 10/02/03 on 10/08/02 due to SSME flowliner crack repairs. - Postponed to NET 01/22/04 on 03/13/03 due to Columbia accident. - Postponed to NET 03/30/04 on 04/17/03 due to Columbia accident. - Postponed to NET 07/29/04 on 05/28/03 due to Columbia accident. - Postponed to NET 12/15/04 on 07/29/03 due to Columbia accident. - Deleted flight from FDRD on 10/03/03. - Re-baselined STS-117 to NET 05/18/06 on 03/17/05. - Postponed to NET 07/13/06 on 05/28/05. Slip reflected latest manifest constraints. - Postponed to NET 12/07/06 on 11/10/05. Slip reflected latest manifest constraints. - Postponed to NET 02/22/07 on 04/04/06. Slip reflected latest manifest constraints. - Postponed to NET 03/16/07 on 11/02/06. Slip due to ET delivery/processing schedule. - Launch date "under review" due to ET hail damage during 02/26/07 storm at the PAD. (ET sustained over 4,000 dings.) - Postponed to 06/08/07 on 04/16/07 due to rollback for ET repairs. LAUNCH SCRUBS: None LAUNCH WINDOW: - Total launch window was 6 minutes 29 seconds with window open at 159:23:34:53Z and close at 159:23:41:22Z. Preferred Launch Time was 159:23:38:04Z (In-Plane Time) for a launch window of 3m18s. LAUNCH DELAYS: - None. Launch occurred on time at 159:23:38:04Z, 7:38:04 PM EDT on Friday, 06/08/07. Continued…

iss015e11705 S3 S4 PL: The 17.8 ton S3/S4 truss to be added to the station is shown berthed in the Shuttle payload bay.

FLT NO.	ORBITER	CREW (7) TITLE, NAMES & EVA'S	LAUNCH SITE, LIFTOFF TIME, LANDING SITES, ABORT TIMES	LANDING SITE/ RUNWAY, CROSSRANGE LANDING TIMES FLT DURATION, WINDS	SSME-TL NOM-ABORT EMERG THROTTLE PROFILE ENG. S.N.	SRB RSRM AND ET	INC	ORBIT HA/HP	FSW	PAYLOAD WEIGHTS, PAYLOADS/ EXPERIMENTS	MISSION HIGHLIGHTS (LAUNCH SCRUBS/DELAYS, TAL WEATHER, ASCENT I-LOADS, FIRSTS, SIGNIFICANT ANOMALIES, ETC.)
STS-117/ ISS 13A Continued...		Continued... SS EVA 105 DOCKED QUEST EVA 28 EMU/TETHERED EVA 98 SCHEDULED EVA 97 DURATION 6:29 MCC WHITE FCR (48) FLIGHT DIRECTORS: SHUTTLE: A/E - N. D. Knight LD/O1 - C. A. Koerner O2 - B. C. Lunney O3/PLNG - R. S. Jones MOD - P. L. Engelauf Team 4 - M. L. Sarafin ISS: LD/O2 - K. B. Beck O1 - A. P. Hasbrook O3/PLNG - H. E. Ridings Team 4 - S. P. Davis CAPCOMS: SHUTTLE: A/E - D. A. Antonelli - T. W. Virts (Wx) LD/O1 - T. W. Virts O2 - K. A. Ford O3/Plng - R. S. Kimbrough Team 4 - N/A ISS: LD/O2 - K. M. McArthur O1 - S. G. Bowen O3/PLNG - R. M. Davis Team 4 - N/A	Continued... VI: 25819.0 25818.5 OMS-2: 37:46 38:30 98.7 FPS 96.8 FPS	Continued... WINDS: 1.9T/0.5R KTS OFFICIAL: 08002P06 KTS 5T/3L KTS DENS ALT: 5169 FT FLT DURATION: 13:20:11:33 S/T: 1096:16:06:06 OV-104: 245:12:44:01 DISTANCE: 5,809,363 sm TOTAL SHUTTLE DISTANCE: 444,524,399 sm							Continued... TAL WEATHER: Launch Day Synopsis: "Showers and thunderstorms will develop during the daylight hours on Friday across Spain and France but are expected to diminish rapidly after sunset. TAL landing times are well after sunset." ZZA and FMI TAL Sites were forecast and observed GO. ZZA was selected as Prime TAL Site. MRN was not available. PERFORMANCE ENHANCEMENTS: - Include the standard set plus: (1) PE Operational High Q SUM/JUN, (2) OMS Assist, (3) 52 nm MECO, Del Psi, and (4) Non-standard Consumables Reduction. FLIGHT DURATION CHANGES/LANDING: STS-117 was planned as an 11+2+2 duration flight. - FD4: The MMT concurred with the recommendation to repair the Port OMS Pod thermal blanket damage incurred during ascent. An additional 2 days, docked to the ISS, and a 4th EVA were added to conduct the repair. - FD14: Two KSC landing attempts (12:55 pm & 2:30 pm CDT) were waved due to weather. After wave-off, an Orbit Adjust Maneuver was added to the timeline. This 11 FPS burn brought in an additional landing opportunity (total of 3) for Edwards AFB on Friday, FD15. - FD15: KSC landing attempt at 1:18 pm CDT was waved due to weather. Landing site was switched to Edwards AFB for a successful landing on Orbit Rev 219 at 2:49 pm CDT (12:49 pm PDT). (PAO: "It's a good day to land in California...") FIRSTS/LASTS: - First flight of 2007. - First Launch from PAD 39A since final flight of Columbia. - First flight of Advanced Health Monitoring System (AHMS) on all three Sesame's. One flew in Active Mode. Two flew in Monitor Mode. In active mode, AHMS provides safe engine shutdown for excessive turbopump vibrations. - Sunita Williams sets new female long duration spaceflight record of 195 Days 18 Hours 58 Min, breaking Shinned Lucid's record of 188 Days 4 Hours. Williams surpassed Lucid's record on Saturday, 06/16/07, at 12:47 a.m. CDT - First EVA repair of Shuttle thermal blanket. - Last flight for James Reilly. Reilly flew to two space stations and clocked more than 853 hours in space, with five space walks totaling over 31 hours. He left NASA in June 2008. RENDEZVOUS #66: Rendezvous and dock with ISS EVENTS: - OMS 2 ignition at 160:00:16:34Z resulted in a 123.7 by 84.7 nm orbit. - SRMS OBSS/LDRI survey of nosecap, port and starboard wing RCC (WLE's) was completed. At 160:03:50Z, the crew reported damage to a thermal blanket on the Port OMS POD. Continued...

S117-E-07686 (16 June 2007) — STS-117 & Exp 15 crewmembers portrait in Destiny Lab. From the left (front row): Anderson/FE Exp 15, Williams/MS/STS-117, Exp 15 CDR Yurchikhin (Russia), & Kotov/FE Exp 15 (Russia). From the left (middle row): PLT Archambault/STS-117 and STS-117 CDR Sturckow. From the left (back row) Forrester, Reilly, Swanson and Olivas, all MS/STS-117.

S117E07686

FLT NO.	ORBITER	CREW (7) TITLE, NAMES & EVA'S	LAUNCH SITE, LIFTOFF TIME, LANDING SITES, ABORT TIMES	LANDING SITE/ RUNWAY, CROSSRANGE LANDING TIMES, FLT DURATION, WINDS	SSME-TL NOM-ABORT EMERG THROTTLE PROFILE ENG. S.N.	SRB RSRM AND ET	ORBIT	INC	HA/HP	FSW	PAYLOAD WEIGHTS, PAYLOADS/ EXPERIMENTS	MISSION HIGHLIGHTS (LAUNCH SCRUBS/DELAYS, TAL WEATHER, ASCENT I-LOADS, FIRSTS, SIGNIFICANT ANOMALIES, ETC.)
STS-117/ ISS 13A Continued…											

S117-E-06886 --- Reilly/EV1(center) Olivas/EV2 (right) connect power, data & cooling cables to S1 & S3, and deploy solar array blanket boxes on S4.

iss015e12948 -- EVA Repair: Anchored to a foot restraint on the RMS robotic arm, astronaut John "Danny" Olivas moves toward port OMS pod thermal blanket damage during EVA 3. Skin stapler and pins were used to make the repair.

S117-E-07789 Forrester/EV3 (left) Swanson/EV4, participate in 4th EVA as construction continues on ISS. Among other tasks, Forrester and Swanson continued activation of the station's new starboard 3 and 4 (S3/S4) truss segments.

Continued…

EVENTS (continued):
- TI maneuver at 161:17:00:57Z: Resultant orbit was 181.2 by 179.4 nm orbit
- Rbar Pitch Maneuver was performed. Photos of Atlantis' tile surfaces and the damaged OMS POD thermal blanket were taken by ISS crew. The thermal blanket damage was later determined to be from ET foam/ice shedding from LO2 line bracket during ascent.
- Docking Capture occurred at 161:19:36:10Z
- Hard Docking occurred at 161:19:47:48Z.
- ISS Hatch open 161:21:20:00Z, 4:20 pm CDT, Sunday, June 10, 2007, ISS crew welcoming
- IELK Seat Liner transfer at 162:00:55Z (7:55 PM CDT, June 10, 2007). At that time, Sunita Williams became a member of STS-120 and Daniel Tani joined the ISS Expedition 16 as Flight Engineer.
- STS-117 delivered new set of solar arrays on 21st flight to ISS; P6 Starboard array was retracted for over 3 days.
- "Suni" Williams was replaced by Clay Anderson on Expedition 15 and returned home on STS-117 with long duration space record for a female (see Firsts above).
- FD4 - Station robotic arm used to install S3/S4 truss on S1 truss.
- FD4 EVA 1: Reilly/EV1 & Olivas/EV2 completed the following tasks for S3/S4 Power Generation work: connected 13 power & data umbilicals, unstowed & deployed 1A & 3A solar arrays, and uncinched/unwinched photovoltaic radiator (PVR) for deployment. SARJ work included: installing 4 alpha joint I/F structure (AJIS) struts, installing drive lock assembly (later, EVA 2 determined a problem, see below), removed 6 SARJ locks, and released all swing bolts along SARJ. EVA 1 duration: 6h16m.
- FD4 - MMT Management Decisions Summary: On 06/11/07, the MMT concurred: (1) that the Port OMS Pod TPS Blanket is considered [to be] suspect in case of a contingency deorbit, (2) with performing a repair of the OMS Pod Blanket, and (3) with adding 2 extension days and a 4th EVA.
- FD5: Activities completed nominally. Solar Array deployment - 8 bays retracted. Array behavior similar to 4B retraction on STS-116 (sticking grommets, asymmetric folding).
- FD6: Russian central and terminal computers failed during docked operations at GMT 164:15:15:00Z and were restored with jumper cables bypassing power monitoring devices.
- FD6 EVA 2: Forrester/EV3 & Swanson/EV4 conducted partial retraction of P6 2B Solar Array (including cut leader). Inspected P6 aft radiator starboard PIP pin (only one confirmed). SARJ work included: Installed 4 SARJ brace beams, installed DLA 1 (discovered DLA's were cross wired on the ground), removed 10 SARJ launch locks, and broke torque on 3 SARJ launch restraints. EVA 2 duration: 7h16m.

Continued…

FLT NO.	ORBITER	CREW (7) TITLE, NAMES & EVA'S	LAUNCH SITE, LIFTOFF TIME, LANDING SITES, ABORT TIMES	LANDING SITE/ RUNWAY, CROSSRANGE LANDING TIMES, FLT DURATION, WINDS	SSME-TL NOM-ABORT EMERG THROTTLE PROFILE ENG. S.N.	SRB RSRM AND ET	ORBIT 	INC	HA/HP	FSW	PAYLOAD WEIGHTS, PAYLOADS/ EXPERIMENTS	MISSION HIGHLIGHTS (LAUNCH SCRUBS/DELAYS, TAL WEATHER, ASCENT I-LOADS, FIRSTS, SIGNIFICANT ANOMALIES, ETC.)
STS-117/ ISS 13A Continued...											

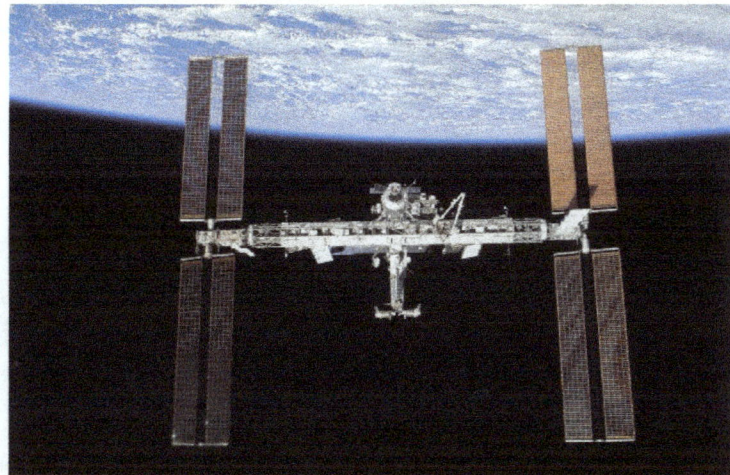

s117e08006ISS-earth.jpg: Back-dropped by the blackness of space and Earth's horizon, the new ISS configuration is viewed from the departing Atlantis.

----- IN THE JSC MISSION CONTROL CENTER -----
LEFT: JSC2007-E-31063 -- Orbit 1 FCT. FD/Cathy Koerner (left) & CAPCOM Terry W. Virts Jr. hold STS-117 logo.
CENTER: JSC2007-E-28303 --- A "fish-eye" perspective of MOCR activity: (lt to rt) CAPCOMs Terry Virts & Tony Antonelli; & FDs Norm Knight & Steve Stich.
RIGHT: JSC2007-E-29876 --- Orbit 2 FCT. FD/Bryan Lunney (wearing business suit) is in foreground.

Continued...

EVENTS (Continued):
- FD8 EVA 3: Conducted by Reilly/EV1 & Olivas/EV2: Removed Lab H2O Vent & installed Lab H_2 Vent, repaired OMS POD thermal blanket with skin stapler and pins, relocated 1 of 3 APFR's for 13A.1, and finished retraction of P6 2B Solar Array. This was unscheduled EVA added by MMT. EVA 3 duration: 7h58 m.
- FD10 EVA 4: Conducted by Forrester/EV3 & Swanson/EV4: Activated SARJ for rotation, cleared S3 Mobile Transporter path, relocated 2 of 3 APFR's for 13A.1, released torque on S4 MMOD Shield bolts, moved VSSA to Camera Port 1, cleared Node 1 Port for 10A Node 2 temporary stowage, and opened Lab H2 Vent. EVA duration: 6h 29m.
- Transfers:
  - Mid-deck resupply cargo transfer to ISS from Atlantis was 1277 lbs.
  - Mid-deck return cargo transfer to Atlantis from ISS was 1528 lbs.
  - Supply Water total to ISS was 751 L (1,656 lbm)
  - Oxygen (net) to ISS was 89 lbm
  - Nitrogen to ISS: to A/L tanks 17.3 lbm; into stack for repress 16 lbm
  - Lithium Hydroxide (LiOH): STS [used] to ISS = 3, ISS (new) to STS = 3
- Undocked at 170:14:42:00Z followed by a fly-around (1/2 lap).
- Sep 1 & Sep 2 maneuvers resulted in orbit of 185.0 x 177.1 nm
- Micrometeoroid Orbital Debris late inspection was completed.
- No communications blackout during Entry.

SIGNIFICANT ANOMALIES:
Orbiter:
- MDM OA2 CARD 5 Failed - Invalid Data
- MADS Recorder Tape Speed Went To 120 IPS (Nom is 15) at Nose Wheel TD
- E3 LH_2 Inlet Pressure Transducer Went OSH at T+ 3.5 Min
SRB: None.
RSRM:
- Gas Penetration Through Nozzle Joint 2 RTV, RSRM-96A&B
SSME: None.
ET:
- Post-Launch Camera & Film Rev. - Loss of LH2 Acreage Foam at Stations 1160, 1623 & 1871
MOD:
- GDR Data Dropouts During Ascent
- Ascent LOC Push Button Inoperative
- LCC Activation Turning Off WLES PGSC
Integration:
- Tile Piece Liberated From Aft Fuselage Body Flap I/F During Ascent
- FOD Found In Aft Compartment
- Port OMS Pod Blanket Damage During Ascent
- Rope-Like Material Noted Moving In Umbilical Well Imagery
- Propellant Use During FDS Extended Shuttle Attitude - Hold Approx 3 Times Higher Than Predicted

FLT NO.	ORBITER	CREW (7) TITLE, NAMES & EVA'S	LAUNCH SITE, LIFTOFF TIME, LANDING SITES, ABORT TIMES	LANDING SITE/ RUNWAY, CROSSRANGE LANDING TIMES FLT DURATION, WINDS	SSME-TL NOM-ABORT EMERG THROTTLE PROFILE ENG. S.N.	SRB RSRM AND ET	INC	ORBIT HA/HP	FSW	PAYLOAD WEIGHTS, PAYLOADS/ EXPERIMENTS	MISSION HIGHLIGHTS (LAUNCH SCRUBS/DELAYS, TAL WEATHER, ASCENT I-LOADS, FIRSTS, SIGNIFICANT ANOMALIES, ETC.)
STS-118/ ISS 13A.1 SEQ FLT# 119 KSC-119 PAD 39A-42 MLP-1 22ND SHUTTLE FLIGHT TO ISS	OV-105 (Flight 20) ENDEAVOUR OMS PODS: LPO3-31 RPO4-27 FRC5-20	CDR: Scott J. Kelly (Flt 2 - STS-103) P733/R253/V187/M220 PLT: Charles O. Hobaugh (Flt 2 - STS-104) P734/R268/V188/M234 MS 1/R: Tracy E. Caldwell P735/R311/F43 MS 2/EV1: Richard A. Mastracchio (Flt 2 - STS-106) P736/R257/V189/M224 MS 3/EV2: David R. Williams (Canada) P737/R312/M269 MS 4: Barbara R. Morgan P738/R313/F44 MS 5: B. Alvin Drew P739/R314/M270 SS EVA 106 DOCKED QUEST EVA 29 EMU/TETHERED EVA 99 SCHEDULED EVA 98 DURATION 6:17 SS EVA 107 DOCKED QUEST EVA 30 EMU/TETHERED EVA 100 SCHEDULED EVA 99 DURATION 6:28 SS EVA 108 DOCKED QUEST EVA 31 EMU/TETHERED EVA 101 SCHEDULED EVA 100 DURATION 5:28 SS EVA 109 DOCKED QUEST EVA 32 EMU/TETHERED EVA 102 SCHEDULED EVA 101 DURATION 5:02 Continued…	KSC 39A 220:22:36:42Z 6:36:42 PM EDT (P) 6:36:42 PM EDT (A) Wednesday (14) 8/8/07 (8) LAUNCH WINDOW: 4M 14S (PLT IN-PLANE) EOM PLS: KSC TAL: ZZA TAL WX: MRN, FMI SELECTED: RTLS: KSC 15 TAL: ZZA 30L (FMI): NO-GO) AOA: KSC 15 1ST DAY PLS: EDW 22 TDEL: 0:000(P) 0.312(A) MAX Q NAV: 707.47(P) 699.34(A) SRB STG: 2:02.56(P) 2:03.04(A) PERF: NOMINAL 2 ENG TAL (MRN*): 2:34 (P) 2:40(A) *ZZA prime TAL site; Call made off MRN (GO site) NEG RETURN: 3:53 3:56 PTA (U/S 167 FPS): 5:04 5:10 SE TAL (ZZA 104): 5:58 6:08 PTM (U/S 179 FPS): 6:16 6:23 SE PRESS 104 6:56 6:58 MECO CMD: 8:25.0 8:25.4 Continued…	KSC 15 (KSC 65) 233:16:32:17Z 12:32:17 PM EDT Tuesday (21) 08/21/07 (8) DEORBIT BURN: 233:15:25:12Z XRANGE 697 NM ORBIT DIR: A/L 37 AIM PT: NOMINAL MLGTD: 1628 FT 233:16:32:17Z VEL: 210 KGS 212 KEAS HDOT: -3.1 FPS TD NORM 205: 2302 FT DRAG CHUTE DEPLOY: 163 KEAS 233:16:32:30Z NLGTD: 5619 FT 233:16:32:29Z VEL: 169 KGS 165 KEAS HDOT: -6.3 FPS BRK INIT: 123 KGS DRAG CHUTE JETTISON: 54 KGS 233:16:32:59Z BRK DECEL FPS^2: AVE 6.1 PK 9.1 WHEELS STOP: 233:16:33:16Z 11862 FT ROLLOUT: 10234 FT 46 SEC Continued…	104/104/109% PREDICTED: 100/104.5/ 104.5/72/ 104.5 ACTUAL: 100/104.5/ 104.5/74/ 104.5 1 = 2047 (10) 2 = 2051 (6) 3 = 2045 (9) M 3 EOM: WEIGHT: 221740 LBS X CG: 1078.1 IN LANDING: WEIGHT: 221660 LBS X CG: 1079.8 IN	BI-130 RSRM 97 ET-117 SLWT 27 ET IMPACT MET 1:14:03 LAT: 36.9S LONG: 159.2W	51.6 (22)	DIRECT INSERTION POST OMS-2: 172.2X124.2 NM DEORBIT: HA 187.2 NM HP 22.8 NM ENTRY VELOCITY: 25860 FPS ENTRY RANGE: 4343 NM	OI-30 (6)	CARGO: 37390 LBS PAYLOAD CHARGEABLE: 23899 LBS DEPLOYED: 11830 LBS NON-DEPLOYED: 11740 LBS MIDDECK: 329 LBS SHUTTLE ACCUMULATED WEIGHTS: DEPLOYED: 1329454 LBS NON-DEPLOYED: 1597761 LBS CARGO TOTAL: 3781165 LBS PERFORMANCE MARGINS (LBS): FPR: 2651 FUEL BIAS: 1063 FINAL TDDP: 1913 RECON: 2435 PAYLOADS: PLB: ISS 13A.1-ITS S5 SPACEHAB SM, ESP-3 MIDDECK: ISS 13A.1 RAMBO MAUI 5 CRYO TK SETS RMS 76 ODS, OBSS RMS USED FOR RMS/OBSS SURVEYS AND GRAPPLE/ UNBERTH S5, HANDOFF TO SSRMS	***BRIEF MISSION SUMMARY:*** *STS-118/13A (22nd ISS mission) continued the assembly and resupply of the International Space Station and fulfilled a long-standing teacher's legacy. The new assembly included the delivery of the S5 Truss segment, installation of a spare parts platform, and changeout of a failed gyroscope. This was the last shuttle resupply mission using the SPACEHAB module. In addition, Barbara R. Morgan, who had served as backup to Christa McAuliffe in the Teacher in Space Project 21 years earlier, flew as the first Educator Mission Specialist. McAuliffe was a member of the crew that lost their lives in the 1986 Challenger accident.* KSC W/D: OPF 1332+64+63+18 = 1477, VAB 9, PAD 25 = 1511 Total Work Days (OPF Processing occurred over a total time period of 1665 days.) LAUNCH POSTPONEMENTS: - Added STS-118 to FDRD - launch date of 10/09/03 on 08/01/02. - Postponed to NET 11/13/03 on 10/08/02 due to engine flowliner crack repairs. - Postponed to NET 05/06/04 on 03/13/03 due to Columbia accident. - Postponed to NET 06/01/04 on 04/17/03 due to Columbia accident. - Deleted flight from FDRD on 05/28/03. - Re-baselined to NET 09/14/06 on 07/14/05. - Revised to "TBD" on 11/10/05. Slip reflected latest manifest constraints. - Postponed to NET 06/11/07 on 04/04/06. Slip reflected latest manifest constraints. - Postponed to NET 06/28/07 on 11/02/06. Slip due to ET delivery/processing schedule. - Postponed to NET 08/09/07 on 04/16/07. Slip due to STS-117 rollback. - Advanced to 08/07/07 on 06/28/07. Provide an adequate number of launch opportunities before a range conflict. - Launch delayed to 08/08/07 on 08/03/07 due to "cabin leak checks and other processing work." LAUNCH SCRUBS: None LAUNCH WINDOW: - Total launch window was 8 minutes 11 seconds with window open at 220:22:32:45Z and close at 220:22:40:56Z. Preferred Launch Time was 220:22:36:42Z (In-Plane Time) for a launch window of 4m14s. LAUNCH DELAYS: - None. Launch occurred on time at 220:22:36:42Z, 6:36:42 PM EDT on Wednesday, 08/08/07. TAL WEATHER: Forecast: Pressure gradient between a surface high over northern Spain and low over northern Italy will keep NW winds at FMI and ZZA Wednesday through Friday. Peak winds at FMI are forecast to be above headwind limits all 3 days, but remain within limits at ZZA. MRN weather is forecast "GO" all 3 days. Continued…

ISS015-E-21711 - Endeavour delivers a new S5 stbd truss segment, cargo inside the SPACEHAB module (in center of bay), and the external stowage platform 3 to ISS.

FLT NO.	ORBITER	CREW (7) TITLE, NAMES & EVA'S	LAUNCH SITE, LIFTOFF TIME, LANDING SITES, ABORT TIMES	LANDING SITE/ RUNWAY, CROSSRANGE LANDING TIMES FLT DURATION, WINDS	SSME-TL NOM-ABORT EMERG THROTTLE PROFILE ENG. S.N.	SRB RSRM AND ET	INC	ORBIT HA/HP	FSW	PAYLOAD WEIGHTS, PAYLOADS/ EXPERIMENTS	MISSION HIGHLIGHTS (LAUNCH SCRUBS/DELAYS, TAL WEATHER, ASCENT I-LOADS, FIRSTS, SIGNIFICANT ANOMALIES, ETC.)
STS-118/ ISS 13A.1 Continued...		Continued... MCC WHITE FCR (49) FLIGHT DIRECTORS: SHUTTLE: A/E - J. S. Stich LD/O1 - M. R. Abbott O2 (FD1-FD6) - R. S. Jones O2 (FD7-EOM) - M. L. Sarafin O3/PLNG (FD1-Undock) - M. P. Moses O3/PLNG/Prelaunch/Post-Undock - P. F. Dye MOD - P. L. Engelauf Team 4 - R. E. LaBrode ISS: LD/O2 - J. R. Montalbano O1 - K. L. Alibaruho O3/PLNG - G. Kerrick Team 4 - J. D. Hassmann CAPCOMS: SHUTTLE: A/E - C. J. Ferguson - J. P. Dutton (Wx) LD/O1 - S. K. Robinson O2 - R. S. Kimbrough O3/PLNG - S. W. Lucid Team 4 - N/A ISS: LD/O2 - S. Walker O1 - D. A. Antonelli O3/PLNG - L. McCullough Team 4 - N/A	Continued... VI: 25819.0 25817.4 OMS-2: 37:00 37:00.7 253.9 FPS 252.6 FPS	Continued... WINDS: 6H 4L KTS OFFICIAL: 11909P13 KTS 10H 8L KTS DENS ALT: 1973 FT FLT DURATION: 12:17:55:35 S/T: 1109:10:01:41 OV-105: 219:08:07:41 DISTANCE: 5,274,977 sm TOTAL SHUTTLE DISTANCE: 449,799,376 sm							Continued... PERFORMANCE ENHANCEMENTS: Include the standard set plus: 1) PE Operational High Q WIN/DEC, 2) OMS Assist, 3) a 52 nm MECO, and 4) Del Psi FLIGHT DURATION CHANGES/LANDING: On 8/12/07, FD5, the MMT concurred with extending the Mission to 14+2 days and adding EVA 4. FIRSTS/LASTS: - First flight of Endeavour in 5 years - First flight test of new system to monitor ECO circuit voltage to fuel sensors. System allows Flight Controllers to recommend manual engine shutdown by the crew if sensor voltage has failed. - First flight of Automated Meteorological Profiling System (AMPS) High Resolution (HR) as primary system for DOLILU wind measurements - replacement for Jimspheres. - First flight that Station Shuttle Power Transfer System (SSPTS) available to provide extended duration capability to shuttle - First flight that three-string Global Positioning System (GPS) was used to replace landing TACAN System - previously flown single string only. - First flight of SRB Command Receiver/Decoder (CRD). Replaced Integrated Receiver/Decoder (IRD) and Range Safety Distributor (RSD) due to obsolescence concerns - Last flight of SPACEHAB resupply module. - First and last flight of Educator Mission Specialist Barbara R. Morgan. She left NASA and returned to Boise State University in 2008. NIGHT LAUNCH - N/A RENDEZVOUS #67: Rendezvous and dock with ISS NINTH SHUTTLE CREWMEMBER REPLACEMENT - Clay Anderson was replaced by Drew in August 2007. (8th Shuttle crewmember replacement occurred on STS-121.) EVENTS: - OMS 2 ignition at 220:22:47:15Z resulted in a 172.2 by 124.7 nm orbit. - SRMS OBSS/LDRI survey of nosecap and port and starboard wing RCC (WLE's) was completed. - TI maneuver at 222:15:15:19Z - resultant orbit was 186.5 by 180.4 nm - During R-Bar Pitch Maneuver, a gouge in the heat shield below the right wing (site 3) was identified. - Docking contact occurred at 222:18:01:54Z. - Hard Dock occurred at 222:18:29:44Z. Continued...

ISS015-E-23031 --- Exp 15 & STS-118 crews in ISS Destiny Lab: Front row, from left: Clayton C. Anderson/FE Exp15, CDR Exp15 Fyodor Yurchikhin (RSA), & Oleg Kotov/FE (RSA). STS-118 crew: middle row, from left: Drew/MS, Morgan/MS, Williams/MS (CSA), & CDR Kelly. Back row, from left: PLT Hobaugh, Mastracchio/MS, & Caldwell/MS.

FLT NO.	ORBITER	CREW (7) TITLE, NAMES & EVA'S	LAUNCH SITE, LIFTOFF TIME, LANDING SITES, ABORT TIMES	LANDING SITE/ RUNWAY, CROSSRANGE LANDING TIMES FLT DURATION, WINDS	SSME-TL NOM-ABORT EMERG THROTTLE PROFILE ENG. S.N.	SRB RSRM AND ET	ORBIT		FSW	PAYLOAD WEIGHTS, PAYLOADS/ EXPERIMENTS	MISSION HIGHLIGHTS (LAUNCH SCRUBS/DELAYS, TAL WEATHER, ASCENT I-LOADS, FIRSTS, SIGNIFICANT ANOMALIES, ETC.)
							INC	HA/HP			
STS-118/ ISS 13A.1 Continued...											Continued...

S118e06114 - Barbara R. Morgan flew as first Educator Mission Specialist

S118-E-06998 - Anchored to the foot restraint on the Canadarm2, Williams, and Mastracchio (out of frame), R&R a faulty control moment gyroscope (CMG-3) into the Z1 truss during EVA 2.

Continued...

EVENTS (Continued):
- ISS Hatch open 222:20:04:00Z, 3:04 pm CDT, Friday, August 10, 2007, ISS crew welcoming
- FD4: MMT, per Flight Rule 13A.1_A2-6 concurred that TPS was considered to be damaged.
- FD4, EVA 1: EV1 and EV2 installed S5 on S4, relocated S5 PVRGF to S5 Keel (ground strap bolt would not seat again, like P5), retracted and cinched P6 Forward PVR, and retrieved EVA ratchet from STBD Z1 toolbox. EVA 1 duration 6h17m.
- FD5: MMT concurred that TPS was considered to be damaged and authorized focused TPS inspection. Mission was extended to 14+2 and EVA4 (preplanned) was added.
- FD6, EVA 2: EV1 and EV2 completed R&R of faulty CMG 3 into ISS Z1 truss, installed old CMG3/FSE/FRAM on nadir ESP-2 FRAM Site #5 with MLI cover (no straps), and retrieved EVA ratchet from PORT Z1 toolbox. The failed CMG will remain at its temporary stowage location until it is returned to Earth on a later shuttle mission. The new gyroscope is one of four CMG's used to control Station attitude on orbit. EVA 2 duration 6h28m.
- FD8, EVA 3: EV1 and EV3 (Exp 15/16) relocated P6 SASA to P1 zenith, installed P1 S-band BSP and Xpdr, moved CETA cart 1 to STBD of MT (connected to MT), moved CETA cart 2 to STBD of MT (connected to CETA 1), and removed P6 S-band Xpdr (dummy box plate installed). EV1 EVA terminated early to EMU glove damage at EVA Phase Elapsed Time (PET) 4:20. The damage did not cause leakage; the suit pressure was unaffected. Due to the early termination, the S-band Antenna Structural Assembly (SASA) Spare Gimbal Locks and Materials International Space Station Experiment (MISSE) 3 and 4 tasks were not completed. EVA 3 duration 5h28m.
- FD8: EVA 4 delayed from FD9 to FD11 by MMT for potential tile repair.
- FD9: MMT decided that the TPS repair issue required a Programmatic assumption of risk and that the MMT was willing to assume that risk. The preponderance of data (including ground analysis and arc jet testing) indicated acceptable margins to fly as is. MMT decided that no TPS repair would be performed on Endeavour and that the nominal planned EVA 4 would be executed on FD11.
- FD11, EVA 4: EV2 and EV3 (EXP 15/16) installed OBSS OSE (2) on S1 zenith trunnions, re-torqued Z1 SASA gimbal bolts, removed MISSE 3 and MISSE 4 from A/L and returned on Shuttle, Lab EWIS antenna handrails and cable installed (Lab fwd endcone nadir - got 3 of 3 DZU's installed), and retrieved tools from A/L toolboxes. Did not perform Lab or Node MMOD shield cleanup or S3 WETA installation. EVA 4 duration 5h 2m.
- FD12: MOD contingency plans for Hurricane Dean Preparedness included decreasing the flight control support to two teams and evacuation on military aircraft if required. The plan was not required to be implemented.

Continued...

FLT NO.	ORBITER	CREW (7) TITLE, NAMES & EVA'S	LAUNCH SITE, LIFTOFF TIME, LANDING SITES, ABORT TIMES	LANDING SITE/ RUNWAY, CROSSRANGE LANDING TIMES FLT DURATION, WINDS	SSME-TL NOM-ABORT EMERG THROTTLE PROFILE ENG. S.N.	SRB RSRM AND ET	INC	ORBIT HA/HP	FSW	PAYLOAD WEIGHTS, PAYLOADS/ EXPERIMENTS	MISSION HIGHLIGHTS (LAUNCH SCRUBS/DELAYS, TAL WEATHER, ASCENT I-LOADS, FIRSTS, SIGNIFICANT ANOMALIES, ETC.)
STS-118/ ISS 13A.1 Continued...											

RIGHT: S118-E-07918 - Category 4 Hurricane Dean, viewed from Endeavour, was moving westerly in the Caribbean nearing Jamaica with sustained winds of 150 mph. MOD contingency evacuation plans were prepared, but not needed.

TOP: JSC2007-E-42079 -- In MCC Lead FD Matt Abbott follows the in-space ops. MIDDLE: JSC2007-E-41693 --- In MCC FD Richard Jones follows launch preps at KSC. BOTTOM: JSC2007-E-42074 --- In MCC Shannon Walker ISS CAPCOM, ISS Lead FD Joel Montalbano (right), & Steven W. Lindsey (standing), Chief of Astronaut Office, keep up with in-space ops.

JSC2007-E-46429 (17 Sept. 2007) --- The STS-118 Ascent/Entry flight control team and crewmembers pose for a group portrait in the space shuttle flight control. Flight director Steve Stich holds mission logo with CDR Kelly (left), & CAPCOM Chris Ferguson (right). Additional crewmembers pictured are PLT Hobaugh, Morgan/MS, Caldwell/MS, & Mastracchio/MS.

Continued...

EVENTS (Continued):
- Transfers:
  - Hardware transferred to ISS (outside and inside): 14,740 lbs
  - Hardware/supplies returned from ISS: 3,297 lbs
  - Water delivered to ISS: 918.6 lbm
  - Oxygen to ISS: 77 lbm
  - Nitrogen to ISS: 33.8 lbs
  - Lithium Hydroxide (LiOH) cans from ISS to STS: 12 cans (9 old, 3 used)
  - LiOH new cans from STS to ISS: 30 cans
  - Power transferred from ISS to orbiter using the SSPTS was 1186 kWh.
- Undocked at 170:14:42:00Z followed by a flyaround (1/2 lap)
- Sep 1 and Sep 2 maneuvers resulted in orbit 185.2 by 183.5 nm.
- Micrometeoroid Orbital Debris late inspection was completed. No issues.
- No communications blackout during Entry.

SIGNIFICANT ANOMALIES:
Orbiter:
- A Magenta Hue Appeared On Camera (GFE).
- STS-118 Drag Chute Reefing Line Cutter Failure to Cut (GFE).
SRB:
- None.
RSRM:
- Gas Penetrations through Nozzle Joint 2 RTV, RSRM-97A&B
SSME:
- 3 Com Card/Cable Failed (GFE).
ET:
- 2007 ET-117 Film Review Found TPS Loss at Sta. 1623 Outboard LO2
- Feedline Support Bracket and TPS Orb Impact
- XT 1973 Inboard LO2 Feedline Bracket Base Fitting TPS Crack on ET-117
- Post-Launch Camera and Film Review Showed Loss of LH2 Acreage Foam
MOD:
- B30M Power Failure B-C Power Feeds
- Margi Output Error
- ET Umbilical Door Closure Timing
- SSRMS Movement Prior To Shuttle Ku Mask
- OBSS Sensor Mode Change From 6 to 2 per MCC
- Procedure Error on PGSC Setup
Integration:
- Partial Tyvek Cover Release
- SSRMS Movement Prior to Shuttle Ku Mask
- BFS Loss of Class III Alert from Spacehab E

SPACE SHUTTLE MISSIONS SUMMARY

FLT NO.	ORBITER	CREW (6+1 UP/6+1 DN) / TITLE, NAMES & EVA'S	LAUNCH SITE, LIFTOFF TIME, / LANDING SITES, ABORT TIMES	LANDING SITE/ RUNWAY, CROSSRANGE / LANDING TIMES FLT DURATION, WINDS	SSME-TL NOM-ABORT EMERG / THROTTLE PROFILE ENG. S.N.	SRB RSRM AND ET	INC	ORBIT HA/HP	FSW	PAYLOAD WEIGHTS, / PAYLOADS/ EXPERIMENTS	MISSION HIGHLIGHTS (LAUNCH SCRUBS/DELAYS, TAL WEATHER, ASCENT I-LOADS, FIRSTS, SIGNIFICANT ANOMALIES, ETC.)
STS-120/ ISS 10A SEQ FLT# 120 KSC-120 PAD 39A-43 MLP-2 23RD SHUTTLE FLIGHT TO ISS	OV-103 (Flight 34) DISCOVERY OMS PODS: LPO1-37 RPO3-35 FRC3-34	CDR: Pamela A. Melroy (Flt 3 - STS-92, STS-112) P740/R261/V175/F34 PLT: George Zamka P741/R315/M271 MS 1/EV1: Scott E. Parazynski (Flt 5 - STS-66, STS-86, STS-95, STS-100) P742/R187/V144/M165 MS 2/R: Stephanie D. Wilson (Flt 2 - STS-121) P743/R298/V190/F39 MS 3/EV2: Douglas H. Wheelock P744/R316/M272 MS 4/R: Paolo A. Nespoli (ESA) P745/R317/M273 MS 5 UP/EXP 16 FLT ENG: Daniel M. Tani (Flt 2 - STS-108) P746/R272/V191/M238 MS 5 DN/EXP 15/16 FLT ENG: Clayton C. Anderson (UP on STS-117, Stay on ISS) P747/R310/M268 SS EVA 110 DOCKED QUEST EVA 33 EMU/TETHERED EVA 103 SCHEDULED EVA 102 DURATION 6:14 Continued…	KSC 39A 296:15:38:19Z 11:38:19 PM EDT (P) 11:38:19 PM EDT (A) Tuesday (16) 10/23/07 (12) LAUNCH WINDOW: 7M 17S (PLT IN-PLANE) EOM PLS: KSC TAL: MRN TAL WX: FMI SELECTED: RTLS: KSC 15 N/N TAL: MRN 20 N/N (ZZA: NO-GO) AOA: NOR 35 N/N 1ST DAY PLS: EDW 04 CI/N TDEL: 0:000(P) 0.162(A) MAX Q NAV: 719.02(P) 701.56(A) SRB STG: 2:02.56(P) 2:03.20(A) PERF: NOMINAL 2 ENG TAL (MRN): 2:37 (P) 2:45(A) NEG RETURN: 3:51 3:55 PTA (U/S 167 FPS): 5:16 5:26 SE TAL (ISTRES 104): 6:04 6:12 PTM (U/S 181 FPS): 6:16 6:27 Continued…	KSC 33 (KSC 66) 311:18:01:17Z 01:01:17 PM EST Wednesday (15) 11/07/07 (12) DEORBIT BURN: 311:16:58:49Z XRANGE: 196 NM ORBIT DIR: D/R 21 AIM PT: CLOSE IN MLGTD: 1247 FT 311:18:01:17Z VEL: 204 KGS 220 KEAS HDOT: -5.4 FPS TD NORM 195: 3249 FT DRAG CHUTE DEPLOY: 189 KEAS 311:18:01:26Z Continued…	104/104/109% PREDICTED: 100/104.5/ 104.5/72/ 104.5 ACTUAL: 100/104.5/ 104.5/72/ 104.5 1 = 2050 (6) 2 = 2048 (7) 3 = 2058 (2) M 3 EOM: WEIGHT: 203067 LBS X CG: 1081.0 IN LANDING: WEIGHT: 202989 LBS X CG: 1083.0 IN	BI-131 RSRM 98 ET-120 SLWT 28 ET IMPACT MET 1:14:06 LAT: 36.749S LONG: 158.983W	51.6 (23)	DIRECT INSERTION POST OMS-2: 169.9X123.8 NM DEORBIT: HA 188.0 NM HP 12.1 NM ENTRY VELOCITY: 25850 FPS ENTRY RANGE: 4436 NM	OI-32 (1)	CARGO: 40872 LBS PAYLOAD CHARGEABLE: 33813 LBS DEPLOYED: 33474 LBS NON-DEPLOYED: 280 LBS MIDDECK: 59 LBS SHUTTLE ACCUMULATED WEIGHTS: DEPLOYED: 1362928 LBS NON-DEPLOYED: 1598100 LBS CARGO TOTAL: 3822037 LBS PERFORMANCE MARGINS (LBS): FPR: 2651 FUEL BIAS: 1063 FINAL TDDP: 2091 RECON: 1880 PAYLOADS: PLB: ISS 10A (NODE 2), PDGF, MBSU, SASA MIDDECK: ISS 10A RAMBO MAUI 5 CRYO TK SETS RMS 77 ODS, OBSS	**BRIEF MISSION SUMMARY:** STS-120/10A (23rd ISS mission) provided for expansion of the ISS with delivery of the Italian-built U.S. multi-port Node 2 connecting module named Harmony. Installation of Harmony allows for attachment of research labs from the European Space Agency (Columbus) and the Japan Aerospace Exploration Agency (Kibo) to be delivered on subsequent flights. The P6 truss segment and solar arrays were replaced from a temporary location (on Z1) to a permanent location on P5 truss. In this new location, the solar arrays were redeployed to maximize needed power generation for inclusion of the future research labs. Also on this mission, a 1-day extension was added to extend EVA 4 for starboard SARJ inspections, but the EVA was later reworked for a successful repair of P6 4B solar power array damaged during deploy. KSC W/D: OPF 234, VAB 7, PAD 23 = 264 Total Work Days (OPF Processing occurred over a total time period of 273 days.) LAUNCH POSTPONEMENTS: - Added STS-120 to FDRD - launch date of 02/19/04 on 01/23/03. - Postponed to NET 09/23/04 on 03/13/03 due to Columbia accident. - Deleted flight from FDRD on 05/28/03. - Re-baselined to NET 08/09/07 on 06/01/06. - Postponed to NET 09/07/07 on 11/02/06. Slip due to ET delivery/processing schedule - Advanced to 08/26/07 on 02/08/07 to avoid spacing problem with Soyuz and ATV. - Postponed to 10/20/07 on 04/16/07. Slip due to STS-117 rollback. - Postponed to 10/23/07 on 08/07/07. Slip to maintain standard minimum interval between Soyuz undocking (changed for landing opportunities) and orbiter docking to the ISS. LAUNCH SCRUBS: None LAUNCH WINDOW: - Total launch window was 11 minutes 19 seconds with window open at 296:15:34:17Z and close at 296:15:45:36Z. Preferred Launch Time was 296:15:38:19Z (In-Plane Time) for a launch window of 7m17s. LAUNCH DELAYS: - None. Launch occurred on time at 296:15:38:19Z, 11:38:19 AM EDT on Tuesday, 10/23/07. (PAO: "It's a nice day in Florida…") Continued…

S120-E-006397 (25 Oct. 2007) --- Historical first space meeting of female Women Commanders. Peggy Whitson (right), ISS EXP 16 CDR, greets Pam Melroy, STS-120 CDR.

FLT NO.	ORBITER	CREW (6+1 UP/6+1 DN) TITLE, NAMES & EVA'S	LAUNCH SITE, LIFTOFF TIME, LANDING SITES, ABORT TIMES	LANDING SITE/ RUNWAY, CROSSRANGE LANDING TIMES FLT DURATION, WINDS	SSME-TL NOM-ABORT EMERG THROTTLE PROFILE ENG. S.N.	SRB RSRM AND ET	INC	ORBIT HA/HP	FSW	PAYLOAD WEIGHTS, PAYLOADS/ EXPERIMENTS	MISSION HIGHLIGHTS (LAUNCH SCRUBS/DELAYS, TAL WEATHER, ASCENT I-LOADS, FIRSTS, SIGNIFICANT ANOMALIES, ETC.)
STS-120/ ISS 10A Continued...		Continued... SS EVA 111 DOCKED QUEST EVA 34 EMU/TETHERED EVA 104 SCHEDULED EVA 103 DURATION 6:33 SS EVA 112 DOCKED QUEST EVA 35 EMU/TETHERED EVA 105 SCHEDULED EVA 104 DURATION 7:08 SS EVA 113 DOCKED QUEST EVA 36 EMU/TETHERED EVA 106 SCHEDULED EVA 105 DURATION 7:19	Continued... SE PRESS 104 7:06 6:57 MECO CMD: 8:25.6 8:25.8 VI: 25819 25817 OMS-2: 37:22 37:19.6 232.8 FPS 230.9 FPS	Continued... NLGTD: 5419 FT VEL: 150 KGS 163 KEAS HDOT: -5.9 FPS BRK INIT: 109 KGS DRAG CHUTE JETTISON: 52 KGS 311:18:01:53Z BRK DECEL FPS^2: AVE 6.3 PK 10.5 WHEELS STOP: 311:18:02:11Z 9593 FT ROLLOUT: 8346 FT 54 SEC WINDS: 10.6H 2.8R KTS OFFICIAL: 35013P22 KTS 21H 6R KTS DENS ALT: 771 FT FLT DURATION: 15:02:22:58 S/T: 1124:12:24:39 OV-103: 291:13:57:03 DISTANCE: 6,249,432 sm TOTAL SHUTTLE DISTANCE: 456,048,808 sm							Continued...

MCC WHITE FCR (50)

FLIGHT DIRECTORS:
SHUTTLE:
A/E - N. D. Knight
LD/O1 - R. E. LaBrode
O2 (FD2-FD13) - M. P. Moses
O2 (FD1, FD14 and Waveoff) - M. R. Abbott
O3/PLNG (FD1-FD13) - M. L. Sarafin
PLNG (Prelaunch, FD1, FD14, and Waveoff) - A. J. Ceccacci
ENT - B. C. Lunney
MOD - P. L. Engelauf
Team 4 - P. F. Dye

ISS:
LD/O2 - J. D. Hassmann
O1 - D. J. Weigel
O3/PLNG - H. L. Rarick
Team 4 - G. Kerrick

CAPCOMS:
SHUTTLE:
A/E - T. W. Virts
      - L. J. Archambault (Wx)
LD/O1 - C. J. Ferguson
O2 - D. A. Antonelli
O3/Plng - S. W. Lucid
Team 4 - N/A

ISS:
LD/O2 - K. A. Ford
O1 - H. Getzelman
O3/PLNG - Z. Jones
Team 4 - N/A

JSC2007-E-095788 --- In MCC, FDs, Knight (left) & Lunney, monitor EVA repair of ISS solar panel shown in photos at right & bottom.

ISS016-E-008875 --- Close-up view of the repaired solar array.

S120-E-007608 --- STS-120 & Exp16 crews ISS Harmony node. From left (bottom): Anderson/MS (DN) , CDR Peggy A. Whitson, Yuri I. Malenchenko/FE/Exp16 (RSA) & PLT Zamka. From left (center): Wilson/MS, CDR Pam Melroy, & Nespoli/MS (ESA). From left (top): Daniel Tani/FE/Exp16 (UP), Parazynski/MS, & Wheelock/MS.

ISS016-E-009207 (3 Nov. 2007) --- While anchored to a foot restraint on the end of the OBSS, Parazynski/EV1 assesses his repair work as the solar array is fully deployed during EVA 4.

TAL WEATHER:
The weather model data for Europe continued to show an area of low pressure near Italy, with high pressure over central France. Windy conditions at ZZA and FMI were expected to contribute to pockets of turbulence in the region. Weakening high pressure was forecast over southern Spain, with partly cloudy skies and southwest winds at MRN Tuesday. All three TAL sites were forecast and observed GO. Moron was selected as Prime TAL Site.

PERFORMANCE ENHANCEMENTS:
Include the standard set plus: 1. PE Operational High Q TRN/OCT, 2. OMS Assist, 3. 52 nautical mile MECO, and 4. Del Psi.

FLIGHT DURATION CHANGES/LANDING:
On FD7, MMT concurred with adding a docked extension day to the mission to extend EVA 4 for starboard SARJ inspections for cause of vibrations and drag.

FIRSTS:
- Historical first meeting of two spacecrafts commanded by women: Peggy Whitson, the first woman to command the ISS, and Pamela A. Melroy, the second woman space shuttle commander.
- Successful first time operation of OV-103 Station-to-Shuttle Power Transfer System (SSPTS).
- First ET LO2 IFR bracket pockets filled with BX (replaces PDL in pockets) to minimize void formation.
- First flight of OI-32 Flight Software. Standard capability release included changes for enhanced crew safety and situational awareness, improved mated control of ISS, and other enhancements for ground and flight operations and safety.
- First High-definition TV coverage of Launch (by CNN)

NIGHT LAUNCH: (N/A)

RENDEZVOUS #68: Rendezvous and dock with ISS

EVENTS
- OMS 2 ignition at 296:15:48:44Z resulted in a 159.9 by 123.8 NM orbit.
- SRMS OBSS/LDRI survey of nosecap and port and starboard wing RCC (WLE's) was completed.
- TI maneuver at 298:09:55:25Z resulted in a 188.7 by 179.7 NM orbit.
- R-Bar Pitch Maneuver was performed. No significant issues
- Docking Capture occurred at 298:12:39:57Z.
- Hard Dock occurred at 298:12:52:50Z.

Continued…

FLT NO.	ORBITER	CREW (6+1 UP/6+1 DN) TITLE, NAMES & EVA'S	LAUNCH SITE, LIFTOFF TIME, LANDING SITES, ABORT TIMES	LANDING SITE/ RUNWAY, CROSSRANGE LANDING TIMES, FLT DURATION, WINDS	SSME-TL NOM-ABORT EMERG THROTTLE PROFILE ENG. S.N.	SRB RSRM AND ET	INC	ORBIT HA/HP	FSW	PAYLOAD WEIGHTS, PAYLOADS/ EXPERIMENTS	MISSION HIGHLIGHTS (LAUNCH SCRUBS/DELAYS, TAL WEATHER, ASCENT I-LOADS, FIRSTS, SIGNIFICANT ANOMALIES, ETC.)

STS-120/ ISS 10A
Continued…

ABOVE: In JSC MCC, Ed Gonzalez/Ascent Trajectory Officer monitors prelaunch data.   CENTER: JSC2007-E-095148 --- In JSC MCC, FD Mike Moses (standing) escorted former President George H.W. Bush and former First Lady Barbara Bush shown talking to Shuttle & ISS crews on-orbit. AT RIGHT: JSC2007-E-097963---- On Nov.8 at Ellington Field, President George W. Bush greets returning CDR Melroy (pictured) and other crew members (out of frame) with JSC Director Mike Coats in the background.

S120-E-008531 (5 Nov. 2007) --- Back-dropped by the blackness of space and Earth's horizon, the new ISS configuration is viewed from the departing STS-120 Discovery.

SIGNIFICANT ANOMALIES:
Orbiter:
- V070-396376-201, Blanket R&R
- Protrusion on the Arrowhead Plate (H-0.38)
- Protruding Ames Gap Filler (H=0.21 & H=0.29)
- Blanket is lifted off left (Port) OMS Pod
- The MPS Engine #1 LO$_2$ Inlet Temperature failed off scale high at 15:41:15GMT during STS-120 Ascent.
- On STS-120/OV-103, Measurement V62T0519A was erratic, diverged from approximately 184 degrees F
- Missing debris
SRB:
- Nonlinear separation on LH SRB of the Frustrum/Forward Skirt Ordnance Ring for STS-120/BI-131
- STS-120/ET-120 launched on 10/23/07: Post Launch camera and film review showed loss of foam at two locations.
RSRM:
- Gas penetrations through Nozzle Joint 2 RTV, RSRM-98A&B
- Gas penetration through RTV, Nozzle Joint 5, RSRM-98B
SSME: None
ET: None
MOD:
- Missing step in PDRS STBD survey procedure
- Typo - IMU align in Orb Ops Checklist
- RMS Joint Angle Ground Display Error
INTEGRATION:
- LH$_2$ Umbilical ice noted prelaunch
- GUCP ice bridged to ET Intertank Foam
- ET LH$_2$ Tank foam acreage losses
- Unexpected debris/expected debris exceeding mass allowable prior to pad clearance (liftoff debris)
- Debris release on outboard side of LO$_2$ Feedline at ~277 sec MET

Continued…
- ISS Hatch opened at 9:39 AM (CDT) on10/25/07 (298:14:39:00Z) - Shuttle Crew welcomed by ISS Crew - Historical first meeting of two spacecrafts commanded by women.
- IELK Seat Liner Transfer at 298:16:12Z (11:12 AM CDT, Oct. 25, 2007). At that time Clayton Anderson became a member of STS-120 and Daniel Tani joined the ISS Expedition 16 as Flight Engineer.
- FD4 EVA 1: (EV1 and EV2) Removed the failed SASA from Z1; installed SASA in PLB sidewall carrier; prepped Node 2 (Harmony) for removal from bay; demated P6/Z1 fluid QD's; used Station robot arm (PDGF) to install Node 2 to temporary location on Node 1 (Unity). [NOTE: Node 2 was moved to its permanent location at the front of the U.S. Lab using the ISS robotic arm after shuttle departure.] EVA1 duration 6h14m
- FD6 EVA 2: EV1 and EV3 conducted P6 truss demate from temporary location on Z1; EV3 performed inspection of suspected sharp edge on S1 CETA rail; Initial stbd SARJ inspection; Node 2 Outfitting (EV1 completed all of this solo); structurally installed the Node 2 PDGF; successfully deployed the two outboard S1 radiators between EVA 2 and EVA 3 (so all three are now deployed). EVA 2 duration 6h33m
- FD7: MMT concurred with adding a docked extension day to the mission to extend EVA 4 for starboard SARJ inspections for cause of vibrations and drag.
- FD8 EVA 3: EV1 and EV2 attached P6 truss to P5 (permanent location). The 2B solar array was 100% deployed. The 4B array was aborted at 25 bays, with a tear in the right blanket (guide wire snag). EVA 3 duration 7h 8m
- FD11: MMT concurred with new plan for EVA4 to repair the Solar Array Wing (SAW) 4B repair. The Tile Ablator Dispenser DTO was postponed.
- FD12 EVA 4: (EV1 & EV2) EV1 repaired the P6 4B array using the OBSS on the SSRMS with a WIF-E. As reported by the Rocky Mountain News: "Parazynski…performed what NASA is calling on e of the greatest 'space saves' in the history of manned spaceflight. …[He] floated outside with wire cutters, pliers, and homemade tools to fix the torn wing" [restoring maximum power capability to the ISS.] EVA 4 duration 7h 19m

- Transfers:
  - Hardware transferred ISS (outside and inside): 33,834 lbs
  - Hardware/supplies returned from ISS: 2,020 lbs
  - Water delivered to ISS: 939.1 lbm
  - Oxygen transferred to ISS: 30 lbm
  - Nitrogen transferred to ISS: 31.6 lbs
  - Power from ISS to Orbiter using SSPTS: 1186 kWh.
- FD14: Undocking from ISS: 309:10:32:03Z (4:32 am CST, 11/05/07)
- Sep 1 & Sep 2 maneuvers resulted in orbit 189.6 by 181.9 nm.
- Micrometeoroid Orbital Debris late inspection was completed. No issues.
- Anderson returned home after 152 days in space.
- Communications blackout time during Entry: 1m

FLT NO.	ORBITER	CREW (6+1 UP/6+1 DN) TITLE, NAMES, & EVA'S	LAUNCH SITE, LIFTOFF TIME, LANDING SITES, ABORT TIMES	LANDING SITE/ RUNWAY, CROSSRANGE LANDING TIMES FLT DURATION, WINDS	SSME-TL NOM-ABORT EMERG THROTTLE PROFILE ENG. S.N.	SRB RSRM AND ET	INC	ORBIT HA/HP	FSW	PAYLOAD WEIGHTS, PAYLOADS/ EXPERIMENTS	MISSION HIGHLIGHTS (LAUNCH SCRUBS/DELAYS, TAL WEATHER, ASCENT I-LOADS, FIRSTS, SIGNIFICANT ANOMALIES, ETC.)
STS-122/ ISS 1E SEQ FLT# 121 KSC-121 PAD 39A-44 MLP-1 24TH SHUTTLE FLIGHT TO ISS	OV-104 (Flight 29) ATLANTIS OMS PODS: LPO4-29 RPO1-36 FRC4-29	CDR: Stephen N. Frick (Flt 2 - STS-110) P748/R276/V192/M242 PLT: Alan G. Poindexter P749/R318/M274 MS 1/R: Leland D. Melvin P750/R319/M275 MS 2/EV1: Rex J. Walheim (Flt 2 - STS-110) P751/R277/V193/M243 MS 3/EV2: Hans Schlegel (Germany) (Flt 2 - STS-55) P752/R163/V194/M143 MS 4/EV3: Stanley G. Love P753/R320/M276 MS 5 UP/EXP 16 FLT ENG: Leopold Eyharts (ESA) (also flew on MIR Feb 1998) P754/R321/M277 MS 5 DN/EXP 16 FLT ENG: Daniel M. Tani (Flt 2 - STS-108, STS-120 up) P755/R272/V191/M238 SS EVA 114 DOCKED QUEST EVA 37 EMU/TETHERED EVA 107 SCHEDULED EVA 106 DURATION 7:58 SS EVA 115 DOCKED QUEST EVA 38 EMU/TETHERED EVA 108 SCHEDULED EVA 107 DURATION 6:45 Continued…	KSC 39A 038:19:45:30Z 2:45:30 PM EST (P) 2:45:30 PM EST (A) Tuesday (35) 2/07/08 (9) LAUNCH WINDOW: 5M1S (PLT IN-PLANE) EOM PLS: KSC TAL: ZZA TAL WX: MRN, BEN SELECTED: RTLS: KSC 15 N/N TAL: ZZA 30L N/N AOA: NOR 23 N/N 1ST DAY PLS: EDW 04 N/N TDEL: 0:000(P) 0.212(A) MAX Q NAV: 756.21(P) 755.17(A) SRB STG: 2:04.16(P) 2:04.16(A) PERF: NOMINAL 2 ENG TAL (MRN): 2:35(P) 2:38(A) NEG RETURN: 3:51 3:54 PTA (U/S 161 FPS): 5:04 5:05 SE TAL (ZZA 104): 6:04 6:082 PTM (U/S 167 FPS): 5:58 6:02 Continued…	KSC 15 (KSC 67) 051:14:07:09Z 9:07 AM EST Thursday (11) 02/21/08 (7) DEORBIT BURN: 051:12:59:52.0Z XRANGE: 408 NM ORBIT DIR: A/L 38 AIM PT: NOMINAL MLGTD: 2344 FT 051:14:07:09Z VEL: 197 KGS 194 KEAS HDOT: -2.1 FPS TD NORM 195: 2200 FT DRAG CHUTE DEPLOY: 188 KEAS 051:14:07:10Z NLGTD: 5175 FT 051:14:07:17Z VEL: 157 KGS 155 KEAS HDOT: -4.9 FPS BRK INIT: 91 KGS DRAG CHUTE JETTISON: 54 KGS 051:14:07:46Z BRK DECEL FPS^2: AVE 4.6 PK 6.9 WHEELS STOP: 051:14:08:07Z 10911 FT ROLLOUT: 8567 FT 58 SEC Continued…	104/104/109% PREDICTED: 100/104.5/ 104.5/72/ 104.5 ACTUAL: 100/104.5/ 104.5/74/ 104.5 1 = 2059 (2) 2 = 2052 (6) 3 = 2057 (3) M 3 EOM: WEIGHT: 207295 LBS X CG: 1078.2 IN LANDING: WEIGHT: 207215 LBS X CG: 1080.4 IN	BI-132 RSRM 99 ET-125 SLWT 29 ET IMPACT MET 1:14:07 LAT: 36.619S LONG: 158.796W	51.6 (24)	DIRECT INSERTION POST OMS-2: 124.0X118.8 NM DEORBIT: HA 187.6 NM HP 23.1 NM ENTRY VELOCITY: 25860 FPS ENTRY RANGE: 4403 NM	OI-32 (2)	CARGO: 40296 LBS PAYLOAD CHARGEABLE: 32941 LBS DEPLOYED: 30657 LBS NON-DEPLOYED: 2162 LBS MIDDECK: 122 LBS SHUTTLE ACCUMULATED WEIGHTS: DEPLOYED: 1393585 LBS NON-DEPLOYED: 1600348 LBS CARGO TOTAL: 3862333 LBS PERFORMANCE MARGINS (LBS): FPR: 2651 FUEL BIAS: 1063 FINAL TDDP: 2402 RECON: 3435 PAYLOADS: PLB: ISS 1E (COLUMBUS MODULE) ICC-LITE ECSH PDGF MIDDECK: ISS 1E MAUI 5 CRYO TK SETS RMS 78 ODS OBSS SSPTS	**_BRIEF MISSION SUMMARY:_** _STS-122/1E (24th ISS mission) delivered the European Space Agency's Columbus research laboratory module to the ISS. Columbus, measuring 23 ft in length and 15 ft in diameter, is ESA's largest contribution to the expansion of the ISS. Also delivered were ESA experiments and two ESA astronauts with one of them to join the ISS crew for operation of Columbus research. This mission also saw the Columbus Control Center in Oberpfaffenhofen, near Munich, Germany, brought on-line for initial checkout and future operations of the laboratory._ KSC W/D: OPF: 121, VAB HB-3: 7, PAD A: 76 = 204 Total Work Days (+1 holiday @ OPF Processing + 10 holidays + 4 contingency days @ PAD) LAUNCH POSTPONEMENTS: - Added STS-122 to FDRD - launch date of 10/17/07 on 10/05/06. - Postponed to 12/06/07 on 04/16/07 due to STS-117 rollback. - After 12/06/07 scrub, see LAUNCH SCRUBS below, launch was reset for 24-hr turnaround on Friday, 12/07/07. - Later, on 12/06/07, during MMT Scrub Turnaround Meeting, it was decided to extend to a 48-hr turnaround for Saturday, 12/08/07 launch to allow additional time to address all concerns. At Friday, 12/07/07 MMT, it was determined that necessary discussion could not be finished in time for Saturday 12/08/07 launch attempt. The launch was moved to Sunday 12/09/07 with a new Launch Commit Criteria (for this launch only) requiring four of four valid ECO sensor readings (rather than three of four) prior to launch. In addition, the following two conditions were added: 1) Launch Window was limited to in-plane +1 minute (to provide additional ascent fuel margin), and 2) utilization of new in-flight ECO circuit voltage readings (successfully tested on STS-118 and STS-120 by ground flight controllers to recommend manual engine shutdown by the crew, if required. - After second scrub on 12/09/07, see LAUNCH SCRUBS below, launch was rescheduled to NET 01/02/08 contingent on development and implementation of fuel ECO sensor system troubleshooting plan. - Postponed to 01/10/08 on 12/13/07 dependent on resolution of the problem with the fuel sensor system. Slip was to allow "as many people as possible to have time with family and friends at the time of year when it means the most." Tanking test using add-on Time Domain Reflectivity (TDR) instrumentation on 12/18/07 isolated ECO Sensor System failures to open circuit in the three-part "pass-through connector". TPS removal on the tank was authorized at the pad to begin moving toward removal of the hardware, if required, to solve the problem. Launch date remained unchanged. - Postponed to TBD on 01/03/08; however, PRCB established a "work to" launch date of 02/02/08 dependent on testing of removed ECO connector, installation of replacement connector, and replacement and retesting procedures of Ascent Thrust Vector Control (ATVC) unit. Continued…

S122-E-007873 (11 Feb. 2008) --- Photographed from ISS, the station's robotic Canadarm2 moves the Columbus laboratory from Atlantis' payload bay to the starboard side of the Harmony module.

FLT NO.	ORBITER	CREW (6+1 UP/6+1 DN) TITLE, NAMES, & EVA'S	LAUNCH SITE, LIFTOFF TIME, LANDING SITES, ABORT TIMES	LANDING SITE/ RUNWAY, CROSSRANGE LANDING TIMES FLT DURATION, WINDS	SSME-TL NOM-ABORT EMERG THROTTLE PROFILE ENG. S.N.	SRB RSRM AND ET	ORBIT INC	ORBIT HA/HP	FSW	PAYLOAD WEIGHTS, PAYLOADS/ EXPERIMENTS	MISSION HIGHLIGHTS (LAUNCH SCRUBS/DELAYS, TAL WEATHER, ASCENT I-LOADS, FIRSTS, SIGNIFICANT ANOMALIES, ETC.)
STS-122/ ISS 1E Continued...		Continued... SS EVA 116 DOCKED QUEST EVA 39 EMU/TETHERED EVA 109 SCHEDULED EVA 108 DURATION 7:25 MCC WHITE FCR (51) FLIGHT DIRECTORS: SHUTTLE: ASC - N. D. Knight LD/O1 - M. L. Sarafin O2 - A. J. Ceccacci PLNG - P. F. Dye ENT - B. C. Lunney MOD - P. L. Engelauf Team 4 - M. R. Abbott ISS: LD/O2 - S. P. Davis O1 - R. C. Dempsey O3 - J. R. Spencer Team 4 - K. L. Alibaruho IP FD - A. P. Hasbrook (I/F w/Columbus CC, Oberpfaffenhofen, Germany)	Continued... SE PRESS 104 6:55 6:55 MECO CMD: 8:22.9 8:22.8 VI: 25819 25818 OMS-2: 37:46 37:40 159.6 FPS158.1 FPS	Continued... WINDS: 1.9T 0.6R KTS OFFICIAL: 31003P05 KTS 5H 2L KTS DENS ALT: 77 FT FLT DURATION: 12:18:21:39 S/T: 1137:06:46:18 OV-104: 258:07:05:40 DISTANCE: 5,296,842 sm TOTAL SHUTTLE DISTANCE: 461,345,650 sm							Continued...

CAPCOMS:
SHUTTLE:
A/E - J. P. Dutton
- T. W. Virts (Wx)
LD/O1 - K. A. Ford
O2 - S. K. Robinson
PLNG - S. W. Lucid
Team 4 - N/A

ISS:
O1 - H. Getzelman
LD/O2 - C. J. Cassidy
O3/PLNG - C. E. Zajac
Team 4 - N/A

S122-E-008923 (15 Feb. 2008) --- Mission Specialist, Rex Walheim, performs work on the outside of the Columbus laboratory. Mission Specialist, Stanley Love (out of frame), shared this EVA with Walheim.

S122-E-009694-- STS-122 & EXP 16 crews in ISS Zvezda SM: STS CDR Frick (bottom left), Walheim/MS (bottom center), Melvin/MS (bottom right), Exp 16 CDR Peggy Whitson, Love/MS (above Whitson), STS PLT Poindexter (top right), Tani/MS (top left), Leopold Eyharts EXP FE (ESA) (left middle), Schlegel/MS (Germany), Yuri I. Malenchenko/EXP FE (RSA) is above Walheim.

S122-E-008911--- Schlegel/MS (ESA Germany) continues work aimed toward readying the new Columbia lab for duty

Continued...

- New "work to" launch date of NET 02/07/08 established on 01/14/08. Testing of removed ECO connector confirmed problem in the connector.
- Officially postponed launch to 02/07/08 on 01/28/08. Slip was due to ECO sensor problems experienced during December launch attempt and implementation of ECO sensor connector soldered mod. (Also, LCC went back to the standard three of four valid ECO sensor readings.)

LAUNCH SCRUBS:
- Thursday, 12/06/07 launch attempt was terminated 2 hours into tanking when two of four engine cutoff (ECO) low-level LH2 fuel sensors failed wet/dry test. (The 5% sensor also failed wet during drain-back.) The ECO sensors are required for backup engine shutdown command to avoid catastrophic failure in the event of early fuel depletion. Launch was scrubbed at 8:56 am CST. Technical Scrub.
- Sunday, 12/09/07 launch attempt was terminated when one of previously failed sensors failed again during tanking, a couple of minutes into fast-fill. Engineers stated that the ET feedthrough and connector assembly was the most likely source of the problems. The 12/06/07 and 12/09/07 launch attempts produced previously unavailable time trending data that showed sensor faults occurring shortly before and after the feedthrough and connector were immersed in the super-cold propellants. Technical Scrub.

LAUNCH WINDOW:
- Total launch window was 10m1s with window open at 038:19:40:29Z and close at 038:19:50:30Z. Preferred Launch Time was 038:19:45:30Z (In-Plane Time) for a launch window of 5m1s.

LAUNCH DELAYS:
- None. Launch occurred on time at 038:19:45:30Z, 1:45:30 PM CST on Thursday 02/07/08.

TAL WEATHER:
Weather for the Transoceanic Abort Landing (TAL) sites during launch was benign. High pressure at the surface and aloft produced clear skies and light winds for Moron, Spain (MRN), Zaragoza, Spain (ZZA), and Istres, France (ISTRES). All three TAL sites were forecast GO throughout the launch count.

Continued...

FLT NO.	ORBITER	CREW (6+1 UP/6+1 DN) TITLE, NAMES, & EVA'S	LAUNCH SITE, LIFTOFF TIME, LANDING SITES, ABORT TIMES	LANDING SITE/ RUNWAY, CROSSRANGE LANDING TIMES, FLT DURATION, WINDS	SSME-TL NOM-ABORT EMERG THROTTLE PROFILE ENG. S.N.	SRB RSRM AND ET	ORBIT INC	HA/HP	FSW	PAYLOAD WEIGHTS, PAYLOADS/ EXPERIMENTS	MISSION HIGHLIGHTS (LAUNCH SCRUBS/DELAYS, TAL WEATHER, ASCENT I-LOADS, FIRSTS, SIGNIFICANT ANOMALIES, ETC.)
STS-122/ ISS 1E Continued…											

S122-E-011027 (18 Feb. 2008) --- The new ISS configuration seen from Atlantis post sep.

S122E011027

Continued…

PERFORMANCE ENHANCEMENTS:
Include the standard set plus: 1) PE Operational High Q WIN/FEB, 2) OMS Assist, 3) a 52 nm MECO, and 4) Del Psi.

FLIGHT DURATION CHANGES/LANDING:
On FD4, MMT concurred with formally changing mission duration from 11+1+2 to 12+0+2 to honor ISSP request for extra docked day for commissioning Columbus. (Activity did not fit 11-day mission.)

On FD7, MMT concurred with extending the mission duration to 13+0+2 to provide additional time needed to complete the activation of the Columbus module. Landing day was moved to 02/20/08.

FIRSTS/LASTS:
- First flight ECO sensor connector soldered mod
- First flight of new RSRM Nozzle-to-Case J-leg Joint insulation configuration
- New Annex Flight Rule in place to outline operational use of ECO sensor voltage measurements
- Addition of the Modified Adjustable Protective Mitten Assemblies (APMA's) or Overgloves
- First operational support from the Columbus Control Center in Oberpfaffenhofen, Germany
- First reboost of ISS since December 2002
- Last Shuttle Mission for Shuttle Program Manager N. Wayne Hale, Jr., a 30-year veteran of NASA who helped lead the space agency's recovery from the 2003 Columbia Disaster.

MEMENTOS:
- Mementos carried aboard STS-122 included three green starter flags celebrating the 50th anniversary of NASA and the 50th running of the Daytona 500 NASCAR Race, a dried red rose to be woven into a NASA-themed 50th anniversary float for the Tournament of Roses Parade, and 20 ESA flags whose use will be to commemorate the addition of Columbus to the ISS.

NIGHT LAUNCH: N/A

RENDEZVOUS #69: Rendezvous and dock with ISS

EVENTS:
- OMS 2 ignition at 038:20:23:09.9Z resulted in a 124.4 by 118.7 nm orbit.
- SRMS OBSS/LDRI survey of nosecap and port and starboard wing RCC (WLE's) was completed.
- TI maneuver at 040:14:37:28Z resulted in a 184.0 by 176.0 nm orbit.
- R-Bar Pitch Maneuver was performed. No significant issues
- Docking Capture occurred at 040:17:17:20Z.
- Hard Dock occurred at 040:17:30:22Z (above the South Australian coast - Columbus reached its permanent home).
- ISS Hatch Open 12:40 PM CST, Saturday, 02/09/08 - welcomed by ISS Crew.

Continued…

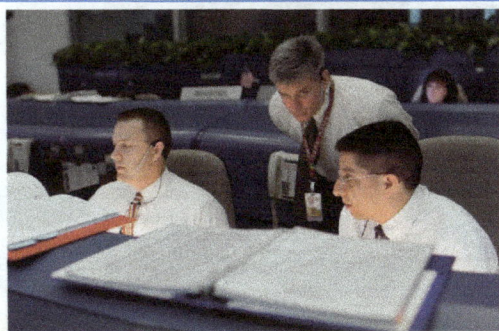

JSC2008-E-010344 --- FD's Norm Knight (left), Bryan Lunney, & Richard Jones monitor data in the Space Shuttle FCR of JSC's MCC during launch countdown activities a few hundred miles away at KSC.

JSC2008-E-010460 (8 Feb. 2008) --- John Shannon (right), Deputy Shuttle Program Manager; and Mike Sarafin, Lead Shuttle Flight Director, participate in an STS-122 press briefing with news media representatives at JSC. Rob Navias, PAO, (left) serves as moderator for the briefing.

FLT NO.	ORBITER	CREW (6+1 UP/6+1 DN) TITLE, NAMES, & EVA'S	LAUNCH SITE, LIFTOFF TIME, LANDING SITES, ABORT TIMES	LANDING SITE/ RUNWAY, CROSSRANGE LANDING TIMES FLT DURATION, WINDS	SSME-TL NOM-ABORT EMERG THROTTLE PROFILE ENG. S.N.	SRB RSRM AND ET	INC	ORBIT HA/HP	FSW	PAYLOAD WEIGHTS, PAYLOADS/ EXPERIMENTS	MISSION HIGHLIGHTS (LAUNCH SCRUBS/DELAYS, TAL WEATHER, ASCENT I-LOADS, FIRSTS, SIGNIFICANT ANOMALIES, ETC.)
STS-122/ ISS 1E Continued...		ABOVE: JSC2008-E-012993 ---- The STS-122 Orbit 1 Flight Control Team pose for a portrait in the Space Shuttle FCR at the JSC MCC. Flight Director Mike Sarafin (center right) holds the STS-122 mission logo. BELOW: JSC2008e020392 --STS-122 Ascent FCT poses with the crew in JSC MCC. FD Norm Knight (left) & CAPCOM Jim Dutton hold the mission logo. Crew pictured are CDR Frick, PLT Poindexter, Melvin/MS, Walheim/MS, & Schlegel/MS. (Not pictured was Love/MS.) 	**NOTES:** - Landing occurred at KSC on Wednesday 02/20/08 at 9:07 AM EST, 46 years to the day after the first American, John Glenn, orbited the Earth. - Daniel Tani returned home after 120 days. **SIGNIFICANT ANOMALIES:** Orbiter: - Overexposed video due to suspect AVIU - Fuel Cell 3 O_2 flowmeter is erratic. - During flight, Port AFT MPM Pedestal Stow indications came on approximately 11 hours after actual stow. - SSOR #1 intermittent comm dropouts - Suspect indication of possible IML crack on noted tile - CCTV black and white video shows intermittent color. - Mid Port Payload Bay Floodlight not illuminating SRB: - One of the three main parachutes on BI-132 LH showed significant damage in the canopy. RSRM: - Missing piece of forward factory joint weather seal, RSRM-99B SSME: None ET: - ET-124 - Post Launch camera and film review showed LH2 acreage foam loss at Sta. 1160 during Launch. - A crack in the +Y SRB Pal Ramp was observed prior to the ET-125 tanking test on 12/18/07. - A crack in the +Y Longeron Closeout was observed during the post-drain walkdown after the ET-125 tanking test on 12/18/07. - During the first launch attempt of ET-125 on 12/06/07, ECO/S #3 and #4 failed wet. - STS-122/ET-125 launched on 02/07/08. Post Launch camera and film review showed LH$_2$ acreage foam loss at Sta. 1145 during Launch. - STS-122/ET-125 - Post Launch camera and film review showed TPS losses at the intertank to Lh2 flange closeout at two locations. MOD: - High-speed data dropouts during Launch - Trajectory Server GPS time misconfiguration Integration: - Stinger tile observed falling after SSME startup - Ku-Band radiated in Hi Power - Unexpected debris/expected debris exceeding mass allowable prior to pad clearance (liftoff debris) - I/T to LH$_2$ Flange closeout foam loss - 2 locations of red foreign material located on SRB - LO$_2$ Umbilical Cable Tray foam loss (aft of Xt-2058) - STS-122 LH$_2$ ECO failure - LH2 acreage loss adjacent to Xt 1129 LO$_2$ Feedline base closeout - LH2 acreage loss aft of +Y bipod - Missing/peeled SF-EPDM on RH Forward Segment Factory Joint	Continued... - IELK Seat Liner Transfer at 040:23:20Z (5:20 PM CST, Feb. 9, 2008). At that time Daniel Tani became a member of STS-122 and Leopold Eyharts/ESA joined the ISS Expedition 16 as Flight Engineer. - Due to crew health issue, EVA1 postponed from FD4 to FD5 - FD5 EVA 1: EV1 and EV3 (sub for EV2, health issue) performed Columbus prep activities: connected data, power, and communications lines; removed LTA cable and CBM seal cover; installed PDGF; performed NTA prep activities; and stowed OTSD. Columbus second stage bolting completed at 3:44 PM CST Monday, 02/11/08. EVA1 duration 7h58m - FD7 EVA 2: EV1 and EV2 completed primary task to R&R a spent Nitrogen Transfer Assembly, outfit Columbus with trunnion covers, and repair Lab MMOD shield. EVA 2 duration 6hr45m - The OMS Pod stinger tile was cleared for entry. - FD9 EVA 3: EV1 and EV3 transferred SOLAR to Columbus, installed Columbus keel pin cover and handrail, transferred CMG to PLB, transferred EuTEF, and performed Airlock handrail damage swatch test. EVA 3 duration 7h25m - EVA NOTE: One EMU glove from STS-122, S/N 6197, had a 3/16-inch hole in the Vectran of left thumb that wasn't seen until postflight inspections on the ground. S/N 6197 was Rex Walheim's left glove worn on all three EVA's (per STS-123 03/11/08 MMT notes). - European Flight Controllers told the crew they had successfully completed initial activation of Columbus with the module's computer systems. German Chancellor Angela Merkel called to congratulate the crew. - FD9: To clear the path to shoot down a crippled spy satellite, NASA agreed to open its California landing strip on Wednesday, 02/20/08 so Atlantis could land that day, even if weather was bad at KSC. "The reason is to give the military the biggest possible window and maximum flexibility to ensure the success of the satellite intercept" per Lead Shuttle Flight Director Sally Davis. - Transfers: • Hardware transferred to ISS (outside and inside): 30404 lbs • Columbus - ESA Laboratory: 26627 lbs • Hardware/supplies transferred from ISS: 3585 lbs • H$_2$O transferred to ISS: 1386 lbs • O$_2$ transferred to ISS: 95 lbs • N$_2$ transferred to ISS: 27 lbs - FD10: Reboost at 047:12:17:00.0Z resulted in 187.8 by 177.6 nm orbit (first reboost since December 2002). ISSP estimated prop savings to get 400 lbs of logistics gains. - Undocked at 049:09:24:40Z followed by a flyaround (1/2 lap) - Separation Burn 1 at 049:10:34:02.0Z resulted in 188.1 by 175.8 nm orbit - Separation Burn 2 at 049:11:01:30.0Z resulted in 187.9 by 175.5 nm orbit - No communications blackout during Entry.

SPACE SHUTTLE MISSIONS SUMMARY

FLT NO.	ORBITER	CREW (6+1 UP/6+1 DN) TITLE, NAMES & EVA'S	LAUNCH SITE, LIFTOFF TIME, LANDING SITES, ABORT TIMES	LANDING SITE/ RUNWAY, CROSSRANGE LANDING TIMES FLT DURATION, WINDS	SSME-TL NOM-ABORT EMERG THROTTLE PROFILE ENG. S.N.	SRB RSRM AND ET	INC	ORBIT HA/HP	FSW	PAYLOAD WEIGHTS, PAYLOADS/ EXPERIMENTS	MISSION HIGHLIGHTS (LAUNCH SCRUBS/DELAYS, TAL WEATHER, ASCENT I-LOADS, FIRSTS, SIGNIFICANT ANOMALIES, ETC.)
STS-123/ ISS 1JA	0V-105	CDR: Dominic L. Gorie	KSC 39A	KSC 15 (KSC 68)	104/104/109%	BI-133	51.6 (25)	DIRECT INSERTION	OI-32 (3)	CARGO: 38915 LBS	BRIEF MISSION SUMMARY: STS-123/1JA (25th ISS mission)...

STS123-S-009 (11 March 2008) — Overcast clouds at 6500 ft provided a spectacular night image as the clouds glowed from the Shuttle's exhaust.

FLT NO.	ORBITER	CREW (6+1 UP/6+1 DN) TITLE, NAMES & EVA'S	LAUNCH SITE, LIFTOFF TIME, LANDING SITES, ABORT TIMES	LANDING SITE/ RUNWAY, CROSSRANGE LANDING TIMES FLT DURATION, WINDS	SSME-TL NOM-ABORT EMERG THROTTLE PROFILE ENG. S.N.	SRB RSRM AND ET	ORBIT INC	HA/HP	FSW	PAYLOAD WEIGHTS, PAYLOADS/ EXPERIMENTS	MISSION HIGHLIGHTS (LAUNCH SCRUBS/DELAYS, TAL WEATHER, ASCENT I-LOADS, FIRSTS, SIGNIFICANT ANOMALIES, ETC.)
STS-123/ ISS 1JA Continued...		Continued... SS EVA 119 DOCKED QUEST EVA 42 EMU/TETHERED EVA 112 SCHEDULED EVA 110 DURATION 6:47 SS EVA 120 DOCKED QUEST EVA 43 EMU/TETHERED EVA 113 SCHEDULED EVA 111 DURATION 6:24 SS EVA 121 DOCKED QUEST EVA 44 EMU/TETHERED EVA 114 SCHEDULED EVA 112 DURATION 6:02 MCC WHITE FCR (52) FLIGHT DIRECTORS: SHUTTLE: ASC - B. C. Lunney LD/O1 - M. P. Moses O2 - R. E. LaBrode PLNG - M. R. Abbott ENT - R. S. Jones MOD - P. L. Engelauf Team 4 - R. S. Jones/ A. J. Ceccacci ISS: LD/O2 - D. J. Weigel O1 - K. L. Alibaruho O 3 - G. Kerrick Team 4 - H. L. Rarick IP FD - E. J. Nelson (I/F w/CSA & JAXA) CAPCOMS: SHUTTLE: A/E - J. P. Dutton K. A. Ford (Wx) LD/O1 - T. W. Virts O2 - N. J. Patrick PLNG - B. A. Drew Team 4 - N/A Continued...	Continued... MECO CMD: 8:23.6 (P) 8:22.6(A) VI: 25819 (P) 25817.6(A) OMS-2: 38:15 (P) 38:30 (A) 97.4 FPS 96.1 FPS	Continued... WINDS: 1.5T 1.3L KTS OFFICIAL: 01002P03 KTS 2H 2R KTS DENS ALT: -336 FT FLT DURATION: 15:18:10:52 S/T: 1153:00:57:10 OV-105: 235:02:18:33 DISTANCE: 6,577,857 sm TOTAL SHUTTLE DISTANCE: 467,923,507 sm							

ISS016-E-032598 (12 March 2008) --- The Canadian-built Dextre robotic system and the Japanese K bo laboratory (JLP) are vis ble in Endeavour's cargo bay on approach to ISS.

ISS016-E-033684-- Crews: STS-123 (green shirts) & ISS Exp 16 (blue shirts), ISS CDR Peggy Whitson (second right, rear), Yuri Malenchenko/FSA FE (left, front), and Garrett Reisman/FE (left rear). Also in green shirt is Leopold Eyharts/ESA (right rear), former Exp16 FE, who has moved over to the STS-123 crew. Leaving ISS with Eyharts are the Endeavour crew CDR Dominic Gorie (second left, rear), PLT Gregory H. Johnson (behind Malenchenko), Takao Doi/JAXA MS (right front), Rick Linnehan/MS (behind Doi); Mike Foreman/MS (second right, center row); Robert L. Behnken/MS (far left, center row).

Continued...

TAL WEATHER:
Weather at the TAL sites was tricky as showers were monitored near Zaragoza, Spain and Istres, France during the launch countdown. Post cold front low level wind flow from the northwest brought showers to the windward sides of the Pyrenees and central French mountains. These showers dissipated as they crossed the high terrain. TAL weather was GO.

PERFORMANCE ENHANCEMENTS:
Include the standard set plus: 1) PE Operational High Q WIN/MAR, 2) OMS Assist, 3) A 52 nm MECO, and 4) Del Psi

FLIGHT DURATION CHANGES/LANDING:
Deorbit burn was planned for 086:21:58:14Z. Due to low clouds moving in at KSC, the deorbit burn was delayed to second opportunity at 086:23:33:13.9Z. Landing occurred at 087:00:39:06Z, Wednesday, 03/26/08, at 8:39:06 PM EDT.

FIRSTS/LASTS:
- First 16-day Space Station Assembly Mission, 12 days docked. (Longest mission is STS-67 - Spacelab, 16D 21H 47M 35S.)
- Tied the current mission record of five spacewalks held by the HST Servicing Missions (STS-61, STS-82, and STS-109). Most EVA's docked to ISS.
- A redesign to RSRM Nozzle Joints 2 and 5, the latter with an additional bolt enhancement, follows up the new Nozzle-to-Case J-leg Joint insulation configuration that debuted on STS-122's motors.
- First flight of a lighting system derived from an off-the-shelf flash (Nikon SB800) was added to a digital camera (in orbiter umbilical well) to capture photos of ET after separation for about 130 ft away.
- This is the last modified tank (before Columbia) and the next will be a tank built with all mods done in line.
- First on-orbit test of orbiter tile repair technique.
- First time the OBSS was left on the Station so that the next flight can deliver the large JAXA Kibo module.
- This mission marks a significant milestone with the inauguration of the JAXA IP support to real-time operations, adding them to the fold with ESA, CSA, and Russia. "We have reached a new pinnacle in the 'international' part of the Space Station operations."
- Spacelab Logistics Pallet (SLP) used by Dextre made its fourth and final flight to space, "concluding a long history that can be traced back before the first shuttle left the launch pad." - PAO.
- First flight with John Shannon as Shuttle Program Manager.
NOTE: The unmanned cargo ship Jules Verne, the ESA's first Automated Transfer Vehicle (ATV), launched toward ISS on March 7. It was parked well away from ISS at a safe distance until Endeavour's departure.

Continued...

FLT NO.	ORBITER	CREW (6+1 UP/6+1 DN) TITLE, NAMES & EVA'S	LAUNCH SITE, LIFTOFF TIME, LANDING SITES, ABORT TIMES	LANDING SITE/ RUNWAY, CROSSRANGE LANDING TIMES FLT DURATION, WINDS	SSME-TL NOM-ABORT EMERG THROTTLE PROFILE ENG. S.N.	SRB RSRM AND ET	INC	ORBIT HA/HP	FSW	PAYLOAD WEIGHTS, PAYLOADS/ EXPERIMENTS	MISSION HIGHLIGHTS (LAUNCH SCRUBS/DELAYS, TAL WEATHER, ASCENT I-LOADS, FIRSTS, SIGNIFICANT ANOMALIES, ETC.)
STS-123/ ISS 1JA Continued…		Continued… ISS: O1 - Z. Jones LD/O2 - S. K. Robinson O3 - M. T. Vande Hei Team 4 - R. C. Dempsey									Continued… NIGHT LAUNCH #30: Shannon: "We are launching in the dark." NIGHT LANDING KSC #16: (#22 in Shuttle history) RENDEZVOUS #70: Rendezvous and dock with ISS EVENTS: - OMS2 ignition at 071:07:06:44.0Z resulted in a 124.9 by 84.8 nm orbit. - SRMS OBSS/LDRI survey of nosecap and port and starboard wing RCC (WLE's) was completed. - TI maneuver at 073:00:42:21.9Z resulted in a 186.3 by 180.6 nm orbit. - R-Bar Pitch Maneuver was performed. No issues - Docking contact occurred at 073:03:46:54Z. - Hard Dock occurred at 073:04:02:11Z - ISS Hatch opened at 073:05:36:00Z, 12:36 AM CDT, Thursday, March 13, 2008, ISS crew welcoming - IELK Seat Liner Transfer at 073:07:50Z (2:50 AM CDT, March 13, 2008). At that time Leopold Eyharts/ESA became a member of STS-123 and Garrett Reisman joined the ISS Expedition 16/17 as Flight Engineer. - The first transfer item after hatch opening was swapping Garrett Reisman/MS for Leopold Eyharts (ESA)/Expedition 16 FE. The transfer was official when the form-fitting Soyuz seatliners were swapped. Eyharts spent 33 days as a member of ISS Expedition 16. With the on-time landing of March 26, Eyharts spent a total of 48 days in space. - FD4/5: EVA 1: EV1 & EV4: JLP prepped for unberthing, shuttle robot arm grappled JLP, Orbital Replacement Unit (ORU) and Tool Changeout Mechanism installed on the Canadian Special Purpose Dexterous Manipulator (SPDM or Dextre) arm 2 and arm 1, shuttle arm unberthed JLP, and shuttle arm installed JLP onto Harmony zenith port (temporary location until Kibo delivery on STS-124). Unable to provide keep-alive power to SPDM (later determined to be flawed cable in pallet). EVA 1 duration 7:01 - FD6: While Expedition 16 and STS-123 crewmembers brought the Kibo logistics module to life, Dextre's power supply unit was brought to life via the SSRMS. - FD6: EVA 2: EV1 & EV3: EVA ran long due to problems with the SPDM Arm Expandable Diameter Fasteners (EDF's) not releasing per procedure. Crew ended up using a pry bar. Time didn't permit removing some of the SPDM blankets. EV3 experienced RTV delamination. Per Rule {1JA_C2-105}, EMU OVERGLOVE EXCEPTIONS, crew continued the SPDM assembly task without donning overgloves due to the thermal constraints on SPDM. EV3 donned overgloves once the thermal critical tasks were complete. ISS multimeter was repaired and would later be swapped with shuttle multimeter prior to hatch closure. Installed the Node 2/JLP vestibule barrier assembly. EVA 2 duration 7:09 Continued…

S123-E-006403 --- Linnehan & Foreman assemble the stick-figure Dextre including attaching its two arms during EVA 2.

S123-E-007088 (18 March 2008) --- Canada's two armed robot, Dextre, is shown in the grasp of the station's robotic Canadarm2.

S123-E-006089 --- Reisman, Exp 16 & Linnehan (out-of-frame) prepare tool change out mechanisms on Dextre during EVA 1.

Foreman & Behnken install a spare arms on Dextre during EVA

SPACE SHUTTLE MISSIONS SUMMARY

FLT NO.	ORBITER	CREW (6+1 UP/6+1 DN) TITLE, NAMES & EVA'S	LAUNCH SITE, LIFTOFF TIME, LANDING SITES, ABORT TIMES	LANDING SITE/ RUNWAY, CROSSRANGE LANDING TIMES FLT DURATION, WINDS	SSME-TL NOM-ABORT EMERG THROTTLE PROFILE ENG. S.N.	SRB RSRM AND ET	INC	ORBIT HA/HP	FSW	PAYLOAD WEIGHTS, PAYLOADS/ EXPERIMENTS	MISSION HIGHLIGHTS (LAUNCH SCRUBS/DELAYS, TAL WEATHER, ASCENT I-LOADS, FIRSTS, SIGNIFICANT ANOMALIES, ETC.)

STS-123/ ISS 1JA

Continued...

S123-E-009262 (24 March 2008) --- The ISS latest configuration is viewed from Endeavour post-separation.

JSC2008-E-025177 --- Flight Controller Bill Foster in JSC MCC during launch countdown activities.

JSC2008-E-025187 --- Astronaut George Zamka, Spacecraft Communicator (CAPCOM), monitors data during launch countdown.

Flight Directors Bryan Lunney & Norm Knight in JSC MCC

SIGNIFICANT ANOMALIES:
Orbiter:
- Sensor Unit S/N 1150 on the port wing had excessive triggers (quantity 4452) during the first hour of MMOD monitoring for Late Inspection.
- Integrated Sensor Inspection System Sensor Pack 1 Pan Tilt Unit 10 degrees offset
- DCS OI1 card 1 failure
- FES shutdown on Primary A Controller
- GG Chamber pressure indicated a shift upward
- APU 1 fuel tank pressure decay
- LH OMS Pod mid surface temperature
- Sensor Unit 1150 (Ref Des: 65V08A01) on the port wing
- APU 3 seal cavity drain line pressures indicate slow decay.
- Body Flap tile damage
- Aft arrowhead damage
- STBD FWD RAD Retract Flexhose did not fully retract into RRSC (ref SPC# 205181853).
- APU 1 Gas Generator Chamber Pressure Transducer shift
- Cabin Temp Controller 1 noisy
- MPS E-3 LOX Inlet pressure showed a shift of 30 psi at Liftoff.
- MADS PCM MSRMNT gradually and abruptly moved to OSH throughout the MADS and MMU1/SSR1 recording phase.
- Lost OMS POD (RH OMS024) putty repair
- Damage to the V070-391044-174 (BRI-18) tile
- Damage to the V070-191101-043 (BRI-18) tile
SRB:
- Loss of data from SRB RH ET Observation Camera during Ascent
RSRM: None
SSME: None
ET: None
MOD:
- White-VTS-Servers hung
Integration:
- Unexpected debris/expected debris exceeding mass allowables prior to pad clearance (Liftoff debris)
- Stub Tile damage during SSME ignition
- Tile chips on orbiter stingers during SSME ignition

Continued...
- FD8: RTV Loss in EVA Gloves: EV3's gloves were NO-GO for subsequent EVA's. First spare set used on EVA 4.
- FD8: EVA3: EV1 & EV2: Finished assembly of Dextre, including installation of tool holder assembly and a Camera Light Pan Tilt Assembly (CLPA) which serves as Dextre's eyes. Also, the Spacelab Logistics Pallet used for assembly was prepared for return to shuttle cargo bay. Attempted to install MISSE-6 experiment (unsuccessful - moved to EVA5). EVA 3 duration 6:53
- FD10: Japanese Prime Minister called to congratulate the crew.
- FD10: During press interview, asked to describe the fast-growing Space Station, Reisman said the crew was struck by the view during final approach and similarities with the famous Space Station scene in the movie "2001: A Space Odyssey" by Stanley Kubrick and Arthur C. Clarke. Clarke died during this mission on 3/19/08 at the age of 90. Clarke in "First on the Moon" stated, "The inspirational value of the space program is probably of far greater importance to education than any input of dollars...a whole generation is growing up which has been attracted to the hard disciplines of science and engineering by the romance of space."
- FD11: EVA4: EV2 & EV3: Tasks were Remote Power Control Module removal and replacement, and the Tile Repair Ablator Dispenser (T-RAD) detailed test objective worksite setup and demonstration. The demonstration was considered a "huge" success, but needs results from post-landing detailed analysis. EVA 4 duration 6:24
- FD13: EVA 5: EV2 & EV3: Primary tasks completed were positioning of OBSS to temporary home on ISS truss, installation of MISSE-6 experiment, and inspection of SARJ. EVA 5 duration 6:02
- FD14: Conducted Rigidizable Inflatable Gas Experiment (RIGEX) funded by the Air Force. RIGEX was designed to test how well ground models and computer simulations predict what happens to the inflated structures in weightlessness. Once rigid, the sample tubes aboard Endeavour were blasted with vibrations to test their structural integrity. The experiment was returned to Earth aboard the shuttle for further scientific analysis.
- Transfers:
  • Hardware transferred to Station (outside and inside: 25839 lbs
  • Hardware transferred to Station (outside): 23776 lbs
  • Hardware transferred to Station (inside): 1432 lbs
  • Japanese pressurized logistics module: 18377 lbs
  • Dextre - Special Purpose Dexterous Manipulator: 3431 lbs
  • Middeck items returned from ISS aboard Endeavour: 1565 lbs
  • Water transferred to Station: 608 lbs
  • Oxygen transferred to Station: N/A
  • Nitrogen transferred to Station: 23 lbs
- Undocked at 085:00:25:00Z followed by a flyaround (1/2 lap). (Undocking was delayed 29 minutes due to two ISS Beta Gimbal Assembly (BGA) latch aborts.)
- Communications blackout time during Entry: 6m
NOTE: Currently, 590826 lbs mass in space of the ISS and ISS assembly 70% complete.

FLT NO.	ORBITER	CREW (6+1 UP/6+1 DN) / TITLE, NAMES & EVA'S	LAUNCH SITE, LIFTOFF TIME, / LANDING SITES, ABORT TIMES	LANDING SITE/ RUNWAY, CROSSRANGE / LANDING TIMES FLT DURATION, WINDS	SSME-TL NOM-ABORT EMERG / THROTTLE PROFILE ENG. S.N.	SRB RSRM AND ET	INC	ORBIT HA/HP	FSW	PAYLOAD WEIGHTS, / PAYLOADS/ EXPERIMENTS	MISSION HIGHLIGHTS (LAUNCH SCRUBS/DELAYS, TAL WEATHER, ASCENT I-LOADS, FIRSTS, SIGNIFICANT ANOMALIES, ETC.)
STS-124/ ISS 1J SEQ FLT# 123 KSC-123 PAD 39A-46 MLP-3 26TH SHUTTLE FLIGHT TO ISS	OV-103 (Flight 35) DISCOVERY OMS PODS: LPO1-38 RPO3-36 FRC3-35	CDR: Mark E. Kelly (Flt 3 - STS-108, STS-121) P763/R271/V181/M237 PLT: Kenneth T. Ham P764/R326/M282 MS 1/Robotics: Karen L. Nyberg P765/R327/F45 MS 2/EV2: Ronald J. Garan P766/R328/M283 MS 3/EV1: Michael E. Fossum (Flt 2 - STS-121) P767/R296/V196/M259 MS 4/Robotics: Akihiko Hoshide (Japan) P768/R329/M284 MS 5 UP/Stay as EXP 17/18 FLT ENG: Gregory E. Chamitoff P769/R330/M285 MS 5 DN/EXP 16/17 FLT ENG: Garrett E. Reisman (Up on STS-123, stay ISS) P770/R325/M281 SPECIAL EDUCATOR "Buzz" Lightyear (UP/EXP 18) See "Firsts" Continued…	KSC 39A 152:21:02:12Z 5:02:12 PM EDT (P) 5:02:12 PM EDT (A) Saturday (7) 05/31/08 (7) LAUNCH WINDOW: 6M 47S (PLT IN-PLANE) EOM PLS: KSC TAL: MRN TAL WX: FMI SELECTED: RTLS: KSC 15 N/N TAL: MRN 20 N/N (ZZA NO-GO) AOA: KSC 15 N/N 1ST DAY PLS: EDT 22 N/N TDEL: 0:000(P) -0.508(A) MAX Q NAV: 715.16(P) 701.98(A) SRB STG: 2:03:36(P) 2:02.56(A) PERF: NOMINAL 2 ENG TAL (ZZA): 2:48(P) 2:47(A) NEG RETURN: 3:48 3:55 PTA (U/S 159 FPS): 5:19 5:23 SE TAL (FMI 104): 6:08 6:13 PTM (U/S 180 FPS): 6:18 6:29 SE PRESS 104 7:01 7:07 MECO CMD: 8:24 8:26.3 Continued…	KSC 15 (KSC 69) 166:15:15:18Z 11:15:18 AM EDT Saturday (22) 06/14/08 (8) DEORBIT BURN: 166:14:10:12Z XRANGE: 270.2 NM ORBIT DIR: A/L 39 AIM PT: NOMINAL MLGTD: 2100 FT 166:15:15:17Z VEL: 209 KEAS 208 KEAS HDOT: -2.1 FPS TD NORM 195: 3172 FT DRAG CHUTE DEPLOY: 194 KEAS 166:15:15:20Z NLGTD: 5601 FT 166:15:15:28Z VEL: 155 KGS 148 KEAS HDOT: -7.0 FPS BRK INIT: 77 KGS DRAG CHUTE JETTISON: 54 KGS 166:15:15:59Z BRK DECEL FPS²: AVE 4.8 PK 6.3 WHEELS STOP: 166:15:16:19Z 11421 FT ROLLOUT: 9321 FT 1:02 M:S Continued…	104/104/109% PREDICTED: 100/104.5/ 104.5/72 104.5 ACTUAL: 100/104.5/ 104.5/72 104.5 1 = 2051 (7) 2 = 2048 (8) 3 = 2058 (2) M 3 EOM WEIGHT: 203604.5 LBS X CG: 1088.03 IN LANDING: WEIGHT: 203558.5 LBS X CG: 1090.00 IN	BI-134 RSRM 102 ET-128 SLWT 31 ET IMPACT MET 1:14:18 LAT: 36.362S LONG: 158.449W	51.6 (26)	DIRECT INSERTION POST OMS-2: 170.3x125.0 NM DEORBIT: HA 190.6 NM HP 23.3 NM ENTRY VELOCITY: 25866 FPS ENTRY RANGE: 4396 NM	OI-32 (4)	CARGO: 41997 LBS PAYLOAD CHARGEABLE: 33969 LBS DEPLOYED: 33890 LBS NON-DEPLOYED: 0 LBS MIDDECK: 79 LBS SHUTTLE ACCUMULATED WEIGHTS: DEPLOYED: 1456917 LBS NON-DEPLOYED 1601747 LBS CARGO TOTAL: 3943245 LBS PERFORMANCE MARGINS (LBS): FPR: 2651 FUEL BIAS: 1063 FINAL TDDP: 1308 RECON: 2513 PAYLOADS: PLB: ISS 1J MIDDECK: ISS 1J MAUAI 5 CRYO TK SETS SRMS (80) ODS, OBSS (Return Only) SSPTS	**BRIEF MISSION SUMMARY:** STS-124/1J (26th ISS mission) delivered the second and main segment of the Japanese (JAXA) Station Kibo (Hope) Laboratory. This segment known as the Japanese Pressurized Module (JPM) is the ISS's largest laboratory measuring 14.4 feet in diameter and 36.7 feet long. The Kibo complex also includes: An airlock and two robotic arms also delivered on this flight; the Japanese Experiment Logistics Module Pressurized Section (launched on STS-123); and an exterior platform for experiments exposed to space, scheduled for delivery on STS-127. The STS-124 mission is the first in which the JAXA Flight Control Team activated and controlled a module from Kibo Mission Control in Tsukuba, Japan. Also, as the STS-124 launch countdown got underway, a special Russian pump was added to Discovery's manifest to fix "a balky toilet" on the ISS. KSC W/D: OPF: 157, VAB HB-1: 7, PAD A: 29 = 193 Total Work Days (+ 13 Holidays @ OPF) LAUNCH POSTPONEMENTS: - Added STS-124 to FDRD - launch date of 02/28/08 on 02/20/07. - Ppd. to 04/24/08 on 04/16/08. Slip due to STS-117 rollback. - Ppd. to 05/25/08 on 03/07/08. Slip due to ET delivery delay and Beta Angle restriction. - Ppd. to 05/31/08 on 04/03/08. Slip due to adverse weather conditions affected on dock delivery date of ET-128. LAUNCH SCRUBS: None LAUNCH WINDOW: Total launch window was 7 minutes 45 seconds with window open at 152:21:01:14Z and close at 152:21:08:59Z. Preferred Launch Time was 152:21:02:12Z (In-Plane Time) for a launch window of 6m47s. LAUNCH DELAYS: None. Launch occurred on time at 152:21:02:12Z, 5:02:12 p.m. EDT, Saturday, May 31, 2008. On launch day, the sea breeze pushed across KSC with showers just west of the launch pad several hours before launch time. However, the sea breeze had pushed west of KSC by early afternoon with near ideal conditions for launch. Thunderstorms were occurring over central Florida but were well outside the 20 nautical mile thunderstorm flight rule limit. "Nice day to send 'Hope' to the ISS" – PAO. Cain: "If you watched today, you saw a flawless countdown." Continued…

080531—"Shuttle launch exhaust thrust damaged flame trench and blasted bricks and other debris beyond a perimeter fence some 1,800 feet from pad. No damage to Shuttle."

FLT NO.	ORBITER	CREW (7) TITLE, NAMES & EVA'S	LAUNCH SITE, LIFTOFF TIME, LANDING SITES, ABORT TIMES	LANDING SITE/ RUNWAY, CROSSRANGE LANDING TIMES FLT DURATION, WINDS	SSME-TL NOM-ABORT EMERG THROTTLE PROFILE ENG. S.N.	SRB RSRM AND ET	ORBIT INC HA/HP	FSW	PAYLOAD WEIGHTS, PAYLOADS/ EXPERIMENTS	MISSION HIGHLIGHTS (LAUNCH SCRUBS/DELAYS, TAL WEATHER, ASCENT I-LOADS, FIRSTS, SIGNIFICANT ANOMALIES, ETC.)
STS-124/ ISS 1J Continued...		Continued… SS EVA 122 DOCKED QUEST EVA 45 EMU/TETHERED EVA 115 SCHEDULED EVA 113 DURATION 6:48 SS EVA 123 DOCKED QUEST EVA 46 EMU/TETHERED EVA 116 SCHEDULED EVA 114 DURATION 7:11 SS EVA 124 DOCKED QUEST EVA 47 EMU/TETHERED EVA 117 SCHEDULED EVA 115 DURATION 6:33 MCC WHITE FCR (53) FLIGHT DIRECTORS: SHUTTLE: ASC - N. D. Knight LD/O1 - M. R. Abbott O2 - M. L. Sarafin PLNG - P. F. Dye/ A. J. Ceccacci ENT - R. S. Jones MOD - J. A. Mccullough Team 4 - R. E. LaBrode ISS: LD/O2 - A. P. Hasbrook O1 - R. C. Dempsey O3 - E. J. Nelson Team 4 - B. T. Smith IP FD - H. E. Ridings (I/F w/JAXA) Continued…	Continued… VI: 25819 25820 OMS-2: 37:20 37:21 250.7 FPS 249.1 FPS	Continued... WINDS: 0 KT 5 L KTS OFFICIAL: 07007P12 KTS 1H 12L KTS DENS ALT: 1748 FT FLT DURATION: 13:18:13:06 S/T: 1166:19:10:16 OV-103: 305:08:10:09 DISTANCE: 5,735,643 sm TOTAL SHUTTLE DISTANCE: 473,659,150 sm						Continued… TAL WEATHER: The TAL weather conditions were rather challenging. An upper low had been spinning over Spain for several days, drifting slowly to the northwest. Timing differences in the models made forecasting where precipitation would develop difficult. Initially on L-2 day, NO-GO forecasts were issued for Moron and Zaragoza, Spain with a GO forecast for Istres, France. Shuttle launches require only one of the three TAL sites have GO weather. As the upper low began to finally move to the northwest, forecasts were updated to GO for Moron, but a NO-GO for Istres. On launch day, Moron weather remained favorable and conditions at Istres improved and were GO. Zaragoza was observed NO-GO at TAL landing time. PERFORMANCE ENHANCEMENTS: Include the standard set plus: 1) PE Operational High Q TRN/JUN, 2) OMS Assist, 3) A 52 nautical mile MECO, and 4) Del Psi. FLIGHT DURATION CHANGES/LANDING: None FIRSTS/LASTS: - First flight of an ET built from scratch with all of the safety modifications stemming from the 2003 Columbia accident. "This essentially is the completed return-to-flight tank," Shannon. - First docking of Shuttle while ATV also docked to ISS. - First OBSS transfer from ISS to Orbiter. - First Post-Undock Inspection (Orbiter heat shield) will be the full "FD2 Inspection" done on previous missions. - First flight of Modified EMU gloves: includes addition of Turtleskin™ patches to thumb and index finger – to provide increased protection against cuts. - A first: NASA and Disney joined forces for education. "Buzz Lightyear," a 12-inch tall action doll, based on the cartoon character from the Pixar Studios Toy Store animated movies was delivered to the ISS for a 6-month stay. While on ISS, Lightyear will demonstrate zero gravity to elementary school children. NIGHT LAUNCH: N/A RENDEZVOUS: #71 - Rendezvous and dock with ISS Continued…

S124-E-005921 --- In the grasp of ISS robotic Canadarm2, the Kibo Japanese Pressurized Module (JPM) is moved from Discovery's payload bay to the port side of the Harmony node.

S124-E-006361 --- Fossum & Garan outfitted the outside of the JPM, installing covers and external television equipment and removing thermal covers and insulation on the JAXA RMS and top hatch.

FLT NO.	ORBITER	CREW (7) TITLE, NAMES & EVA'S	LAUNCH SITE, LIFTOFF TIME, LANDING SITES, ABORT TIMES	LANDING SITE/ RUNWAY, CROSSRANGE LANDING TIMES FLT DURATION,	SSME-TL NOM-ABORT EMERG THROTTLE PROFILE	SRB RSRM AND ET	ORBIT INC HA/HP	FSW	PAYLOAD WEIGHTS, PAYLOADS/ EXPERIMENTS	MISSION HIGHLIGHTS (LAUNCH SCRUBS/DELAYS, TAL WEATHER, ASCENT I-LOADS, FIRSTS, SIGNIFICANT ANOMALIES, ETC.)
STS-124/ ISS 1J Continued…		Continued… CAPCOMS: SHUTTLE: A/E - T. W. Virts - K. A. Ford (Wx) LD/O1 - N. J. Patrick O2 - B. A. Drew PLNG - S. W. Lucid Team 4 - N/A ISS: O1 - M. T. Vande Hei LD/O2 - C. J. Cassidy O3/PLNG - M. C. Jensen Team 4 - N/A							Continued…

AT RIGHT: S124-E-005615 --- STS-124 & Exp 17 crews greet each other shortly after docking. Left Foreground: EXP17 CDR Sergei Volkov (RSA). Left, partially obscured CDR Kelly & PLT Ham; Fossum/MS (center left), Reisman/MS (center right); Oleg Kononenko/FE EXP17/RSA (right), Garan/MS, Chamitoff/MS, & Nyberg/MS. BELOW: Hoshide/MS (JAXA), not in photo at right, works in newly installed Kibo JPM.

ABOVE: S124-E-009982 (11 June 2008) --- View of ISS configuration post Shuttle sep shows Kibo attached to Harmony at bottom center with first ESA ATV Docked at top center. AT LEFT: S124-E-010186 --- The Kibo laboratory (center left) is shown after attachment to port side of Harmony Node with: Kibo logistics module at bottom left, Columbus lab at center right, and at top center is Dextre along with two docked Russian spacecrafts.

Continued…

EVENTS:
- Shuttle launch sent asbestos 1,800 feet from pad. The 6 million pounds of thrust from Discovery's engines, channeled by the flame trench, blasted bricks, concrete rubble, and asbestos beyond a perimeter fence some 1,800 feet away. Bricks and some asbestos landed in a retention pond behind the fence. No damage to Shuttle.
- OMS2 ignition at 152:21:39:32.5Z resulted in a 170.3 by 125.0 NM orbit.
- NOTE: SRMS OBSS/LDRI survey of nosecap and port and starboard wing RCC (WLE's) was not performed until post undocking (no OBSS on Shuttle).
- FD2: TI Maneuver at 154:15:16:26.0Z resulted in a 183.9 by 182.2 NM orbit.
- R-Bar Pitch Maneuver was performed. No issues
- FD3: Docking Contact occurred at 154:18:03:20Z.
- Hard Dock occurred at 154:18:16:30Z.
- ISS Hatch opened at 154:19:30:00Z, 2:30 PM CDT, Monday, June 02, 2008; welcomed by ISS crew.
- IELK Seat Liner Transfer at 154:22:35Z (5:35 PM CDT, June 2, 2008). At that time Garrett Reisman became a member of STS-124 and Greg Chamitoff joined the ISS Expedition 17 as Flight Engineer.
- FD4: EVA 1: Egress was delayed by about 1 hour to reconnect Fossum's comm cap - lost comm during pre-breathe. Fossum & Garan, prepared the Kibo (JPM) for its removal from the Shuttle payload bay, disconnecting cables and removing covers. JAXA MS/Hoshide and MS/Nyberg robotically removed Kibo from the Shuttle P/L bay and latched it to Harmony, Node 2. Hoshide noted: "We have a new 'Hope' on the ISS." EV1 & EV2 assisted in the transfer of the OBSS from its ISS stored position (since STS-123) back to the Shuttle. The OBSS would be used with the shuttle robotic arm on FD12 to inspect the Orbiter heat shield. EV1& EV2 also demonstrated a technique that could be used to clean the starboard SARJ, which has had limited capability for several months. EV2 installed a new bearing and EV1 verified by inspection that a spot on earlier EVA's was a divot. This will feed into further analysis of the origin of the damage. EVA 1 duration 6:48.
- FD4: Based on review of launch imagery, the MMT decided that the focused inspection of the Orbiter heat shield was not required.
- FD6: EVA 2 - Fossum & Garan outfitted the outside of the JPM, installing covers and external television equipment and removing thermal covers and insulation on the JAXA RMS and top hatch. They also loosened bolts holding two Nitrogen Tank Assemblies in place on the Station's truss. Those tanks will be swapped during EVA 3. They also retrieved a failed external television camera from the port truss. In addition, Fossum inspected the left SARG, which had been performing perfectly. No shavings or debris were found, but photos were taken to be sent to the ground for review. EVA 2 duration 7:11.

Continued…

SPACE SHUTTLE MISSIONS SUMMARY

FLT NO.	ORBITER	CREW (7) TITLE, NAMES & EVA'S	LAUNCH SITE, LIFTOFF TIME, LANDING SITES, ABORT TIMES	LANDING SITE/ RUNWAY, CROSSRANGE LANDING TIMES FLT DURATION, WINDS	SSME-TL NOM-ABORT EMERG THROTTLE PROFILE ENG. S.N.	SRB RSRM AND ET	ORBIT INC	HA/HP	FSW	PAYLOAD WEIGHTS, PAYLOADS/ EXPERIMENTS	MISSION HIGHLIGHTS (LAUNCH SCRUBS/DELAYS, TAL WEATHER, ASCENT I-LOADS, FIRSTS, SIGNIFICANT ANOMALIES, ETC.)

STS-124/ ISS 1J Continued...

Mosaic of the Zenith and Aft Sides of the ISS During Flyaround 1J/STS-124

(Labels: SSRMS, JLP, JEM Remote Manipulator System, JEM, Node 2, Columbus, P5 Truss, ESP3, S3 Truss, S5 Truss, P6 Truss, P4 Truss, P3 Truss, P1 Truss, S0 Truss, Z1 Truss, S1 Truss, S4 Truss, ESP2)

JSC2008-E-043220 --- John McCullough (left), chief of the Flight Director Office, part of the Mission Operations Directorate at JSC, and Bryan Lunney, Flight Director and a mission manager observe KSC launch from MCC.

STS124-S-072 --- A close look at Discovery post landing at KSC. From left: KSC Director Bill Parsons and Bill Gerstenmaier, NASA Associate Administrator for Space Operations. At right: JAXA Director of Program Management & Integration Yuichi Yamaura & VP Kaoru Mamiya.

SIGNIFICANT ANOMALIES:
Orbiter:
- TCS Dropouts during Rendezvous
- Engine # 2 Dome Heat C/P Tile Damage
- Imagery Showed F3D (V070-421558-024) and F44 (V070-421558-025) Tyvek Rain Covers Released Late
- IMU 1 Z Gyro excessive drift
- The Left Hand ET Door BRI-18 Tile V070-395055-255
- Rudder Speed Brake Thermal Tab found dislodged and floating
- A buildup of ceramic adhesive identified under the Thermal Barrier
- Closed 2 Indication failed to Transfer On when door was closed
- Crew reported difficulty latching the External Airlock Upper Hatch prior to Undocking
KSC:
- STS-124 Pad debris items
SRB:
- STS-124/BI-134rh Data Acquisition System failed to record video and obtained erroneous Accelerometer data
RSRM: None. SSME: None. MOD: None
ET:
- STS-124/ET-128 Post-Launch Camera Film Review showed two foam losses (80971008428-510) on Xt 1129 LO2 Feedline Support Fitting Closeout
Integration:
- Unexpected Debris/Expected Debris Exceeding Mass Allowable prior to Pad clearance (Liftoff Debris)
- Late Tyvek partial cover releases
- Roll Moment during SRB Tail-off
- Liberated Refractory Brick, NE Flame Trench Wall Pad A
- ET TPS loss at ~Xt 1129, near LO2 Feedline Bracket

Continued...

- FD9: EVA 3: Fossum & Garan began the EVA 30 minutes ahead of schedule. The EVA was highlighted by Garan's dramatic robot ride some 80 feet over the top of the ISS to replace a 550 lb nitrogen tank on the starboard truss. The ride was dubbed the "windshield wiper maneuver" or as Mark Carreau (Houston Chronicle) headlined it: "Wild robot-arm ride caps workday at Space Station." Fossum returned to the port SARJ (inspected on EVA 2) taking particulate matter from inside the joint, using a strip of tape that was returned to Earth for analysis. He also removed thermal insulation from the Kibo robotic arm's wrist and elbow cameras and launch locks from one of the Kibo windows and deployed debris shields on Kibo. Other tasks by the pair included: The repaired video camera retrieved on EVA 2 was re-installed and several extra tasks (installation of thermal cover on Harmony, relocation of foot restraint aid, and removal of SARJ launch lock) were conducted. EVA 3 duration 6:33.
- Transfers:
  - Hardware transferred to ISS (outside & inside): 34,353 lbs
  - Hardware transferred to ISS (inside): 1,787 lbs
  - Hardware transferred to shuttle (outside – OBSS): 536 lbs
  - Hardware/supplies transferred from ISS (inside): 1,807 lbs
  - H2O delivered to ISS: 569 lbs
  - O2 used for the 3 EVA's: 92 lbs
  - O2 used for "stack maintenance:" 29 lbs
  - N2 transferred to ISS: 15 lbs
- FD12: Undocked at 163:11:41:54Z followed by a fly-around (1/2 lap).
- Conducted the late inspection of the Shuttle's heat shield using the OBSS. No issues.
- FD14: Rudder/Speedbrake thermal spring tab was seen floating away from the vehicle during the FCS checkout. The function of the tab is to prevent a flow path for ascent heating and is not required for entry. The TPS was cleared for entry.
- [Post-flight, this issue was presented to 08/07/08 PRCB; decision was made to continue to fly as is. PRCB directed a new ascent thermal environmental assessment to consider flying without the tabs.]
- No communications blackout during Entry

FLT NO.	ORBITER	CREW (6+1 UP/6+1 DN) TITLE, NAMES & EVA'S	LAUNCH SITE, LIFTOFF TIME, LANDING SITES, ABORT TIMES	LANDING SITE/ RUNWAY, CROSSRANGE LANDING TIMES FLT DURATION, WINDS	SSME-TL NOM-ABORT EMERG THROTTLE PROFILE ENG. S.N.	SRB RSRM AND ET	INC	HA/HP	FSW	PAYLOAD WEIGHTS, PAYLOADS/ EXPERIMENTS	MISSION HIGHLIGHTS (LAUNCH SCRUBS/DELAYS, TAL WEATHER, ASCENT I-LOADS, FIRSTS, SIGNIFICANT ANOMALIES, ETC.)
STS-126/ ISS- ULF2 SEQ FLT # 124 KSC-124 PAD 39A (47) MLP- 3 27th SHUTTLE FLIGHT TO ISS	OV-105 (Flight 22) ENDEAVOUR OMS PODS: LPO3-33 RPO4-29 FRC5-22	CDR: Chris Ferguson Flt 2 (STS-115) P771/R300/V197/M179 PLT Eric Boe P772/R331/M286 MS1 Donald Pettit Flt 2 (STS-113 Up – Soyuz TMA-1 Dn) P773/R289/V198/M253 MS2 Steve Bowen P774/R332/M287 MS3 Heidemarie Stefanyshyn-Piper Flt 2 (STS-115) P775/R301/V199/F40 MS4 Shane Kimbrough P776/R333/M288 MS5 UP Stay ISS EXP 18/FLT ENG Sandra Magnus Flt 2 (STS-112) P777/R284/V200/F36 MS5 DN EXP 17/Flt ENG Greg Chamitoff (UP ON STS-124, stay ISS) P778/R330/M285 SS EVA 125 DOCKED QUEST EVA 48 EMU/TETHERED EVA 118 SCHEDULED EVA 116 DURATION 6:52 Continued…	KSC 39A 320:00:55:39Z 7:55:39 PM EST (P) 7:55:39 PM EST (A) Friday (26) 11/14/08 (15) LAUNCH WINDOW: 4M 39S (PLT in-plane) EOM PLS: KSC TAL: ZZA TAL WX: FMI SELECTED: RTLS: KSC15 CI/NOM TAL: ZZA30L N/N AOA: KSC15 CI/N 1ST DAY PLS: EDT22 N/SFD Continued…	EDT04 CONC EDW 52 CONC 33 335:21:25:09Z 1:25:09 PM PST Sunday (15) 11/30/08 (13) DEORBIT BURN: 335:20:19:29Z XRANGE: 169.6 NM ORBIT DIR: A/L (40) AIM PT: Close-In MLGTD: 2040 FT 335:21:25:09Z VEL: 219 KGS 211 KEAS HDOT: -1.1 FPS TD NORM 205: 2482 F Continued…	104/104/109% PREDICTED: 100/104.5/104.5/ 72/104.5 ACTUAL: 100/104.5/104.5/ 72/104.5 1 = 2047 (12) 2 = 2052 (7) 3 = 2054 (8) M 3 EOM: WEIGHT: 221787 LBS X CG: 1087.2 IN LANDING: WEIGHT: 221712 LBS X CG: 1089.0 IN	BI-136 RSRM 104 ET-129 SLWT 32 ET IMPACT MET 1:14:18 LAT: 36.202 S LONG: 158.215W	51.6 (27)	DIRECT INSERTION POST OMS-2: 125.7x 84.6NM DEORBIT: HA 193.1 NM HP 21.9 NM ENTRY VELOCITY: 25863 FPS ENTRY RANGE: 4400NM	OI-33 (1)	CARGO: 39471 LBS PAYLOAD CHARGEABLE: 32403 LBS DEPLOYED: 30432 LBS NON-DEPLOYED: 1760 LBS MIDDECK: 211 LBS SHUTTLE ACCUMULATED WEIGHTS: DEPLOYED: 1487349 LBS NON-DEPLOYED: 1603708 LBS CARGO TOTAL: 3982716 LBS PERFORMANCE MARGINS (LBS): FPR: 2651 FUEL BIAS: 1063 FINAL TDDP: 1682 RECON: 2329 PAYLOADS: PLB: ISS-ULF2 (MPLM, LMC),SSPL/PSSC MIDDECK: ISS-ULF2, MAUI SEITE 5 CRYO TANK SETS RMS (81) SRMS, ODS, OBSS, SSPTS	***Brief Mission Summary:*** *"Extreme Home Improvements" STS-126/ULF2 (27th ISS mission) outfitted the ISS to increase accommodations from a crew of three to six. Life support and habitability additions included: an advanced resistive exercise device, a second toilet, a galley, two sleep stations and an integrated water recycling system. The mission also included EVA's for lubricating the sluggish Solar Alpha Rotary Joints (SARJ) and installation of other external systems.* *Endeavour was originally rolled to Launch Pad 39B as the Launch on Need (LON) vehicle in support of STS-125 HST servicing mission. Last minute complications with HST caused an indefinite delay for STS-125. Endeavour was rolled to Launch Complex 39A and prepared for the STS-126 November launch date. (Shuttles have only moved from one spaceport launch pad to another twice before in the program's history, in 1990 and 1993.)* **KSC W/D** The Orbiter prep days are 162 workdays (W/D) + 3 holidays + 3 weather days in the OPF. VAB ops = 7 W/D + 1 weather day Pad B ops = 19 W/D + 15 contingency days Pad A ops = 18 W/D + 5 contingency days Total W/D = 206 **LAUNCH POSTPONEMENTS** - Added STS-126 to FDRD - launch date of 09/18/08 on 08/15/07. - Ppd. to 10/16/08 on 02/14/08. Slip due to ECO sensor problems experienced during December launch attempt of STS-122. - Ppd. to 11/10/08 on 05/27/08. Slip due to delays in delivery of ET-127 & ET-129 for STS-125 & STS-400, respectively. - Ppd. to 11/12/08 on 09/08/08. Slip due to Hurricane Faye impacts to HST payload readiness. - Ppd. to 11/16/08 on 09/24/08. Slip due to STS-125 slip to from 10/10/08 to 10/14/08 caused by Hurricane ke. - Launch moved forward to 11/14/08 on 10/19/08. Move due to critical path adjustment. STS-126/ULF2 now "prime crew" as STS-125 postponed to NET Mid-Feb 2009 on 10/02/08. **LAUNCH SCRUBS**: None. Continued…

STS-125 (HST Service) & LON Vehicle on Pads 39A & 39B. LON Vehicle became STS-126 when STS-125 was ppd to 2009. Picture courtesy of Rod Ostoski/KSC-USA.

FLT NO.	ORBITER	CREW (7) TITLE, NAMES & EVA'S	LAUNCH SITE, LIFTOFF TIME, LANDING SITES, ABORT TIMES	LANDING SITE/ RUNWAY, CROSSRANGE LANDING TIMES FLT DURATION, WINDS	SSME-TL NOM-ABORT EMERG THROTTLE PROFILE ENG. S.N.	SRB RSRM AND ET	ORBIT INC	HA/HP	FSW	PAYLOAD WEIGHTS, PAYLOADS/ EXP	MISSION HIGHLIGHTS (LAUNCH SCRUBS/DELAYS, TAL WEATHER, ASCENT I-LOADS, FIRSTS, SIGNIFICANT ANOMALIES, ETC.)
STS-126/ ISS- ULF2 Continued ...		Continued... SS EVA 126 DOCKED QUEST EVA 49 EMU/TETHERED EVA 119 SCHEDULED EVA 117 DURATION 6:45 SS EVA 127 DOCKED QUEST EVA 50 EMU/TETHERED EVA 120 SCHEDULED EVA 118 DURATION 6:57 SS EVA 128 DOCKED QUEST EVA 51 EMU/TETHERED EVA 121 SCHEDULED EVA 119 DURATION 6:07 MCC WHITE FLIGHT FCR (54) FLIGHT DIRECTORS: SHUTTLE: ASC- Bryan Lunney LD/O1- Mike Sarafin O2- Tony Ceccacci FD 1-12 - Paul Dye FD 13-EOM Planning- Paul Dye FD 1-3 - Kwatsi Alibarufo FD 4-EOM ENT- Bryan Lunney MOD – John Mccullough Team 4- Richard Jones ISS O1 – Holly Ridings LD/O2- Ginger Kerrick O3 – Brian Smith Team 4- Courtenay McMillan Continued...	Continued... TDEL: 0.000 (P) 0.192 (A) MAX Q NAV: 757.6 (P) 750.2 (A) SRB STG: 2:04.32(P) 2:06.24(A) PERF: NOMINAL 2 ENG TAL (MRN): 2:38 (P) 2:39 (A) NEG RETURN: 3:52 3:54 PTA (U/S 157 FPS): 5:08 5:14 SE TAL (ZZA 104): 6:01 6:04 PTM (U/S 168 FPS): 6:07 6:18 SE PRESS 104 6:54 6:59 MECO CMD: 8:22.1 8:23.0 VI: 25819.0 25818.8 OMS-2: 38:20 38:19.3 97.4 FPS 95.9 FPS	Continued... DRAG CHUTE DEPLOY: 193 KEAS 335:21:25:12Z NLGTD: 6761 FT 335:21:25:20Z VEL: 154 KGS 146 KEAS HDOT: -6.2 FPS BRK INIT: 124 KGS DRAG CHUTE JETTISON: 53 KGS 335:21:25:42Z BRK DECEL FPS²: AVE 6.2 PK 9.3 WHEELS STOP: 335:21:26:02Z 11180 FT ROLLOUT: 9140 FT 0:53 M:S WINDS: 4H KT 0 KTS OFFICIAL: 04004P06 KTS 6H 0CROSS KTS DENS ALT: 3234 FT FLT DURATION: 15:20:29:30 S/T: 1183:15:39:46 OV-105: 274:03:35:10 DISTANCE: 6,615,109 sm TOTAL SHUTTLE DISTANCE: 480,274,259 sm							Continued... **LAUNCH WINDOW:** Total launch window was 9 minutes 26 seconds with window open at 320:00:50:52Z and close at 320:01:00:18Z. Preferred Launch Time was 320:00:55:39 (In-Plane Time) for a launch window of 4m39s. **LAUNCH DELAYS:** None. Launch occurred on time at 320:00:55:39Z, 7:55:39 p.m. EST, Friday, November 14, 2008. Weather on launch day was acceptable. Isolated afternoon showers were observed at 60 miles south of KSC along the sea breeze late in the day. The showers diminished by sunset - not a threat for the evening launch time or RTLS. **TAL WEATHER** Weather at the TAL sites was forecast/observed GO. **PERFORMANCE ENHANCEMENTS:** Include the standard set plus: 1) PE Operational High Q TRN/NOV, 2) OMS Assist, 3) a 52 nautical mile MECO, and 4) Del Psi **FLIGHT DURATION CHANGES/LANDING:** - FD 11 MMT decision made for a one-day extension for additional on-orbit time for the Urine Processing Assembly (UPA) troubleshooting & processing or possible Distillate Assembly (DA) return. - Weather for landing was quite complex. Both KSC and EAFB were activated on Sunday, November 30, 2008, as possible landing sites. A large upper level low pressure system over the eastern US with a cold front moving across FL were concerns for landing at KSC on Sunday (EOM) & Monday (EOM+1). Spaceflight Meteorology Group (SMG) weather forecasts were "NO GO" for KSC with crosswind, ceiling, precipitation, and thunderstorm flight rule violations. Also, two Tornado Watches were issued for central FL and a third Watch included KSC. A squall line moving east at 20 kts combined with an unstable air mass across south and central FL generated numerous thunderstorms and isolated tornadoes by mid day. The weather continued to deteriorate across central FL, prompting the MMT to assess the possibility of staying on orbit and attempting EOM+1 landing at KSC. The SMG forecasts for that day indicated marginal conditions for a safe return to KSC. After waving off the first opportunity to KSC and with weather conditions deteriorating through the day at KSC, the decision was made to land at EAFB. Weather conditions at EAFB were nearly ideal with light northeast surface winds and mostly clear skies. Endeavour touched down at 335:21:25:09Z (3:25 PM CST, November 30, 2008) on temporary runway 04. This runway was built due to construction and resurfacing of the primary runway. Continued...

Parade of storms during STS-125 & STS-126 launch preps as seen on Sep. 04, 2008: Gustav (inland remnants, upper left) followed by Hanna, Ike, & Josephine. (From:Robert Harvey/DA8)

IKE08-notrack.gif: Hurricane IKE tracking. Category 2 landfall at 2:10 a.m. CDT near Galveston Sep. 13, 2008. (From: JSC Roundup Nov. 2008) Damage from hurricanes cost NASA $50M this season.

At Left: STS126-S-044 --- NASA Administrator Michael Griffin (front) & Associate Administrator for Space Operations Bill Gerstenmaier watch the launch of the Space Shuttle Endeavour from KSC Launch Control Center on Nov. 14, 2008.

FLT NO.	ORBITER	CREW (7) TITLE, NAMES & EVA'S	LAUNCH SITE, LIFTOFF TIME, LANDING SITES, ABORT TIMES	LANDING SITE/ RUNWAY, CROSSRANGE LANDING TIMES FLT DURATION, WINDS	SSME-TL NOM-ABORT EMERG THROTTLE PROFILE ENG. S.N.	SRB RSRM AND ET	ORBIT INC / HA/HP	FSW	PAYLOAD WEIGHTS, PAYLOADS/ EXP	MISSION HIGHLIGHTS (LAUNCH SCRUBS/DELAYS, TAL WEATHER, ASCENT I-LOADS, FIRSTS, SIGNIFICANT ANOMALIES, ETC.)
STS-126/ ISS- ULF2 Continued…		Continued… **CAPCOMS:** **SHUTTLE** A/E – Alan Poindexter - Greg (Box) Johnson (Wx) LD/01 – Steve Robinson O2 – Jim Dutton Planning – Shannon Lucid Team 4 - N/A **ISS** O1- Terry Virts LD/O2- Mark Vande Hei O3 – Robert Hanley Team 4 - N/A								Continued…

S126-E-012247 --- Endeavour & Exp 18 crews shared a Thanksgiving meal on middeck: At top Center, Magnus /STSUp/FE Exp18. Clockwise from her: Kimbrough/MS, PLT Boe, Yury Lonchakov/FE Exp 18, Bowen/MS (partially visible behind Lonchakov), Pettit/MSDn, Exp 18 CDR Michael Fincke, Chamitoff/MS, Stefanyshyn-Piper/MS,CDR Ferguson (partially visible top Lt).

STS126-S-024 --- After STS-126 successful launch Launch Director Mike Leinbach (right) performs tie-cutting ceremony on KSC Center Director Bob Cabana in LCC Firing Room. Cabana experienced his first shuttle launch as Center Director.

S126-E-008741 (20 Nov. 2008) --- Stefanyshyn-Piper (left) and Kimbrough during EVA2 continue removing debris and applying lubrication around starboard SARJ.

FIRSTS/SECONDS:
- First water regeneration system to recycle urine into drinking water delivered and installed on ISS.
- First flight OI-33 Flight Software. Several minor changes made to improve Post MECO attitude control and reduce the risk of recontact with the ET.
- First flight of new SSME controller S/W to downlink Advanced Health Management System (AHMS) data on-orbit - provides backup to MADS data.
- First flight of redesigned EVA Prime Flight Glove TMG, a Turtleskin® reinforcement layer sandwiched between molded palm and RTV on thumb and index finger and new RTV-3145.
- First flight of ET redesigned LO2-to-Intertank Flange closeout per RTF B/L Plan
- First flight of ATK BSMs in both forward and aft positions.
- First Flight of BSM Forward Segment Grain Redesign - eliminated waiver.
- First flight of SRB Installed Enhanced Data Acquisition System (EDAS) Units and Instrumentation.
- First flight of SRB Redesigned Frangible Nut with Pyrotechnic Crossover Assembly to help prevent stud hang-up.
- A Second: "World Toilet Organization (WTO) is a global nonprofit organization committed to improving toilet and sanitation conditions worldwide. World Toilet Day November 19th - During this mission the crew did their bit for WTD with installation of a new second toilet facility on ISS."

NIGHT LAUNCH: # 31 NASA Test Director Charlene Blackwell-Thompson, "Endeavour is ready to go. And we're really excited to share our version of a sunrise with you ..."

RENDEZVOUS: #71 Rendezvous and dock with ISS.

EVENTS:
- At L-1 hr NASA Security was informed of an inbound threat to the Shuttle about two miles off shore. Security sweeps came up all clear. At L-5 min officials determined no threat and cleared Shuttle for launch. The perpetrator of the hoax was later arrested, found guilty and sentenced to jail in November 2010.
- FD1: OMS2 ignition at 320:01:33:58.3Z resulted in a 125.7 by 84.6 NM orbit.
- FD2: RCC inspection found no areas of concern - focused inspection cancelled on FD4.
- T1 maneuver at 321:19:26:48.0Z resulted in a 192.4 by 184.3 NM orbit
- FD3: R-Bar Pitch Maneuver was performed. No issues.
- Docking Contact occurred at 321:22:01:17Z
- Hard Dock occurred at 321:22:44:35Z
- ISS Hatch opened at 321:24:16:00Z (6:16PM CST, Nov 16, 2008) welcomed by ISS crew.

Continued…

FLT NO.	ORBITER	CREW (7) TITLE, NAMES & EVA'S	LAUNCH SITE, LIFTOFF TIME, LANDING SITES, ABORT TIMES	LANDING SITE/ RUNWAY, CROSSRANGE LANDING TIMES FLT DURATION, WINDS	SSME-TL NOM-ABORT EMERG THROTTLE PROFILE ENG. S.N.	SRB RSRM AND ET	ORBIT		FSW	PAYLOAD WEIGHTS, PAYLOADS/ EXPERIMENTS	MISSION HIGHLIGHTS (LAUNCH SCRUBS/DELAYS, TAL WEATHER, ASCENT I-LOADS, FIRSTS, SIGNIFICANT ANOMALIES, ETC.)
						INC	HA/HP			

| STS-126/ ISS- ULF2 Continued ... | | | | | | | | | | |

S126-E-008178 (18 Nov. 2008) --- Pettit installs the Water Recovery System (WRS) rack in Destiny lab.

296595main_ED08-0306-131c_946-710.jpg: STS-126 Ferry Flight in route to KSC

Pawel-Warchal-EndISS281108_1227890243.jpg: Impressive photo taken by Polish astronomer just after Shuttle/ISS undocking.

SIGNIFICANT ANOMALIES:

Orbiter:
-The Fuel Cell 1 S/N P760106 Hydrogen Flowmeter Measurement Began Drifting High And Erratic At 320/12:36 GMT.
- MER-02, LV57 E2 GH_2 FCV, After Engine Throttle up E2 GH_2 Line Shows a Drop of 200 Psi
- MPS Helium Bottle Lost 140 Psi During Ascent, OMRSD Allows 60 Psi Max. (MER-10)
- GNC Bypass of Ku-Band Radar Data
- Tile Damage on Edge .65l × .23w × .05d
KSC:
- RDUnassigned - Column parity errors on all ME FEPs.
- IRAMS Failed at GMT Rollover.
SRB:
- STS126/Bi136 Squawk 126-001: HDP 3 Blast Container Debris Containment Failure
RSRM, SSME, & ET: None.
MOD:
- Updating Minimum EPS Consumables
- Loss of Crewlock Bag during Eva #1
- Over Torque of Trundle Bearing Assembly Mount
- Middeck Return Item Weights Missing
- Debris Released Near the LH2 T-0 Plate
Integration:
- SM GPC Failure to Send GCIL Commands
- Unexpected Debris/Expected Debris Exceeding Mass Allowable Prior to Pad Clearance (Liftoff Debris)

EVENTS: Continued…

- IELK Seat Liner Transfer at 322:02:50:00Z (8:50 PM CST, Nov 16, 2008). At that time Greg Chamitoff became a member of STS-126 and Sandra Magnus joined the ISS Expedition 18 as Flight Engineer
- FD5: Based on review of launch imagery, the MMT decided that the focused inspection of the Orbiter heat shield was not required.
-FD5: EVA 1: Piper & Bowen transferred the Nitrogen Tank Assembly (NTA) from the External Stowage Platform (ESP)-3 to Lightweight MPESS Carrier (LMC), followed by the Flex Hose Rotary Coupler (FHRC) transfer from LMC to ESP-3. JEM EFBM Multi-Layered Insulation (MLI) Cover was removed in prep for c/o of EFBM (to be installed on 2JA later in 2009). Stbd SARJ trundle bearing assembly (TBA) #10 and #6 were replaced, and the stbd race ring was partially cleaned and lubed. A crew equipment bag was inadvertently released during the EVA, but there was sufficient redundant cleaning and lube equipment to finish scheduled tasks. EVA 1 duration 6:52.
- FD6: Home improvements continued aboard ISS with installation of two new bedrooms and preparations to activate the water recycling facility.
- FD7: EVA2: Piper & Kimbrough relocated the CETA carts in prep for 15A install of S6 solar array upcoming in Feb. 2009; SSRMS Latching End Effector (LEE) A snares were lubricated; all stbd SARJ cleaning and lube objectives were completed except for cleaning under covers 11 and 12; & 4 more trundle bearing assemblies were replaced. EVA was terminated slightly early due to high CO2 readings in Kimbrough's suit. EVA2 duration 6:45. **[During this EVA the ISS marked the 10th Anniversary of launching its first element - the Russian-built Zarya control module. "It's hard to believe it's been 10 years," said Kirk Shireman, NASA's Deputy Manager for ISS, who remembers it being a cold day on the steppes of Kazakhstan.]**
- FD9: UPA anomalous shutdown due to centrifuge speed below limits & high motor current.
- FD9: EVA3: Piper & Bowen continued cleaning of ISS stbd SARJ; R&R'ed the remaining TBA; and cleaned area around SARJ's drive lock assemblies. EVA3 duration 6:57.
- FD11:EVA4: Bowen & Kimbrough completed stbd and port SARJ lube tasks; P1 lower inboard camera installed in camera port 7; external facility berthing mechanism latch bolt retracted via EVA override and cover reinstalled; JEM GPS A installed and heaters checked out ok, JEM GPS B deferred to stage or next flight; and, no get-ahead radiator imagery was taken. EVA4 duration 6:07.
- SARJ put back in autotrack at 330/00:35 GMT (post-EVA).
- FD12: UPA processing was completed for the docked mission.
Transfers:
   - 16,390 lbs of hardware transferred to ISS (Leonardo & middeck)
   - 3,642 lbs of hardware returned from ISS to Endeavour (inside)
   - 25 lbs O2 transferred to ISS
- FD15: Undocked at 333:14:47:26Z followed by Sep-1, Sep-2 and Sep-3; OBSS surveys on starboard, nose cap and port; and LDRI downlink.
- Communications blackout during Entry: "There [were] a few drop outs but nothing big around GMT 335:21:09 d:h:m."

FLT NO.	ORBITER	CREW (6+1 UP/6+1 DN) TITLE, NAMES & EVA'S	LAUNCH SITE, LIFTOFF TIME, LANDING SITES, ABORT TIMES	LANDING SITE/ RUNWAY, CROSSRANGE LANDING TIMES FLT DURATION, WINDS	SSME-TL NOM-ABORT EMERG THROTTLE PROFILE ENG. S.N.	SRB RSRM AND ET	INC	ORBIT HA/HP	FSW	PAYLOAD WEIGHTS, PAYLOADS/ EXPERIMENTS	MISSION HIGHLIGHTS (LAUNCH SCRUBS/DELAYS, TAL WEATHER, ASCENT I-LOADS, FIRSTS, SIGNIFICANT ANOMALIES, ETC.)
STS-119/ ISS- 15A SEQ FLT # 125 KSC-125 PAD 39A (48) MLP-1 28th SHUTTLE FLIGHT TO ISS	OV-103 (Flight 36) DISCOVERY OMS PODS LPO1-39 RPO3-37 FRC3-36	CDR: Lee Archambault Flt 2 (STS-117) P779/R307/V201/M265 PLT Tony Antonelli P780/R334/M289 MS1 Joseph Acaba P781/R335//M290 MS2 Steve Swanson Flt 2 (STS-117) P782/R308/V202/M266 MS3 Richard Arnold P783/R336/M291 MS4 John Phillips Flt 2 (STS-100) P784/R266/V203/M232 MS5 UP Stay ISS EXP 18FLT ENG Koichi Wakata (JAXA) Flt 3 (STS-72, STS-92) P785/R208/V164/M181 MS5 DN EXP 18/Flt ENG Sandra Magnus Flt 2 (STS-112) (UP ON STS-126, stay ISS) P786/R284/V200/F36 SS EVA 129 DOCKED QUEST EVA 52 EMU/TETHERED EVA 122 SCHEDULED EVA 120 DURATION 6:07 SS EVA 130 DOCKED QUEST EVA 53 EMU/TETHERED EVA 123 SCHEDULED EVA 121 DURATION 6:30 Continued...	KSC 39A 074:23:43:44Z 7:43:44 PM EDT (P) 7:43:44 PM EDT (A) Sunday (12) 03/15/09 (10) LAUNCH WINDOW: 4M 14S (PLT in-plane) EOM PLS: KSC TAL: ZZA TAL WX: MRN SELECTED: RTLS: KSC15 CI/NOM TAL: ZZA30L N/N AOA: KSC15 CI/N 1ST DAY PLS: EDW22 N/N TDEL: 0.000 (P) -0.008 (A) MAX Q NAV: 739.4 (P) 722.9 (A) SRB STG: 2:04.00 (P) 2:05.12 (A) PERF: NOMINAL 2 ENG TAL (MRN): 2:35 (P) 2:37 (A) NEG RETURN: 3:54 3:55 PTA (U/S 166 FPS): 5:12 5:15 SE TAL (ZZA 104): 6:00 6:00 PTM (U/S 181 FPS): 6:13 6:16 SE PRESS 104 6:56 6:57 Continued...	KSC 15 (KSC 70) 087:19:13:26Z 2:13:26PM CDT Saturday (23) 03/28/09 (10) DEORBIT BURN: 087:18:08:14Z XRANGE: 222.2 NM ORBIT DIR: A/R (14) AIM PT: Close-In MLGTD: 2705 FT 087:19:13:26Z VEL: 188 KGS 203 KEAS HDOT: -2.7 FPS TD NORM 195: 3473 FT DRAG CHUTE DEPLOY: 194 KEAS 087:19:13:29Z NLGTD: 5369 FT 087:19:13:34Z VEL: 152 KGS 167 KEAS HDOT: -6.7 FPS BRK INIT: 40 KGS DRAG CHUTE JETTISON: 60 KGS 087:19:13:59Z BRK DECEL FPS2: AVE 3.1 PK 4.2 WHEELS STOP: 87:19:14:43Z 12050 FT ROLLOUT: 10345 FT 1:17 M:S Continued...	104/104/ 109% PREDICTED: 100/104.5/ 104.5/72/104.5 ACTUAL: 100/104.5/ 104.5/72/104.5 1 = 2048 (9) 2 = 2051 (8) 3 = 2058 (3)	BI-135 RSRM 103 ET-127 SLWT- 33	51.6 (28)	DIRECT INSERTION POST OMS-2: 126.0x84.9 NM DEORBIT: HA 184.8 NM HP 21.6 NM ENTRY VELOCITY: 25849 FPS ENTRY RANGE: 4377 NM	OI-33 (2)	CARGO: 39088 LBS PAYLOAD CHARGEABLE: 32546 LBS DEPLOYED: 32489 LBS NON-DEPLOYED: 0 LBS MIDDECK: 57 LBS SHUTTLE ACCUMULATED WEIGHTS: DEPLOYED: 1517781 LBS NON-DEPLOYED: 1603765 LBS CARGO TOTAL: 4021804 LBS PERFORMANCE MARGINS (LBS): FPR: 2651 FUEL BIAS: 1063 FINAL TDDP: 1746 RECON:2016 PAYLOADS: PLB: ISS 15A (S6) MIDDECK: ISS 15A, MAUI SEITE, SIMPLEX 5 CRYO TANK SETS RMS (82) SRMS, ODS, OBSS, SSPTS	*Brief Mission Summary:* ISS United States Operational Segment (USOS) assembly was completed with installation of S6 truss with final set of power generating Solar Arrays on Shuttle's 28th ISS Mission. This additional power prepares the ISS with the capability of housing six member crews in the near future. **KSC W/D:** OPF = 191+13H+3Wx, VAB = 6 + 0C, PAD = 47 + 14C: Total Work Days = 244 (OPF Processing occurred over a total time period of 207 days.) **LAUNCH POSTPONEMENTS** - Added STS-119 to FDRD - launch date of 01/15/04 on 01/23/03 - Ppd. to NET 06/10/04 on 03/13/03 due to Columbia accident. - Ppd. to NET 06/30/04 on 04/17/03 due to Columbia accident. - Deleted from FDRD on 05/28/03 pending Columbia accident investigation outcome. - Re-Baselined in FDRD - Launch date of 11/06/08 on 10/04/07 - Ppd. to 12/04/08 on 02/14/08. Slip due to ECO Sensor problems during STS-122 launch attempt. - Ppd. to 02/12/09 on 07/03/08. Slip due to ET delivery schedule. - Ppd. to NET 02/19/09 on 02/04/09. Slip due to additional testing & analysis required to resolve MPS flow control valve issue - Ppd. to NET 02/22/09 on 02/09/09. Slip due to additional testing & analysis required to resolve MPS flow control valve issue - Ppd. to 02/27/09 on 02/14/09. Slip due to additional testing & analysis required to resolve MPS flow control valve issue - Ppd. to TBD at STS-119 "Continuation" FRR on 02/20/09. Managers could not reach a consensus. - Ppd. to tentative date of 03/12/09 on 02/25/09. MPS flow control valve U/R. - Launch date set for NET 03/11/09 on 03/04/09. MPS flow control valve U/R. - Launch date set for 03/11/09 at Delta FRR on 03/06/09. - Officially ppd. launch to 03/15/09 on 03/12/09 after Scrub on 03/11/09. Scrub was due to gaseous hydrogen leak in vent line. **LAUNCH SCRUB:** **Mar.11, 2009**, Wednesday, with fewer than 20 minutes left in tanking process launch was scrubbed due to a gaseous hydrogen vent line leak. This line connects the Ground Umbilical Carrier Plate (GUCP), attached to ET, to the "flare stack" for burn-off of vented gaseous hydrogen. Launched scrubbed at 1:37 PM CDT. Technical Scrub. **LAUNCH WINDOW:** Total launch window was 8M 27S with window open at 074:23:39:31Z and close at 074:23:47:58Z. Preferred Launch Time was 074:23:43:44Z (In-Plane Time) for a launch window of 4M 14S. Continued...

M 3 EOM: WEIGHT: 201795 LBS X CG: 1082.8 IN

LANDING: WEIGHT: 201713 LBS X CG: 1084.7 IN

ET IMPACT: 1:14:30 MET

LAT: 35.725 S

LONG: 157.56 W

STS-119 - Waiting for GO! Moon - Waiting for Constellation! (It will be a long wait - President directed cancellation of Constellation in 2010.) 317861main_image_1301946-710STS119Moon.jpg :

FLT NO.	ORBITER	CREW (7) TITLE, NAMES & EVA'S	LAUNCH SITE, LIFTOFF TIME, LANDING SITES, ABORT TIMES	LANDING SITE/ RUNWAY, CROSSRANGE LANDING TIMES FLT DURATION, WINDS	SSME-TL NOM-ABORT EMERG THROTTLE PROFILE ENG. S.N.	SRB RSRM AND ET	INC	ORBIT HA/HP	FSW	PAYLOAD WEIGHTS, PAYLOADS/ EXPERIMENTS	MISSION HIGHLIGHTS (LAUNCH SCRUBS/DELAYS, TAL WEATHER, ASCENT I-LOADS, FIRSTS, SIGNIFICANT ANOMALIES, ETC.)
STS-119/ ISS-15A Continued …		Continued… SS EVA 131 DOCKED QUEST EVA 54 EMU/TETHERED EVA 124 SCHEDULED EVA 122 DURATION 6:27 MCC WHITE FCR (55) FLIGHT DIRECTORS: SHUTTLE: ASC/ENT- Richard Jones LD/O1- Paul Dye O2- Mike Sarafin (FD1- FD12) O2-Tony Ceccacci (FD13-EOM) O3- Richard LaBrode (Prelaunch – FD1) O3- Norman Knight (FD2-FD8) O3- Bryan Lunney (FD9-EOM) Planning- Norm Knight - Bryan Lunney MOD – John Mccullough Team 4 - Tony Ceccacci Continued…	Continued… MECO CMD: 8:23.6 8:23.8 VI: 25819.0 25819.6 OMS-2: 38:00 38:30.0 97.7 FPS 96.1 FPS	Continued… WINDS: 15H KT 0.3L KTS OFFICIAL: 15017P23 KTS X1P1H17P23 KTS DENS ALT: 1718 FT FLT DURATION: 12:19:29:42 S/T: 1196:11:09:28 OV-103: 318:03:39:51 DISTANCE: 5,304,106 sm TOTAL SHUTTLE DISTANCE: 485,578,259 sm							

ABOVE: STS-119 launch panorama into twilit sky. Photo by Ryan R. Smith (KSC-BOE-K2)
http://www.ryansmithphotography.com/

BELOW: S119-E-007747 --- STS-119 & Exp18 crews in ISS Harmony. From left (bottom row): PLT Antonelli, CDR Archambault, & Acaba/MS. From left (middle row): Magnus/MS, Exp 18 CDR Michael Fincke, Yury Lonchakov/Exp18FE(RSA), & Koichi Wakata/Exp18FE (JAXA). From left (top row) Swanson/MS, Arnold/MS, & Phillips/MS.

S119E007747

Continued…

LAUNCH DELAYS: None. Launch occurred on time at 074:23:43:44Z, 7:43:44 p.m. EST, Sunday, March 15, 2009. Launch weather was relatively benign at KSC. A sea breeze developed at KSC and moved west of the Banana River about 3 hours prior to launch. The movement of the sea breeze inland produced favorable weather conditions with widely scattered clouds.

TAL WEATHER
TAL sites at both Zaragoza and Moron, Spain were acceptable for launch due to a high pressure system. Winds at Istres were out of limits following the passage of a cold front the day prior to launch, but launch proceeded with two acceptable TAL sites.

PERFORMANCE ENHANCEMENTS:
Include the standard set plus: 1) PE Operational High Q WIN/MAR, 2) OMS Assist, 3) 52 nautical mile MECO, & 4) Del Psi

FLIGHT DURATION CHANGES/LANDING:
- When STS-119 launch was slipped to March 15, 2009, (due to earlier scrub) the mission duration was reduced from 14 to 13 days to accommodate a Russian Soyuz mission to ISS later in the month. This also reduced number of EVA's from 4 to 3.
- For first KSC landing opportunity weather was no go with cloud decks building in at lower than anticipated broken (5/8) at 3000. Weather improved as did the wind direction. Discovery was given "Go" to land on second KSC opportunity. Landing occurred at 087:19:13:26Z (2:13:26 PM CDT Saturday, 03/28/09).

FIRSTS/SECONDS/LASTS:
- SSME ECP 1514 – LPOTP Bearing Ball Process Change
- SRB Hold Down Post Debris Containment mod
- S&MA: Orbiter LH_2 T-0 Umbilical Ice: Update to IDBR-01 and NSTS-60559 to reflect new expected debris source.
- Last to be installed on ISS, the 45-foot S6 aluminum girder weighing more than 31,000 pounds was the first truss segment built (stored at KSC for six years).
- Second time a bat attempted to fly into space on Space Shuttle ET; coincidentally Koichi Wakata was on both flights.
- Discovery served as a hypersonic test bed during entry for new heat shield tiles in development for NASA's next-generation spacecraft.

Continued…

FLT NO.	ORBITER	CREW (7) TITLE, NAMES & EVA'S	LAUNCH SITE, LIFTOFF TIME, LANDING SITES, ABORT TIMES	LANDING SITE/ RUNWAY, CROSSRANGE LANDING TIMES FLT DURATION, WINDS	SSME-TL NOM-ABORT EMERG THROTTLE PROFILE ENG. S.N.	SRB RSRM AND ET	INC	ORBIT HA/HP	FSW	PAYLOAD WEIGHTS, PAYLOADS/ EXPERIMENTS	MISSION HIGHLIGHTS (LAUNCH SCRUBS/DELAYS, TAL WEATHER, ASCENT I-LOADS, FIRSTS, SIGNIFICANT ANOMALIES, ETC.)
STS-119/ ISS- 15A Continued …		Continued… ISS LD/O1 - Kwatsi Alibaruho O2 - Heather Ranick O3 - David Korth Team 4 - Robert dempsey CAPCOMS: SHUTTLE A/E – George Zamka Asc (Wx)- C. Hobaugh Ent (Wx)- Al Poindexter LD/01 – George Zamka O2 – Greg (Box) Johnson Planning – Shannon Lucid Team 4 - N/A ISS LD/O1 – Rick Davis O2- Lucia McCullough O3 – Jay Marschke Team 4 - N/A									

S119E007312

S119-E-006673 --- Swanson (center) and Arnold (partially obscured above Swanson) during EVA 1 connected bolts to attach S6 truss to S5, plugged in power and data connectors, prepared a radiator for cooling, and readied new solar arrays.

S119-E-009765 (25 March 2009) --- ISS USOS assembly complete as seen during Shuttle fly-around [labeled the "$100 Billion Picture" by ISS Lead Flight Director Kwatsi Alibaruho]. The ISS truss backbone measures 361 feet - longer than a football field.

"$100 BILLION PICTURE"
However, NASA estimates its total direct cost of building ISS at $58.5 billion since 1985.

SEE: MCC ROSES: Above right under Mission Highlights Column

Continued… **FIRSTS/SECONDS/LASTS:**

- March 27, 2009: In a rare example of overlapping space missions, a U.S. space shuttle [STS-119] is set to return to Earth on Saturday just a few hours after a Russian Soyuz arrives at the ISS. Together the crews of the three craft total 13 people, tying the record for humans in space, first set 14 years ago this month. [Robert Pearlman - collectSPACE.com]

MCC ROSES:
This was the 100th flight since the Challenger accident that a beautiful bouquet of roses was delivered to the Houston MOCR to celebrate each mission since the landing of STS-26 in 1988. In 1989 it was determined that the roses were sent by the Shelton family (Mark, MacKenzie & Terry) of Bedford, TX. On March 27, 2009, the Sheltons personally delivered their 100th bouquet in recognition of STS-119. They received a warm welcome in the MOCR, led by James "Milt" Heflin, JSC Associate Director, Technical. They also received several JSC mementos for their kindness and dedication to the Space Program.

NIGHT LAUNCH: # 32 (Into twilit sky)

RENDEZVOUS: #72 Rendezvous and dock with ISS.

EVENTS:
- FD1: OMS2 ignition at 075:00:22:14Z resulted in a 126.0 by 84.9 NM orbit.
- FD2: RCC inspection found no areas of concern
- T1 maneuver at 076:18:35:39.0Z resulted in a 196.8 by 183.3 NM orbit
- FD3: R-Bar Pitch Maneuver was performed. No issues.
- Docking Contact occurred at 076:21:19:49Z, St. Patrick's Day
- Hard Dock, hooks closed, occurred at 076:21:33:59Z
- ISS Hatch opened at 076:23:22:59Z (6:09 PM CDT, March17, 2009) welcomed by ISS crew.
- IELK Seat Liner Transfer at 077:02:00Z (9:00 PM CDT) March 17, 2009. At that time Sandra Magnus became a member of STS-119 and Koichi Wakata joined the ISS Expedition 18 as Flight Engineer.
- FD5: Based on review of launch imagery, MMT cancelled FD6 focused inspection of Orbiter heat shield.
- FD5: EVA 1: Steve Swanson & Ricky Arnold: Activities included: S6 Connected to ISS, SABB Unstow, PCDF-PU Transfer, PVR Deploy, and 1B & 3B solar arrays deployed EVA1 duration 6:07.

Continued…

FLT NO.	ORBITER	CREW (7) TITLE, NAMES & EVA'S	LAUNCH SITE, LIFTOFF TIME, LANDING SITES, ABORT TIMES	LANDING SITE/ RUNWAY, CROSSRANGE LANDING TIMES FLT DURATION, WINDS	SSME-TL NOM-ABORT EMERG THROTTLE PROFILE ENG. S.N.	SRB RSRM AND ET	ORBIT		FSW	PAYLOAD WEIGHTS, PAYLOADS/ EXPERIMENTS	MISSION HIGHLIGHTS (LAUNCH SCRUBS/DELAYS, TAL WEATHER, ASCENT I-LOADS, FIRSTS, SIGNIFICANT ANOMALIES, ETC.)
							INC	HA/HP			
STS-119/ ISS- 15A Continued ...											

Two of FCT's That Participated In ISS USOS Complete

JSC2009-E-060960 (20 March 2009) --- Group portrait of Shuttle STS-119 Orbit 1 Flight Control Team in JSC MCC. FD Paul Dye (left) is visible on the front row.

JSC2009-E-060959 (20 March 2009) --- Group portrait of STS-119/15A ISS Orbit 1 Flight Control Team in JSC MCC. FD Kwatsi Alibaruho (right) is visible on the front row.

In JSC MCC at Landing Support Officer (LSO) console: On left, Marty Linde/USA , Lt. Col. Dave Impiccini/USAF (standing), Wayne Hensley/USA (on phone), & Brenton Hartung (student observer in rear). Laughter caused by photographer always catching Wayne on telephone.

Continued... **SIGNIFICANT ANOMALIES:**

- Ground Imagery Showed That When Thruster F4D's Tyvek Rain Cover Released at 5:28 Sec Met (~93fps Or 63 Mph), A ~21 Inches × ~7.4 Inches Piece Remained Attached to the Thruster Lip as Shown In Figures 1 and 2.
KSC:
- STS-119 Post Launch Debris
SRB: RSRM: SSME: None.
ET:
-During Initial Launch Attempt of STS-117/Et-127, a GH2 Leak was Detected at Approximately One Minute After Start of LH2 Topping
MOD:
-Inadvertent Abort Light Command Sent from FDO
Integration:
-Unexpected Debris/Expected Debris Exceeding Mass Allowable Prior to Pad Clearance (Liftoff Debris)
-High GH2 Concentrations at the Ground Umbilical Carrier Plate (GUCP)
-MPS LH2 ORB Umbilical Plate Gap Pressure LCC Violation
-Stub Tile Damage

Continued... **EVENTS:**

Downlinked, P3 UCCAS Deploy unsuccessful, temporary tethers installed, S3 PAS Deploy deferred to EVA3, and Z1 Patch Panel Reconfig unsuccessful. EVA2 duration 6:30.
- FD8: CDR Lee Archambault maneuvered the Shuttle-ISS "stack" to avoid a 9-year-old piece of Chinese space junk (4" fragment) that could have been a close encounter during upcoming EVA3. (A 4' fragment from a Russian satellite had previously passed at a safe distance prior to Shuttle/ISS docking.)
- FD9: EVA3: Joe Acaba & Ricky Arnold: Activities included: UCCAS troubleshooting; tethered in place, CETA cart relocation and SSRMS LEE B lube completed. Numerous get aheads accomplished: CETA coupler, S1/S3 SSAS panel BBC reconfig, S1 FHRC outboard p-clamps released 2 of 6 (#5, #6), and retrieved bungee caddy from Nadir STBD A/L toolbox. EVA3 duration 6:27.
- Transfers:
   - 32,962 lbs of hardware transferred to ISS (S6 Truss & Middeck)
   - 1963 lbs of hardware returned from ISS to Discovery (middeck)
   - 1142 lbs of water transferred to ISS
- FD11: Undocked at 084:19:53:26Z
- Flyaround initiated 084: 20:19Z
- Communications blackout during Entry occurred at GMT 87:18:47 to 87:18:52 d:h:m due to plasma effect.

SIGNIFICANT ANOMALIES:
Orbiter:
- Galley Water Leakage.
- WLES Group 2 Sensor S/N# 1033 Time Slip
- During MM/OD Monitoring With Group 2 Sensors, Sensor S/N 1024 On The Port Wing Unexpectedly Dropped Out Of On-Orbit Mode After 5-6 Hrs Of Monitoring.
- AVIU S/N 1031 Failure
- Failed Camera Shutter Actuation.
- Incorrect SORG Needle Installed
- V07P9379A Dropped To Lower Limit (Unit Step) During STS-119 Ascent
- Aft Stub Tile on the Upper Body Flap Was Suspect to be Damaged During FD3 On-Orbit Inspection. During Post-Flight Inspection the V070-395018-144 Tile Was Verified As Damaged.

Continued at left...

FLT NO.	ORBITER	CREW (7) TITLE, NAMES & EVA'S	LAUNCH SITE, LIFTOFF TIME, LANDING SITES, ABORT TIMES	LANDING SITE/ RUNWAY, CROSSRANGE LANDING TIMES FLT DURATION, WINDS	SSME-TL NOM-ABORT EMERG THROTTLE PROFILE ENG. S.N.	SRB RSRM AND ET	INC	ORBIT HA/HP	FSW	PAYLOAD WEIGHTS, PAYLOADS/ EXPERIMENTS	MISSION HIGHLIGHTS (LAUNCH SCRUBS/DELAYS, TAL WEATHER, ASCENT I-LOADS, FIRSTS, SIGNIFICANT ANOMALIES, ETC.)
STS-125 SEQ FLT # 126 KSC-126 PAD 39A (49) MLP-2 5TH & Final HST Service Flight	OV-104 (Flight 30) ATLANTIS OMS PODS LPO4-30 RPO1-37 FRC4-30	CDR: Scott Altman (Flt 4 - STS-90,STS-106, STS-109) P787/R237/V161/M207 PLT Gregory C. Johnson P788/R337/M292 MS1 Michael Good P789/R338//M293 MS2 Megan McAuthur P790/R339/F46 MS3 John Grunsfeld (Flt 5-STS-67, STS-81, STS-103, STS-109) P791/R191/V133/M167 MS4 Mike Massimino (Flt 2 - STS-109) P792/R275/V204/M241 MS5 Andrew Feustel P793/R340/M294 SS EVA 132 EMU/TETHERED EVA 125 SCHEDULED EVA 123 DURATION 7:20 SS EVA 133 EMU/TETHERED EVA 124 SCHEDULED EVA 124 DURATION 7:56 Continued...	KSC 39A 131:18:01:56Z 2:01:56 PM EDT (P) 2:01:56 PM EDT (A) Monday (14) 05/11/09 (8) LAUNCH WINDOW: 59M 45S (Total) 41M 50S (Preferred) EOM PLS: KSC TAL: MRN TAL WX: None. SELECTED: RTLS: KSC15 N/N TAL: MRN20 CI/N AOA: KSC15 N/N 1ST DAY PLS: NOR17 N/N TDEL: 0.000 (P) -0.448 (A) MAX Q NAV: 740.95 (P) 734.75 (A) SRB STG: 2:04.16(P) 2:04.32(A) PERF: NOMINAL 2 ENG TAL (MRN): 2:48 (P) 2:55 (A) NEG RETURN: 3:53 (P) 3:56 (A) PTA (U/S 483 FPS): 4:11 (P) 4:12 (A) PTM (U/S 500 FPS): 5:09 (P) 5:12 (A) Continued...	EDW22 CONC EDW 53 CONC 34 144:15:39:04Z 10:39:04 AM CDT Sunday (16) 05/24/09 (11) DEORBIT BURN: 144:14:24:41.0Z XRANGE. 405.6 NM ORBIT DIR: D/L (50) AIM PT: Nominal MLGTD: 3863 FT 144:15:39:04Z VEL: 192 KGS 200 KEAS HDOT: -2.5 FPS Continued...	104/104/109% PREDICTED: 100/104.5/104.5/ 72/104.5 ACTUAL: 100/104.5/94/ 72/104.5 1 = 2059 (3) 2 = 2044 (11) 3 = 2057 (4) M 3 EOM: WEIGHT: 225509.5 LBS X CG: 1078.3 IN LANDING: WEIGHT: 225898 LBS X CG: 1080.9 IN	BI-137 RSRM 105 ET-130 SLWT 34 ET IMPACT 1:27:55 MET LAT: 16.699 N LONG: 147.375 W	28.45 (51)	DIRECT INSERTION POST OMS2: 298.1 NM X 106.8 NM DEORBIT: HA 294.3 NM HP 26.4 NM ENTRY VELOCITY: 26046 FPS ENTRY RANGE: 4267 NM	OI-32 (5)	CARGO: 32418 LBS PAYLOAD CHARGEABLE: 22254 LBS DEPLOYED: 4694 LBS NON-DEPLOYED: 17560 LBS MIDDECK: 0 LBS SHUTTLE ACCUMULATED WEIGHTS: DEPLOYED: 1524432 LBS NON-DEPLOYED 1621371 LBS CARGO TOTAL: 4054222 LBS PERFORMANCE MARGINS (LBS): FPR: 2651 FUEL BIAS: 1063 FINAL TDDP: 1689 RECON:2499 PAYLOADS: PLB: HST SM4, ICBC 3D MIDDECK: HST SM4 5 CRYO TANK SETS RMS (83) SRMS, OBSS	*Brief Mission Summary:* STS-125 was the 5th and final service mission (SM) visit to the 19 year old Hubble Space Telescope (HST) deployed on STS-31 in 1990. This was the 4th planned SM for HST. (The 3rd SM was conducted in two parts, 3A on STS-103 & 3B on STS-109.) HST improvements included a new camera, a new spectrograph, repair of two other instruments, and replacement of six batteries and six gyroscopes. These improvements resulted in a higher definiton view of the universe and HST life extension into the next decade. A launch- on-need (LON) vehicle, STS-400, was readied on Pad B for potential crew rescue since there was no ISS safe haven on this misssion. STS-400 release from rescue duty occurred on May 21st , 2009, as the STS-125 crew prepared for the first deorbit/landing opportunity. **KSC W/D:** OPF Run 1: 178+2H+3Wx OPF Run 2: 120+11H VAB Run 1: 12+0C VAB Run 2: 8+0C PAD Run 1: 40+2C PAD Run 2: 38+4C Total Work Days = 396 (OPF Processing occurred over a total time period of 314 days.) **POSTPONEMENTS:** - Added STS-125 to FDRD - launch date of 08/07/08 on 06/29/07. - Ppd. to 08/28/08 on 02/14/08. Slip due to ECO sensor problems experienced during December launch attempt of STS-122. - Ppd. to 10/08/08 on 05/27/08. Slip due to delays in delivery of ET 127 & ET-129 (STS-400). - Ppd. to 10/10/08 on 09/08/08. Slip due to Hurricane Faye impacts to HST payload readiness. - Ppd. to 10/14/08 on 09/24/08. Slip due primarily to training time lost in the aftermath of Hurricane Ike. - Ppd. to NET Mid-Feb 2009 on 10/02/08. Slip due to HST on-orbit failure of A-side of Control Unit Science Data Formatter. - Ppd. to NET Mid-May 2009 on 10/30/08. Slip due to checkout problems with HST spare control unit. - Selected May 12, 2009 launch date on 12/04/08. - Advanced from 05/12/09 to 05/11/09 on 05/01/09. Advancing one day provided a 3rd launch opportunity before range conflicts. **LAUNCH SCRUBS:** None. Continued...

S125-E-012154 --- HST Service Crew pose on middeck . Front row (left to right): PLT Johnson, CDR Altman, and McArthur/MS. Back row (left to right): Good/MS, Massimino/MS, Grunsfeld/MS, and Feustel/MS.

FLT NO.	ORBITER	CREW (7) / TITLE, NAMES & EVA'S	LAUNCH SITE, LIFTOFF TIME, / LANDING SITES, ABORT TIMES	LANDING SITE/ RUNWAY, CROSSRANGE / LANDING TIMES FLT DURATION, WINDS	SSME-TL NOM-ABORT EMERG / THROTTLE PROFILE ENG. S.N.	SRB RSRM AND ET	ORBIT / INC	HA/HP	FSW	PAYLOAD WEIGHTS, / PAYLOADS/ EXPERIMENTS	MISSION HIGHLIGHTS (LAUNCH SCRUBS/DELAYS, TAL WEATHER, ASCENT I-LOADS, FIRSTS, SIGNIFICANT ANOMALIES, ETC.)

STS-125 Continued …		Continued…
		
		 SS EVA 134
		 EMU/TETHERED EVA 127
		 SCHEDULED EVA 125
		 DURATION 6:36
		
		 SS EVA 135
		 EMU/TETHERED EVA 128
		 SCHEDULED EVA 126
		 DURATION 8:02
		
		 SS EVA 136
		 EMU/TETHERED EVA 129
		 SCHEDULED EVA 127
		 DURATION 7:02
		
		 MCC WHITE FLIGHT FCR (56)
		
		 FLIGHT DIRECTORS:
		 ASC/ENT- Norm Knight
		 LD/O1- Tony Ceccacci
		 O2- Rick LaBrode
		 Planning- Paul Dye
		 MOD – John Mccullough
		 Team 4- Bryan Iunneyi
		
		 CAPCOMS:
		 A/E - Greg (Box) Johnson
		 - Eric Boe (Wx)
		
		 LD/01 – Dan Burbank
		 O2 – Alan poindexter
		 Planning – Janice Voss
		 Team 4 - N/A
			Continued…
			
			 SE BYD 104
			 5:39 (P) 5:46 (A)
			
			 NEG MRN (2@104)
			 5:59 (P) 6:02 (A)
			
			 SE PRESS 109
			 6:22 (P) 6:29 (A)
			
			 MECO CMD:
			 8:23.4 (P) 8:24.3 (A)
			
			 VI:
			 26088.0 (P)
			 26086.0 (A)
			
			 OMS-2:
			 43:46 (P) 43:45.0 (A)
			 142.5 (P) 139.7 (A)
			 FPS
			
			 -------------------- 	
			
			 Continued from col @ right…
			
			 FLT DURATION:
			 12:21:37:18
			 S/T:
			 1196:08:46:46
			
			 OV-104:
			 271:04:42:58
			
			 DISTANCE:
			 5,276,106 sm
			
			 TOTAL SHUTTLE DISTANCE:
			 490,854,365 sm
				Continued…
				
				 TD NORM 205:
				    3201 FT
				
				 DRAG CHUTE DEPLOY:
				    189 KEAS
				 144:15:39:06Z
				
				 NLGTD: 7134 FT
				 144:15:39:15Z
				 VEL: 137 KGS
				    141 KEAS
				 HDOT: -6.3 FPS
				
				 BRK INIT: 96 KGS
				
				 DRAG CHUTE JETTISON:
				    55 KGS
				 144:15:39:40Z
				
				 BRK DECEL FPS^2:
				 AVE 2.8 PK 7.4
				
				 WHEELS STOP:
				 144:15:40:13 Z
				    12367 FT
				
				 ROLLOUT:
				    8504 FT
				    1:09 M:S
				
				 WINDS:
				   16H KT 0 KTS
				 OFFICIAL:
				 23016P20 (X 2 PK 2
				 HD 16 PK 20)
				
				 DENS ALT: 3848 FT
				
				 Continued @ left …							

S125-E-007221 (14 May 2009)-- Grunsfeld & Feustel and mirrored reflection during first HST EVA. Activities included installation of a new WFC3 and SI C&DH unit.

Continued…

LAUNCH WINDOW:
Total launch window was 59M 45S with window open at 131:17:44:01Z and close at 131:18:43:46Z. Preferred Launch Time was 131:18:01:56Z (In-Plane Time) for a launch window of 41M 50S.

LAUNCH DELAYS: None. Launch occurred on time at 131:18:01:56Z, 2:01:56 p.m. EDT, Monday, May 11, 2009. The Spaceflight Meteorology Group (SMG) forecast no flight rule violations for launch or RTLS. The SMG also tracked a large wildfire 18nm northwest of KSC that stayed north of the orbiter track for an RTLS if needed.

TAL WEATHER
At Moron, the only TAL site for the HST low inclination orbit, a trough of low pressure initially resulted in a "NO GO" with a slight chance of showers within 20nm. Balloon data showed the atmosphere was too dry for showers and the forecast was updated to "GO" at 1636Z. Peak crosswinds of 15.5 kts surpassed the 15kt limit for a brief time at TAL landing, however, the FD had previouly stated a peak crosswind of 17kts was acceptable.

PERFORMANCE ENHANCEMENTS:
Include the standard set plus: PE Operational High Q TRN/MAY

FLIGHT DURATION CHANGES/LANDING:
- For both KSC landing opportunities on Friday, May 22nd the unstable weather was no go with low ceilings and thunderstorms expected. Landing was postponed to Saturday (EOM + 1).
- KSC weather was no go for EOM+1 with broken low ceilings and thunderstorms. Little change was expected for Sunday (EOM+2) and Monday (EOM+3) as moisture remained abundant over KSC.
- KSC landing for Sunday (EOM+2) waived off due to weather. Next opportunity to EDW's was selected on EOM +2 with typical summer weather and mostly clear skies. Landing occurred at 144:15:39:04Z (10:39:04 AM CDT Sunday, 05/24/09).

FIRSTS/LASTS:
- First mission post-STS-107 incident without ISS safe haven. LON STS-400 mission was on standby on PAD 39B. "First time since 2001 that two such birds have simultaneously perched on NASA's twin shuttle launch pads" - Todd Halvorson, Florida Today.
- 116 new EVA tools (GSFC) were developed to meet unique demands of this HST SM.
- First flight of food bars and Metamucil wafers
- First ET build with elimination of "Hand Pack Ablator (SLA)"

Continued…

FLT NO.	ORBITER	CREW (7)	LAUNCH SITE, LIFTOFF TIME,	LANDING SITE/ RUNWAY, CROSSRANGE	SSME-TL NOM-ABORT EMERG	SRB RSRM	ORBIT		FSW	PAYLOAD WEIGHTS,	MISSION HIGHLIGHTS (LAUNCH SCRUBS/DELAYS,
		TITLE, NAMES & EVA'S	LANDING SITES, ABORT TIMES	LANDING TIMES FLT DURATION, WINDS	THROTTLE PROFILE ENG. S.N.	AND ET	INC	HA/HP		PAYLOADS/ EXPERIMENTS	TAL WEATHER, ASCENT I-LOADS, FIRSTS, SIGNIFICANT ANOMALIES, ETC.)
STS-125 Continued …										

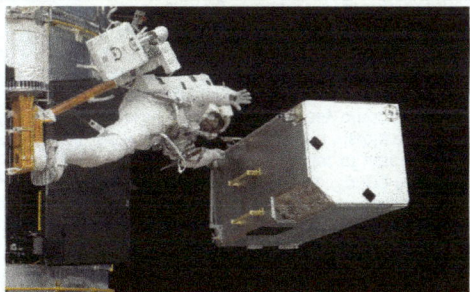

S125-E-008120 (16 May 2009)-- Andrew Feustel moves Corrective Optics Space Telescope Axial Replacement (COSTAR) in 3rd EVA to upgrade HST.

JSC2009-E-120479 --- In MCC: Members of the STS-125 Hubble Space Telescope Planning and Orbit Flight Control Team.

S125-E-009918 (18 May 2009) "Hugging the Hubble!" - Grunsfeld, on end of RMS, and Feustel, conduct mission's fifth and final HST service EVA: Replaced batteries, a Fine Guidance Sensor, and three thermal blankets (NOBL).
NOTE: Dr. John M. Grunsfeld was later appointed Deputy Director of the Space Telescope Science Institute (STScI) in Baltimore, Md. effective January 4, 2010.

JSC2009-E-118819 --- In MCC: John McCullough (seated foreground), Chief Flight Directors Office; Brent Jett (seated right), Director, Flight Crew Operations; Lead Flight Director Tony Ceccacci (standing, left);and Asc/Des Flight Director Norm Knight (standing, right).

Continued… **FIRSTS/LASTS:**

- First flight of ATK BSM's in both forward and aft positions
- SRB Frangible nut redesigned with pyrotechnic crossover assembly
- Mike Massimino first to 'Tweet' from space, through email to JSC to his Twitter.
- First job offer in space: John Grunsfeld, while flying high in space, was named an adjunct professor at the University of Colorado at Boulder
- Fifth & last HST Service mission.

NIGHT LAUNCH: N/A

RENDEZVOUS: #73 Rendezvous with HST.

EVENTS:

- FD1: OMS2 ignition at 131:18:45:40.9Z resulted in a 298.1 by 106.6 NM orbit.
- T1 maneuver at 133:14:41:56.0Z resulted in a 303.2 by 302.9 NM orbit
- FD2: RCC inspection found no areas of concern - no requirement for Focused Inspection.
- FD3: HST Grapple by McArthur occurred at 133:17:14Z. Timeline was about 20 min. behind schedule due to a comm. problem with HST that delayed HST prep for capture.
- FD4: EVA 1: Grunsfeld & Feustel: Activities included installing and completing good aliveness tests for new WFC3 and SI C&DH unit. The HST can now see farther into space and across a wider spectrum of colors. EVA ran 50 min longer than planned as the crew encountered difficult (aging) latches and bolts. EVA1 duration 7:20.
- FD5: EVA 2: Massimino & Good: Activities included Rate Sensor Unit changeouts & Bay 2 Battery checkout. EVA ran long due to the challenges for seating and bolting of RSU's. EVA2 duration 7:56.
- FD6: EVA 3: Grundsfeld & Feusel: Activities included replacement of the COSTAR instrument with the Cosmic Origins Spectrograph and repair of the Advanced Camera for Surveys. EVA3 duration 6:36.
- FD7: EVA 4: Massimino & Good: Activities included refurbishment of Space Telescope Imaging Spectrograph and replacement of 6 Gyros. EVA 4 duration 8:02 (6th longest in program history).
- FD8: EVA 5:Grundsfeld & Feustel: Activities included Bay 3 battery changeout and FGS 2 changeout . On way back to A/L crew found debris liberated from carrier and head under HST. On retrieving the debris, PLSS contact damaged the TPS cover on the Low Gain Aantenna (LGA). The LGA cover was reinstalled. The HST was in a good configuration for long term exposure to space. EVA5 duration 7:02.
- On departing the telescope, astronaut Grunsfeld called the week a "tour de force of tools and human ingenuity." He also added: "'Hubble Isn't Just a Satellite, It Is About Mankind's Quest for Knowledge".
- FD9: HST was released at 139:12:57:00Z. This was followed shortly by OBSS late inspection of Atlantis TPS.
- During Entry comm blackout occurred at GMT 144/1513 - 1517 due to plasma effect.

FLT NO.	ORBITER	CREW (7) TITLE, NAMES & EVA'S	LAUNCH SITE, LIFTOFF TIME, LANDING SITES, ABORT TIMES	LANDING SITE/ RUNWAY, CROSSRANGE LANDING TIMES FLT DURATION, WINDS	SSME-TL NOM-ABORT EMERG THROTTLE PROFILE ENG. S.N.	SRB RSRM AND ET	ORBIT INC HA/HP	FSW	PAYLOAD WEIGHTS, PAYLOADS/ EXPERIMENTS	MISSION HIGHLIGHTS (LAUNCH SCRUBS/DELAYS, TAL WEATHER, ASCENT I-LOADS, FIRSTS, SIGNIFICANT ANOMALIES, ETC.)

STS-125
Continued ...

After Shuttle release, the HST orbital observatory returns to its cosmic duties, see photos at right and below.

Pillar and Stellar Jet in the Carina Nebula
HST WFC3/UVIS
2 light-years
0.61 pasec
HST WFC3/IR
F164N [Fe II]
F108N [Fe II]

HST Program released the above photos on 09/10/09 taken by the "Refurbished Hubble" (using WFC3). At upper right are two views of: Stars Bursting to Life in Chaotic Carina Nebula. These two images of a huge pillar of star birth demonstrate how observations taken in visible and in infrared light by HST reveal dramatically different and complementary views of an object. Above left is cauldrons of gas at 36K Deg F tearing across space at 600K mph resembling a "butterfly". Above center is NGC 6302 Stephan's Quintet Galactic Wreckage - a clash among members of the quintet revealing stars from young blue stars to aging red stars. See: http://www.nasa.gov/hubble Credit: NASA, ESA, and the Hubble SM4 ERO Team.

Hubble Ultra Deep Field 2009–2010
Hubble Space Telescope • WFC3/IR
UDF-39546284

SIGNIFICANT ANOMALIES:
Orbiter:
- FWD STBD PLB FLOODLIGHT (#2) FAILED DURING STS-125
- DURING SSME IGNITION, AN ELECTRICAL ANOMALY OCCURRED THAT CAUSED ASA 1 TO BE LOST.
- AFTER CARRIER PANEL REMOVAL AN IN-PLANE CRACK WAS DETECTED AT THE DENSIFICATION LAYER INTERFACE WITH BASE MATERIAL ON TILES V070-395018-143 (SERIAL S83057) AND V070-395018-151 (SERIAL 7HB1DR)
- THE CREW DISCOVERED CARRYOVER OR UNPROCESSED CONDENSATE IN THE IMMEDIATE AREA OF THE HUMIDITY SEPARATORS IN THE LOWER EQUIPMENT BAY.
- THE IMU FAN DELTA PRESSURE (V61P2869A) WAS OBSERVED TO SLOWLY INCREASE ON FD 12, WITH THE FIRST INCIDENCE OF TOGGLING ABOVE THE FLIGHT RULE LIMIT OF 4.71 PSI OCCURRING AT GMT 142/18:22:37.
- DURING SSME IGNITION AN ELECTRICAL SHORT OCCURRED ON THE 26VAC EXCITATION CIRCUIT BETWEEN AEROSURFACE SERVOAMPLIFIER 1 (ASA-1) AND THE RIGHT HAND INBOARD ELEVON ACTUATOR PRIMARY DELTA PRESSURE TRANSDUCER.
- MDU CRT 4 REPORTED 'MSG COM 1553B ERROR', 'MESSAGE 1553B FAIL' AND 'MEDS I/O ERROR' IN DOWNLIST AT NOSE GEAR TOUCHDOWN.
KSC:
- Fondu-Fyre Liberated from SRB Main Flame Deflector, STS-125, Pad A
- Brick Liberated from East Flame Trench Wall, SSME Side, STS-125, Pad A
SRB: None. SSME: None. ET: None. MOD: None.
RSRM:
- MISSING STIFFENER RING FOAM WITH DISCOLORATION, STIFFENER RINGS, RSRM-105B
Integration:
- Aerosurface Servo Amplifier-1 (ASA-1) Power Supply Failed
- Unexpected Debris/Expected Debris Exceeding Mass Allowable Prior to Pad Clearance (Liftoff Debris)
- Ice Internal and External to the LH2 T-0 Umbilical
- Gap Filler Releases From Port OMS Pod

At Left: **HUBBLE DETECTS - MOST ANCIENT OBJECT**
On Jan 26, 2011, NASA reported that Hubble using its new camera, discovered a faint red blob (see ultra-deep-field exposure insert above right) thought to be the most distant object ever seen: a small proto galaxy some 13.2 billion light years away (faint optical image in insert below right). This galaxy existed 480 million years after the "Big Bang". These exposures were taken in 2009 & 2010. Credit NASA, ESA, G.Illingworth (U. of Calif Santa Cruz & R. Bouwens (U. of Calif, Santa Cruz & Leiden U.), & HUDF09 Team.

FLT NO.	ORBITER	CREW (6+1 UP/6+1 DN) / TITLE, NAMES & EVA'S	LAUNCH SITE, LIFTOFF TIME, LANDING SITES, ABORT TIMES	LANDING SITE/ RUNWAY, CROSSRANGE / LANDING TIMES FLT DURATION, WINDS	SSME-TL NOM-ABORT EMERG THROTTLE PROFILE ENG. S.N.	SRB RSRM AND ET	INC	ORBIT HA/HP	FSW	PAYLOAD WEIGHTS, PAYLOADS/ EXPERIMENTS	MISSION HIGHLIGHTS (LAUNCH SCRUBS/DELAYS, TAL WEATHER, ASCENT I-LOADS, FIRSTS, SIGNIFICANT ANOMALIES, ETC.)
STS-127/ ISS-2JA SEQ FLT # 127 KSC-127 PAD 39A (50) MLP-3 29th SHUTTLE FLIGHT TO ISS	OV-105 (Flight 23) ENDEAVOUR OMS PODS LPO3 -33 RPO4 29 FRC5-22	CDR: Mark Polansky (Flt 3 - STS-98,STS-116) P794/R262/V185/M228 PLT Doug Hurley P795/R341/M295 MS 1 Christopher Cassidy P796/R342/M296 MS 2 Julie Payette (Canada) (Flt 2 -STS-96) P797/R249/V205/F33 MS 3 Tom Marshburn P798/R343/M297 MS 4 Dave Wolf (Flt 4 - STS-58, Up to Mir on STS-86, Dn on STS-89, STS-112) P799/R173/V147/M151 MS 5 UP Stay ISS EXP20/FLT ENG T1m Kopra P800/R344/M298 Continued…	KSC 39A 196:22:03:09Z 6:03:10 PM EDT (P) 6:03:10 PM EDT (A) Wednesday (15) 07/15/09 (10) LAUNCH WINDOW: 10M 0S (Total) 5M 0S (Preferred) EOM PLS: KSC TAL: MRN TAL WX: ZZA. SELECTED: RTLS: KSC15 N/N TAL: MRN20 N/N AOA: NOR 17 N/SFD S B 1ST DAY PLS: EDW 22L N/N TDEL: 0.000 (P) -0.308 (A) MAX Q NAV: 722.7 (P) 705.3 (A) SRB STG: 2:04.2 (P) 2:03.8 (A) PERF: NOMINAL 2 ENG TAL (MRN): 2:29 (P) 2:35 (A) NEG MRN (2@ 104): 3:53 (P) 3:58(A) PTA (U/S 158 FPS): 5:02(P) 5:10(A) Continued…	KSC 15 (KSC 71) 212:14:48:07Z 09:48:07 AM CDT Friday (15) 07/31/09 (12) DEORBIT BURN: 212:13:41:09.9Z XRANGE: 672.5 NM ORBIT DIR: A/L (41) AIM PT: Nominal MLGTD: 1797 FT 212:14:48:07Z VEL: 208 KGS 209 KEAS HDOT: -2.8 FPS TD NORM 195: 2865 FT DRAG CHUTE DEPLOY: 186 KEAS 212:14:48:13Z NLGTD: 5842 FT 212:14:48:19Z VEL: 152 KGS 150 KEAS HDOT: -5.0 FPS BRK INIT: 71 KGS DRAG CHUTE JETTISON: 56 KGS 12:14:48:52Z BRK DECEL FPS2: AVE 4.8 PK 6.3 Continued…	104/104/109% PREDICTED: 100/104.5/104.5/ 72/104.5 ACTUAL: 100/104.5/100// 72/104.5 1 = 2045 (10) 2 = 2060 (1) 3 = 2054 (9) M 3 EOM WEIGHT: 215899.5 LBS X CG: 1089.8 IN LANDING: WEIGHT: 215816.5 LBS X CG: 1091.7 IN	BI-138 RSRM 106 ET-131 SLWT 35 ET IMPACT 1:14:27 MET LAT: 35.889 S LONG: 157.79 W	51.6 (29)	DIRECT INSERTION POST OMS-2: 123.8x32.3 NM DEORBIT: HA 184.5 NM HP 22.2 NM ENTRY VELOCITY: 25855 FPS ENTRY RANGE: 4334 NM	OI-33 (3)	CARGO: 36253LBS PAYLOAD CHARGEABLE: 24682 LBS DEPLOYED: 24266 LBS NON-DEPLOYED: 290 LBS MIDDECK: 126 LBS SHUTTLE ACCUMULATED WEIGHTS: DEPLOYED: 1548698 LBS NON-DEPLOYED: 1621661 LBS CARGO TOTAL: 4090475 LBS PERFORMANCE MARGINS (LBS): FPR: 2651 FUEL BIAS: 1059 FINAL TDDP: 2553 RECON:2734 PAYLOADS: PLB: ISS-2J/A, ANDRE-2, DRAGONSAT MIDDECK: ISS-2A,MAUI, SEITE,SIMPLEX 5 CRYO TANK SETS ODS, SRMS (84), OBSS,SSPTS, ECSHS(2),ROEU, PPSUS(2)	*Brief Mission Summary:* STS-127 (29th mission to ISS) was a "16 day marathon construction mission". The final pieces of the Japanese Kibo Complex including an Experiment Exposed Facility ("Porch in Space" - PAO) and the unpressurized Experiment Logistics Module were delivered along with spare equipment intended to keep ISS operational long after Shuttle is retired. Five EVA's and operations of three robotic arms were conducted for completion of all objectives. KSC W/D: OPF: 109 + 9H VAB: 7 + 0C PAD B: 32 + 10C + 1 SD (STS-125 launch) + 1 CR (Crew Rest Day) PAD A: 42 + 3C + 1H Total Work Days = 190 (OPF processing occurred over a total time period of 118 days.) POSTPONEMENTS: - Added STS-127 to FDRD - launch date of 04/23/09 on 04/24/08. - Ppd. to 05/15/09 on 07/03/08. Slip due to ET deliveries. - Ppd. to 06/13/09 on 03/10/09. Slip due to interim changes while Cx and SSP schedules were assessed and prioritized. LAUNCH SCRUBS: - Launch scrubbed officially on Saturday, 06/13/09 at 12:26 a.m. EDT due to GH2 leak at the GUCP – the same type of leak that scrubbed STS-119 in March. Launch rescheduled for 06/17/09. Technical Scrub. - Launch scrubbed officially on Wednesday 06/17/09 at 1:55 EDT with the reoccurrence of the same type of GUCP leak as previous scrub. Launch rescheduled for 07/11/09. Technical Scrub. - Launch officially scrubbed during L-11 Hour Hold at MMT meeting on Saturday morning, 07/11/09, due to unstable weather and lightning strikes overnight in KSC area. Seven strikes hit the lightning protection system, but none hit the vehicle. Launch rescheduled for 07/12/09. Weather Scrub. - Launch scrubbed during a final hold at T-9 minute mark on Sunday 07/12/09 due to predicted thunderstorms within 20 nm limit of SLF. Launch rescheduled for 07/13/09. Weather Scrub. - Launch scrubbed at 6:39 PM EDT on Monday 07/13/09 due to weather violations in KSC area. Launch rescheduled for 07/15/09. Weather Scrub. Continued…

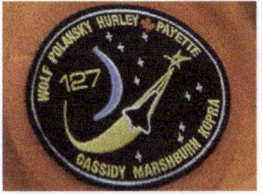

Gaseous hydrogen vent line leak caused STS-119 scrub in March 2009, also caused two scrubs on STS-127. This line connects the Ground Umbilical Carrier Plate (GUCP), attached to ET, to "flare stack" for burn-off of vented gaseous hydrogen.

SPACE SHUTTLE MISSIONS SUMMARY

Page 2-210 - STS-127/2JA

FLT NO.	ORBITER	CREW (6+1 UP/6+1 DN) TITLE, NAMES & EVA'S	LAUNCH SITE, LIFTOFF TIME, LANDING SITES, ABORT TIMES	LANDING SITE/ RUNWAY, CROSSRANGE LANDING TIMES FLT DURATION, WINDS	SSME-TL NOM-ABORT EMERG THROTTLE PROFILE ENG. S.N.	SRB RSRM AND ET	INC	HA/HP	FSW	PAYLOAD WEIGHTS, PAYLOADS/ EXPERIMENTS	MISSION HIGHLIGHTS (LAUNCH SCRUBS/DELAYS, TAL WEATHER, ASCENT I-LOADS, FIRSTS, SIGNIFICANT ANOMALIES, ETC.)
STS-127/ ISS-2JA Continued …		Continued… MS 5 DN EXP 18/19/20 FLT ENG (Japan) Koichi Wakata (Flt 3 - STS-72,STS-92, Up on STS-119 stay ISS) P801/R208/V164/M181 SS EVA 137 DOCKED QUEST EVA 55 EMU/TETHERED EVA 130 SCHEDULED EVA 128 DURATION 5:32 SS EVA 138 DOCKED QUEST EVA 56 EMU/TETHERED EVA 131 SCHEDULED EVA 129 DURATION 6:53 SS EVA 139 DOCKED QUEST EVA 57 EMU/TETHERED EVA 132 SCHEDULED EVA 130 DURATIO N 5:59 SS EVA 140 DOCKED QUEST EVA 58 EMU/TETHERED EVA 133 SCHEDULED EVA 131 DURATIO N 7:12 SS EVA 141 DOCKED QUEST EVA 59 EMU/TETHERED EVA 134 SCHEDULED EVA 132 DURATION 4:54 Continued…	Continued… SE TAL (ZZA 104): 6:03(P) 6:08(A) PTM (U/S 181 FPS): 6:01(P) 6:14(A) SE PRESS 104 6:52(P) 7:01(A) MECO CMD: 8:22.4(P) 8:24.9(A) VI: 25819(P) 25820(A) OMS-2: 35:45 (P) 38:30(A) 98.7(P) 96.9(A) FPS	Continued… WHEELS STOP: 212:14:49:13Z 11856 FT ROLLOUT: 10059 FT 1:06 M:S WINDS: 7H KT 6R KTS OFFICIAL: 19008P13KT (X5P7 H7P11) DENS ALT: 1916 FT FLT DURATION: 15:16:44:58 S/T: 1212:01:31:44 OV-105: 266:15:33:01 DISTANCE: 6,547,853 sm TOTAL SHUTTLE DISTANCE: 497,402,218 sm							Continued… **LAUNCH WINDOW:** Total launch window was 10M 5S with window open at 196:21:58:10Z and close at 196:22:08:10Z. Preferred Launch Time was 196:22:03:10Z (In-Plane Time) for a launch window of 5M 0S. **LAUNCH DELAYS:** - None. Launch occurred on time at 196:22:03:10Z, 6:03:10 p.m. EDT, Wednesday, July 15, 2009. The Spaceflight Meteorology Group (SMG) forecast was challenged by thunderstorms along the east coast breeze throughout the day. However, the weather improved at the SLF and within the 20nm limit prior to launch for a "Go". **TAL WEATHER:** TAL weather also cooperated for a Go for launch. A high pressure system produced dry and stable conditions across southern Spain. The two Spanish TAL sites were forecast for clear skies and winds within flight rule limits. Istres was forecasting a slight chance of a ceiling below flight rule limits for launch day. **PERFORMANCE ENHANCEMENTS:** Include the standard set plus: 1) PE Operational High Q SUM/JUL, 2) OMS Assist, 3) a 52 nautical mile MECO, and 4) Del Psi **FLIGHT DURATION CHANGES**: NONE - Planned landing at KSC on orbit 248. Landed at KSC Runway 15 on orbit 248 at 212:14:48:07Z on Friday, July 31, 2009. **FIRSTS/SECONDS/LASTS:** - Five launch scrubs is second highest number: STS-73 in 1995 & STS-61C in 1986 had six. - Koichi Wakata, first Japanese astronaut to have engaged in long-duration on-orbit, returned to Earth after 4 1/2 months. - First flight of SSME controller constant updates, an updated MPS propellant inventory, and an updated CMR. - Record-size space crew of thirteen (ISS & Shuttle). **NIGHT LAUNCH**: N/A **RENDEZVOUS: #74** Rendezvous and dock with ISS. Continued…

JSC2009-E-143033 --- Retired NASA Launch Director Robert Sieck, right, talks with Associate Administrator for Space Operations Bill Gerstenmaier in KSC Firing Room during a built-in launch countdown hold.

ISS020-E-022626 (20 July 2009) --- Endeavour's crew cabin, along with the ISS's Kibo laboratory and Harmony node are shown during 2nd EVA.

FLT NO.	ORBITER	CREW (6+1 UP/6+1 DN) TITLE, NAMES & EVA'S	LAUNCH SITE, LIFTOFF TIME, LANDING SITES, ABORT TIMES	LANDING SITE/ RUNWAY, CROSSRANGE LANDING TIMES FLT DURATION, WINDS	SSME-TL NOM-ABORT EMERG THROTTLE PROFILE ENG. S.N.	SRB RSRM AND ET	ORBIT INC / HA/HP	FSW	PAYLOAD WEIGHTS, PAYLOADS/ EXPERIMENTS	MISSION HIGHLIGHTS (LAUNCH SCRUBS/DELAYS, TAL WEATHER, ASCENT I-LOADS, FIRSTS, SIGNIFICANT ANOMALIES, ETC.)
STS-127/ ISS-2JA Continued …		Continued… MCC WHITE FLIGHT FCR (57) FLIGHT DIRECTORS: SHUTTLE: A/E- Bryan Lunney LD/O1- Paul Dye O2- Kwatsi Alibaruho Planning- Gary Horlacher - Mike Sarafin MOD – John Mccullough Team 4- Richard Jones ISS O1 - Brian Smith LD/O2 – Holly Ridings O3 – Derek Hassmann Team 4 - Ron Spencer CAPCOMS: SHUTTLE A/E – Alan Poindexter - Eric Boe (Wx) LD/O1 – Greg (Box) Johnson O2 - Janice Voss Planning - Stan Love - Shannon Lucid Team 4 - N/A ISS O1 – Hal Getzelman LD/O2-Akihiko Hoshide O3 – Jason hutt Team 4 – N/A								

S127-E-009733 (28 July 2009) --- **Record Size Space Crew:** The STS-127 and Expedition 20 crew members pose for a group portrait in ISS Harmony Node. From left (front row) are NASA astronauts Michael Barratt, Exp 20 FE; Mark Polansky, STS-127 CDR; cosmonaut Gennady Padalka, Exp 20 CDR; and NASA astronaut Dave Wolf, STS-127 MS. From left (middle row) are JAXA astronaut Koichi Wakata, STS-127 MS; Canadian astronauts Julie Payette, STS-127 MS and Robert Thirsk, Exp 20 FE; and NASA astronaut Tom Marshburn, STS-127 MS. From left (back row) are cosmonaut Roman Romanenko, Exp 20 FE; NASA astronauts Christopher Cassidy, STS-127 MS; Doug Hurley, STS-127 Pilot; Tim Kopra, Exp 20 FE; and ESA astronaut Frank De Winne, Exp 20 FE.

Continued…

EVENTS:

- During liftoff several pieces of foam insulation came off the ET. Shuttle was hit two or three times, said Bill Gerstenmaier. Some scuff marks were spotted on the belly, but that probably was coating loss and considered minor, he said. That was later determined to be the case.
- FD1: OMS2 ignition at 196:22:41:40.0.9Z resulted in a 125.4 by 85.1 NM orbit.
- FD2: RCC inspection found no areas of concern
- T1 maneuver at 198:15:17:25.9Z resulted in a 188.7 by184.0 NM orbit
- FD3: R-Bar Pitch Maneuver was performed. No issues.
- Hard Dock, hooks closed, occurred at 198:15:47:10Z (12:47 CDT, July 17, 2009)
- ISS Hatch opened at 198:17:48:10Z (2:48 PM CDT, July 17, 2009) welcomed by ISS crew.
- IELK Seat Liner Transfer at 198:19:22:10Z (9:00 PM CDT March 17, 2009). At that time Koichi Wakata became a member of STS-127 and Tim Kopra joined the ISS Expedition 20 as Flight Engineer.
- Reboost - ~2.5 fps posigrade delta V. Increased altitude approx 4700 ft . Cleared vehicles of conjunction with Object 84180.
- FD4: Based on review of launch imagery, MMT cancelled FD5 focused inspection of Orbiter heat shield.
- FD4: EVA 1: David Wolf & Tim Kopra: Activities included: JPM berthing mechanism prep and install, CETA cart mods, and the P3 Nadir UCCAS deploy. EVA was shortened due to suit consumables. The PAS deploy was ppd. EVA1 duration 5:32.
- Using the SSRMS and SRMS the JEM Exposed Facility (JEF) was successfully unberthed from the Shuttle P/B and captured on the Japanese Experiment Module (JEM).
- FD6: EVA2: Dave Wolf & Tom Marshburn: Activities included: Transfer of ORU's (Space-to-Ground Antenna, Linear drive Unit & Pump Module) from the Integrated Cargo Carrier (ICC) to the External Stowage Platform. Installation of the JEF forward Vision Equipment [VE] was deferred. EVA2 duration 6:53.
- FD8: EVA3: Dave Wolf & Chris Cassidy: Activities included: Node 2 WIF 14 removal and installation to COL WIF 2, JLE payload prep, completion of 2 Lab FPP grounding sleeves, changeout of 2 of 6 batteries on P6 (batts A & B from the ICC-VLD) and positioning of ICC-VLD in overnight parking configuration. EV2's LiOH performance caused early termination. EVA3 duration 5:59.
- FD10: EVA4: Chris Cassidy & Tom Marshburn: Activities included: successful R&R of all batteries and successful latching of the ICC-VLD back into the Shuttle P/L bay for return. EVA4 duration 7:12.

Continued…

FLT	ORBITER	CREW (6+1 UP/6+1 DN)	LAUNCH SITE, LIFTOFF TIME,	LANDING SITE/ RUNWAY, CROSSRANGE	SSME-TL NOM-ABORT EMERG	SRB RSRM	ORBIT		FSW	PAYLOAD WEIGHTS,	MISSION HIGHLIGHTS (LAUNCH SCRUBS/DELAYS,
NO.		TITLE, NAMES & EVA'S	LANDING SITES, ABORT TIMES	LANDING TIMES FLT DURATION, WINDS	THROTTLE PROFILE ENG. S.N.	AND ET	INC	HA/HP		PAYLOADS/ EXPERIMENTS	TAL WEATHER, ASCENT I-LOADS, FIRSTS, SIGNIFICANT ANOMALIES, ETC.)

STS-127/ ISS-2JA

Continued …

JSC2009-E-145586 --- Orbit 1 Lead FD Paul Dye (foreground) on console during docking of STS-127 Endeavour to ISS. In background are CAPCOM's Dominic Gorie (far left) and Greg Johnson.

BELOW: S127-E-009372 (27 July 2009) Marshburn (left) & Cassidy, STS-127 MS's, participate in fifth and final EVA as construction and maintenance continue on the ISS.

S127E009372

S127E011200

S127-E-011200 (28 July 2009) --- The ISS is seen from Space Shuttle Endeavour as the two spacecraft begin their relative separation.

SIGNIFICANT ANOMALIES: Continued…

ET:
- POST-LAUNCH CAMERA AND FILM REVIEW SHOWED LOSS OF FOAM AT SEVERAL LOCATIONS ON THE INTERTANK.
- POST-LAUNCH CAMERA & FILM REVIEW SHOWED LOSS OF FOAM IN THE AFT INBOARD CORNER OF THE LO_2 ICE FROST RAMP AT STATION 718
- ET TPS Loss Outboard Section of the -Y Bipod Closeout
MOD: None.
Integration:
- Unexpected Debris/Expected Debris Exceeding Mass Allowable Prior to Pad Clearance (Liftoff Debris)
- LH_2 Leak at ET Ground Umbilical Carrier Plate (GUCP)
- Ice Internal and External to the LH_2 T-0 Umbilical

EVENTS: Continued…

- FD13: EVA5: Chris Cassidy & Tom Marshburn: Activities included: completion of Z1 patch panel reconfig, SPDM covers, JEF Vision Equipment installation and several get-aheads (JEM handrail and WIF installation, Lab cable tiedowns, Node 2 Gap Spanner installation, and relocating two APFR's for STS-128). The S3 Zenith Outboard PAS task was not performed due to lack of time based on predicted METOX capability. EVA5 duration 4:54.
-Transfers:
24,638 Pounds of hardware transferred to ISS (inside & out)
10,479 Pounds of hardware returned aboard Endeavour
2,175 Pounds of middeck items delivered to ISS aboard Endeavour
1,980 Pounds of middeck items returned from ISS to Endeavour
1,225 Pounds of water transferred to ISS
45 Pounds of Oxygen used for "stack maintenance"
12 Pounds of Nitrogen transferred to ISS
- ISS Mass in space 685,986 mass - pounds
- FD14: Undocked at 209:17:26:00Z (12:26 PM CDT, July 28, 2009)
- After undocking, Hurley initiated Endeavour fly-around at a distance of 400 feet from ISS and completed Sep-maneuver at 209:19:09:00Z (2:09 PM CDT, July 28, 2009)
- During Entry comm blackout occurred at 212:14:34:05Z - 212:14:36:24Z due to plasma effect.

SIGNIFICANT ANOMALIES:
Orbiter:
- MICROBIAL REMOVAL ASSEMBLY LEAKAGE
- FUEL CELL 3 SN 121 SUSTAINING HEATER TURNED ON WHEN THE FC STACK OUT TEMPERATURE REACHED A VALUE OF 185 DEG F
- DURING THE RCS HOTFIRE TEST, FORWARD RCS THRUSTER F2F EXHIBITED LOW PC (V42P1542A) OF APPROXIMATELY 16 PSI. F2F WAS DECLARED FAILED OFF AND AUTO DESELECTED BY RCS RM AT MET 14/10:45:40 (GMT 211/08:48:50).
KSC:
- The Istres Backup Azimuth system is in a Hard Overscan Alarm
- STS-127 Post Launch Debris
SRB:
- TOP LAYERS OF MSFC CONVERGENT COATING (MCC-1) MISSING ON AFT SKIRT TPS ACREAGE (BOTH LEFT & RIGHT HAND)POST FLIGHT OF STS-127/BI-138
- LEFT-HAND SOLID ROCKET BOOSTER ENHANCED DATA ACQUISITION SYSTEM (EDAS) ASSEMBLY CHANNEL 4 DID NOT RECORD NOMINAL STRAIN RESPONSE.
RSRM: None.
SSME: None.

Continued at left…

FLT NO.	ORBITER	CREW (6+1 UP/6+1 DN) TITLE, NAMES & EVA'S	LAUNCH SITE, LIFTOFF TIME, LANDING SITES, ABORT TIMES	LANDING SITE/ RUNWAY, CROSSRANGE LANDING TIMES FLT DURATION, WINDS	SSME-TL NOM-ABORT EMERG THROTTLE PROFILE ENG. S.N.	SRB RSRM AND ET	INC	ORBIT HA/HP	FSW	PAYLOAD WEIGHTS, PAYLOADS/ EXPERIMENTS	MISSION HIGHLIGHTS (LAUNCH SCRUBS/DELAYS, TAL WEATHER, ASCENT I-LOADS, FIRSTS, SIGNIFICANT ANOMALIES, ETC.)
STS-128 (17A) SEQ FLT # 128 KSC-128 PAD 39A (51) MLP-2 30th SHUTTLE FLIGHT TO ISS	OV-103 (Flight 37) DISCOVERY OMS PODS LPO1-40 RPO3-38 FRC3-37	CDR: Rick Sturckow (Flt 4 - STS-88,STS-105 STS-117) P802/R247/V173/M215 PLT Kevin Ford P803/R345/M259 MS 1 Patrick Forrester (Flt 3 - STS-105, STS-117) P804/R269/V186/M235 MS 2 Jose Hernandez P805/R346/M300 MS 3 Danny Olivas (Flt 2-STS-117) P806/R309/V207 /M267 MS 4 Christer Fuglesang (ESA) (Flt 2 - STS-116) P807/R304/V208/M264 MS 5 UP Stay ISS EXP20/FLT ENG Nicole Stott P808/R347/F47 MS 5 DN EXP 20 FLT ENG Tim Kopra Up on STS-127 stay ISS) P809/R344/M298 Continued...	KSC 39A 241:03:59:37Z 11:59:37 PM EDT (P) 11:59:37 PM EDT (A) Friday (27) 08/28/09 (9) LAUNCH WINDOW: 9M 36S (Total) 4M 48S (Preferred) EOM PLS: KSC TAL: MRN TAL WX: FMI.(NO GO) ZZA (NO GO) SELECTED: RTLS: KSC33 N/N TAL: MRN20 N/N AOA: NOR 17 N/SFD S B 1ST DAY PLS: EDW 22L N/SFD TDEL: 0.000 (P) -0.078 (A) MAX Q NAV: 752.76 (P) 738.70 (A) SRB STG: 2:02.2 (P) 2:02.6 (A) PERF: NOMINAL 2 ENG TAL (MRN): 2:38 (P) 2:41 (A) NEG MRN (2@ 104): 3:52 (P) 3:53(A) PTA (U/S 157 FPS): 5:09(P) 5:12(A) Continued...	EDW22 CONC EDW 54 CONC 35 255:00:53:20Z 7:53:20 PM CDT Friday (16) 09/11/09 (12) DEORBIT BURN: 254:23:47:37Z XRANGE: 374.6NM ORBIT DIR: A/L (42) AIM PT: Nominal MLGTD: 1515 FT 255:00:53:20Z VEL: 220 KGS 199 KEAS HDOT: -4.3 FPS TD NORM 195: 1753 FT DRAG CHUTE DEPLOY: 155 KEAS 255:00:53:32Z NLGTD: 4854 FT 255:00:53:29Z VEL: 185 KGS 161 KEAS HDOT: -6.3 FPS BRK INIT: 113 KGS DRAG CHUTE JETTISON: 54 KGS 255:00:54:06Z BRK DECEL FPS²: AVE 4.8 PK 7.4 Continued...	104/104/ 109% PREDICTED: 100/104.5/104.5/ 72/104.5 ACTUAL: 100/104.5/100// 72/104.5 1 = 2052 (8) 2 = 2051 (9) 3 = 2047 (13) M 3 EOM: WEIGHT: 222200 LBS X CG: 1088.4 IN LANDING: WEIGHT: 222271 LBS X CG: 1090 IN	BI-139 RSRM 107 ET-132 SLWT 36 ET IMPACT 1:14:26 MET LAT: 35.875 S LONG: 157.761 W	51.6 (30)	DIRECT INSERTION POST OMS-2 127.5x84.4 NM DEORBIT: HA 192.1 NM HP 22.5 NM ENTRY VELOCITY: 25863 FPS ENTRY RANGE: 4399.1 NM	OI-34 (1)	CARGO: 40605LBS PAYLOAD CHARGEABLE: 33056 LBS DEPLOYED: 30572 LBS NON-DEPLOYED: 2331 LBS MIDDECK: 153 LBS SHUTTLE ACCUMULATED WEIGHTS: DEPLOYED: 1579270 LBS NON-DEPLOYED: 1623992 LBS CARGO TOTAL: 4131080 LBS PERFORMANCE MARGINS (LBS): FPR: 2908 FUEL BIAS: 1059 FINAL TDDP: 1707 RECON: 2077 PAYLOADS: PLB: ISS-17A (MPLM,LMC), MISSE 6, TRIDAR AR&D SENSOR,DTO-701A Continued...	*Brief Mission Summary:* The STS-128 (30th mission to ISS), dubbed "Racking Up New Science" by PAO, main objective was to deliver science and environmental racks to dramatically enhance the scientific capability of the ISS. These racks were carried in the Leonardo MPLM. Included in the cargo was the highly publicized Combined Operational Load Bearing External Resistance Treadmill (COLBERT) named after TV comedian Stephen Colbert. Three EVA's were conducted and included replacement of the massive ammonia tank used by the ISS Thermal Control System. **KSC W/D** OPF: 117+ 2H VAB: 9 +0C PAD A: 25 + 0C Total Work Days = 151 (OPF processing occurred over a total time period of 119 days.) **POSTPONEMENTS:** - Added STS-128 to FDRD - launch date of 07/30/09 on 06/23/08. - Ppd. to 08/06/09 on 12/10/08. Interim manifest while HST final placement is considered. - Ppd. to 08/07/09 on 06/08/09. Slip due to MA direction. - Ppd. to 08/18/09 on 06/30/09. Slip due to STS-127 GUCP delays. - Ppd. to 08/25/09 on 08/20/09. Slipped to support KSC processing. **LAUNCH SCRUBS:** - 08/25/09 weather did not cooperate, systems looked good. Setting up for the next opportunity, window open at 12:05am CDT tomorrow with the in-plane time at 12:10am. Weather Scrub. - 08/25/09 the 2nd launch attempt was scrubbed officially at 4:52 p.m. CDT (5:52 Eastern) by Launch Director Pete Nickolenko due to stuck "fill & drain valve during ET loading. Based on the results of a technical review of the MPS Hydrogen Fill & Drain Valve data, a 48 hour scrub turnaround was initiated. Technical scrub. - 08/27/09 Official no go for launch today. Launch postponed to allow engineers additional time to develop flight rationale based on testing of F&D valve. Moses, "Will try tomorrow night if we get there." Next opportunity is Friday at 10:59 pm CDT (11:59 Eastern). -08/28/09 MMT Summary at 12:55 PM: Reviewed LH₂ valve (PV12) and agreed to plan for tonight's launch attempt. MMT is go to proceed for launch. Continued...

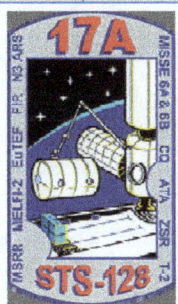

STS128-S-011 (28 Aug. 2009) ---Viewed from the Banana River Viewing Site, the Space Shuttle heads toward Earth orbit and rendezvous with ISS. Night launch #33.

FLT NO.	ORBITER	CREW (6+1 UP/6+1 DN) TITLE, NAMES & EVA'S	LAUNCH SITE, LIFTOFF TIME, LANDING SITES, ABORT TIMES	LANDING SITE/ RUNWAY, CROSSRANGE LANDING TIMES, FLT DURATION, WINDS	SSME-TL NOM-ABORT EMERG THROTTLE PROFILE ENG. S.N.	SRB RSRM AND ET	INC	ORBIT HA/HP	FSW	PAYLOAD WEIGHTS, PAYLOADS/ EXPERIMENTS	MISSION HIGHLIGHTS (LAUNCH SCRUBS/DELAYS, TAL WEATHER, ASCENT I-LOADS, FIRSTS, SIGNIFICANT ANOMALIES, ETC.)

STS-128 (17A)

Continued…

At FRR News Conf: News 13 Flordia: "Buzz Lightyear doing okay?" Suffredini: "There are big plans for him. He's been stowed, so I didn't talk to him."

Continued…

SPECIAL EDUCATOR "Buzz" Lightyear (DN/EXP 20, see below left)

SS EVA 142
DOCKED QUEST EVA 60
EMU/TETHERED EVA 135
SCHEDULED EVA 133
DURATION 6:35

SS EVA 143
DOCKED QUEST EVA 61
EMU/TETHERED EVA 136
SCHEDULED EVA 134
DURATION 6:39

SS EVA 144
DOCKED QUEST EVA 62
EMU/TETHERED EVA 137
SCHEDULED EVA 135
DURATIO N 7:01

MCC WHITE FLIGHT FCR (58)

FLIGHT DIRECTORS:
SHUTTLE:
A/E- Richard Jones
LD/O1- Tony Ceccacci
O2- Kwatsi Alibaruho
Planning- Gary Horlacher
MOD – John Mccullough
Team 4- Mike Sarafin

ISS
O1 - Ron Spencer
LD/O2 – Heather Rarick
O3 – Royce Renfro
Team 4 - Derek Hassmann

Continued…

STS-128: Weather scrubs launch. Xenon lights over Launch Pad 39A compete with the lightning strike. Photo source: Not identified.

Continued…

SE TAL (FMI 104):
6:05(P)   6:08(A)

PTM (U/S 181 FPS):
6:09(P)   6:16(A)

SE PRESS 104
6:57(P)   6:58 (A)

MECO CMD:
8:24.0(P) 8:24.7 (A)
VI:
25819(P) 25820(A)

Continued…

<---- Water Tower Strike

Google Earth Plots of StrikeNet and CGLSS Coordinates. From: Aug 15, 2009 Daily PRCB, John Apfelbaum/KSC PHI10

Continued…

MIDDECK:
ISS-17A,MAUI, SEITE,SIMPLEX

5 CRYO TANK SETS
ODS, SRMS (85), OBSS,SSPTS

Continued…

LAUNCH WINDOW: Total launch window was 9M 36S with window open at 241:03:54:49Z and close at 241:04:04:25Z. Preferred Launch Time was 241:03:59:37Z (In-Plane Time) for a launch window of 4M 48S.

LAUNCH DELAYS:
- None. Launch occurred on time at 241:03:59:37Z, 11:59:37 PM EDT, Friday, August 28, 2009. The Spaceflight Meteorology Group (SMG) gave a "Go" for weather.

TAL WEATHER: SMG Forecast: A frontal system is approaching Istres and a upper level shortwave is dropping into northern Spain and southern France. Result in very windy conditions at Istres and breezy conditions at Zaragozal. Istres winds will be violating flight rule limits while Zaragoza will be very near the headwind limit. Moron weather is looking very favorable with clear skies and relatively light winds.

PERFORMANCE ENHANCEMENTS:
Include the standard set plus: 1) PE Operational High Q - SUM/AUG, 2) OMS Assist, 3) a 52 nautical mile MECO, and 4) Del Psi

FLIGHT DURATION CHANGES:
- Thursday, Sep 10, 2009, first deorbit opportunity waved off for violations of showers within 30nm & crosswind violations at 17 kts. Second opportunity also waved off; showers, instability, broken cloud deck and crosswind violation. Flight extended for EOM +1 day to Friday, 4 opportunities available. First & second opportunities at KSC were again waved off due to weather. EDW had no violations and low winds, first opportunity shows winds 230 8p12 kts. GO for EDW given. Landed on EDW Runway 22 at 255:00:53:20Z, Friday, Sep 11, 2009.

FIRSTS:
RSRM Improved Resiliency O-rings, Nozzle-to-Case Joint. Fly with higher margins.
RSRM Inactive Stiffener Stub Removal - Eliminated four debris liberation/debris impact causes

NIGHT LAUNCH: #33

Continued…

| FLT NO. | ORBITER | CREW (6+1 UP/6+1 DN) TITLE, NAMES & EVA'S | LAUNCH SITE, LIFTOFF TIME, LANDING SITES, ABORT TIMES | LANDING SITE/ RUNWAY, CROSSRANGE LANDING TIMES FLT DURATION, WINDS | SSME-TL NOM-ABORT EMERG THROTTLE PROFILE ENG. S.N. | SRB RSRM AND ET | ORBIT | | FSW | PAYLOAD WEIGHTS, PAYLOADS/ EXPERIMENTS | MISSION HIGHLIGHTS (LAUNCH SCRUBS/DELAYS, TAL WEATHER, ASCENT I-LOADS, FIRSTS, SIGNIFICANT ANOMALIES, ETC.) |
							INC	HA/HP			
STS-128 (17A) Continued...		Continued... CAPCOMS: SHUTTLE A/E – Eric Boe - Chris Ferguson (Wx) LD/O1 - Chris Ferguson - Tony Antonelli O2 - Stan Love Planning - Shannon Lucid Team 4 - N/A ISS O1 - Chris Zajac LD/O2- Robert Hanley O3 – Mike Jensen Team 4 – N/A	Continued... OMS-2: 39:00 (P) 39:00(A) 95.1(P) 94.5(A) FPS	Continued... WHEELS STOP: 255:00:54:33Z 13109 FT ROLLOUT: 11594 FT 1:13 M:S WINDS: -6.5T KT -2.5L KTS OFFICIAL: 09007P08KT (X4P4 T6P7) DENS ALT: 5489 FT FLT DURATION: 13:20:53:43 S/T: 1225:22:25:27 OV-103: 332:00:33:34 DISTANCE: 5,702,716 sm TOTAL SHUTTLE DISTANCE: 503,104,934 sm							Continued... RENDEZVOUS: #75 Rendezvous and dock with ISS. EVENTS: - FD1: OMS2 ignition at 241:04:38:36.9Z resulted in a 127.5 by 84.4 nm orbit. - FD2: RCC inspection found no areas of concern - T1 maneuver at 242:22:26:17Z resulted in a 193.2 by181.6 NM orbit - FD3: R-Bar Pitch Maneuver was performed. No issues. - Docking Contact occurred at 243:00:53:56Z - Hard Dock, hooks closed, occurred at 243:01::07:23Z - ISS Hatch opened at (9:32 PM CDT, Aug 30, 2009) welcomed by ISS crew. - IELK Seat Liner Transfer at (10:50 PM CDT, Aug 30). At that time Tim Kopra became a member of STS-128 and Nicole Stott joined ISS EXP 20. - MMT FD3 reported VRCS jet F5R experienced a jet fail leak at 00/4:37 MET. ISS to perform all attitude control & maneuvers during the docked mission. - MMT FD5 concurred that no Focused Inspection of Orbiter was required. - FD5: "Leonardo" MPLM transferred to ISS, Zero-G stowage rack t "Harmony" node & COLBERT treadmill transferred. - EVA 1: Olivas & Stott successfully completed: Prep of P1 truss Ammonia Tank Assembly (ATA) for removal, EuTEF & MISSE experiment removal from Columbus module. EVA1 duration 6:35. - FD7: EVA2: Olivas & Fuglesang: EVA was about 51 min late due to Olivas' comm. cap chin strap came undone while in pre-breathe. The ATA task was completed early & 3 get ahead tasks were completed: CLA cover installation, APFR 4 tool stanchion relocation, & CLPA cover installation. EVA2 duration 6:39. - FD9: EVA3: Olivas & Fuglesang: Activities included: Deploy S3 Truss Payload Attach System, Rate Gyro Assembly 2 R&R, S0 Truss Remote Power Control Unit R&R, Global Positioning System 4 installation, "Tranquility" Node 3 avionics cable routing (full), & Oxygen Generator Assembly water filter R&R. A lens became mechanically detached from Fuglesang's helmet at the end of the EVA. Without intact helmet lights he headed to the A/L before sunset . His PET was 6:22. Olivas performed cleanup. EVA3 duration (PET) 7:01. Continued...

In the JSC MCC: JSC2009-E-155032 --- FDs Richard Jones (left) & Tony Ceccacci on console during 2nd launch attempt. The launch was later postponed due to a valve issue in Discovery's main propulsion system. FD Bryan Lunney is in the background.

S128E007229

Construction and maintenance continued on the ISS.
ABOVE: S128-E-007229 (1 Sept. 2009) --- Nicole Stott/EXP 20 FE, during EVA 1with Danny Olivas/MS3 (out of frame). Activities included removal of an empty ammonia tank from ISS truss .
BELOW: S128-E-007720 (5 Sept. 2009) --- Olivas/MS3 (left) & Christer Fuglesang//ESA/MS4, participate in EVA3 activites.

S128E007720

FLT NO.	ORBITER	CREW (6+1 UP/6+1 DN) TITLE, NAMES & EVA'S	LAUNCH SITE, LIFTOFF TIME, LANDING SITES, ABORT TIMES	LANDING SITE/ RUNWAY, CROSSRANGE LANDING TIMES FLT DURATION, WINDS	SSME-TL NOM-ABORT EMERG THROTTLE PROFILE ENG. S.N.	SRB RSRM AND ET	INC	ORBIT HA/HP	FSW	PAYLOAD WEIGHTS, PAYLOADS/ EXPERIMENTS	MISSION HIGHLIGHTS (LAUNCH SCRUBS/DELAYS, TAL WEATHER, ASCENT I-LOADS, FIRSTS, SIGNIFICANT ANOMALIES, ETC.)
STS-128 (17A) Continued...											Continued...

ISS020-E-038322 --- STS-128 & Exp 20 crew in-flight portrait on ISS. STS-128 red-clad crew are: front row, from left, CDR Sturckow, Hernandez, & Forrester; middle row in red, PLT Ford, Olivas, & Fuglesang (ESA). EXP 20 crew (in blue) are: bottom left, Kopra, who joined ISS crew in July, now scheduled to return to Earth with STS-128. Clockwise from him are: Stott, Robert Thirsk/CSA, Roman Romanenko/RSA, Frank De Winne/ESA, Gennady Padalka/RSA, and Michael Barratt.

S128-E-009998 (8 Sept. 2009) --- Back-dropped by Earth's horizon and the blackness of space, ISS as seen from Discovery as the two spacecraft begin their relative separation.

S128E009968

STS128-S-047 (11 Sept. 2009) --- Shuttle Discovery's main landing gear touchdown at EAFB. Landing was diverted from KSC due to marginal weather.

Continued...

-Transfers:
  18,548 Lbs of hardware transferred to ISS
   1,705 Lbs "New" ATA (with 600 lbs of ammonia) to ISS
   1,295 "Old" ATA to Discovery
   5,223 Lbs hardware returned to Discovery
   1,705 Lbs of middeck items transferred to ISS
    861 Lbs of middeck items returned from ISS to Discovery
   1,243 Lbs of water transferred to ISS
710,966 Mass in space of the ISS (lbs)
     84 Percentage complete of ISS assembly
- FD12: Undocked at 251:19:26:22Z
- During Entry comm blackout occurred at 255:00:38:39Z - 255:00:39:02Z due to plasma effect.
- FD15: Deorbit burn on orbit 219 for EDW landing.

SIGNIFICANT ANOMALIES:
Orbiter:
- EV2 UNACCEPTABLE COMM DURING EVA 2.
- Vernier Thruster F5R Indicates Leak In Flight
- APU 3 EGT 2 R&R
- Vernier Thruster F5R Indicates Leak In Flight
KSC:
- HANDLES ON BULK HEAD PLATES ARE LIBERATING
- STS-128 Post Launch Debris
SRB:
- DEBRIS OBSERVED NEAR HOLD DOWN POST (HDP-4) DURING ASCENT.
- RH MAIN CHUTE CANOPY DAMAGED WITH A VERTICAL TEAR EXTENDING FROM THE TOP VENT BAND TO THE CANOPY BOTTOM SKIRT BAND DURING STS-128 ON BI-139
RSRM: None.
SSME: None.
ET:
- STS-128/ET-132 REVIEW SHOWED FOAM LOSS BETWEEN +Y JACKPAD/-Y BIPOD CLOSEOUTS AT LH2/IT FLANGE
MOD: None.
Integration:
- LH₂ PV-12 Inboard Fill and Drain valve did not indicate closed when commanded
- Debris Observed Near RH SRB Aft Skirt HDP #4 Foot
- LH₂ PV-12 Inboard Fill and Drain valve did not indicate closed when commanded

FLT NO.	ORBITER	CREW (6 UP/6+1 DN) TITLE, NAMES & EVA'S	LAUNCH SITE, LIFTOFF TIME, LANDING SITES, ABORT TIMES	LANDING SITE/ RUNWAY, CROSSRANGE LANDING TIMES FLT DURATION, WINDS	SSME-TL NOM-ABORT EMERG THROTTLE PROFILE ENG. S.N.	SRB RSRM AND ET	INC	ORBIT HA/HP	FSW	PAYLOAD WEIGHTS, PAYLOADS/ EXPERIMENTS	MISSION HIGHLIGHTS (LAUNCH SCRUBS/DELAYS, TAL WEATHER, ASCENT I-LOADS, FIRSTS, SIGNIFICANT ANOMALIES, ETC.)
STS-129/ ULF3 SEQ FLT # 129 KSC-129 PAD 39A (52) MLP-3 31th SHUTTLE FLIGHT TO ISS	OV-104 (Flight 31) ATLANTIS OMS PODS LPO4-30 RPO1-38 FRC4-31	CDR: Charles O. Hobaugh (Flt 3 - STS-104, STS-118) P810/R268/V188/M234 PLT Barry E. Wilmore P811/R348/M301 MS 1 Leland Melvin (Flt-2 - STS-122) P812/R319/V209/M275 MS 2 Randy Bresnik P813/R349/M302 MS 3 Mike Foreman (Flt 2 -STS-123) P814/R324/V210/M280 MS 4, EV2 Robert Satcher, Jr. P815/R350/M303 MS 5 DN EXP20/21 FLT ENG Nicole Stott (UP STS-128) P816/R347/F47 SS EVA 145 DOCKED QUEST EVA 63 EMU/TETHERED EVA 138 SCHEDULED EVA 136 DURATION 6:37 SS EVA 146 DOCKED QUEST EVA 64 EMU/TETHERED EVA 139 SCHEDULED EVA 137 DURATION 6:08 Continued…	KSC 39A 320:19:28:10Z 1:28:01 PM CST (P) 2:28:01 PM EST (A) Monday (15) 11/16/09 (15) LAUNCH WINDOW: 9M 01S (Total) 4M 28S (Preferred) EOM PLS: KSC TAL: ZZA TAL WX: MRN, FMI (Cloud Ceiling) SELECTED: RTLS: KSC33N/N TAL: ZZA 30L N/SFD AOA: KSC 33 N/N 1ST DAY PLS: EDW 22L N/N TDEL: 0.000 (P) -0.072 (A) MAX Q NAV: 760.9 (P) 733.8 (A) SRB STG: 2:03.0 (P) 2:04.0 (A) PERF: NOMINAL 2 ENG TAL (ZZA): 2:36 (P) 2:43 (A) NEG ZZA (2@ 104): 3:52 (P) 3:57(A) PTA (U/S 157 FPS): 5:08(P) 5:09(A) SE TAL (ZZA 104): 5:57(P) 6:13(A) Continued…	KSC 33 KSC (72) 331 / 14:44:21Z 8:44:21 AM CST Saturday (24) 11/7/09 (14) DEORBIT BURN: 331:13:37:09Z XRANGE: 344.1NM ORBIT DIR: A/L (43) AIM PT: (Close-In) MLGTD: 2971 FT 331:14:44:20Z VEL: 184 KGS 197 KEAS HDOT: -2.1 FPS TD NORM 195: 2989 FT DRAG CHUTE DEPLOY: 189 KEAS 331:14:44:24Z Continued…	104/104/ 109% PREDICTED: 100/104.5/104.5/ 72/104.5 ACTUAL: 100/104.5/100/ 72/104.5 1 = 2048 (10) 2 = 2044 (12) 3 = 2058 (4) M 3 EOM: WEIGHT: 206917 LBS X CG: 1083.8 IN LANDING: WEIGHT: 207200 LBS X CG: 1084.6 IN	BI-140 RSRM 108 ET-133 SLWT 37 ET IMPACT 1:14:13 MET LAT: 36.434 S LONG: 158.531 W	51.6 (31)	DIRECT INSERTION POST OMS-2 125.0x84.8 NM DEORBIT HA 191.9 NM HP 23.3 NM ENTRY VELOCITY: 25867 FPS ENTRY RANGE: 4390.31 NM	OI-34 (2)	CARGO: 38893LBS PAYLOAD CHARGEABLE: 29372 LBS DEPLOYED: 27615 LBS NON-DEPLOYED 1404 LBS MIDDECK: 353 LBS SHUTTLE ACCUMULATED WEIGHTS: DEPLOYED: 1606885 LBS NON-DEPLOYED 1625396 LBS CARGO TOTAL: 4131080 LBS PERFORMANCE MARGINS (LBS): FPR: 2908 FUEL BIAS: 1059 FINAL TDDP: 2228 RECON: 2041 PAYLOADS: PLB: ISS-ULF3 (ELC 1, ELC 2, SASA, MISSE 7A, MISSE 7B) Continued…	Brief Mission Summary: The STS-129 (31th mission to ISS), dubbed "Stocking the Station" by PAO, main objective was to deliver nearly 14 tons of ISS systems spares. The most critical spares being transferred were two 600 lb. control moment gyros. "They've done a tremendous job of really outfitting station with all the spares that are going to be needed, essentially through its lifetime," Bill Gerstenmaier, NASA Associate Administrator for Space Operations. KSC W/D: OPF: 113 days + 10 non-workdays + 1 holiday VAB: 7 days +1 contingency day PAD A: 32 days + 2 contingency days Total Work Days = 152 (OPF processing occurred over a total time period of 124 days.) POSTPONEMENTS: - Baselined STS-129 to FDRD - launch date of 10/15/09 on 10/06/08. - Ppd. to 11/12/09 on 12/04/08. Interim manifest while HST final placement is considered. - Ppd. to 11/16/09 at10/29/09 FRR. Slip due to latest SSP planning. LAUNCH SCRUBS: None. LAUNCH WINDOW: Total launch window was 9M 01S with window open at 320:19:23:37Z and close at 320:19:32:38Z. Preferred Launch Time was 320:19:28:10Z (In-Plane Time) for a launch window of 4M 28S. LAUNCH DELAYS: - None. Launch occurred on time at 320/19:28:10Z, 2:28:10 PM EST, Monday, November 16, 2009. A cloud ceiling below 5000 feet developed early in the morning, violating flight rule limits. The ceiling lifted to above flight rule limits about 5 hours prior to launch, but continued to violate US Air Force Range Safety cloud criteria. Astronaut Steve Lindsey, flying weather reconnaissance, provided measurements of the cloud thickness for the 45th Space Wing's Launch Weather Officer and found the thickness to be acceptable about 3 hours prior to launch.. (Courtesy NWS SMG Post-Mission Summary.) Continued…

KSC-2009-5945 (28 Oct. 2009) --- NASA's new Ares I-X test rocket launches from PAD 39B as STS-129 readies for Nov. 16, 2009 launch at PAD 39A.

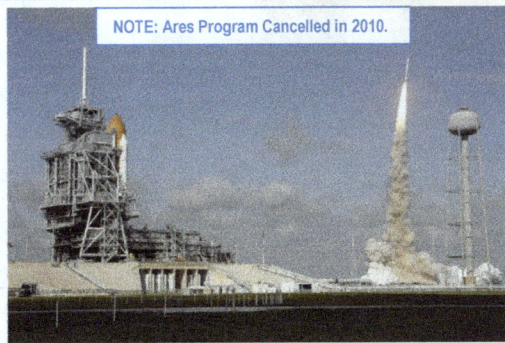

NOTE: Ares Program Cancelled in 2010.

SPACE SHUTTLE MISSIONS SUMMARY

FLT NO.	ORBITER	CREW (6 UP/6+1 DN) TITLE, NAMES & EVA'S	LAUNCH SITE, LIFTOFF TIME, LANDING SITES, ABORT TIMES	LANDING SITE/ RUNWAY, CROSSRANGE LANDING TIMES FLT DURATION, WINDS	SSME-TL NOM-ABORT EMERG THROTTLE PROFILE ENG. S.N.	SRB RSRM AND ET	INC	ORBIT HA/HP	FSW	PAYLOAD WEIGHTS, PAYLOADS/ EXPERIMENTS	MISSION HIGHLIGHTS (LAUNCH SCRUBS/DELAYS, TAL WEATHER, ASCENT I-LOADS, FIRSTS, SIGNIFICANT ANOMALIES, ETC.)
STS-129/ ULF3 Continued...		Continued... SS EVA 147 DOCKED QUEST EVA 65 EMU/TETHERED EVA 140 SCHEDULED EVA 138 DURATION 5:42	Continued... PTM (U/S 181 FPS): 6:11(P) 6:13(A) SE PRESS 104 6:56(P) 6:56 (A) MECO CMD: 8:24.2(P) 8:24.3 (A) VI: 25819(P) 25819(A) OMS-2: 37:55 (P) 38:15(A) 98.8(P) 96.7(A) FPS	Continued... BRK DECEL FPS²: AVE 6.4 PK 7.9 NLGTD: 5810 FT 331:14:44:30Z VEL: 140 KGS 150 KEAS HDOT: -5.1 FPS BRK INIT: 100 KGS DRAG CHUTE JETTISON: 51 KGS 331:14:44;25Z BRK DECEL FPS²: AVE 6.4 PK 7.9 WHEELS STOP: 331:14:45:04Z 9557 FT ROLLOUT: 6586 FT 0:44 M:S WINDS: 11H KTS -1L KTS OFFICIAL: 33011P17KTS (X1P2H11P17) DENS ALT: - 473 FT FLT DURATION: 10:19:16:14 S/T: 1236:17:41:41 OV-104: 281:23:59:12 DISTANCE: 4,490,138 sm TOTAL SHUTTLE DISTANCE: 507,595,072 sm						Continued... PAYLOADS: MIDDECK: ISS-ULF3, MAUI, SEITE, SIMPLEX, RAMBO-2 5 CRYO TANK SETS ODS, SRMS (86), OBSS	Continued... **TAL WEATHER:** Weather on launch day caused a couple minor issues at back-up site, Istres. Weather conditions at Zaragoza, the prime TAL site, and Moron were observed and forecast acceptable throughout the countdown. However, a cloud ceiling developed at Istres 2 hours prior to launch limiting the use of that landing site. (Courtesy NWS SMG Post-Mission Summary.) Istres became GO close to launch update. **PERFORMANCE ENHANCEMENTS:** Include the standard set plus: 1) PE Operational High Q - TRN/NOV, 2) OMS Assist, 3) a 52 nautical mile MECO, and 4) Del Psi **FLIGHT DURATION CHANGES:** None. Landed on KSC Runway 33 at 331:14:44:21Z, Friday, November 27, 2009, at8:24:21 CST. **FIRSTS/SECONDS/LASTS:** Second child born while astronaut dad in space. Randy Bresnik's wife, Rebecca, gave birth to Abigail Mae Bresnik, 6 lbs 13 oz, at 11:04 p.m. Saturday, Nov. 21st, in Houston. First "dad while in space" was Mike Fincke in 2004 on ISS during a 6 mo tour- a girl. First Orthopedic Surgeon in space: Dr. Robert Satcher, Jr. First flight of new variable Alt DAP First flight ET replaced LH2 ice Frost Ramp (IFR) base TPS with NCFI at 14 locations First Flight SSME Nozzle Corrosion Inhibitor Application Change First Monarch Butterflies delivered to ISS. Butterflies took flight on 12/09/09 as monitored by thousands of students back on Earth. Super Bowl XLIV opening-toss coin flown to ISS & returned. **NIGHT LAUNCH:** N/A **RENDEZVOUS: #76** Rendezvous and dock with ISS. **EVENTS:** - FD1: OMS2 ignition at 320:20:06:25Z resulted in a 125.0 by 84.8 NM orbit. - FD2: RCC inspection found no areas of concern - T1 maneuver at 322:14:05:57Z resulted in a 185.6 by179.5 NM orbit - FD3: R-Bar Pitch Maneuver was performed. No issues. - Docking Contact occurred at 322:16:51:16Z Continued...

STS129-S-027 (16 Nov. 2009) --- NASA mission managers monitor Atlantis launch in KSC FR 4.   Bill Gerstenmaier, NASA Assoc. Administrator for Space Operations is at bottom left. Photo Credit: NASA/Bill Ingalls.

ISS021-E-029824 (18 Nov. 2009) --- Atlantis loaded with spares is photographed on approach to ISS by an Expedition 21 crew member. The Russian Progress 35P spacecraft is docked at left.

MCC WHITE FLIGHT FCR (59)

FLIGHT DIRECTORS:
SHUTTLE:
A/E- Bryan Lunney
LD/O1- Mike Sarafin
O2- Gary Horlacher
Planning- PaulDye
MOD – John Mccullough
Team 4- Kwatsi Alibaruho

ISS

O1 - Emily Nelson

LD/O2 – Brian Smith

O3 – Jerry Jason

Team 4 - Heather Rarick

Continued...

FLT NO.	ORBITER	CREW (6 UP/6+1 DN) TITLE, NAMES & EVA'S	LAUNCH SITE, LIFTOFF TIME, LANDING SITES, ABORT TIMES	LANDING SITE/ RUNWAY, CROSSRANGE LANDING TIMES FLT DURATION, WINDS	SSME-TL NOM-ABORT EMERG THROTTLE PROFILE ENG. S.N.	SRB RSRM AND ET	INC	ORBIT HA/HP	FSW	PAYLOAD WEIGHTS, PAYLOADS/ EXPERIMENTS	MISSION HIGHLIGHTS (LAUNCH SCRUBS/DELAYS, TAL WEATHER, ASCENT I-LOADS, FIRSTS, SIGNIFICANT ANOMALIES, ETC.)
STS-129/ ULF3 Continued…		Continued… CAPCOMS: SHUTTLE A/E - Chris Ferguson - Steve Frick (Wx) LD/O1 - Stan Love O2 - Megan McArthur Planning - Aki Hoshide Team 4 - N/A ISS O1 - Drew Feustel LD/O2- Steve Swanson O3 – Ryan Lien Team 4 - N/A									Continued… EVENTS: Continued… - Hard Dock, hooks closed, occurred at 322:17:03:49 - ISS Hatch opened at 12:28 PM CST, Nov. 18, 2009, welcomed by ISS crew. At that time Stott ended her stay as EXP 21 FE and became an STS-129 MS. - FD4: EVA 1: Foreman & Satcher successfully completed all ISS maintenance and spares transfer tasks ahead of schedule. A get-ahead task was the most difficult. In releasing a cargo platform, a spring loaded device jammed and had to be manhandled to achieve release. EVA1 duration 6:37. - MMT concurred that no Focused Inspection of Orbiter was required. - FD6: EVA2: Russian false depress event overnight, but EVA2 was conducted on time. Foreman & Bresnik completed all nominal tasks plus the following get-aheads: S3 Nadir/Inboard PAS Deploy, SGANT Y-cable check (CHIT 8025) , Tool stanchion relocation to P1 WIF 3, & APFR 5 retrieve. EVA2 duration 6:08. - FD8: EVA3: Satcher & Bresnik: EVA-3 started one hour late due to EV2's drink bag valve coming loose. All tasks successfully completed included: transfer of HPGT & MISSE & from ExPRESS Logistics Carrier 2 to Quest airlock. Towards the end of the EVA two [unknown] items were lost overboard at 327:17:37Z. All tools were accounted for. EVA3 duration (PET) 5:42. - Hard Dock, hooks closed, occurred at 322:17:03:49 - ISS Hatch opened at (12:28 PM CST, Nov. 18, 2009) welcomed by ISS crew. At that time Stott ended her stay as EXP 21 FE and became an STS-129 MS. -Transfers: 31,789 Pounds of hardware transferred to station (inside & out) 40 Pounds of Oxygen "transferred" (pumped) into ISS cabin 11 Pounds of Nitrogen transferred into ISS tanks 2,211 Pounds of middeck items delivered to ISS 2,110 Pounds of middeck items returned from ISS ~1,400 Pounds of water transferred to ISS - Mass in space of the ISS 759,222 pounds - ISS assembly: 86 Percentage complete - FD10: Undocked at 329:09:53:02Z - During Entry there was no RF blackout. It was avoided by a handover to the Eastern TDRS early, then a handover to the ground station. Continued…

ISS021-E-032724-– (24 Nov. 2009) Portrait Time: Twelve internationally-represented astronauts and cosmonauts spend time together in space. The group includes the seven STS-129 astronauts CDR Hobaugh, PLT Wilmore; & Mission Specialists Stott, Foreman, Melvin, Satcher, & Bresnik, plus the five ISS crewmembers: Jeffrey Williams, Frank De Winne/ESA, Robert Thirsk/CSA and Russia's FSA Roman Romanenko & Maxim Suraev.

------------------ SPACEMEN AT WORK ------------------

ISS021-E-030165 (19 Nov. 2009) Foreman installing a spare S-band antenna structural assembly to the Z1 segment of the station's truss. EVA 1.

S129-E-007762 " New Dad In Space", Bresnik, installing a Grappling Adaptor to On-Orbit Railing Assembly (GATOR) on Columbus Lab. EVA 2. (21 Nov. 2009)

S129-E-008103 (23 Nov. 2009) Satcher moves debris shields from Quest airlock to the External Stowage Platform #2. EVA 3.

SPACE SHUTTLE MISSIONS SUMMARY

FLT NO.	ORBITER	CREW (6 Up/6+1 DN) TITLE, NAMES & EVA'S	LAUNCH SITE, LIFTOFF TIME, LANDING SITES, ABORT TIMES	LANDING SITE/ RUNWAY, CROSSRANGE LANDING TIMES FLT DURATION, WINDS	SSME-TL NOM-ABORT EMERG THROTTLE PROFILE ENG. S.N.	SRB RSRM AND ET	INC	ORBIT HA/HP	FSW	PAYLOAD WEIGHTS, PAYLOADS/ EXPERIMENTS	MISSION HIGHLIGHTS (LAUNCH SCRUBS/DELAYS, TAL WEATHER, ASCENT I-LOADS, FIRSTS, SIGNIFICANT ANOMALIES, ETC.)

STS-129/ ULF3
Continued…

S129-E-009497 (24 Nov. 2009) --- Nicole Stott/MS takes one of her final "strolls" through the ISS modules on the eve of her departure from the orbital outpost.

JSC2009e240939 --- In MCC, Tim Oram with the Space Flight Meteorology Group gathers data for weather forecast.

JSC2009-E-244757 --- In MCC, Joshua Byerly/PAO narrates mission post undocking activities.

against the background of Earth's horizon and the blackness of space.

JSC2009-E-243548--- The members of the STS-129 Ascent Flight Control Team pose for a group portrait in MCC at JSC. Flight Director Bryan Lunney and Flight Controller Christi Worstell hold the STS-129 logo.

Continued…

SIGNIFICANT ANOMALIES:
Orbiter:
- WASTE DUMP STOPPED PREMATURELY. THE WASTE WATER DUMP INITIATED POST-UNDOCK AT APPROX. 329/12:07:38 GMT, EXHIBITED A NOMINAL WASTE DUMP RATE (APPROX. 2.0 %//MIN) UNTIL APPROX. 329/12:19:36 GMT WHEN THE WASTE DUMP RATE DEGRADED TO 0.3/ %/MIN. WASTE DUMP WAS TERMINATED BY CLOSING THE DUMP VALVE AND NOZZLE WAS REHEATED TO APPROX. 258 DEG F. DUMP VALVE WAS THEN OPENED AT 329/12:35:34 GMT FOR CONTINUATION OF THE DUMPING OPERATION. THE OBSERVED DUMP RATE CONTINUED OFF-NOMINALLY AT NEAR 0 %/MIN AND THE WASTE DUMP WAS TERMINATED AFTER 19 MINUTES. THIS IFA IS CONSIDERED A CONSTRAINT TO STS-132/ULF4 (NEXT FLIGHT OF OV-104), BUT IS EXPECTED TO BE RESOLVED WITH A DUMP LINE FILTER CHANGE.
- APU WATER TANK HEATER A (50V46HR01A) DID NOT OPERATE AT EXPECTED TEMP. APU WATER TANK TEMP
- LRCS BFS FUEL AND OXIDEZER QUANTITIES INCREASED OFF NOMINAL
KSC: None.
SRB:
RH SOLID ROCKET BOOSTER AFT SKIRT FOAM ON THE OUTBOARD SIDE OF HOLDDOWN POST M2 NEAR THE GN2 PURGE LINE IS OBSERVED TO CRACK DURING LIFTOFF
RSRM: None.
SSME: None.
ET: None.
MOD: None.
Integration:
- Unexpected Debris/Expected Debris Exceeding Mass Allowable Prior to Pad Clearance (Liftoff Debris)
- Single Transient SRB I/O Error at Liftoff

FLT NO.	ORBITER	CREW (6) TITLE, NAMES & EVA'S	LAUNCH SITE, LIFTOFF TIME, LANDING SITES, ABORT TIMES	LANDING SITE/ RUNWAY, CROSSRANGE LANDING TIMES FLT DURATION, WINDS	SSME-TL NOM-ABORT EMERG THROTTLE PROFILE ENG. S.N.	SRB RSRM AND ET	INC	ORBIT HA/HP	FSW	PAYLOAD WEIGHTS, PAYLOADS/ EXPERIMENTS	MISSION HIGHLIGHTS (LAUNCH SCRUBS/DELAYS, TAL WEATHER, ASCENT I-LOADS, FIRSTS, SIGNIFICANT ANOMALIES, ETC.)
STS-130/ 20A	OV-105 (Flight 24) ENDEAVOUR	CDR: George D. Zamka (Flt 2 - STS-120) P817/R315/V211/M271	KSC 39A 39:09:14:07Z 4:14:07 AM EST (P) 4:14:07 AM EST (A) Monday (16) 02/08/10 (10)	KSC15 KSC (73) 053:03:20:29Z 9:20:29 PM CST Sunday (17) 02/21/10 (8)	104/104/ 109%	BI-141	51.6 (32)	DIRECT INSERTION	OI-34 (3)	CARGO: 40956 LBS	**Brief Mission Summary:** The STS-130 (32nd mission to ISS) main objectives were to deliver and assemble the final U.S. module (Tranquility) and the Italian built Cupola Node plus delivery of ISS equipment, supplies, and experiments. Tranquility provides additional room for the ISS crew and life support systems. The Cupola is a robotic control station and provides a panoramic view of earth through 7 windows, "A Room With a View" - PAO. The mission included 3 EVA's.
SEQ FLT # 130	OMS PODS LPO3 -34 RPO4 30 FRC5-23	PLT Terry W. Virts, Jr. P818/R351/M304	LAUNCH WINDOW: 11M 57S (Total) 7M 32S (Preferred)	DEORBIT BURN: 053:02:14:47Z XRANGE: 336.9NM	PREDICTED: 100/104.5/104.5/ 72/104.5 ACTUAL: 100/104.5/100/ 74/104.5	RSRM 109 ET-134 SLWT 38		POST OMS-2 124.0x110.08 NM DEORBIT HA 190.3 NM HP 23.3 NM		PAYLOAD CHARGEABLE: 34931 LBS DEPLOYED: 34648 LBS	**KSC W/D** OPF-2: 130 days + 3 holidays VAB-1: 9 days + 5 contingency days +11 holidays PAD A: 31 days + 3 contingency days Total Work Days = 170 (OPF processing occurred over a total time period of 133 days.)
KSC-130 PAD 39A (53)		MS 1 Kathyrn P. Hire (Flt 2 - STS-90) P819/R238/V212/F31	LAUNCH WINDOW: EOM PLS: KSC TAL: ZZA TAL WX: MRN (NO GO), FMI (NO GO)	ORBIT DIR: A/L (44) AIM PT: (Close-In)	1 = 2059 (4) 2 = 2061 (1) 3 = 2057 (5)	ET IMPACT 1:13:54 MET		ENTRY VELOCITY: 25866 FPS		NON-DEPLOYED: 0 LBS MIDDECK: 283 LBS	**POSTPONEMENTS:** - Baselined STS-130 to FDRD - launch date of 12/10/09 on 11/17/08. - Ppd. to 02/04/10 on 03/10/09. Interim change while Cx and SSP schedules were assessed and prioritized. - Ppd. to 02/07/10 on 12/17/09. Launch date change supports efficient use of KSC ground operation resources.
MLP-2 32nd SHUTTLE FLIGHT TO ISS		MS 2 Stephen K. Robinson (Flt 4 - STS-85, STS-95, STS-114) P820/R222/V152/M196	SELECTED: RTLS: KSC15 N/N TAL: ZZA 30L N/N AOA: KSC 15 N/N 1ST DAY PLS: EDW 22R N/N TDEL: 0.000 (P) 0.232 (A)	MLGTD: 2760 FT 053:03:20:29Z VEL: 188 KGS 190 KEAS HDOT: -1.9 FPS TD NORM 195: 2405 FT	M 3 EOM: WEIGHT: 201138 LBS X CG: 1082.8 IN LANDING: WEIGHT: 201084 LBS X CG: 1084.8 IN	LAT: 37.192 S LONG: 159.603 W		ENTRY RANGE: 4367.5 NM		SHUTTLE ACCUMULATED WEIGHTS: DEPLOYED: 1641533 LBS NON-DEPLOYED: 1626311 LBS	**LAUNCH SCRUBS:** Sunday, 02/07/10 launch attempt was terminated about an hour before scheduled launch of 4:40 AM EST. Launch scrub was due to a massive area of low cloud ceilings that blanked the northern half of Florida.. launch was reset for 02/08/10. WEATHER SCRUB.
		MS 3 Nicholas J. M. Patrick (Flt 2 - STS-105, STS-116) P821/R303/V186/M263	MAX Q NAV: 757.6 (P) 756.6 (A) SRB STG: 2:05.9 (P) 2:07.2 (A) PERF: NOMINAL	DRAG CHUTE DEPLOY: 185 KEAS 053:03:20:31Z NLGTD: 5219 FT 053:03:20:36Z VEL: 157 KGS 158 KEAS HDOT: -6.2 FPS	iss022e062672					CARGO TOTAL: 4210929 LBS PERFORMANCE MARGINS (LBS): FPR: 2908 FUEL BIAS: 1059 FINAL TDDP: 1188 RECON: 2828	**LAUNCH WINDOW:** Total launch window was 11M 57S with window open at 39:09:09:42Z and close at 39:09:21:39Z. Preferred Launch Time was 39:09:14:07Z (In-Plane Time) for a launch window of 7M32S. **LAUNCH DELAYS:** None. Launch occurred on time at 39:09:14:07Z on Monday 02/08/10.
		MS 4 Robert L. Behnken (Flt 2 - STS-123) P822/R323/V213/M279	2 ENG TAL (ZZA): 2:42 (P) 2:43 (A) NEG ZZA (2@ 104): 3:52 (P) 3:54(A) PTA (U/S 160 FPS): 5:08(P) 5:06(A) SE TAL (ZZA 104): 6:02(P) 6:00(A) Continued…	BRK INIT: 113 KGS DRAG CHUTE JETTISON: 54 KGS 255:00:54:06Z BRK DECEL FPS²: AVE 2.7 PK 10.1 WHEELS STOP: 053:03:22:00Z 12966 FT Continued…	Shuttle approaches ISS with Node 3/Cupola. iss022-e-068832					PAYLOADS: PLB: ISS-20A (NODE 3 W/CUPOLA) MIDDECK: ISS-20A, MAUI, SEITE, SIMPLEX, RAMBO-2 5 CRYO TANK SETS ODS, SRMS (87), OBSS, SSPTS, SPDUS	**TAL WEATHER:** Spaceflight Meteorology Group (SMG) reported "quite challenging" weather for TAL sites: low clouds & showers at Moron & showers in 20 circle at ZZA. Recon aircraft at ZZA reported moisture (not rain droplets) so TAL "rain shower rule " was invoked for "GO". Istres changed form "GO" to "NO GO" (Low cloud ceiling) late in launch count. **PERFORMANCE ENHANCEMENTS:** Include the standard set plus: 1) PE Operational High Q - WIN/FEB, 2) OMS Assist, 3) a 52 nautical mile MECO, and 4) Del Psi Continued…
		SS EVA 148 DOCKED QUEST EVA 66 EMU/TETHERED EVA 141 SCHEDULED EVA 139 DURATION 6:32 SS EVA 149 DOCKED QUEST EVA 67 EMU/TETHERED EVA 142 SCHEDULED EVA 140 DURATION 5:53 SS EVA 150 DOCKED QUEST EVA 68 EMU/TETHERED EVA 143 SCHEDULED EVA 141 DURATION 5:48 Continued…									

FLT NO.	ORBITER	CREW (6) TITLE, NAMES & EVA'S	LAUNCH SITE, LIFTOFF TIME, LANDING SITES, ABORT TIMES	LANDING SITE/ RUNWAY, CROSSRANGE LANDING TIMES FLT DURATION, WINDS	SSME-TL NOM-ABORT EMERG THROTTLE PROFILE ENG. S.N.	SRB RSRM AND ET	INC	ORBIT HA/HP	FSW	PAYLOAD WEIGHTS, PAYLOADS/ EXPERIMENTS	MISSION HIGHLIGHTS (LAUNCH SCRUBS/DELAYS, TAL WEATHER, ASCENT I-LOADS, FIRSTS, SIGNIFICANT ANOMALIES, ETC.)
STS-130/ 20A Continued…		Continued… MCC WHITE FLIGHT FCR (60) CAPCOMS: SHUTTLE A/E - Rick Sturckow - Steve Frick (Wx) LD/O1 - Danny Olivas - Rick Sturckow (Flt Days 3 & 12) O2 - Mike Massimino Planning - Shannon Lucid Team 4 - N/A ISS O1 - Robert Hanley LD/O2- Hal Getzelman O3 – Kathy Bolt FLIGHT DIRECTORS: SHUTTLE: A/E- Norm knight LD/O1- Kwatsi Alibaruho O2- Gary Horlacher Planning- Chris Edelen MOD – John Mccullough Team 4- Paul Dye ISS O1 - Royce Renfrew LD/O2 - Bob Dempsey O3 - Mike Lammers Team 4 - Dana Weigel	Continued… PTM (U/S 181 FPS): 6:10(P) 6:12(A) SE PRESS 104 6:57(P) 6:56 (A) MECO CMD: 8:22.5 (P) 8:21.4 (A) VI: 25819(P) 25817(A) OMS-2: 37:44 (P) 37:42(A) 143.4(P) 142.1(A) FPS								Continued… **FLIGHT DURATION CHANGES**: On FD6 *MMT agreed to add +1 day to nominal flight plan to facilitate complete transfer of the regen ECLSS racks to Node 3 as well as assist with accomplishing other flight objectives.* Landed on KSC Runway 15 at 053:03:20:29Z, Sunday, February 21, 2010 at 9:20:29 CST. **FIRSTS/LASTS:** - Shuttle's last night launch. - Last U.S. on-orbit Segment (Node 3) installed on ISS. - Orbiter: First flight of Main Engine Ignition Overpressure Acoustic Instrumentation. - First lunar rock returned to space. The sample was collected on Apollo 11 by Neil Armstrong in 1969 and carried by Scott Parazynski (Shuttle astronaut) in 2009 on his climb of Mt. Everest. Now on ISS, it orbits Earth once again. **NIGHT LAUNCH**: # 34 **NIGHT LANDING KSC #17: (#23 in Shuttle history)** **RENDEZVOUS: #77** Rendezvous and dock with ISS. **EVENTS:** - FD1: OMS2 ignition at 039:09:51:49Z resulted in a 124.0 by 110.0 NM orbit. - FD2: During RCC surveys the crew downlinked some views of pulled up portion of port wing upper surface flapper door seal area. Area was cleared. - T1 maneuver at 041:02:28:25Z resulted in a 187.4 by180.7 NM orbit - FD3: R-Bar Pitch Maneuver was performed. No issues. MMT concurred no focus inspection required. - Docking Contact occurred at 041:05:05:56Z - Hard Dock, hooks closed, occurred at 041/05:54:12Z - ISS Hatch opened at 1:16 AM CST Wednesday, Feb. 10, 2010, welcomed by ISS crew. - FD4: EVA 1: Behnken & Patrick successfully completed preparations for unberthing Tranquility (Node 3). ISS arm unberthed Node 3 & installed it on Node 1 port side followed by crew activation. EVA1 duration 6:32. - FD7: EVA2: Behnken & Patrick All planned activities were completed including installation of the ammonia jumpers, integrating Node 3 to EATCS Loop A, and installing the Node 3 port center disc cover (CDC). Cupola was successfully relocated. EVA2 duration 5:53. Continued…

Prelaunch in JSC MOCR, Flight Dynamics Officer (FDO) Mark McDonald works on abort landing site plannning.

Endeavour launch as seen in time lapse photo from top of the Intracoastal Waterway Bridge in Ponte Vedra, FL, 115 Miles from the launch site, Monday, February 8, 2010 @ 4:14 am EST. Photo by: James Vernacotola, copyright 2010: www.jamesvernacotola.com

FLT NO.	ORBITER	CREW (6) TITLE, NAMES & EVA'S	LAUNCH SITE, LIFTOFF TIME, LANDING SITES, ABORT TIMES	LANDING SITE/ RUNWAY, CROSSRANGE LANDING TIMES FLT DURATION, WINDS	SSME-TL NOM-ABORT EMERG THROTTLE PROFILE ENG. S.N.	SRB RSRM AND ET	INC	ORBIT HA/HP	FSW	PAYLOAD WEIGHTS, PAYLOADS/ EXPERIMENTS	MISSION HIGHLIGHTS (LAUNCH SCRUBS/DELAYS, TAL WEATHER, ASCENT I-LOADS, FIRSTS, SIGNIFICANT ANOMALIES, ETC.)
STS-130/ 20A Continued…				Continued…							Continued…

ROLLOUT:
10206 FT
1:31 M:S

WINDS:
5H KTS 0.3R KTS
OFFICIAL:
16007P10KT
(X1P1 H7P10)

DENS ALT: 410 FT

FLT DURATION:
13:18:06:22

S/T:
1250:11:48:03

OV-105:
280:09:39:23

DISTANCE:
5,738,991 sm

TOTAL SHUTTLE DISTANCE:
513,386,662 sm

ISS construction and maintenance continue.
Above: ISS022-E-062844 -- Patrick during EVA1.
Below: ISS022-E-065750 -- Behnken during EVA 2

ISS022-E-067727 --- Crews for STS-130 (red) & Exp 22 (blue) in Harmony node. Front row (lt to rt): Exp 22 CDR Jeffrey Williams, Patrick/MS, CDR Zamka, & Behnken/MS. Middle row: Exp 22 Soichi Noguchi/FE (JAXA), Hire/MS, & Exp 22 T.J. Creamer/FE. Back row: Maxim Suraev & Oleg Kotov, both Exp 22/FE (RSA); along with Robinson/MS & PLT Virts.

2010-02-17-0001Hq --- U.S. President Barack Obama, with members of Congress and middle school pupils, waves goodbye to Shuttle crew from the White House.

EVENTS: Continued…
- FD8: Cupola unberthed and moved from forward end to nadir port of Tranquility.
- FD10: EVA3: Behnken & Patrick All planned and a number of get ahead tasks were completed including Loop B QD opening (integration of EATCS Loop B with Node 3 heat exchanger), PMA-3 cable installation, Cupola MLI removal, and VSC video cable routing. EVA3 duration (PET) 5:48.

-Transfers:
36,130 Pounds of hardware transferred to ISS (inside & out)
29,788 Tranquility Node 3 weight in pounds (as installed)
3,594 Cupola
757 Integrated Stowage Platform cargo
24 Pounds of Oxygen transferred into ISS Airlock tanks
0 Pounds of Nitrogen transferred (N2 was used to repress the stack)
1,991 Pounds of middeck items delivered to ISS aboard Endeavour
1,803 Pounds of middeck items returned from ISS to Endeavour
~1,095 Pounds of water transferred to ISS

799,045 Mass in space of the International Space Station (in pounds)

- FD13: Undocked at 051:00:53:52Z

- During entry a manual handover to TDRS-46 early avoided rolling on to a lower antenna and prevented a comm blackout period.

Continued…

FLT NO.	ORBITER	CREW (6) TITLE, NAMES & EVA'S	LAUNCH SITE, LIFTOFF TIME, LANDING SITES, ABORT TIMES	LANDING SITE/ RUNWAY, CROSSRANGE LANDING TIMES FLT DURATION, WINDS	SSME-TL NOM-ABORT EMERG THROTTLE PROFILE ENG. S.N.	SRB RSRM AND ET	INC	ORBIT HA/HP	FSW	PAYLOAD WEIGHTS, PAYLOADS/ EXPERIMENTS	MISSION HIGHLIGHTS (LAUNCH SCRUBS/DELAYS, TAL WEATHER, ASCENT I-LOADS, FIRSTS, SIGNIFICANT ANOMALIES, ETC.)
STS-130/ 20A Continued…			Quoting Oscar Wilde's "Life imitates art far more than art imitates life", Dave Zani - CinemaBlend.com, sees the Cupola window as the inside window of a Star Wars TIE Fighter.								Continued… **SIGNIFICANT ANOMALIES:** Orbiter: - During STS-130 Ascent monitoring, WLE Sensor Unit S/N 1155 experienced two (2) off-scale high data spikes. - MUX bypass switch will not switch to Bypass front for OCA 48Mbps downlinks. - Audio drop-out during EVA 1. - Trajectory Control Sensor (TCS) had trouble transitioning to CW mode. CW data became ratty and unusable. KSC: 12 IFA's entitled "STS-130 Post Launch Debris" SRB: None. RSRM: None. SSME: None. ET: - POST-FLIGHT REV. IDENT. 2 FOAM LOSSES +Z SIDE INTERTANK NCFI 24-124 ACREAGE, 19 FOAM LOSSES ?Z SIDE OF THE INTERTANK NCFI 24-12 ACREAGE MOD: - INCORRECT TAL RUNWAY SURFACE IN FLIGHT RULE Integration: None.

ISS022-E-067184 --- Behnken (left) & Patrick removing insulation blankets & launch bolts from Cupola's windows.

S130-E-010380--- Soichi Noguchi/ JAXA/FE ISS Exp 22, takes earth photo from a window in Cupola.

ISS022-E-068724 -- CDR Zamka tries out view from Cupola.

JSC2010-E- 017955 ---- Flight Directors in JSC MCC: From left: Chris Edelen, Norm Knight, Kwatsi Alibaruho and Gary Horlacher.

S130-E-012188 --- ISS as seen by Endeavour post-undocking and separation. Tranquility & Cupola are located just left of center.

STS130-S-128 --- Drag chute is deployed at MLGTD on KSC Runway 15 at 10:20:29 PM EST on Feb. 21, 2010. It was the 23rd night landing in Shuttle history and the 17th at KSC.

FLT NO.	ORBITER	CREW (7) TITLE, NAMES & EVA'S	LAUNCH SITE, LIFTOFF TIME, LANDING SITES, ABORT TIMES	LANDING SITE/ RUNWAY, CROSSRANGE LANDING TIMES FLT DURATION, WINDS	SSME-TL NOM-ABORT EMERG THROTTLE PROFILE ENG. S.N.	SRB RSRM AND ET	INC	ORBIT HA/HP	FSW	PAYLOAD WEIGHTS, PAYLOADS/ EXPERIMENTS	MISSION HIGHLIGHTS (LAUNCH SCRUBS/DELAYS, TAL WEATHER, ASCENT I-LOADS, FIRSTS, SIGNIFICANT ANOMALIES, ETC.)
STS-131/19A SEQ FLT # 131 KSC-131 PAD 39A (54) MLP-3 33rd SHUTTLE FLIGHT TO ISS	OV-103 (Flight 38) DISCOVERY OMS PODS LPO1 -41 RPO3-39 FRC3-38	CDR: Alan G. Poindexter (Flt 2- STS-122) P823/R318/V214/M274 PLT James P. Dutton, Jr. P824/R352/M305 MS 1 Rick Mastracchio (Flt 3 - STS-106, STS-118) P825/R257/V189/M224 MS 2 Dorthy Metcalf-Lindenburger P826/R353/F48 MS 3 Stephanie Wilson (Flt 3 - STS-121, STS-120) P827/R298/V190/F39 MS 4 Naoko Yamazaki (JAXA) P828/R354/F49 MS5 Clayton Anderson (Flt 2-UP ON STS-117STAY ISS, DN ON STS-120) P829/R310/V215/ M268 SS EVA 151 DOCKED QUEST EVA 69 EMU/TETHERED EVA 144 SCHEDULED EVA 142 DURATION 6:27 SS EVA 152 DOCKED QUEST EVA 70 EMU/TETHERED EVA 145 SCHEDULED EVA 143 DURATION 7:26 SS EVA 153 DOCKED QUEST EVA 71 EMU/TETHERED EVA 146 SCHEDULED EVA 144 DURATION 6:24 Continued…	KSC 39A 95:10:21:25Z 6:21:25 AM EDT (P) 6:21:25 AM EDT (A) Monday (17) 04/05/10 (16) LAUNCH WINDOW: Dual pane day with window open at 95:10:18:40Z and close at 95:10:27:17Z 5M 52S (Preferred) EOM PLS: KSC TAL: ZZA TAL WX: MRN FMI (NO GO) SELECTED: RTLS: KSC33 N/N TAL: MRN20 N/N AOA: KSC33 N/N 1ST DAY PLS KSC15 N/N TDEL: 0.000 (P) 0.142 (A MAX Q NAV: 708.0 (P) 700.5 (A) SRB STG: 2:04.8 (P) 2:05.8 (A) PERF: NOMINAL 2 ENG TAL (MRN): 2:36 (P) 2:41 (A) NEG MRN (2@ 104): 3:47 (P) 3:54(A) PTA (U/S 157 FPS): 5:17(P) 5:206(A) SE TAL (ZZA 104): 6:02(P) 6:03(A) Continued…	KSC33 KSC (74) 110:13:08:34Z 8:08:34 AM CDT Tuesday (18) 04/20/10 (12) DEORBIT BURN: 110:12:02:59Z XRANGE: 20.4 NM ORBIT DIR: D/L (50) AIM PT: NOMINAL MLGTD: 3559 FT 110:13:08:34Z VEL: 198 KGS 198 KEAS HDOT: -1.6 FPS TD NORM 195: 2955 FT DRAG CHUTE DEPLOY: 191 KEAS 110:13:08:36Z NLGTD: 6398 FT 110:13:08:43Z VEL: 157 KGS 160 KEAS HDOT: -4.4 FPS BRK INIT: 107 KGS DRAG CHUTE JETTISON: 58 KGS 110:13:09:31Z BRK DECEL FPS²: AVE 5.2 PK 7.0 WHEELS STOP: 110:13:09:32Z 11886 FT ROLLOUT: 8327 FT 0:58 M:S Continued…	104/104/ 109% PREDICTED: 100/104.5/104.5/ 72/104.5 ACTUAL: 100/104.5/100/ 72/104.5 1 = 2045 (11) 2 = 2060 (2) 3 = 2054 10) M 3 EOM: WEIGHT: 224257 LBS X CG: 1089.0 IN LANDING: WEIGHT: 224206 LBS X CG: 1090.7 IN	BI-142 RSRM 110 ET-135 SLWT 39 ET IMPACT 1:13:55 MET LAT: 37.233 S LONG: 159.667 W	51.6 (33)	DIRECT INSERTION POST OMS-2 140.0x123.8 NM DEORBIT HA 190.6 NM HP 14.2 NM ENTRY VELOCITY: 25862 FPS ENTRY RANGE: 4480 NM	OI-34 (4)	CARGO: 39516 LBS PAYLOAD CHARGEABLE: 32131 LBS DEPLOYED: 30512 LBS NON-DEPLOYED: 1388 LBS MIDDECK: 231 LBS SHUTTLE ACCUMULATED WEIGHTS: DEPLOYED: 1672045 LBS NON-DEPLOYED: 1627930 LBS CARGO TOTAL: 4250445 LBS PERFORMANCE MARGINS (LBS): FPR: 2908 FUEL BIAS: 1059 FINAL TDDP: 1133 RECON: 1491 PAYLOADS: PLB: ISS-19A (MPLM,LMC), TRIDAR AR&D SENSOR DTO-701A MIDDECK: ISS-19A, MAUI, SEITE, SIMPLEX, RAMBO-2 5 CRYO TANK SETS ODS, SRMS (88), OBSS, SSPTS,	Brief Mission Summary: The STS-131 (33rd mission to ISS), dubbed "Experiment Express" by PAO, main objectives were to bring some 8 tons of supplies and scientific equipment to ISS, remove & replace a depleted Ammonia tank, and return a large load of experiments and no longer useful gear back to earth. KSC W/D OPF: 142 days + 11 holidays VAB: 9 days + 0 contingency days PAD A: 32 days + 2 contingency days Total Work Days = 183 (OPF processing occurred over a total time period of 153 days POSTPONEMENTS: - Baselined STS-131 to FDRD - launch date of 03/18/10 on 02/05/09. - Ppd. to 04/05/10 on 03/09/10. Due to cold weather conditions, Orbiter rollover from the OPF to VAB was delayed such that the March 18, 2010 launch date could not be met. LAUNCH SCRUBS: None LAUNCH WINDOW: Dual pane day with window open at 95:10:18:40Z and close at 95:10:27:17Z. Preferred Launch Time was 95:10:21:25Z (In-Plane Time) for a launch window of 5M52S. LAUNCH DELAYS: None. Launch occurred on time at 95:10:21:25Z on Monday 04/05/10. TAL WEATHER: Spaceflight Meteorology Group (SMG) reported a pressure gradient between a high & a departing low contributed to winds at Istres above headwind limits. Only high cirrus clouds prevailed at both Zaragoza & Moron with winds well within flight rule limits. Weather was "GO". PERFORMANCE ENHANCEMENTS: Include the standard set plus: 1) PE Operational High Q - TRN/APR, 2) OMS Assist, 3) a 52 nautical mile MECO, and 4) Del Psi FLIGHT DURATION CHANGES: - FD 4: MMT approved plan for conducting a docked late inspection using +1 day - extended mission from 12 to 13 days. - Landing postponed 1 day due to unstable weather. Weather was still unsatisfactory next day with fog and area showers for first opportunity. Weather cleared for "Go" on 2nd opportunity at KSC. Landing occurred at 110:13:08:34Z, Tuesday, April 20, 2010, at 8:08:34 AM CDT Continued…

ISS023-E-020718 --- ISS robotic Canadarm2 relocates Leonardo (MPLM) from Discovery's PLB to port on Harmony node.

FLT NO.	ORBITER	CREW (7) / TITLE, NAMES & EVA'S	LAUNCH SITE, LIFTOFF TIME, LANDING SITES, ABORT TIMES	LANDING SITE/ RUNWAY, CROSSRANGE / LANDING TIMES, FLT DURATION, WINDS	SSME-TL NOM-ABORT EMERG / THROTTLE PROFILE ENG. S.N.	SRB RSRM AND ET	INC	ORBIT HA/HP	FSW	PAYLOAD WEIGHTS, PAYLOADS/ EXPERIMENTS	MISSION HIGHLIGHTS (LAUNCH SCRUBS/DELAYS, TAL WEATHER, ASCENT I-LOADS, FIRSTS, SIGNIFICANT ANOMALIES, ETC.)
STS-131/19A Continued…		Continued… MCC WHITE FLIGHT FCR (61) CAPCOMS: SHUTTLE A/E - Rick Sturckow - George Zamka/Wx LD/O1 - Rick Sturckow O2 - Aki Hoshide Planning - Megan McArthur - Chris Cassidy Team 4 - N/A ISS O1 - Mike Jensen LD/O2 - Stan Love O3 - Marcus Reagant Team 4 – N/A FLIGHT DIRECTORS: SHUTTLE: A/E- Bryan Iunney LD/O1- Richard Jones O2- Mike Sarafin Planning- Ginger Kerrick MOD – John Mccullough Team 4- Gary Horlacher ISS O1 - Courtenay McMillan LD/O2 - Roy Spencer O3 - Ed Van Cise Team 4 - Brian Smith	Continued… PTM (U/S 180 FPS): 6:20(P) 6:29(A) SE PRESS 104 6:58(P) 7:01 (A) MECO CMD: 8:22.5 (P) 8:23.5 (A) VI: 25819(P) 25816(A) OMS-2: 37:16 (P) 37:14(A) 197.2(P) 196.5(A) FPS	Continued… WINDS: 2.1H KTS 2.2R KTS OFFICIAL: 02003P05KT (X0P0 H5P6) DENS ALT: 908 FT FLT DURATION: 15:02:47:09 S/T: 1265:14:35:12 OV-103: 347:03:20:09 DISTANCE: 6,232,235 sm TOTAL SHUTTLE DISTANCE: 519,613,765 sm						Continued… FIRSTS/LASTS: - Last return trip for MPLM Leonardo. After STS-133 it will remain on ISS as a permanent fixture. - First time for four women living in space. - First time for two Japanese astronauts in space together. - First special cookies from the Italian Café in Seabrook, TX, requested originally by Col. Timothy Creamer after a 6-month ISS tour, were delivered to ISS. The sand tarts passed NASA tests with the request to go light on the powdered sugar. NIGHT LAUNCH: N/A RENDEZVOUS: #78 Rendezvous and dock with ISS. EVENTS: - FD1: OMS2 ignition at 095:10:58:39Z resulted in a 140.0 by 123.8 NM orbit. - FD2: During RCC surveys showed no areas of concern. - T1 maneuver at 097:05:06:44Z resulted in a 189.3 by181.7 NM orbit - Ku Band failed. - FD3: R-Bar Pitch Maneuver was performed. Four areas of interest were identified: 1) RSB Trailing Edge Tile, 2) FWD Gap Filler, 3) Port ET Door Tile Chip, 4) three closely grouped OMS POD tile damage sites. The Damage Assessment Team later cleared these areas for entry and MMT concurred no focus inspection required. - Crew executed the radar fail procedures for rendezvous after the system failed to respond to a last attempt early in the rendezvous. - Docking Contact occurred at 097:07:44:09Z - Hard Dock, hooks closed, occurred at 097:07:58:52Z - ISS Hatch opened at 4:11 AM CDT April 7, 2010, welcomed by ISS crew. - FD4: MPLM was grappled, unberthed, and installed on the Node 2 Nadir without issue. - FD5: EVA 1:Mastracchio & Anderson remove old ATA and handover new ATA to SSRMS, retrieve JEM SEED, & R&R RGA. EVA1 duration 6:27. Continued…

STS131-S-050 --- NASA Commentator Mike Curie and astronaut Kathryn (Kay) Hire discuss launch in LCC Firing Room 4 at KSC.

S...-E-010002 --- STS-131 & EXP 23 crews ... in ... ISS ... Lab STS-131 crew members pictured (light blue shirts) are CDR Poindexter, PLT Dutton; Anderson/MS, Mastracchio/MS, Metcalf-Lindenburger/MS, Wilson/MS, & Yamazaki/MS (JAXA). EXP 23 crew members are CDR Oleg Kotov (RSA), Mikhail Kornienko/FE (RSA), Alexander Skvortsov/FE (RSA); Soichi Noguchi/FE (JAXA), T.J. Creamer/FE (USA), & Tracy Caldwell Dyson/FE (USA).

In JSC MCC, Carson Sparks/FDO (Flight Dynamics Officer) in foreground & Tom Schmidt/GPO (Guidance & Procedures Officer), in rear, working launch data updates one hour prior to launch.

FLT NO.	ORBITER	CREW (7) TITLE, NAMES & EVA'S	LAUNCH SITE, LIFTOFF TIME, LANDING SITES, ABORT TIMES	LANDING SITE/ RUNWAY, CROSSRANGE LANDING TIMES FLT DURATION, WINDS	SSME-TL NOM-ABORT EMERG THROTTLE PROFILE ENG. S.N.	SRB RSRM AND ET	INC	ORBIT HA/HP	FSW	PAYLOAD WEIGHTS, PAYLOADS/ EXPERIMENTS	MISSION HIGHLIGHTS (LAUNCH SCRUBS/DELAYS, TAL WEATHER, ASCENT I-LOADS, FIRSTS, SIGNIFICANT ANOMALIES, ETC.)
STS-131/19A Continued…											

AT LEFT:
S131-E-008710 -- Mastracchio (left) & Anderson conduct 2nd EVA during which they unhooked and removed depleted ammonia tank and installed a 1,700-pound ammonia tank on ISS Starboard 1 truss. Crew had problems with bolting down the new ATA tank on S1. They eventually got all 4 bolts secured, however, the time required to do this resulted in several tasks dropping off this EVA.

S131E008710

S131-E-009456 --- Mastracchio (right) & Anderson conduct 3rd & final session of EVA. Activities included fluid lines hookup of new 1,700-pound ammonia tank and prepared cables on the Zenith 1 truss for a spare Space to Ground Ku-Band antenna.

S131-E-007954 --- First time four women in space shown in the Zvezda Service Module: clockwise from lower left: are Tracy Caldwell Dyson/FE EXP 23, Metcalf-Lindenburger/MS, Yamazaki/MS(JAXA), & Wilson/MS.

Continued…

EVENTS: Continued
- FD6&7: EVA2: Mastracchio & Anderson had difficulty installing new ATA onto S1 truss due to sticky plungers on bolt 4. Numerous workarounds were employed and eventually the bolt did cooperate. Alignment of the bolts and soft dock mechanisms are orientation sensitive and the task took much more time than booked. Several tasks were not completed & were rescheduled to EVA 3. EVA2 duration 7:26.
- FD9: EVA3: Mastracchio & Anderson completed: S1 ATA Fluid connectors (from EVA 2), Retrieve A/L MMOD shields (from EVA 2), Old ATA transfer to the LMC in Shuttle payload bay (all 4 bolts were engaged, though the last bolt required extra time due to some alignment challenges), & S1 ATA FGB install. EVA3 duration (PET) 6:24.
- **FD9; Monday, April 12th celebrated the 49th Anniversary of the Soviet cosmonaut, Yuri Gagarin, first human to orbit the earth in 1961 and the 29th Anniversary of the first U.S. Space Shuttle launch in 1981.**
- Transfers:
  - ~15,222 Lbs of hardware transferred to ISS (inside & out)
  - 12,060 Lbs of MPLM supplies & logistics transferred to ISS
  - 4,109 Lbs of MPLM supplies & logistics returned from ISS
  - 1,702 Lb Ammonia Tank Assembly (ATA) delivered to ISS
  - 1,295 Lb ATA (old) returned from ISS
  - 94.5 Lbs of O_2 used to repress the stack
  - 1,460 Lbs middeck items delivered to ISS
  - 1,235 Lbs of middeck items returned from ISS to Discovery
  - 6,639 Lbs of total hardware returned aboard Discovery
  - 975 Lbs of water transferred to ISS

  - 806,282 Mass (Lbs) of ISS now in space
  - 98 Percentage complete of ISS assembly (pressurized volume)

- FD13: Undocked at 107:12:52:10Z

- During entry comm blackout times were approx 110/12:49:15 to 12:54:34 (~ 5.5 min). Early H/O to TDRS 46 was not an option as TDRS 46 stayed on a lower antenna. INCO prediction of LOS was in error due to DOL PAD error, noted in Significant Anomalies below. Also, see Ascent/Entry Flight Techniques Panel #255 of April 30, 2010.

Continued…

FLT NO.	ORBITER	CREW (7)	LAUNCH SITE, LIFTOFF TIME,	LANDING SITE/ RUNWAY, CROSSRANGE	SSME-TL NOM-ABORT EMERG	SRB RSRM	ORBIT		FSW	PAYLOAD WEIGHTS,	MISSION HIGHLIGHTS (LAUNCH SCRUBS/DELAYS,
		TITLE, NAMES & EVA'S	LANDING SITES, ABORT TIMES	LANDING TIMES FLT DURATION, WINDS	THROTTLE PROFILE ENG. S.N.	AND ET	INC	HA/HP		PAYLOADS/ EXPERIMENTS	TAL WEATHER, ASCENT I-LOADS, FIRSTS, SIGNIFICANT ANOMALIES, ETC.)
STS-131/19A Continued…											

ABOVE: JSC2010-E-045167 --- Flight Directors for the STS-131/19A: From the left are Tony Ceccacci, Bryan Lunney, Paul Dye, Richard Jones, Ginger Kerrick and Mike Sarafin.

BELOW: JSC2010-E-051978 -- STS-131 Orbit 2 Flight Control Team pose in JSC MCC. FD Mike Sarafin holds mission logo.

Discovery's planned approach and landing track across the continental U.S. Photo courtesy JSC/PAO.

SHUTTLE051109 049(KSC)--- Discovery on approach to KSC Runway 33 on April 20, 2010, after weather waveoffs on April 19th and again on first opportunity of April 20th.

Continued…

SIGNIFICANT ANOMALIES:
Orbiter:
- CCTV Camera C zoom not functioning
- DURING STS-131, KU-BAND FAILED FROM POWER UP FOR BOTH COMM AND RADAR OPERATIONS.
- NIRD 131-005, D-131-RPM-410-001: DEBRIS EVENT DURING ACCENT AT 42SEC MET FROM PORT UPPER RSB TRAILING EDGE. TILE HAS BROKEN AWAY, APPEARS TO BE PARTIAL LIBERATION. VISIBLE CHARRING ALONG THE AFT EDGE.
- LRCS fuel helium ISO B valve slow to close during post wave off system reconfigure.
- FRCS fuel helium ISO A valve slow to close during post entry valve test.
KSC:
- STS-131 Post Launch Debris
SRB:
- UPLOADED ACCELEROMETER DATA FROM THE S/N 2000003 DAS SHOWED 446 SECONDS OF PREFLIGHT TESTING FOLLOWED BY THE FIRST 94 SECONDS OF FLIGHT DATA
RSRM: None.
SSME:
- ME-2 HPFTP 21 DEGREE ACCEL DISQUALIFIED @ T+7:19
ET: None.
MOD:
- INCORRECT COMM PREDICTS DUE TO PADS ERROR
Integration:
- Base Heat Shield TPS Liberation
- Windows 5, 6 Missing/Protruding Ceramic Plugs
- Rudder Speedbrake TPS Liberation

FLT NO.	ORBITER	CREW (6) TITLE, NAMES & EVA'S	LAUNCH SITE, LIFTOFF TIME, LANDING SITES, ABORT TIMES	LANDING SITE/ RUNWAY, CROSSRANGE LANDING TIMES FLT DURATION, WINDS	SSME-TL NOM-ABORT EMERG THROTTLE PROFILE ENG. S.N.	SRB RSRM AND ET	INC	ORBIT HA/HP	FSW	PAYLOAD WEIGHTS, PAYLOADS/ EXPERIMENTS	MISSION HIGHLIGHTS (LAUNCH SCRUBS/DELAYS, TAL WEATHER, ASCENT I-LOADS, FIRSTS, SIGNIFICANT ANOMALIES, ETC.)
STS-132/ULF4 SEQ FLT # 132 KSC-132 PAD 39A (55) MLP-2 34th SHUTTLE FLIGHT TO ISS	OV-104 (Flight 32) ATLANTIS' LAST SCHEDULED FLIGHT OMS PODS LPO4-31 RPO1-39 FRC4-31	CDR: Kenneth T. Ham (Flt 2- STS-124) P830/R326/V216 /M282 PLT Dominic A. Antonelli (Flt 2 - STS-119) P831/R334/V217M289 MS 1 Garrett Reisman (Flt 2 -Up on STS-123, stay ISS, DN STS-124) P832/R325/V 218/M281 MS 2 Michael Good (Flt 2 STS-125) P833/R338/V219/M293 MS 3 Steve Bowen (Flt 2 - STS-126) P834/R332/V220M287 MS 4 Piers Sellers (Flt 3 (STS-112, STS-121)) P835/R285/V182/M249 SS EVA 154 DOCKED QUEST EVA 72 EMU/TETHERED EVA 147 SCHEDULED EVA 145 DURATION 7:25 SS EVA 155 DOCKED QUEST EVA 73 EMU/TETHERED EVA 148 SCHEDULED EVA 146 DURATION 7:09 SS EVA 156 DOCKED QUEST EVA 74 EMU/TETHERED EVA 149 SCHEDULED EVA 147 DURATION 6:46 Continued...	KSC 39A 134:18:20:09Z 2:20:09 PM EDT (P) 2:20:09 PM EDT (A) Friday (28) 05/14/10 (9) LAUNCH WINDOW: 10M 01S (Total) 5M 01S (Preferred) EOM PLS: KSC TAL: ZZA TAL WX: MRN FMI (NO GO) SELECTED: RTLS: KSC33 N/N TAL:ZZA30 CI/N AOA: KSC33 N/N 1ST DAY PLS EDW22 N/N TDEL: 0.000 (P) 0.162 (A) MAX Q NAV: 722.4 (P) 708.3 (A) SRB STG: 2:02.7 (P) 2:05.0 (A) PERF: NOMINAL 2 ENG TAL (MRN): 2:42 (P) 2:36 (A) NEG MRN (2@ 104): 3:56(P) 3:58(A PTA (U/S 157 FPS): 4:45(P) 4:56(A) Continued...	KSC33 KSC (75) 146:12:48:08Z 7:48:08 AM CDT Saturday (25) 05/26/10 (12) DEORBIT BURN: 146:11:41:59Z XRANGE: 611.3 NM ORBIT DIR: A/L (45) AIM PT: Close-In MLGTD: 2919 FT 146:12:48:08Z VEL: 193 KGS 194 KEAS HDOT: -1.9 FPS TD NORM 195: 3174 FT Continued...	104/104/ 109% PREDICTED: 100/104.5/104.5/ 72/104.5 ACTUAL: 100/104.5/72/ 104.5 1 = 2052 (9) 2 = 2051 (8) 3 = 2047 (14) M 3 EOM WEIGHT: 210434 LBS X CG: 1081.0 IN LANDING WEIGHT: 210370 LBS X CG: 1082.9 IN	BI-143 RSRM 111 ET-136 SLWT 40 ET IMPACT 1:14:24 MET LAT: 35.906S LONG: 157.809W	51.6 (34)	DIRECT INSERTION POST OMS-2 125.1x85.2 NM DEORBIT HA 195.4 NM HP 23.6 NM ENTRY VELOCITY: 25877 FPS ENTRY RANGE: 4334 NM	OI-34 (5)	CARGO: 35963 LBS PAYLOAD CHARGEABLE: 26740 LBS DEPLOYED: 26619 LBS NON-DEPLOYED: 0 LBS MIDDECK: 121 LBS SHUTTLE ACCUMULATED WEIGHTS: DEPLOYED: 1698664 LBS NON-DEPLOYED: 1628051 LBS CARGO TOTAL: 4286408 LBS PERFORMANCE MARGINS (LBS): FPR: 2908 FUEL BIAS: 1059 FINAL TDDP: 5074 RECON:4326 PAYLOADS: PLB: ISS-ULF4 (MRM1,ICC-VLD, ICAPC/PDGF0 MIDDECK: ISS-ULF4, MAUI, SEITE, SIMPLEX, RAMBO-2 5 CRYO TANK SETS, ODS, SRMS (89), OBSS, SSPTS	*Brief Mission Summary:* The STS-132 (34th mission to ISS), dubbed "Finishing Touches" by PAO, main objectives were to conduct three Eva's, deliver & install the 2nd Russian Mini-Research Module, a complement of batteries, a backup Ku-band antenna, and other ISS supplies. *This was the last scheduled flight of Atlantis; however ,Congress later approved one more flight, see STS-135.* **KSC W/D** OPF: 127 days + 9 holidays VAB: 7 days + 2 Wx days PAD A: 22 days + 1 contingency day Total Work Days = 156 (OPF processing occurred over a total time period of 136 days. **POSTPONEMENTS:** - Baselined STS-132 to FDRD - launch date of 05/13/10 on 04/02/09. - Ppd. to 05/14/10 on 05/04/09. ISS request to de-conflict dynamic vehicle events of a Soyuz undocking and Orbiter docking on the same day. **LAUNCH SCRUBS:** None **LAUNCH WINDOW:** Window open at 134:18:15:09Z and close at 134:18:25:10Z. Preferred Launch Time was 134:18:20:09Z (In-Plane Time) for a launch window of 5M01S. **LAUNCH DELAYS:** None."It's a beautiful day in Florida to bid *'Bon Voyage'* to the good ship *Atlantis* on its sunset cruise. " -KjH (Space Shuttle Program Public Affairs). Launch occurred on time at 134:18:20:09Z on Friday 05/14/10. **TAL WEATHER:** Spaceflight Meteorology Group (SMG) reported a high pressure ridge provided benign weather at KSC for launch and RTLS. Things were trickier for TAL Sites with low pressure system resulting in breezy conditions at ZZA & MRN. By launch winds decreased below Flight Rule limits. At Istres rains remained outside of 20 NM watch area . Weather was "GO". **PERFORMANCE ENHANCEMENTS:** Include the standard set plus: 1) PE Operational High Q - TRN/MAY, 2) OMS Assist, 3) a 52 nautical mile MECO, and 4) Del Psi **FLIGHT DURATION CHANGES:** None. Continued...

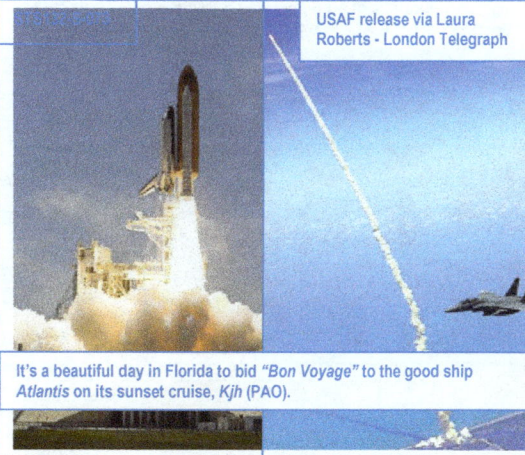

USAF release via Laura Roberts - London Telegraph

It's a beautiful day in Florida to bid *"Bon Voyage"* to the good ship *Atlantis* on its sunset cruise, *Kjh* (PAO).

SPACE SHUTTLE MISSIONS SUMMARY

FLT NO.	ORBITER	CREW (7) TITLE, NAMES & EVA'S	LAUNCH SITE, LIFTOFF TIME, LANDING SITES, ABORT TIMES	LANDING SITE/ RUNWAY, CROSSRANGE LANDING TIMES, FLT DURATION, WINDS	SSME-TL NOM-ABORT EMERG THROTTLE PROFILE ENG. S.N.	SRB RSRM AND ET	INC	ORBIT HA/HP	FSW	PAYLOAD WEIGHTS, PAYLOADS/ EXPERIMENTS	MISSION HIGHLIGHTS (LAUNCH SCRUBS/DELAYS, TAL WEATHER, ASCENT I-LOADS, FIRSTS, SIGNIFICANT ANOMALIES, ETC.)

STS-132/ULF4
Continued…

OV-104

Continued…

MCC WHITE FLIGHT FCR (62)

CAPCOMS:
SHUTTLE
A/E - Charlie Hobaugh
 - Steve Frick (Wx)
LD/O1 - Chris Cassidy
O2 - Stan Love
Planning - Shannon Lucid
Team 4 - N/A
ISS
O1 - Zach Jones
LD/O2 - Steve Swanson
O3 – Rob Hayhurst
Team 4 - N/A

FLIGHT DIRECTORS:
SHUTTLE:
Ascent- Richard Jones
LD/O1- Mike Sarafin
O2- Chris Edelen
Planning- Ginger Kerrick
Entry - Tony Ceccacci
MOD – John Mccullough
Team 4 - Paul Dye
ISS
O1 - Holly Ridings
LD/O2 - Emily Nelson
O3 - Dina Contella
Team 4 - Royce Renfrew

Continued…

SE TAL (ZZA 104):
6:02(P)   6:02(A)

PTM (U/S 181 FPS):
5:48(P)   5:59(A)

SE PRESS 104
6:51(P)   6:53 (A)

MECO CMD:
8:24.1 (P) 8:25.6 (A)

VI:
25819(P) 25819(A)

OMS-2:
37:47 (P) 38:15(A)
98.8(P) 97.4(A) FPS

Continued…

DRAG CHUTE DEPLOY: 190 KEAS
146:12:48:10Z

NLGTD: 6227 FT
146:12:48:19Z
VEL: 135 KGS
141 KEAS
HDOT: -5.0 FPS

BRK INIT: 59 KGS

DRAG CHUTE JETTISON:
57 KGS
146:12:48:47Z

BRK DECEL FPS²:
AVE 2.7 PK 4.1

WHEELS STOP:
146:12:49:27Z
12019 FT
ROLLOUT:
9100 FT
1:19 M:S
WINDS:
8 H KTS 2 L KTS

OFFICIAL:
31508P11 (X 3p4
HD 8p10)

DENS ALT: 1652 FT

FLT DURATION:
11:18:27:59

S/T:
1277:09:03:11

OV-104:
294:18:27:11

Continued…

ABOVE: ISS023-E-050819 --- STS-132 (blue shirts) and Exp 23 crews (red shirts) on ISS. Front: Exp CDR Oleg Kotov/RSA (ctr lt), CDR Ham (ctr rt), with (from lt) T.J. Creamer/FE Exp 23, Good/MS, Alexander Skvortsov (RSA)/FE Exp 23, & Reisman/MS. Back (from lt): Bowen/MS, Tracy Caldwell Dyson/FE Exp 23, Piers Sellers/MS, Mikhail Kornienko (RSA)/FE, PLT Antonelli, & Soichi Noguchi (JAXA)/FE Exp 23.

BELOW: --- While preparing for the routine inspection of Atlantis' TPS on FD2, crew discovered a pinched cable preventing the sensor package pan and tilt unit from moving properly. During EVA 2 the crew successfully unsnagged & tied off cable. Photo courtesy K. Herring/PAO.

Possible contributor to snag

Continued…

FIRSTS/LASTS:
- **Last scheduled flight of Atlantis.**
- The Mini Research Module 1 (MRM1), aka Rassvet, is first & only major piece of Russian H/W that U.S. hauled to ISS.
- First evaluation of Commercial Compression Garments to prevent post-spaceflight Orthostatic Intolerance.
- First RSRM incorporation of V1288 fluorocarbon O-rings in nozzle joints 4 and 5.

TENTH SHUTTLE CREWMEMBER REPLACEMENT
- Karen Nyberg (medical condition) was replaced by Michael Good in August 2009. (9th Shuttle crewmember replacement occurred on STS-118.)

NIGHT LAUNCH: N/A

RENDEZVOUS: #79 Rendezvous and dock with ISS.

EVENTS:
- **Gerst**:"The entire team gave us a great launch...nice ET… [only] one small piece of foam late in ascent."
- FD1: OMS2 ignition at 134:18:58:24Z resulted in a 125.1 by 85.2 NM orbit.
- FD2: During RCC surveys a camera cable was wedged between camera & OBSS structure limiting tilt capability. This left gaps in RCC survey. Ops team developed plan to get docked imagery and cable assess during EVA.[Post mission: It was determined that the snag was attributed to cable S/N unique memory characteristics. Cable was replaced with a different S/N cable.]
- T1 maneuver at 136:11:40:09Z resulted in a 189.7 by184.8 NM orbit
- FD3: R-Bar Pitch Maneuver was performed.
- Docking Contact occurred at 136:14:28:25Z.
- Hard Dock, hooks closed, occurred at 136:14:40:49Z.
- ISS Hatch opened at 11:18 AM CDT May 16, 2010, welcomed by ISS crew.
- FD4: EVA 1: Reisman & Bowen installed SGANT & EOPT EVA1 duration 7:25.
- FD5 Russian MRM1 successfully unberthed and docked to ISS.
- FD6: EVA2: Bowen & Good successfully completed all tasks: cleared cable from the Orbiter LDRI tilt axis, installed 4 new batteries in truss 3 old batteries into pallet, & stowed a temp. EVA2 duration 7:09.
- FD8: EVA 3: Good & Garrett activities included: completion of batteries R&R's, P6 cleanup, & PDGF trial. EVA3 duration (PET) 6:46.

Continued…

ISS023E044569 -- Atlantis on 'Final Approach' to ISS with Russian MRM1.

FLT NO.	ORBITER	CREW (7)	LAUNCH SITE, LIFTOFF TIME,	LANDING SITE/ RUNWAY, CROSSRANGE	SSME-TL NOM-ABORT EMERG	SRB RSRM		ORBIT	FSW	PAYLOAD WEIGHTS,	MISSION HIGHLIGHTS (LAUNCH SCRUBS/DELAYS,
		TITLE, NAMES & EVA'S	LANDING SITES, ABORT TIMES	LANDING TIMES FLT DURATION, WINDS	THROTTLE PROFILE ENG. S.N.	AND ET	INC	HA/HP		PAYLOADS/ EXPERIMENTS	TAL WEATHER, ASCENT I-LOADS, FIRSTS, SIGNIFICANT ANOMALIES, ETC.)

STS-132/ULF4
Continued...

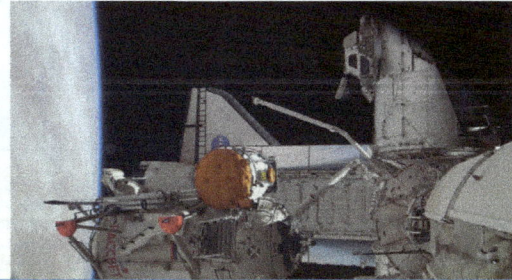

ISS023-E-047488 --- In the grasp of ISS Canadarm2, Russian-built Mini-Research Module 1 (MRM-1) is moved for permanent attachment to ISS FGB. Named Rassvet, Russian for "dawn," the module is the second in a series of new pressurized cargo storage components for Russia. Rassvet also gives ISS an additional docking port.

Continued...

DISTANCE:
4,879.978 sm

TOTAL SHUTTLE DISTANCE:
524,493,743 sm

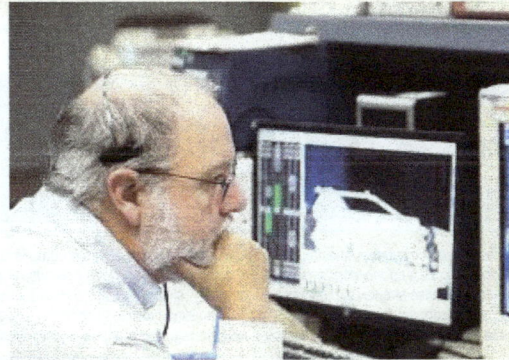

William "Bill" Miller/DX/USA honored "In the [MOD] Spotlight" for significant contributions to Shuttle Ops. On STS-132 he provided great effort in development of alternate survey procedures when the laser sensor package camera was snagged prohibiting it from tilting correctly, see previous page.

Continued...

EVENTS: Continued...
- Transfers:
  - 28,792 Lbs H/W transferred to ISS (inside & out) includes MRM1 "Rassvet" - loaded (17,670 Lbs)
  - 7,573 Lbs ICC with supplies to ISS
  - 6,466 Lbs ICC with supplies from ISS
  - 42 Lbs Oxygen to ISS
  - 30 Lbs Oxygen to ISS (stack repress)
  - 10.5 Lbs Nitrogen to ISS
  - 1,325 Lbs water to ISS
  - 2,192 Lbs middeck items to ISS aboard Atlantis
  - 1,763 Lbs middeck items returned from ISS aboard Atlantis
  - 8,229 Lbs total H/W returned aboard Atlantis includes ICC
- 816,349 Mass (Lbs) of ISS now in space
- Undocked at 143:15:22:04Z
- During entry comm outage time due to blackout was 146/12:32:00Z - 12:34:30Z (~ MET 011/18:12 - 18:14:30). S/W handover to TS 46 was not available as TS 46 was on a Lower Antenna resulting in plasma blackout. This was well advertised. At 12:34:30Z due to Roll Reversal, TS 46 satellite works over to upper antenna and regains comm. Comm through Mila was available at 12:36:00Z with hand down to Mila at 12:37:00Z.

Continued...

LEFT: ISS023-E-032398 --- Soichi Noguchi (JAXA) ISS EXP 23 FE, photographed the Mississippi Delta showing the BP oil slick in the Gulf of Mexico on May 4, 2010. Part of the river delta and nearby Louisiana coast appear dark in the sunlight. Location of oil rig is out of frame to the left. USGS Comment: "Worst oil spill in U.S. history."

S132-E-008106 -- Bowen during first EVA with Reisman (out of frame), continues construction and maintenance on the ISS, with battery replacements & installation of a 2nd Ku-band antenna.

S132-E008900 -- Good (foreground) & Reisman, are surrounded by ISS hardware during the flight's final EVA.

FLT NO.	ORBITER	Atlantis was named after the primary research vessel for the Woods Hole Oceanographic Institute in Massachusetts from 1930 to 1966. The two-masted, 460-ton ketch was the first U.S. vessel to be used for oceanographic research. Such research was considered to be one of the last bastions of the sailing vessel as steam-and-diesel-powered vessels dominated. [From STS-132 Press Kit by PAO]	MISSION HIGHLIGHTS (LAUNCH SCRUBS/DELAYS, TAL WEATHER, ASCENT I-LOADS, FIRSTS, SIGNIFICANT ANOMALIES, ETC.)
STS-132/ULF4 Continued			

STS-132/ULF4 Continued

S132-E-012208 -- Alantis bids final farewell to ISS!

Shuttle Legacy Mural - Hanging in LCC Firing Room at KSC

ATLANTIS TRIBUTE: From Mike Leinbach/Launch Director/KSC

Continued…

SIGNIFICANT ANOMALIES:
Orbiter:
During Flight, a FES Shutdown Occurred While Operating on the Primary B Controller. Reference: MER-09
KSC:
- STS-132 Post Launch Debris
SRB:
- LEFT-HAND SRB FRUSTUM UPPER RIGHT BSM ROOM TEMPERATURE VULCANIZATION (RTV) 133 IS MISSING, MEASURING 5? LONG
RSRM: None.
SSME: None.
ET:
- STS-132/ET-136 FOAM LOSS ON THE +Z SIDE OF THE INTERTANK
MOD: None.
Integration:
- Unexpected Debris/Expected Debris Exceeding Mass Allowable Prior to Pad Clearance (Liftoff Debris)
- Ice Observed on the T-0 Umbilical at Retraction

ABOVE RIGHT: KSC-2010-4450 (http://mediaarchive.ksc.nasa.gov/index.cfm). This Tribute Display features *Atlantis* soaring above the earth. *Atlantis* flew seven missions to space station Mir. In addition to its many assembly, construction, and resupply missions to the International Space Station, *Atlantis* also flew the last Hubble Space Telescope servicing mission on STS-125. The planet Venus represents the Magellan probe deployed during STS-30, and the planet Jupiter represents the Galileo probe deployed during STS-34. Threaded through the design are the mission patches for each of *Atlantis'* flights. The inset photos illustrate various aspects of space shuttle processing as well as significant achievements such as the "glass cockpit" and the first shuttle docking with Mir during STS-71. The inset photo in the upper left corner shows a rainbow over *Atlantis* on Pad A and *Endeavour* on Pad B. *Endeavour* was the assigned vehicle had *Atlantis'* STS-125 mission needed rescue, and this was the last time both launch pads were occupied simultaneously. The stars in the background represent the many people who have worked with *Atlantis* and their contributions to the vehicle's success.

CONGRESS ALLOWS ATLANTIS TO FLY AGAIN - SEE STS-135

----------- ATLANTIS NOW HEADS TO STS-335 RESCUE MISSION PREP THEN TO THE BARN/MUSEUM! -----------

"Space Shuttle Atlantis comes home to the Kennedy Space Center for the final time, 25 years, 32 flights, and more than 120 million miles traveled; the legacy of Atlantis, now in the history books," Commentator Josh Byerly remarked from his console in Houston. NASA Photos courtesy: *Susan Phipps Multimedia Librarian/AP3 JSC*

FLT NO.	ORBITER	------------------------------ SOME OF THE OPERATIONS SUPPORT TEAM ------------------------------
STS-132/ULF4 Continued…	OV-104 Atlantis	

STS132-S-012 (14 May 2010) --- Secretary of Defense Dr. Robert M. Gates, right, NASA Associate Administrator for Space Operations William H. Gerstenmaier, center, and other NASA mission managers monitor the last scheduled launch of Space Shuttle Atlantis from Firing Room 4 at KSC.

JSC2010-E-086698 -- Orbit 1 FCT: Flight Director Mike Sarafin (center) on front row.

JSC2010-E-086451-- Orbit 2 FCT: Flight Director Chris Edelen (second left) on front row.

JSC2010-E-086504 -- Orbit 3 FCT: Flight Director Ginger Kerrick (right) holds mission logo.

JSC2010-E-087358 -- Entry FCT Flight Director Tony Ceccacci holds mission logo.

**Lonnie Schmitt -
First "Century Club" Controller**

(From: collectSPACE.com - Robert Pearlman) - CDR Ken Ham joined in with past and present members of MCC Thursday morning [May 20, 2010] to recognize Lonnie Schmitt as the first Flight Controller to reach his 100th shuttle mission. *"This is truly a momentous occasion," radioed Ham from onboard Atlantis. "We were just kicking this around on the flight deck here between us who have spent a lot of time in MCC as Capcom and know a lot of the flight controllers and offhand, we can't come up with any other individual that we know of that has been around as a flight controller since STS-1."*

JSC2010-E-080436 ---Kyle J. Herring (left) & Joshua Byerly, both PAO commentators, on JSC MCC consoles during launch countdown.

JSC2010-E-063832-- ISS FD's: Left (front row) Emily Nelson & Scott Stover. Back row: Royce Renfrew & Holly Ridings.

JSC2010-E-045162 --- STS FD's: From left: Chris Edelen, Richard Jones, Mike Sarafin, Ginger Kerrick & Tony Ceccacci.

JSC2010-E-090665-- Ascent FCT: FD Richard Jones (right) & STS-132 CDR Ken Ham hold the mission logo.

FLT NO.	ORBITER	CREW (6) TITLE, NAMES & EVA'S	LAUNCH SITE, LIFTOFF TIME, LANDING SITES, ABORT TIMES	LANDING SITE/ RUNWAY, CROSSRANGE LANDING TIMES FLT DURATION, WINDS	SSME-TL NOM-ABORT EMERG THROTTLE PROFILE ENG. S.N.	SRB RSRM AND ET	INC	ORBIT HA/HP	FSW	PAYLOAD WEIGHTS, PAYLOADS/ EXPERIMENTS	MISSION HIGHLIGHTS (LAUNCH SCRUBS/DELAYS, TAL WEATHER, ASCENT I-LOADS, FIRSTS, SIGNIFICANT ANOMALIES, ETC.)
STS-133/ULF5 SEQ FLT # 133 KSC-133 PAD 39A (56) MLP-3 35th SHUTTLE FLIGHT TO ISS	OV-103 (Flight 39) Discovery's **LAST FLIGHT** OMS PODS LPO1 -42 RPO3-40 FRC3-39	CDR Steven W. Lindsey (Flt 5 - STS-87, STS-95, STS-104, STS-121) P836/R229/V131/M200 PLT Eric A. Boe (Flt 2 - STS-126) P837/R331/V 221/M286 MS 1 Alvin Drew (Flt 2 STS-118) P838/R314/V221/M270 MS 2 Steve Bowen (Flt 3 - STS-126, STS-131) P839/R332/V220M287 MS 3 Michael Barratt (TMA-14 ISS EXP 19 & 20) P840/R355/M306 MS 4 Nicole Stott (Flt 2 - Up STS-128 stay ISS Dn STS-129) P841/R347/V223/F47 SPECIAL PASSENGER Robonaut 2 First dexterious humanoid robot in space - stay ISS Continued...	KSC 39A 055:21:53:24Z 4:50:27 PM EST (P) 4:53:24 PM EST (A) Thursday (35) 02/24/11 (11) LAUNCH WINDOW: 6M 02S (Total) 3M02S (Preferred) EOM PLS: KSC TAL: ZZA TAL WX: MRN , FMI SELECTED: RTLS: KSC15 CI/N TAL: ZZA30 CI/N AOA: KSC15 CI/N 1ST DAY PLS EDW22 CI/N (Briefed to crew) KSC15 CI/N (Go Wx) TDEL: 0.000 (P) 0.092 (A) MAX Q NAV: 714.8 (P) 710.4 (A) SRB STG: 2:05.9 (P) 2:06.9 (A) PERF: NOMINAL 2 ENG TAL (MRN) 2:41 (P) 2:44 (A PERF: NOMINAL NEG MRN (2@ 104): 3:54(P) 3:56(A PTA (U/S 160 FPS): 5:24(P) 5:15(A) Continued...	KSC15 KSC (76) 068:16:57:15Z 10:57:15 AM CST Wednesday (17) 03/09/11 (11) DEORBIT BURN: 068:15:52:04Z XRANGE: 24.8 NM ORBIT DIR: A/R (15) AIM PT: Close-In MLGTD: 2446 FT 068:16:57:15Z VEL: 180 KGS 197 KEAS HDOT: -1.4 FPS TD NORM 195: 2645 FT Continued...	104/104/ 109% PREDICTED: 100/104.5/104.5/ 72/104.5 ACTUAL: 100/104.5/72/ 104.5 1 = 2044 (13) 2 = 2048 (9) 3 = 2058 (5) M 3 EOM WEIGHT: 205011 LBS X CG: 1082.4 IN LANDING: WEIGHT: 205022 LBS X CG: 1084.3 IN	BI-144 RSRM 112 ET-137 SLWT 41 w/Stringer Mod ET IMPACT 1:14:20 MET LAT: 35.535S LONG: 158.000W	51.6 (35)	DIRECT INSERTION POST OMS-2 125.5x84.9 NM DEORBIT HA 192.9 NM HP 23.2 NM ENTRY VELOCITY: 25868 FPS ENTRY RANGE: 4387 NM	OI-34 (6)	CARGO: 40108 LBS PAYLOAD CHARGEABLE: 31802 LBS DEPLOYED: 30576 LBS NON-DEPLOYED 818 LBS MIDDECK: 408 LBS SHUTTLE ACCUMULATED WEIGHTS: DEPLOYED: 1729240 LBS NON-DEPLOYED 1629277 6LBS CARGO TOTAL: 4326516 LBS PERFORMANCE MARGINS (LBS): FPR: 2821 FUEL BIAS: 954 FINAL TDDP: 1481 RECON: 394 PAYLOADS: PLB: ISS-ULF 5 (ELC 4,PMM), LWAPA MIDDECK: ISS-ULF 5, MAUI, SEITE, SIMPLEX, RAMBO-2 4 CRYO TANK SETS, ODS, SRMS (89), OBSS, SSPTS	*Brief Mission Summary:* The STS-133 (35th mission to ISS) delivered two key components to ISS – the Italian-built Permanent Multipurpose Module (PMM) and Express Logistics Carrier 4 (ELC4) – for spare parts and storage capacity. Also delivered was Robonaut 2, the first dexterous humanoid robot in space. *This was the final flight of the most flown Orbiter, Discovery (39 flights) - The Beginning of the END!* **KSC W/D** OPF: 138 days + 3 holidays VAB HB3 (part 1):10 days + 2 contingency days PAD A (part 1): 82 days + 8 contingency days = 2 holidays (rolled back for ET repairs) VAB (part 2): 35 days + 5 holidays PAD A (part 2): 19 days + 5 contingency Total Work Days = 284 (OPF processing occurred over a total time period of 141 days **POSTPONEMENTS:** - Baselined STS-133 to FDRD - launch date of 07/29/10 on 06/30/09. - Ppd. to 09/16/10 on 09/30/09. Adjustments needed for flight product planning. - Ppd to 11/01/10 on 07/01/10. Slip was required to complete preparations of critical spares that will be launched in the Permanent Multi-Purpose Module (PMM). **LAUNCH SCRUBS:** - Launch scrubbed on 10/29/10 due to helium & nitrogen leaks discovered in the right OMS pod. Launch rescheduled for 11/02/10. On 10/30/10 launch rescheduled to 11/03/10 to allow additional time for reloading the helium tank after repair in the right OMS pod. Technical scrub. - Launch scrubbed on 11/02/10 at L-1 MMT meeting due to problem with center SSME controller. Launch rescheduled for 11/04/10. Technical scrub. - Launch scrubbed on 11/04/10 at tanking MMT meeting due to predictions of bad weather. Launch rescheduled for 11/05/10. Weather scrub. - Launch scrubbed on Friday, 11/05/10 when a liquid hydrogen leak was detected about 6:30 a.m. CDT in the Ground Umbilical Carrier Plate (GUCP). Mike Moses, MMT Chair stated: "This is not a stranger to us – we saw this on STS-119 and STS-127." In addition to the leak, a crack was detected on the flange of the ET intertank near the oxygen tank. To allow time for engineering analyses of these issues, for compatibility with on orbit sun angles, and for avoidance of other space traffic to/from ISS, the launch was reset for NET 11/30/10. Continued...

ROBONAUT 2

Delivered to ISS on STS-133. Flight Controllers successfully "awakened" Robonaut 2 on August 23, 2011.

Giant Crawler Carries Shuttle To Pad
ABOVE: STS-133 rolls to PAD on 09/20/10 for first launch attempt - scrubbed on 10/29/10. The KSC crawler-transporters (two) carried all Apollo Saturn V's and all Shuttle vehicles on the gravel path from the VAB to Launch Complex 39.

FLT NO.	ORBITER	CREW (7) TITLE, NAMES & EVA'S	LAUNCH SITE, LIFTOFF TIME, LANDING SITES, ABORT TIMES	LANDING SITE/ RUNWAY, CROSSRANGE LANDING TIMES FLT DURATION, WINDS	SSME-TL NOM-ABORT EMERG THROTTLE PROFILE ENG. S.N.	SRB RSRM AND ET	INC	ORBIT HA/HP	FSW	PAYLOAD WEIGHTS, PAYLOADS/ EXPERIMENTS	MISSION HIGHLIGHTS (LAUNCH SCRUBS/DELAYS, TAL WEATHER, ASCENT I-LOADS, FIRSTS, SIGNIFICANT ANOMALIES, ETC.)
STS-133/ULF5 Continued,,,	OV-103	Continued… SS EVA 157 DOCKED QUEST EVA 75 EMU/TETHERED EVA 150 SCHEDULED EVA 148 DURATION 6:34 SS EVA 158 DOCKED QUEST EVA 76 EMU/TETHERED EVA 151 SCHEDULED EVA 149 DURATION 6:14	Continued… SE TAL (ZZA 104): 6:03(P) 6:01(A) PTM (U/S 180 FPS): 6:25(P) 6:28(A) SE PRESS 104 6:58(P) 7:00 (A) MECO CMD: 8:22.6 (P) 8:23.8 (A) VI: 25819(P) 25818(A) OMS-2: 37:46 (P) 38:30(A) 98.8(P) 96.4(A) FPS	Continued… DRAG CHUTE DEPLOY: 191 KEAS 068:16:57:18Z NLGTD: 5439 FT 68:16:57:26Z VEL: 129 KGS 141 KEAS HDOT: -6.2 FPS BRK INIT: 56 KGS DRAG CHUTE JETTISON: 58 KGS 68:16:57:47Z BRK DECEL FPS^2: AVE 4.4 PK 6.3 WHEELS STOP: 68:16:58:11Z 9641 FT ROLLOUT: 7195 FT 0:56 M:S WINDS: 18 H KTS 2 L KTS OFFICIAL: 15018P25KT (X2P2H18P25) DENS ALT: 1266 FT FLT DURATION: 12:19:03:53 S/T: 1290:04:07:04 OV-103: 359: 22: 24:02 Continued…						

DISCOVERY'S FINAL LIFT-OFF

STS133-S-039 (24 Feb. 2011)

STS-133 / ET-137 Intertank Stringer Crack Observation

- During post drain inspections a crack was noted on LO$_2$-Intertank Flange Closeout
  - Dissection of foam revealed a crack on both sides of stringer at S7-2 (~9.0° L) and a crack at the adjacent stringer (S6-2, ~3.0°L)
- 1st observation in history of program
  - Design unchanged since SLWT (ET-96)

PRCB Briefing Chart for ET-137 Intertank Stringer Crack Issue found after fourth launch scrub on 11/05/10 when a liquid hydrogen leak was detected.

CAPCOMS:
SHUTTLE
Asc - Charlie Hobaugh
 - Barry Wilmore (Wx)
LD/O1 - Steve Robinson
O2 - Megan McArthur
Planning - Mike Massimino
Ent - Charlie Hobaugh
 - Terry Virts (Wx)
Team 4 - N/A
ISS
O1 - Hal Getzelman
LD/O2 - Stan Love
O3 - Ricky Arnold
Team 4 - N/A

MCC WHITE FLIGHT FCR (63)
FLIGHT DIRECTORS:
SHUTTLE:
Ascent- Richard Jones
LD/O1 - Bryan Lunney
O2- Ginger Kerrick
O3 - Rick LaBrode
Entry - Tony Ceccacci
MOD – John Mccullough
Team 4 & Prelaunch:
 - Paul Dye
ISS
O1 - David Korth
LD/O2 - Royce Renfrew
O3 - Chris Edelen
Team 4 - Kwatsi Alibaruho

External Tank Foam Loss 3 min, 51 sec into Ascent - No Severe Damage
External Tank Foam Loss (3 min, 51 sec)

Continued…

LAUNCH SCRUBS: (Continued)
On 11/18/10 launch rescheduled for NET 12/03/10 due to identified analysis and ET repairs required for safe launch. On 11/24/10 launch rescheduled to NET 12/17/10 to allow analysts additional time to determine likelihood of additional ET stringer cracks during ascent. "This is turning out to be a little more complicated from an analysis standpoint," NASA's associate administrator Bill Gerstenmaier. On 12/03/10 launch rescheduled to NET 02/03/11 to validate repairs and to support engineering analysis with instrumented ET Tanking Test. On 01/08/11 launch rescheduled to NET 02/24/11 to allow engineers additional time to assess new cracks resulting from tanking test. And, on 01/20/11 launch date was established as 02/24/11. This date allowed for completion of all stringer work.   Technical scrub.

LAUNCH WINDOW: Window open at 055:21:47:25Z and close at 055:21:53:27Z. Preferred Launch Time was 055:21:53:27Z (In-Plane Time) for a launch window of 3M02S.

LAUNCH DELAYS: 2M 57S due to Range Safety Central Command Computer anomaly. "We had about two seconds of hold time remaining, which is about one second more than Mike [Launch Director Leinbach] needed to get the job done, so we had plenty of margin," quipped Launch Integration Mgr Mike Moses.

TAL WEATHER: Spaceflight Meteorology Group (SMG) reported high pressure across Spain and France for generally acceptable weather at the TAL sites. ZZA was selected as prime TAL site at crew briefing [however, earliest TAL call was based on MRN]. Winds were gusting to 30 kts prior to crew brief, but headwinds dropped within limits at time of briefing. Isolated showers in Eastern France were never a threat and strong winds at Istres weakened enough for forecast to be amended GO.

PERFORMANCE ENHANCEMENTS:
Include the standard set plus: 1) PE Operational High Q -WIN/ FEB, 2) OMS Assist, 3) a 52 nautical mile MECO, & 4) Del Psi

FLIGHT DURATION CHANGES: Plus 1 day added for PMM outfitting was **approved** by MMT on FD 5. The IMMT/MMT added a 2nd extra day on FD 8 to allow the six member shuttle crew to further help unload the new PMM storage unit.

Continued…

FLT NO.	ORBITER	CREW (7)	LAUNCH SITE, LIFTOFF TIME,	LANDING SITE/ RUNWAY, CROSSRANGE	SSME-TL NOM-ABORT EMERG	SRB RSRM	ORBIT		FSW	PAYLOAD WEIGHTS,	MISSION HIGHLIGHTS (LAUNCH SCRUBS/DELAYS,
STS-133/ULF5		TITLE, NAMES & EVA'S	LANDING SITES, ABORT TIMES	LANDING TIMES FLT DURATION, WINDS	THROTTLE PROFILE ENG. S.N.	AND ET	INC	HA/HP		PAYLOADS/ EXPERIMENTS	TAL WEATHER, ASCENT I-LOADS, FIRSTS, SIGNIFICANT ANOMALIES, ETC.)

ISS026-E-030282 (26 Feb. 2011) --- Backdropped by a blue and white part of Earth, Discovery approaches ISS for its last visit.

Continued...

DISTANCE:
  5,304,140 sm

TOTAL OV-103 DISTANCE
  148,221,675 sm

TOTAL SHUTTLE DISTANCE
  529,797,883 sm

S133-E-007808 --- ISS's Canadarm2 grasps the Italian-built Permanent Multipurpose Module (PMM) for transfer from Discovery's payload bay to be permanently attached to the Unity node.

S133-E-007375 --- Bowen (top) and Drew, conduct EVA 1 as construction and maintenance continues on ISS.

S133-E-008627 --- In ISS, Lab Destiny, crews pose for a joint STS-133/Exp 26 group portrait. The STS-133 crew in red shirts (from left) are Stott, Drew, PLT Boe, CDR Lindsey, Barratt & Bowen. In dark blue Exp 26 crew, from left, are Paolo Nespoli/ESA, Oleg Skripochka/RSA, Dmitry Kondratyev/RSA (below), Alexander Y. Kaleri/ RSA and CDR Scott Kelly and Cady Coleman (below).

Continued...

FIRSTS/LASTS:
- Last flight of Discovery - 1st vehicle to be retired.
- Robonaut 2 is first dexterous humanoid robot in space
- First flight of SRB Thrust Vector Control (TVC) Auxiliary Power Unit (APU) Phase II fuel pump
- All six existing major spacecraft from Japan, Europe, Russia and the US that service ISS were simultaneously docked for first and last time. (Proposed Soyuz fly around of ISS for historic photo of the 6 vehicles - ruled out by Russia's FSA as safety risk.)
- Last NASA module (Italian-built), the Permanent Multipurpose Module (PMM), a storage room, was attached to ISS.
- Steve Bowen is first NASA astronaut to fly on back-to-back Shuttle missions (see below).
- FD13: First "Live" Wakeup Call! Performed by Big Head Todd & the Monsters playing "Blue Sky" from MCC, Tuesday, March 8, at 3:23 a.m. CST.

11th SHUTTLE CREWMEMBER REPLACEMENT
- Tim Kopra (injury) was replaced by Bowen in Jan. 2011. (10th Shuttle crewmember replacement occurred on STS-132..)

NIGHT LAUNCH: N/A

RENDEZVOUS: #80 Rendezvous and dock with ISS.

EVE NTS:
- FD1: OMS2 ignition at 55:22:31:54Z resulted in a 125.5 by 84.9 NM orbit.
- FD2: No Focus Inspection required for TPS/RCC
- T1 maneuver at 57:16:33:24Z resulted in a 192.4 by184.9 NM orbit.
- FD3: Performed R-Bar Pitch Maneuver.
- Docking Contact occurred at 057:19:14:18Z
- Hard Dock, hooks closed, occurred at 057:20:04:09Z
- ISS Hatch opened at 3:16 PM CST Feb. 26, 2011.
- Reboost (26 mins) at 62:14:29:36Z resulted in a 194.6 by 184.8 NM orbit.
- FD5: EVA 1: Bowen & Drew completed all planned tasks: J612 extension cable install, Pump module retrieval from POA, Pump module install on ESP-2, CP3 camera wedge install, and Message in a Bottle Experiment. During pump installation task the cupola robotic workstation had a "loss of comm." resulting in Bowen holding the 800 lb (but now weightless) pump for 25 min. He reported "I'm fine as long as it's not too much longer." Then added "How much longer?" Operations were transferred to the Lab robotics and task completed. EVA1 duration 6:34 Continued...

FLT NO.	ORBITER	CREW (7) TITLE, NAMES & EVA'S	LAUNCH SITE, LIFTOFF TIME, LANDING SITES, ABORT TIMES	LANDING SITE/ RUNWAY, CROSSRANGE LANDING TIMES FLT DURATION, WINDS	SSME-TL NOM-ABORT EMERG THROTTLE PROFILE ENG_S N	SRB RSRM AND ET	INC	ORBIT HA/HP	FSW	PAYLOAD WEIGHTS, PAYLOADS/ EXPERIMENTS	MISSION HIGHLIGHTS (LAUNCH SCRUBS/DELAYS, TAL WEATHER, ASCENT I-LOADS, FIRSTS, SIGNIFICANT ANOMALIES, ETC.)

S133-E-007866 --- CDRs Scott Kelly (left) Exp 26 & Steve Lindsey STS-133 are shown in the hatch leading to the newly-installed PMM.

Discovery's planned final approach and landing track to KSC. Chart courtesy Kyle Herring/JSC-PAO.

STS-133/ULF5   Discovery departs ISS for last time!

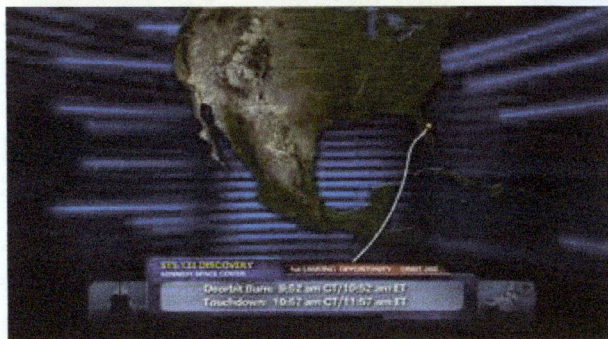

DISCOVERY'S FLAWLESS FINALE
MLGTD @ KSC March 09, 2011, 10:57:15 CST
201103090001HQ. - Courtesy: Rob Navias/JSC-PAO

Continued...

EVE NTS: (Continued)
 - FD5 MMT Decision: Based on FD2 inspection and RPM data, the TPS was cleared for entry per Flight Rule A2-142
 - FD6: PMM, an extra storage room/closet, was installed and hatch opened.
 - FD7: EVA2: Bowen & Drew successfully completed all tasks: Vent Ops/QD bag cleanup, Light Weight Adapter Plate Assembly (LWAPA) Retrieval & Install, P3 CETA Light Install, SPDM Camera Light Pan/Tilt Assy 1 Install and EP1 MLI Removal, and P1 Grapple Beam re-torque bolts down, plus several get-aheads. EVA2 duration 6:14.
 - Transfers:
       31,459 Pounds of H/W to ISS (inside & out)
          110 Pounds of Oxygen to ISS (Quest tanks)
           72 Pounds of Oxygen to ISS (stack repress)
           26 Pounds of Nitrogen to ISS
          931 Pounds of water to ISS
        2,031 Pounds of middeck items to ISS
        2,599 Pounds of H/W (middeck only) returned to Discovery
 - ISS Mass in space 919,964 Pounds
 - 100 Percent ISS complete (pressurized volume)
 - FD12: Undock from ISS complete at 066:12:00:10Z
 - FD14:During entry comm outage times due to blackout were:
   - 1st outage 068:16:39:25Z. INCO cmds H/O from TDRS 174 to TDRS 46 prior to roll cmd - at 068:16:30:25Z 1st outage ends.
   - 2nd outage at 068:16:37:53Z. INCO cmds H/O back to TDRS 174 prior to 1st roll reversal - at 068:16:37:58Z 2nd outage ends. MILA AOS at 68:16:45:00Z good return link and UHF.

SIGNIFICANT ANOMALIES:
Orbiter
- TPS Anomalies
- ATVC Ch 1 Power Supply Failed to Restart
- Ammonia Spray Boiler Sys B Unexpected Switchover
- KSC, RSRM, SSME, MOD, SRB - None.
- VIDEO FROM RH ET OBSERVATION CAMERA NOT RECORDED BY DAS DURING FLIGHT
ET: (See Integration issues below)
Integration:
- ET Intertank Stringer Cracks
- Hydrogen Leak at ET Ground Umbilical Carrier Plate (GUCP)
- Unexpected Debris/Expected Debris Exceeding Mass Allowable Prior to Pad Clearance (Liftoff Debris)
- Debris Released From LH2 Flange Area Near the Bipod

FLT NO.	ORBITER	-------------------------- SOME OF THE OPERATIONS SUPPORT TEAM --------------------------
STS-133/ULF5	OV-103	

JSC2011-E-021930 - STS-133 Lead FD Bryan Lunney monitors rendezvous data. His last flight.

JSC2011-E-023001 --- STS-133 Orbit 1 FCT - Flight Director Bryan Lunney (center left) on 2nd row.

Jsc2011e023002 --- ISS Orbit 3 FCT - Orbit 3 - FD Chris Edelen (lt) & CAPCOM Richard Arnold with STS-133 logo.

Pat Ryan/PAO   Ginger Kerrick/FD O2

JSC2011-E- 021648 -- Rt to Lt: FDs Tony Ceccacci & Richard Jones, & CAPCOMs Charlie Hobaugh & Barry Wilmore.

JSC2011-E-024279 --- STS-133 Ascent and Entry FCT in shuttle FCR in JSC. Flight Directors Tony Ceccacci (left) and Richard Jones hold the STS-133 mission logo.

IN KSC LCC: ABOVE: NASA Ctr Directors: (lt to rt) are Patrick Scheuermann/Stennis, Bob Cabana/KSC, Mike Coats/JSC, & Robert Lightfoot/MSFC. BELOW: We have lift-off! (lt to rt) Stephanie Stilson/Discovery Flow Director, Charlie Blackell-Thompson/Lead Test Director, & Mike Leinbach/ Launch Director.

------- SALUTE -------

STS-133/ULF5 "A MIXTURE OF SADNESS AND PRIDE"

JSC Center Director: "I am proud to have been the Pilot on the first flight of Discovery in 1984. I also flew Discovery on my two missions as Commander." **- Mike Coats**

Shuttle Program Manager/JSC: "Discovery's landing yesterday was an outstanding end to an amazing mission. I was really struck by the 'business as usual' attitude of the dedicated team that takes care of our Orbiters. ... To those team members that have flown their last flight with us – You should walk away with your head held very high. You have built and kept safe a unique capability in the most extreme of environments. I can only hope that others that come after us will look back at the Space Shuttle team and emulate the dedication, perseverance, and excellence that this team represents. If they do, we will have an outstanding human spaceflight program. For those team members remaining - Let's go finish this program strong." **- John Shannon**

STS-133 Crew: Nearing the end of the shuttle's final mission, the crew sentiments were a mixture of sadness and pride. "When you look out the Cupola window, times like that, I really reflect on what a great vehicle it's been – 39 missions, nearly one year on orbit, thinking about all the things the vehicle has done, it's kind of bittersweet." And later, "Houston for the last time, Wheels Stop!" **- CDR Steve Lindsey**.
"She retires with all of the honors and dignity due any of those ships that made great discoveries. So I think we salute Discovery in that way, with all the accolades she deserves. But it also lays out a challenge. What will be the next ship named Discovery? The next ship to bear this name hopefully will go farther than this one and make every bit as much of a contribution to history and to discovery as this ship." **- Michael Barratt/MS**

Launch Director/KSC: "I'm going to take away the attitude of the team on the ground that safed the vehicle. They did that today just like they've done every mission. They didn't skip a beat today and that's a true testament to their work ethic. It was heartwarming. ... Proud of the people that put the vehicle together and the flight controllers in Houston that executed the mission." **- Mike Leinbach**

Lead Flight Director/JSC: "Discovery represents the ingenuity, creativity and diligence of the teams who originally designed and built Discovery and also the teams who operated and evolved the capabilities of Discovery across three decades. Discovery evolved from a short duration LEO delivery vehicle to a much more capable delivery and service spacecraft staying on orbit more than twice as long as originally intended. The engineering teams and operations teams expanded Discovery's capabilities well beyond the original designers intentions enabling scientists to learn more and more about the world and universe around us." **- Bryan Lunney/Onyx Flight**

NASA Assoc Admin. for Space Ops: "I don't really know what to say other than to thank the Discovery team. I think of all the processing work, the folks throughout the history of this vehicle back to Downey and Palmdale who gave us a phenomenal vehicle. It's legacy is the future with station in great shape and that's only possible because Discovery performed so well. That extra work sets us up so well for the research period aboard station." **- Gerst**

DISCOVERY NOW HEADS TO THE SMITHSONIAN NATIONAL AIR AND SPACE MUSEUM's UDVAR-HAZY CENTER IN CHANTILLY, VA.

Shuttle Legacy Mural - Hanging in LCC Firing Room at KSC

DISCOVERY TRIBUTE: From Mike Leinbach/Launch Director/KSC

KSC-2010-4453 (http://mediaarchive.ksc.nasa.gov/index.cfm). This Tribute Display features Discovery demonstrating the renowned Rendezvous Pitch Maneuver on approach to the International Space Station (ISS) during STS-114. Having accumulated the most space shuttle flights, Discovery's 39 mission patches are shown encircling the vehicle. The background image was taken from the Hubble Space Telescope, which was launched aboard Discovery on STS-31 and serviced by Discovery on STS-82 and STS-103. The prominent American flag and eagle represent Discovery's two "Return to Flight" missions, STS-26 and STS-114, and symbolize Discovery's heroic role in returning American astronauts to spaceflight. Discovery's significant accomplishments include the first female Shuttle pilot (Eileen Collins on STS-63), John Glenn's legendary STS-95 mission, and the celebration of the 100th space shuttle mission with STS-92. In addition, Discovery supported numerous DOD programs, satellite deploy/repair missions, and 13 flights for construction and operation of the ISS.

Cabana & Crew

Leinbach & Stott

Gerst - Lindsey - Moses

Bolden & Crew

FLT NO.	ORBITER	CREW (6) TITLE, NAMES & EVA'S	LAUNCH SITE, LIFTOFF TIME, LANDING SITES, ABORT TIMES	LANDING SITE/ RUNWAY, CROSSRANGE LANDING TIMES FLT DURATION, WINDS	SSME-TL NOM-ABORT EMERG THROTTLE PROFILE ENG. S.N.	SRB RSRM AND ET	INC	ORBIT HA/HP	FSW	PAYLOAD WEIGHTS, PAYLOADS/ EXPERIMENTS	MISSION HIGHLIGHTS (LAUNCH SCRUBS/DELAYS, TAL WEATHER, ASCENT I-LOADS, FIRSTS, SIGNIFICANT ANOMALIES, ETC.)
STS-134/ULF6 SEQ FLT # 134 KSC-134 PAD 39A (57) MLP-2 36th SHUTTLE FLIGHT TO ISS	OV-105 (Flight 25) **ENDEAVOUR FINAL FLIGHT** OMS PODS LPO3 -35 RPO4 -31 FRC5 -24	CDR: Mark E. Kelly (Flt 4 - STS-124,STS-108, STS-121) P842/R271/V181/M237 PLT Gregory H. Johnson (Flt 2 - STS-123) P843/R322/V224/M278 MS 1 Michael Finke (ISS EXP 9 & 19) P844/R356//M307 MS 2 Greg Chamitoff (Flt 2 UP ON STS-124, stay ISS, DN ON STS-126) P845/R330/V225/M285 MS 3 Andrew Feustel (ESA) (Flt 2 - STS-125) P845/R340/V226/M294 MS 4 Roberto Vittori (ESA) (ISS "Marco Polo" & "Eneide") P846/R357/M308 SS EVA 159 DOCKED QUEST EVA 77 EMU/TETHERED EVA 152 SCHEDULED EVA 150 DURATION 6:19 SS EVA 160 DOCKED QUEST EVA 78 EMU/TETHERED EVA 153 SCHEDULED EVA 151 DURATION 8:07 SS EVA 161 DOCKED QUEST EVA 79 EMU/TETHERED EVA 154 SCHEDULED EVA 152 DURATION 6:54 Continued…	KSC 39A 136:12:56:28Z 8:56:28 AM EDT (P) 8:55:42 AM EDT (A) Monday (18) 05/16/11 (10) LAUNCH WINDOW: 5M 46S (Total) 5M01S (Preferred) EOM PLS: KSC TAL: ZZA TAL WX: MRN , FMI SELECTED: RTLS: KSC15 N/N TAL: ZZA30L N/N AOA: NOR17 N/N 1ST DAY PLS EDW22 N/N TDEL: 0.000 (P) 0.122 (A) MAX Q NAV: 733.5 (P) 733.1 (A) SRB STG: 2:04.8 (P) 2:04.8 (A) PERF: NOMINAL 2 ENG TAL (MRN): 2:36 (P) 2:36 (A NEG RET (2@ 104): 3:53(P) 3:54(A PTA (U/S 156 FPS): 5:00(P) 4:57(A) Continued…	KSC15 KSC (77) 152:06:34:50Z 01:34:50 AM CDT Wednesday (18) 06/01/11 (9) DEORBIT BURN: 152:05:29:03Z XRANGE:141.1 NM ORBIT DIR: A/L (46) AIM PT: Nominal MLGTD: 3138 FT 152:06:34:50Z VEL: 196 KGS 191 KEAS HDOT: -1.0 FPS TD NORM 195: 2931 FT Continued…	104/104/ 109% PREDICTED: 100/104.5/104.5/ 72/104.5 ACTUAL: 100/104.5/72/ 104.5 1 = 2059 (5) 2 = 2061 (2) 3 = 2057 (6) M 3 EOM: WEIGHT: 204532 LBS X CG 1080.4 IN LANDING: WEIGHT: 204463 LBS X CG: 1082.3 IN	BI-145 RSRM-113 ET-122 SLWT 42 w/Stringer Mod ET IMPACT 1:14:11 MET LAT: 36.436S LONG: 158.531W	51.6 (36)	DIRECT INSERTION POST OMS-2 175.9x124.7 NM DEORBIT HA 188.7 NM HP 23.1 NM ENTRY VELOCITY: 25860 FPS ENTRY RANGE: 4419 NM	OI-34 (6)	CARGO: 39210 LBS PAYLOAD CHARGEABLE: 31693 LBS DEPLOYED: 30721 LBS NON-DEPLOYED: 811 LBS MIDDECK: 161 LBS SHUTTLE ACCUMULATED WEIGHTS: DEPLOYED: 1759961 LBS NON-DEPLOYED: 1630249LBS CARGO TOTAL: 4365726 LBS PERFORMANCE MARGINS (LBS): FPR: 2821 FUEL BIAS: 954 FINAL TDDP: 1968 RECON: 3211 PAYLOADS: PLB: ISS-UL F6 (AMS-02 ELC 3) MISSE 7 Return , MISSE 8, STORRM DTO-703 MIDDECK: ISS-UL65, MAUI, SEITE, SIMPLEX, RAMBO-2 Continued…	**Brief Mission Summary:** *The STS-134 (36th mission to ISS) delivered the $2 billion Alpha Magnetic Spectrometer-2 (AMS-02) to the ISS. AMS-02 is a particle physics detector designed to search for dark matter and for antimatter (first discovered by British physicist Paul Dirac in 1920's) in the universe. MIT Prof. Sam Ting is the AMS Principal Investigator. ISS spare parts and a suite of DoD Experiments were also delivered to orbit. Four EVA's were conducted for ISS maintenance and the Orbiter OBSS was transferred to ISS as a permanent fixture. This was the final flight of Endeavour (25 flights).* **KSC W/D** OPF: 263 days+ 89 Non-work days + 17 holidays + 2 safety days VAB: 9 +1C (Contingency) day + 1Wx PAD A: 53+14C Total Work Days = 325 (OPF processing occurred over a total time period of 371 days) **POSTPONEMENTS:** - Baselined STS-134 to FDRD - launch date of 07/29/10 on 06/30/09. - Ppd. to 11/26/10 on 07/01/10. Delayed to late November after a decision to replace the magnet at the heart of the Alpha Magnetic Spectrometer payload. - Ppd. to 02/26/11 on 07/01/10. A late-November/early December launch was ruled out because of conflicts with other planned station launches. Temperature constraints related to the station's orbit prevented a launch in January and range conflicts with other unmanned missions pushed the approved launch date to Feb. 26. - Ppd. to NET 04/01/11 on 12/03/10 due to STS-133 slip for ET stringer problems. - Ppd. to 04/19/11 on 01/26/11. This date was driven by the launch pad turnaround time required after STS-133 launch. - Ppd. to 04/29/11 on 04/04/11 due to conflicts with Russian Progress vehicle flight to ISS. **LAUNCH SCRUBS:** - Launch scrubbed on 04/29/11 due to failed APU fuel line heater. Launch rescheduled for NET 05/02/11. On 05/02/11 launch was initially rescheduled to NET 05/08/11, then later to 05/10/11 to allow time to R&R faulty Load Control Assembly (LCA) box. On 05/06/11 managers announced earliest launch date was now 05/16/11 pending resolution of additional electrical testing. Mike Moses at 05/09/11 Press Brief: "Our confidence to fly is we've done functional on all the components and know we have good power to those heaters and they're performing just fine. ... we have confidence in launching Monday the 16th at 8:56 am EDT". Technical scrub. Continued…

STS-134 - (16 May 2011)

ENDEAVOUR'S FINAL LIFT-OFF

FLT NO.	ORBITER	CREW (7) TITLE, NAMES & EVA'S	LAUNCH SITE, LIFTOFF TIME, LANDING SITES, ABORT TIMES	LANDING SITE/ RUNWAY, CROSSRANGE LANDING TIMES FLT DURATION, WINDS	SSME-TL NOM-ABORT EMERG THROTTLE PROFILE ENG. S.N.	SRB RSRM AND ET	ORBIT INC	HA/HP	FSW	PAYLOAD WEIGHTS, PAYLOADS/ EXPERIMENTS	MISSION HIGHLIGHTS (LAUNCH SCRUBS/DELAYS, TAL WEATHER, ASCENT I-LOADS, FIRSTS, SIGNIFICANT ANOMALIES, ETC.)
STS-134/ULF6	OV-105	Continued...	Continued...	Continued...					Continued...	Continued...

Crew column (continued):

Continued...

SS EVA 162
DOCKED QUEST EVA 80
EMU/TETHERED EVA 155
SCHEDULED EVA 153
DURATION 7:24:15

MCC WHITE FLIGHT FCR (63)
FLIGHT DIRECTORS:
SHUTTLE:
Ascent- Richard Jones
LD/O1 - Gary Horlacher
O2- Paul Dye
O3 & Prelaunch - Kwatsi Alibaruho
O4 - Rick LaBrode
Entry - Tony Ceccacci
MOD – John Mccullough
Team 4 - Richard Jones
ISS
O1 - Dana Weigel
LD/O2 - Derek Hassmann
O3 - Dina Contella
Team 4 - Dave Korth

CAPCOMS:
SHUTTLE
Ascent - Barry Wilmore
- Lee Archambault/Wx
LD/O1 - Megan McArthur
O2 - Steve Robinson
Planning:
- Shannon Lucid
Entry - Barry Wilmore
- Terry Virts/Wx
Team 4 - N/A
ISS
O1 - Rob hayhurst
LD/O2 - Lucia McCullough
O3 - Dan Tani

Launch Site column (continued):

Continued...

SE TAL (ZZA 104):
6:00(P)   5:56(A)

PTM (U/S 181 FPS):
6:01(P)   5:56(A)

SE PRESS 104
6:54(P)   6:55 (A)

MECO CMD:
8:21.8 (P) 8:21.5 (A)
VI:
25819(P) 25818(A)

OMS-2:
37:00 (P) 36:57(A)
260(P) 259(A) FPS

Landing Site column (continued):

Continued...

DRAG CHUTE DEPLOY: 184 KEAS
152:06:34:53Z

NLGTD: 6034 FT
152:06:34:54Z
VEL: 154 KGS
149 KEAS
HDOT: - 4.5 FPS

BRK INIT: 119 KGS

DRAG CHUTE JETTISON:
152:06:35:19Z
47 KGS

BRK DECEL FPS²:
AVE 8.3 PK 11.7

WHEELS STOP:
152:06:35:32Z
9712 FT

ROLLOUT:
6574 FT 0:42 M:S

WINDS:
-2 H KTS 0 KTS
OFFICIAL:
32002P03KT(X1P1 T2P3)

DENS ALT: 844 FT

FLT DURATION:
15:17:38:22

S/T:
1305:21:45:26

OV-105:
296:03:17:45
Continued...

Payload column (continued):

Continued...

5 CRYO TANK SETS, ODS, SRMS (90) OBSS Remains on ISS, SSPTS

ABOVE: Jsc2011e036650 --STS-134 was first flight controlled from JSC Mission Control Center after it was renamed in honor of Christopher C. Kraft, Jr. on April 14, 2011.
BELOW: Jsc2011e036646 – Chris speaks at the ceremony. He was NASA's 1st Flight Director for manned spaceflight. He served on all Mercury & several Gemini flights, was one of the designers & implementers of the MCC, and was JSC Center Director from 1972 to 1982. Call Name - Red Flight.

S134-E- 7189 - AMS In the grasp of the Orbiter's robotic Canadarm for transfer to ISS.

MIT Professor Sam Ting the Principal Investigator for the $2 Billion AMS-02 in search of cosmic dark matter & antimatter. (AMS-01 was flown & tested on STS-91.)

Mission Highlights column (continued):

Continued...

LAUNCH WINDOW:
Window open at 136:12:55:43Z and close at 136:13:01:29Z Preferred Launch Time was 136:12:56:28 (In-Plane Time) for a launch window of 5M01S (preferred).

LAUNCH DELAYS: None.

TAL WEATHER: Spaceflight Meteorology Group (SMG) reported at least one site would have favorable weather. High pressure gave mostly clear skies at ZZA & FMI, but head winds near limit at FMI. Moron reported concerns for thunderstorms. ZZA was selected as prime TAL site for launch. The slight chance of thunderstorms at MRN was removed one hour before launch giving Mission Control Team two acceptable TAL sites with FMI marginally acceptable.

PERFORMANCE ENHANCEMENTS
Include the standard set plus: 1) PE Operational High Q - TRN/APR, 2) OMS Assist, 3) a 52 nautical mile MECO, & 4) Del Psi

FLIGHT DURATION CHANGES: None.

FIRSTS/LASTS/MOSTS:
- Last flight of Endeavour.
- First flight controlled from JSC MCC renamed for Dr. Christopher C. Kraft, Jr. on April 14, 2011.
- First Papal call to space. On Saturday, May 21, 2011 Pope Benedict XVI commended crews for their courage and blessed them with prayers.
- First undock of Soyuz while Shuttle is docked to ISS. Leagcy photo by Soyuz of ISS with Docked Shuttle.
- Last EVA's of Shuttle crew. Feustel, Chamitoff, & Fincke rotated
through 4 EVA's. "We will be traversing from one end of the station to the other," said Feustel.
- Most time in space by an American: Mike Fincke surpassed Peggy Whitson's record of 377cumulative days finishing with 382 days.

NIGHT LAUNCH: N/A

NIGHT LANDING KSC #18: (#24 in Shuttle history)

RENDEZVOUS: #80 Rendezvous and dock with ISS.

Continued...

FLT NO.	ORBITER	CREW (7) TITLE, NAMES & EVA'S	LAUNCH SITE, LIFTOFF TIME, LANDING SITES, ABORT TIMES	LANDING SITE/ RUNWAY, CROSSRANGE LANDING TIMES, FLT DURATION, WINDS	SSME-TL NOM-ABORT EMERG THROTTLE PROFILE ENG. S N	SRB RSRM AND ET	INC	ORBIT HA/HP	FSW	PAYLOAD WEIGHTS, PAYLOADS/ EXPERIMENTS	MISSION HIGHLIGHTS (LAUNCH SCRUBS/DELAYS, TAL WEATHER, ASCENT I-LOADS, FIRSTS, SIGNIFICANT ANOMALIES, ETC.)
STS-134/ULF6 Continued...		Crew "Star Trek" Connection S134-E-009631-- EVA 4 Fincke (Below) Last Shuttle Crew EVA May 25, 2011 S134-E-009647 -- EVA 4 Chamitoff (Above)	Continued... DISTANCE: 6,510,221 sm TOTAL OV-105 DISTANCE 122,883,151 sm TOTAL SHUTTLE DISTANCE: 536,308,104 sm								S134-E-009265 -- EVA-1 Feustel (rt) & Chamitoff (lt) ISS027-E-035698 --- Crews STS-134 (in Black) & EXP 27 (in Blue) pose in ISS Kibo: lt to rt (front row) are Paolo Nespoli/ESA, CDR Dmitry Kondratyev/RSA, CDR Kelly & Vittori/ESA; and (back row), Cady Coleman, Andrey Borisenko/RSA, Alexander Samokutyaev/RSA, Ron Garan, Fincke, Feustel/ESA, Chamitoff & PLT Johnson.	Continued... **VIP** CDR Mark Kelly's wife & U.S. Representative Gabrielle Giffords severely wounded in a shooting at a public event in Tucson, Arizona on Jan. 8, 2011, was able to attend the launch. **EVE NTS** - FD1: OMS2 ignition at 136:13:33:25Z resulted in a 175.9 by 124.7 NM orbit. - FD2: RCC surveys showed some areas of concern. Focus Inspection required on FD6 - T1 maneuver at 138:07:38:13Z resulted in a 186.1 by182.8 NM orbit. - FD3: Performed R-Bar Pitch Maneuver. - Docking Contact occurred at 138:10:13:52Z - Hard Dock, hooks closed, occurred at 138:10:25:15Z - ISS Hatch opened at 6:38 AM CDT May 18, 2011. - FD4: AMS handed off from Shuttle arm to ISS arm and installed on ISS. Scientists immediately began detecting "thousands and thousands" of subatomic particles from deep space. FD5: - DAT team cleared ascent RCC damage, but recommended a Focused Inspection of area between MLGD & ET door. MMT approved for FD6. - FD5: EVA 1: Feustel & Chamitoff completed Installation & retrieval of MISSE experiments, & installations of: S3 CETA light, SARJ cover 7, P3/P4 ammonia jumper on ISS. Chamitoff's ppCO2 sensor dropped out during EWC antenna task. Flight rule required termination of the EVA. EVA1 duration 6:19 - FD6: Focused Inspection was completed. DAT team analysis using these images cleared TPS for safe entry. - FD7: EVA 2: Feustel & Fincke completed all tasks, however, duration was 1:30 longer than planned due to H/W issues. During port SARJ lube task some loose bolts prevented removal of 2 covers & reinstallation of another. Also, after filling P6 truss PVTCS one ammonia flake was seen near Fincke's suit. Inspections revealed no visible contamination. Other tasks included SPDM LEE lube & S1 Radiator Stowage Beam installation. EVA2 duration 8:07. - FD8: GMT 143/21:35 Soyuz TMA-20 undocking from ISS & imagery operations of Shuttle docked to ISS. - FD10: EVA3: Feustel & Fincke completed all tasks for servicing of ISS, installing cables for the power system & completion of work on a wireless communications system. EVA3 duration 6:54. - FD10 the OBSS will be left behind to serve as an extension for station use if needed in the future. Continued...

STS-134/ULF6

FD 7 - JSC2011-E-046603 (21 May 2011) - This overall view of the space Shuttle FCR in the Christopher C. Kraft, Jr. Mission Control Center was taken during a special call from Pope Benedict XVI (upper left) in the Vatican to the STS-134 and Expedition 27 crews (center screen) on the ISS.

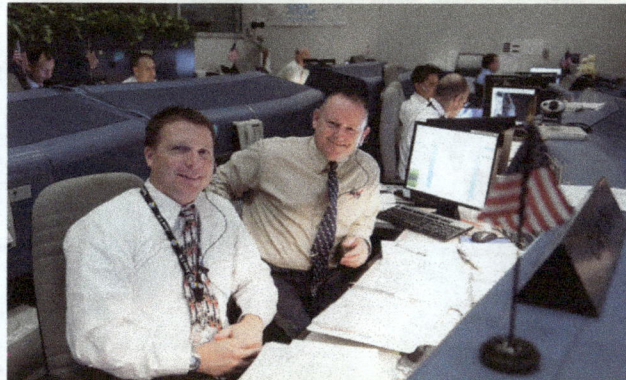

JSC2011-E-050144 --- CAPCOMs Terry Virts (left) and Barry Wilmore on console in CCK-MCC.

FD 11: May 25, 2011 the 50th Anniversary of President John F. Kennedy's historic space message to a joint session of Congress, on May 25, 1961.

"...I believe this nation should commit itself to achieving the goal, before this decade is out, of landing a man on the Moon and returning him safely to the Earth."
And ending with… *"We have a long way to go in this space race. But this is the new ocean, and I believe that the United States must sail on it and be in a position second to none."*

America has sailed this ocean for the past 50 years, and grabbed the lead on July 20, 1969. The question now is: will she still be the lead ship on this ocean for the next 50 years?

Continued…

EVENTS: Continued
- FD12: EVA 4: Fincke & Chamitoff completed all major objectives including: OBSS Stow on ISS, P6 PGDF Retrieve, & OBSS EFGF/PDGR Swap. Only items not completed: relocation of APFR from P6 and stowing EFGF on Tool Stowage Assembly (TSA) in Shuttle P/L Bay. EVA4 duration 7:24:15 . **This was last EVA of the Shuttle era.**
- FD14: ISS reboost with 14-min Orbiter RCS Verniers completed for 1.8 fps burn.
- **Transfers:**
    29,370 pounds of H/W to ISS
       17 Pounds of Oxygen to ISS (Quest tanks)
      278 Pounds of Oxygen to ISS (stack repress)
      18 Pounds of Nitrogen to ISS
    2,266 Pounds of middeck items to ISS
    2,235 Pounds of middeck returned to Endeavour
- ISS Mass in space 904,991 Pounds
- FD15: Undock from ISS complete at 150:03:55:12Z
- FD15: Completed DTO for new docking system - Sensor Test for Orion Relative Navigation Risk Mitigation (STORRM) - "Went exceptionally well." Anthony Ceccacci/FD.
- FD16: No significant comm outage during blackout timeframe. Start of RF Blackout : MET 15/17:09, good comm on TDRS-174 & Orbiter upper antennas. 1st Roll Maneuver: MET 15/17:11, still on upper antenna & good comm. 1st Roll Reversal: MET 15/17:20, INCO handoff to TDRS-46, still on uppers & good comm. Hand down to MILA: MET 15/17:25.

SIGNIFICANT ANOMALIES:
Orbiter:
- RH NLG P2 Pressure Measurement(V51P0373A) Erratic/Off Scale High During Ascent
- DURING PRE-LAUNCH OPERATIONS ON APRIL 29, AUXILIARY POWER UNIT (APU) 1 FUEL TEST LINE AND FUEL SERVICE LINE B HEATERS FAILED TO ACTIVATE (TEMPERATURE TRENDED BELOW THE LCC LIMIT OF 45 DEG F IN LCC APU-14) WITH BOTH GROUND COMMAND AND PANEL SWITCH ACTIVATION.
SRB:
- RH SRB MAIN CHUTE FAILURE ? GORE 26 FAILED FROM THE SKIRT BAND THROUGH THE VENT BAND.
KSC, RSRM, SSME, MOD.& ET - None.
Integration:
- Unexpected Debris/Expected Debris Exceeding Mass Allowable Prior to Pad Clearance (Liftoff Debris)
- Cylindrical Debris Observed Near +Y Thrust Panel During SRB Separation

SPACE SHUTTLE MISSIONS SUMMARY

ABOVE: STS134-070 (1 June 2011) --- After 19 years of service, xenon lights illuminate Endeavour's drag chute during it's last landing & Shuttle's last night landing.

BELOW: 2011060100044hq (1 June 2011) --- Crew poses in front of Endeavour post- landing: (Lt to Rt) Vittori, Johnson, CDR Kelly, Fincke, Chamitoff, & Feustel.

STS-134/ULF7

FD 8: iss027e036679 (May 23, 2011) --- One of first legacy photos taken from Soyuz TMA-20 of a Shuttle (Endeavour, left of center) docked to ISS.

STS-134/ULF7 ----- SALUTE TO ENDEAVOUR AND SOME OF ITS OPERATIONS SUPPORT TEAM -----

TOP: JSC2011-E-048881 --- STS-134 Orbit 3 FCT. FD Kwatsi Alibaruho (left) on the front row.

BOTTOM: JSC2011-E-048941 --- Entry FCT. FD Tony Ceccacci (third from left) on the front row with CAPCOM Barry Wilmore holding STS-134 mission logo.

Shuttle Legacy Mural - Hanging in LCC Firing Room at KSC

ENDEAVOUR: From Mike Leinbach/Launch Director/KSC

KSC-2010-4454 (http://mediaarchive.ksc.nasa.gov/index.cfm). This Tribute Display features Endeavour soaring into orbit above the sailing vessel HMS Endeavour for which the orbiter was named. The Cupola, delivered to the International Space Station by Endeavour on STS-130, is shown framing various images of Endeavour. The images represent the phases of mission processing and execution for the Space Shuttle Program. The first ever use of a drag chute during orbiter landing (STS-49) is depicted in the top window and moving clockwise the images symbolize the following: Rollout to the Pad, Ferry Flight return to Kennedy Space Center, Orbiter Processing Facility Roll-in, Docking at the International Space Station, and Lifting Operations for Orbiter Mate in the Vehicle Assembly Building. The background image was captured by the Hubble Space Telescope and signifies the first servicing mission which was performed by the Endeavour crew on STS-61. Crew-designed patches from Endeavour's maiden voyage through her final mission are shown ascending toward the stars.

FD LD/O1 Gary Horlacher (left) & Chief FD John McCullough

FD O4 Rick LaBrode

Kelly Humphries/PAO

ATLANTIS COMING HOME TO KSC

An unprecedented view, as seen by the ISS Exp 28 crew, of Space Shuttle Atlantis on its way home with its plasma trail generated during the heat of entry. Airglow over Earth and stars can be seen in the background.
(ISS028-E-018214)

SPACE SHUTTLE MISSIONS SUMMARY

FLT NO.	ORBITER	CREW (4) TITLE, NAMES & EVA'S	LAUNCH SITE, LIFTOFF TIME, LANDING SITES, ABORT TIMES	LANDING SITE/ RUNWAY, CROSSRANGE LANDING TIMES FLT DURATION, WINDS	SSME-TL NOM-ABORT EMERG THROTTLE PROFILE ENG. S.N.	SRB RSRM AND ET	INC	ORBIT HA/HP	FSW	PAYLOAD WEIGHTS, PAYLOADS/ EXPERIMENTS	MISSION HIGHLIGHTS (LAUNCH SCRUBS/DELAYS, TAL WEATHER, ASCENT I-LOADS, FIRSTS, SIGNIFICANT ANOMALIES, ETC.)

STS-135/ULF7

OV-104 (Flight 33) ATLANTIS LAST FLIGHT OF SHUTTLE PROGRAM

SEQ FLT # 135
KSC-135
PAD 39A (58)
MLP-3
37th SHUTTLE FLIGHT TO ISS

CDR: Chris Ferguson (Flt 3 - STS-115, STS-126) P847/R300/V197/M179

PLT Doug Hurley (Flt 2 - STS-127) P848/R358/V227/M295

MS 1 Sandy Magnus Flt3- STS-112, (UP ON STS-126, stay ISS, DN on STS-119) P849/R284/V200/F36

MS 2 Rex Walheim (Flt3 - STS-110, STS-122) P850/R277/V193/M243

SS EVA's No SS EVAs were scheduled for this flight. (There was an ISS Crew EVA by Michael Fossum & Ronald Garan during this mission for a duration of 6:31 hr:min)

The "Katrina Tank" recovered from Hurricane Katrina and is ready to fly. (Courtesy Lockheed Martin – Michoud)

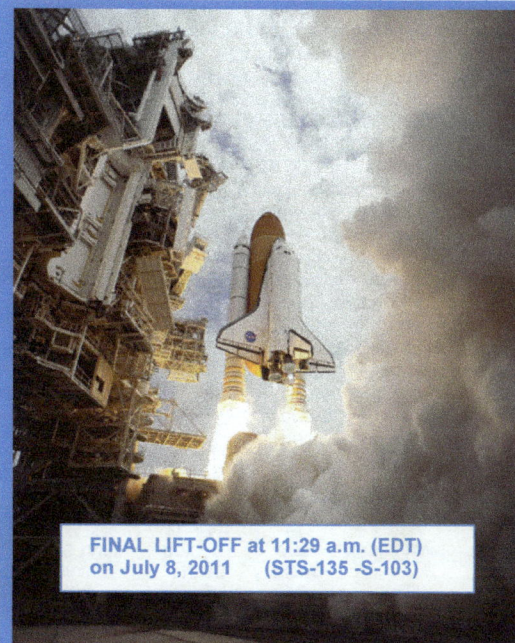

FINAL LIFT-OFF at 11:29 a.m. (EDT) on July 8, 2011 (STS-135 -S-103)

FLT NO.	ORBITER	CREW (7) TITLE, NAMES & EVA'S	LAUNCH SITE, LIFTOFF TIME, LANDING SITES, ABORT TIMES	LANDING SITE/ RUNWAY, CROSSRANGE LANDING TIMES FLT DURATION, WINDS	SSME-TL NOM-ABORT EMERG THROTTLE PROFILE ENG. S.N.	SRB RSRM AND ET	INC	ORBIT HA/HP	FSW	PAYLOAD WEIGHTS, PAYLOADS/ EXPERIMENTS	MISSION HIGHLIGHTS (LAUNCH SCRUBS/DELAYS, TAL WEATHER, ASCENT I-LOADS, FIRSTS, SIGNIFICANT ANOMALIES, ETC.)
STS-135/ULF7 Continued...	OV-104	Continued... MCC WHITE FLIGHT FCR (64) **FLIGHT DIRECTORS:** SHUTTLE: Ascent - Richard Jones LD/O1 - Kwastsi Alibaruho O2 - Rick LaBrode Planning - Paul Dye Entry - Tony Ceccacci MOD - John Mccullough Team 4 - N/A ISS O1 - Jerry Jason LD/O2 - Chris Edelen O3 - Courtenay Team 4 - N/A	Continued... PTM (U/S 180 FPS): 6:14(P) 6:23(A) SE PRESS 104 6:52(P) 6:55 (A) MECO CMD: 8:23.5 (P) 8:23.8 (A) VI: 25819(P) 25817(A) OMS-2: 37:46 (P) 37:45(A) 98.7(P) 96.8(A) FPS	Continued... MLGTD: 1649 FT 202:09:56:58Z VEL: 221 KGS 216 KEAS HDOT: -2.1 FPS TD NORM 205: 2809 FT DRAG CHUTE DEPLOY:192 KEAS 202:09:57:03Z NLGTD: 6713 FT 202:09:57:11Z VEL: 156 KGS 146 KEAS HDOT: - 6.6 FPS BRK INIT: 118 KGS DRAG CHUTE JETTISON: 52 KGS 202:09:57:35Z BRK DECEL FPS^2: AVE 6.8 PK 8.0 WHEELS STOP: 202:09:57:49Z 11,361 ROLLOUT: 9712 FT 0:51 M:S WINDS: -1 H KTS 0 KTS OFFICIAL: 33001P02KT (X0P0 T1P2) DENS ALT: 1239 FT _Continued..._	Continued... M 3 EOM: WEIGHT: 226333 LBS Xcg: 1090.7 IN LANDING: WEIGHT: 226270 LBS Xcg:1092.4 IN						

BELOW: S135-E-008061 --- In what has become a tradition for Shuttle & ISS crews, STS-135 & ISS Exp 28 crews formed a microgravity circle for a portrait aboard JAXA Kibo Laboratory. STS-135 crew consists of NASA astronauts Chris Ferguson, Doug Hurley, Sandy Magnus and Rex Walheim; Exp 28 crewmembers are JAXA astronaut Satoshi Furukawa, NASA astronauts Ron Garan and Mike Fossum, and Russian cosmonauts Andrey Borisenko, Alexander Samokutyaev and Sergei Volkov. Shuttle & ISS CDRs Ferguson and Borisenko are in the 12 o'clock & six o'clock positions, respectively, on the circle. The U.S. flag pictured was flown on the first Space Shuttle mission, STS-1, and flew on this mission to be presented to the ISS crew. It will remain on board until the next crew launched [commercially] from the U.S. will retrieve it for return to Earth. It will fly from Earth again, with the crew that launches from the U.S. on a journey of exploration beyond Earth orbit.

ISS028-E-015565 Atlantis as seen from ISS brings supplies & spare parts to ISS packed in MPLM at rear of P/L Bay. Last flight of the "Banana Truck"! [Atlantis was happily called the "Banana Truck" on STS-71 by Cosmonaut Strehalov, see page 2-84.]

CAPCOMS: SHUTTLE Ascent - Barry Wilmore - Lee Archambault/Wx LD/O1 - Steve Robinson O2 - Megan McArthur Planning: - Shannon Lucid Entry - Barry Wilmore - Charlie Hobaugh/Wx Team 4 - N/A ISS O1 - Dan Tani LD/O2 - Rob Hayhurst O3 - Kathy Bolt Team 4 - N/A

Continued...

FLIGHT DURATION CHANGES: On FD4, MMT agreed to add 1 day to mission, stating: "Additional mission content would benefit ISS transfer and utilization".

FIRSTS/LASTS/MOSTS:
- Last flight of Atlantis & Space Shuttle Program.
- Sandra Magnus is "Last Woman to Blast Off" in Space Shuttle.
- First iPhone launched into space to run an experimental app designed by Odyssey Space Research.

NIGHT LAUNCH: N/A

NIGHT LANDING KSC #19: (#25 in Shuttle history)

RENDEZVOUS: #81 Rendezvous and dock with ISS.

EVE NTS:
- FD1: OMS2 ignition at 189:16:06:49Z resulted in a 124.3 by 84.9 NM orbit.
- FD2 Wakeup: "Viva la Vida" by Coldplay for Doug Hurley (w/greeting from MSFC employees)
- RCC survey data collected for DAT. Go to MMT on FD4.
- T1 maneuver at 191:12:29:04Z resulted in a 210.5 by 207.3 nm orbit.
- FD3: Performed R-Bar Pitch Maneuver.
- Docking Contact occurred at 191:15:07:15Z
- Hard Dock, hooks closed, occurred at 191:15:19:32Z
- ISS Hatch opened at11:47 AM CDT July 10, 2011.
- FD4: MMT agreed to add 1 day to mission. "Additional mission content would benefit ISS transfer and utilization."
- MPLM installed on ISS
- FD5: MMT concurred with DAT assessment that Orbiter TPS was cleared for deorbit, entry, & landing.
- ISS crewmembers Mike Fossum & Ron Garan conduced EVA completing the following activities: Readied Pump Module in P/L Bay for return, RRM installed on SPDM EOTP, MISSE 8 ORMATE installed, FGB PDGF exposed grounding wire corrected, & PMA3 cover installed. PET duration 6 hrs 31 minutes
- FD7: GPC 4 (SMGPC) failed - most likely cause was a transient single event upset (radiation hit). Same event occurred on STS-71 (same vehicle & same GPC).
- FD 8: GPC 2 was reconfigured as SM GPC.
- FD 10: GPC 4 reconfigured to SM and treated as fully functional for Entry.

Continued...

FLT NO.	ORBITER	CREW (7)	LAUNCH SITE, LIFTOFF TIME,	LANDING SITE/ RUNWAY, CROSSRANGE	SSME-TL NOM-ABORT EMERG	SRB RSRM	ORBIT		FSW	PAYLOAD WEIGHTS,	MISSION HIGHLIGHTS (LAUNCH SCRUBS/DELAYS,
STS-135/ ULF7		TITLE, NAMES & EVA'S	LANDING SITES, ABORT TIMES	LANDING TIMES FLT DURATION, WINDS	THROTTLE PROFILE ENG. S.N.	AND ET	INC	HA/HP		PAYLOADS/ EXPERIMENTS	TAL WEATHER, ASCENT I-LOADS, FIRSTS, SIGNIFICANT ANOMALIES, ETC.)

S135-E-006297 --- Chris Ferguson, left, and Doug Hurley are pictured at the Commander's Station and Pilot's Station, respectively.

Continued...

FLT DURATION:
12:18:27:56

S/T:
1318:16:13:22

OV-104:
307:12:55:0 7

DISTANCE:
5,284,862 sm

TOTAL OV-104 DISTANCE
125,935,769 sm

TOTAL SHUTTLE DISTANCE:
541,592,966 sm

ISS028-E-017042 --- Sandy Magnus/ MS enjoys view in the panoramic Cupola, an ISS addition since her last visit.

Continued...

- **Transfers:**
  30,576 lbs of h/w to ISS – MPLM, LMC, RRM & Picosat
  28,100 lbs of h/w from ISS MPLM LMC, & PM
  2,977 LtWt MPESS Carrier (up mass)
  1,409 LtWt MPESS Carrier (down mass)
  550 Robotic Refueling Mission P/L mass
  8 PicoSat mass (final deplow from a Shuttle - the 180th)
  65 lbs of Oxygen transferred to ISS (stack repress)
  111 lbs of Nitrogen transferred to ISS (stack repress)
  1,652 lbs of water transferred to ISS
  1,283 lbs of Atlantis middeck items transferred to ISS
  723 lbs of Atlantis middeck items returned from ISS
- ISS Mass in space 901,745 Pounds
- ISS assembly (pressurized volume) 100% complete
- FD12: Undock from ISS complete at 200:06:27:58Z
- FD13 did not have an LOS due to black out but if we had been on a lower antenna we estimated blackout to be GMT 202/09:28 – 202/09:41.

SIGNIFICANT ANOMALIES:
Orbiter:
- WLEIDS SENSOR UNIT 1111 DROPPED OUT OF OOM PREMATURELY
- WLEIDS SENSOR UNIT 1080 HAD COMMUNICATION DROP OUTS
- GPC 4 (SM) AR5450
SRB, KSC, RSRM, SSME, ET, Integration & MOD: None.

S135-E-007515 --- With his feet secured on a restraint on the ISS robotic arm or Canadarm2, Ron Garan/ Exp 28 Flight Engineer, carries the pump module,

S135-E-007457: -- Rex Walheim, Mission Specialist, works on the aft flight deck of Space Shuttle Atlantis.

S135-E-007637 --- Close-up of Mike Fossum/ Exp 28 Flight Engineer, as he participates in the July 12 six and a half hour spacewalk on ISS.

A Shuttle Goodbye to ISS

S135-E-011914

------- SALUTE -------

Space Shuttle's Last Landing Path (From PAO)

ATLANTIS STAYS AT KSC

FINAL LANDING at 5:57 a.m. (EDT) July 21, 2011

CDR Ferguson has a big smile for the thermal tiles (JSC2011-E-067990)

"God Bless America" sung by the unmistakable Kate Smith signaled the start of landing prep. FD Paul Dye, CapCom Shannon Lucid (in rear center) and the rest of the team in the CCK MOCR stood during the song, which was played for the crew and all those who have worked for the Space Shuttle Program.
(From PAO)

Entry Flight Control Team in Shuttle FCR in CCK MCC. FD Tony Ceccacci (center front) holds STS-135 mission logo.
(JSC2011-E-067253)

ROSES FOR CCK MCC

Seated on INCO console with Atlantis model is a bouquet of roses sent once again by the Shelton & Murphy families of North Texas. See history of "MCC Roses" given on flight STS-119/14A (page 2-203).
(JSC2011-E-063867)

------- SALUTE -------

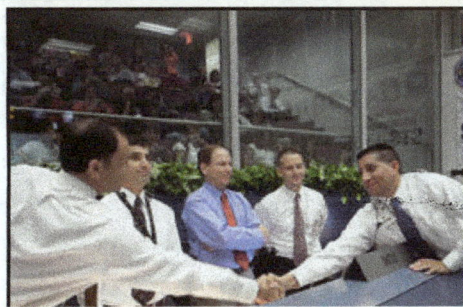

JSC2011-E-067473-- In MOCR in CCK MCC Michael Lopez-Algeria (left) Director/Flight Crew Ops for ISS shakes hands with Ascent FD Richard Jones. In middle (l to r) are Paul Hill/ Director/MOD, John Mccullough/Chief FD Office & Norm Knight/Dep Ch FD Office.

JSC2011-E-070840 --- STS-135 Ascent flight control team and flight crew (black shirts) in shuttle flight control room in the CCK MCC. Flight Director Richard Jones (left) and CDR Chris Ferguson, STS-135 Commander, hold the mission logo.

JSC2011-E-064806 --- STS-135 Shuttle & ISS FD's in the shuttle FCR in CCK MCC at JSC. From left (front row) are Tony Ceccacci, Courtenay McMillan, Chris Edelen, Kwatsi Alibaruho, Gary Horlacher, and Rick LaBrode. Back row are Paul Dye, Royce Renfrew, Richard Jones and Jerry Jason.

JSC2011-E-062692 --- STS-135 Orbit 1 FCT group portrait. Flight Director Kwatsi Alibaruho (center) stands on the front row.

JSC2011-E-063635 --- STS-135 Orbit 2 FCT group portrait. Flight Director Rick LaBrode holds the STS-135 mission logo.

JSC2011-E-064789 --- STS-135 Orbit 3 FCT group portrait. FD Paul Dye is in front near Shuttle model & MCC Roses.

JSC2011-E-063846 --- Shannon Lucid, STS-135 Planning Shift CAPCOM. She was one of NASA's first six women astronauts.

------- SALUTE -------

On behalf of the Astronaut Office...

Now that Atlantis and the final Shuttle crew have safely returned to planet Earth, we are all feeling the finality of 30 years of Space Shuttle flights. ... While the Shuttle is an incredible, one-of-a-kind flying machine, the most important thing that this program has given us is wrapped up in all the people and expertise that turned a concept into something real. ... We are exceptionally honored to have flown with all of you as part of the Shuttle Program, and look forward to the continuation of our journey on board the International Space Station and beyond.

Peggy A. Whitson
STS-111/Exp 5/STS-113, ISS Exp 16 CDR
Chief, Astronaut Office

30 YEARS OF FLIGHT

This plaque flew on the final Space Shuttle Mission in July, 2011. From the fortunate few who have served in space to the thousands who make spaceflight a reality, thank you for keeping the dream alive. Your passion for these amazing space ships will always stand as proof of what this country can do when it dares to be bold!

STS-135 crew left this plaque in the cockpit of Atlantis as a tribute to all of the people who have worked on the Space Shuttle Program.

SPACE SHUTTLE PROGRAM MANAGERS

Robert Thompson
February 1970 - June 1981
Glynn Lunney
June 1981 - June 1985
Richard Kohrs (Interim Mgr)
June 1985 - August 1985
Arnold Aldrich
August 1985 - November 1986
Richard Kohrs
November 1986 - June 1989
Leonard Nicholson
June 1989 - March 1993
Brewster Shaw
March 1993 - November 1995
Tommy Holloway
November 1995 - April 1999
Ronald Dittemore
April - 1999 - July 2003
William Parsons
July 2003 - September 2005
Wayne Hale
September 2005 - February 2008
John Shannon
February 2008 - To End of Shuttle Program 2011

Above:(jsc2011e071116): Lead INCO Heidi Brewer hangs final Shuttle plaque for STS-135 in White FCR in CCK MCC at JSC.

Below: Atlantis Shuttle Legacy Mural [UPDATED for STS-135] - Hanging in LCC Firing Room at KSC - From Mike Leinbach/Launch Director

------- SALUTE -------

A LARGE WELOME HOME & A SUPER WELL DONE!

ABOVE: (JSC2011-E-068785) --- A large crowd welcomes home the crew of STS-135 on July 22, 2011 at Ellington Field near JSC.   AT RIGHT: (JSC2011-E-070276) --- JSC Director Michael L. Coats (left), Houston Mayor Annise Parker, U.S. Senator Kay Bailey Hutchison (R.-Texas) and STS-135 Commander Chris Ferguson enjoy the crew return ceremony.   Poster reads: "HOUSTON! Always the first word in Space. Thank You!" The Mayor was also presented a flown flag by the STS-135 crew.

NOTES: From STS-135 (ULF7) Post Landing News Conference
- July 21, 2011 (From PAO)

Gerst – I really want to thank the Space Shuttle team and Program for today and the entire history of the Program. I can't say enough about meeting the challenges and finishing strong. Today they met all the objectives. I'd also like to thank the nation for supporting this vehicle. It is a true marvel and allowed us to do amazing things. It's going to allow us to move forward and utilize the station and commercial cargo providers come online later this year. We need to go forward and explore.
I recognize that change is very hard, but huge improvement comes from change, so this team can accomplish great things in the future. I wish them the best. They will be successful in the future.

Moses – It's been a heck of a day and heck of a Program. I'm representing a team across the country today and the vehicle performed perfectly. The team here and in Houston are world class. The Marshall team put together a propulsion system that also finished strong. It's been a nice ride.

Cabana – It is great to have Atlantis home to stay after this mission. I can't say enough about the teams here at KSC and how they performed the last few flights. The folks that knew they were going to be out of work performed flawlessly and were dedicated to what they were doing. That is what they do. I am proud to be part of this program. We've achieved the goal of flying out the shuttle safely and we'll celebrate what we've accomplished over the last 30 years. But when that's done, we'll move on to the future.

Leinbach – Thanks to the KSC workforce. I've worked here all my career – 27 years. They did their job just like always. The workers here and across the country that dedicated their lives to this are my friends and I wish them well. I want to thank the press as well. You guys have been good friends of the space program as well. It was a good day. Mission complete and we're looking forward to new challenges.

------- SALUTE -------

Space Shuttle's "Final Four"

STS135-S-214 (21 July 2011) -- Space Shuttle's "Final Four" stand proudly in front of Atlantis after landing at KSC. From right, are CDR Chris Ferguson, PLT Doug Hurley; Sandy Magnus/MS and Rex Walheim/MS.

U.S. SENATE RESOLUTION 233 -

HONORING THE MEN AND

WOMEN OF THE NATIONAL

AERONAUTICS AND SPACE

ADMINISTRATION SPACE

SHUTTLE PROGRAM ON

REACHING THE HISTORIC

MILESTONE OF THE 135TH AND

FINAL FLIGHT OF THE SPACE

TRANSPORTATION SYSTEM

Passed U.S. Senate on July 13. 2011

No, they are not rolling out for launch! Discovery & Endeavour are rolling to storage locations at KSC where they will remain until ready for transport to museums , see below.
(Photo from Herirng/PAO)

Safely Home...
...Mission Accomplished!

At Left:

Safely Home...
...Mission Accomplished!

"What a privilege to be on the scene for the last Apollo splashdown AND the last Space Shuttle landing ... and, what a privilege for each of us to have been associated with such talented and dedicated people ..."

Milt Heflin
Apollo Recovery Engineer- Primary Recovery Ship for Apollo 8, 10, 16, 17, Skylab 2,3,4, & ASTP
Space Shuttle EPS, Thermal, EGIL, EECOM & Flight Director
JSC Associate Director (Technical)

[That's Milt with "hands on hips". Yes, he was there & there. Well Done!]

Where will they go?

Announced April 12, 2011:

OV-101 Enterprise Test Vehicle
- To New York City's Intrepid Museum

OV-103 Discovery
- To Smithsonian National Air & Space Museum in Chantilly, Virginia

OV-104 Atlantis
- To Kennedy Space Center, Florida

OV-105 Endeavour
- To California Science Center in Los Angeles

NASA officially closed the books on the Space Shuttle Program on August 31, 2011.

APPENDIX A- SPACE SHUTTLE FLIGHT WEIGHT SUMMARY

A summary table of weight data for each shuttle element and payloads for each mission is provided in herein. The data for flights STS-1 through STS-57 was extracted from the SODB, Volume II. Effective with STS-51, the SODB data was no longer updated as flown. Therefore, the data has been obtained from the Day-of-Launch (DOL) Trajectory Design Data Package (TDDP), with Mach 3 EOM (End of Mission) and Landing Weights/CG's from the Postflight Prop 30 Reports. The Performance Margin data was extracted from the RI/Boeing Postflight Trajectory Reconstruction Reports.

Starting with STS-75, the Shuttle Program agreed to a 900-lb Ascent Performance Margin (APM) gain for all missions. STS-75 and STS-76 have 900 lbs of inert weight adjustment (-450 lbs inert weight discrepancy adjustment and -450 lbs

performance discrepancy adjustment, which were subtracted from the STS Operator Chargeable Cargo). Effective with STS-77, the -450 lbs was subtracted from the STS Operator Chargeable Cargo and the -450 lbs performance discrepancy is included in the MPS Prop Inventory. Effective with STS-79, the performance adjustment was changed to -200 lbs which is subtracted from the STS Operations Chargeable Cargo. Finally, beginning with STS-97 the TDDP included an item for "RECONSTRUCTED ASCENT PERFORMANCE COLLECTOR" in the "Shuttle Total Weight at SRB Igniton".

The P/L Deployed Weights for MIR flights reflect the weights of hardware transferred to the MIR (does not include consumables transferred to MIR). DOD mission weight data was not available for this document.

APPENDIX A - SPACE SHUTTLE FLIGHT WEIGHT SUMMARY

(SOURCES: SODB, VOL II Thru STS-57 & DOL TDDP for STS-51 and Beyond)

FLIGHT	TAIL NO. OV-	W/O CONS.	NON-PROP CONS @ SRB IGN	OMS PROP	RCS PROP FWD	RCS PROP AFT	ORB TOTAL @ SRB IGN	PRI DPLY/ NON-DPLY	DPLY AND RETR	RETR ONLY	ANCIL OR * MID-DECK	CHARGE-ABLE PYLD/ STS	FLIGHT CARGO TOTAL	ACCUM PYLD DPLY/ NON-DPLY	ACCUM CARGO TOTAL	WT @ SRB IGN	WT @ ORBIT INSERT	ACCUM WT @ ORBIT INSERT	ET TOTAL @ SRB IGN	SRB TOTAL LEFT & RIGHT	SHUTTLE TOTAL @ SRB IGN	PERF MARG FINAL TDDP & RECON	MACH 3 EOM WT	MACH 3 EOM X CG	LANDING WT	LANDING X CG
STS-1	102 (1)	172425	5197	18408	2461	5371	208437	0 / 10823	0	0			10823	0 / 10823	10823	219260	208415	208415	1664455	1295940 / 1298160	4459280	NOT AVAILABLE	195943	1096.7	195473	1098.1
STS-2	102 (2)	175211	5922	18011	2469	5383	212161	0 / 18778	0	0			18778	0 / 29601	29601	230939	219844	428259	1647514	1296747 / 1296784	4471984	2049 / 275	204356	1096.6	204263	1098.1
STS-3	102 (3)	175374	6560	17919	2446	5384	212846	0 / 22710	0	0			22710	0 / 52311	52311	235556	222985	651244	1643507	1296696 / 1296915	4470555	5343 / 2278	207349	1095.4	207073	1096.9
STS-4	102 (4)	175581	6588	22155	2446	5344	217280	0 / 9800	0	0	1844	11644 / 12848	24492	0 / 63955	76803	241772	228442	879696	1644745	1298213 / 1299253	4483983	4038 / 1195	209141	1092.9	208947	1094.4
STS-5	102 (5)	176729	5507	19804	2448	5379	215033	14585 / 5167	0	0	1078	20830 / 12596	32080	14585 / 70200	108883	247113	231213	1110869	1644995	1298256 / 1298714	4489078	822 / -1017	202643	2094.8	202480	1096.3
STS-6	099 (1)	172837	5364	19242	1964	5384	209957	37546 / 6853	0	0	2263	46662 / 1794	46971	52131 / 79316	155854	256928	241325	1352224	1644495	1295364 / 1296180	4488967	4755 / 2463	190627	1099.7	190330	1101.2
STS-7	099 (2)	172822	5415	21015	2449	5372	212239	14949 / 13002	3192	0	3942	31893 / 5448	37124	67080 / 96260	192978	249363	233619	1585843	1644631	1295695 / 1294346	4484035	2940 / 2021	204340	1089.8	204043	1091.2
STS-8	099 (3)	172879	5363	22011	2456	4962	212837	7445 / 13179	0	0	5166	25790 / 4440	30076	74525 / 114605	223054	242913	227365	1813208	1656386	1297016 / 1297508	4493822	14863 / 15735	204141	1090.4	203945	1091.9
STS-9	102 (6)	179369	6184	16000	2446	5384	214549	0 / 32261	0	0	MIDDECK 0 CRYO TK 870	33131 / 1708	33264	74525 / 147736	256318	247813	235793	2049001	1662238	1298367 / 1297983	4505505	841 / -411	220288	1085.8	220027	1087.1
STS 41-B (STS-11)	099 (4)	173041	6210	24704	2446	4970	216537	15073 / 10198	0	0	2981	28252 / 5598	33868	89598 / 160915	290186	250405	234108	2283109	1662570	1295569 / 1296187	4500237	12062 / 6961	201529	1087.9	201239	1089.3
STS 41-C	099 (5)	173207	5285	25096	2449	5012	216215	21396 / 12394	0	0	41	33831 / 4446	38266	110994 / 173350	328452	254481	245167	2528276	1661790	1295828 / 1296378	4508234	995 / -3322	197170	1100.0	196976	1101.6
STS-41-DR	103 (1)	173911	5748	23864	2446	4970	216105	30086 / 10122	0	0	1174	41382 / 6521	47516	141080 / 184646	375968	263621	246903	2775179	1662823	1296101 / 1298244	4518538	-1611 / -1564	202317	1090.7	201675	1091.7
STS 41-G	099 (6)	175411	6236	25088	2465	4970	219326	4949 / 11986	0	0	657	17592 / 5772	23465	146029 / 197289	399433	242791	226344	3001523	1662451	1296571 / 1296300	4495592	2194 / 3375	202829	1083.7	202266	1084.8
STS 51-A	103 (2)	174036	6311	25107	2446	4970	218016	22764 / 15052	0	2381	187	38003 / 7717	45306	168793 / 212528	444739	263352	247014	3248537	1662369	1299428 / 1299700	4522111	281 / 1003	207983	1081.4	207506	1082.6
STS 51-C	103 (3)	---------- DOD WEIGHT DATA NOT AVAILABLE ----------																				-- / -1457		1091.8	197700	1096.8
STS 51-D	103 (4)	174756	6272	21464	2446	4970	214855	22576 / 5092	0	0	1079	28747 / 7265	35794	191369 / 218699	480513	250679	239298	3487835	1661830	1297460 / 1296665	4504439	1243 / 1957	198167	1092.7	198014	1094.3
STS 51-B	099 (7)	174968	5397	22900	2446	4970	215847	105 / 30341	0	0	302	30748 / 1727	31377	191474 / 249342	511910	247254	230944	3718779	1661509	1296246 / 1296969	4501978	2536 / 3609	213795	1084.1	213499	1085.4
STS 51-G	103 (5)	174862	6272	18600	2446	4970	212316	22832 / 14866	2217	0	560	38258 / 6299	44477	214306 / 264762	556387	256823	243779	3962558	1661726	1297968 / 1298704	4518845	160 / -1664	204321	1082.1	204169	1083.7
STS-51-F	099 (8)	175260	5397	25064	2446	4970	218303	0 / 31257	0	0	1755	33012 / 1492	34400	214306 / 297780	590787	252733	237931	4200489	1661338	1300211 / 1300031	4514313	NOT AVAILABLE	216894	1079.8	216735	1081.3

APPENDIX A - SPACE SHUTTLE FLIGHT WEIGHT SUMMARY
(SOURCES: SODB, VOL II Thru STS-57 & DOL TDDP for STS-51 and Beyond)

FLIGHT	TAIL NO. OV-	ORBITER W/O CONS.	NON-PROP CONS @ SRB IGN	OMS PROP	RCS PROP FWD	RCS PROP AFT	ORB TOTAL @ SRB IGN	PRI DPLY/ NON-DPLY	DPLY AND RETR	RETR ONLY	MID-DECK	CHARGEABLE PYLD/ STS	FLIGHT CARGO TOTAL	ACCUM PYLD DPLY/ NON-DPLY	CARGO TOTAL	WT @ SRB IGN	WT @ ORBIT INSERT	ACCUM WT @ ORBIT INSERT	ET TOTAL @ SRB IGN	SRB TOTAL LEFT & RIGHT	SHUTTLE TOTAL @ SRB IGN	PERF MARG FINAL TDDP & RECON	MACH 3 EOM WT	MACH 3 EOM X CG	LANDING WT	LANDING X CG
STS 51-I	103 (6)	174785	6272	24646	2446	4970	218285	30289 / 8221	0	0	374	38884 / 5223	43988	244595 / 306375	634775	262303	249479	4449968	1661662	1297697 / 1298536	4520198	176 / -1145	196856	1092.4	196674	1094.2
STS 51-J	104 (1)	--------- DOD WEIGHT DATA NOT AVAILABLE ---------																							190765	1101.2
STS 61-A	099 (9)	175531	5397	18300	2446	4970	211810	150 / 27330	0	0	2164	30519 / 1587	31911	244745 / 335869	666686	243751	227797	4677765	1665455	1298021 / 1297886	4505113	6222 / 6219	214325	1083.8	214171	1085.2
STS 61-B	104 (2)	175615	6272	20000	1882	4970	213905	27465 / 13986	0	0	1337	42788 / 5236	47509	272210 / 351192	714195	261444	250836	4928601	1661470	1296606 / 1296018	4515538	874 / 2332	205880	1084.4	205732	1085.9
STS 61-C	102 (7)	185529	5692	22700	2096	4970	223153	12351 / 15837	0	0	437	28625 / 5547	32733	284561 / 367466	746928	255916	238764	5167365	1665325	1295611 / 1295702	4512534	10754 / 11127	210325	1083.6	210161	1085.1
STS 51-L	099 (10)	175760	5397	21500	2446	4970	215239	37636 / 10167 PROJECTED	0	0	830	48633 / 4171	52655			267929			1665170	1297828 / 1297848	4528770	NOT AVAILABLE	199585 PROJECTED	1089.7 PROJECTED	199438 PROJECTED	1091.3 PROJECTED
STS-26	103 (7)	176680	5409	14000	1914	4970	208139	37514 / 5928	0	0	1159	44601 / 3073	46448	322075 / 374553	793376	254617	243158	5410523	1664857	1301509 / 1301424	4522407	1546 / 624	194347	1096.6	194184	1098.3
STS-27	104 (3)	--------- DOD WEIGHT DATA NOT AVAILABLE ---------																					2905 / -286			190956	1095.1
STS-29	103 (8)	177365	5409	13984	1914	4973	208811	37640 / 6727	0	0	949	45316 / 3303	47394	359715 / 382226	840770	256235	244949	5655472	1664872	1300254 / 1300916	4522277	3772 / 2995	194940	1093.7	194790	1095.3
STS-30	104 (4)	177163	5415	18916	1906	4977	213543	40118 / 5540	0	0	165	45823 / 3140	17783	399833 / 387934	888553	261356	245051	5900523	1664743	1300247 / 1300880	4527426	4709 / 2650	192558	1097.4	192460	1099.1
STS-28	102 (8)	--------- DOD WEIGHT DATA NOT AVAILABLE ---------																					409 / 158			200214	1089.4
STS-34	104 (5)	177407	5479	14007	1926	4987	208972	38323 / 6696	0	0	886	45905 / 3871	48613	438156 / 395516	937166	257615	246268	6146791	1664981	1300812 / 1300165	4523573	2103 / -132	196112	1093.1	195954	1094.7
STS-33	103 (9)	--------- DOD WEIGHT DATA NOT AVAILABLE ---------																					1157 / 653			194282	1094.8
STS-32	102 (9)	184982	7165	25187	2224	4951	229575	15316 / 1962	0	21398	1039	18317 / 8141	26458	453472 / 398517	963624	256063	244557	6391348	1664843	1299175 / 1299406	4510498	1956 / 992	228523	1078.2	228335	1079.6
STS-36	104 (6)	--------- DOD WEIGHT DATA NOT AVAILABLE ---------																					881 / 930			187200	1096.4
STS-31	103 (10)	177516	5556	25045	2219	4966	220468	23095 / 960	0	0	652	25517 / 3126	28643	476567 / 400129	992267	249141	231665	6623013	1665069	1300241 / 1300214	4514665	2861 / 1352	189309	1087.9	189118	1089.7
STS-41	103 (11)	177599	5406	14509	1861	4961	209502	38604 / 6732	0	0	837	46173 / 3796	49969	515171 / 407698	1042236	259501	248128	6871141	1664877	1301372 / 1301388	4527138	1270 / -152	196982	1089.4	196869	1091.2
STS-38	104 (7)	--------- DOD WEIGHT DATA NOT AVAILABLE ---------																					863 / 474			191091	1098.6

APPENDIX A - SPACE SHUTTLE FLIGHT WEIGHT SUMMARY
(SOURCES: SODB, VOL II Thru STS-57 & DOL TDDP for STS-51 and Beyond)

FLIGHT	TAIL NO. OV-	W/O CONS.	NON-PROP CONS @ SRB IGN	OMS PROP	RCS PROP FWD	RCS PROP AFT	ORB TOTAL @ SRB IGN	PRI DPLY/NON-DPLY	DPLY AND RETR	RETR ONLY	MID-DECK	CHARGEABLE PYLD/STS	CARGO FLIGHT TOTAL	ACCUM PYLD DPLY/NON-DPLY	ACCUM CARGO TOTAL	WT @ SRB IGN	WT @ ORBIT INSERT	ACCUM WT @ ORBIT INSERT	ET TOTAL @ SRB IGN	SRB TOTAL LEFT & RIGHT	SHUTTLE TOTAL @ SRB IGN	PERF MARG FINAL TDDP & RECON	MACH 3 EOM WT	MACH 3 EOM X CG	LANDING WT	LANDING X CG
STS-35	102 (10)	184580	7156	19339	2232	4971	223444	0 / 25968	0	0	1792	27760 / 5277	33037	515171 / 435458	1075273	256511	243474	7114615	1664775	1300088 / 1300140	4521514	4131 / 3812	225531	1079.1	225329	1080.5
STS-37	104 (8)	177763	5379	20053	1835	4971	215167	34442 / 1615	0	0	743	36800 / 3761	40561	549613 / 437816	1115834	255758	240809	7355424	1664803	1300130 / 1299254	4519945	1116 / 525	190266	1087.4	190098	1089.2
STS-39	103 (12)	179611	6257	22553	2451	4974	221012	827 / 16046	4046 / 3955	0	494	21413 / 4881	26294	550440 / 454356	1142128	247336	236623	7592047	1664494	1299733 / 1301485	4513048	1054 / 2768	211673	1080.3	211512	1082.0
STS-40	102 (11)	185755	7111	13265	1919	4968	218184	0 / 26237	0	0	1877	28114 / 5593	33707	550440 / 482470	1175835	251921	241175	7833222	1664845	1301303 / 1301723	4519792	3137 / 4212	226737	1079.6	226535	1080.9
STS-43	104 (9)	177623	6245	14126	1860	4972	209992	37575 / 8146	0	0	991	46712 / 2613	49325	588015 / 491607	1225160	259347	247964	8081186	1664898	1299653 / 1299220	4523118	2656 / 2593	196353	1087.4	196088	1089.7
STS-48	103 (13)	178149	5466	22643	2061	4970	218455	14388 / 2066	0	0	690	17144 / 4425	21569	602403 / 494363	1246729	240054	224141	8305327	1665078	1298959 / 1298580	4502671	510 / -562	192925	1096.0	192780	1097.8
STS-44	104 (10)	177916	6245	16390	1893	4976	212586	37588 / 5809	0	0	1240	44637 / 2598	47235	639991 / 501415	1293964	259851	247087	8552414	1664283	1298356 / 1300086	4522576	565 / 1025	195047	1090.8	194818	1092.5
STS-42	103 (14)	178203	6341	14469	1908	4974	211062	0 / 26453	0	0	2210	28663 / 3701	32364	639991 / 530075	1326328	243456	231497	8783911	1664527	1300167 / 1299324	4507474	2511 / 2716	218159	1080.6	218089	1082.2
STS-45	104 (11)	177732	6337	16894	2180	4970	213279	0 / 15538	0	0	2145	17683 / 2658	20341	639991 / 547758	1346669	233650	222086	9005997	1664861	1298457 / 1298957	4496035	11017 / 10427	205672	1085.4	205588	1087.2
STS-49	105 (1)	180161	6197	19916	2448	4971	218859	23346 / 8766	0	0	697	32809 / 4635	37444	663337 / 557221	1384113	256333	246008	9252005	1664838	1299195 / 1298788	4519154	3351 / 3206	201400	1084.4	201235	1086.2
STS-50	102 (12)	186622	9760	16830	1903	4967	225218	0 / 22126	0	0	2179	24305 / 8142	32447	663337 / 581526	1416560	257695	245902	9497907	1664945	1298413 / 1299050	4520103	2940 / 3276	225865	1077.7	225615	1079.1
STS-46	104 (12)	178089	6380	24887	2451	4968	221941	9901 / 16094	1486 / 1396	0	1104	28585 / 5475	34060	673238 / 598724	1450620	256031	241797	9739704	1664720	1297746 / 1298292	4516789	2825 / 1942	209851	1078.2	209532	1079.6
STS-47	105 (2)	179161	6286	14559	1917	4979	212058	0 / 26247	0	0	1845	28092 / 4388	32480	673238 / 626816	1483100	244568	232661	9972365	1664720	1298225 / 1299291	4506804	1348 / 2887	220325	1083.7	220195	1085.3
STS-52	102 (13)	186650	7127	17398	2163	4974	223478	5577 / 12475	0	0	2080	20132 / 6730	26862	678815 / 641371	1509962	250370	239178	10211543	1664613	1299187 / 1300395	4514565	10788 / 9801	216043	1082.6	215935	1084.3
STS-53	103 (15)	179035	5874	18600	1912	4964	215551	20789 / 4299	0	0	1030	26118 / 2198	28316	699604 / 646700	1538278	243897	230731	10442274	1664985	1299174 / 1298531	4506587	1368 / 2844	194028	1089.5	193851	1091.3
STS-54	105 (3)	178558	5895	14278	1925	4980	210802	37497 / 7991	0	0	1052	46540 / 2499	49039	737101 / 655743	1587317	259871	248338	10690612	1664458	1299819 / 1299151	4523299	2659 / 3421	197481	1091.6	197353	1093.4
STS-56	103 (16)	179811	6287	17526	2456	4967	216223	0 / 12568	2840 / 2798	0	1031	16439 / 4561	21003	737101 / 669342	1608317	237253	225597	10916209	1664388	1299765 / 1300514	4501920	9521 / 10714	208052	1084.6	207946	1086.3
STS-55	102 (14)	186929	7345	15687	1928	4967	222022	0 / 24599	0	0	2282	26881 / 6535	33416	737101 / 696223	1641733	255468	244156	11160365	1664456	1299515 / 1300561	4519000	6248 / 7559	227484	1078.4	227209	1079.7
STS-57	105 (4)	179410	6412	25147	2450	4969	223554	132 / 18244	0	9424	1254	19630 / 9489	29119	737233 / 715721	1670852	252703	239319	11399684	1664332	1300548 / 1300983	4518566	2030 / 2162	224752	1081.1	224468	1082.5

FLIGHT	TAIL NO. OV-	W/O CONS.	NON-PROP CONS @ SRB IGN	OMS PROP	RCS PROP FWD	RCS PROP AFT	ORB TOTAL @ SRB IGN	PRI DPLY/NON-DPLY	DPLY AND RETR	RETR ONLY	MID-DECK	CHARGEABLE PYLD/STS	FLIGHT CARGO TOTAL	ACCUM PYLD DPLY/NON-DPLY	ACCUM CARGO TOTAL	WT @ SRB IGN	WT @ ORBIT INSERT	ACCUM WT @ ORBIT INSERT	ET TOTAL @ SRB IGN	SRB TOTAL LEFT & RIGHT	SHUTTLE TOTAL @ SRB IGN	PERF MARG FINAL TDDP & RECON	MACH 3 EOM WT	MACH 3 EOM X CG	LANDING WT	LANDING X CG
STS-51	103 (17)	179422	6003	16743	2451	4978	214763	26889 / 7305	7321	0	1122	42637 / 4048	46685	764122 / 724148	1717537	261487	250023	11649708	1664649	1298328 / 1298670	4523125	1358 / 1273	207043	1084.8	206932	1086.5
STS-58	102 (15)	187669	9789	14520	1945	4975	224062	0 / 21754	0	0	1373	23127 / 8884	32011	764122 / 747275	1749548	256103	244860	11894568	1664337	1298196 / 1298502	4517138	767 / 1114	229481	1078.8	229369	1080.4
STS-61	105 (5)	181308	7000	24989	2451	4971	225885	2308 / 14428	0	2148	665	17401 / 6962	24363	766430 / 762368	1773911	250279	236670	12131238	1664521	1298559 / 1298436	4511794	927 / 554	212947	1078.9	212836	1080.6
STS-60	103 (18)	179635	6510	18045	2450	4972	216778	171 / 21015	*0	0	1110	22296 / 6661	28957	766601 / 784493	1802868	245765	233290	12364528	1664515	1298776 / 1298783	4508839	110 / 306	216663	1079.6	216595	1081.3
STS-62	102 (16)	187779	9733	16797	2091	4968	226533	0 / 18512	0	0	1280	19792 / 10224	30016	766601 / 804285	1832884	256579	245457	12609985	1664370	1299668 / 1299184	4519801	871 / 1795	228360	1082.6	228250	1084.1
STS-59	105 (6)	180488	7220	13287	1924	4976	213061	0 / 26002	0	0	1445	27447 / 6311	33758	766601 / 831732	1866642	246849	237048	12847033	1664202	1300061 / 1300299	4511411	2856 / 1731	221981	1079.6	221865	1081.2
STS-65	102 (17)	188398	9567	16385	1898	4975	226389	0 / 22521	0	0	1761	24282 / 8598	32880	766601 / 856014	1899522	259296	247778	13094811	1664460	1299585 / 1300097	4523441	2169 / 3531	229368	1078.6	229261	1080.1
STS-64	103 (19)	180122	6286	16789	2451	4969	215783	0 / 16212	2842 / 2800	0	1363	20417 / 5204	25621	766601 / 873589	1925143	241439	230743	13325554	1664420	1298946 / 1299121	4503921	6409 / 9639	212294	1082.3	212180	1083.9
STS-68	105 (7)	180520	7225	13321	1913	4976	213121	0 / 25997	0	0	1643	27640 / 6612	34252	766601 / 901229	1959395	247404	237742	13563296	1664393	1299294 / 1299523	4510613	1721 / 2071	221784	1078.7	221673	1080.4
STS-66	104 (13)	180096	7163	20801	2448	4974	220648	0 / 9901	7154 / 7011	0	1080	18135 / 5426	23560	766601 / 912210	1982955	244238	232278	13795574	1664386	1299860 / 1300231	4508715	3284 / 3158	211562	1084.4	211411	1086.1
STS-63	103 (20)	179828	6285	23979	2454	4980	222692	23 / 15249	2651 / 2617	0	1128	19051 / 5852	24903	766624 / 928587	2007858	247630	235671	14031245	1664161	1299714 / 1300130	4511630	1830 / 3476	212775	1079.5	212693	1081.2
STS-67	105 (8)	180588	10610	24154	2447	4972	227937	0 / 18303		0	1764	20067 / 8461	28528	766624 / 948654	2036386	256495	243809	14275054	1664446	1299857 / 1299389	4520187	4099 / 6754	217646	1083.5	217437	1085.0
STS-71	104 (14)	180545	7390	21956	2452	4972	222481	0 / 17251	0	476	690	17941 / 8636	26577	766624 / 966595	2062963	249089	238682	14513736	1664561	1299083 / 1298854	4511586	1040 / 1398	216527	1079.7	216352	1081.3
STS-70	103 (21)	179039	5537	15110	1921	4982	211755	37774 / 5585	0	0	1086	44445 / 2354	46799	804398 / 973266	2109762	258584	247141	14760877	1664631	1299218 / 1299339	4521772	3789 / 5299	194267	1097.2	194190	1099.1
STS-69	105 (9)	180072	7149	24993	2452	4973	224805	0 / 16739	7306 / 7258	0	1301	25346 / 6203	31549	804398 / 991306	2141311	256385	243328	15004205	1664169	1299385 / 1299176	4519114	5409 / 7966	219395	1080.7	219298	1082.3
STS-73	102 (18)	188174	10734	12653	1883	4972	223592	0 / 23302	0	0	2008	25310 / 8395	33705	804398 / 1016616	2175016	257321	246718	15250923	1664190	1299554 / 1300510	4521581	1906 / 4902	230603	1080.7	230479	1082.3
STS-74	104 (15)	179624	7175	25155	2453	4976	224549	10015 / 9623	0	690	914	14064	23687	814413 / 1020665	2198703	248266	237141	15488064	1664354	1299872 / 1299903	4512395	1823 / 3689	202767	1078.7	202718	1080.6
STS-72	105 (10)	181188	7149	25038	2452	4970	225963	0 / 10546	2643	10459	898	14087 / 6931	21018	814413 / 1032109	2219721	247011	238498	15726562	1664138	1302278 / 1301220	4514647	11447 / 13346	218496	1081.7	218345	1083.3
STS-75	102 (19)	188372	9386	19109	2452	4970	229455	1494 / 20490	*0	0	1369	23353 / 8653	32006	815907 / 1053968	2251727	261491	250226	15976788	1663825	1300542 / 1300635	4526493	1594 / 638	226443	1079.4	226287	1080.9

* NOTE: DEPLOYED, NON-DEPLOYED, AND DEPLOYED/RETRIEVED REFLECT ACTUALS, E.G., WSF WAS NOT DEPLOYED AND RETRIEVED ON STS-60; TSS WAS LEFT IN SPACE ON STS-75.

APPENDIX A - SPACE SHUTTLE FLIGHT WEIGHT SUMMARY
(SOURCES: SODB, VOL II Thru STS-57 & DOL TDDP for STS-51 and Beyond)

FLIGHT	TAIL NO. OV-	W/O CONS.	NON-PROP CONS @ SRB IGN	OMS PROP	RCS PROP FWD	RCS PROP AFT	ORB TOTAL @ SRB IGN	PRI DPLY/ NON-DPLY	DPLY AND RETR	RETR ONLY	MID-DECK	CHARGEABLE PYLD/ PYLD/ STS	FLIGHT CARGO TOTAL	ACCUM PYLD DPLY/ NON-DPLY	ACCUM CARGO TOTAL	WT @ SRB IGN	WT @ ORBIT INSERT	ACCUM WT @ ORBIT INSERT	ET TOTAL @ SRB IGN	SRB TOTAL LEFT & RIGHT	SHUTTLE TOTAL @ SRB IGN	PERF MARG FINAL TDDP & RECON	MACH 3 EOM WT	MACH 3 EOM X CG	LANDING WT	LANDING X CG
STS-76	104 (16)	180112	7216	21664	2451	4976	221585	2814 / 10578	0	736	760	14152 / 10453	24605	818721 / 1065306	2276332	246222	238531	16215319	1664159	1299899 / 1299353	4509631	3140 / 3563	211913	1082.8	211805	1084.5
STS-77	105 (11)	180204	7235	19483	2453	4976	219518	1104 / 23586	1837 / 1820	0	866	27393 / 7812	35205	819825 / 1089758	2311537	254753	243818	16459137	1664470	1300764 / 1299175	4519162	5381 / 8528	222399	1080.5	222276	1082.0
STS-78	102 (20)	188422	10876	13227	1940	4979	224611	0 / 21598	0	0	2066	23666 / 8188	31854	819825 / 1113422	2343391	256495	245723	16704860	1664859	1297868 / 1298255	4517477	3683 / 4245	229134	1081.9	228986	1083.4
STS-79	104 (17)	180241	7286	21473	2450	4971	221598	3170 / 15151	0	2126	718	19039 / 8773	27812	822995 / 1129291	2371203	249440	241776	16946636	1664353	1297828 / 1298848	4510469	462 / 716	215990	1081.3	215904	1083.0
STS-80	102 (21)	187805	9760	20528	2451	4975	230676	0 / 7575	12524 / 12427	0	1109	21208 / 9903	31111	822995 / 1137975	2402314	261817	248721	17195357	1663927	1299137 / 1299854	4524735	487 / 1103	227815	1079.1	227670	1080.6
STS-81	104 (18)	180533	7284	21574	2452	4978	221988	4019 / 14492	0	2842	810	19321 / 8828	28149	827014 / 1153277	2430463	250167	242178	17437535	1663879	1298753 / 1298212	4511011	1286 / 2118	215403	1081.4	215337	1083.1
STS-82	103 (22)	182897	6572	25010	2448	4971	227065	6941 / 9921	0	6638	512	17374 / 7517	24891	833955 / 1163710	2455354	251986	239583	17677118	1663879	1299604 / 1298386	4513855	3503 / 4235	213949	1077.8	213869	1079.6
STS-83	102 (22)	187924	10876	15000	1912	4970	225849	0 / 23536	0	0	2020	25566 / 8817	34373	833955 / 1189266	2489727	259963	248526	17925644	1663889	1299392 / 1299392	4522925	4820 / 3741	235510	1078.5	235421	1080.0
STS-84	104 (19)	179665	7163	21674	2455	4973	221097	3902 / 14605	0	2576	1136	19643 / 8854	28497	837857 / 1205007	2518224	249624	241827	18167471	1663879	1298206 / 1298123	4509832	938 / 1103	216169	1081.0	216021	1082.6
STS-94	102 (23)	187901	10876	15058	1918	4968	225890	0 / 23536	0	0	2032	25568 / 8791	34359	837857 / 1230575	2552583	260279	248956	18416427	1664630	1297078 / 1297346	4519333	2845 / 4193	230818	1078.4	230773	1080.1
STS-85	103 (23)	181354	7072	17089	2450	4978	218082	0 / 15666	7726 / 7587	0	1590	24982 / 6977	31959	837857 / 1247831	2584542	250101	238142	18654569	1664460	1298435 / 1299129	4512125	1446 / 3065	221335	1082.0	221264	1083.6
STS-86	104 (20)	180477	7283	21682	2451	4975	222037	6058 / 14379	0	2859	602	21039 / 8689	29728	843915 / 1262812	2614270	251795	241773	18896342	1664491	1297660 / 1298078	4512024	1756 / 81	215387	1081.3	215303	1083.0
STS-87	102 (24)	188297	10459	16179	2188	4978	227270	0 / 17496	2998 / 2998	0	1452	21946 / 12448	34394	843915 / 1281760	2648665	261664	250693	19147035	1664353	1297733 / 1298120	4521900	4384 / 6115	232930	1081.0	232849	1082.6
STS-89	105 (12)	182187	7059	20679	2450	4972	222513	4596 / 16699	0	3508	868	22163 / 5877	28040	848511 / 1299327	2676705	250583	239584	19386619	1664543	1298227 / 1298526	4511879	2309 / 3544	217475	1086.5	217422	1088.2
SYS-90	102 (25)	187562	10884	15763	1841	4972	226191	0 / 23865	0	0	2340	26205 / 9844	36049	848511 / 1325532	2712754	262270	247955	19634574	1663992	1298901 / 1298520	4523683	3162 / 1999	233031	1080.3	232979	1081.9
STS-91	103 (24)	182624	7273	21882	2450	4976	224374	2419 / 22315	0	2964	891	25625 / 9944	35549	850933 / 1348738	2748303	259973	249580	19884154	1658766	1298618 / 1297292	4514649	631 / 403	226968	1079.5	226872	1081.1
STS-95	103 (25)	182647	7085	25032	2294	4980	227207	125 / 24108	2973 / 2945	0	1314	28520 / 10098	38618	851055 / 1374160	2786921	265855	247947	20132101	1658996	1297332 / 1298008	4520191	1587 / 2740	228455	1076.8	228388	1079.5
STS-88	105 (13)	182065	6997	24612	2451	4971	226265	26791 / 3073	0	335	1122	30986 / 6745	37731	877846 / 1378355	2824652	264026	251336	20383437	1658691	1297827 / 1297945	4518489	2365 / 1043	201538	1084.3	201492	1086.2
STS-96	103 (26)	183197	7174	25007	2450	4977	227974	4228 / 17994	0	213	1034	22707 / 11101	33808	882074 / 1397383	2858460	261812	245256	20628693	1658803	1297048 / 1296568	4514231	4435 / 4306	222366	1080.2	222299	1081.8

* NOTE: STS-91 WAS FIRST FLIGHT OF SLWT, 59212 LBS. STS-95 WAS SECOND FLIGHT OF SLWT, 59942 LBS. STS-88 WAS THIRD FLIGHT OF SLWT, 59137 LBS. STS-89 ET WEIGHED 66353 LBS.

APPENDIX A - SPACE SHUTTLE FLIGHT WEIGHT SUMMARY
(SOURCES: SODB, VOL II Thru STS-57 & DOL TDDP for STS-51 and Beyond)

FLIGHT	TAIL NO. OV-	ORBITER W/O CONS.	NON-PROP CONS @ SRB IGN	OMS PROP	RCS PROP FWD	RCS PROP AFT	ORB TOTAL @ SRB IGN	PRI DPLY/NON-DPLY	DPLY AND RETR	RETR ONLY	MID-DECK	CHARGEABLE PYLD/STS	FLIGHT CARGO TOTAL	ACCUM PYLD DPLY/NON-DPLY	CARGO TOTAL	WT @ SRB IGN	WT @ ORBIT INSERT	ACCUM WT @ ORBIT INSERT	ET TOTAL @ SRB IGN	SRB TOTAL LEFT & RIGHT	SHUTTLE TOTAL @ SRB IGN	PERF MARG FINAL TDDP & RECON	MACH 3 EOM WT	MACH 3 EOM X CG	LANDING WT	LANDING X CG
STS-93	102 (26)	185743	4820	14814	2473	4976	217975	43080 5171	0	0	1538	49789 2593	52382	925154 1404092	2910842	270387	258911	20887604	1658826	1297760 1297999	4524972	2081 -3981	202872	1097.5	202796	1099.4
STS-103	103 (27)	183199	7065	24990	2451	4979	227853	5423 6451	0	5351	1334	13208 7068	20276	930577 1411877	2931118	248159	236285	21123889	1658784	1299709 1299767	4506419	13576 13388	212288	1080.6	212217	1082.4
STS-99	105 (14)	182260	6989	19605	2308	4968	221299	260 26987	0	0	1822	29069 6341	35410	930837 1440686	2966528	256739	242322	21366211	1664331	1299767 1299817	4520450	1085 395	225092	1078.5	225030	1080.2
STS-101	104 (21)	183166	7235	23891	2453	4980	226894	3371 20159	0	1391	1262	24733 10871	35604	934208 1462107	3002132	262528	252056	21618267	1658873	1299223 1298831	4519455	1480 988	226277	1081.2	226212	1082.9
STS-106	104 (22)	183426	7235	23786	2449	4978	227032	5399 17935	0	948	1172	23967 11024	34991	939607 1481214	3037123	262053	253389	21871656	1658741	1299561 1298823	4519178	1940 317	222835	1080.1	222774	1081.7
STS-92	103 (28)	183363	7235	24629	2447	4968	227808	21998 4678	0	293	1333	28009 7241	35250	961605 1487225	3072373	263088	253459	22125115	1658781	1299531 1299149	4520549	1532 2330	205188	1080.0	205129	1081.8
STS-97*	105 (15)	181992	6989	22156	2452	4971	223736	36376 719	0	227	1021	37496 5308	42804	997981 1488965	3115177	266570	253646	22378761	1658695	1299246 1300085	4524795 *	1920 2032	197829	1085.9	197781	1087.7
STS-98	104 (23)	182605	7055	22904	2227	4978	224935	32270 583	0	872	987	33286 5876	39162	1030251 1490535	3154339	264127	251033	22629794	1658647	1298270 1298137	4519380	2138 1538	197909	1083.1	197854	1082.0
STS-102	103 (29)	182881	7055	24940	2452	4975	227469	9649 3517	0	1086	472	28739 8559	37328	1039900 1494524	3191667	264797	253436	22883230	1658484	1299774 1298555	4521809	2847 3031	218094	1083.2	218031	1084.9
STS-100	105 (16)	182943	7301	24075	2451	4972	226908	6346 4282	0	1608	781	29472 8858	38330	1046246 1499587	3229997	265268	253063	23136293	1658593	1298945 1299241	4522246	2670 2296	220623	1083.8	220556	1085.5
STS-104	104 (24)	182862	7301	25033	2452	4975	227787	19782 6060	0	626	582	26424 8711	35135	1066028 1506229	3265132	262952	254358	23390651	1658552	1298897 1299559	4520159	2884 2990	209142	1083.8	209097	1085.6
STS-105	103 (30)	182831	7055	23428	1886	4974	225340	9657 4654	0	3802	475	29305 7802	37107	1075685 1511358	3298239	262477	253897	23644548	1658085	1298852 1298417	4518170	705 631	222682	1081.0	222620	1085.6
STS-108	105 (17)	182106	7058	25057	2452	4972	226711	6454 8635	0	4156	690	31393 6784	38177	1082139 1520683	3336416	264918	252854	23897402	1657831	1298263 1298521	4519872	2381 1182	220623	1083.8	220556	1085.5
STS-109	102 (27)	188444	6969	25066	2451	4975	233071	8256 10672	0	6409	1216	20144 7420	27564	1090395 1532571	3363980	260665	250970	24148372	1658065	1298219 1298358	4515646	3309 4170	222447	1082.9	222366	1084.6
STS-110	104 (25)	184160	7060	25072	2451	4975	228854	30600 0	0	2607	757	28379 7470	35849	1120995 1533328	3399829	264763	253486	24401858	1658030	1298947 1298885	4520964	1256 2670	201513	1085.3	201463	1087.2
STS-111	105 (18)	183220	7060	25059	2454	4976	227935	9512 906	0	6342	288	29712 6370	36082	1130507 1534522	3435911	264047	253522	24655380	1657969	1297561 1298161	4518077	2484 1870	220234	1083.6	220279	1085.3
STS-112	104 (26)	183924	7060	25043	2179	4869	228341	29543 0	0	1839	381	29502 7939	37441	1160050 1534904	3473352	265812	254269	2490949	1658013	1298072 1299078	4521314	2744 3860	202688	1087.1	202621	1088.9
STS-113	105 (19)	183037	7060	25064	2254	4970	227551	29672 46	0	2250	288	30217 8176	38393	1189722 1535238	3511745	265974	250282	25159931	1658011	1298806 1298119	4521249	1736 2486	200993	1087.6	200939	1089.5
STS-107	102 (28)	189487	10160	17619	2180	4976	229588	0 23515	0	0	801	24316 11147	35463	1189722 1559554	3547208	265081	250270	25410201	1663352	1298648 1298614	4526034	1335 1348	234495 **	1078.5 **	234167 **	1077.9 **

* Beginning with STS-97 the TDDP included an item for "RECONSTRUCTED ASCENT PERFORMANCE COLLECTOR" in the "Shuttle Total Weight at SRB Igniton".

** WT & CG ARE AT EI AND EI+15 MINUTES.

APPENDIX A - SPACE SHUTTLE FLIGHT WEIGHT SUMMARY
(SOURCES: SODB, VOL II Thru STS-57 & DOL TDDP for STS-51 and Beyond)

FLIGHT	TAIL NO. OV-	W/O CONS.	NON-PROP CONS @ SRB IGN	OMS PROP	RCS PROP FWD	RCS PROP AFT	ORB TOTAL @ SRB IGN	PRI DPLY/NON-DPLY	DPLY AND RETR	RETR ONLY	MID-DECK	CHARGEABLE PYLD/STS	FLIGHT CARGO TOTAL	ACCUM PYLD DPLY/NON-DPLY	ACCUM CARGO TOTAL	WT @ SRB IGN	WT @ ORBIT INSERT	ACCUM WT @ ORBIT INSERT	ET TOTAL @ SRB IGN	SRB TOTAL LEFT & RIGHT	SHUTTLE TOTAL @ SRB IGN	PERF MARG FINAL TDDP & RECON	MACH 3 EOM WT	MACH 3 EOM X CG	LANDING WT	LANDING X CG
STS-114	103 (31)	184906	7076	24931	2174	4972	229219	26413 3231	0	6600	163	29807 8845	38652	1216135 1562948	3585860	267901	253950	25664151	1657242	1298074 1298565	4523083	2111 3792	225792	1086.6	225727	1088.2
STS-121	103 (32)	184902	7076	24922	2174	4968	229235	23696 5426	0	8456	158	29280 8456	37736	1239831 1568532	3623596	267001	253267	25917418	1657055	1299220 1299312	4523889	2290 N/A (Sensor Fail)	226063	1084.6	225972	1086.3
STS-115	104 (27)	184260	6986	24926	2449	4970	228734	35552 0	0	993	206	35758 6090	41848	1275383 1568738	3665444	270612	252240	26169658	1657088	1298957 1298678	4526580	1749 349	199711	1086.0	199642	1087.0
STS-116	103 (33)	185153	7189	24959	2134	4977	229555	5748 1652	0	806	182	22502 13188	35690	1281131 1585492	3701134	265275	250980	26420638	1657123	1298200 1298501	4520334	3768 4559	224053	1077.5	223986	1079.2
STS-117	104 (28)	184487	7018	24298	1926	4974	227846	36393 0	0	857	200	36593 6048	42641	1317524 1585692	3743775	270517	255388	26676026	1657157	1298138 1298472	4525519	1306	199418	1084.6	199305	1086.8
STS-118	105 (20)	185133	7189	24899	2030	4975	229369	11830 11740	0	316	329	23899 13491	37390	1329354 1597761	3781165	266789	250805	2692831	1657180	1298333 1297781	4521318	1913 2435	221740	1078.1	221660	1079.8
STS-120	103 (34)	185405	7108	22763	1885	4971	227275	33474 280	0	1577	59	33813 7059	40872	1362828 1598100	3822037	268177	251790	2944621	1657012	1298906 1298777	4524107	2091 1880	203069	1078.3	202989	1083.0
STS-122	104 (29)	184885	7042	20823	1914	4979	226743	30657 2162	0	2162	122	32941 7355	40296	1393485 1600384	3862333	267069	252667	3197288	1657253	1298675 1299004	4523236	2402 3435	207013	1078.2	207215	1080.4
STS-123	105 (21)	185393	7108	22763	1928	4981	227316	29442 1132	0	4891	188	30762 8153	38915	1422927 1601704	3901248	266261	253348	3450636	1657249	1298163 1298480	4521388	2109 5128	208916	1079.7	208762	1081.8
STS-124	103 (35)	185476	6868	22771	1923	4971	227152	33890 0	0	1608	79	33969 8028	41997	1456817 1601783	3943245	269179	251247	3701883	1656958	1299147 1298621	4525140	1308 2513	203605	1088.0	203755	1089.3
STS-126	105 (22)	185343	7108	22761	2187	4971	227513	30432 1760	0	19436	211	32403 7068	39471	1487249 1603754	3982716	267014	254431	3956314	1657112	1298611 1299270	4523242	1682 2329	221787	1087.2	221712	1089.0
STS-119	103 (36)	185710	6808	22762	2162	4973	227558	32489 0	0	1279	57	32546 6542	39088	1519738 1603811	4021804	266676	254546	4210860	1656990	1298197 1298799	4521897	1746 2016	201795	1082.8	201713	1084.7
STS-125	104 (30)	186902	7087	24984	2450	4982	231548	4694 17560	3893	0		22254 10164	32418	1524432 1621371	4054222	231548	254376	4465236	1657233	1297936 1298774	4519550	1689 2499	225509	1078.3	225469	1080.1
STS-127	105 (23)	185510	7108	22762	2204	4973	227700	24266 290	0	9756	126	24682 11571	36253	1548698 1621787	4090475	263983	252658	4717894	1657094	1298273 1298296	4518787	2553 2734	215900	1089.8	215817	1091.7
STS-128	103 (37)	185683	6586	22762	1934	4970	227078	30572 2331	0	19130	153	33056 7549	40605	1579270 1624271	4131080	267713	254672	4972566	1657188	1298511 1298323	4522876	1707 2077	222200	1088.4	222148	1090.2
STS-129	104 (31)	185268	7042	22762	2205	4967	227387	27615 1404	0	1176	353	29372 9521	38893	1606885 1626028	4169973	266310	254734	5227300	1657082	1298893 1298843	4522269	2228 2041	206917	1083.8	207200	1084.6
STS-130	105 (24)	185488	6397	22763	1918	4974	226683	34648 0	0	1262	283	34931 6025	40956	1641533 1626311	4210929	267669	252838	5480138	1657227	1298385 1297738	4522160	1188 2828	201138	1082.8	201084	1084.8
STS-131	103 (38)	186007	6392	22762	1931	4976	227212	30512 1388	0	21764	231	32131 7385	39516	1672045 1627930	4250445	266758	251459	5731597	1657053	1298230 1298461	4521643	1133 1491	224257	1089.0	224206	1090.7

APPENDIX A - SPACE SHUTTLE FLIGHT WEIGHT SUMMARY
(SOURCES: SODB, VOL II Thru STS-57 & DOL TDDP for STS-51 and Beyond)

	ORBITER							CARGO								ORBITER TOTALS			ET TOTAL @ SRB IGN	SRB TOTAL LEFT & RIGHT	SHUTTLE TOTAL @ SRB IGN	PERF MARG FINAL TDDP & RECON	ORBITER AT MACH 3 EOM		ORBITER AT LANDING	
FLIGHT	TAIL NO. OV-	W/O CONS.	NON-PROP CONS @ SRB IGN	OMS PROP	RCS PROP FWD	RCS PROP AFT	ORB TOTAL @ SRB IGN	PRI DPLY/NON-DPLY	DPLY AND RETR	RETR ONLY	MID-DECK	CHARGEABLE PYLD/STS	FLIGHT CARGO TOTAL	ACCUM PYLD DPLY/NON-DPLY	ACCUM CARGO TOTAL	WT @ SRB IGN	WT @ ORBIT INSERT	ACCUM WT @ ORBIT INSERT					WT	X CG	WT	X CG
STS-132	104 (32)	185064	7042	22762	2166	4974	227151	26619 / 0	0	7564	121	26740 / 9223	35963	1698664 / 1628051	4286408	263144	251170	5982767	1657088	1299411 / 1299029	4519813	5074 / 4326	210434	1081.0	210370	1082.9
STS-133	103 (39)	185336	6084	24861	1927	4971	228323	30576 / 818	0	1949	408	31802 / 8306	40108	1729240 / 1629277	4326516	268461	254067	6236834	1657403	1299112 / 1299345	4525061	1431 / 394	205075	1082.4	205022	1084.2
STS-134	105 (25)	185638	7007	24860	1907	4973	229529	30721 / 811	0	1609	161	31693 / 7517	39210	1759961 / 1630249	4365726	268769	256331	6493165	1657445	1298824 / 1299313	4525091	1968 / 3211	204532	1080.4	204463	1082.3
STS-135	104 (33)	184276	7072	24861	2171	4962	228486	27997 / 2137	0	24175	291	30425 / 7109	37534	1787958 / 1632677	4403260	266050	254325	6747490	1657525	1298160 / 1298628	4521103	1987 / N/A*	226333	1090.7	226270	1092.4

*Reconstruction analysis was not available (N/A) for STS-135 due to lack of funding.

The authors would like to acknowledge the following individuals for their contributions to the preparation of this book. Data Sources are also provided.

ACKNOWLEDGEMENTS - LEGLER INFORMAL BOOK

To: Brewster H. Shaw, while COO of United Space Alliance, for his sponsorship of Legler's informal book.

To: Mary C. Thomas/DA8 for her dedicated services as Book Manager for Revisions and Change Notices to Bob Legler's informal book through flight STS-115.

To: Karen J. Chisholm/DA8 for her dedicated services as editor and typist for Revisions and Change Notices to Bob Legler's informal book through flight STS-115.

To: All those who helped Bob Legler gather data through flight STS-115.

DATA SOURCES - LEGLER INFORMAL BOOK

This document provides "as flown" operational mission data and has been compiled from many sources including the following:
- Flight Logs
- Flight Rules
- Flight Anomaly Logs
- MOD Post-Flight Reports (Ascent, On-Orbit and Descent)
- Post Flight Analysis of MPS propellants
- FDRD - Flight Definiton Requirements Document
- FRD - Flight Requirements Document
- SODB - Shuttle Operational Data Book
- MER (Mission Evaluation Room) Shuttle Flight Data.
- Orbit Distance traveled is taken from the PAO Mission Statistics.

ACKNOWLEDGEMENTS - BENNETT (STS-116 Through STS-135)

To: James M. Heflin/AB111, Associate Center Director Technical, for his leadership role to publish the informal "Legler Book" as an official NASA Technical Memorandum.

To: USA's continued sponsorship to finalize the NASA Technical Memorandum.

To: Michael Curie, NASA HQs PAO Specialist & Commentator for his many responses to requests for information from Floyd Bennett.

To: M. Cathleen (Cat) Buehrer/DA32 (REDE CRITIQUE NSS JV) for her invaluable tutorial assistance and knowledge provided to Floyd Bennett for navigating through Microsoft Word software for preparation of this final book.

To: Edward P. Gonzalez/DM321 and John A. Fields/ DM111 for excellent technical review.

To: Dale H. Ward/IS4 (Tessada) and Sharon Hecht/IS4 (DB Consulting Group, Inc.) for their excellent editorial comments and final document preparation.

DATA SOURCES - FOR BENNETT

And finally, thanks to all the Data Source Contributors who helped Floyd Bennett find his way to the correct mission data for flights STS-116 through STS-135.
See the listing to follow:

This listing provides the data sources and Point(s) of Contact (POC's) used in preparing the portion of the Space Shuttle Mission Summary Book for missions STS-116 through STS-135. My thanks to all these contributors and many others who helped this author find his way to the correct mission data.

Floyd V. Bennett

ITEM	DATA SOURCES

COLUMN 1: FLT NUMBER

FLT NO.	FDRD: https://sspweb.jsc.nasa.gov/webdata/pdcweb/07700.htm
SEQ FLT #	Calculated from previous missions
KSC- #	Calculated from previous missions at KSC
PAD # PAD # (#)	FDRD: https://sspweb.jsc.nasa.gov/webdata/pdcweb/07700.htm Calculated from previous missions same pad
MLP-3	FDRD: https://sspweb.jsc.nasa.gov/webdata/pdcweb/07700.htm
# SHUTTLE FLIGHT TO ISS	Calculated from previous missions to ISS

COLUMN 2: ORBITER

ORBITER Vehicle Designation (Number of Flights) Vehicle Name	FDRD: https://sspweb.jsc.nasa.gov/webdata/pdcweb/07700.htm
OMS PODS Left POD #, Right POD #, & FWD RCS POD # # of Flights of each POD	MV/Orbiter Project Office, POC's: Dwyer, Kenneth J. (JSC-MV6) and Storm, Michael D. (KSC-USA) # of flights calculated from previous flight of the POD's

ITEM   DATA SOURCES

COLUMN 3: CREW SIZE: TITLE, NAMES, AND EVA'S

FLIGHT CREW SIZE   FDRD: https://sspweb.jsc.nasa.gov/webdata/pdcweb/07700.htm

FLIGHT CREW,
FLIGHT DIRECTORS,
& CAPCOMS
TITLES & NAMES   DA8/Lead Flight Directors Memos & JSC PAO Mission Press Kit: http://www.shuttlepresskit.com

EVA's
 Type and Duration   JSC PAO Shuttle Status Reports: http://www.nasa.gov/centers/johnson/news/shuttle/index.html
                                        MMT Briefings: John A. Mccullough, Annette P. Hasbrook, Norm Knight DA8
 # of EVA's   Calculated from previous mission EVAs

COLUMN 4: LAUNCH SITE, LIFTOFF TIME, LANDING SITES, ABORT TIMES

LAUNCH SITE
Launch Pad   FDRD: https://sspweb.jsc.nasa.gov/webdata/pdcweb/07700.htm
Launch Date & Time   Ascent FDO Post Flight Report/DM, POC's: Gonzalez,Edward P. Sparks, Carson W. (USA), & McDonald, Mark A. (USA)
Day of Week (#)   Refers to # of launches that day of week calculated from previous missions same day of week
Date (#)   Refers to # of launches that month calculated from previous missions same month

LAUNCH WINDOW   Real-time data, POC: Sparks, Carson W. and Mark McDonald (JSC-DM) [USA]

EOM PLS   Planned landing site: FDRD: https://sspweb.jsc.nasa.gov/webdata/pdcweb/07700.htm

TAL   Ascent FDO Post Flight Report/DM, POC's: Gonzalez, Edward P. Sparks, Carson W. (USA), McDonald, Mark A. (USA),
                                        & Kalil, Jose G. (JSC-DM)[USA]

TAX WX   Spacecraft Meteorology Group Post Mission Summary, POC: Oram, Timothy D. (JSC-WS8) [NOAA]

SELECTED:
RTLS, TAL, AOA, PLS   LSO Post-Flight Summary, POC: Linde, Martin G. (JSC-DM) [USA] & Hensley,Doyle W. (JSC-DM461)

MAX Q NAV   STS-XXX GNC First Stage Reconstruction: https://sspweb.jsc.nasa.gov/w ebdata/sei/t_Post%20Flight%20Reports/
                                        POC's: Cooper, Carling C. (Boeing), and Biskup, Bruce A., (Boeing)

SRB STG: [MET]   STS-XXX Ascent Performance Trajectory Reconstruction letter, POC: Stephen P. Brod/The Boeing Company (HM5-20)

ALL REMAINING DATA   Ascent FDO Post Flight Report/DM, POC's: Gonzalez,Edward P. Sparks, Carson W. (USA), & McDonald, Mark A. (USA)
THIS COLUMN

ITEM   DATA SOURCES

COLUMN 5: LANDING SITE/RUNWAY, CROSSRANGE, LANDING TIMES, FLIGHT DURATION, WINDS

ALL ITEMS EXCEPT   Descent Postflight Summary & Quicklook Reports: http://usa1.unitedspacealliance.com/usahou/orgs/48-20/dsct/pf/
ENTRIES LISTED BELOW: POC: Barbara Schill (USA) & Chris Re (USA), Chris Lessman (USA), Rosalyn Mark

LANDING EVENTS
  Time of Landing   Image Science & Analysis group: http://isal-web1.jsc.nasa.gov/Shuttle/ShowPage.pl?template=default.htm
                            Ascent/Descent Flight Design, POC: Lessmann, Christopher F. (USA)
  Site (#)   Site (#) refers to # of landings at a site, calculated from previous landing at that site
  Surface (#)   Surface (#) refers to # of landings on surface from previous landings on same surface
  Landing Day of Week (#)   (#) refers to # of landings on that particular weekday, calculated from landings on same weekday
  Landing Date (#)   (#) refers to # of landings in a particular calendar month, calculated from landings in the same calendar month

DEORBIT BURN   GMT (e.g., 051:12:59:52.0Z)-DM Trajectory Server - Legler Report, POC's:Propst, Carolyn A. (USA) & Deboeck, Toni F
                            (USA)

ORBIT DIR   (#) refers to # of landings from the same direction, calculated from # of last mission at same direction

TIME OF EVENTS
DURING LANDING   LLIMS Events: http://isal-web1.jsc.nasa.gov/llims/ObservationPublic.aspx?Mode=screening&mission=STS-XXX

ROLLOUT
  Distance (ft)   Calculated: wheels stop position - MLGTD position
  Time (sec)   Calculated: wheels stop GMT - MLGTD GMT

WINDS: OFFICIAL
and DENS ALT (ft)   Spaceflight Meteorology Post Flight Mission Summary, POC: Oram, Timothy D. (JSC-WS8) [NOAA]

FLT DURATION
  S/T   Shuttle total flight time, calculated: mission duration + sum of previous missions
  OV-XXX: Total flight time for specific orbiter vehicle, calculated: mission duration + sum of previous missions

DISTANCE   Statute miles traveled this mission: PAO Missions Stats Report, POC: Herring, Kyle J. (JSC-AP311)

TOTAL SHUTTLE DISTANCE   Calculated: distance traveled this mission + sum of previous missions
                            PAO Missions Stats Report, POC: Herring, Kyle J. (JSC-AP311)

ITEM   DATA SOURCES

<u>COLUMN 6: SSME-TL, NOM-ABORT, EMERG THROTTLE PROFILE</u>

SSME THROTTLE LEVELS
   PREDICTED   FDRD: https://sspweb.jsc.nasa.gov/webdata/pdcweb/07700.htm
ACTUAL   Ascent FDO Post Flight Report/DM, POC's: Gonzalez,Edward P. Sparks, Carson W. (USA), & McDonald, Mark A. (USA)

ENG. S.N.   FDRD: https://sspweb.jsc.nasa.gov/webdata/pdcweb/07700.htm
(#)   Refers to # of flights by engine serial number - calculated from previous flight by that SSME

M 3 EOM and LANDING
   WEIGHT and X CG   IDP Cycle/Prop30 Aerosciences Report/Version 01, POC: Schill, Barbara C. (USA)

<u>COLUMN 7: SRB, RSRM, AND ET</u>

SRB, RSRM, and ET   FDRD: https://sspweb.jsc.nasa.gov/webdata/pdcweb/07700.htm

ET IMPACT: MET, LAT, LONG   STS-XXX Nominal ET Disposal Chart and ET Summary Table, POC: Dulski, Matthew B. (USA) & Strach, Daniel P (USA)

<u>COLUMN 8: ORBIT INCLINATION</u>

INC   FDRD: https://sspweb.jsc.nasa.gov/webdata/pdcweb/07700.htm

<u>COLUMN 9: ORBIT HA/HP</u>

INSERTION (type)   FDRD: https://sspweb.jsc.nasa.gov/webdata/pdcweb/07700.htm

POST OMS-2 (nm) and
DEORBIT HA/HP   DM Trajectory Server - Legler Report Request, POC's: Propst, Carolyn A. (USA), Deboeck, Toni F (USA), and Leleux, Darrin P. (JSC-DM411)

ENTRY VELOCITY (fps) and
ENTRY RANGE (nm)   Descent Post Flight Summary: http://usa1.unitedspacealliance.com/usahou/orgs/48-20/dsct/pf/, POC: Hill, Trudy D. (Debbie) (USA)

ITEM   DATA SOURCES

COLUMN 10: FLIGHT SOFTWARE (FSW)

OI-XX   Orbit Insertion Flight Software version # - FDRD: https://sspweb.jsc.nasa.gov/webdata/pdcweb/07700.htm
(#)   (#) refers to # of flights flown - calculated from last flight of that FSW version

COLUMN 11: PAYLOAD WEIGHTS; PAYLOADS, EXPERIMENTS

PAYLOAD WEIGHTS   Day of Launch (DOL) Trajectory Design Data Package (TDDP), POC: Bhula, Jayantilal (Jay) (USA)
  TOTAL, MIDDECK,
  DEPLOYED, and
  NON-DEPLOYED

SHUTTLEACCUMULATED
WEIGHTS   Calculated (summed) from previous missions
  DEPLOYED,
  NON-DEPLOYED,
  and CARGO TOTAL

PERFORMANCE MARGIN (LBS)
  FPR and FUEL BIAS,   Day of Launch (DOL) Trajectory Design Data Package (TDDP), POC's: Bhula, Jayantilal (Jay) (USA)
  FINAL TDDP   Provided by Mike . L. Scott/USA/FDD POC
  RECON   STS-XXX Ascent Performance Trajectory Reconstruction, POC:Steven P. Brod/Boeing

ASSIGNMENTS   FDRD: https://sspweb.jsc.nasa.gov/webdata/pdcweb/07700.htm
  PAYLOADS: PLB and MIDDECK
  # CRYO TANK SETS
  STS OPERATOR SELECTIONS

RMS (#)   -# of flights RMS flown - calculated from previous missions with RMS

ITEM   DATA SOURCES

COLUMN 12: MISSION HIGHLIGHTS

BRIEF MISSION SUMMARY   JSC PAO Mission Press Kit: http://www.shuttlepresskit.com/
                              MMT Minutes: https://sspweb.jsc.nasa.gov/mmt/

KSC W/D (Work Days)   KSC Milestone Interface Chart, POC: Overton, Thomas L. (KSC) [ASRC AEROSPACE] & Clark D. Ford (KSC PHO00)

LAUNCH POSTPONEMENTS   SSPO PRCB Directives: https://sspweb.jsc.nasa.gov/meeting/mtgdata.cfm

LAUNCH SCRUBS   MMT Minutes: https://sspweb.jsc.nasa.gov/mmt/

LAUNCH WINDOW   Real-time Data, POC: Sparks, Carson W. (JSC-DM) [USA]

LAUNCH DELAYS   MMT Minutes: https://sspweb.jsc.nasa.gov/mmt/

TAL WEATHER   Spaceflight Meteorology Group Post Mission Summary,
                              http://www.srh.noaa.gov/smg/XXX_Postmission_Summary.pdf (XXX is STS Flight #)
                              POC: Oram, Timothy D. (JSC-WS8) [NOAA]

PERFORMANCE
ENHANCEMENTS   Day of Launch (DOL) Trajectory Design Data Package (TDDP), POC: Bhula, Jayantilal (Jay) (USA)

FLIGHT DURATION
CHANGES/LANDING   MMT Minutes: https://sspweb.jsc.nasa.gov/mmt/
                              Spaceflight Meteorology Group Post Mission Summary, POC: Oram, Timothy D. (JSC-WS8) [NOAA]

FIRSTS/LASTS   JSC PAO Mission Press Kit: http://www.shuttlepresskit.com/
                              Flight Readiness Reviews: https://sspweb.jsc.nasa.gov/webdata/launch/

NIGHT LAUNCH (#)   Number of night launches, calculated from previous night launch mission

NIGHT LANDING (Site, #)   Number of night landings at specified site, calculated from previous night landing mission at that site

RENDEZVOUS   Number of rendezvous missions, calculated from previous rendezvous mission

Continued…

ITEM   DATA SOURCES

COLUMN 12 MISSION HIGHLIGHTS (Continued)

EVENTS

Time of on-orbit maneuver
events (OMS 2, IT, etc.)   DM Trajectory Server - Legler Report, POC's: Propst, Carolyn A. (USA) and Deboeck, Toni F (USA)

Time of docking/undocking
events   APDS sensor Data from the ODRC, POC: Dake, Janna J., Murphy, Rachel & Haskovec, Doug (JSC-DS421)

Time of ISS hatch opening
and crew welcome   JSC PAO Shuttle Status Reports: http://www.nasa.gov/centers/johnson/news/shuttle/index.html

EVA descriptions and
durations   Post flight EVA notes (provided by DX POC)
      JSC PAO Shuttle Status Reports: http://www.nasa.gov/centers/johnson/news/shuttle/index.html

Transfers (hardware and
consumables weights)   STS-XXX Final Customer Support Room (CSR) Report and STS-XXX Mission by the Numbers (provided by MO POC's)

SIGNIFICANT ANOMALIES   PCASS In-flight Anomalies: https://usa93.usa-spaceops.com:4443/adamvweb/ifa.ifa_search2.wp_execfind

ENTRY BLACKOUT   INCO Electronic Flight Log (Provided by DS POC Steve Sides & Mark Williamson)

WEIGHT SUMMARY
All entries except entries below: Day of Launch (DOL) Trajectory Design Data Package (TDDP): POC: Bhula, Jayantilal (Jay)/USA

Orbiter Tail No.   FDRD: https://sspweb.jsc.nasa.gov/webdata/pdcweb/07700.htm
Shuttle /PL Accumulated WTs   Calculated from previous missions
Weight at Orbit Insertion   Ascent Post Flight Data (provided by Gonzalez, Edward P./JSC-DM)
Performance Margin
  Final TDDP   Provided by Mike . L. Scott/USA/FDD POC
  Reconstructed   STS-XXX Ascent Performance Trajectory Reconstruction (Kristin Smaltz & Stephen Brod/Boeing)
Orbiter weight at Mach 3 EOM
  and at Landing   IDP Cycle/Prop30 Aerosciences Report (provided by Barbara Shill & Rosalyn Mark/USA/FDD/SDM)

PHOTOS (All Missions)   Identified by NASA Number, unless otherwise noted. POC: Jody Russell/JSC-AP (Tessada)

APPENDIX C - FLIGHT DIRECTOR LOG

This appendix provides the JSC Flight Director Log initially compiled and kept updated by Bob Legler, "History Flight". Since his death the log has been maintained by the Flight Director Office staff. This is a listing of Flight Directors beginning with Christopher C. Kraft, Jr. "Red Flight" in 1960 with Project Mercury flights, and ending with the completion of the Space Shuttle Program in 2011.

Note: Names listed in blue denote photo available from electronic copy by "control-hold-click".

FLIGHT DIRECTOR LOG
(August 2010)

Compiled by HISTORY FLIGHT and updated by DA8 Staff

#	COLOR	NAME	CLASS	STATUS	First Shift on Console	Retired as a Flight Director
1	Red Flight	Christopher C. Kraft	1960	Retired		1967
2	Blue Flight	John Hodge	1963	Retired		
3	White Flight	Eugene F. Kranz	1963	Retired		1974
4*	Black Flight	Glynn S. Lunney	1963	Retired		
5	Green Flight	Clifford E. Charlesworth	1966	Deceased - 2001		1970
6	Gold Flight	Gerald D. Griffin	1968	Retired		1973
7	Maroon Flight	Milton L. Windler	1968	Retired		
8	Orange Flight	M. P. (Pete) Frank	1968	Deceased - 2005		1983
9	Purple Flight	Phillip C. Shaffer	1971	Deceased - 2007		1974
10	Crimson Flight	Donald R. Puddy (STS-1)	1971	Deceased - 2007		1981
11	Silver Flight	Neil B. Hutchinson (STS-1)	1971	Retired		1984
12	Bronze Flight	Charles R. Lewis (STS-1)	1971	Retired		1984
13	Ivory Flight	Tommy W. Holloway	1979	Retired		1984
14	Crystal Flight	Harold M. Draughon	1979	Retired		1984
15	Gray Flight	Gary E. Coen	May-81	Retired		1995
16	Granite Flight	John T. Cox	May-81	Retired		1988
17	Emerald Flight	Jay H. Greene	May-81	Retired		1987
18	Amber Flight	Brock (Randy) Stone	Nov-81	Retired		1993
19	Indigo Flight	Lawrence S. Bourgeois	Nov-81	Retired		1991
20	Aquila Flight	A. (Lee) Briscoe	Mar-83	Retired		1991
21	Orion Flight	T. Cleon Lacefield	Mar-83	Retired		1986
22	Polaris Flight	Granvil A. Pennington	Mar-83	Retired		2007
23	Alpha Flight	William D. Reeves	Mar-83	Retired		2001
24	Altair Flight	Charles W. Shaw	Mar-83	Retired		2003
25	Sirius Flight	J. Milton Heflin, Jr.	Mar-83	Retired		2005
26	Rigel Flight	Charles R. Knarr	Mar-83	Retired		1991
27	Phoenix Flight	Ronald D. Dittemore	Nov-85	Retired		1992
28	Turquoise Flight	N. Wayne Hale, Jr.	Feb-88	Retired		2004
29	Antares Flight	Robert E. Castle, Jr.	Feb-88	Retired		2003
30	Falcon Flight	Robert M. Kelso	Feb-88	Retired		2000
31	Regulus Flight	Philip L. Engelauf	Dec-89	Retired		2008
32	Aurora Flight	Jeffrey W. Bantle	Dec-89	Retired		2001
33	Corona Flight	Linda J. (Hautzinger) Ham	Jan-91	Retired		2000
34	Burgundy Flight	Richard D. Jackson, Jr.	Jan-91	Retired		1997
35	Kitty Hawk Flight	John F. Muratore	Jan-92	Retired		1994
36	Iron Flight	Paul F. Dye	Nov-93			
37	Perseus Flight	Bryan P. Austin	Nov-93	Retired		2003
38	Midnight Flight	John P. Shannon	Nov-93	Retired		2004
39	Argon Flight	Andrew F. Algate	Oct-94	Retired		2008

Continued…

#	COLOR	NAME	CLASS	STATUS	First Shift on Console	Retired as a Flight Director
40	*Atlas Flight*	Paul S. Hill	Jun-96	Retired		2005
41	*Ares Flight*	Jeffrey M. Hanley	Jun-96	Retired		2005
42	*Cardinal Flight*	Mark A. Kirasich	Jun-96	Retired		2006
43	*Cassini Flight*	Sally P. Davis	Jun-96	Retired		2008
44	*Azure Flight*	Mark J. Ferring	Jun-96	Retired		2007
45	*Arcturus Flight*	John M. Curry	Jun-98	Retired		2007
46	*Pegasus Flight*	Richard E. La Brode, Jr.	Jun-98			
47	*Chromium Flight*	Leroy E. Cain	Jun-98	Retired		2005
48	*Sapphire Flight*	Kelly B. Beck	Jun-98	Retired		2008
49	*Flash Flight*	Joel R. Montalbano	Oct-00	Retired		2008
50	*Eagle Flight*	John A. McCullough	Oct-00			
51	*Amethyst Flight*	Norman D. Knight	Oct-00			
52	*Fuchsia Flight*	Annette P. Hasbrook	Oct-00	Retired		2009
53	*Titanium Flight*	J. Derek Hassmann	Oct-00			
54*	*Onyx Flight*	Bryan C. Lunney	Oct-00	[Retired]		[2011]
55	*Aquarius Flight*	Matthew R. Abbott	Oct-00			
56	*Topaz Flight*	Catherine A. Koerner	Oct-00	Retired		2007
57	*Intrepid Flight*	Anthony J. Ceccacci	Oct-00			
58	*Garnet Flight*	Steven J. Stich	Oct-00	Retired		2007
59	*Defiant Flight*	Kwatsi Alibaruho	Feb-05			
60	*Vega Flight*	Ginger Kerrick	Feb-05			
61	*Galileo Flight*	Robert Dempsey	Feb-05			
62	*Viking Flight*	Holly Ridings	Feb-05			
63	*Mercury Flight*	Dana Weigel	Feb-05		01/16/06	
64	*Liberty Flight*	Brian Smith	Feb-05		02/13/06	
65	*Sigma Flight*	Richard Jones	Feb-05		06/30/06	
66	*Kodiak Flight*	Michael Sarafin	Feb-05		07/13/06	
67	*Apex Flight*	Michael Moses	Feb-05	Retired	09/05/06	2008
68	*Sequoia Flight*	Heather Rarick	June-06		02/26/07	
69	*Gemini Flight*	Ron Spencer	June-06		04/23/07	
70	*Peridot Flight*	Emily J. Nelson	2007		12/03/07	
71	*Tranquility*	Courtenay McMillan	2007		12/07/07	
72	*Odyssey Flight*	David Korth	2007		3/31/08	
73	*Venture Flight*	J. Chris Edelen	2007		4/11/08	
74	*Tungsten Flight*	Royce J. Renfrew	2008		10/31/08	
75	*Raptor Flight*	Jerry P. Jason	2008		4/14/09	
76	*Viper Flight*	Gary C. Horlacher	2008		7/16/09	
77	*Saturn Flight*	Michael L. Lammers	2008		7/11/09	
78	*Carbon Flight*	Edward A. Van Cise	2009		1/20/10	
79	*Keystone Flight*	Scott Stover	2009		3/29/10	
80	*Steel Flight*	Dina Contella	2009		5/10/10	

* Second generation FDs, #4 Glynn Lunney and #54 Bryan Lunney

NOTE: There were two additional individuals that were selected as flight directors but elected to not continue: Rick Fitts and Michele Brekke.
Continued…

	HONORARY FLIGHT DIRECTORS		
	COLOR	*NAME*	*STATUS*
1	*Grey Flight*	Howard W. Tindall, Jr.	Deceased - 1995
2	*Pink Flight*	Lois Ransdell	Deceased - 1996
3	*Diamond Flight*	Alene Ganzer	
4	*Scarlet Flight*	John W. O'Neill	
5	**History Flight**	**Robert D. Legler**	**Deceased - 2007**

THE FLIGHT DIRECTOR OFFICE: *"Provides leadership and direction for conducting human space flight operation. Our mission is to ensure excellence in mission operations for Human Space Flight." (DA8 Home Page)*

IN MEMORIAM

Bob Legler, the originator of the informal Space Shuttle Missions Summary Book, was born a natural Corn Husker and lived a full life. His true love was serving his country in the US Coast Guard, Merchant Marines, United Nations, US Army, and the NASA Space Programs as an aerospace engineer. As one of a handful of people to ever support the Mercury, Gemini, Apollo, Skylab, Space Shuttle, and International Space Station missions, Bob was an icon to his peers. He spent 44 years in this noble endeavor called manned space flight. In the memorial service for Bob, Milt Heflin, JSC Associate Director and former JSC Chief Flight Director, provided the following insight:

"Bob was about making things happen, no matter what his position or rank, in whatever the enterprise was at that time…it might have been dodging bullets and bombs while establishing communication systems for United Nations outposts in crazy places…it might have been while riding the Coastal Sentry Quebec Tracking ship in the Indian Ocean…watching over the Lunar Module electrical power system or the operation of the Apollo Telescope Mount…serving as a SPAN Manager in the MCC (where a lot of really good stories were told during crew sleep)…or even while serving as the Chairman of the Annual FOD Chili Cook-off or his beloved Chairmanship of the Apollo Flight Operations Association [for reunions]…`in each case he gave of himself so that the "mission," no matter what it was, could be successful…Bob might not have been the most efficient chairman…story telling could get in the way from time to time…but he made up for it by being a catalyst, causing the team to rise to the occasion…

(Continued)

Bob Legler
April 4, 1927 - March 16, 2007

IN MEMORIAM

(Continued)

And, we all know quite well his love of capturing the history of manned space flight…Apollo reunions and producing the Space Shuttle Missions Summary Book are two of his legacies…events and things with Bob's hands that were done for the enjoyment of all…he took great pride in keeping the "official" Flight Director Log, a listing of those that have served as a Flight Director in Mission Control…the Log today lists 69 Flight Directors beginning with Red Flight, Chris Kraft…even I had a hard time in convincing Bob that I would not abuse my electronic copy of this list, if he would just send it to me…this list also contains the names of only five individuals designated as an Honorary Flight Director…Bob is number 5, known as 'History Flight,' given that honor upon his retirement…"

From Randy Stone, former JSC Chief Flight Director and former JSC Deputy Director: "Bob mentored all of the new Flight Controller's with his wisdom, knowledge, but more importantly his passion for human space flight."

Others commented: "Bob was a walking encyclopedia of space knowledge and also had a great sense of humor." "Bob was a rarity in the annuals of human space flight – a joyful cheerleader [with] unabashed love of the space program." "I could always rely on Bob for hard to find info. His enthusiasm for his work was obvious." "Bob was good natured and enjoyed a good joke, even if it was on him. I love Bob and will miss him."

And, shortly before Bob died, he received the following note from Bob Cabana, KSC Center Director and former Astronaut: "Bob, I look forward to your Summary Shuttle Book after the last [final] Shuttle mission. I think it's the only way I'm ever going to remember what missions I CAPCOM'ed on and who was on console with me."

The detail, the accuracy, the completeness of this Space Shuttle Missions Summary Book are a testament to Bob Legler's "passion and knowledge" for human space flight. We will finish this book for him with the same dedication.

Floyd Bennett
Friend & Colleague

ABOUT THE SECOND AUTHOR - FLOYD V. BENNETT

After Bob Legler's death in 2007, Floyd Bennett asked for and was given the task of completing Bob Legler's Space Shuttle Missions Summary Book, beginning with flight STS-116 and ending with the final Space Shuttle Mission. He was a friend and colleague of Bob's during the Apollo and Space Shuttle Programs. He also worked with Bob as a member of the Apollo Flight Operations Association for reunion events and was a co-author of Bob's 35th [and last] Apollo Anniversary Reunion Book.

Floyd has 57 years of technical and managerial experience in the field of Aerospace Engineering. After graduation from Virginia Tech University in 1954, he joined the National Advisory Committee for Aeronautics (which became NASA in 1958) at Langley Research Center in Hampton, VA. As a research engineer he published several NACA/NASA

Technical Reports on aircraft aeroelasticity. In 1962 he transferred with the Space Task Group to the Manned Spacecraft Center (now Johnson Space Center) in Houston, TX. Here he performed and managed analyses for manned spaceflight in engineering development, mission planning, flight operations, systems integration, and finally as a Space Shuttle Missions historian.

He performed key roles during the Apollo Program in establishing the Lunar Module Spacecraft landing and ascent operational trajectory strategies, lunar landing site selection, mission planning and real-time mission support for all Apollo manned lunar landing missions. During the Space Shuttle Program he performed a key role in systems integration for establishing program control of vehicle weight and performance for initial Space Shuttle manned development flights.

After NASA retirement in 1982 he continued making contributions in Space Shuttle Systems Integration for resolution of Payload, SSME, and Orbiter technical issues while working for three different NASA contractors, retiring from United Space Alliance in 2006.

Floyd is an Associate Fellow & Emeritus Lifetime Member American Institute of Aeronautics & Astronautics. He has received numerous NASA and USA awards for exceptional service during the Apollo and Space Shuttle Programs including an Apollo 15 Astronaut's Lunar Landmark named "Bennett Hill".